Statistical Mechanics and Applications in Condensed Matter

This innovative and modular textbook combines classical topics in thermodynamics, statistical mechanics and many-body theory with the latest developments in condensed matter physics research. Written by internationally renowned experts and logically structured to cater for undergraduate and postgraduate students and researchers, it covers the underlying theoretical principles, and includes numerous problems and worked examples to put this knowledge into practice.

Three main streams provide a framework for the book: beginning with thermodynamics and classical statistical mechanics, including mean field approximation, fluctuations and the renormalization group approach to critical phenomena. The authors then examine quantum statistical mechanics, covering key topics such as normal Fermi and Luttinger liquids, superfluidity and superconductivity. Finally, they explore classical and quantum kinetics, Anderson localization and quantum interference, and disordered Fermi liquids.

Unique in providing a bridge between thermodynamics and advanced topics in condensed matter, this textbook is an invaluable resource for all students of physics.

Carlo Di Castro is a member of the Accademia Nazionale dei Lincei and Emeritus Professor of Theoretical Physics at the Sapienza University of Rome, where he has been at the forefront of teaching and research in statistical mechanics and many-body theory for over 40 years. His current interests include strongly correlated electron systems, quantum criticality, high temperature superconductivity and non-Fermi-liquid metals.

Roberto Raimondi is Associate Professor of Condensed Matter Physics at Roma Tre University, where he has made important contributions to the understanding of the transport properties of disordered and mesoscopic systems. His current research interests include spintronics, especially the spin Hall effect and topological insulators.

Statistical Mechanics and Applications in Condensed Matter

CARLO DI CASTRO
Università degli Studi di Roma 'La Sapienza', Italy

ROBERTO RAIMONDI
Università Roma Tre, Italy

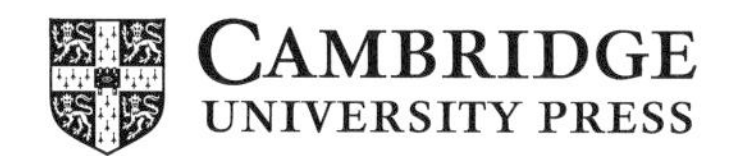

CAMBRIDGE
UNIVERSITY PRESS

University Printing House, Cambridge CB2 8BS, United Kingdom

Cambridge University Press is part of the University of Cambridge.

It furthers the University's mission by disseminating knowledge in the pursuit of education, learning and research at the highest international levels of excellence.

www.cambridge.org
Information on this title: www.cambridge.org/9781107039407

First published 2015

Printed in the United Kingdom by TJ International Ltd. Padstow Cornwall

A catalogue record for this publication is available from the British Library

ISBN 978-1-107-03940-7 Hardback

Contents

Preface

D. Pines in his Editor's Foreword to the important series "Frontiers in Physics, a Set of Lectures" of the sixties and seventies of the past century (W. A. Benjamin, Inc.) was suggesting as a possible solution to "the problem of communicating in a coherent fashion the recent developments in the most exciting and active fields of physics" what he called "an informal monograph to connote the fact that it represents an intermediate step between lecture notes and formal monographs."

Our aim in writing this book has been to provide a coherent presentation of different topics, emphasizing those concepts which underlie recent applications of statistical mechanics to condensed matter and many-body systems, both classical and quantum. Our goal has been indeed to reach an up to date version of the book *Statistical Mechanics. A Set of Lectures* by R. P. Feynman, one of the most important monographs of the series mentioned above. We felt, however, that it would have been impossible to give to a student the full flavor of the recent topics without putting them in the classical context as a continuous evolution. For this reason we introduced the basic concepts of thermodynamics and statistical mechanics. We have also concisely covered topics that typically can be found in advanced books on many-body theory, where usually the apparatus of quantum field theory is used.

In our book we have kept the technical apparatus at the level of the density matrix with the exception of the last four chapters. Up to Chapter 17 no particular prerequisite is needed except for standard courses in Classical and Quantum Mechanics. Chapter 18 provides an introduction to statistical quantum field theory, which is used in the last chapters. Chapters 20 and 21 cover topics which, although covered in recent monographs, are not commonly found in classical many-body books.

In our book then the student will find a bridge from thermodynamics and statistical mechanics towards advanced many-body theory and its applications. In our attempt to give a coherent account of several topics of condensed matter physics, we have at the same time preserved the personal point of view of the notes of our courses. Our bibliography is for this reason far from complete and the presentation of some topics is somewhat informal and partial. Many important contributions and fundamental references have been left out.

Some fundamentals topics of modern statistical mechanics for condensed matter physics have been left out, e.g. quantum criticality (except for the two examples of Luttinger liquid and Anderson localization), the quantum Hall effect and spin glasses. For them, however, there are recent devoted books.

Suggestions on the usage of the book

Although this book is conceived and organized as an organic self-contained whole, it can also be used as a modular text. We provide here a few suggestions for planning courses on the great amount of the covered material at undergraduate, graduate and postgraduate level. In this way we hope also to facilitate consultation by teachers and researchers.

1. **Introductory statistical mechanics [Chapters 1–4, 6–7, (5)]**[1]
   Chapters 1–4 and 6–7 are a self-contained introduction to basic statistical mechanics. Chapter 1 is a concise review of thermodynamics in order to put statistical mechanics in context. In particular we emphasize the concepts of thermodynamic stability with respect to deviations and fluctuations from average by means of the introduction of the thermodynamic function availability as the maximum work available. The aim is to alert the reader from the start of the statistical interpretation of the law that the entropy of an isolated system can never decrease. Fluctuations play an important role in phase transitions. Both these two issues are at the heart of the foundations of statistical mechanics. The logical sequence of the thermodynamic potentials and the related pairs of conjugate variables prepares the basis of the thermodynamics for ensembles. Chapters 2 and 3 cover the basic concepts of statistical mechanics from the fundamental insights by Boltzmann to the Gibbs formulation. Chapter 4 is about the consequences of the basic axioms and discusses the relation between the different statistical ensembles. Chapter 5, which is not included in the above list, concludes the exposition of classical statistical mechanics by presenting the physical and mathematical significance of the thermodynamic limit especially with reference to the problem of phase transitions. This chapter can be used as supplementary material at undergraduate level or included in a more advanced course at graduate level.
   Chapter 6 covers elementary quantum statistical mechanics. It first uses an heuristic point of view by showing how the ideas of classical kinetics can be modified to include the quantum nature of particles. Fermi and Bose statistics emerge then in a transparent way. In the rest of the chapter the basic description of a quantum system is introduced in terms of the density matrix, a tool used extensively throughout the book.
   Chapter 7 finally presents the derivation of the Bose and Fermi statistics from the Grand Canonical ensemble. The Fermi and Bose non-relativistic gases are discussed in detail. Given the current experimental relevance of the so-called cold atoms, presented in Chapter 15, we have included the discussion of Fermi and Bose gases in a confining harmonic potential. The important application of Bose statistics to photons and phonons is also included.
2. **Classical mean-field theory and critical phenomena [Chapters 8, 13, 14, 16, 17]**
   Chapters 8, 13, 14, 16, 17 offer a self-contained description of classical and modern theory of critical phenomena. Whereas in Chapter 8, two classical mean-field theories is discussed in detail (van der Waals and Curie–Weiss), in Chapter 13 we introduce the unifying description by Landau. The relevance of the fluctuations and of the symmetry properties in critical phenomena and in limiting the validity of the mean-field theories is discussed extensively. The more general Landau description is therefore presented in preparation for the scaling theory of critical phenomena, the use of renormalization group

[1] Chapters in round brackets contain supplemental topics which can be covered partially and/or optionally.

with the Gaussian approximation and the formulation of the Landau–Wilson model. The modern theory of critical phenomena develops through Chapters 14, 16 and 17. The presentation of the renormalization group approach in Chapter 17 strongly relies on physical concepts like the various aspects of the universality principle and even the discussion of the field-theoretic formulation does not require any technical knowledge of field theory. The present program is suitable for graduate students up to Chapter 17. This last chapter could be included in an advanced or postgraduate course.

A standard graduate one-year course on classical statistical mechanics and critical phenomena could be organized by assembling together modules 1 and 2.

3. **Mean-field theory for quantum systems and a unified presentation of superfluids and superconductors [Chapters 9, 12, 15, (8, 11, 21)]**
   Chapter 9 is about second quantization, reduced density matrices and the Hartree–Fock approximation. It could also be included together with the the classical mean-field theory of Chapter 8, in the first module as the simplest one-body approximation of a many-particle system. Chapter 9 introduces the following Chapters 12 and 15, where the reader becomes acquainted with the normal and superfluid phases of interacting Bose and Fermi systems, starting from the phenomenology and adopting the reduced-density-matrix approach as a unifying framework at the technical level and the quasiparticle gas at the conceptual level. The last one was the paradigmatic interpretation of the whole of condensed matter physics of the last century. All together this is a comprehensive course on superfluids and on the normal and superconductive phases of fermions for graduate and postgraduate students. For the last more advanced level, after the Fermi liquid theory of Chapter 12, one could include Chapter 11 on the transport in disordered systems and the first five sections of Chapter 21 on the Anderson localization.
4. **Dissipative phenomena in classical and quantum systems [Chapters 2, 6, 10, 11, (21)]**
   Chapters 2, 6, 10 and 11 may be used for a one-semester course on dissipative phenomena in classical and quantum systems. In particular, in Chapter 10 we present the theory of linear response for quantum systems and show how it reduces to the simpler classical case, already discussed in Chapter 8. The fluctuation–dissipation theorem is also generalized to non-equilibrium and its relation with the availability function and the maximum entropy production is given. Chapter 11 introduces Brownian and diffusive motion in the classical case. It also shows how these ideas can be adapted to the quantum case of a disordered Fermi gas, which plays an important role in the description of transport phenomena in condensed matter systems such as metals and semiconductors.

   It is also preparatory to the discussion of Anderson localization as a phase arising from the normal Fermi liquid system in the presence of disorder, discussed in the first two sections of Chapter 21, which could also be included in this module.
5. **Modern trends in quantum statistical mechanics [Chapters 6, 7, 9, 12, 15 (20, 11, 21)]**
   The last two modules (3 and 4) can be included in a larger one-year, graduate or postgraduate, course on quantum statistical mechanics and the density matrix (Chapter 6), quantum gases (Chapter 7), second quantization and the Hartree–Fock method

(Chapter 9), normal Fermi liquid (Chapter 12), superfluidity and superconductivity (Chapter 15), the Luttinger liquid (20 (the first three sections)) as an example of non-Fermi-liquid behavior, diffusive motion and the metal–insulator transition (Chapters 11 and 21 (the first three sections)).

6. **Advanced topics and techniques in quantum statistical mechanics [Chapters 10, 18, 19, 20, 21]**
   Chapters 10, 18, 19, 20 and 21 may be used, together with related appendices for an advanced course in quantum statistical mechanics. In particular, in Chapter 18 we provide a concise self-contained introduction to the thermal Green function and its diagrammatic associated perturbation expansion technique, which is the tool most used by researchers when applying statistical mechanics to quantum systems. As a first application, Chapter 19 shows how to set a microscopic foundation for the phenomenological theory of Fermi liquids of Chapter 12. Chapter 20 uses the Green function technique to study the renormalization group treatment of the one-dimensional Tomonoga–Luttinger model, which provides the first example of non-Fermi-liquid behavior. This model, introduced in the nineteen-fifties mostly as an interesting theoretical model, has found an increasingly widespread application in the description of several systems of experimental interest. Its theoretical interest was recently rekindled by the non-Fermi-liquid behavior of the normal phase of the high temperature superconductors discussed in Chapter 15. Finally, Chapter 21 deals with the problem of disordered interacting Fermi gases and liquids and the so-called Anderson metal–insulator transition and its generalization to include interaction among the electrons. The last two chapters describe also a non-trivial application of the renormalization group approach to quantum systems somehow differing from its use in critical phenomena. In the last case the symmetry properties of the system (gauge invariance) have been exploited to select the proper renormalizations to be carried out in this most complex problem. In the Luttinger one-dimensional model, singularities plague the calculation of the physical response functions in perturbation theory. Singularities cannot be present in the final measurable responses of a stable liquid phase and must disappear in the final finite answers instead of being resummed to a power behavior as in critical phenomena. The renormalization group together with the proper symmetries of this one-dimensional system keeps the singularities under control and provides the exact final answer.

We would like also to indicate a few general references, which we have found useful when preparing our lecture notes and writing the institutional part of each chapter. Below we group them in correspondence with chapters.

- Pippard (1957) for Chapter 1.
- Huang (1963); Thompson (1972); Landau and Lifshitz (1959) for Chapters 2–5, 8, 13.
- Feynman (1972) for Chapters 6, 15.
- Rammer (2007); Peliti (2011) for Chapters 10, 11.
- Nozières and Pines (1966); Nozières (1964) for Chapters 12, 19.
- Patashinskij and Pokrovskij (1979); Ma (1976); Chaikin and Lubensky (1995); Amit and Martín-Mayor (2005) for Chapters 13, 14, 16, 17.

- London (1959); Atkins (1959); Feynman (1972); De Gennes (1966); Tinkham (1975); Schrieffer (1999); Leggett (2006) for Chapter 15.
- Abrikosov et al. (1963); Fetter and Walecka (1971); Mahan (2000); Rammer (2007) for Chapters 18, 20, 21.

Acknowledgments

This volume is based on a series of lectures we have given along the decades at the undergraduate, graduate, postgraduate and postdoctoral level. We therefore first of all thank all our students who have been the primary actors of this work. We should like to express our gratitude to our colleagues G. Jona Lasinio, C. Castellani, W. Metzner, S. Caprara, L. Benfatto, P. Schwab, C. Gorini and G. Vignale for continuous stimulating discussions on the topics presented in this volume and in particular to M. Grilli and A. Varlamov who also read parts of it, suggesting important corrections, and to J. Lorenzana who tested one of the suggested modules providing us with an important feedback. One of us (R. R.) would also like to thank his present students J. Borge and D. Guerci for "testing" parts of the book during its development.

Last but not least, we are tremendously grateful to E. Gasbarro for a careful check of all the bibliographical records.

1 Thermodynamics: a brief overview

Thermodynamics is a phenomenological theory based on empirical observations. The thermodynamic laws are a coherent account of the experimental analysis and provide a universally valid description of the behavior of macroscopic systems without referring to their detailed structure. In this chapter we introduce thermodynamics as a prerequisite scenario for statistical mechanics and we do not give an extended exposition of the theory, but only recall the basic concepts as an introduction to the statistical approach.[1]

1.1 Equilibrium states and the empirical temperature

A thermodynamic system is any portion of matter, solid, liquid, mixtures, ..., which can be described in terms of a small number of parameters, the macroscopic thermodynamic variables, as for example the pressure P and the volume V for a fluid or the magnetic moment and the magnetic field for a paramagnet. These variables are called extensive or intensive, depending whether their value does or does not depend on the amount of matter present in the system. Energy, volume and magnetization are extensive variables, whereas pressure, temperature and magnetic field are intensive.

Thermodynamics deals with macroscopic properties of macro-systems at equilibrium. A thermodynamic equilibrium state must then be characterized as a macroscopic phenomenon defined via macroscopic variables. A system is in equilibrium if the values of the observables we take into consideration and of the parameters describing its state do not change with time. The thermodynamic equilibrium is never fully static. The characterization of a thermodynamic state depends on the observation time, usually at least of the order of fractions of a second, which in any case must always be much longer than any characteristic time scale of the molecular motion ($\sim 10^{-11}$ s).

We use a simple example of common experience to illustrate this point. Soon after hot water is poured into a container, it appears to be at rest. If we ask whether the water is in a state of equilibrium, the answer depends obviously on the observation time and on the sensitivity of the measurement. Within the interval of time of a few seconds we do not feel changes in the temperature and in the volume of the water, which during this observation time can be considered at equilibrium by itself. If we prolong the observation, we start to appreciate that the temperature of the water is changing and the water cannot be considered

[1] For a more detailed exposition we suggest the book by A. B. Pippard (1957).

at equilibrium. After several minutes the temperature of the water will be equal to the room temperature and will not change anymore. Even if the molecules of the water continually evaporate, within a few hours also the volume of the water does not change appreciably. The water at the same temperature of the room in this interval of time can be considered in equilibrium. In a few days, the water will eventually evaporate and the equilibrium state of the water in the container is meaningless. Unlike mechanical equilibrium, thermodynamic equilibrium is never static. A thermodynamic state does not refer, as in mechanics, to instantaneous positions and velocities of all the microscopic constituents, the molecules of the water in the previous example, which define a micro-state or configuration. All the observation times considered above are macroscopic time scales much longer than the microscopic time characteristic of molecular motion. Provided the observation time considered remains within certain limits, the observables that we consider to characterize the thermodynamic equilibrium do not change their values appreciably. During this time the system goes through a large number of microscopic configurations and the proper macroscopic observables of thermodynamic equilibrium can be considered as averages quantities in the complicated evolution of the motion of all molecules during this interval of time. A state of absolute thermodynamic equilibrium from which the system never departs cannot be considered, strictly speaking, as an experimental truth. Its occurrence, as we have seen, depends on the observation time and on the precision that we require or that we have in the experimental apparatus. In reality we can assume that for a given small set of external intensive parameters, such as temperature and pressure when the system is conditioned by the external world, or the volume and the energy for a system in complete isolation, a macroscopic system has definite equilibrium properties. With the caution discussed above, macroscopic violations of this working hypothesis, as we shall see later, would require observation times so long as to be considered infinite for any practical purpose.

If two systems are in equilibrium, their states and, therefore, their thermodynamic variables cannot be specified arbitrarily. Moreover, we can postulate the transitive property of equilibrium, zeroth law, that if two systems A and B are in equilibrium with a reference system C, then A is in equilibrium with B. For single component systems, as for instance fluids, whose states are specified by P and V,[2] their states must be conditioned in the sense that if two systems A and B are in equilibrium with another system C, specified by the two conditions

$$f_{\mathrm{AC}}(P_1, V_1, P_3, V_3) = 0, \quad f_{\mathrm{BC}}(P_2, V_2, P_3, V_3) = 0, \tag{1.1}$$

then A and B are in equilibrium with one another:

$$f_{\mathrm{AB}}(P_1, V_1, P_2, V_2) = 0. \tag{1.2}$$

Here (P_1, V_1), (P_2, V_2) and (P_3, V_3) refer to systems A, B and C, respectively. The functions f_{AC}, f_{BC} and f_{AB} depend on the nature of the systems A, B and C.

[2] It can be experimentally verified that, independently from the process by which the pressure and the volume of the gas are fixed, its properties (viscosity, thermal conductivity, ...) are the same.

The above conditions allow us to introduce the notion of the empirical temperature. By solving the two equations in (1.1) with respect to P_3 there must be a function such that

$$\theta_A(P_1, V_1, V_3) = \theta_B(P_2, V_2, V_3). \tag{1.3}$$

Since the relation (1.3) must have the same content of the thermal equilibrium condition (1.2) between the two systems A and B, the functional dependence on V_3 of θ_A and θ_B must be the same and can then be dropped by fixing the volume V_3 of the system C assumed as a reference system. The resulting function $\theta(P, V)$ defines, then, the empirical temperature and its form depends on the chosen reference system. This argument can be generalized to any macrosystem and the empirical temperature θ will depend on all the parameters necessary to characterize the state of the system.

Once the above definition is assumed, the equilibrium implies a condition among the variables specifying a thermodynamic state and the expression of any one of them in terms of the other variables defines the equation of state. For a perfect gas the equation of state is given by the empirical observation for a very dilute gas, which can be summarized by Boyle's law, according to which the product $PV = f(\theta)$. Using a mercury-in-glass thermometer, the function $f(\theta)$ is linear over a wide range of temperatures. The coefficient of proportionality, per mole of the gas, is identified with the gas constant R once the unity of the scale coincides with the degree Celsius equal to the Kelvin (per one mole $R =$ 8.3145 J/molK). The above temperature, for which from now on we use the standard notation T, also coincides with the absolute temperature T to be defined later according to the second law.

The condition of mechanical and thermal equilibrium requires that the system pressure, P, and temperature, T, coincide with those of the environment. In the case of a multi-component system, in addition to the standard thermodynamic variables, the specification of the relative concentration of each component is also required. Furthermore a system may or may not be homogeneous. If a system is homogeneous it is said to consist of a single phase. Hence the state of a system is given, in general, by specifying, besides P and T, the number of phases f and components n. However, not all of these are independent parameters. The actual number of degrees of freedom of a thermodynamic system, i.e., the number υ of independent parameters, is given by the Gibbs rule (which will be proved later on)

$$\upsilon = 2 + n - f. \tag{1.4}$$

For instance, an ideal gas is a single-component single-phase system, and its state is entirely determined by two variables, pressure and temperature.

Depending on the value of the thermodynamic parameters, a single-component system may be in different phases. The information about the possible phases of a system is graphically represented in a phase diagram, which is a Cartesian diagram whose coordinates correspond to any two of the thermodynamic parameters P, T and V. A generic thermodynamic state corresponds to a point in the phase diagram. According to Gibbs' rule, the coexistence of two phases, for a single-component system, implies a number of degrees of freedom, $\upsilon = 1$, and hence different phases may coexist only along curves in the phase

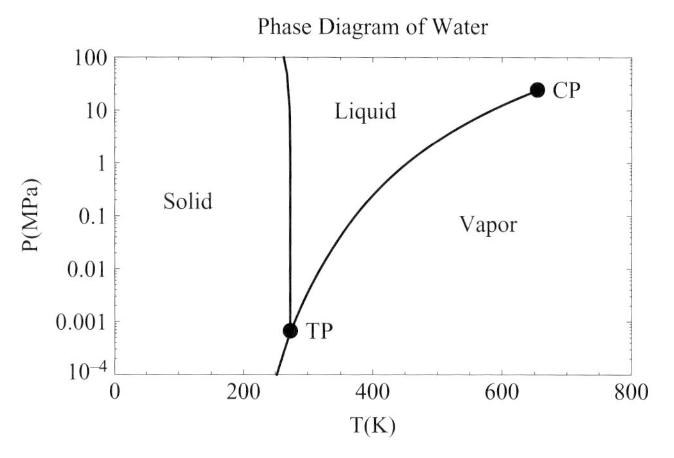

Fig. 1.1 Phase diagram of a one-component system (water) in the $T-P$ plane. The critical point (CP) is the end point of the liquid–vapor coexistence curve $T_c = 647.096$ K, $P_c = 27.064$ MPa. The triple point (TP) is the coexistence of three phases $T_{tr} = 273.16$ K, $P_{tr} = 6.11 \times 10^2$ Pa.

diagram. These curves are the phase boundaries. For the coexistence of three phases, $v = 0$, which corresponds to the triple point of the phase diagram. The phase diagram of water is shown in Fig. 1.1 as an example.

Within thermodynamics we consider also processes during which the state of a system changes. If a system is in equilibrium, its macroscopic state cannot change unless the external constraints are modified. If the external constraints are suddenly modified, the system will find itself, in general, in a non-equilibrium state. After some time, the system will reach a new equilibrium state corresponding to the modified constraints. A process from a non-equilibrium state to an equilibrium one is called *natural* or *irreversible*. To make our discussion more precise, let us suppose that τ_{rel} is the time over which the system relaxes to a new equilibrium state and τ_{con} the time over which the value of the external constraints is varied. If $\tau_{con} \ll \tau_{rel}$, the system will find itself, suddenly, in a non-equilibrium state. On the contrary, when $\tau_{con} \gg \tau_{rel}$ the system readjusts almost instantaneously with respect to the modification of the external constraints. In this sense the system passes from an equilibrium state to another one. Suppose that we change the external constraints in several steps and at each step we change them by a small amount such that the system is always arbitrarily close to an equilibrium state. The succession of these steps yields a *quasi-static* process. This process is called *reversible*, if by reversing back to the old values of the external constraints, the system returns to the original equilibrium state without any modification of the external environment. An isolated system (fully constrained by isolating walls which forbid any contact with any other system) eventually will reach equilibrium. A *thermally* isolated system is enclosed within adiabatic walls, which are defined by the following property. Any change of the system may only occur by a displacement of the walls or by forces acting at distance.

1.2 The principles of thermodynamics

First principle and the internal energy

When a system which is thermally isolated changes its thermodynamic state, we say that the transformation is *adiabatic* and we denote by W the work exchanged between the system and the environment during the transformation. By convention, we define $W > 0$ when the system makes work on the environment. We can state in general that for each thermally isolated system there is a state function U named internal energy such that exchanged work corresponds to the variation of the energy U of the system in going from an initial to a final state and does not depend on the intermediate steps of the transformation. For an adiabatic transformation we have

$$\Delta U = -W, \tag{1.5}$$

where ΔU represents the variation of internal energy between the final and the initial state.

In general the energy U of a system can be changed by external factors which can perform work on the system, e.g. by compressing a fluid. In analogy with the work done by a force f to move an object a distance $\mathrm{d}x$, the generalized displacements are introduced as η_i modifications of the internal parameters of the system to adjust to the variation of the external factors. This work for an infinitesimal quasi-static change can then be expressed by the product of the generalized force $f_i = \partial U/\partial \eta_i$ times the generalized displacement $\mathrm{d}\eta_i$. For instance, on a macroscopic fluid a compression by a pressure P determines a volume contraction $\mathrm{d}V$ with a work $\mathrm{d}W = P\mathrm{d}V$, where P assumes the role of generalized force and V is the displacement η. Similarly, for a magnetic body the magnetic field B and the magnetic moment variation $\mathrm{d}M$ along the direction of B determine an increase of energy with a "work" $\mathrm{d}W = B\mathrm{d}M$. Another example consists of the energy variation $\mathrm{d}W = \mu\mathrm{d}N$ necessary to change the number N of particles of a macroscopic system where the chemical potential μ is the "force" which produces the change in N.[3] The variables representing the external conditions of the system, like the pressure, the magnetic field and the chemical potential are intensive. Their effect is on extensive variables, like the volume, the magnetic moment and the number of particles characterizing the macroscopic state of the system. The pairs of variables specified by the generalized forces and the generalized displacements of the system define the conjugate variables with respect to the energy U.

For a non-thermally isolated system undergoing a transformation, the equality (1.5) is not valid. The thermodynamic state and the energy of a system can also be changed without any mechanical work, for instance by radiation, by lighting a fire underneath or by running a current in a resistor immersed in it. The energy flow into and out of the system is called heat, Q. By convention, we define $Q > 0$ when the system receives heat from the environment.

[3] This can be taken as the definition of the chemical potential which will be clarified later.

In general the first principle of thermodynamic states that *in any thermodynamic process the balance of exchanged work and heat equals the variation of the internal energy* U, in the formula

$$\Delta U = Q - W, \tag{1.6}$$

where ΔU is fully determined by the parameters specifying the initial and final states. The first principle may be expressed in differential form for infinitesimal changes in the thermodynamic parameters by

$$\mathrm{d}U = \delta Q - \delta W, \tag{1.7}$$

where the symbol δ implies that the corresponding quantities δQ and δW are *not* exact differentials, in contrast to $\mathrm{d}U$. The first principle is then a formulation of the conservation of energy, generalized to include thermal exchange. For a one-component fluid system with $\delta W = P\mathrm{d}V$ an adiabatic process is such that

$$\mathrm{d}U + P\mathrm{d}V = 0 \quad \text{or} \quad \left[P + \left(\frac{\partial U}{\partial V}\right)_P\right]\mathrm{d}V + \left(\frac{\partial U}{\partial P}\right)_V \mathrm{d}P = 0, \tag{1.8}$$

while an isothermic process is determined by

$$\mathrm{d}T = \left(\frac{\partial T}{\partial V}\right)_T \mathrm{d}V + \left(\frac{\partial T}{\partial P}\right)_V \mathrm{d}P = 0 \quad \text{or} \quad \left(\frac{\partial P}{\partial V}\right)_T = -\left(\frac{\partial T}{\partial V}\right)_P \Big/ \left(\frac{\partial T}{\partial P}\right)_V . \tag{1.9}$$

As is easy to verify for a perfect gas, in the V–P plane the curves describing the isothermic and adiabatic processes are decreasing functions of V with the slope in the adiabatic case being larger. The proof is provided in the answer to Problem 1.1.

The first principle distinguishes between the two different forms of energy which can be exchanged by a system with the environment. The addition of heat to a body is not due to the application of a generalized force but by conduction, convection or radiation. In all these cases the body changes its thermodynamic state as a result of a change of motion at the microscopic level. In thermodynamics this change must be characterized by an appropriate pair of global conjugate variables, the intensive variable temperature and the corresponding extensive entropy, which will be defined next by means of the second principle.

Second principle and the entropy

Before introducing the second principle of thermodynamic we must discuss the concept of hotness and how it is related to the temperature. When two systems being out of thermal equilibrium (at different empirical temperatures) are put in contact with one another, we define as the hotter system the one which loses heat to reach equilibrium and the same final empirical temperature together with the other system. By the zeroth law, we can unequivocally define a correspondence between hotness and empirical temperature: if we say that the system at higher empirical temperature is hotter, then it will be hotter than all the other systems at the same lower temperature.

The second principle essentially establishes the empirical observation that there are two different forms of energy transmission and that, via the state function entropy, the heat cannot be transformed into work without losses allowing for a thermodynamic characterization of irreversibility. It can be stated as follows:

no process is possible, the sole result of which is that heat is transferred from a body to a hotter one (Clausius's formulation)

or

no process is possible, the sole result of which is that a body is cooled and work is done (Kelvin's formulation)

or

in any neighborhood of any thermodynamic state there are states that cannot be reached from it by an adiabatic process (Carathéodory's formulation).[4]

The logical sequence that leads to the thermodynamic characterization of the irreversibility is first the introduction of the absolute temperature T, through the universal efficiency of a Carnot reversible cycle; then the existence of a state function $S(T, V)$ called entropy, via the Clausius theorem. This theorem states that for an arbitrary reversible cycle C_R

$$\oint_{C_R} \frac{\delta Q}{T} = 0, \tag{1.10}$$

from which the definition of the variation of the entropy function follows:

$$S(V_2, T_2) - S(V_1, T_1) = \int_{\Gamma_{rev}} \frac{\delta Q}{T}, \tag{1.11}$$

where Γ_{rev} represents a reversible process from the state V_1, T_1 to V_2, T_2. The difference (1.11) is independent of the path connecting the initial and final states; the entropy is therefore a state function completely determined for each state once its value is fixed for one state. Finally, the function $S(V, T)$ quantifies the irreversibility by stating that for an infinitesimal generic transformation

$$\frac{\delta Q}{T} \leq dS. \tag{1.12}$$

In the above the (in)equality applies to (ir)reversible processes. Clearly the entropy cannot diminish for a thermally isolated system for which $\delta Q = 0$.

We now follow more closely the logical sequence sketched above. The Carnot reversible cycle starts from a state (P_1, V_1) and returns to it via an isothermal expansion from (P_1, V_1) to (P_2, V_2) (the empirical temperature θ_1, with the old notation for the time being, is maintained fixed by absorbing the heat Q_1 from a heat bath or reservoir),[5] followed by an adiabatic expansion from (P_2, V_2) to (P_3, V_3), an isothermal compression from (P_3, V_3) to (P_4, V_4) (the empirical temperature θ_2 is maintained fixed by giving the heat Q_2 to a heat bath), and finally an adiabatic compression back to (P_1, V_1). The balance of work on the basis of the first principle and the efficiency η defined as $\eta = W/Q_1$ are

$$W = \oint P dV = Q_1 - Q_2, \quad \eta = 1 - \frac{Q_2}{Q_1}. \tag{1.13}$$

[4] For a discussion of this less common formulation, one may consult the book by Zemansky (1968).

[5] A heat bath is a sufficiently large system capable of absorbing and yielding a finite amount of heat without changing its natural temperature.

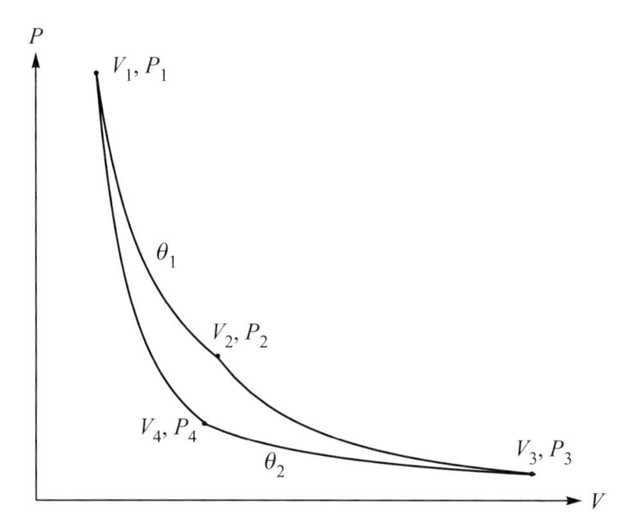

Fig. 1.2 Carnot cycle in the $V-P$ plane. The isothermic expansion (compression) at temperature $\theta_1(\theta_2)$ goes from the state $V_1(V_3), P_1(P_3)$ to the state $V_2(V_4), P_2(P_4)$. The adiabatic expansion (compression) goes from the state $V_2(V_4), P_2(P_4)$ to the state $V_3(V_1), P_3(P_1)$.

Consider now two systems R and R' undergoing two Carnot cycles between the temperatures θ_1 and θ_2. If for simplicity we choose $Q'_1 = Q_1$, then $Q'_2 = Q_2$ or $W' = W$. If it were not so, by combining the Carnot cycle for R' with the reverse one for R, one would violate the Kelvin statement of the second principle. See the answer to Problem 1.2 for the proof. The heat ratio must be an universal function solely depending on θ_1 and θ_2, $Q_2/Q_1 = f(\theta_1, \theta_2)$. Moreover, by considering two cycles working between θ_1 and θ_2 and between θ_2 and θ_3 and the combined cycle working between θ_1 and θ_3, one has

$$Q_3/Q_1 = f(\theta_1, \theta_3) = f(\theta_1, \theta_2) f(\theta_2, \theta_3).$$

From the last equation one sees that the function f factorizes:

$$f(\theta_1, \theta_2) = \frac{\phi(\theta_2)}{\phi(\theta_1)}.$$

The absolute temperature T is then defined as the universal function of the empirical temperature $T = \alpha\phi(\theta)$. The efficiency η is then

$$\eta = 1 - \frac{Q_2}{Q_1} = 1 - \frac{T_2}{T_1}$$

and therefore for a Carnot cycle

$$\frac{Q_1}{T_1} + \frac{-Q_2}{T_2} = 0,$$

for any reversible closed cycle the Clausius theorem (1.10) holds, as can be proved by covering the cycle by a net of Carnot cycles and going to the limit of finer and finer networks. The entropy function (1.11) is then defined. The scale α can be determined by the equation of state of a perfect gas.

For any cycle acting between the temperatures T_1 and T_2, to avoid contradiction with the second principle the Clausius theorem is modified into

$$\frac{Q_1}{T_1} + \frac{-Q_2}{T_2} \leq 0,$$

where the equality holds for reversible cycles. In general, the corresponding algebraic sum for all the transformations occurring in a cyclical process can only be less than zero, leading to the inequality for the Clausius theorem:

$$\oint_C \frac{\delta Q}{T} \leq 0$$

and to (1.12) in differential form for an infinitesimal change.

We remark that the existence of the entropy as state function is related to the fact that there exists an integrating factor for the infinitesimal heat exchange δQ and that such an integrating factor is the inverse absolute temperature. By considering T and V as independent variables, from the first principle one has

$$\delta Q = \left(\frac{\partial U}{\partial T}\right)_V \mathrm{d}T + \left[\left(\frac{\partial U}{\partial V}\right)_T + P\right] \mathrm{d}V, \tag{1.14}$$

and the condition for δQ being an exact differential is

$$\frac{\partial^2 U}{\partial V \partial T} = \frac{\partial}{\partial T}\left[\left(\frac{\partial U}{\partial V}\right)_T + P\right]. \tag{1.15}$$

By considering, for instance, the perfect gas ($U = C_V T$ with the heat capacity at constant volume C_V constant, and $PV = nRT$, n being the number of moles and R the gas constant), one soon sees that the above equation cannot be satisfied. On the other hand, by considering $\delta Q/T$, the condition for being an exact differential becomes

$$\frac{\partial}{\partial V}\frac{C_V}{T} = \frac{\partial}{\partial T}\left[\frac{1}{T}\left(\frac{\partial U}{\partial V}\right)_T + \frac{P}{T}\right], \tag{1.16}$$

and is obviously satisfied.

The definition of the state function entropy is the missing quantity to complete the set of thermodynamic variables arranged in pairs of conjugated variables. For a reversible transformation the entropy appears as the generalized displacement variable whose conjugate generalized force is the temperature.

The first principle establishes how a process can take place without violating energy conservation. The second principle fixes the direction of these processes. For example if two bodies at different temperatures are put in thermal contact the energy flows from the hotter to the colder and eventually the final temperature is the same. Generally speaking, however, we could imagine that the energy flows in the opposite direction as long as the total energy is conserved in agreement with the first principle. We know that such a process which would correspond to a diminishing of the entropy of the total isolated system never occurs. The second principle may be considered to be empirical and up to now there have been no known violations of it.

Third principle and the unattainability of the absolute zero

The third principle of thermodynamics, also known as Nernst's principle, is about the behavior of the entropy when the temperature decreases towards absolute zero. As for the second principle, there exist different formulations, starting from the one originally put forward by Nernst (1906), all logically equivalent.

The simplest formulation, which has a direct interpretation within the statistical mechanics, as we shall see later (see the discussion in Section 4.2), is the following.

The entropy of a system at the absolute zero of temperature is a universal constant, which can be set to zero.

In mathematical terms one has

$$\lim_{T\to 0} S(T) = 0. \tag{1.17}$$

Alternatively, the third principle may be stated as *the impossibility of attaining the absolute zero of temperature in a finite number of reversible thermodynamic transformations*.

Finally, the original formulation of Nernst is usually enunciated in the Nernst–Simon form (see the book by Zemansky (1968)): *the entropy variation of any isothermic reversible transformation of a condensed system (liquid or solid) tends to zero when the temperature decreases indefinitely*. This can be mathematically expressed by

$$\lim_{T\to 0} (S(T, X_\mathrm{f}) - S(T, X_\mathrm{i})) = 0. \tag{1.18}$$

In the above X indicates (besides the temperature) any other thermodynamic parameter (such as pressure or magnetic field), which can be varied reversibly during an isothermic transformation. Clearly, if one assumes the first formulation of Eq. (1.17), there can be no dependence on the value of X, and Eq. (1.18) follows. The logical connection of the first and third formulations with the impossibility of attaining absolute zero in a finite number of transformations requires the mathematical development of the consequences of the first two principles. This development will be carried out in the next two sections and the equivalence between the unattainability formulation and Eq. (1.18) is shown in Problem 1.4. We will not then embark here in a detailed discussion of the equivalence of the various formulations and, for the time being, we limit ourselves to a graphical illustration following Zemansky (1968). To this end we refer to Fig. 1.3, where the behavior of the entropy as a function of temperature is shown for two different values of a parameter X. Both curves go to zero when $T \to 0$ according to Eq. (1.17). We can now imagine performing an isothermic reversible transformation (see thin arrow (1) in Fig. 1.3) starting on a point of the upper curve corresponding to $X = X_\mathrm{i}$ and letting X vary from X_i to X_f. Then, we perform an isoentropic transformation during which the parameter X is reversibly changed back to the original value X_i (see thin arrow (2) in Fig. 1.3). The net result of the two transformations is a decrease of the temperature of the system. In experimental practice, when the system is a magnetic salt and the parameter X is the external magnetic field, such a sequence of transformations is called adiabatic demagnetization. Clearly, one can iterate the two-step transformation (see thin arrows (3) and (4)) and reach a further lower temperature. However, the amount of decrease of temperature in this second iteration is lower than in

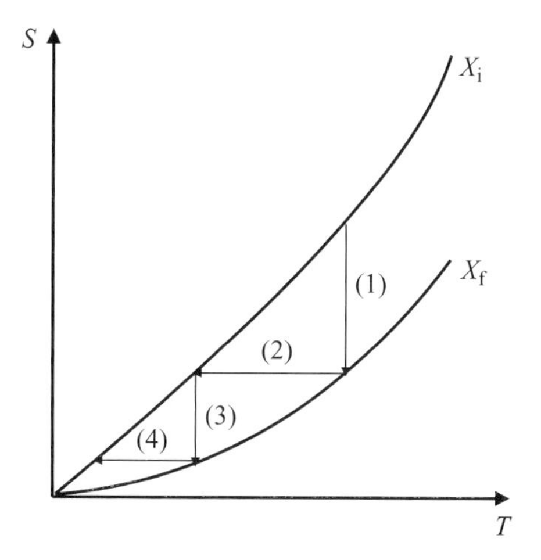

Fig. 1.3 The entropy as function of temperature for two different values of a parameter X. Vertical thin arrows ((1) and (3)) indicate isothermic reversible transformations obtained by changing the parameter X from X_i to X_f. Horizonal thin arrows ((2) and (4)) indicate isoentropic transformations. The temperature decrease in the transformation made up of steps (1) and (2) is smaller than the decrease of the transformation with steps (3) and (4), the change in the parameter X being the same.

the first because of the evident reason that both curves must go to zero, despite the fact that the change in the parameter X is the same. This means that a third iteration, although still lowering the temperature, will produce an even smaller decrease of the temperature. As long as the number of iterations or transformations remains finite, it is impossible to reach absolute zero. We notice that this would not be the case if the two entropy curves had a different zero-temperature limit.

1.3 Thermodynamic potentials and equilibrium

The thermodynamic potentials are special thermodynamic state functions that satisfy the following properties: (i) their variations are given by exact differentials; (ii) they are homogeneous among each other with dimensions of energy; (iii) their derivatives define pairs of conjugate thermodynamic variables; (iv) they are extensive quantities; and (v) if the system is homogeneous, the thermodynamic potentials are additive quantities. The last property specifies the general property that a thermodynamic description of a system and of a macroscopic part of it are identical.

For a reversible and infinitesimal quasi-static process the first two thermodynamic principles imply

$$dU = T dS - P dV. \tag{1.19}$$

Because U is a state function, Eq. (1.19) implies the definition of the two pairs of conjugate variables

$$T = \left(\frac{\partial U}{\partial S}\right)_V, \quad P = -\left(\frac{\partial U}{\partial V}\right)_T. \tag{1.20}$$

The entropy and the temperature are conjugate variables, the first being the generalized displacement and the second the generalized force.

A number of useful relations among partial derivatives, named the Maxwell relations, exist. The first Maxwell relation

$$\left(\frac{\partial T}{\partial V}\right)_S = -\left(\frac{\partial P}{\partial S}\right)_V \tag{1.21}$$

follows from the equality of the mixed second derivatives.

Equation (1.20) leads to the natural choice of $U = U(S, V)$. By solving $U(S, V)$ with respect to S, S can be expressed in terms of U and V with

$$\left(\frac{\partial S}{\partial U}\right)_V = \frac{1}{T}, \quad \left(\frac{\partial S}{\partial V}\right)_U = \frac{P}{T}. \tag{1.22}$$

For an irreversible process from Eq. (1.12) one has

$$\mathrm{d}U < T\mathrm{d}S - P\mathrm{d}V. \tag{1.23}$$

For a system undergoing a process while being thermally ($\mathrm{d}S = 0$) and mechanically ($\mathrm{d}V = 0$) isolated, the approach to equilibrium implies a decrease of the internal energy, which then must be minimum at equilibrium. This fact is well known in mechanics, where the equilibrium conditions correspond to the minima of the potential. Due to this analogy, the internal energy U is a first example of a thermodynamic potential. The use of U as a thermodynamic potential comes from the choice of S and V as independent variables. Suppose we want to use T and V as variables, i.e., we consider a system in contact with a reservoir at a temperature T and at constant volume V. By using $\mathrm{d}(TS) = S\mathrm{d}T + T\mathrm{d}S$, we get

$$\mathrm{d}U \leq \mathrm{d}(TS) - S\mathrm{d}T - P\mathrm{d}V$$

from which we derive the existence of another state function F such that

$$\mathrm{d}(U - TS) \equiv \mathrm{d}F \leq -S\mathrm{d}T - P\mathrm{d}V. \tag{1.24}$$

The state function F is called free energy or Helmholtz free energy which is minimum at T and V constant in the equilibrium state. By the same reasoning as for the internal energy, one gets again a pair of conjugate variables and the second of the Maxwell relations

$$S = -\left(\frac{\partial F}{\partial T}\right)_V, \quad P = -\left(\frac{\partial F}{\partial V}\right)_T, \quad \left(\frac{\partial S}{\partial V}\right)_T = \left(\frac{\partial P}{\partial T}\right)_V. \tag{1.25}$$

This leads to the natural choice of the independent variables T and V. Only in this case the thermodynamic potential defines the conjugacy relation, e.g.

$$S \neq \left(\frac{\partial F}{\partial T}\right)_P.$$

At constant T and V, $F(T, V)$ is minimum at equilibrium. By using instead S and P as independent variables, we get, by means of similar steps, the enthalpy state function H, for which

$$\mathrm{d}(U + PV) \equiv \mathrm{d}H \leq T\mathrm{d}S + V\mathrm{d}P, \tag{1.26}$$

from which the third pair of conjugacy relations and a third Maxwell relation follow:

$$T = \left(\frac{\partial H}{\partial S}\right)_P, \quad V = \left(\frac{\partial H}{\partial P}\right)_S, \quad \left(\frac{\partial T}{\partial P}\right)_S = \left(\frac{\partial V}{\partial S}\right)_P. \tag{1.27}$$

For S and P constant the enthalpy is minimum at equilibrium. Finally, by adopting T and P as independent variables, one obtains the Gibbs free energy G for which

$$\mathrm{d}(U - TS + PV) \equiv \mathrm{d}G \leq -S\mathrm{d}T + V\mathrm{d}P, \tag{1.28}$$

together with the fourth set of relations

$$S = -\left(\frac{\partial G}{\partial T}\right)_P, \quad V = \left(\frac{\partial G}{\partial P}\right)_T, \quad \left(\frac{\partial S}{\partial P}\right)_T = -\left(\frac{\partial V}{\partial T}\right)_P, \tag{1.29}$$

where G is minimum at T and P constant.

Until now we have considered the number N of the particles in the system to be fixed. We now consider N as an independent variable together with S and V, T and V, or T and P. The conjugate variable of N, the *chemical potential* μ, is

$$\mu = \left(\frac{\partial U}{\partial N}\right)_{S,V} = \left(\frac{\partial F}{\partial N}\right)_{T,V} = \left(\frac{\partial G}{\partial N}\right)_{T,P} = \left(\frac{\partial H}{\partial N}\right)_{S,P} = -T\left(\frac{\partial S}{\partial N}\right)_{U,V}. \tag{1.30}$$

We first generalize Eq. (1.19) as $\mathrm{d}U = T\mathrm{d}S - P\mathrm{d}V + \mu\mathrm{d}N$, thus $\mu = (\partial U/\partial N)_{S,V}$, then the other relations follow. Let us use as variables T, V and μ. This leads us, in the usual way, to the introduction of the state function grand potential Ω defined by

$$\Omega = U - TS - \mu N = F - \mu N, \tag{1.31}$$

with

$$\mathrm{d}\Omega \leq -S\mathrm{d}T - P\mathrm{d}V - N\mathrm{d}\mu \tag{1.32}$$

and

$$S = -\left(\frac{\partial \Omega}{\partial T}\right)_{V,\mu}, \quad P = -\left(\frac{\partial \Omega}{\partial V}\right)_{T,\mu}, \quad N = -\left(\frac{\partial \Omega}{\partial \mu}\right)_{T,V}. \tag{1.33}$$

We observe that, for the Gibbs free energy, since T and P are intensive variables, we must have

$$G(T, P, N) = Ng(T, P),$$

since G is extensive and $g(T, P)$ is the Gibbs free energy per particle. Then it follows that the chemical potential is nothing but the Gibbs free energy per particle

$$\mu = g(T, P), \tag{1.34}$$

and the grand potential is then

$$\Omega = -PV. \tag{1.35}$$

The thermodynamic description of any system can be equivalently given in terms of any one of the above thermodynamic potentials provided they are expressed as a function of their natural variables and each one can be obtained from any other one by a simple Legendre transformation.

In general, given a function of two variables $f = f(x, y)$ with

$$\mathrm{d}f = \left(\frac{\partial f}{\partial x}\right)_y \mathrm{d}x + \left(\frac{\partial f}{\partial y}\right)_x \mathrm{d}y \equiv u\mathrm{d}x + v\mathrm{d}y,$$

$h = h(u, y)$, the Legendre transform of f with respect to the variable $x \to u$, is

$$h(u, y) = f(x(u, y), y) - x(u, y)u,$$

where $x = x(u, y)$ is defined by inverting $u = (\partial f/\partial x)_y$. Then it follows that

$$\left(\frac{\partial h}{\partial u}\right)_y = \left(\frac{\partial f}{\partial x}\right)_y \left(\frac{\partial x}{\partial u}\right)_y - x - u\left(\frac{\partial x}{\partial u}\right)_y = -x; \quad v = \left(\frac{\partial h}{\partial y}\right)_u.$$

All the thermodynamic potentials are equivalent and are obtained from each other. For example, given $F(T, V)$, $U(S, V)$ is obtained as

$$U(S, V) = F(T(S, V), V) + T(S, V)S = -T^2\left(\frac{\partial}{\partial T}\left(\frac{F}{T}\right)\right)_V, \tag{1.36}$$

which is the Gibbs–Helmholtz equation. We stress that if F is not expressed in terms of the natural variables T and V, U will not be directly obtained via the Gibbs–Helmholtz equation. Among the equivalent potentials the choice of the thermodynamic potential to use in the various physical situations depends on the variables conditioning the macro-state of the system.

The extensivity property of the thermodynamic potentials implies their homogeneity, for instance

$$U(S, V, N) = Nu\left(\frac{S}{N}, \frac{V}{N}\right) \quad \text{or} \quad U(\lambda S, \lambda V, \lambda N) = \lambda U(S, V, N).$$

Then the Euler equation holds:

$$U(S, V, N) = \left(\frac{\partial U}{\partial S}\right)_{V,N} S + \left(\frac{\partial U}{\partial V}\right)_{S,N} V + \left(\frac{\partial U}{\partial N}\right)_{S,V} N = TS - PV + \mu N.$$

The Legendre transform of $U(S, V, N)$ in all its extensive variables vanishes.

1.4 Thermodynamic stability and availability function

At first sight it may seem impossible to discuss the problem of the fluctuations from equilibrium within a thermodynamic context. As we have said at the outset, a thermodynamic state is the equilibrium macroscopic state compatible with the external constraints. A change of the constraints leaves the system in a non-equilibrium state with respect to the constraints. For instance, consider a box divided into two parts by a wall. Suppose that only one of the two parts is filled with gas. A sudden removal of the wall does not, at least for some initial period of time, change the fact that the gas occupies only half the volume of the box. Hence, the state of the gas is characterized by a value of the volume which is not the one required by the external constraints. Clearly, as time goes by, the gas will, eventually, fill all the available volume, reaching a new equilibrium state with uniform density over the larger volume. There is no reason, at least from the point of view of physical possibility, that the gas at some later time will remain confined again in one half of the box. Such a situation is a giant deviation from equilibrium and experience tells us that such an occurrence has practically no chance of happening spontaneously. Nevertheless, minute density deviations, and occasionally even large deviations, will occur spontaneously as fluctuations. It may be useful to investigate whether it is possible to quantify the cost of a deviation and therefore of a fluctuation from equilibrium. After all, such a fluctuation is the opposite process of the approach to equilibrium, which has been described in terms of the thermodynamic potentials. Thermodynamic arguments, however, can be applied only to equilibrium states. We can consider a situation of fluctuation as an equilibrium state for a system with an additional constraint, as for instance the gas confined in half the volume of the box. The entropy of a gas at equilibrium is associated with all configurations compatible with those constraints, e.g., in the previous example, the entropy increases as soon as the separating wall is taken off. Indeed, the second principle of thermodynamics can be formulated in the sense that the constraints of an isolated system cannot be varied with a simultaneous decrease of entropy.

Small deviations from equilibrium and related small fluctuations of a system can be studied via the introduction of a particular thermodynamic function named availability. We consider the possibility that a small portion of an isolated system finds itself in a state where the temperature T and the pressure P may be different from T_0 and P_0 of the remaining part at an equilibrium state of a constrained system. Both parts are macroscopic (sufficiently large to define thermodynamic variables), but the second one is assumed to be much larger than the first. The small portion plays the role of the system in the following discussion, while the remaining part acts like a reservoir. In this way we may consider the situation that locally the system is not in equilibrium with the rest or has a fluctuation from the global equilibrium. The system and the reservoir may exchange heat and work. We assume that such heat and work exchanges are infinitesimal when referred to the reservoir and do not affect its temperature and pressure. The system and the reservoir considered together are an isolated system and the irreversible process which leads to reciprocal equilibrium, after eliminating the constraints of the system, implies an increase of the total entropy ΔS_{tot} of the reservoir and the system:

$$\Delta S_{\mathrm{tot}} = \Delta S_0 + \Delta S > 0. \tag{1.37}$$

We indicate with ΔQ the heat absorbed or given by the system. Since, by assumption, such an amount of heat, while affecting the state of the system, only represents an infinitesimal change for the reservoir, the entropy change of the latter is $\Delta S_0 = -\Delta Q/T_0$. For the total entropy change we then have

$$\Delta S_{\text{tot}} = -\frac{\Delta Q}{T_0} + \Delta S. \tag{1.38}$$

By using the first principle relative to the system

$$\Delta Q = \Delta U + P_0 \Delta V, \tag{1.39}$$

where ΔU is the variation of the internal energy and $P_0 \Delta V$ represents the work done by the system against the pressure of the reservoir. By combining Eqs. (1.37)–(1.39) we get

$$0 < \Delta S_{\text{tot}} = -\frac{1}{T_0}(\Delta U - T_0 \Delta S + P_0 \Delta V) \equiv -\frac{1}{T_0}\Delta A, \tag{1.40}$$

where the thermodynamic function *availability* A expressed in terms of the mixed variables of the system and the reservoir has been introduced.

At equilibrium A coincides with the Gibbs free energy and gets its minimum value when $T = T_0$ and $P = P_0$ since to first order $\Delta U = T\Delta S - P\Delta V$ and therefore for small deviations

$$\Delta A = (T - T_0)\Delta S - (P - P_0)\Delta V \leq 0. \tag{1.41}$$

The two terms of ΔA represent the thermal and mechanical cost of bringing the system out of equilibrium. The decrease of A is therefore the maximum work *available* in the process of reaching equilibrium. This is also the minimum work necessary to trigger the process.

The minimum value of A is the most general statement of the condition for thermodynamic equilibrium. Indeed, for an isolated system ($\Delta Q = 0$) $\Delta A = -T_0 \Delta S \leq 0$ and the entropy is maximum; when T and V are constant, $\Delta A = \Delta F \leq 0$ and the Helmholtz free energy is minimum; finally, when T and P are constant, A coincides with G.

Having established the conditions for equilibrium, we investigate the thermodynamic stability. Starting from equilibrium, the variation of A, for a non-equilibrium state, must be positive:

$$-\Delta S_{\text{tot}} = \frac{1}{T_0}\Delta A = \frac{1}{T_0}\Delta U^{(2)} > 0, \tag{1.42}$$

where $\Delta U^{(2)} > 0$ is the second order expansion of U in terms of the variations of the variables S and V,

$$\Delta U^{(2)}(V, S) = \frac{1}{2}\left[\left(\frac{\partial^2 U}{\partial V^2}\right)_S \Delta V^2 + \left(\frac{\partial^2 U}{\partial S^2}\right)_V \Delta S^2 + 2\left(\frac{\partial^2 U}{\partial V \partial S}\right)\Delta V \Delta S\right], \tag{1.43}$$

where all the derivatives are taken at equilibrium. The expansion has been limited to second order, because only small fluctuations from equilibrium are considered.

For thermodynamic stability, we must impose that the quadratic form in Eq. (1.43) be positive definite, i.e.,

$$\left(\frac{\partial^2 U}{\partial V^2}\right)_S > 0, \quad \left(\frac{\partial^2 U}{\partial V^2}\right)_S \left(\frac{\partial^2 U}{\partial S^2}\right)_V - \left(\frac{\partial^2 U}{\partial V \partial S}\right)^2 > 0, \quad \left(\frac{\partial^2 U}{\partial S^2}\right)_V > 0. \tag{1.44}$$

The first and the last conditions correspond to the adiabatic compressibility:

$$\kappa_S = -\frac{1}{V}\left(\frac{\partial V}{\partial P}\right)_S \tag{1.45}$$

and the specific heat per unit volume at constant V,

$$c_V = \frac{T}{V}\left(\frac{\partial S}{\partial T}\right)_V \tag{1.46}$$

being positive. To discuss the second condition, it is useful to recall a few mathematical properties about coordinate transformations and Jacobians. The Jacobian of a transformation $u = u(x, y)$, $v = v(x, y)$ is

$$J = \begin{vmatrix} \left(\frac{\partial u}{\partial x}\right)_y & \left(\frac{\partial u}{\partial y}\right)_x \\ \left(\frac{\partial v}{\partial x}\right)_y & \left(\frac{\partial v}{\partial y}\right)_x \end{vmatrix} = \left(\frac{\partial u}{\partial x}\right)_y\left(\frac{\partial v}{\partial y}\right)_x - \left(\frac{\partial u}{\partial y}\right)_x\left(\frac{\partial v}{\partial x}\right)_y \equiv \frac{\partial(u, v)}{\partial(x, y)}. \tag{1.47}$$

Let us consider the two successive transformations

$$t = t(x, y), \quad s = s(x, y) \ \text{(I)}$$

and

$$u = u(t, s), \quad v = v(t, s) \ \text{(II)}.$$

The Jacobian of the transformation $x, y \to u, v$ is obtained as

$$J = \frac{\partial(u, v)}{\partial(x, y)} = \frac{\partial(u, v)}{\partial(t, s)}\frac{\partial(t, s)}{\partial(x, y)} = J_{\text{II}} J_{\text{I}}. \tag{1.48}$$

We now go back to the thermodynamic stability conditions of Eq. (1.44). By recalling Eq. (1.20) we may write

$$\left(\frac{\partial^2 U}{\partial V^2}\right)_S\left(\frac{\partial^2 U}{\partial S^2}\right)_V - \left(\frac{\partial^2 U}{\partial V \partial S}\right)^2 = \frac{\partial(P, T)}{\partial(S, V)} = \frac{\partial(P, T)}{\partial(V, T)}\frac{\partial(V, T)}{\partial(S, V)} = \frac{T}{V^2 c_V \kappa_T}, \tag{1.49}$$

where

$$\kappa_T = -\frac{1}{V}\left(\frac{\partial V}{\partial P}\right)_T \tag{1.50}$$

is the isothermal compressibility. Hence the thermodynamic stability conditions (1.44) become in terms of measurable quantities

$$c_V > 0, \quad \kappa_T > 0, \quad \kappa_S > 0, \quad c_P > c_V > 0, \tag{1.51}$$

where we introduced the specific heat at constant pressure c_P. The last condition follows from

$$c_P = c_V + \frac{T}{V^2 \kappa_T}\left(\frac{\partial V}{\partial T}\right)_P^2 > 0, \tag{1.52}$$

which can be derived by using the Maxwell relation (1.29) $(\partial S/\partial P)_T = -(\partial V/\partial T)_P$. See the answer to Problem 1.3 for a proof. The stability conditions imply convexity or concavity properties of the thermodynamic potentials. For instance, for the free energy

$$\left(\frac{\partial F}{\partial T}\right)_V = -S < 0, \quad c_V = -\frac{T}{V}\left(\frac{\partial^2 F}{\partial T^2}\right)_V > 0, \tag{1.53}$$

i.e., F is a decreasing concave function of T. Similarly, F is a decreasing convex function of V.

1.4.1 Probability distribution of small fluctuations

So far our discussion about the fluctuations from equilibrium has been purely on thermodynamic grounds and has been confined to the evaluation of the cost and to establish the thermodynamic stability conditions. To estimate the chance of small fluctuations from equilibrium we must then connect the variation of the availability with the probability for their occurrence. In so doing we go beyond pure thermodynamics. Einstein (1910) was the first to point out that such a connection is indeed provided by reversing the Boltzmann relation, to be discussed in Chapter 3, which expresses the entropy of a state in terms of the probability of that state:

$$S = k_{\mathrm{B}} \ln W. \tag{1.54}$$

In the above W is the number of micro-states compatible with the constraints of a given macroscopic state and k_{B} is the Boltzmann constant, defined in terms of the gas constant R and Avogadro's constant N_{A} as $k_{\mathrm{B}} = R/N_{\mathrm{A}} = 1.38064 \times 10^{-23}\,\mathrm{J} \cdot \mathrm{K}^{-1}$. A micro-state of a system is determined by the positions and momenta of all its particles. Clearly the probability of the macro-state, p, is related to W by some suitable normalization factor. By reversing Eq. (1.54) one may assume that the probability of a state is

$$p = \mathrm{const} \cdot \mathrm{e}^{S/k_{\mathrm{B}}}, \tag{1.55}$$

where the constant can be determined by requiring a normalized distribution. Einstein pointed out that, although the exact evaluation of W, and hence of p, would require a complete theory at microscopic level, nonetheless the apparent irreversibility of natural phenomena implies that the probabilities of different macro-states may differ by orders of magnitude.

Suppose now that a small portion of the entire system, which initially is in equilibrium with the rest, is driven out of equilibrium by a modification of the original constraints or by a small fluctuation with a subsequent increase of A. Then using Eq. (1.42) we have

$$p_{\mathrm{fluct}} = \mathrm{e}^{\Delta S_{\mathrm{tot}}/k_{\mathrm{B}}} = \mathrm{e}^{-\beta \Delta A}, \tag{1.56}$$

where we have defined (as is customary) $\beta = (k_{\mathrm{B}} T)^{-1}$. The quadratic form (1.43) expressed in terms of the two extensive variables V and S is not suitable for evaluating the variance of the fluctuations directly due to the presence of the mixed term in the two variables. Clearly one can consider thermodynamic fluctuations also in terms of, say, T and V. It is enough to express ΔA in the following way:

$$\begin{aligned}\Delta A &= \frac{1}{2}\left[\left(\frac{\partial^2 U}{\partial V^2}\right)_S (\Delta V)^2 + \left(\frac{\partial^2 U}{\partial S^2}\right)_V (\Delta S)^2 + 2\left(\frac{\partial^2 U}{\partial V \partial S}\right)\Delta V \Delta S\right] \\ &= \frac{1}{2}\left[-\Delta V\left(\left(\frac{\partial P}{\partial V}\right)_S \Delta V + \left(\frac{\partial P}{\partial S}\right)_V \Delta S\right) + \Delta S\left(\left(\frac{\partial T}{\partial V}\right)_S \Delta V + \left(\frac{\partial T}{\partial S}\right)_V \Delta S\right)\right] \\ &= \frac{1}{2}\left[-\Delta V \Delta P + \Delta S \Delta T\right]. \end{aligned} \tag{1.57}$$

Equation (1.57) is symmetric in all the variables and we can choose any two as independent variables, say T and V. From

$$\Delta P = \left(\frac{\partial P}{\partial T}\right)_V \Delta T + \left(\frac{\partial P}{\partial V}\right)_T \Delta V,$$

$$\Delta S = \left(\frac{\partial S}{\partial T}\right)_V \Delta T + \left(\frac{\partial S}{\partial V}\right)_T \Delta V$$

we get

$$\begin{aligned}\Delta A &= \frac{1}{2}\left[-\left(\frac{\partial P}{\partial V}\right)_T (\Delta V)^2 + \left(\frac{\partial S}{\partial T}\right)_V (\Delta T)^2 + \Delta V \Delta T\left(\left(\frac{\partial S}{\partial V}\right)_T - \left(\frac{\partial P}{\partial T}\right)_V\right)\right] \\ &= \frac{1}{2}\left[\frac{1}{V\kappa_T}(\Delta V)^2 + \frac{V c_V}{T}(\Delta T)^2\right], \end{aligned} \tag{1.58}$$

where we used the second Maxwell relation Eq. (1.25). Notice that ΔA is extensive, i.e., depends linearly on the volume.

Inserting Eq. (1.58) in Eq. (1.56) one obtains that small fluctuations from equilibrium obey a Gaussian distribution. This is a physical manifestation of the central limit theorem of probability theory according to which the properly normalized sum of a large number of identically distributed random variables follows a Gaussian distribution. A sketch of the proof by using the Fourier transform is shown in Appendix A. The mean square roots of the fluctuations of ΔV and ΔT from equilibrium are given by

$$\langle(\Delta V)^2\rangle = V k_{\mathrm{B}} T \kappa_{\mathrm{T}}, \quad \langle(\Delta T)^2\rangle = \frac{k_{\mathrm{B}} T^2}{V c_V}, \tag{1.59}$$

where we omitted the subscript 0 for the equilibrium temperature.

The particle density $n = N/V$ has fluctuations $\langle(\Delta n)^2\rangle = (n^2/V) k_{\mathrm{B}} T \kappa_T$. This expression can be generalized to include smooth dependence of the density on the position $\mathbf{x}$ with the substitution

$$\Delta A = \frac{1}{2}\left[\frac{V}{n^2\kappa_T}(\Delta n)^2\right] \to \int d\mathbf{x} \frac{1}{2}\left[\frac{1}{n^2\kappa_T}(\Delta n(\mathbf{x}))^2 + u^{(2)}(\nabla n(\mathbf{x}))^2\right], \tag{1.60}$$

where $u^{(2)} = \partial^2 u/\partial(\nabla n)^2|_{\nabla n=0}$ is the coefficient of the expansion in terms of the ∇n and u is the energy density. From the last equation we obtain the famous Ornstein–Zernike response function (Ornstein and Zernike (1914)) for the fluctuations of the density in the Fourier transformed $\mathbf{k}$ space. The Fourier transform of the local fluctuations and the associated antitransform are defined as

$$\delta n(\mathbf{k}) = \frac{1}{\sqrt{V}} \int d\mathbf{x}\, e^{-i\mathbf{k}\cdot\mathbf{x}} \Delta(\mathbf{x}), \quad \Delta n(\mathbf{x}) = \frac{1}{\sqrt{V}} \sum_{\mathbf{k}} e^{i\mathbf{k}\cdot\mathbf{x}} \delta n(\mathbf{k}). \tag{1.61}$$

By using the antitransform in Eq. (1.60) and the integral representation of the Kronecker delta

$$\delta_{\mathbf{k},-\mathbf{k}'} = \frac{1}{V} \int d\mathbf{x}\, e^{i(\mathbf{k}+\mathbf{k}')\cdot\mathbf{x}} \tag{1.62}$$

we get

$$\Delta A = \sum_{\mathbf{k}} \frac{1}{2} \left[\frac{1}{n^2 \kappa_T} + u^{(2)} \mathbf{k}^2 \right] |\delta n(\mathbf{k})|^2, \tag{1.63}$$

so that the statistical average gives

$$\langle |\delta n(\mathbf{k})|^2 \rangle = \frac{k_B T}{n^{-2} \kappa_T^{-1} + u^{(2)} k^2}. \tag{1.64}$$

By using S and P as independent variables, the variation of availability from equilibrium is

$$\Delta A = \frac{1}{2} \left[\frac{T}{V c_P} (\Delta S)^2 + V \kappa_S (\Delta P)^2 \right]$$

and

$$\langle (\Delta S)^2 \rangle = k_B V c_P, \quad \langle (\Delta P)^2 \rangle = \frac{k_B T}{V \kappa_S}. \tag{1.65}$$

The stability conditions also guarantee the exponential decrease of the probability of large fluctuations. At fixed density n, the particle number is given in terms of the volume $N = nV$. It is clear that the variance of the fluctuations of an extensive variable scales as N, whereas the variance of the intensive variables scales as $1/N$. For all the variables the mean square root of the fluctuations with respect to its average scales as $1/\sqrt{N}$. When the number of particles is large, the fluctuations are negligible compared to the averaged quantities.

There is an important exception to this rule, when the second order thermodynamic derivatives become singular as happens at the critical point of the second order phase transitions. For instance, for a critical point of the liquid–gas transition where $(\partial P/\partial V)_T \to 0$, the density fluctuations become singular. We will have an example of this in Section 1.6 where we will discuss the van der Waals equation for a real gas.

1.5 Phase equilibrium

Let us consider the phase diagram of a generic substance by using T and P as thermodynamic parameters. The boundary between two phases is referred to as the coexistence curve. Along this curve, the Gibbs free energy is the sum of the contributions of the two phases:

$$G = N_a g_a + N_b g_b, \tag{1.66}$$

where N_a and N_b are the particle numbers in the two phases indicated as a and b. The total number of particles is fixed:

$$N = N_a + N_b. \tag{1.67}$$

The two functions g_a and g_b are the thermodynamic potentials per particle, i.e. the chemical potentials μ_a and μ_b, of the two phases and depend on the intensive variables T and P only. The equilibrum condition requires G to be a minimum, which implies that, at a given transition point (T, P) on the coexistence line, we must have

$$\mathrm{d}G = \mathrm{d}N_a g_a + \mathrm{d}N_b g_b = 0$$

from which

$$g_a = g_b, \tag{1.68}$$

e.g. along the gas–liquid $T - P$ coexistence curve $\mu_{\text{liquid}} = \mu_{\text{gas}}$. To derive Eq. (1.68) we used the fact that $\mathrm{d}N_a = -\mathrm{d}N_b$ as required by Eq. (1.67).

We are now ready to prove the Gibbs phase rule (Eq. (1.4)). Consider a system made up of n components at constant T and P. In each of the f phases we have to specify pressure, temperature and the relative amount of each component so that we get in total $2 + f(n-1)$ parameters. For each component we have $f-1$ phase-coexistence conditions like Eq. (1.68), i.e., $n(f-1)$ in total. Then the number of independent parameters v is obtained by considering the fact that

$$v = [2 + f(n-1)] - n(f-1) = 2 + n - f.$$

If we move along the coexistence line, we change the value of temperature and pressure by $\mathrm{d}T$ and $\mathrm{d}P$, but the condition (1.68) does not change:

$$\mathrm{d}g_a = \mathrm{d}g_b. \tag{1.69}$$

By recalling Eqs. (1.29), we introduce the entropy and volume per particle in the two phases as

$$s_{a,b} = -\left(\frac{\partial g_{a,b}}{\partial T}\right)_P, \tag{1.70}$$

$$v_{a,b} = \left(\frac{\partial g_{a,b}}{\partial P}\right)_T. \tag{1.71}$$

The Gibbs potentials are decreasing functions of T and increasing functions of P. In each region of the $P - T$ plane where a phase is stable its Gibbs potential must be minimum, i.e. lower than the Gibbs potential of the other phase. To make an example, let us consider the liquid–gas or liquid–solid transition with $a = \mathrm{L}$ and $b = \mathrm{G}$ or $a = \mathrm{S}$ and $b = \mathrm{L}$, respectively. Whereas the entropy of the phase stable at greater temperature is always greater, $s_\mathrm{G} > s_\mathrm{L}$ or $s_\mathrm{L} > s_\mathrm{S}$ at fixed P, at fixed T it is $v_\mathrm{G} > v_\mathrm{L}$ for the liquid–gas transition, but both possibilities $v_L > v_S$ or $v_L < v_S$ arise for the liquid–solid one. The latter situation is for instance realized in water, see Fig. 1.1.

By combining Eqs. (1.69), (1.70) and (1.71), one gets the Clausius–Clapeyron equation for the coexistence curve:

$$\frac{dP}{dT} = \frac{s_a - s_b}{v_a - v_b}. \quad (1.72)$$

We know that the liquid occupies a smaller volume with respect to the gas so that $v_L < v_G$. Also we know that we must supply heat to the liquid, at constant temperature, to evaporate it. This amount of heat is called the *latent heat* of vaporization, λ, and is related to the variation of entropy by

$$\lambda = T(s_G - s_L), \quad (1.73)$$

so that $s_G > s_L$. Hence the positive jumps of specific entropy and volume imply a positive slope for the coexistence curve. As is well known, the case of the solid–liquid transition of water is anomalous in the fact that the solid, more ordered phase has a bigger specific volume, implying a negative slope of the coexistence curve.

The phase transitions along the coexistence curve are then characterized, in general, by a finite jump or, mathematically speaking, by a discontinuity, in the first derivatives of the thermodynamic Gibbs free energy. Due to this fact, they are called *first order* phase transitions. Phase transitions where the first derivatives are continuous are called *higher order* phase transitions. In general one would have an n-th order phase transition, when the corresponding derivative of the thermodynamic potential is discontinuous (Ehrenfest (1933)). *Second order* phase transitions are the most commonly studied and occur, for instance, at the end critical point of the liquid–gas coexistence line. From the discussions of the previous section, we have learned that second derivatives of the thermodynamic potentials are associated with measurable response functions such as specific heat and compressibility. These response functions, furthermore, control the thermodynamic stability and the fluctuations from equilibrium. Hence the behavior at the critical point, associated with singularities in the second derivatives, is characterized by the presence of strong fluctuations.

1.6 The liquid–gas transition

The equation of state derived by van der Waals[6] for real gases provides the basis of a first example of a classical theory of *critical phenomena*, which we will discuss in detail in this book. In this section we limit ourselves to a phenomenological derivation of the van der Waals equation and of its solution. This will allow us to introduce a few concepts useful for the following discussions.

The aim of the van der Waals equation is to take into account the interaction among the molecules, in oder to generalize the equation of state of the perfect gas. The potential

[6] Johannes Diderik van der Waals wrote this in 1873 in his doctoral thesis (van der Waals (1873)).

between two molecules is generically characterized by a strong repulsive behavior at small distances and by an asymptotically vanishing weak attractive tail at large distances. The first feature leads to the idea of *excluded* volume, i.e., each molecule cannot use the volume occupied by the other molecules. On average this is taken into account by the following replacement in the equation of state of a perfect gas:

$$V \to V - b.$$

The attractive character of the potential at large distances implies a reduction of the pressure that the molecules are able to exercise on the walls of the container of the gas. This is taken into account, on average, by adding to the pressure an additional pressure via the following replacement:

$$P \to P + \frac{a}{V^2}.$$

What matters here is the inverse square dependence on the volume. This can be justified in the following manner. Since the pressure depends on the attractive interaction among the particles it must depend on the square of the number of particles. The pressure, on the other hand, is an intensive quantity and must depend on N only via an intensive combination of extensive quantities. Hence the additional pressure will depend on the square of the density N/V. At fixed number of particles, this leads to the inverse square dependence on the volume.

In Chapter 8 we will give a microscopic justification of these phenomenological assumptions, which lead to the van der Waals equation for one mole:

$$\left(P + \frac{a}{V^2}\right)(V - b) = RT, \tag{1.74}$$

where a and b depend on the specific substance. At fixed T and P, the equation of state of the perfect gas gives only one value of V, whereas Eq. (1.74) yields three values. This is seen by writing Eq. (1.74) as a cubic equation in V:

$$V^3 - V^2\left(b + \frac{RT}{P}\right) + \frac{a}{P}V - \frac{ab}{P} = 0. \tag{1.75}$$

In general there will be three solutions. When only one is real, the system is in a single phase, as above the critical point. The presence of distinct real solutions indicates phase coexistence among the liquid and gas phases with different specific volumes $v_L < v_G$. The merging of the three real solutions identifies the critical point, where the transition becomes of second order as the end point of the coexistence line with the divergence of the compressibility.

A useful way to analyze Eq. (1.74) or (1.75) is to write it in the following form:

$$V - b = \frac{RT}{P}\frac{V^2}{V^2 + a/P} \equiv f(V), \tag{1.76}$$

so that the solution can be found graphically by looking at the intersection of the two functions of V in the two sides of Eq. (1.76). We observe that the function $f(V)$ is quadratic at small values of V and goes to a constant for $V \to \infty$ with an inflection point, while the straight line has positive slope and a negative intercept. Therefore there must be, at

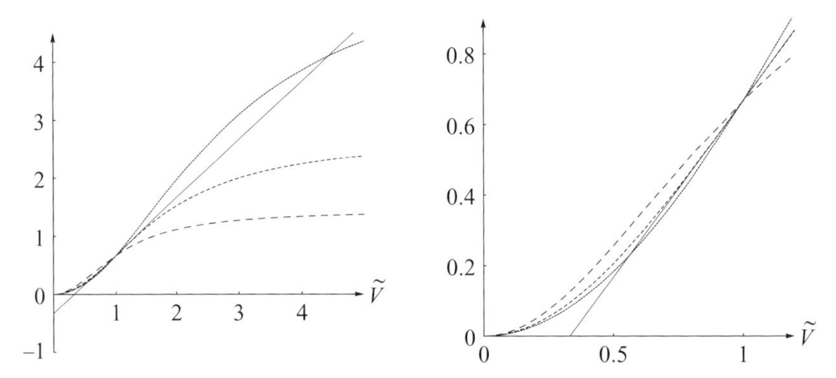

Fig. 1.4 The graphical solution of Eq. (1.84) for three different sets of values of the reduced variables $\tilde{T}$ and $\tilde{P}$: $(1.4, 2.6)$ (long-dashed), $(1.0, 1.0)$ (dashed), $(0.85, 0.4)$ (full line). The plot on the right represents an enlarged vision of the range of values of $\tilde{V}$ where further solutions appear.

least, one intersection. The possibility of more intersections arises when the function $f(V)$ is tangent to the straight line $V - b$ at the inflection point, where two, initially coincident, new intersections develop. The conditions to be imposed for such a *critical point* are given by

$$V_c - b = f(V_c), \tag{1.77}$$

$$1 = f'(V_c), \tag{1.78}$$

$$0 = f''(V_c), \tag{1.79}$$

which yield

$$V_c = 3b, \quad P_c = \frac{a}{27b^2}, \quad RT_c = \frac{8a}{27b}. \tag{1.80}$$

By introducing reduced variables as

$$T \to \tilde{T}T_c, \tag{1.81}$$

$$P \to \tilde{P}P_c, \tag{1.82}$$

$$V \to \tilde{V}V_c, \tag{1.83}$$

Eq. (1.76) becomes

$$\tilde{V} - \frac{1}{3} = \frac{8\tilde{T}}{3\tilde{P}} \frac{\tilde{V}^2}{\tilde{V}^2 + 3/\tilde{P}} \equiv f(\tilde{V}). \tag{1.84}$$

Its graphical solution is shown in Fig. 1.4 for three different regimes. Notice that in order to have the three solutions, the values of $\tilde{T}$ and $\tilde{P}$ must lie on the coexistence line. Near the critical point the coexistence line can be evaluated analytically, as will be shown explicitly in Chapter 8.

Finally, to make contact with the concluding remarks of the previous section, we show that the compressibility evaluated with the van der Waals equation becomes infinite at the

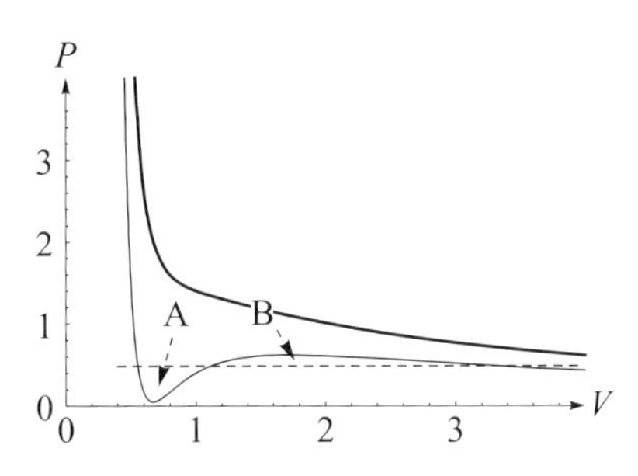

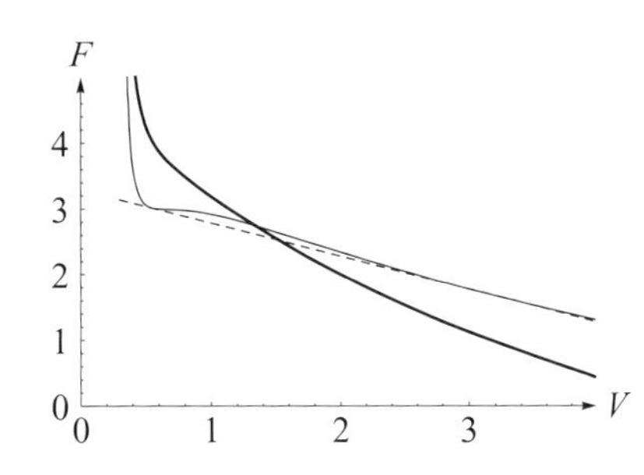

Fig. 1.5 The pressure, at constant temperature, as a function of the volume (left) and the free energy (right), obtained by integrating the relation $P = -(\partial F/\partial V)_T$. Thick and thin lines are above and below the critical point ($T = 1.1\,T_c$ and $T = 0.85\,T_c$). In the right panel, the dashed line represents the linear interpolation for the free energy corresponding to a combination of the free energies of the liquid and gas phases. This yields the Maxwell construction in the left panel with equal areas A and B.

critical point. To this end we rewrite Eq. (1.74) in terms of the reduced variables to read

$$\left(\tilde{P} + \frac{3}{\tilde{V}^2}\right)\left(\tilde{V} - \frac{1}{3}\right) = \frac{8}{3}\tilde{T}. \tag{1.85}$$

By resolving it with respect to $\tilde{P}$, we have

$$\tilde{P} = \frac{8}{3}\frac{\tilde{T}}{\tilde{V} - 1/3} - \frac{3}{\tilde{V}^2} \tag{1.86}$$

and, by taking the derivative with respect to $\tilde{V}$ in both sides,

$$\kappa_{\tilde{T}} = -\frac{1}{\tilde{V}}\left(\frac{\partial \tilde{V}}{\partial \tilde{P}}\right)_{\tilde{T}} = -\frac{1}{\tilde{V}}\left[-\frac{8}{3}\frac{\tilde{T}}{(\tilde{V} - 1/3)^2} + \frac{6}{\tilde{V}^3}\right]^{-1}, \tag{1.87}$$

which becomes infinite at the critical point, $\tilde{T} = 1$, $\tilde{V} = 1 = \tilde{V}_G = \tilde{V}_L$. It is clear from the left panel in Fig. 1.5 that the compressibility, above the critical point, obeys the thermodynamic stability condition and that the free energy is accordingly a convex decreasing function of the volume. Below the critical point, there exists a range of values of volume with negative compressibility and correspondingly a concave free energy. Maxwell (1875) proposed how to cure this unphysical situation via the equal area construction for the van der Waals isotherms as shown in Fig. 1.5. It amounts to saying that no work can be obtained by a cycle performed with a source at a single temperature. The idea is based on the observation that the free energy in the range where it has the wrong curvature is replaced by a straight line interpolating between two points with equal derivative, i.e., equal pressure for the two phases. This pressure is determined by

$$P = P(V_L) = P(V_G) = -\frac{F(V_L) - F(V_G)}{V_L - V_G} = \frac{1}{V_G - V_L}\int_{V_L}^{V_G} dV\,P(V), \tag{1.88}$$

where the integral has to be taken along the van der Waals isotherm.

This construction, now called after Maxwell, determines the onset of phase separation between gas and liquid along the coexistence curve in the P–T plane.

1.7 Problems

1.1 Show that in the V–P plane, the isothermic and adiabatic transformations are decreasing functions of V and the adiabatic curve decreases more rapidly.

1.2 Show that
- for two Carnot cycles R and R' operating between θ_1 and θ_2, the equality $Q'_1 = Q_1$ implies $Q'_2 = Q_2$;
- for any cycle

$$\frac{Q_1}{T_1} + \frac{-Q_2}{T_2} \leq 0,$$

where T_1 and T_2 are the absolute temperatures corresponding to the empirical temperatures θ_1 and θ_2.

1.3 Prove Eq. (1.52).

1.4 By making reference to Fig. 1.3, show that the vanishing of the zero-temperature limit of the entropy variation upon a isothermic reversible transformation implies the unattainability of absolute zero and vice versa.

2 Kinetics

According to the second law of thermodynamics, the evolution towards an equilibrium state corresponds to the increase of the entropy of a thermally isolated system, or, depending on the external conditions, by the decrease of an appropriate thermodynamic potential. Statistical mechanics must explain the origin of such a law starting from the laws which control the motion of the microscopic constituents of matter. At a microscopic level, however, all the basic laws are symmetric with respect to time reversal. This suggests that a purely mechanical derivation of the second law of thermodynamics is conceptually impossible and a more subtle reasoning, based on probability considerations, is necessary to explain irreversibility and the thermodynamics of a system in terms of the motion of its elementary constituents. This has been clearly explained by Boltzmann after several years of intense thinking. Towards the end of his life, Boltzmann published (in 1896) *Lectures on Gas Theory* (Boltzmann (1964)). This is the topic of this chapter and the next.

2.1 The birth of kinetic theory

Kinetic theory makes its first appearance in the book *Hydrodynamics* published by Daniel Bernoulli in 1738 (Bernoulli and Bernoulli (2005)). The assessment of the kinetic approach of Bernoulli was undertaken by Clausius (1857). Bernoulli's argument explains the pressure exercised by a gas on the walls of the container in terms of the collisions of its constituent molecules. Let us imagine a molecule approaching one of the walls of the gas container as shown in Fig. 2.1. We assume elastic scattering and the wall perpendicular to the x-axis. Let $v_x > 0$ be the component along the x-axis of the initial velocity. Since upon bouncing on the wall the molecule reverses the sign of its perpendicular velocity component, the transferred momentum to the wall is $2mv_x$, m being the mass of the molecule. The transferred momentum per unit time is the force exercised by the molecule on the wall during a single collision. The total force on the wall is the cumulative effect of the collisions occurring in unit time. The number of collisions is the number of molecules which during the unit time, $\mathrm{d}t$, will collide with the wall, i.e. those which are $v_x\mathrm{d}t$ far away from the wall. Let us consider a cylinder of base area A and height $v_x\mathrm{d}t$. Clearly the number of molecules in such a cylinder is $nv_x\mathrm{d}tA$, where $n = N/V$ is the average density of the N molecules in a container of volume V. We may then conclude that the momentum transferred per unit time and per unit area of the wall, i.e. the pressure, is $2mv_x^2n$.

To make the above argument independent of the particular value of the velocity, we may replace v_x^2 with its average value. In so doing we must keep in mind that in evaluating the

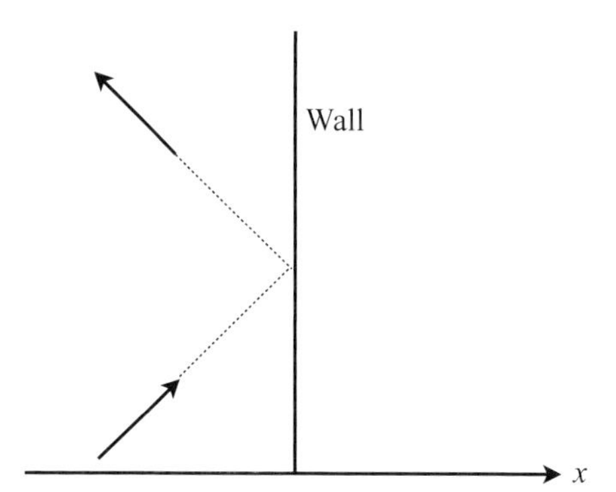

Fig. 2.1 Elastic scattering of molecules on a rigid wall. The velocity component parallel to the wall is conserved, whereas the perpendicular one reverses its sign after bouncing on the wall. The momentum transferred to the wall is twice the perpendicular velocity times the mass of the molecules.

pressure exercised on the wall we considered only the positive values of the x-component of the velocity. In the average value, clearly, the negative ones are included as well. If we assume complete symmetry in the probability of occurrence of positive and negative values of the velocity, we conclude that upon averaging we must replace $2v_x^2$ with $\langle v_x^2 \rangle$, as was envisaged by Boltzmann with the introduction of the distribution function of the velocity. A further generalization of the argument arises if we assume space isotropy so that the average value of the square of the x-component of the velocity is the same of the average value of the square of the other two velocity components, i.e. $\langle v_x^2 \rangle = \langle v_y^2 \rangle = \langle v_z^2 \rangle = \langle v^2 \rangle /3$, v being the absolute value of the velocity. As a result the pressure exercised by the gas reads

$$P = \frac{n}{3} m \langle v^2 \rangle = \frac{2}{3} \frac{N}{V} \left\langle \frac{mv^2}{2} \right\rangle = \frac{2}{3} \frac{E}{V}, \tag{2.1}$$

i.e. the pressure is directly proportional to the average kinetic energy E of the molecules of the gas and inversely proportional to the volume occupied, which is the Boyle law for a perfect gas. By recalling the state equation for perfect gases $PV = RT$, assuming one mole for the sake of simplicity so that $N = N_A$ is the Avogadro number, one concludes that

$$T = \frac{2}{3} \frac{N_A}{R} \left\langle \frac{mv^2}{2} \right\rangle = \frac{2}{3k_B} \left\langle \frac{mv^2}{2} \right\rangle. \tag{2.2}$$

This basic relation unveils the meaning of the absolute temperature as a measure of the average kinetic energy (Krönig (1856)) and shows the origin of the Boltzmann constant k_B.[1]

Clausius (1858) noted that the mean free path of the colliding molecules should be taken into account in order to make the time necessary for the molecules of a gas to travel from one side to the opposite side of a room compatible with the observed propagation e.g. of the smell of a volatile substance (see Problem 2.4).

[1] From the principle of least action applied to the time averaged kinetic and potential energies in a periodic motion, Boltzmann and Clausius, independently, showed that the averaged kinetic energy, i.e. the absolute temperature, is the integrating factor of δQ, the Clausius theorem (1.12) for a reversible process.

The arguments given above, which contain the basic physical ingredients of the kinetic theory, do not specify how the average over the values of the velocity is performed in the sense that no explicit form of the velocity probability distribution is given. Neither say anything about how the equilibrium is reached and what aspects the equilibrium distribution function depends on. For that it was necessary to wait till the fundamental work of Maxwell and Boltzmann. The rest of the chapter will deal with their work.

2.2 The Boltzmann equation

Consider a diluted gas of N classical particles whose mechanical state for each time t is defined by the values of position $\mathbf{q}$ and momentum $\mathbf{p}$ of all the particles. The six components $\mathbf{q}$ and $\mathbf{p}$ of each particle play the role of coordinates in the so-called μ phase-space. The thermodynamic properties of the gas are determined by how the particles are distributed in the μ phase-space rather than on the specific motion of each particle. We will assume with Boltzmann that the gas is out of equilibrium and *show* how equilibrium is reached via the binary collisions among the particles. The object of the kinetic theory is the determination of the distribution function, f, defined such that

$$\mathrm{d}N(\mathbf{q},\mathbf{p},t) = f(\mathbf{q},\mathbf{p},t)\mathrm{d}^3q\,\mathrm{d}^3p, \tag{2.3}$$

where $\mathrm{d}N(\mathbf{q},\mathbf{p},t)$ represents the number of particles in the μ phase-space volume element centered around the point $(\mathbf{q},\mathbf{p})$ at time t. If we neglect the interactions among the particles and consider the evolution according to the Hamilton equations of motion during a time interval $\mathrm{d}t$, we must have

$$f(\mathbf{q},\mathbf{p},t)\mathrm{d}^3q\,\mathrm{d}^3p = f(\mathbf{q}+\dot{\mathbf{q}}\mathrm{d}t,\mathbf{p}+\dot{\mathbf{p}}\mathrm{d}t,t+\mathrm{d}t)\mathrm{d}^3q'\mathrm{d}^3p', \tag{2.4}$$

since the particles that were in the volume element around the point $(\mathbf{q},\mathbf{p})$ at time t will have moved, after the time $\mathrm{d}t$, into the volume element around the point $(\mathbf{q}',\mathbf{p}')$, with $\mathbf{q}'=\mathbf{q}+\dot{\mathbf{q}}\mathrm{d}t$ and $\mathbf{p}'=\mathbf{p}+\dot{\mathbf{p}}\mathrm{d}t$, where $\dot{\mathbf{p}}=\mathbf{F}$ with $\mathbf{F}$ the external force. Owing to the conservation of the phase-space volume $\mathrm{d}^3q\,\mathrm{d}^3p$,[2] by expanding to linear order in $\mathrm{d}t$, Eq. (2.4) becomes

$$\frac{\partial f}{\partial t}+\dot{\mathbf{q}}\cdot\frac{\partial f}{\partial \mathbf{q}}+\dot{\mathbf{p}}\cdot\frac{\partial f}{\partial \mathbf{p}}=0. \tag{2.5}$$

If we add the interactions among the particles we must include in the equation for f the number of particles that, as a result of collisions, enter and leave the phase-space volume element centered around the point $(\mathbf{q},\mathbf{p})$. We now define the quantities $I_{\mathrm{in(out)}}$ in such a way that

$$I_{\mathrm{in(out)}}\mathrm{d}t\,\mathrm{d}^3q\,\mathrm{d}^3p \tag{2.6}$$

[2] Under the action of the force F and for each component $i=x,y,z$, the rectangular element $\mathrm{d}q_i\,\mathrm{d}p_i$ transforms in the time $\mathrm{d}t$ into a parallelogram of equal area $\mathrm{d}q_i'\,\mathrm{d}p_i'$.

is the number of collisions occurring in the time dt and in the phase-space volume element $d^3q\, d^3p$ where one of the final (initial) colliding particle is in the phase-space volume element $d^3q\, d^3p$, around the point $(\mathbf{q}, \mathbf{p})$. Then Eq. (2.5) becomes

$$\frac{\partial f}{\partial t} + \dot{\mathbf{q}} \cdot \frac{\partial f}{\partial \mathbf{q}} + \dot{\mathbf{p}} \cdot \frac{\partial f}{\partial \mathbf{p}} = I \equiv I_{\text{in}} - I_{\text{out}}. \tag{2.7}$$

From this last equation, Boltzmann obtained his famous equation for the time evolution of the distribution function $f(\mathbf{q}, \mathbf{p}_1, t)$ whose derivation will follow shortly:

$$\frac{\partial f_1}{\partial t} + \frac{\mathbf{p}_1}{m} \cdot \frac{\partial f_1}{\partial \mathbf{q}} + \mathbf{F} \cdot \frac{\partial f_1}{\partial \mathbf{p}_1} = \int d\Omega d^3 p_2 \frac{d\sigma}{d\Omega} \frac{|\mathbf{p}_1 - \mathbf{p}_2|}{m} (f_{1'} f_{2'} - f_1 f_2), \tag{2.8}$$

where

$$f_1 \equiv f(\mathbf{q}, \mathbf{p}_1, t), \ldots, \text{etc.},$$

$d\sigma/d\Omega$ is the differential cross section for a binary collision, and $\mathbf{p}_1$, $\mathbf{p}_2$ and $\mathbf{p}_1'$, $\mathbf{p}_2'$ are the initial (final) and final (initial) momenta of the colliding particles for "out" and "in" processes, respectively. In the center-of-mass frame a binary collision reduces to a deflection of the relative velocity $\mathbf{p}_1 - \mathbf{p}_2$ with angles (θ, ϕ), θ being the angle with respect to the initial direction of $\mathbf{p}_1 - \mathbf{p}_2$ and ϕ the azimuthal angle. The elementary solid angle is then $d\Omega = \sin(\theta) d\theta d\phi$.

The derivation is based on the two simplifying assumptions of binary collisions and molecular chaos. The first one can always be satisfied for sufficiently high dilution of the system, whereas the second is non-dynamical in nature and will be discussed at length in the following. Under these assumptions, as we will discuss in the following sections, the Maxwell–Boltzmann distribution is a stationary solution for f to which every solution of the Boltzmann equation evolves in time.

We begin the derivation of the Boltzmann equation by showing how to obtain the expressions for I_{in} and I_{out}. Let us consider a collision where an initial state of two particles with momenta $\mathbf{p}_1$ and $\mathbf{p}_2$ goes into a final state with momenta $\mathbf{p}_1'$ and $\mathbf{p}_2'$. Furthermore, we denote by $f^{(2)}(\mathbf{q}, \mathbf{p}_1, \mathbf{p}_2, t) d^3p_1 d^3p_2 d^3q$ the number of pairs of particles with momenta $\mathbf{p}_1$, $\mathbf{p}_2$ that, at time t, are found in the space volume-element d^3q, and in the momentum volume elements d^3p_1, d^3p_2, respectively. Clearly, then, we must have, by considering the definition Eq. (2.6), that

$$I_{\text{out}} = \int d^3p_2 f^{(2)}(\mathbf{q}, \mathbf{p}_1, \mathbf{p}_2, t) \gamma_{\text{coll}}, \tag{2.9}$$

where γ_{coll} is the number of collisions per unit time at fixed $\mathbf{p}_2$. We stress that the function $f^{(2)}(\mathbf{q}, \mathbf{p}_1, \mathbf{p}_2, t)$ conveys information about the statistical distribution of the particles, whereas γ_{coll} depends on the mechanism of interaction. In the original derivation of Boltzmann, the quantity γ_{coll} was obtained by using classical mechanics. Let us, first, discuss the classical case in the simple case of an homogeneous system in the absence of external forces ($f^{(2)}$ does not depend on $\mathbf{q}$). We later consider the quantum extension.

We notice first that a binary collision can be reduced to the problem of a particle scattered by a center of force. This corresponds to using a reference frame attached to the center of mass. The center-of-mass and relative momenta before and after the collision are defined

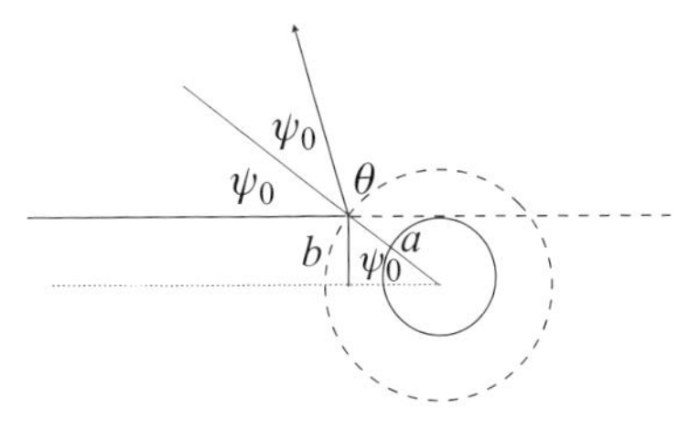

Fig. 2.2 Geometrical description of the scattering by a hard-core potential of a particle arriving from the left with impact parameter b. The full-line circle represents the hard particle of diameter a; the dashed one represents the action sphere of radius a.

as

$$\mathbf{P} = \frac{\mathbf{p}_1 + \mathbf{p}_2}{2}, \quad \mathbf{p} = \mathbf{p}_1 - \mathbf{p}_2, \quad \mathbf{P}' = \frac{\mathbf{p}_1' + \mathbf{p}_2'}{2}, \quad \mathbf{p}' = \mathbf{p}_1' - \mathbf{p}_2'.$$

During the collision both the total momentum and the energy are conserved so that the collision only produces a rotation of the relative momentum. Such a rotation is described by the scattering angle θ, defined by the angle between $\mathbf{p}$ and $\mathbf{p}'$, and ϕ is the azimuth angle about the direction of $\mathbf{p}$.

Let $N\gamma_{\text{coll}}\mathrm{d}t$ be the number of particles with relative momentum $\mathbf{p}$ which enter in collision in the time $\mathrm{d}t$ with the center of force. These are the particles contained in the cylinder of length $(p/m)\mathrm{d}t$ and base area σ, the scattering cross section. As a result the number of collisions per particle per unit time reads

$$\gamma_{\text{coll}} = \frac{p}{Vm}\sigma. \tag{2.10}$$

To be simple, let us consider the case of particles interacting via a hard-core potential

$$U(\mathbf{x}) = \begin{cases} \infty, & x < a, \\ 0, & x > a, \end{cases} \tag{2.11}$$

where a is the diameter of the hard spheres. In this case the differential cross section can be obtained by geometrical considerations as depicted in Fig. 2.2. The trajectory is a straight line until it touches the surface of a sphere of radius a located at the origin. Energy conservation requires the angle of incidence ψ_0 to be equal to that of reflection. As a result the angle of incidence is related to the scattering angle θ by the relation $\psi_0 = \pi/2 - \theta/2$. For a particle to be deflected in the time $\mathrm{d}t$ by an angle between θ and $\theta + \mathrm{d}\theta$ and between ϕ and $\phi + \mathrm{d}\phi$, its impact parameter, the minimal distance between the undeflected trajectory and the center of force (see Fig. 2.2), $b = a\cos(\theta/2)$, must lie between b and $b + \mathrm{d}b$. As a result the differential cross section, which is the portion of scattering surface of differential area $b\mathrm{d}\phi\mathrm{d}b$, is $\mathrm{d}\sigma = b(\mathrm{d}b/\mathrm{d}\theta)\mathrm{d}\Omega/\sin(\theta)$. The differential cross section then reads $\mathrm{d}\sigma/\mathrm{d}\Omega = a^2/4$ and therefore

$$\gamma_{\text{coll}} = \frac{p}{Vm}\int \mathrm{d}\Omega \frac{\mathrm{d}\sigma}{\mathrm{d}\Omega}. \tag{2.12}$$

We then obtain for $I_{\rm out}$

$$I_{\rm out}\mathrm{d}^3p_1\mathrm{d}^3q = \int \mathrm{d}\Omega \mathrm{d}^3p_2 \frac{\mathrm{d}\sigma}{\mathrm{d}\Omega}\frac{|\mathbf{p}_1-\mathbf{p}_2|}{Vm} f^{(2)}(\mathbf{p}_1,\mathbf{p}_2,t)\mathrm{d}^3p_1\mathrm{d}^3q. \tag{2.13}$$

By reasoning in the same way, one may calculate the rate at which particles enter the phase-space volume element at $\mathbf{p}_1$ and $\mathbf{q}$. It is enough to exchange the initial and the final momenta of the scattering to obtain

$$I_{\rm in}\mathrm{d}^3p_1'\mathrm{d}^3q = \int \mathrm{d}\Omega \mathrm{d}^3p_2' \frac{\mathrm{d}\sigma}{\mathrm{d}\Omega}\frac{|\mathbf{p}_1'-\mathbf{p}_2'|}{Vm} f^{(2)}(\mathbf{p}_1',\mathbf{p}_2',t)\mathrm{d}^3p_1'\mathrm{d}^3q. \tag{2.14}$$

In obtaining Eq. (2.14) we used the fact that the differential cross section may depend only on the scattering angle. Due to momentum ($\mathbf{p}_1+\mathbf{p}_2=\mathbf{p}_1'+\mathbf{p}_2'$) and energy ($p_1^2+p_2^2=(p_1')^2+(p_2')^2$) conservation, it follows that $|\mathbf{p}_1-\mathbf{p}_2|=|\mathbf{p}_1'-\mathbf{p}_2'|$ and $\mathrm{d}^3p_1\mathrm{d}^3p_2=\mathrm{d}^3p_1'\mathrm{d}^3p_2'$. Equation (2.7) then becomes

$$\frac{\partial f_1}{\partial t} = \int \mathrm{d}\Omega \int \mathrm{d}^3p_2 \frac{\mathrm{d}\sigma}{\mathrm{d}\Omega}\frac{|\mathbf{p}_1-\mathbf{p}_2|}{Vm}(f^{(2)}(\mathbf{p}_1',\mathbf{p}_2',t) - f^{(2)}(\mathbf{p}_1,\mathbf{p}_2,t)). \tag{2.15}$$

Up to now we have used the equation of motion under the hypothesis of binary collisions only, that, as we have already mentioned, can be made always valid by a sufficient dilution. More importantly, in order to close Eq. (2.15) $f^{(2)}$ must be expressed in terms of f. Here Boltzmann introduced the basic assumption of molecular chaos. The function $f^{(2)}(\mathbf{p}_1,\mathbf{p}_2,t)$, as the distribution function f, depends on the specific microscopic state of the gas and in turn on the time-dependent multi-particle distribution functions. One should in principle consider the full hierarchy of equations or, equivalently, follow the phase-space trajectories of all the particles. The description of the system in terms of f only, i.e. the evolution in μ-space rather than in the phase-space consisting of all particles' coordinates cannot be done in general on a purely dynamical level. Here comes the brilliant intuition of Boltzmann. At a microscopic level, the gas may find itself in a state in which, given a particle at space point $\mathbf{q}$ and momentum $\mathbf{p}_1$, the nearest particle must have a momentum $\mathbf{p}_2$ which depends on $\mathbf{p}_1$. Boltzmann calls this a *molecular ordered* state. In a diluted gas, however, the mean free path between collisions is large compared with the distance over which the collisions occur, so that it may be reasonable to expect that for the majority of the microscopic configurations, the two events of finding, at the space point $\mathbf{q}$, two particles with momenta $\mathbf{p}_1$ and $\mathbf{p}_2$ are statistically independent.[3] Microscopic states of this type are called by Boltzmann *molecular disordered*. Boltzmann then formulates the *molecular chaos* hypothesis as

$$f^{(2)}(\mathbf{p}_1,\mathbf{p}_2,t) \simeq Vf(\mathbf{p}_1,t)f(\mathbf{p}_2,t). \tag{2.16}$$

[3] This situation is essentially realized, and f evolves within short times according to the Boltzmann equation in the so-called Boltzmann–Grad limit ($N\to\infty$, $a\to 0$, $Na^2\to$ constant), an extremely diluted limit which maintains the interaction (see e.g. Lanford (1975)).

Molecular chaos is equivalent to saying that the probability for a pair of particles $(\mathbf{p}_1, \mathbf{p}_2)$, is the product of probability for each particle $\mathbf{p}_1$ and $\mathbf{p}_2$ at any time t and in any region of space.[4]

Equation (2.15) can be easily generalized to the case of non-homogeneous systems in the presence of external forces by allowing the $\mathbf{q}$-dependence in f and $f^{(2)}$ and by replacing the time derivative of f with the total derivative

$$\frac{\partial f}{\partial t} + \dot{\mathbf{q}} \cdot \frac{\partial f}{\partial \mathbf{q}} + \dot{\mathbf{p}}_1 \cdot \frac{\partial f}{\partial \mathbf{p}_1} = \int \mathrm{d}\Omega \int \mathrm{d}^3 p_2 \frac{\mathrm{d}\sigma}{\mathrm{d}\Omega} \frac{|\mathbf{p}_1 - \mathbf{p}_2|}{Vm} (f^{(2)}(\mathbf{q}, \mathbf{p}'_1, \mathbf{p}'_2, t) - f^{(2)}(\mathbf{q}, \mathbf{p}_1, \mathbf{p}_2, t)). \quad (2.17)$$

One then obtains the celebrated Boltzmann equation (2.8).

We now express the Boltzmann equation (2.8) in a form more suitable to be generalized to the quantum case. Due to the total momentum and energy conservations, one can write

$$\begin{aligned} &\int \mathrm{d}^3 p'_1 \mathrm{d}^3 p'_2 \delta^{(3)}(\mathbf{p}_1 + \mathbf{p}_2 - \mathbf{p}'_1 - \mathbf{p}'_2)\delta\left(\frac{p_1^2}{2m} + \frac{p_2^2}{2m} - \frac{(p'_1)^2}{2m} - \frac{(p'_2)^2}{2m}\right) \\ &= \int \mathrm{d}^3 P' \mathrm{d}^3 p' \delta^{(3)}(\mathbf{P} - \mathbf{P}')\delta\left(\frac{P^2}{m} + \frac{p^2}{4m} - \frac{(P')^2}{m} - \frac{(p')^2}{4m}\right) \\ &= \int \mathrm{d}^3 p' \delta\left(\frac{p^2}{4m} - \frac{(p')^2}{4m}\right) \\ &= 2mp \int \mathrm{d}\Omega, \end{aligned}$$

where Ω indicates the solid angle of the relative momentum $\mathbf{p}'$ after the collision. In going from the second to the third line we used the fact that the energy of center of mass is conserved because of the conservation of the total momentum. The Boltzmann equation (2.8) can now be rewritten as

$$\frac{\partial f_1}{\partial t} + \frac{\mathbf{p}_1}{m} \cdot \frac{\partial f_1}{\partial \mathbf{q}} + \mathbf{F} \cdot \frac{\partial f_1}{\partial \mathbf{p}_1} = \int_2 \int_{1'} \int_{2'} W_{\text{class}}(\mathbf{p}'_1, \mathbf{p}'_2; \mathbf{p}_1, \mathbf{p}_2)(f_{1'} f_{2'} - f_1 f_2), \quad (2.18)$$

where

$$\int_2 \equiv \int \mathrm{d}^3 p_2, \ldots, \text{etc.}$$

and

$$W_{\text{class}}(\mathbf{p}'_1, \mathbf{p}'_2; \mathbf{p}_1, \mathbf{p}_2) = \frac{1}{2m^2} \frac{\mathrm{d}\sigma}{\mathrm{d}\Omega} \delta^{(3)}(\mathbf{p}_1 + \mathbf{p}_2 - \mathbf{p}'_1 - \mathbf{p}'_2)\delta(E'_1 + E'_2 - E_1 - E_2). \quad (2.19)$$

We notice that the scattering kernel $W_{\text{class}}(\mathbf{p}'_1, \mathbf{p}'_2; \mathbf{p}_1, \mathbf{p}_2)$ obeys the principle of detailed balance, according to which the collisions $(\mathbf{p}_1, \mathbf{p}_2) \to (\mathbf{p}'_1, \mathbf{p}'_2)$ and $(\mathbf{p}'_1, \mathbf{p}'_2) \to (\mathbf{p}_1, \mathbf{p}_2)$

[4] A direct way to obtain the molecular chaos hypothesis is to say that the molecules $f(\mathbf{q}, \mathbf{p}_1, t)\mathrm{d}^3 q \mathrm{d}^3 p_1$ will experience in a time dt a number of collisions which corresponds to the molecules $f(\mathbf{q}, \mathbf{p}_2, t)\mathrm{d}^3 p_2 \mathrm{d}\sigma \mathrm{d}t |\mathbf{p}_1 - \mathbf{p}_2|/m$ within a cylinder of basis area $\mathrm{d}\sigma$ and height $\mathrm{d}t|\mathbf{p}_1 - \mathbf{p}_2|/m$.

occur with the same probability. This fact is the result of time reversal invariance and parity, as shown in Problem 2.1.

2.3 Extension of the Boltzmann equation to quantum systems

We now see how the previous calculation must be modified in quantum mechanics. In practice, the quantum mechanical calculation is relevant because when particles collide, their wave functions overlap and quantum mechanical effects become important, even though their motion between two successive collisions can be approximated by a classical description. The possibility of using the distribution function for a quantum system, for which the concept of trajectory loses its meaning, is discussed in Chapter 6. Here we assume that this is possible under certain circumstances. Once this possibility is accepted, the molecular-chaos hypothesis can be correspondingly formulated. The quantum Boltzmann equation is then obtained by substituting the classical scattering kernel $W_{\rm class}(\mathbf{p}'_1, \mathbf{p}'_2; \mathbf{p}_1, \mathbf{p}_2)$ with its quantum analog $W_{\rm qu}(\mathbf{p}'_1, \mathbf{p}'_2; \mathbf{p}_1, \mathbf{p}_2)$. In the classical case the scattering kernel integrated over the final states $d^3p'_1 d^3p'_2$ gives the collision rate $\gamma_{\rm coll}$. Once the the quantum expression for $\gamma_{\rm coll}$ is known, the quantum version of Eq. (2.18) follows. To evaluate $\gamma_{\rm coll}$ one must use the Fermi Golden Rule according to which the collision rate is obtained by integrating over the final states the transition probability $T(\mathbf{p}'_1, \mathbf{p}'_2; \mathbf{p}_1, \mathbf{p}_2)$ between the final and initial states, taking into account the conservation of energy:

$$\gamma_{\rm coll} = \frac{2\pi}{\hbar}\int \frac{V d^3p'_1}{(2\pi\hbar)^3}\int \frac{V d^3p'_2}{(2\pi\hbar)^3}\delta(E'_1 + E'_2 - E_1 - E_2)T(\mathbf{p}'_1, \mathbf{p}'_2; \mathbf{p}_1, \mathbf{p}_2), \tag{2.20}$$

where

$$T(\mathbf{p}'_1, \mathbf{p}'_2; \mathbf{p}_1, \mathbf{p}_2) = \frac{(2\pi\hbar)^3}{V^3}\delta^{(3)}(\mathbf{p}'_1 + \mathbf{p}'_2 - \mathbf{p}_1 - \mathbf{p}_2)|U(\mathbf{p}_1 - \mathbf{p}'_1)|^2. \tag{2.21}$$

In Eq. (2.21) T is the modulus square of the matrix element for a transition from the initial state $(\mathbf{p}_1, \mathbf{p}_2)$ to the final state $(\mathbf{p}'_1, \mathbf{p}'_2)$ and $U(\mathbf{k})$ is the Fourier transform of the interaction potential:

$$U(\mathbf{k}) = \int d^3q\ U(\mathbf{q})e^{-i\mathbf{k}\cdot\mathbf{q}/\hbar}. \tag{2.22}$$

The integrals over $\mathbf{p}'_1$ and $\mathbf{p}'_2$ are required since one must sum over all possible final states. The delta function in Eq. (2.20) expresses the energy conservation, while the one in Eq. (2.21) expresses the momentum conservation and originates from taking the matrix element of the interaction potential among plane waves for initial and final states. To understand the origin of the volume V and $(2\pi\hbar)^3$ factors, it is useful to remember that

$$\left|\int d^3x e^{i(\mathbf{p}'_1 + \mathbf{p}'_2 - \mathbf{p}_1 - \mathbf{p}_2)\cdot\mathbf{x}/\hbar}\right|^2 = V(2\pi\hbar)^3\delta^{(3)}(\mathbf{p}'_1 + \mathbf{p}'_2 - \mathbf{p}_1 - \mathbf{p}_2). \tag{2.23}$$

The factor V^3 in the denominator of Eq. (2.21) is then the result of the factor V in Eq. (2.23) and a factor V^4 due to the normalization of the wave functions for the initial and final plane wave states. The square of the transition matrix element has then the correct dimensions of an energy squared.

By the expression (2.20) of $\gamma_{\rm coll}$ we obtain the quantum scattering kernel as

$$W_{\rm qu}(\mathbf{p}_1', \mathbf{p}_2'; \mathbf{p}_1, \mathbf{p}_2) = \frac{V^2}{(2\pi\hbar)^6} \frac{2\pi}{\hbar} \delta(E_1' + E_2' - E_1 - E_2) T(\mathbf{p}_1', \mathbf{p}_2'; \mathbf{p}_1, \mathbf{p}_2). \quad (2.24)$$

From the explicit expression (2.21), one sees that $W_{\rm qu}$ also obeys the principle of detailed balance and this allows us, as for the classical case, to collect together the direct and inverse scattering rate. As emphasized already, this is valid also beyond the Fermi Golden Rule approximation. From now on we will omit the subscript in the scattering kernel $W(\mathbf{p}_1', \mathbf{p}_2'; \mathbf{p}_1, \mathbf{p}_2)$, since its classical or quantum origin does not play a relevant role in the following discussion. When the system reaches the equilibrium state the distribution f must become independent of time. A sufficient condition for equilibrium is $f_1 f_2 = f_{1'} f_{2'}$. In Section 2.5 the famous Maxwell distribution for a gas in equilibrium will be derived from this condition. We first show that the condition is also necessary for equilibrium by introducing the H-theorem in the next section.

2.4 The H-theorem and the approach to equilibrium

In a diluted gas in the absence of external fields, the distribution function does not depend on the coordinates $\mathbf{q}$. Boltzmann defines the H functional

$$\mathrm{H} = \int \mathrm{d}^3 p_1 f(\mathbf{p}_1, t) \ln f(\mathbf{p}_1, t). \quad (2.25)$$

The H-theorem states that if f satisfies the Boltzmann equation, then

$$\mathrm{dH}/\mathrm{d}t \leq 0. \quad (2.26)$$

We now proceed to the proof. The time derivative of H with respect to time is

$$\frac{\mathrm{dH}}{\mathrm{d}t} = \int \mathrm{d}^3 p_1 \frac{\partial f(\mathbf{p}_1, t)}{\partial t} \left[\ln f(\mathbf{p}_1, t) + 1\right]. \quad (2.27)$$

If $f(\mathbf{p}_1, t)$ satisfies the Boltzmann equation, by replacing in the above equation the time derivative of the distribution function as given by the Boltzmann equation (2.18) we get

$$\frac{\mathrm{dH}}{\mathrm{d}t} = \int_1 \int_2 \int_{1'} \int_{2'} W(1', 2'; 1, 2) \left[f_{1'} f_{2'} - f_1 f_2\right] \left[\ln f_1 + 1\right]. \quad (2.28)$$

In Eq. (2.28) the momenta $\mathbf{p}_1$ and $\mathbf{p}_2$ are integration variables and can be interchanged without affecting the value of the integral which can be rewritten as

$$\frac{\mathrm{dH}}{\mathrm{d}t} = \int_1 \int_2 \int_{1'} \int_{2'} W(1', 2'; 2, 1) \left[f_{1'} f_{2'} - f_1 f_2\right] \left[\ln f_2 + 1\right]. \quad (2.29)$$

We note that $W(1', 2'; 2, 1) = W(1', 2'; 1, 2)$ since the collision probability cannot change, the two particles 1 and 2 being identical. By taking half the sum of (2.28) and (2.29) one obtains

$$\frac{\mathrm{dH}}{\mathrm{d}t} = \frac{1}{2}\int_1\int_2\int_{1'}\int_{2'} W(1', 2'; 1, 2)\,[f_{1'}f_{2'} - f_1 f_2]\,[\ln f_1 f_2 + 2]\,. \tag{2.30}$$

By using again the principle of detailed balance we may interchange the role of the initial and final states and rewrite Eq. (2.30) as

$$\begin{aligned}\frac{\mathrm{dH}}{\mathrm{d}t} &= \frac{1}{2}\int_1\int_2\int_{1'}\int_{2'} W(1, 2; 1', 2')\,[f_1 f_2 - f_{1'}f_{2'}]\,[\ln f_{1'}f_{2'} + 2] \\ &= \frac{1}{2}\int_1\int_2\int_{1'}\int_{2'} W(1', 2'; 1, 2)\,[f_1 f_2 - f_{1'}f_{2'}]\,[\ln f_{1'}f_{2'} + 2]\,.\end{aligned} \tag{2.31}$$

By considering again half the sum of Eqs. (2.30) and (2.31), we finally arrive at

$$\frac{\mathrm{dH}}{\mathrm{d}t} = \frac{1}{4}\int_1\int_2\int_{1'}\int_{2'} W(1', 2'; 1, 2)\,[f_{1'}f_{2'} - f_1 f_2]\,[\ln f_1 f_2 - \ln f_{1'}f_{2'}]\,. \tag{2.32}$$

By defining $f_1 f_2 \equiv x > 0$ and $f_{1'}f_{2'} \equiv y > 0$, one sees that, since $W(1', 2'; 1, 2) > 0$, the integrand of Eq. (2.32) contains a factor

$$-(x - y)(\ln x - \ln y) \leq 0, \quad (= 0 \Leftrightarrow x = y).$$

As a result the H functional satisfies

$$\frac{\mathrm{dH}}{\mathrm{d}t} \leq 0, \tag{2.33}$$

where the equality holds when

$$f_1 f_2 = f_{1'}f_{2'}. \tag{2.34}$$

Equations (2.33) and (2.34) are the H-theorem, i.e., that the extremum condition for H is a necessary and sufficient condition for the stationarity of the distribution function. Indeed, if $\partial f/\partial t = 0$, then from Eq. (2.27) it follows that $\mathrm{dH}/\mathrm{d}t = 0$. On the other hand, if $\mathrm{dH}/\mathrm{d}t = 0$, then (2.34) holds and via the Boltzmann equation (2.8) one gets $\partial f/\partial t = 0$.

The distribution function, $f_0(\mathbf{p})$, that satisfies Eq. (2.34) is the Maxwell–Boltzmann equilibrium distribution function. The equilibrium condition (2.34) is such as to make the collision integral vanish. This fact makes its physical meaning transparent: at equilibrium in a given volume element of the phase space the number of particles that, as a result of collisions, enter the volume element is exactly matched by that of those leaving the volume element. We further notice that the condition (2.34) makes the collision integral vanish irrespective of the value of the scattering kernel W. This corresponds physically to the fact that, in a diluted gas, the collisions, although essential for the system to go to equilibrium, do not determine the form of the equilibrium distribution function. If at some initial time $t = 0$ the system is not at equilibrium and has a distribution function $f(\mathbf{p}, t) \neq f_0(\mathbf{p})$, then the time variation of f is such as to make H diminish as indicated by (2.33). Besides, since the normalization condition due to the conservation of particle number and energy is valid,

$$\frac{N}{V} = n = \int d^3 p f(\mathbf{p}, t), \quad \epsilon = \frac{E}{V} = \int d^3 p \frac{p^2}{2m} f(\mathbf{p}, t), \tag{2.35}$$

it follows that H has a finite lower bound and the Maxwell–Boltzmann distribution gives this lower bound. This can be easily seen by looking for the extrema of H with the constraints (2.35), as shown in Problem 2.2. This lower bound is independent of the hypothesis of molecular chaos. However, only if the Boltzmann equation is satisfied at any time, $f(\mathbf{p}, t)$ evolves smoothly towards the equilibrium distribution.

Boltzmann emphasizes two aspects. (1) Given a microscopic configuration of the gas, one can, at least in principle, evaluate the corresponding distribution function and hence the H functional. Let us consider such a configuration as the starting one at a given time t_0. We can evaluate at a later time $t > t_0$ the distribution function $f(\mathbf{p}, t)$ from the microscopic configuration at the time t. In general the value of the H functional may have increased or decreased. (2) According to the H-theorem, if at the time t_0, the molecular chaos hypothesis holds, then the following time evolution is such that H must decrease. One expects that, taking a microscopic configuration for a system with a large number of particles at random, the fact that it satisfies the molecular chaos hypothesis is an event with a very high probability. Hence for a given non-equilibrium initial configuration, one concludes that, almost always, the evolution towards the equilibrium implies a diminishing value for the H functional.

Following Boltzmann, one may consider this example. Let us evisage the unlikely microscopic configuration of the gas in which all the particles have the same magnitude of the velocity v but random directions at an initial time t_0. Clearly, for such an unlikely microscopic state the molecular-chaos hypothesis holds. This is clear by observing that, if one of the two colliding particles has velocity v, the probability that the other has the same velocity v is independent of the velocity of the first particle. After some time, at $t = t_1$, the majority of the particles have experienced at least one collision. Let us focus on those particles that, as a result of the collision, have acquired a velocity v_1. It is clear that there must be some other particles with velocity $\sqrt{2v^2 - v_1^2}$. During the time interval (t_0, t_1), H has decreased. Let us now reverse, at the time t_1, all particle velocities so that the gas, at microscopic level, retraces back all the configurations it came through before. In such a way H must necessarily increase in going from t_1 to $t = 2(t_1 - t_0)$. This does not contradict the H-theorem. Indeed, the microscopic configuration of the gas at time t_1 (which is the initial time for the reversed evolution) is such that a particle with velocity v_1 must necessarily collide with a particle with velocity $\sqrt{2v^2 - v_1^2}$. This configuration, clearly, does not satisfy the molecular-chaos hypothesis.

The collisions responsible for the time evolution of f can therefore bring the system to a molecular chaos state as well as take it out. The Boltzmann equation cannot be true at all times. The Boltzmann equation can only give the evolution of f on average. If it were valid at any time, once the equilibrium is reached, the system would remain in this state for ever and no fluctuation could exist.

2.5 The Maxwell–Boltzmann distribution

We have already noted that the relation (2.34) defines the equilibrium Maxwell–Boltzmann distribution function $f_0(\mathbf{p})$.[5] In fact by considering the logarithm of (2.34), one gets

$$\ln f_0(\mathbf{p}_1) + \ln f_0(\mathbf{p}_2) = \ln f_0(\mathbf{p}_{1'}) + \ln f_0(\mathbf{p}_{2'}), \tag{2.36}$$

which has the typical form of a conservation law. This suggests that the quantity $\ln f_0(\mathbf{p})$ must be a linear function of a conserved quantity. In a collision of two particles, the total kinetic energy is conserved:

$$\frac{1}{2m}\mathbf{p}_1^2 + \frac{1}{2m}\mathbf{p}_2^2 = \frac{1}{2m}\mathbf{p}_1'^2 + \frac{1}{2m}\mathbf{p}_2'^2,$$

so that a solution of (2.36) reads

$$\ln(h^3 f_0(\mathbf{p})) = a - b\mathbf{p}^2. \tag{2.37}$$

We have introduced a constant h with the dimension of an action in such a way as to make the argument of the logarithm dimensionless. We will see later on that quantum statistical mechanics requires this constant being the Planck constant and thus we have used the same symbol from the outset for simplicity. The two quantities a and b must be determined by requiring

$$\frac{N}{V} = \int \mathrm{d}^3 p f_0(\mathbf{p}) \tag{2.38}$$

and

$$\epsilon = \frac{\int \mathrm{d}^3 p f_0(\mathbf{p})(\mathrm{p}^2/2m)}{\int \mathrm{d}^3 p f_0(\mathbf{p})}, \tag{2.39}$$

where N/V and ϵ are the density and average energy of the particles. One then finds, by using the properties of the Gaussian integrals,

$$\frac{N}{V} = \frac{e^a \pi^{3/2}}{h^3 b^{3/2}} \tag{2.40}$$

and

$$\epsilon = \frac{3}{2}\frac{1}{2mb}. \tag{2.41}$$

By recalling the Bernoulli relation (2.1), which relates the average kinetic energy to the pressure P of the gas as $\epsilon = (3/2)(V/N)P$ and the equation of state for the perfect gases, $PV = Nk_\mathrm{B}T$, one obtains

$$b = \frac{1}{2mk_\mathrm{B}T}. \tag{2.42}$$

[5] Before the Boltzmann derivation via the H-theorem, the distribution that satisfies (2.34) was derived by J. C. Maxwell by exploiting the symmetry conditions defining the equilibrium (Maxwell (1860b, a)), as shown in Problem 2.3.

Then the Maxwell–Boltzmann distribution reads

$$f_0(\mathbf{p}) = \frac{N}{V}\frac{1}{(2\pi m k_B T)^{3/2}} e^{-p^2/(2mk_B T)}. \tag{2.43}$$

Notice that the constant h does not appear explicitly in the Maxwell–Boltzmann distribution.

By evaluating H for the Maxwell–Boltzmann distribution function, one obtains

$$\begin{aligned} \mathrm{H} &= \int d^3 p f_0(\mathbf{p}) \ln f_0(\mathbf{p}) \\ &= \int d^3 p f_0(\mathbf{p}) \left[\ln \left(\frac{N}{V} \frac{1}{(2\pi m k_B T)^{3/2}} \right) - \frac{\mathbf{p}^2}{2mk_B T} \right] \\ &= \frac{N}{V} \ln \left(\frac{N}{V} \frac{1}{(2\pi m k_B T)^{3/2}} \right) - \frac{1}{k_B T} \frac{N}{V} \epsilon \\ &= -\frac{N}{V} \ln \left(\frac{V}{N} (2\pi m k_B T)^{3/2} \right) - \frac{3}{2} \frac{N}{V}. \end{aligned} \tag{2.44}$$

The function $-k_B V \mathrm{H}[f_0]$, apart from a constant in temperature, coincides with the entropy S of a perfect gas which will be evaluated in Chapter 3, Eq. (3.66).

In this sense, the law of decreasing H generalizes the law of increasing entropy when a system evolves from non-equilibrium to equilibrium. Furthermore, the expression of the entropy clarifies the meaning of the constant a defined by the relation (2.40). By recalling that the chemical potential can be obtained by

$$\mu = -T \left(\frac{\partial S}{\partial N} \right)_{U,V},$$

one sees immediately that a, introduced in Eq. (2.37), is $\mu / k_B T$.

2.6 The mean free path for Maxwell–Boltzmann statistics

An important application of the Maxwell–Boltzmann distribution is the evaluation of the mean free path, a concept introduced by Clausius (1858) even before Boltzmann pointed out the importance of collisions for reaching equilibrium. The reasoning that led Clausius to the mean free path concept was due to the necessity of avoiding a result of the kinetic theory which disagreed with common experience. Suppose we have an odorous gas coming out of a container in the corner of a large room. From experience we know that it will take several minutes or more for the gas to reach the opposite corner in the room. By estimating the velocity of the gas from the temperature by using the Krönig relation (2.2), one would conclude that it will take a very small fraction of a second for the gas to cover the distance in the room. The solution to this apparent failure of the kinetic theory lies in the fact that the gas particles undergo several collisions and the velocity estimated from the Krönig relation corresponds to the velocity between two successive collisions. To cover the distance across the room the gas particle will perform a zig zag motion that will take a much longer time.

To estimate this time we refer to Problem 2.4. We now determine the formula for the mean free path.

Let us consider a particle with velocity $\mathbf{v}_1$ colliding with a *target* particle with velocity $\mathbf{v}_2$. Let us assume a constant cross section σ. In unit time, the first particle will experience a number of collisions given by

$$|\mathbf{v}_1 - \mathbf{v}_2|\sigma f(\mathbf{v}_2)\mathrm{d}^3 v_2,$$

where $f(\mathbf{v}_2)\mathrm{d}^3 v_2$ is the density of particles with velocity $\mathbf{v}_2$, instead of the momentum $\mathbf{p}_2$. The total number of collisions is obtained by integrating over $\mathbf{v}_2$. The average number of collisions per unit time, τ^{-1}, is given by averaging over the velocity $\mathbf{v}_1$ of the first particle:

$$\frac{1}{\tau} = \frac{1}{\int \mathrm{d}^3 v_1 f(\mathbf{v}_1)} \int \mathrm{d}^3 v_1 f(\mathbf{v}_1) \int \mathrm{d}^3 v_2 |\mathbf{v}_1 - \mathbf{v}_2|\sigma f(\mathbf{v}_2)\mathrm{d}^3 v_2. \tag{2.45}$$

To evaluate τ^{-1} we use the center-of-mass and relative velocities of the two colliding particles:

$$\mathbf{V} = \frac{\mathbf{v}_1 + \mathbf{v}_2}{2}, \qquad \mathbf{v} = \mathbf{v}_1 - \mathbf{v}_2. \tag{2.46}$$

In terms of newly defined velocities the expression for τ^{-1} becomes (with $n = N/V$)

$$\frac{1}{\tau} = n\sigma \left(\frac{m}{2\pi k_B T}\right)^3 \int \mathrm{d}^3 V \mathrm{e}^{-mV^2/k_B T} \int \mathrm{d}^3 v \; v \; \mathrm{e}^{-mv^2/4k_B T} = \sqrt{2} n\sigma \bar{v}, \tag{2.47}$$

where we have exploited the normalization (2.38) and the properties of the Gaussian integral. $\bar{v} = 2\sqrt{2k_B T/\pi m}$ is the average value of the modulus of the velocity, which is different from the mean square velocity given in Eqs. (2.41) and (2.42). The mean free path is evaluated as the product of the average velocity times the average time between the collisions:

$$l = \bar{v}\tau = \frac{1}{\sqrt{2} n\sigma}. \tag{2.48}$$

With a typical density at ambient pressure, $P \sim 10^5$ Pa, and room temperature, $T \sim 290$ K, $n =\sim 0.25 \times 10^{26}\,\mathrm{m}^{-3}$ and a cross section, $\sigma = \pi a^2$, determined by a molecular size of the order of the Ångstrom, $\sigma \sim 3.14 \times 10^{-20}\,\mathrm{m}^2$, one estimates $l \sim 0.9 \times 10^{-6}\,\mathrm{m} = 0.9 \times 10^4$ Å.

2.7 The Kac model

After the study of the Boltzmann analysis of the second law of thermodynamics and the introduction of molecular chaos hypothesis, it is useful to study a simple model introduced by Kac (1959), where the mathematics can be followed exactly and the physical conditions are clearly traced. The model is defined in the following way.

On a circle we consider n equidistant points; m of the points are marked and form a set called S. The complementary set (of $n - m$ points) will be called $\overline{S}$. In each interval between a point i ($i = 1, 2, \ldots, n$) and the subsequent point $i + 1$, ($n + 1$ coincides with 1), there is a ball which can be either white (w) or black (b). During an elementary time interval each ball moves clockwise to the nearest point with the following proviso. If the ball crosses into S, it changes color upon completing the move but if it crosses into $\overline{S}$, it performs the move without changing color.

This model mimics a real diluted gas in the following sense. The thermodynamic state of a gas may be specified by giving the number of particles whose velocity lies in a given interval. This defines a distribution function for the velocities. The equilibrium state is characterized by the Maxwell–Boltzmann distribution. At some initial time, the gas can be prepared in a non-equilibrium state, whose distribution function clearly differs from the equilibrium one. Experience tells that, after a sufficiently long time, the distribution function will evolve towards the equilibrium distribution function and will not change from then on. In order to reach the equilibrium distribution function, the molecules of the gas must interact among themselves so as to change their velocities. In the Kac model, the color of the balls corresponds to the dynamic variables the position and velocity of each particle, while the marked points represent the effect of the collisions. We will assume the condition that $m \ll n$ which is a *diluteness* condition. In a diluted gas, the collisions are rare.

We indicate with $N_{\rm w}(t)$ and $N_{\rm b}(t)$ the number of white and black balls at a given instant of time t. The equations of motion must determine the time evolution of $N_{\rm w}(t)$ and $N_{\rm b}(t)$, for a given initial condition $N_{\rm w}(0)$ and $N_{\rm b}(0)$. There is, of course, a conservation

$$n = N_{\rm w}(t) + N_{\rm b}(t). \tag{2.49}$$

Let us call $n_{\rm w}(t)$ the number of white balls that at the next instant of time, $t + 1$, will change their color and will become black. With $n_{\rm b}(t)$ we indicate the corresponding quantity for the black balls. These two quantities clearly depend on the specific distribution of the marked points. The equations of motion are

$$N_{\rm w}(t + 1) = N_{\rm w}(t) - n_{\rm w}(t) + n_{\rm b}(t), \tag{2.50}$$

$$N_{\rm b}(t + 1) = N_{\rm b}(t) - n_{\rm b}(t) + n_{\rm w}(t), \tag{2.51}$$

with the conservation condition (2.49) and

$$m = n_{\rm w}(t) + n_{\rm b}(t). \tag{2.52}$$

A more detailed description can be obtained in terms of microscopic variables rather than $n_{\rm w}$ and $n_{\rm b}$. At each interval we associate a two-valued variable $\eta_i = \pm 1$, $i = 1, \ldots, n$. The plus (minus) sign indicates that the ball is black (white). At each point we also associate a two-valued variable ϵ_i: $\epsilon_i = 1$ if the point belongs to $\overline{S}$ and $\epsilon_i = -1$ otherwise. By moving clockwise on the circle,

$$\eta_i(t) = \epsilon_{i-1}\eta_{i-1}(t - 1) \tag{2.53}$$

and by iteration

$$\eta_i(t) = \epsilon_{i-1} \ldots \epsilon_{i-t}\eta_{i-t}(0). \tag{2.54}$$

The interesting dynamic quantity is the difference

$$\Delta(t) \equiv N_b(t) - N_w(t) = \sum_{i=1}^{n} \eta_i(t) = \sum_{i=1}^{n} \epsilon_{i-1} \ldots \epsilon_{i-t} \eta_{i-t}(0). \tag{2.55}$$

Equation (2.55) gives the detailed microscopic solution in terms of the configuration of the marked points and of the initial conditions. The model is deterministic and reversible. If we take the set of the $\eta_i(0)$s as initial conditions and let the circle evolve *clockwise*, after t time steps we recover the values $\eta_i(0)$. We note also that the model has a "Poincaré cycle"[6] $T = 2n$: after two complete turns of the circle each ball goes through a marked point twice and recovers its initial color. We will limit to times $t \ll n$ as n goes to infinity to mimic the thermodynamic limit.

2.7.1 The statistical solution

If we are not interested in following the result for a specific system, but we want to learn about the average response of the system for $t \gg 1$ but $t \ll n \to \infty$ (which corresponds to the thermodynamic limit), we can make a probabilistic assumption analog of the molecular chaos assumption. We assume that the color of the ball is uncorrelated with having a marked point ahead. As a result $n_b(t)$, ($n_w(t)$) factorizes into the product of the probability of having a marked point in front, $\mu \equiv m/n$, times the total number $N_b(t)$ ($N_w(t)$). In this case $n_w(t) = \mu N_w(t)$ and $n_b(t) = \mu N_b(t)$ and from Eq. (2.50) and Eq. (2.51)

$$\Delta(t+1) = (1 - 2\mu)\Delta(t) = (1 - 2\mu)^t \Delta(0). \tag{2.56}$$

In the "thermodynamic limit" ($n \to \infty$, $(m/n) < 1/2$ finite), as $t \to \infty$, $\Delta(t) \equiv N_b(t) - N_w(t) \to 0$ and reaches equilibrium with equal numbers of white and black balls whatever their initial difference. This solution, based on a probabilistic assumption, presents the same paradox as the H-theorem. It introduces an arrow in time even if the equations of motion are time reversal invariant. This contradiction has to be traced back in the probabilistic hypothesis analog to molecular chaos.

To see how this "molecular chaos" hypothesis arises in the statistical approach of the Kac model we look for a statistical solution.

According to Eq. (2.55), $\Delta(t)$ is expressed as sum of the initial values of the variables $\eta_i(0)$, each weighted by a string $\epsilon_i^t = \epsilon_{i-1} \ldots \epsilon_{i-t}$. The latter takes the values ± 1 and a statistical approach amounts to evaluating the corresponding probabilities. If we assume that the marker positions are equally probable, then the average value of ϵ_i^t is independent of the initial position i and

$$\langle \epsilon_i^t \rangle = \langle \epsilon_1 \ldots \epsilon_t \rangle. \tag{2.57}$$

[6] The concept of the Poincaré cycle arises in the context of the Poincaré theorem which will be proven in the next chapter. For the purpose of the present discussion it suffices to regard the time T of the Poincaré cycle as the first return time of a dynamical system to its initial condition.

Consider that s out of these t intervals are marked. Clearly in this case the value taken by the string variable reads

$$\epsilon_i^t = (-1)^s. \tag{2.58}$$

We are then interested in the probability of the occurrence of s marked points out of t points. This is equivalent to asking the probability of having s *heads* out of t tosses of a coin with p being the probability of having *heads* in a single trial:

$$P_t(s) = \frac{t!}{s!(t-s)!} p^s (1-p)^{t-s}, \tag{2.59}$$

where in this case $p = \mu$. The evaluation of the average value of the string variable is easily carried out and one has

$$\langle \epsilon_i^t \rangle = \sum_{s=1}^{t} (-1)^s P_t(s) = (1-2\mu)^t. \tag{2.60}$$

Then the statistical average value of Δ at time t will be

$$\langle \Delta(t) \rangle = \sum_{i=1}^{n} \langle \epsilon_{i-1} \ldots \epsilon_{i-t} \rangle \eta_i(0) = (1-2\mu)^t \Delta(0). \tag{2.61}$$

Since $1 - 2\mu < 1$, *remarkably*, when $t \to \infty$ (i.e., large but still small compared to n; take for instance $t = 10^6$ and $n = 10^{23}$), the average value of Δ goes to the *equilibrium* value $\Delta = 0$. This means that for most (and overwhelmingly the majority) of the instances of time evolutions considered in the average $\langle \ldots \rangle$ the variable Δ goes to zero. Clearly, there will be instances that will *deviate* from this behavior. This example shows clearly the probabilistic nature of a tendency to equilibrium.

At equilibrium the time evolution of a single system has been substituted with the averaged behavior of an ensemble. This procedure will be carried out in general by introducing the Gibbs ensembles.

2.8 Problems

2.1 Show that $W(\mathbf{p}_1', \mathbf{p}_2'; \mathbf{p}_1, \mathbf{p}_2) = W(\mathbf{p}_1, \mathbf{p}_2; \mathbf{p}_1', \mathbf{p}_2')$.

2.2 Find the extrema of H with the constraints (2.35).

2.3 Derive the Maxwell–Boltzmann distribution by using the following assumptions: (i) the distribution of the i-th component of the velocity is a function $g(v_i)$ identical for all $i = x, y, x$; (ii) the distributions for the different components are independent of each other; and (iii) the distribution of the total velocity only depends on its magnitude.

2.4 Estimate the order of magnitude of the time necessary for gas particle to travel the distance of one meter taking into account the collisions with the other particles.

2.5 Use the Boltzmann equation to derive the law of conservation of mass.

2.6 By assuming that a gas is weakly perturbed with respect to the Maxwell–Boltzmann equilibrium distribution, find the linearized form of the collision integral I.

2.7 A weak temperature gradient, $\partial T(\mathbf{q})/\partial \mathbf{q}$ is applied to a gas of hard spheres. By assuming that the mean free path (2.48) is smaller than the characteristic length over which the temperature varies, find the approximate solution of the Boltzmann equation.

2.8 By defining the thermal energy current density as

$$\mathbf{j}_{\mathrm{Q}} = \int \mathrm{d}^3 p \, \frac{\mathbf{p}}{m} \frac{p^2}{2m} f(\mathbf{q}, \mathbf{p})$$

find the thermal conductivity of a gas of hard spheres.

3 From Boltzmann to Boltzmann–Gibbs

After the H-theorem appeared, two questions arose. Molecular collisions are governed by the time reversal invariant equations of motion and the Poincaré theorem, which we will discuss soon, implies a quasi-periodic H function (Zermelo's recurrence paradox). On the contrary, according to the Boltzmann equation and the H-theorem, a system evolves continuously towards equilibrium (Loschmidt's reversibility paradox).

The two apparent paradoxes can be resolved by stating that the Boltzmann equation, the H-theorem, and the assumption of molecular chaos, valid at any time, are not a direct consequence of the microscopic laws. Firstly, as we have shown explicitly in the Kac model, probabilistic arguments have to be taken into account and, secondly, stating that macro-states of a system will be retraced in the future without mentioning, with Boltzmann, how long it will take for this to happen does not make any physical sense as far as observability is concerned. The Poincaré theorem, a direct consequence of the equations of motion, is obviously correct. It is not correct to apply it to refute macroscopic irreversibility and the laws of thermodynamics. Before proceeding in this discussion we briefly recall the Liouville and Poincaré theorems.

3.1 Liouville and Poincaré theorems

The micro-state of an N-body system is determined by a representative point P in the $6N$-dimensional phase space Γ of the position and momentum coordinates of the N particles $(q_1, \ldots, q_{3N}, p_1, \ldots, p_{3N})$, each particle moving in a three-dimensional ambient space. The point P evolves in time according to the Hamilton equation of motion

$$\dot{q}_i = \frac{\partial H}{\partial p_i}, \quad \dot{p}_i = -\frac{\partial H}{\partial q_i}, \quad i = 1, \ldots, 3N, \tag{3.1}$$

where H is the Hamiltonian of the system (not to be confused with the Boltzmann H function). The Hamilton equations determine uniquely the trajectory of a representative point $P(t)$ in its evolution with velocity $\mathbf{v} = (\dot{q}_1, \ldots, \dot{q}_{3N}, \dot{p}_1, \ldots, \dot{p}_{3N})$ on the energy surface $\sigma(E)$ with

$$H(q_1, \ldots, q_{3N}, p_1, \ldots, p_{3N}) = E, \tag{3.2}$$

actually in a space region $E < H < E + \Delta E$, where $\Delta E \ll E$ is the uncertainty with which the energy is known.

According to the Liouville theorem, the measure of any domain A in Γ and that of its evoluted domain at time t, A_t, are equal:

$$V(A) = \int_A \mathrm{d}\Gamma = V(A_t). \tag{3.3}$$

Here with $\mathrm{d}\Gamma$ we indicate the infinitesimal volume element in the phase space. Equation (3.3) can be proved by observing that the solution of the equation of motion (3.1) defines a transformation in the Γ space from the variables x_i^0 at the starting time to the variables x_i at time t, $x_i = g_i(\{x_j^0\}, t)$, where $x_i = q_i$ for $i = 1, \ldots, 3N$ and $x_i = p_i$ for $i = 3N + 1, \ldots, 6N$ and that the corresponding Jacobian,[1] J, does not depend on time:

$$\begin{aligned} \frac{\partial J}{\partial t} &= \frac{\partial}{\partial t} \frac{\partial(x_1, \ldots, x_{6N})}{\partial\left(x_1^0, \ldots, x_{6N}^0\right)} \\ &= \sum_{i=1}^{6N} \frac{\partial(x_1, \ldots, \dot{x}_i, \ldots, x_{6N})}{\partial\left(x_1^0, \ldots, x_{6N}^0\right)} \\ &= \sum_{i=1}^{6N} \sum_{l=1}^{6N} \frac{\partial \dot{x}_i}{\partial x_l} \frac{\partial(x_1, \ldots, x_{l-1}, x_l, x_{l+1}, \ldots, x_{6N})}{\partial\left(x_1^0, \ldots, x_{6N}^0\right)} \\ &= J \mathrm{div} \cdot \mathbf{v} = 0. \end{aligned} \tag{3.4}$$

The last two equalities follow from the vanishing of the determinant when i is different from l and from the Hamilton equations, as shown in Problem 3.1.

Since during the time evolution the system remains on the energy surface $\sigma(E)$, it is convenient to define a measure for domains on this surface. The measure on $\sigma(E)$ can be derived by the Euclidean measure on Γ by the following argument. Call $\mathrm{d}\sigma$ an element on this surface and $\mathrm{d}n$ the normal distance between $\sigma(E)$ and $\sigma(E + \Delta E)$, then, considering ΔE as a differential,

$$\int_{A_t} \mathrm{d}n \mathrm{d}\sigma = \int_A \mathrm{d}n \mathrm{d}\sigma = \Delta E \int_\Sigma \frac{\mathrm{d}\sigma}{\|\nabla H\|}, \tag{3.5}$$

where Σ is the domain in $\sigma(E)$ corresponding to A in Γ and the norm of ∇H is

$$\|\nabla H\|^2 = \sum_{i=1}^{3N} \left[\left(\frac{\partial H}{\partial q_i} \right)^2 + \left(\frac{\partial H}{\partial p_i} \right)^2 \right]. \tag{3.6}$$

Since ΔE is a prefixed constant, we can define a normalized invariant measure $\mu(A)$ on the surface as

$$\mu(A) = \int_\Sigma \frac{\mathrm{d}\sigma}{\|\nabla H\|} \left(\int_\sigma \frac{\mathrm{d}\sigma}{\|\nabla H\|} \right)^{-1}. \tag{3.7}$$

We now recall the proof of the Poincaré theorem, which states that a system having a finite energy and confined to a finite volume will return, in its evolution in the space Γ, arbitrarily close to any initial point, provided one waits a sufficiently long time. The proof begins by observing that the volume (or the surface at constant energy) in the space Γ

[1] See Eq. (1.47) for the explanation of the notation used for the Jacobian.

available to the system is finite. Consider now a domain A on the constant energy surface and its evolution A_t after some time t. At time $2t$, the initial domain will have evolved to A_{2t}. By considering the sequence of times $0, t, 2t$, and so on, there is a sequence of domains

$$A, \quad A_t, \quad A_{2t}, \ldots, \quad A_{nt}, \ldots. \tag{3.8}$$

By Liouville's theorem all the domains have the same measure and since they are all contained in the total available finite domain, it must occur that

$$A_{kt} \bigcap A_{lt} \neq 0, \quad k > l \geq 0. \tag{3.9}$$

From the unicity of the motion and by tracing the motion backwards, this implies that $B \equiv A_{(k-1)t} \bigcap A \neq 0$. Hence a point contained in B (and also in A) will evolve in a point still in B. Since B is contained in A and A can be taken arbitrarily small, the final point will come back arbitrarily close to the initial point except for a subset of measure zero. The Poincaré theorem does not give any estimate of the recurrence time. Boltzmann pointed out that this time for a macroscopic system is so long that there is no possibility of it being observed.

3.2 Equilibrium statistical mechanics for a gas

Going back to the analysis of the two apparent paradoxes, one can state, more precisely, the H-theorem in a form that is in agreement with time reversal and recurrence. If, at a time $t = 0$, a gas is in a state of molecular chaos, then, at $t = 0^+$, $\mathrm{dH}/\mathrm{d}t \leq 0$, the equality being valid if and only if $f(\mathbf{p}, t)$ is a Maxwell–Boltzmann distribution. Then, by using the time reversal invariance, one can show that, if the gas is in a state of molecular chaos, i.e. $f(\mathbf{p}, t)$ is a solution of the Boltzmann equation, the function H is at a local peak. At $t = 0$ one considers a gas identical to the first one but with all the velocities reversed. The second gas also is in a state of molecular chaos and $\mathrm{dH}/\mathrm{d}t \leq 0$. But the future of this gas is the past of the original one and $\mathrm{dH}/\mathrm{d}t \geq 0$ at $t = 0^-$ and the function H can increase and decrease. Indeed the Boltzmann equation and the molecular chaos cannot be valid at all times, otherwise not only could the improbable big fluctuations related to the Poincaré cycles not be recovered, but also the ordinary observed small fluctuations should be absent. Let us remember that H reaches a minimum (see Problem 2.2) when f is the Maxwell–Boltzmann distribution, independently of the fact that the micro-state of the gas satisfies the hypothesis of molecular chaos. When H corresponds to an equilibrium state a fluctuation must lead to an increase of it. We can then state with Boltzmann that the largest majority of possible states are Maxwellian, i.e. if we choose at random among all possible micro-states compatible with a set of macroscopic conditions specifying a macro-state, in the overwhelming majority of cases we will end up with states corresponding to a Maxwell–Boltzmann distribution. This leads to a probabilistic interpretation of the time evolution towards equilibrium; the inverse process is very improbable. This answers the second paradox. More importantly, this means that if we follow the trajectory of a representative point in Γ space for the longest time the point will meet micro-states corresponding to equilibrium for the given macroscopic conditions.

It is then possible to derive the Maxwell–Boltzmann distribution without referring to its time evolution, but only to equilibrium micro-states. Instead of following a specific system in its time evolution, we can consider a statistical ensemble, i.e. a collection of identical copies of the given system occupying all the compatible micro-states. A statistical ensemble is defined by giving a distribution $\rho(\{q\}, \{p\})$ of representative points in Γ space such that $\rho\, \mathrm{d}^{3N}q\mathrm{d}^{3N}p$ gives the number of points of the micro-states that are in the phase space volume element $\mathrm{d}\Gamma = \mathrm{d}^{3N}q\mathrm{d}^{3N}p$ around the point $(\{q\}, \{p\})$. Representative points in Γ cannot be created or destroyed as a consequence of the Liouville theorem, so that during its evolution ρ must satisfy the continuity equation

$$\frac{\partial \rho}{\partial t} + \nabla \cdot \rho \mathbf{v} = \frac{\partial \rho}{\partial t} + \nabla \rho \cdot \mathbf{v} + \rho \nabla \cdot \mathbf{v} = 0. \tag{3.10}$$

Since by the Hamilton equations $\nabla \cdot \mathbf{v} \equiv 0$ (see Problem 3.1), we can conclude that the total time derivative of ρ is zero,

$$\frac{\mathrm{d}\rho}{\mathrm{d}t} \equiv \frac{\partial \rho}{\partial t} + \nabla \rho \cdot \dot{\mathbf{x}} = 0.$$

For an equilibrium ensemble, the distribution function cannot depend explicitly on time and then must depend on the Γ space coordinates only through the Hamiltonian.

It is clear that the distribution $f(\mathbf{q}, \mathbf{p}, t)$ in the six-dimensional μ-space corresponds to a coarse-grained description of the system: given a point in Γ space, $f(\mathbf{q}, \mathbf{p}, t)$ is determined, and vice versa, given f there are several micro-states compatible with the same $f(\mathbf{q}, \mathbf{p}, t)$. Hence to a given f corresponds a volume in Γ. Again following Boltzmann we can discretize the distribution function by specifying the set of M occupation numbers n_i $(i = 1, \ldots, M)$ of M cells of volume ω_i with the condition that

$$\sum_{i=1}^{M} n_i = N, \quad \sum_{i=1}^{M} \epsilon_i n_i = E, \quad \epsilon_i = \frac{p_i^2}{2m}. \tag{3.11}$$

In this way the ratio n_i/ω_i represents the discretized value of the distribution function in the volume element ω_i of dimension of an action. The volume in Γ corresponding to a given set of occupation numbers, $\{n_i\}$, is

$$V_\Gamma(\{n_i\}) = \frac{N!}{\prod_i n_i!} \prod_i \omega_i^{n_i}, \tag{3.12}$$

where the combinatorial factor $N!/\prod_i n_i!$ comes from the distribution of N distinguishable objects in the M boxes, the i-th box containing n_i objects. We want to determine the set $\{\bar{n}_i\}$ of occupation numbers that makes $V_\Gamma(\{n_i\})$ largest with the constraints (3.11). To this end it is convenient to study the maximum of the logarithm of the volume $V_\Gamma(\{n_i\})$:

$$\begin{aligned} \ln V_\Gamma(\{n_i\}) &\sim N \ln N - N + \sum_i n_i \ln \omega_i - \sum_i (n_i \ln n_i - n_i), \\ &= N \ln N + \sum_i n_i \ln \frac{\omega_i}{n_i}, \end{aligned} \tag{3.13}$$

where we have used the Stirling formula, whose derivation can be found in Appendix B. Notice that the Stirling formula has been applied also to $\ln n_i!$, i.e. we consider the cell's volume much larger than any microscopic characteristic volume, but much smaller than the size of the system ($M \ll N$). We see that the logarithm of the volume $V_\Gamma(\{n_i\})$ has the same functional form as the H function, Eq. (2.25), changed in sign. After introducing two Lagrange multipliers α and β for the particle and energy conservation constraints, respectively, we perform a variation of the occupation numbers $n_i \to n_i + \delta n_i$ and get

$$\delta(\ln V_\Gamma(\{n_i\})) - \alpha \sum_{i=1}^{M} \delta n_i - \beta \sum_{i=1}^{M} \epsilon_i \delta n_i = 0, \tag{3.14}$$

from which we derive

$$\frac{\bar{n}_i}{\omega_i} = C e^{-\beta\epsilon_i}, \tag{3.15}$$

with C a suitable normalization constant. The second variation is

$$\delta^2 \ln V_\Gamma(\{n_i\}) = -\sum_{i=1}^{M} \frac{1}{n_i} \delta n_i^2 < 0, \tag{3.16}$$

which shows that at the set $\bar{n}_i$, $V_\Gamma(\{n_i\})$ has a maximum. This maximum is also strongly peaked. We calculate the average value $\langle n_i \rangle$ and the mean square fluctuations $\langle n_i^2 \rangle - \langle n_i \rangle^2$ by a general procedure which can be applied whenever the quantity we want to average appears multiplied by a "field" h_i in the exponent of a generating functional of the probability. In this case introducing $h_i = \ln \omega_i$ and the generating functional $\ln \sum_{\{n_j\}} V_\Gamma(\{n_j\})$, where the sum is over all the possible sets of $\{n_i\}$, we have the average value

$$\langle n_i \rangle = \frac{\sum_{\{n_j\}} n_i V_\Gamma(\{n_j\})}{\sum_{\{n_j\}} V_\Gamma(\{n_j\})} = \frac{\partial}{\partial h_i} \ln \sum_{\{n_j\}} V_\Gamma(\{n_j\}) \tag{3.17}$$

and the variance

$$\langle n_i^2 \rangle - \langle n_i \rangle^2 = \frac{\sum_{\{n_j\}} n_i^2 V_\Gamma(\{n_j\})}{\sum_{\{n_j\}} V_\Gamma(\{n_j\})} - \left(\frac{\sum_{\{n_j\}} n_i V_\Gamma(\{n_j\})}{\sum_{\{n_j\}} V_\Gamma(\{n_j\})} \right)^2 = \frac{\partial^2 \ln V_\Gamma(\{n_j\})}{\partial h_i^2}. \tag{3.18}$$

By putting (3.17) into (3.18), one has the general result

$$\left\langle n_i^2 \right\rangle - \langle n_i \rangle^2 = \frac{\partial \langle n_i \rangle}{\partial h_i}. \tag{3.19}$$

Assuming $\langle n_i \rangle \approx \bar{n}_i$ i.e. that the distribution is strongly peaked, as can be checked a posteriori, we get from Eq. (3.15)

$$\sqrt{\left\langle \left(\frac{n_i}{\bar{n}_i}\right)^2 \right\rangle - \left\langle \frac{n_i}{\bar{n}_i} \right\rangle^2} = \frac{\sqrt{\bar{n}_i/\bar{n}_i}}{\sqrt{\bar{n}_i}}, \quad \bar{n}_i \gg 1. \tag{3.20}$$

In the large N limit, the cells can be taken large enough such that also each $\bar{n}_i$ is large and the Maxwell–Boltzmann distribution is therefore the most probable distribution. If a system leaves equilibrium then it must, with high probability, return to it. The number of states which deviate from the Maxwell–Boltzmann distribution is evanescently small in the thermodynamic limit. Furthermore, if initially the system is in a non-equilibrium state, the time to come back to this state is enormously large, due to the almost vanishing probability of hitting again the region of the space Γ corresponding to the initial state, thus supporting the Boltzmann hypothesis that a system during its time evolution spends the longest time in an equilibrium state.

All the above arguments support a close connection between the logarithm of the volume $V_\Gamma(\{n_i\})$ and the H function at equilibrium where it reaches its minimum. It is therefore very natural to define the entropy as proportional to the logarithm of $V_\Gamma(\{n_i\})$, at its maximum, i.e. for the equilibrium distribution $\{\bar{n}_i\}$. However, if one requires that the entropy must be extensive, Eq. (3.12) naturally suggests to define the entropy as

$$S = k_{\rm B} \ln \left(\frac{V_\Gamma(\{n_i\})}{\omega_{\rm min}^N N!} \right) = \sum_{i=1}^{M} k_{\rm B} \ln \left(\frac{\omega_i^{n_i}}{\omega_{\rm min}^{n_i} n_i!} \right) \equiv \sum_{i=1}^{M} s_i , \tag{3.21}$$

having associated a natural entropy with each volume element ω_i. The constant $\omega_{\rm min}$, with dimension of the Planck constant cubed, has been introduced to make the argument of the logarithm dimensionless. It will be shown that with the above definition of the entropy, the functional H evaluated with the Maxwell–Boltzmann distribution in Eq. (2.44), coincides, but for a sign, with the entropy (3.66) for a gas.

Notice that Eq. (3.21) is the Boltzmann relation (1.54) (Boltzmann (1872)) between the entropy of a macroscopic system and the number of its micro-states W compatible with given macroscopic conditions, i.e. with the volume V_Γ corresponding to a given set of occupation numbers $\{n_i\}$. The definition (3.21) contains the so-called *correct Boltzmann counting*, which has been often attributed to the ultimate quantum mechanical nature of particles. We see that there is no need to invoke quantum mechanics in order to have a good definition of entropy in classical statistical mechanics. On the other hand it is reasonable to expect that quantum statistical mechanics yields in the classical limit a result in agreement with (3.21). The same correspondence will fix the constant $\omega_{\rm min}$ as the cube of the Planck constant. In this case $\omega_{\rm min} = h^3$ is the minimum cell available for discretization compatible with the uncertainty principle.

In the next section we generalize the concepts discussed so far to introduce the Gibbs approach (Gibbs (1902)) and the equilibrium ensembles.

3.3 Equilibrium statistical mechanics for a generic macroscopic system

Following the idea that a gas can be described by a set of M occupation numbers n_i, we can say, in general, that a macroscopic state of any system is defined by a set of macroscopic variables

$$y_i = f_i(q_1, \ldots, q_{3N}, p_1, \ldots, p_{3N}) \equiv f_i(\{q\}, \{p\}), \quad i = 1, \ldots, M \ll N. \tag{3.22}$$

This coarse-grained description of a macroscopic system is somewhat arbitrary and depends on how detailed the information we require is, just as the definition itself of an equilibrium macro-state depends on the time of observation.

As for the gas, to each set of $\{y_i\}$ corresponds a volume in Γ space, in particular a region on the energy surface σ defined by (3.2). The macroscopic equilibrium is assumed to be given by the set $\{y_i\}$ corresponding to the overwhelmingly largest region Σ^0 in σ as for the Maxwell–Boltzmann distribution.

During the time evolution of a system, a representative point will move along a trajectory on the energy surface. Assuming as a postulate that all the micro-states, i.e., all the representative points on the energy surface, are equally accessible or probable, the time $t(\Sigma^0)$ that the system spends in Σ^0 is proportional to the measure $\mu(\Sigma^0)$ and, if the system is not at equilibrium, very soon it will reach equilibrium, while, if it is initially at equilibrium, it will stay at equilibrium, but for small deviations due to fluctuations within the noise range. Larger and larger fluctuations will be more and more rare. For instance, the Poincaré recurrence time can be evaluated for a generic region Σ of the surface at constant energy $\sigma(E)$. Consider the sum $T_\mathrm{R} = \Sigma_i^M \tau_i^\mathrm{R}$ of M return time τ_i^R. The fraction of time spent in Σ is given by $M\tau_\Sigma / T_\mathrm{R}$, where τ_Σ is the time of permanence in Σ. For M very large one has $M\tau_\Sigma / T_\mathrm{R} = \mu(\Sigma)/\mu(\sigma)$ and the average recurrence time is equal to the inverse normalized measure on the hypersurface in $6N$ dimensions, i.e. it is proportional to (const. $> 1)^N$, N being of the order of the Avogadro number, in agreement with the Boltzmann intuition.

If we consider in general a set $\{y_i\}$ characterizing a macro-state of a system out of equilibrium there will be a region Σ of the energy surface σ with a uniform distribution of representative points $\rho(\{q\}, \{p\}, t) \neq 0$ such that $\mu(\Sigma) \ll \mu(\Sigma^0)$. However, ρ evolves according to the Liouville theorem and, therefore, the measure is conserved. In what sense, then, does a system evolve towards an equilibrium state with the largest measure?

Actually the description in terms of the $\{y_i\}$ is a coarse-grained description in the sense that the set $\{y_i\}$ is compatible with all the micro-states in the region Σ. By the same token, the time evolution also occurs in a coarse-grained sense, and the initial set of points will evolve to occupy uniformly a much larger surface portion to reach equilibrium. The degree of uniformity and the time required to reach it depends on how refined our coarse-grained description of the system is. All the information on the initial conditions is lost during the coarse-grained evolution.

Now it is clear how irreversibility arises. If we reverse the velocity of all the particles of a micro-state belonging to the region Σ^0, we obtain a micro-state which is still within Σ^0. In order to trace back the former evolution, we should pick up, among all possible micro-states, exactly the evolved values of the initial micro-state in Σ. However, since the volume of Σ^0 is overwhelmingly larger than that of Σ, this possibility is evanescently small. Irreversibility, rather than being due to an impossibility, arises from an improbability.

Let us consider the example of the adiabatic expansion of a gas of N particles, initially confined by a wall in one half of a container of volume V. When the wall is removed the gas will find itself initially in a non-equilibrium macroscopic state with respect to the modified constraints. As a macroscopic variable we can consider the gas density, which initially has a value $N/(V/2)$ on one side and zero on the other. After some time, the gas will fill the entire volume of the container with uniform density N/V. Clearly such a macroscopic state could have been also reached starting from the different initial condition in which the gas was confined in the other half of the container or from that in which the gas occupied the two extreme quarters of the container and so on. Since the volume in the space Γ has been increased by a factor 2^N, the number of initial micro-states different from that considered is overwhelmingly larger than the number of the initial micro-states. Starting from a situation in which the gas fills the entire container, in order to observe the reverse process in which the gas contracts itself into one half of the container, we should be sure to pick up, among all the possible micro-states compatible with the uniform distribution over the entire box, one of those relatively few that when reversed in time would end up with a micro-state having all the particles in one half of the container. Since the arrangement of the microscopic state of the gas is not something that can be done in terms of thermodynamic variables, we can only rely on the accidental fact of putting the gas in the right configuration. But the probability for this to occur is of order 2^{-N}. Furthermore, in the space Γ the right configuration will be very close to a large number of bad configurations (i.e. configurations whose future time evolution will produce a different density distribution) and then is extremely sensitive to small perturbations. On the hypothesis that the time a system spends in a region of the phase-space Γ is proportional to its measure, the spontaneous contraction of a gas from a larger volume V_1 to a smaller one V_2 will have a time ratio $T_1/T_2 \sim e^{N \ln(V_1/V_2)}$ between the time of observation of a homogeneous system in V_1 with respect to the observation of an homogeneous system within the smaller volume V_2, which is exponentially large with the Avogadro number as soon as the ratio V_2/V_1 is different from one by any finite amount. In Problem 3.2, the probability of deviation from homogeneity is evaluated explicitly.

The fundamental principle by Boltzmann is that during the time evolution of an N-body system, the representative point $P(\{q\}, \{p\})$ of its state will visit any part of the Γ space compatible with the macroscopic constraints, possibly returning after a long time (possibly infinite) to the starting point. We can then postulate that, over a long time compared to any microscopic characteristic time, the average of the properties of a single system is equal to the average over a uniform distribution of representative points in Γ space. When a macroscopic system is at thermodynamic equilibrium its state (the representative point in Γ space) can be, with equal probability, in any configuration compatible with the specified macroscopic conditions. Therefore, at thermodynamic equilibrium, the average of every observable with respect to the time evolution of a single system according to its molecular dynamics on an observation time much longer than any microscopic scale must be equal to the ensemble average over a uniform distribution of an infinite number of copies of the system. This is essentially the content of the two ergodic theorems. Preliminarily we notice that the observables of a system are uniquely determined by its micro-state and are therefore a set of phase functions y_i of the $6N$ variables of a point in the Γ space specified by Eq. (3.22). When a measurement is performed, the process is never instantaneous, rather the time of measurement is much longer than any microscopic characteristic time, i.e. the result of the measure is not the instantaneous value of the observable $y_i(q_1(t), \ldots, q_{3N}(t), p_1(t), \ldots, p_{3N}(t))$, but rather its time average

$$\bar{y}_i = \lim_{\tau \to \infty} \frac{1}{\tau} \int_0^\tau \mathrm{d}t \; y_i(q_1(t), \ldots, q_{3N}(t), p_1(t), \ldots, p_{3N}(t)).$$

We can now state the two ergodic theorems.

(1) For a bounded mechanical motion, during its time evolution on the energy surface, the time average defined above of any phase function integrable on the surface $E = \text{const.}$ exists almost always and does not depend on the initial point on the trajectory.

(2) If the system is such that the energy surface cannot be partitioned into two disjoint portions with separated evolution of their points (metrically transitive property), the time average of an integrable phase function is almost always equal to the ensemble average over the energy surface with uniform distribution.[2] This entails that the time spent in a region of the energy surface σ is proportional to the measure of this region. Roughly speaking, if the two averages of an observable must be equal, the time average corresponds to the most probable value and the deviations are small.

Transferring the possibility of a macroscopic description of any given system to a mathematical condition such as the metrically transitive property does not make it more acceptable than the Gibbs postulate itself we are going to discuss in the next section.

3.4 The microcanonical ensemble and the entropy

The above analysis makes plausible the use, for an isolated macro-system, of a uniform distribution over all the accessible and equally probable micro-states on the energy surface to obtain, according to the Gibbs postulate, its macroscopic equilibrium:

$$\bar{\rho}(\{q\},\{p\}) = \begin{cases} 1, & E \leq H(\{q\},\{p\}) \leq E + \Delta E, \\ 0, & \text{otherwise.} \end{cases} \tag{3.23}$$

The volume $V_\Gamma(E)$ in the Γ space occupied by the microcanonical distribution is

$$V_\Gamma(E) = \int_{E<H<E+\Delta E} \mathrm{d}\Gamma \; \bar{\rho}(\{q\},\{p\}) = \int_{H=E} \frac{\mathrm{d}\sigma}{\|\nabla H\|} \Delta E \equiv \omega(E)\Delta E, \tag{3.24}$$

where $\mathrm{d}\sigma$ is the infinitesimal surface element on the constant energy surface $\sigma(E)$ and the norm of the gradient in phase space of the Hamiltonian H is defined by Eq. (3.6). $\omega(E)$ coincides with the density of states of the system on the energy surface. To see this we define $\Xi(E)$, the volume of the entire hypersphere in Γ by introducing the theta function

$$\Xi(E) = \int_{H<E} \mathrm{d}\Gamma = \int \theta(E - H)\mathrm{d}\Gamma, \tag{3.25}$$

from which it follows that

$$V_\Gamma(E) = \Xi(E + \Delta E) - \Xi(E) \tag{3.26}$$

and since $\Delta E \ll E$,

$$\omega(E) = \frac{\partial \Xi}{\partial E} = \int \delta(E - H)\mathrm{d}\Gamma. \tag{3.27}$$

[2] A discussion of the ergodic problem may be found in the book by Khinchin (1949).

The normalized distribution ρ is

$$\begin{aligned}\rho &= \frac{\bar{\rho}}{\int \mathrm{d}\Gamma\,\bar{\rho}}\\ &= \frac{\delta(H-E)}{\omega(E)},\end{aligned} \tag{3.28}$$

where $\omega(E)$ is the measure of the energy surface $\sigma(E)$ and we have used the approximation $\bar{\rho} \approx \delta(H-E)\Delta E$ valid for $\Delta E \ll E$. This shows that the energy uncertainty ΔE, provided that the condition $\Delta E \ll E$ is satisfied, does not play any role as far as ensemble averages are concerned.

In the thermodynamic limit we can define the function

$$S = k_{\mathrm{B}} \ln V_\Gamma \sim k_{\mathrm{B}} \ln \Xi \sim k_{\mathrm{B}} \ln \omega. \tag{3.29}$$

Indeed the three definitions are equivalent in the large-N limit. The function S coincides with the entropy of the system, i.e. it is additive and cannot decrease for an isolated system, according to the probabilistic setting by Boltzmann. To see the property of additivity, suppose that the system is decomposed into two parts in a such a way that the Hamiltonian can be written as

$$H = H_1(\{q_1\}, \{p_1\}) + H_2(\{q_2\}, \{p_2\}), \tag{3.30}$$

where the indices 1 and 2 label the two subsystems. The energy, the number of particles and the volume of the two subsystems must satisfy the relations

$$E_1 + E_2 = E, \tag{3.31}$$

$$N_1 + N_2 = N, \tag{3.32}$$

$$V_1 + V_2 = V. \tag{3.33}$$

To begin with let us imagine that the two systems can exchange energy and particles through the separation wall dividing them, but maintain their respective volumes fixed. The volume in the Γ space of the entire system can be decomposed in terms of the phase-space volumes of the two subsystems as

$$V_\Gamma(E, N) = \sum_{N_1=0}^{N} \frac{N!}{N_1!(N-N_1)!} \sum_{k=1}^{E/\Delta E} V_{\Gamma,1}(E_k, N_1) V_{\Gamma,2}(E-E_k, N-N_1). \tag{3.34}$$

To keep the notation simple, we do not indicate the volume dependence for the time being. The binomial coefficient in the right-hand side of Eq. (3.34) counts the number of ways in which we can assign N_1 particles to the first subsystem and $N_2 = N - N_1$ to the second. Given the uncertainty with which the energy is known, there are $E/\Delta E$ ways in which we can distribute the energy between the two subsystems.

Among all the terms in the two sums over N_1 and k, the biggest contribution comes from the values $\bar{N}_1$ (and $\bar{N}_2 = N - \bar{N}_1$) and $\bar{E}_1$ (and $\bar{E}_2 = E - \bar{E}_1$) for which the single term

$$\frac{N!}{N_1!(N-N_1)!} V_{\Gamma,1}(E_k, N_1) V_{\Gamma,2}(E-E_k, N-N_1) \tag{3.35}$$

is maximum.

Since (3.34) is a sum of positive terms,

$$\frac{V_{\Gamma,1}(\bar{E}_1,\bar{N}_1)}{\bar{N}_1!}\frac{V_{\Gamma,2}(\bar{E}_2,\bar{N}_2)}{\bar{N}_2!} \le \frac{V_\Gamma(E,N)}{N!} \le \frac{NE}{\Delta E}\frac{V_{\Gamma,1}(\bar{E}_1,\bar{N}_1)}{\bar{N}_1!}\frac{V_{\Gamma,2}(\bar{E}_2,\bar{N}_2)}{\bar{N}_2!}, \tag{3.36}$$

we conclude that

$$\begin{aligned} &k_B \ln\frac{V_{\Gamma,1}(\bar{E}_1,\bar{N}_1)}{\bar{N}_1!} + k_B \ln\frac{V_{\Gamma,2}(\bar{E}_2,\bar{N}_2)}{\bar{N}_2!} \\ &\le k_B \ln\frac{V_\Gamma(E,N)}{N!} \\ &\le k_B \ln\frac{V_{\Gamma,1}(\bar{E}_1,\bar{N}_1)}{\bar{N}_1!} + k_B \ln\frac{V_{\Gamma,2}(\bar{E}_2,\bar{N}_2)}{\bar{N}_2!} + k_B \ln\frac{NE}{\Delta E}. \end{aligned} \tag{3.37}$$

In the thermodynamic limit, $N \to \infty$, the last term on the right-hand side of (3.37) becomes negligible and we obtain

$$S(E,N) = S_1(\bar{E}_1,\bar{N}_1) + S_2(\bar{E}_2,\bar{N}_2) \tag{3.38}$$

where, in order to satisfy the additivity property with respect to the number of particles, the definition of the entropy (3.29) has been modified as

$$S(E,N) = k_B \ln\frac{V_\Gamma(E,N)}{h^{3N}N!}, \tag{3.39}$$

consistently with Eq. (3.21). The conditions fixing the values $\bar{E}_1$, $\bar{E}_2$, $\bar{N}_1$, $\bar{N}_2$ for which the quantity in Eq. (3.35) is a maximum are

$$\frac{\partial}{\partial E_1}(S_1(E_1,N_1))_{E_1=\bar{E}_1,N_1=\bar{N}_1} = \frac{\partial}{\partial E_2}(S_2(E_2,N_2))_{E_2=\bar{E}_2,N_2=\bar{N}_2}, \tag{3.40}$$

$$\frac{\partial}{\partial N_1}(S_1(E_1,N_1))_{E_1=\bar{E}_1,N_1=\bar{N}_1} = \frac{\partial}{\partial N_2}(S_2(E_2,N_2))_{E_2=\bar{E}_2,N_2=\bar{N}_2}. \tag{3.41}$$

From the thermodynamic relation (1.22) we see that the condition (3.40) expresses the thermal equilibrium among the two subsystems,

$$\frac{1}{T_1} = \frac{1}{T_2}, \tag{3.42}$$

and among all the possible energy pairs only one realizes the equilibrium between the two subsystems, assigning to each one its internal energy.

By recalling the other thermodynamic relation (1.30) the condition (3.41) gives the chemical equilibrium among the subsystems:

$$-\frac{\mu_1}{T_1} = -\frac{\mu_2}{T_2} \quad \Rightarrow \mu_1 = \mu_2, \tag{3.43}$$

after using (3.42). The partition of the number of particles at equilibrium is such that each subsystem assumes its Gibbs free energy and their values per particle are equal.

Finally, by allowing the two subsystems to vary their volumes V_1 and V_2 with the condition that the total volume V remains fixed, after the thermodynamic relation (1.22) we have in a similar manner the condition expressing the mechanical equilibrium $P_1 = P_2$.

3.5 The equipartition theorem

One of the most important consequences of the Gibbs postulate is the equipartition theorem, telling how the energy of a system is divided, on average, among all the degrees of freedom of the system itself.

Let us indicate with x_i any of the $6N$ coordinates upon which the Hamiltonian depends. By using the definition of the microcanonical average, we may write

$$\left\langle x_i \frac{\partial H}{\partial x_i} \right\rangle = \frac{1}{\omega(E)} \int \mathrm{d}\Gamma x_i \frac{\partial H}{\partial x_i} \delta(E - H). \tag{3.44}$$

The integral can be transformed into a energy-constant surface integral

$$\left\langle x_i \frac{\partial H}{\partial x_i} \right\rangle = \frac{1}{\omega(E)} \int_{\sigma(E)} x_i \frac{\partial H}{\partial x_i} \frac{\mathrm{d}\sigma}{||\nabla H||}. \tag{3.45}$$

We now observe that ∇H defines a vector orthogonal to the energy-constant surface and $\nabla H / ||\nabla H||$ defines, correspondingly, the unit vector $\hat{\mathbf{n}}$ normal to the surface. Then, by using the generalized divergence theorem,

$$\left\langle x_i \frac{\partial H}{\partial x_i} \right\rangle = \frac{1}{\omega(E)} \int_{\sigma(E)} x_i \hat{n}_i \mathrm{d}\sigma = \frac{1}{\omega(E)} \int_{H<E} \mathrm{d}\Gamma \frac{\partial}{\partial x_i} x_i = \frac{\Xi(E)}{\omega(E)}. \tag{3.46}$$

By Eq. (3.27), we then obtain

$$\left\langle x_i \frac{\partial H}{\partial x_i} \right\rangle = \left(\frac{\partial \ln \Xi}{\partial E} \right)^{-1} = k_{\mathrm{B}} T. \tag{3.47}$$

As an important example, we consider the case when the Hamiltonian depends on x_i quadratically. Hence for an Hamiltonian with a quadratic dependence on f degrees of freedom one has

$$U = \langle H \rangle = \frac{f}{2} k_{\mathrm{B}} T. \tag{3.48}$$

The perfect gas of N particles in three-dimensional space has an Hamiltonian quadratically dependent on $3N$ momenta and hence has average energy $U = (3Nk_{\mathrm{B}}T)/2$. Notice that, on deriving the result (3.47), we used the condition that the system has the only constraint of being on the surface of constant energy E. Additional constraints will yield modifications to the results (3.47).

A further important consequence of the equipartition theorem is the introduction of the *virial* concept, first introduced by Clausius. Let us assume that the Hamiltonian of N particles is the sum of the kinetic energy, K, quadratic in the momenta $p_{i,\alpha}$ ($i = 1, \ldots, N$ and $\alpha = x, y, z$) and of the potential energy $U(\{q\})$ (not to be confused with the thermodynamic internal energy), depending on the $3N$ coordinates of all the particles. Clearly, the derivative of the Hamiltonian with respect to the coordinate $q_{i,\alpha}$ is minus the α-th component of the total force acting on the i-th particle:

$$F_{i,\alpha} = -\frac{\partial U(\{q\})}{\partial q_{i,\alpha}}. \tag{3.49}$$

Clausius defines the virial of the system as the quantity

$$\sum_{i=1}^{N} \mathbf{q}_i \cdot \mathbf{F}_i . \tag{3.50}$$

By using the equipartition theorem we then obtain

$$2\langle K\rangle + \left\langle \sum_{i=1}^{N} \mathbf{q}_i \cdot \mathbf{F}_i \right\rangle = 0. \tag{3.51}$$

This is the virial theorem. It is important to stress that the force $\mathbf{F}_i$ includes all forces acting on the i-th particle, both external and internal to the system under consideration. As a simple and illustrative example we consider the force exercised by the walls of the container where a gas is confined. Let us assume that the container is a cubic box with edge size L and volume $V = L^3$. Let us also assume that the edges of the box are parallel to a system of coordinates and one of the vertices of the box lies at the origin. Then the walls perpendicular to the x axis exercise a force $P\,[\delta(x) - \delta(x-L)]\,\mathrm{d}y\mathrm{d}z$ per unit surface. The contribution to the virial of the two walls is then obtained by integrating over all the surfaces and averaging over all the values of the coordinate x of the generic particle perpendicular to the surface

$$\int \mathrm{d}x\mathrm{d}y\mathrm{d}z\; x\; P\,[\delta(x) - \delta(x-L)] = -L^3 P = -PV. \tag{3.52}$$

The total contribution to the virial of all the walls is three times the contribution of (3.52). Hence, by recalling that the average value of the kinetic energy is given by (3.48) with $f = 3N$, the virial theorem (3.51) yields

$$2\frac{3}{2}Nk_{\mathrm{B}}T - 3PV = 0, \tag{3.53}$$

which reproduces the equation of state of the perfect gas.

3.6 The perfect gas in the microcanonical ensemble

The simplest application of the microcanonical ensemble distribution is the perfect gas. The Hamiltonian for the perfect gas of N identical particles with mass m is

$$H = \sum_{i=1}^{N} \frac{\mathbf{p}_i^2}{2m}. \tag{3.54}$$

The entropy is given by Eq. (3.39), where, according to the spirit of the thermodynamic large N limit, we may use $\Gamma(E)$ or $\Xi(E)$ or $\omega(E)$. For convenience, we use the last. For the Hamiltonian of the perfect gas, the integration over the coordinates is straightforward:

$$\omega = \int \delta(E - H)\mathrm{d}^{3N}q\,\mathrm{d}^{3N}p = V^N \frac{\partial}{\partial E}\int \theta(E - H)\,\mathrm{d}^{3N}p. \tag{3.55}$$

To perform the integration over the momenta, we define the following integral:

$$\mathcal{I}(n,\epsilon) = \int_{x_1^2+\cdots+x_n^2<\epsilon} \mathrm{d}^n x \equiv \int \theta\left(\epsilon - \left(x_1^2+\cdots+x_n^2\right)\right) \mathrm{d}^n x. \tag{3.56}$$

Since $\mathcal{I}$ is defined within an hypersphere, we use generalized spherical coordinates and write

$$\mathcal{I}(n,\epsilon) = \Omega_n \int_0^{\sqrt{\epsilon}} r^{n-1}\mathrm{d}r = \Omega_n \frac{\epsilon^{n/2}}{n},$$

where r is the radius and Ω_n the n-dimensional solid angle. To evaluate the latter we exploit the following identity based on the known properties of the Gaussian integral:

$$\int_{-\infty}^{\infty} \mathrm{d}x_1 \mathrm{e}^{-x_1^2} \cdots \int_{-\infty}^{\infty} \mathrm{d}x_n \mathrm{e}^{-x_n^2} = \pi^{n/2},$$

which may also be written as

$$\int_{-\infty}^{\infty} \mathrm{d}x_1 \cdots \int_{-\infty}^{\infty} \mathrm{d}x_n \mathrm{e}^{-\left(x_1^2+\cdots+x_n^2\right)} = \Omega_n \int_0^{\infty} \mathrm{e}^{-r^2} r^{n-1}\mathrm{d}r = \frac{1}{2}\Omega_n \Gamma(n/2),$$

where the Euler Gamma function has made its appearance. The Gamma function is defined by

$$\Gamma(z) = \int_0^{\infty} \mathrm{e}^{-t} t^{z-1}\mathrm{d}t, \tag{3.57}$$

and a brief summary of its properties is given in the Appendix B.

By using the Gamma function, we then have for the n-dimensional solid angle

$$\Omega_n = \frac{2\pi^{n/2}}{\Gamma(n/2)}. \tag{3.58}$$

We then get

$$\mathcal{I}(n,\epsilon) = \frac{2\pi^{n/2}\epsilon^{n/2}}{n\Gamma(n/2)} = \frac{\pi^{n/2}\epsilon^{n/2}}{\Gamma(n/2+1)}. \tag{3.59}$$

The entropy of the perfect gas reads then

$$S(E,V,N) = k_\mathrm{B} \ln\left(V^N \frac{3N\pi^{3N/2}(2mE)^{3N/2-1}}{N!h^{3N}2\Gamma(3N/2+1)}\right). \tag{3.60}$$

By means of the Stirling formula Eq. (B.5), apart from terms of order $\ln N$, which are negligible in the large N limit,

$$S(E,V,N) = Nk_\mathrm{B} \ln\left(\frac{V}{N}\left(\frac{4\pi mE}{3Nh^2}\right)^{3/2}\right) + \frac{5}{2}Nk_\mathrm{B}. \tag{3.61}$$

By solving with respect to the energy, we arrive at

$$E \equiv U(S,V,N) = \frac{3Nh^2}{4\pi m}\left(\frac{N}{V}\right)^{2/3} \exp\left(\frac{2S}{3Nk_\mathrm{B}} - \frac{5}{3}\right). \tag{3.62}$$

Now, by taking the derivative with respect to the entropy S, we get the expression for the temperature

$$T = \left(\frac{\partial U}{\partial S}\right)_{V,N} = \frac{2}{3}\frac{U}{Nk_{\mathrm{B}}}, \tag{3.63}$$

while by differentiating with respect to the volume we have the pressure

$$P = -\left(\frac{\partial U}{\partial V}\right)_{S,N} = \frac{2}{3}\frac{U}{V}. \tag{3.64}$$

Equation (3.63) yields the known fact that the energy of a perfect gas depends solely on the temperature. Furthermore, the total energy has the form required by the equipartition theorem (3.47): with each momentum degree of freedom is associated the average kinetic energy $k_{\mathrm{B}}T/2$.

By inserting Eq. (3.63) into Eq. (3.64), we have the equation of state

$$P = \frac{Nk_{\mathrm{B}}T}{V}. \tag{3.65}$$

Finally, by inserting Eq. (3.63) back into Eq. (3.61) to replace the energy we have the so-called Sackur–Tetrode equation for the entropy:

$$S(E, V, N) = Nk_{\mathrm{B}} \ln\left(\frac{V}{N}\left(\frac{2\pi m k_{\mathrm{B}} T}{h^2}\right)^{3/2}\right) + \frac{5}{2}Nk_{\mathrm{B}}. \tag{3.66}$$

3.7 Problems

3.1 Show that the velocity of the system representative point in the phase-space has zero divergence.

3.2 Evaluate the probability of deviations from homogeneity in a gas of N distinguishable particles in a volume V.

4 More ensembles

4.1 The canonical ensemble

When we considered the extensive property of the entropy, we saw that the value of the entropy for a composite system of two subsystems with energies E_1 and E_2 such that the total energy is fixed, $E_1 + E_2 = E$, is mostly determined by the values $\overline{E_1}$ and $\overline{E_2}$ which make the product $\Gamma_1(E_1)\Gamma_2(E_2)$ maximum. This maximum condition corresponds to the thermal equilibrium among the two subsystems with $\overline{E_1}$ and $\overline{E_2}$ being the respective internal energies. Let us now imagine that one of the two subsystems, say subsystem 2, is much larger than the other subsystem which, however, is still of macroscopic size to locally define thermodynamic variables. We indicate subsystem 2 as the *reservoir* or thermostat and subsystem 1 as the *system*. From now on we adopt the notation E_s and E_r for the energies of the two subsystems. The subsystems may exchange energy between each other, maintaining the total energy fixed:

$$E = E_r + E_s. \tag{4.1}$$

According to our assumption, the reservoir energy satisfies $E_r \approx E$, implying that the energy exchanges between the system and the reservoir are such that little modification occurs in the state of the latter, while the state of the former can change considerably. Furthermore the state of the system will depend on the energy partition among the system and the reservoir. Our aim is to determine this dependence. To this end, let us indicate with $\{\mathbf{q}_s, \mathbf{p}_s\}$ and $\{\mathbf{q}_r, \mathbf{p}_r\}$ the phase-space coordinates of the system and the reservoir, respectively. We are interested in the distribution function defined in the phase-space of the system, irrespective of the position of the reservoir in its own phase-space. The former can be obtained by integrating out the reservoir coordinates for the microcanonical distribution function of the composite system:

$$\begin{aligned}\rho(\{\mathbf{q}_s, \mathbf{p}_s\}) &= \frac{\int_{E<E_s+E_r<E+\Delta E} d\Gamma_r}{\int_{E<E_s+E_r<E+\Delta E} d\Gamma_r d\Gamma_s} \\ &= \frac{V_{\Gamma,r}(E-E_s)}{\int V_{\Gamma,r}(E-E_s) d\Gamma_s} \\ &= \frac{e^{\ln V_{\Gamma,r}(E-E_s)}}{\int e^{\ln V_{\Gamma,r}(E-E_s)} d\Gamma_s}.\end{aligned} \tag{4.2}$$

Since the equipartition of the energy occurs essentially with a given pair of values, we can assume that the energy of the system is small compared to that of the reservoir, so we can

function is obtained by integrating over the coordinates of the reservoir and we get

$$\rho(\{\mathbf{q}_s, \mathbf{p}_s\}, N_s) = \frac{N!}{N_s! N_r!} \frac{e^{-\beta H(\{\mathbf{q}_s, \mathbf{p}_s\})} \int d\Gamma_r e^{-\beta H(\{\mathbf{q}_r, \mathbf{p}_r\})}}{\int d\Gamma e^{-\beta H(\{\mathbf{q}, \mathbf{p}\})}},$$

where $d\Gamma_r$ and $d\Gamma$ denote the volume elements in the phase-spaces of the reservoir and the composite system, respectively. The combinatorial factor corresponds to the number of ways in which N_s particles can be chosen out of the N present in the composite system. By recalling the expression of the partition function for the canonical ensemble, the distribution function for the system can be written as

$$\rho(\{\mathbf{q}_s, \mathbf{p}_s\}, N_s) = \frac{e^{-\beta H(\{\mathbf{q}_s, \mathbf{p}_s\})}}{h^{3N_s} N_s!} \frac{Z_r(T, V_r, N - N_s)}{Z(T, V, N)}. \tag{4.22}$$

The partition functions $Z_r(T, V_r, N - N_s)$ and $Z(T, V, N)$ can be expressed in terms of the corresponding free energies:

$$\rho(\{\mathbf{q}_s, \mathbf{p}_s\}, N_s) = \frac{e^{-\beta H(\{\mathbf{q}_s, \mathbf{p}_s\})}}{h^{3N_s} N_s!} e^{-\beta(F_r(T, V - V_s, N - N_s) - F(T, V, N))}. \tag{4.23}$$

By exploiting the condition $N_s \ll N$, which again follows from the fact that among all pairs of N_r and N_s, only one pair is of relevance, we expand the free energy F_r in the second exponential of (4.23):

$$F(T, V, N) - F_r(T, V_r, N - N_s) \approx N_s \left(\frac{\partial F_r}{\partial N} \right)_{T,V} + V_s \left(\frac{\partial F_r}{\partial V} \right)_{T,N} = N_s \mu - P V_s$$

and we finally arrive at the expression for the system distribution function:

$$\rho(\{\mathbf{q}, \mathbf{p}\}, N) = \frac{e^{-\beta(H(\{\mathbf{q}, \mathbf{p}\}) - \mu N) - \beta P V}}{h^{3N} N!}, \qquad \sum_{N=0}^{\infty} \int d\Gamma \, \rho(\{\mathbf{q}, \mathbf{p}\}, N) = 1, \tag{4.24}$$

where the subscript s has been suppressed since from now on we will be concerned solely with the system, the reservoir appearing only via β and μ. We note also that, due to the presence of the factorial in the denominator, which guarantees the convergence, the sum over the number of particles has been extended to infinity, in the same spirit in which we extend the integration over energy to infinity in the derivation of the canonical ensemble.

Equation (4.24) defines the grand canonical distribution function. As for the canonical ensemble, we define the partition function $\mathcal{Z}$ as

$$\mathcal{Z} = \sum_{N=0}^{\infty} \int d\Gamma \frac{e^{-\beta(H(\{\mathbf{q}, \mathbf{p}\}) - \mu N)}}{h^{3N} N!}. \tag{4.25}$$

Due to the normalization condition in Eq. (4.24) together with the thermodynamic relation (1.35) $PV = -\Omega(T, V, \mu)$, we may express the grand potential $\Omega(T, V, \mu)$ (see Eq. (1.31)) in the terms of the grand canonical partition function $\mathcal{Z}$:

$$\Omega(T, V, \mu) = -k_B T \ln \mathcal{Z}(T, V, \mu). \tag{4.26}$$

To prove that Eq. (4.26) defines the thermodynamic grand potential correctly, we write the normalization condition in (4.24) as

$$\sum_{N=0}^{\infty} e^{\beta\mu N} Z(T, V, N) e^{\beta\Omega(T,V,\mu)} = 1. \tag{4.27}$$

Then, by deriving both sides of (4.27) with respect to μ, T and V, we get the known thermodynamic relations (1.33) (see Problem 4.3):

$$\langle N \rangle = -\left(\frac{\partial\Omega}{\partial\mu}\right)_{T,V}, \quad S = -\left(\frac{\partial\Omega}{\partial T}\right)_{\mu,V}, \quad P = -\left(\frac{\partial\Omega}{\partial V}\right)_{T,\mu}. \tag{4.28}$$

Finally, by using Eq. (4.26) and the first of the (4.28), keeping in mind the relation $\Omega = -PV$, we have the set of coupled relations

$$P = \frac{k_B T}{V} \ln \mathcal{Z}(T, V, z), \tag{4.29}$$

$$\langle N \rangle = z\frac{\partial}{\partial z} \ln \mathcal{Z}(T, V, z), \tag{4.30}$$

where $z = e^{\beta\mu}$ is the so-called fugacity. The system of Eqs. (4.29) and (4.30) yields the equation of state in an implicit form. Indeed, from (4.30) one gets the fugacity as a function of temperature, volume and number of particles $z = z(T, V, N)$. Such a function is then inserted into (4.29).

By recalling the canonical partition function for a perfect gas, Eq. (4.12), the grand partition function reads

$$\mathcal{Z} = \sum_{N=0}^{\infty} \frac{e^{\beta\mu N}}{N!} Z_0(T, V, N)^N = e^{Z_0(T,V,N)e^{\beta\mu}}, \tag{4.31}$$

where Z_0 is given by Eq. (4.13). The equation of state is obtained by considering the grand potential and the average number of particles:

$$PV = -\Omega = k_B T V (2\pi m k_B T h^{-2})^{3/2} z, \tag{4.32}$$

$$\langle N \rangle = -\frac{\partial\Omega}{\partial\mu} = \beta z \frac{\partial}{\partial z} k_B T V (2\pi m k_B T h^{-2})^{3/2} z. \tag{4.33}$$

From (4.33), one has $\langle N \rangle = Vz(2\pi m k_B T h^{-2})^{3/2}$, which when inserted into (4.32) yields the equation of state of the perfect gas with the average number of particles, and the pressure P as a continuous decreasing function of V characterizing a single phase. The density, $n = \langle N \rangle / V = 1/v$, can be written as

$$n = \left(\frac{\partial P}{\partial\mu}\right)_T = \left(\frac{\partial P}{\partial v}\right)_T \left(\frac{\partial\mu}{\partial v}\right)_T^{-1} = nV\left(\frac{\partial P}{\partial V}\right)_T \left(-n^2\left(\frac{\partial\mu}{\partial n}\right)_T\right)^{-1}, \tag{4.34}$$

which, after recalling the isothermal compressibility defined in Eq. (1.50), leads to

$$\kappa_T = \frac{1}{n^2}\left(\frac{\partial n}{\partial\mu}\right)_T. \tag{4.35}$$

In the last equation the isothermal compressibility for a gas has been expressed in terms of the number density n and we have introduced the thermodynamic density of states $\partial n/\partial\mu$.

4.4 Fluctuations of a gas

As we have seen for the perfect gas, all the ensembles give the same result for the equation of state. This is not by accident. To understand the origin of this fact, let us examine the energy in the microcanonical and canonical ensembles. In the first case, the energy E is fixed and coincides with the thermodynamic energy U. In the second case, the latter corresponds to the average energy in the ensemble. It is then natural to ask how representative this value of the energy is, or in other words, how important the fluctuations about the average energy are. To answer this we consider the mean square fluctuation by taking the derivative of Eq. (4.11) for $\langle H \rangle$ with respect to β:

$$\begin{aligned}\langle (H - \langle H \rangle)^2 \rangle &= \langle H^2 \rangle - \langle H \rangle^2 \\ &= -\frac{\partial}{\partial \beta} \langle H \rangle \\ &= -\frac{\partial}{\partial \beta} \left(F - T \left(\frac{\partial F}{\partial T} \right)_V \right) \\ &= -k_B T^3 \left(\frac{\partial^2 F}{\partial T^2} \right)_V \\ &= k_B T^2 V c_V . \end{aligned} \tag{4.36}$$

where the symbol $\langle \dots \rangle$ stands for the ensemble average and c_V is the specific heat per unit volume. Considering the Krönig relation (2.2) for a gas, the fluctuation for the energy coincides with the second expression in Eq. (1.59) for $\langle (\Delta T)^2 \rangle$.

Equation (4.36) shows that the mean square fluctuation is proportional to the volume or to the number of particles, given the extensive character of the energy and its derivatives with respect to an intensive variable such as the temperature. To evaluate the importance of the fluctuations, we take the ratio of the mean square fluctuation and the average energy:

$$\frac{\sqrt{\langle (H - \langle H \rangle)^2 \rangle}}{\langle H \rangle} = \frac{T\sqrt{k_B V c_V}}{U} \approx \frac{\sqrt{N}}{N} = \frac{1}{\sqrt{N}}, \tag{4.37}$$

and, in the thermodynamic limit with $N \to \infty$, such a ratio goes to zero.

In a similar way, in the equation of state for a gas in the grand canonical ensemble the average number of particles appears instead of N. We may consider the fluctuations of the number of particles in the grand canonical ensemble. The mean square fluctuation reads

$$\begin{aligned}\langle (N - \langle N \rangle)^2 \rangle &= \langle N^2 \rangle - \langle N \rangle^2 \\ &= k_B T \frac{\partial}{\partial \mu} \langle N \rangle \\ &= k_B T V \left(\frac{\partial n}{\partial \mu} \right)_T . \end{aligned} \tag{4.38}$$

From Eq. (4.35), it is immediate to see that

$$\langle (N - \langle N \rangle)^2 \rangle = \langle N \rangle k_B T n \kappa_T , \tag{4.39}$$

which is the analog of Eq. (1.59) for $\langle(\Delta V)^2\rangle$ and where the isothermal compressibility is an intensive quantity. The ratio between the mean square fluctuation and the average value reads

$$\frac{\sqrt{\langle(N-\langle N\rangle)^2\rangle}}{\langle N\rangle} = \frac{1}{\sqrt{\langle N\rangle}}\sqrt{k_{\rm B}Tn\kappa_T}, \tag{4.40}$$

which goes to zero in the thermodynamic limit.

4.5 Problems

4.1 A perfect gas of N classical particles is kept, at constant temperature T, in a cylinder of infinite height and base area A and is immersed in a constant gravitational field with single-particle Hamiltonian given by

$$H = \frac{\mathbf{p}^2}{2m} + mgz,$$

where z is the coordinate along the cylinder axis, m is the mass of the particles and g is the gravitational acceleration. What is the dependence of the pressure on the height?

4.2 Determine the specific heat of a gas of N biatomic molecules with the constituent atoms having masses m_a and m_b. The Hamiltonian, besides the kinetic energy contains a potential term, which is the sum of N intra-molecular potentials $U(|\mathbf{r}_a - \mathbf{r}_b|)$, with $\mathbf{r}_{a,b}$ indicating the positions of the two atoms in the molecule.

4.3 Derive the relation (4.28).

4.4 Derive the equipartition theorem (3.47) using the canonical ensemble.

4.5 Derive the specific heat of N harmonic oscillators with Hamiltonian

$$H = \sum_{i=1}^{N}\left[\frac{\mathbf{p}_i^2}{2m} + \frac{m\omega_i^2\mathbf{r}_i^2}{2}\right].$$

5 The thermodynamic limit and its thermodynamic stability

Historically, the main aim of statistical mechanics was to derive the thermodynamics. As we have seen in the previous chapter, the thermodynamic potentials evaluated in the different ensembles differ by terms proportional to the logarithm of the extension due to the fluctuations either of the energy or of the number of particles. Only in the thermodynamic limit, N, $V \to \infty$, with fixed density $n = N/V$, all the ensembles provide equivalent extensive thermodynamic potentials. It is therefore relevant to show that this limit exists. Moreover, the full knowledge of the macroscopic behavior of a system implies the derivation of the equation of state of a substance and the corresponding phase diagram. This program is fully realized in the case of the perfect gases, where the lack of interparticle interaction makes the problem sensibly easy and provides a description of a single gas phase with the pressure P being a continuous decreasing function of V. When one considers interparticle interaction the problem becomes, in general, more difficult, even though it is certainly much more interesting. In fact, only by taking into account interparticle interaction can one hope to understand how a phase transition occurs and hence to understand the phase diagram of a given substance with different phases.

In order to describe a phase transition, the statistical mechanics must be able to reproduce the singular behavior associated with the transition itself. For instance, at a first order phase transition, we know that the first derivatives of the thermodynamic potentials have a jump. On the other hand, for any system with a given finite number of particles, N, in a given finite volume, V, and total finite energy E, there is no chance of having singular behavior.

For instance let us consider the grand canonical ensemble. Recall that the grand partition function is written as

$$\mathcal{Z} = \sum_{N=0}^{\infty} Z_N z^N \tag{5.1}$$

where $z = \exp(\beta\mu)$ is the fugacity, Z_N is the partition function for N particles and the sum has been extended to infinity assuming the thermodynamic limit.

In general a finite system with a strongly repulsive interaction (hard-core potential) at short distances has a grand partition function with the sum over N limited to the maximum number of particles $N_{\max}(V)$ for each finite volume V, corresponding to a close-packing configuration. The grand partition function (5.1) becomes a polynomial of degree $N_{\max}$ in the fugacity z with positive coefficients Z_N, whose zeros cannot be real. $\mathcal{Z}$ can therefore be written as

$$\mathcal{Z}(z) = \prod_{i=1}^{N_{\max}} \left(1 - \frac{z}{z_i}\right), \tag{5.2}$$

where the z_i are the complex roots. As long as the system remains finite, the elimination of z between the grand potential Ω and the number of particles,

$$P = \frac{k_B T}{V} \ln \mathcal{Z}, \qquad \frac{\langle N \rangle}{V} = z \frac{\partial}{\partial z} \frac{1}{V} \ln \mathcal{Z}, \tag{5.3}$$

leads to an analytic function P of $v = V/\langle N \rangle$ on a domain containing the entire real axis. To see this, we observe that both P and $1/v$ are analytic functions of z in a domain containing the real axis. Moreover,

$$\frac{\partial P}{\partial v} = \frac{\partial P}{\partial z} \frac{\partial z}{\partial v} < 0, \tag{5.4}$$

since

$$\frac{\partial P}{\partial z} = \frac{k_B T}{z v} > 0 \tag{5.5}$$

and

$$\frac{\partial v}{\partial z} = -\frac{V}{\langle N \rangle^2} \frac{\partial \langle N \rangle}{\partial z} = -\frac{V}{\langle N \rangle^2} \frac{1}{z} (\langle N^2 \rangle - \langle N \rangle^2) \leq 0. \tag{5.6}$$

Therefore P and $1/v$ are analytic non-decreasing functions of z on the real axis and $P(v)$ is an analytic non-increasing function on the real axis describing a single phase as for a gas. (This latter result is consistent with the request of thermodynamic stability.) In this case P cannot be a constant on a segment of the real axis, otherwise it would be constant everywhere. No isotherms corresponding to a first order transition, with P constant on a finite segment of the variable v, can be strictly obtained for this finite system. $\partial P/\partial v$ could be extremely small, but finite. On a macroscopic scale this could be equivalent to a constant pressure, but not mathematically.

The possibility of having singular behavior may present itself, however, when considering the thermodynamic limit, in which also the energy density E/V stays finite. In this case the number of zeros $N_{\max}(V)$ increases to infinity. They may become dense on a line crossing the real positive axis and in the limit a pair of zeros can pinch a real positive value, disconnecting the real axis into different regions of analyticity of $P(v)$. The thermodynamic limit is, clearly, a mathematical idealization, which nonetheless well represents the situation of a macroscopic system formed by a large number of elementary constituents. Basically, when one considers the thermodynamic limit, one neglects surface and finite volume effects.

How this singular behavior may arise in the thermodynamic limit is controlled by the two Yang–Lee theorems (Yang and Lee (1952)), which we discuss in the next section.

5.1 The Yang–Lee theorems

The existence of the equation of state in the thermodynamic limit corresponds to the existence of the two limits

$$P = \lim_{V \to \infty} \frac{k_B T}{V} \ln \mathcal{Z}(V, z, T), \tag{5.7}$$

$$\frac{1}{v} = \lim_{V \to \infty} \frac{\partial}{\partial \ln z} \frac{1}{V} \ln \mathcal{Z}(V, z, T) \tag{5.8}$$

with $v = V/N$ kept constant ($\langle N \rangle \to N$). Furthermore, for a single phase, one requires that the two limits define analytic functions. The Yang–Lee theorems are just about the conditions under which the limits (5.7) and (5.8) exist.

First theorem *For all positive real values of z and for a suitable class of interparticle potentials (to be specified below), the limit (5.7) exists. Furthermore, the limit is a continuous, monotonically increasing function of z and is independent of the form of V.*

Second theorem *If in the z-complex plane, a domain R, which contains a segment of the positive real axis, is always free of roots of $\mathcal{Z}(z)$, then, in this domain, as $V \to \infty$ with $N/V =$ constant, the quantity $\ln \mathcal{Z}(z)/V$ is an analytic function of z and its derivative is a monotonically increasing function of z.*

The analyticity of the limits (5.7) and (5.8) allows us to interchange the two operations of taking the derivative and the limit, obtaining the equation of state in the form

$$\frac{1}{v} = \frac{\partial}{\partial \ln z} \frac{P}{k_B T}. \tag{5.9}$$

In the case when R contains the full positive real axis, the situation is equal to that of a finite system and hence P is an analytic function of v. The key point of the second theorem is about whether the domain R does or does not contain the entire positive real axis. If R contains only a segment of the positive real axis, this implies that there is a value $z = z_c$ at which the limit (5.7), though continuous, is not analytic at $z = z_c$. Suppose, for instance, a discontinuity of the first kind with a jump in the value of v. Let us call v_1 and v_2 the left and right limits of v for $z \to z_c$. This in turn implies that for the value $P_c = P(z_c)$ all values of v between v_2 and v_1 are possible, as occurs at a first order phase transition. If $\partial P/\partial z$ is continuous but $\partial^2 P/\partial z^2$ has a discontinuity of the second kind, we get a second order phase transition and the critical isotherm with infinite compressibility. The essential idea to obtain a description of phase transitions in the thermodynamic limit is that in this limit the pairs of complex zeros in the complex z-plane accumulate on a line and in the limit of $V \to \infty$ and therefore $N(V) \to \infty$ a couple of them pinch the real axis of z, determining a separation into two phases.

Before undertaking the proof of the two theorems, we must specify the class of interparticle potentials considered. We limit ourselves to two-body interactions for simplicity. Also we confine our discussion to classical systems with Hamiltonian

$$H = \sum_{i=1}^{N} \frac{\mathbf{p}_i^2}{2m} + \sum_{1 \leq i < j \leq N} \phi(|\mathbf{q}_i - \mathbf{q}_j|), \tag{5.10}$$

where the two-body potential is of the form ($x \geq 0$)

$$\phi(x) = \begin{cases} \infty, & x < r_0, \\ 0, & x > b, \\ -\epsilon < \phi(x) < 0, & r_0 < x < b. \end{cases} \tag{5.11}$$

The potential (5.11) mimics more realistic potentials such as those responsible for the van der Waals interactions. There is a repulsive part at short distances, the so-called hard-core part. As a consequence, the particles have a finite size and only a finite number

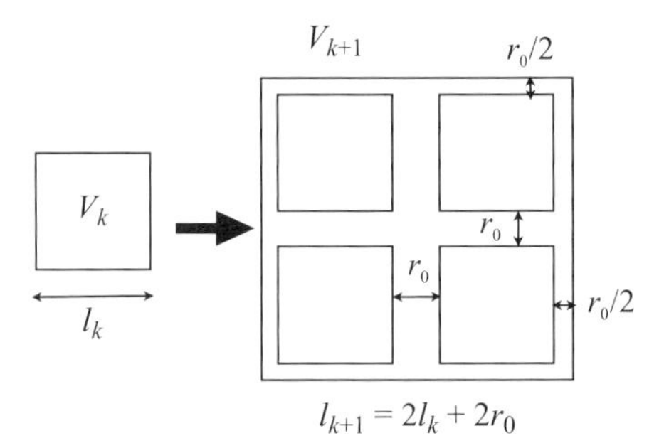

Fig. 5.1 Schematic sketch of the generation of the volume sequence V_k. The size of r_0 has been enlarged with respect to l_k.

of them can be accommodated in a finite volume, $N_{\max}(V)$. The potential has also an attractive part extending over a finite range. This property will be crucial for the proof of the thermodynamic stability in the thermodynamic limit. The long-range Coulomb potential clearly does not satisfy this condition. In the case of Coulomb forces it is necessary to have globally neutral systems so that the effect of long-range forces compensates at large distances.

We remark that the potential (5.11) is by no means the most general potential, maintaining the property of diverging at short distance and going to zero at long distance sufficiently fast, for which one can prove the existence of the thermodynamic limit. Clearly, more general potentials may render the proof more demanding. Since our aim here is to illustrate the meaning of the thermodynamic limit in connection with the existence of phase transitions, we content ourselves with the potential (5.11). In the next section the proof of the first theorem for the existence of the limit and its thermodynamic stability is given. The second theorem is proven in Appendix C.

5.2 Proof of the first theorem

The proof is carried out in two steps. First we demonstrate the existence of the thermodynamic limit for the free energy in the canonical ensemble and the thermodynamic stability of the limit. In the second step we show how this result can be used to establish the existence of the thermodynamic limit in the grand canonical ensemble.

5.2.1 Thermodynamic limit in the canonical ensemble

We begin by defining an increasing sequence of volumes V_k such that $V_k \to \infty$ for $k \to \infty$. In going from k to $k+1$, we put together 2^3 cubes each of volume $V_k = l_k^3$ into a bigger cube of volume $V_{k+1} = (2l_k + 2r_0)^3$ (see Fig. 5.1). The original cubes are at a distance r_0 from one another and a distance $r_0/2$ from the walls of the bigger cube. We also define a sequence of particle numbers N_k. The number of particles at step $k+1$ is simply the sum of the particles contained in all the cubes of volume V_k put into the bigger cube of volume

V_{k+1}. Let N_k be the particle number at step k, then $N_{k+1} = 2^3 N_k$. By defining the density at step k as $n_k = N_k/V_k$, we get

$$n_{k+1} = n_k \frac{1}{(1 + r_0/l_k)^3}. \tag{5.12}$$

Since l_k is the edge of the cube at step k, we have $l_k \to \infty$ as $k \to \infty$. Moreover r_0 has atomic size, hence, for sufficiently large k, the density becomes independent of k. Then we have a suitable sequence of volumes and particle numbers approaching the thermodynamical limit. It is convenient to define, at step k,

$$f_k = \frac{1}{N_k} \ln Z(T, V_k, N_k), \tag{5.13}$$

which is related to the free energy by

$$F(T, V_k, N_k) = -k_B T N_k f_k. \tag{5.14}$$

As a first step in the proof of the existence of the thermodynamic limit, we show that f_k is an increasing sequence as $k \to \infty$. By recalling Eqs. (4.12) and (4.13) for the integration of the kinetic energy term,

$$Z_{k+1} = \frac{(2\pi m k_B T \hbar^{-2})^{3N_{k+1}/2}}{N_{k+1}!} \int_{V_{k+1}^{N_{k+1}}} d^{3N_{k+1}} q \exp\left\{ -\beta \sum_{1 \le i < j \le N_{k+1}} \phi(|\mathbf{q}_i - \mathbf{q}_j|) \right\}, \tag{5.15}$$

where $V_{k+1}^{N_{k+1}}$ stands for the hypervolume in the space of $3N_{k+1}$ coordinates. The integral over V_{k+1} can be decomposed into the eight volumes V_k and the regions between the cubes. Owing to the form of the potential (5.11), the interactions among particles belonging to different cubes of the step k are negative and hence give a positive contribution to the exponent, resulting in a factor greater than one. Then, by neglecting these intercube interactions, we can make (5.15) smaller. By confining ourselves to the intracube interactions, we can decompose the volume integral of (5.15) into the eight volumes V_k. In so doing we must divide the N_{k+1} particles among the eight cubes. There are many ways to do this. For instance, we can put just one particle in seven of the cubes and all the remaining ones in the last cube. If we restrict only to those configurations in which there are N_k particles in each cube, we clearly underestimate the partition function. But this is exactly what we need in order to obtain something smaller than the original partition function. On the other hand the equipartition of the number of particles among the eight cubes is the most probable partition at equilibrium. Once we have decided to have N_k particles in each cube, there are still

$$\frac{N_{k+1}!}{(N_k!)^8}$$

different equivalent ways of doing it. As a result we get

$$Z_{k+1} > \frac{(2\pi m k_B T \hbar^{-2})^{3N_{k+1}/2}}{N_{k+1}!} \frac{N_{k+1}!}{(N_k!)^8} \left[\int_{V_k^{N_k}} d^{3N_k} q \exp\left\{ -\beta \sum_{1 \le i < j \le N_k} \phi(|\mathbf{q}_i - \mathbf{q}_j|) \right\} \right]^8 \tag{5.16}$$

from which one gets the inequality

$$Z_{k+1} > Z_k^8. \tag{5.17}$$

By taking the logarithm, we then obtain

$$f_{k+1} = \frac{1}{N_{k+1}} \ln Z_{k+1} > \frac{1}{N_k} \ln Z_k = f_k, \tag{5.18}$$

which proves that the f_k is an increasing sequence. To prove that this sequence has a limit, we need to show that f_k is bounded from above. To this end, we observe that each particle interacts only with those particles contained in a spherical shell of radii r_0 and b. Given the finite size of the particles, only a finite number of them can be accommodated in this spherical shell. As a result we have the inequality

$$\sum_{1 \leq i < j \leq N_k} \phi(|\mathbf{q}_i - \mathbf{q}_j|) > -BN_k, \tag{5.19}$$

where B is a constant not depending on N_k, but only on the ratio b/r_0 and ϵ. Then we have

$$\exp\left\{-\beta \sum_{1 \leq i < j \leq N_k} \phi(|\mathbf{q}_i - \mathbf{q}_j|)\right\} < \mathrm{e}^{\beta B N_k}$$

and hence the desired upper bound

$$f_k < \ln\left[\frac{V_k}{N_k}(2\pi m k_{\mathrm{B}} T \hbar^{-2})^{3/2}\right] + \beta B + 1. \tag{5.20}$$

Equations (5.18) and (5.20), which depend only on intensive variables, prove the existence of the thermodynamic limit for the sequence f_k. This limit is

$$f(T, v) = -\lim_{N \to \infty} \frac{F(T, V, N)}{N k_{\mathrm{B}} T}, \tag{5.21}$$

where $v = V/N$. It can also be shown that the limit does not depend on the form of the sequence of increasing volumes.

The existence of the limit does not guarantee by itself the thermodymanic stability, i.e., $\kappa_T > 0$. In terms of the function $f(T, v) \equiv f(v)$ we have

$$P = -\frac{\partial F}{\partial V} = k_{\mathrm{B}} T \frac{\partial f}{\partial v} \tag{5.22}$$

and

$$\kappa_T = -\frac{1}{v}\left(k_{\mathrm{B}} T \frac{\partial^2 f}{\partial v^2}\right)^{-1}, \tag{5.23}$$

so that thermodynamic stability implies

$$\frac{\partial^2 f}{\partial v^2} < 0, \tag{5.24}$$

i.e., $f(v)$ must be a concave function (which is equivalent to saying that the free energy must be a convex function). We now show that indeed $f(v)$ is a concave function. This amounts to showing that the function in any intermediate point between v_1 and v_2 is above the secant passing through $f(v_1)$ and $f(v_2)$:

$$f(tv_1 + (1-t)v_2) \geq tf(v_1) + (1-t)f(v_2), \tag{5.25}$$

with $t \in (0, 1)$. Consider the partition function at step $k+1$ with N_{k+1} particles distributed in the volume V_{k+1}. Let us assign $\alpha N_k^{(1)}$ particles to the volume αV_k, while the remaining $(8-\alpha)N_k^{(2)} = N_{k+1} - \alpha N_k^{(1)}$ particles are assigned to the volume $(8-\alpha)V_k$. α is chosen to be an integer < 8. By reasoning as before the partition function at step $k+1$ satisfies the inequality

$$Z_{k+1}(T, V_{k+1}, N_{k+1}) \geq Z_k\big(T, V_k, N_k^{(1)}\big)^{\alpha} Z_k\big(T, V_k, N_k^{(2)}\big)^{8-\alpha}. \tag{5.26}$$

At finite volume and particle number, let us introduce the function

$$g(n) = \frac{1}{V} \ln Z(T, V, N) = nf(v), \quad n = \frac{1}{v}. \tag{5.27}$$

Clearly, if $f(v)$ has a limit when $V, N \to \infty$ at fixed ratio N/V, so does $g(n)$. By using the inequality (5.26), we get

$$g\left(\frac{N_{k+1}}{V_{k+1}}\right) \geq \frac{\alpha}{8} g\left(\frac{N_k^{(1)}}{V_k}\right) + \frac{8-\alpha}{8} g\left(\frac{N_k^{(2)}}{V_k}\right). \tag{5.28}$$

By taking the thermodynamic limit this implies

$$g(n) \equiv g\left(\frac{\alpha}{8} n_1 + \frac{8-\alpha}{8} n_2\right) \geq \frac{\alpha}{8} g(n_1) + \frac{8-\alpha}{8} g(n_2), \tag{5.29}$$

where $n = N_{k+1}/V_{k+1}$ and $n_{1,2} = N_k^{(1,2)}/V_k$. After expressing, finally, $g(n)$ in terms of $f(v)$, we obtain

$$f(v) \geq \frac{\alpha n_1}{8n} f(v_1) + \frac{(8-\alpha)n_2}{8n} f(v_2). \tag{5.30}$$

Equation (5.25) follows by choosing $t = \alpha n_1/8n$. This proves that $f(v)$ is a bounded downwards concave function. Furthermore since v_1 and v_2 can be chosen arbitrarily, the function is also continuous.

5.2.2 Connection between canonical and grand canonical ensembles

As we have already stated, given the hard core part of the potential, only a finite number of particles, $N_{\max}(V)$, can be accommodated into the volume V. Hence the grand partition function can be written as

$$\mathcal{Z}(T, V, z) = \sum_{N=0}^{N_{\max}(V)} e^{V\phi(v,z)} \tag{5.31}$$

where, with $v = V/N$,

$$\phi(v, z) = \frac{1}{v} f(v) + \frac{1}{v} \ln z. \tag{5.32}$$

At fixed volume, by changing N in the various addends of (5.32), v changes. The greatest term in the sum is given by the value v which maximizes $\phi(v, z)$ at fixed z. Let $\phi_0(z) = \phi(v(z), z)$ the maximum over v, where $v(z)$ is the value for which the maximum occurs. Then clearly we have the inequalities

$$\phi_0(z) \leq \frac{1}{V} \ln \mathcal{Z}(T, V, z) \leq \phi_0(z) + \frac{\ln N_{\max}(V)}{V}. \tag{5.33}$$

Then

$$\lim_{V \to \infty} \frac{1}{V} \ln \mathcal{Z}(T, V, z) = \phi_0(z), \tag{5.34}$$

since $N_{\max}(V)$ does not grow exponentially with V. Hence, via (5.34), the existence of the limit (5.7) is related, via (5.32), to the existence of the limit (5.21). One can easily verify that the conditions that $\phi(v, z)$ has a maximum are

$$\frac{\partial \phi}{\partial v} = \frac{1}{v}\left(\frac{\partial f}{\partial v} - \phi\right) = 0 \tag{5.35}$$

and

$$\frac{\partial^2 \phi}{\partial v^2} = -\frac{2}{v}\frac{\partial \phi}{\partial v} + \frac{1}{v}\frac{\partial^2 f}{\partial v^2}, \tag{5.36}$$

that is negative when evaluated at the extremum. The last inequality is guaranteed by the thermodynamic stability condition (5.24). This result is equivalent to saying that $\mathcal{Z} \sim z^{\bar{N}} Z(T, V, \bar{N})$ with $\bar{N}$ which maximizes the weight $W(N) = z^N Z(T, V, N)$ and assuming $\bar{N} = \langle N \rangle$ as a result of the width of the distribution function going to zero in the thermodynamic limit. Equation (5.34) also corresponds to evaluating the pressure P as $P = (G - F)/V$ where the free energy F and the Gibbs potential G are evaluated in the canonical ensemble.

6 Density matrix and quantum statistical mechanics

6.1 Kinetics for quantum gases

In approaching quantum statistical mechanics we start by extending the Boltzmann equation to quantum gases.

As a prelude, we recall few concepts for N identical particles. In quantum mechanics, the knowledge of the state of a system is no longer specified by positions and momenta at any time for all the N indistinguishable particles which evolve according to Hamilton's equations, but rather the maximum information is obtained by the time evolution of the wave function $\Phi_N(\mathbf{x}_1, \mathbf{x}_2, \dots, \mathbf{x}_N, t)$, which gives the probability amplitude of finding the particles at positions $\mathbf{x}_1, \mathbf{x}_2, \dots, \mathbf{x}_N$. We emphasize that the particles are indistinguishable and it clearly makes no sense to specify which particle is found at a particular position. The probability of observing a particle at a point $\mathbf{x}_i$ and a second particle at point $\mathbf{x}_j$ must be equal to the probability in reverse order

$$|\Phi_N(\mathbf{x}_1, \mathbf{x}_2, \dots, \mathbf{x}_i, \dots, \mathbf{x}_j, \dots, \mathbf{x}_N, t)|^2 = |\Phi_N(\mathbf{x}_1, \mathbf{x}_2, \dots, \mathbf{x}_j, \dots, \mathbf{x}_i, \dots, \mathbf{x}_N, t)|^2.$$

The two wave functions differ at most by a phase factor

$$\Phi_N(\mathbf{x}_1, \mathbf{x}_2, \dots, \mathbf{x}_i, \dots, \mathbf{x}_j, \dots, \mathbf{x}_N, t) = \mathrm{e}^{\mathrm{i}\phi}\Phi_N(\mathbf{x}_1, \mathbf{x}_2, \dots, \mathbf{x}_j, \dots, \mathbf{x}_i, \dots, \mathbf{x}_N, t).$$

By repeating the argument one obtains $\mathrm{e}^{2\mathrm{i}\phi} = 1$, i.e., $\Phi_N(\mathbf{x}_1, \mathbf{x}_2, \dots, \mathbf{x}_N, t)$ must be either symmetric or antisymmetric under the interchange of two variables.

If we consider the simple case of two non-interacting independent identical particles with the normalized single-particle wave function $g_{\mathbf{k}}(\mathbf{x})$ labelled by the quantum number $\mathbf{k}$, the state for the two particles is given by the product $g_{\mathbf{k}}(\mathbf{x}_1)g_{\mathbf{k}'}(\mathbf{x}_2)$. Since the two particles are indistinguishable, the wave function $g_{\mathbf{k}}(\mathbf{x}_2)g_{\mathbf{k}'}(\mathbf{x}_1)$ is also representative of the same quantum system. The linear combination $g_{\mathbf{k}}(\mathbf{x}_1)g_{\mathbf{k}'}(\mathbf{x}_2) + cg_{\mathbf{k}}(\mathbf{x}_2)g_{\mathbf{k}'}(\mathbf{x}_1)$, according to the argument given above, must satisfy the identity

$$g_{\mathbf{k}}(\mathbf{x}_1)g_{\mathbf{k}'}(\mathbf{x}_2) + cg_{\mathbf{k}}(\mathbf{x}_2)g_{\mathbf{k}'}(\mathbf{x}_1) = \mathrm{e}^{\mathrm{i}\phi}\left(g_{\mathbf{k}}(\mathbf{x}_2)g_{\mathbf{k}'}(\mathbf{x}_1) + cg_{\mathbf{k}}(\mathbf{x}_1)g_{\mathbf{k}'}(\mathbf{x}_2)\right),$$

i.e. $\mathrm{e}^{\mathrm{i}\phi} = c$, $c\mathrm{e}^{\mathrm{i}\phi} = 1$. The state for the two identical particles is represented by the two symmetric or antisymmetric combinations

$$g^{(2)}_{\mathbf{k}\mathbf{k}'}(\mathbf{x}_1, \mathbf{x}_2) = \frac{1}{\sqrt{2}}\left(g_{\mathbf{k}}(\mathbf{x}_1)g_{\mathbf{k}'}(\mathbf{x}_2) \pm g_{\mathbf{k}}(\mathbf{x}_2)g_{\mathbf{k}'}(\mathbf{x}_1)\right). \tag{6.1}$$

No other possibility is allowed by the probabilistic interpretation of the wave function.

Bose (1924) recognized that the black-body radiation distribution (discussed in Chapter 7) found by Planck in 1900 could be obtained by implementing the indistinguishability of photons with a proper counting of states, different from the Boltzmann counting discussed in Chapter 3. Einstein (1924, 1925) generalized this prescription, now known as Bose–Einstein statistics, to generic non-interacting particles.

Fermi (1926a, b) introduced a further method of counting, now known as Fermi–Dirac statistics, to implement, together with the indistinguishability of particles, the Pauli exclusion principle originally formulated for electrons in atomic clouds.

We will develop these two counting prescriptions in Appendix D.

Heisenberg (1926) and Dirac (1926) related the symmetric and antisymmetric wave functions for identical particles to the Bose–Einstein and Fermi–Dirac statistics, respectively, that describe real particles with either integer or semi-integer spin.

We now proceed to the extension of the Boltzmann equation in the quantum case.

There are two main points to be discussed: (1) the use of the classical distribution function; and (2) the modification of the collision integral. We now see how this can be done. We emphasize that the following treatment is not completely rigorous, although it may be justified by means of a more sophisticated technique as the non-equilibrium quantum field theory. (For a pedagogical introduction see Rammer (2007).)

Concerning the first point, the physical hypothesis is that there is a hierarchy of length and time scales in the problem which allows us to use a semiclassical approximation. For instance, in a Fermi gas (which will be dealt with in the next chapter) in an external potential field ϕ, one assumes that the typical length, l, over which the field varies is much longer than the Fermi wavelength, λ_{F} (see Eq. (7.25)), or in general the De Broglie wavelength $\lambda_{\mathrm{D}} = \hbar/p$, the characteristic wavelength of quantum particles:

$$\lambda_{\mathrm{F}} \ll l. \tag{6.2}$$

In this way we avoid a conflict with the Heisenberg principle. In the presence of ϕ, the chemical potential becomes space dependent as $\mu(\mathbf{q}) = \mu + e\phi(\mathbf{q})$ $(e > 0)$. Hence the idea is to allow for a space dependence of the thermodynamic parameters that characterize the distribution function, and the system is in equilibrium with their local values. Equation (6.2) ensures that the thermodynamic parameters are approximately constant over a region which is large compared to microscopic lengths.

The modification of the collision integral can be understood in the following way. Consider a collision between two particles with the initial state of the two particles labelled by the momenta $\mathbf{p}_1$ and $\mathbf{p}_2$ and the final state by momenta $\mathbf{p}'_1$ and $\mathbf{p}'_2$. Let $W(1, 2; 1', 2') \equiv W$ be the scattering kernel as in the classical case, whose quantum expression (2.24) was derived in Chapter 2 by means of the Fermi Golden Rule. We adopt again the hypothesis of molecular chaos and neglect correlations between the particles. From the point of view of quantum mechanics, the scattering process may be seen as the absorption of the colliding particles with the initial momenta and the emission of two particles with the final momenta. The absorption of two particles with momenta $\mathbf{p}_1$ and $\mathbf{p}_2$ must be then proportional to the probability of having the two particles available for the collision. Having neglected the correlations between two particles, this yields a factor $f_1 f_2$ as in the classical case, where now the distribution functions are normalized as probabilities. Whereas in the classical limit, due to the diluteness condition, there are no limitations on the possible final states compatible with energy and momentum conservation, in quantum mechanics the situation is different. Let us consider first the case of fermions. If we have two fermions in the final state with momenta $\mathbf{p}'_1$ and $\mathbf{p}'_2$, due to the Pauli exclusion principle we must have the corresponding momentum states not previously occupied by other particles. Hence the emission process must be proportional to a factor $(1 - f_{1'})(1 - f_{2'})$. The resulting Boltzmann equation for fermions must then read[1]

$$\frac{\partial f_1}{\partial t} + \dot{\mathbf{q}} \cdot \frac{\partial f_1}{\partial \mathbf{q}} + \mathbf{F} \cdot \frac{\partial f_1}{\partial \mathbf{p}_1} = \int_{2,1',2'} W(f_{1'} f_{2'}(1 - f_1)(1 - f_2) - f_1 f_2 (1 - f_{1'})(1 - f_{2'})). \tag{6.3}$$

By considering the collision between bosons, again we may easily state that the absorption process must yield a factor $f_1 f_2$ as for classical and fermionic particles. The emission process is, however, quite different. Following Einstein, we know that the Planck distribution obeyed by photons in thermal equilibrium with matter is obtained when one postulates, besides the spontaneous emission, also the stimulated emission. This must be true not only for photons, but for all bosons and we may expect that the emission process will be proportional to a factor $(1 + f_{1'})(1 + f_{2'})$, where in each bracket the term with unity refers to spontaneous emission and the term with the distribution function to the stimulated one favored by the presence of other Bose particles. As a result the Boltzmann equation for bosons reads

$$\frac{\partial f_1}{\partial t} + \dot{\mathbf{q}} \cdot \frac{\partial f_1}{\partial \mathbf{q}} + \mathbf{F} \cdot \frac{\partial f_1}{\partial \mathbf{p}_1} = \int_{2,1',2'} W(f_{1'} f_{2'}(1 + f_1)(1 + f_2) - f_1 f_2 (1 + f_{1'})(1 + f_{2'})). \tag{6.4}$$

By requiring the vanishing of the collision integral in Eqs. (6.3) and (6.4), one obtains the equilibrium Fermi–Dirac and Bose–Einstein distribution functions, respectively. This can be seen easily by remembering (2.36), which in the present case becomes

$$\ln \frac{f_1}{1 \pm f_1} + \ln \frac{f_2}{1 \pm f_2} = \ln \frac{f_{1'}}{1 \pm f_{1'}} + \ln \frac{f_{2'}}{1 \pm f_{2'}}. \tag{6.5}$$

In the equation above, the $+(-)$ sign refers to bosons (fermions). The Fermi–Dirac and Bose–Einstein distributions are then obtained by using the same position as in (2.37). Indeed one has

$$f = \frac{1}{e^{b\mathbf{p}^2 - a} \pm 1}, \tag{6.6}$$

with the plus sign referring, in this case, to fermions and the minus one to bosons.

[1] A comment about physical dimensions may be useful. In Eq. (6.3) the scattering kernel W has the meaning of a collision rate as for instance given by the Fermi Golden Rule. It is given by the expression (2.24) without the factor $V^2/(2\pi\hbar)^6$. This is because the integral symbol, which we have kept for analogy with the classical case, must be understood as a sum over states $\int_{\mathbf{p}} \equiv \sum_{\mathbf{p}}$, and the factor $V(2\pi\hbar)^{-3}$ is included when, in the thermodynamic limit, the sum is replaced by the integral over the continuous variables $V(2\pi\hbar)^{-3} \int d^3p$ (see also Eq. (7.10) in Chapter 7.)

As in the classical case, it is also possible to show that Fermi–Dirac and Bose–Einstein distributions are in fact the most probable ones, as will be shown in Appendix D. The derivation of the Fermi–Dirac and Bose–Einstein distributions can also be carried out within the Gibbs ensemble approach, as we will do in Chapter 7.

6.2 Density matrix for pure states

We now turn to equilibrium in quantum statistical mechanics via the density matrix. Each observable is represented by an Hermitian operator which acts on the Hilbert space of the possible states of the system. We say that a system of N identical particles is in a pure state represented by a ket $|\Phi\rangle$, when is completely characterized by a wave function which is the projection of the ket $|\Phi\rangle$ on a representation. If we indicate by $|\{x\}\rangle \equiv |\mathbf{x}_1, \mathbf{x}_2, \ldots, \mathbf{x}_N\rangle$ the autoket of the position operator $\hat{x} \equiv (\hat{\mathbf{x}}_1, \hat{\mathbf{x}}_2, \ldots, \hat{\mathbf{x}}_N)$, then the wave function at any moment is $\Phi_N(\{x\}) = \langle\{x\}|\Phi\rangle$. Knowing the state $\Phi(\{x\})$, the average value of an observable $\hat{O}$ is

$$\begin{aligned} \langle\hat{O}\rangle = \langle\Phi|\hat{O}|\Phi\rangle &= \int \mathrm{d}\{x\} \int \mathrm{d}\{x'\} \langle\Phi|\{x\}\rangle\langle\{x\}|\hat{O}|\{x'\}\rangle\langle\{x'\}|\Phi\rangle \\ &= \int \mathrm{d}\{x\} \int \mathrm{d}\{x'\} \Phi_N^*(\{x\}) O_{\{x\}\{x'\}} \Phi_N(\{x'\}), \end{aligned} \tag{6.7}$$

where we have used the unity decomposition $\int \mathrm{d}\{x\}|\{x\}\rangle\langle\{x\}| = 1$ and $O_{\{x\}\{x'\}}$ is the matrix element of the observable $\hat{O}$ in the x-representation.

We now introduce the density matrix operator $\hat{\rho}$ as the projection operator on the ket $|\Phi\rangle$:

$$\hat{\rho} = |\Phi\rangle\langle\Phi| \tag{6.8}$$

with matrix element in the x-representation

$$\rho(\{x\}, \{x'\}) = \Phi_N^*(\{x'\})\Phi_N(\{x\}). \tag{6.9}$$

The density matrix for any observable has the property

$$\langle\hat{O}\rangle = \mathrm{Tr}\, \hat{\rho}\, \hat{O}, \tag{6.10}$$

with $\mathrm{Tr}\, \hat{\rho} = 1$ and for the pure state $\hat{\rho}^2 = \hat{\rho}$.

The pure state of a system $|\Phi\rangle$ can be in general given as a linear combination, $|\Phi\rangle = \sum_n c_n|n\rangle$, of a set of eigenstates $|n\rangle$ of a dynamical observable which form a complete basis and is specified by a set of quantum numbers n. The coefficients c_n are given by the projection of the state $|\Phi\rangle$ on $|n\rangle$, $c_n = \langle n|\Phi\rangle$. In this representation the matrix element of the density matrix reads

$$\rho_{nm} = c_m^* c_n, \tag{6.11}$$

and the average of the observable is

$$\mathrm{Tr}\, \hat{\rho}\, \hat{O} = \sum_{n,m} \rho_{nm} O_{mn}. \tag{6.12}$$

6.3 Density matrix for systems interacting with the environment

We consider a system X, specified by a set of coordinates $\{x\}$, in weak interaction with the environment Y, characterized by a set of coordinates $\{y\}$, all together forming an isolated system in a pure state with a wave function $\Psi(\{x\}, \{y\})$. The average of any observable $\hat{O}$ of the system X is given by

$$\langle \hat{O} \rangle = \int d\{x\} \int d\{x'\} \int d\{y\}\ \Psi^*(\{x\}, \{y\}) O_{\{x\}\{x'\}} \Psi(\{x'\}, \{y\}) = \text{Tr}\ \hat{\rho}\ \hat{O}, \quad (6.13)$$

where we have defined the density matrix for this mixed state

$$\rho(\{x\}, \{x'\}) = \int d\{y\}\ \Psi^*(\{x'\}, \{y\}) \Psi(\{x\}, \{y\}), \quad (6.14)$$

with $\text{Tr}\ \hat{\rho} = 1$. The density matrix $\rho(\{x\}, \{x'\})$ is reminiscent of the environment through the integration of the $\{y\}$ variables. If, however, the system and the environment are independent, the global wave function can be factorized as a product of two wave functions $\Psi(\{x\}, \{y\}) = \Phi(\{x\})\chi(\{y\})$ and one recovers immediately the definition (6.9) of the pure state.

In general, we can change the representation of the density matrix by considering a complete set of wave functions $\phi_n(\{x\})$ for the system, a complete set $\chi_l(\{y\})$ for the environment and a complete set for the whole system, $\phi_n(\{x\})\chi_l(\{y\})$:

$$\Psi(\{x\}, \{y\}) = \sum_n c_n(\{y\})\phi_n(\{x\}) = \sum_{n,l} c_{n,l}\phi_n(\{x\})\chi_l(\{y\}). \quad (6.15)$$

By plugging the expression (6.15) into the definition of the density matrix in the x-representation, we obtain the density matrix in the generic representation

$$\rho_{mn} = \int d\{y\}\ c_n^*(\{y\})c_m(\{y\}) = \sum_l c_{nl}^* c_{ml}. \quad (6.16)$$

Since the density matrix is Hermitian, it can always be diagonalized by a complete orthonormal set of eigenfunctions $\phi_s(\{x\})$ with real eigenvalues w_s. As an operator $\hat{\rho}$ can be written as

$$\hat{\rho} = \sum_s w_s |\phi_s\rangle\langle\phi_s|, \quad \sum_s w_s = 1 \quad (6.17)$$

and

$$\hat{\rho}^2 = \sum_s w_s^2 |\phi_s\rangle\langle\phi_s|. \quad (6.18)$$

The condition $\hat{\rho}^2 = \hat{\rho}$ is not only a necessary condition, but also a sufficient condition for a system to be in a pure state. In fact, from (6.18) it follows that there must be a state ϕ_k for which $w_k = 1$ and $w_s = 0$ for any s different from k. In general then $\text{Tr}\ \hat{\rho}^2 \leq 1$, while $\text{Tr}\ \hat{\rho} = 1$ always. If the system is not in a pure state, the full quantum state of the system is not known. The probabilities w_n for every state in the superposition of the density matrix are known, not their phases which are necessary to build up the state.

To illustrate the difference between the density matrices in the two cases discussed so far, we consider the following simple example. Let us imagine linearly polarized light in a plane. The most general state, $|\phi\rangle$, can be written as a linear combination of the two polarizations along the x- and y-directions

$$|\phi\rangle = c_1|\phi_1\rangle + c_2|\phi_2\rangle, \quad |\phi_1\rangle = \begin{pmatrix} 1 \\ 0 \end{pmatrix}, \quad |\phi_2\rangle = \begin{pmatrix} 0 \\ 1 \end{pmatrix}, \tag{6.19}$$

where $|c_1|^2 + |c_2|^2 = 1$. In this case, the density matrix reads

$$\hat{\rho} = \begin{pmatrix} c_1 c_1^* & c_1 c_2^* \\ c_2 c_1^* & c_2 c_2^* \end{pmatrix}. \tag{6.20}$$

Since all the amplitudes, c_1 and c_2, are known and constrained by the normalization condition, $\hat{\rho}^2 = \hat{\rho}$.

If instead we know only that the light beam is made of a fraction w_1 polarized along x and a fraction w_2 polarized along y, with $w_1 + w_2 = 1$, the density matrix reads

$$\hat{\rho} = w_1\hat{\rho}_x + w_2\hat{\rho}_y = \begin{pmatrix} w_1 & 0 \\ 0 & w_2 \end{pmatrix} \tag{6.21}$$

and

$$\hat{\rho}^2 = \begin{pmatrix} w_1^2 & 0 \\ 0 & w_2^2 \end{pmatrix} \neq \hat{\rho}.$$

Clearly, in this case we do not know the amplitudes of the states, but only the probabilities, and the interference terms are missing.

Finally, we derive the equation of motion for the density matrix. By writing the density matrix in its diagonal form (6.17), the state $|\phi_s\rangle$ can change with time and the density matrix evolves in time $\hat{\rho}(t) = \sum_s w_s|\phi_s(t)\rangle\langle\phi_s(t)|$. Using the Schrödinger equation for $|\phi_s(t)\rangle$ and $\langle\phi_s(t)|$,

$$i\hbar\frac{\partial|\phi_s(t)\rangle}{\partial t} = H|\phi_s(t)\rangle, \quad i\hbar\frac{\partial\langle\phi_s(t)|}{\partial t} = -\langle\phi_s(t)|H, \tag{6.22}$$

we obtain the time evolution of the density matrix $\hat{\rho}$:

$$i\hbar\frac{\partial\hat{\rho}}{\partial t} = [H, \hat{\rho}]\,. \tag{6.23}$$

Notice the opposite sign with respect to the Heisenberg equation of motion for the time evolution of an operator. If $\hat{\rho}$ has the same set of eigenstates as the Hamiltonian, it will not depend on time as for the statistical ensembles we are going to discuss.

6.4 Quantum statistical ensembles

In the above formulation the knowledge at any instant of the quantum average of any observable $\hat{O}$ is given by the time evolution of the density matrix (6.11) for a pure state and (6.16) for a mixed state. As in classical statistical mechanics, we are not interested in following instant by instant the complete mechanical evolution of the system, but, for a system assumed in thermodynamic equilibrium, we are only interested in its time averaged behavior. In order to obtain the thermodynamic quantities, we need to know the time average, $\overline{\hat{\rho}}$, of the density matrix over times much longer than any characteristic microscopic time of the dynamical evolution:

$$\overline{\langle \hat{O} \rangle} = \sum_{nm} \bar{\rho}_{nm} O_{mn}. \tag{6.24}$$

In analogy with the classical case, we can replace the time average with an ensemble average. In particular for the microcanonical ensemble we assume the following two postulates for the averaged density matrix $\overline{\hat{\rho}}$.

1. Random phase postulate:

$$\overline{\rho}_{nm} = \delta_{nm} w_n, \tag{6.25}$$

n labelling the eigenstates of the Hamiltonian with eigenvalues E_n. This postulate implies the disappearing of the interference terms in the process of time averaging during a measure.

2. Equal a priori probabilities of all the allowed microscopic states:

$$w_n = \begin{cases} \text{constant}, & E < E_n < E + \Delta E, \\ 0, & \text{otherwise}, \end{cases} \tag{6.26}$$

where ΔE is the uncertainty in the knowledge of the energy and the constant ≤ 1 must be determined by the normalization condition $\sum_n w_n = 1$. In quantum systems the uncertainty in the energy has an intrinsic minimum value governed by the Heisenberg principle.

On the basis of the above postulates, in general, from Eq. (6.25) the operational form of the density matrix for the statistical ensemble $\hat{\rho}$ is equal to $\overline{\hat{\rho}}$:

$$\hat{\rho} = \overline{\hat{\rho}} = \sum_n |\phi_n\rangle w_n \langle \phi_n|, \tag{6.27}$$

with $\text{Tr}\overline{\hat{\rho}} = 1$.

These two postulates define the microcanonical ensemble for a quantum system. Before deriving the thermodynamics we introduce the operator counting the allowed states, defined as

$$\hat{\rho}_c = \sum_{E<E_n<E+\Delta E} |\phi_n\rangle\langle\phi_n|.$$

The thermodynamics of the microcanonical ensemble is obtained by replacing the volume in the phase space $\Gamma(E)$ defined in Eq. (3.24) with

$$\Gamma(E) = \text{Tr}\hat{\rho}_c. \tag{6.28}$$

For the allowed states $w_n = 1/\Gamma(E)$. The natural corresponding thermodynamic potential is the entropy:

$$S(E, V) = k_B \ln \Gamma(E) = -k_B \sum_{E<E_n<E_n+\Delta E} w_n \ln w_n \qquad (6.29)$$

and all the thermodynamic relations follow. Notice that we have found the Gibbs expression for the entropy given in Eq. (4.19) again.

At equilibrium the system can be considered to be represented by an incoherent superposition of states, each one appearing with probability w_n. Effectively, this is equivalent to saying that the density matrix and the Hamiltonian of the system can be diagonalized simultaneously. A system being in a mixed state corresponds to a situation where there is a certain lack of information. The interaction of a system with the environment is in many cases the origin of such loss of information. The measured value of any macroscopic observable is not represented by its istantaneous value, but rather by

$$\langle\langle\hat{O}\rangle\rangle = \mathrm{Tr}\, \hat{\rho}\, \hat{O}, \qquad (6.30)$$

where the double average means quantum and statistical averages.

From now on the introduction of the canonical and grand canonical ensembles follows the same route as in the classical case of Chapter 4 with

$$\overline{\hat{\rho}} = \frac{1}{Z}\mathrm{e}^{-\beta H}, \quad Z = \mathrm{Tr}\,\mathrm{e}^{-\beta H} = \mathrm{e}^{-\beta F} \qquad (6.31)$$

and

$$\overline{\hat{\rho}} = \frac{1}{\mathcal{Z}}\mathrm{e}^{-\beta(H-\mu\hat{N})}, \quad \mathcal{Z} = \mathrm{Tr}\,\mathrm{e}^{-\beta(H-\mu\hat{N})} = \mathrm{e}^{-\beta\Omega}, \qquad (6.32)$$

where μ is the chemical potential and $\hat{N}$ the total number operator. The symbol Tr includes the summation over all possible numbers of particles. Z and $\mathcal{Z}$ are the canonical and grand canonical partition functions, to which correspond the two thermodynamical potentials $F(T, V, N)$ and $\Omega(T, V, \mu)$.

6.5 Problem

6.1 Show that non-interacting indistinguishable identical Bose particles tend, on average, to be closer than distinguishable particles. Fermi particles, instead, tend to be further apart.

7 The quantum gases

7.1 The statistical distribution of the quantum gases

The equilibrium Fermi–Dirac and Bose–Einstein distributions can be obtained as the most probable ones (see Appendix D). We here develop the Fermi–Dirac and Bose–Einstein distributions as ensemble averages. We evaluate the grand potential Ω for the quantum gases, N non-interacting particles, either fermions or bosons, distributed over a set of levels with energy ϵ_i, where n_i is the occupation number of the level ϵ_i. The grand canonical partition function reads

$$\mathcal{Z} = \sum_{\{n_i\}} \mathrm{e}^{-\beta \sum_i (\epsilon_i - \mu) n_i}, \tag{7.1}$$

where $\{n_i\}$ indicates all possible sets of occupation numbers without any restriction on the total number of particles. Given the separability of the Hamiltonian, the summation over the sets $\{n_i\}$ can be written as the product of the summations over each occupation number:

$$\mathcal{Z} = \prod_i \sum_{n_i} \mathrm{e}^{-\beta(\epsilon_i - \mu) n_i}. \tag{7.2}$$

For fermions, there are only two possible values for the occupation number, $n_i = 0$ and $n_i = 1$ due to the Pauli exclusion principle. This leads soon to

$$\mathcal{Z}_{\mathrm{F}} = \prod_i \left(1 + \mathrm{e}^{-\beta(\epsilon_i - \mu)}\right). \tag{7.3}$$

For bosons, there is no restriction on the occupation numbers, i.e., $n_i = 0, 1, 2, \ldots$ and we get

$$\mathcal{Z}_{\mathrm{B}} = \prod_i \frac{1}{1 - \mathrm{e}^{-\beta(\epsilon_i - \mu)}}, \tag{7.4}$$

where by performing the summation of Eq. (7.2) we have used the geometrical series. In order for the latter to be convergent, we must require that $\epsilon_i - \mu > 0$ for each value of the index i. This leads to the condition that the chemical potential for the boson gas must satisfy $\mu \leq 0$.

The grand potential for fermions and bosons then reads

$$\Omega = \mp k_{\mathrm{B}} T \sum_i \ln\left(1 \pm \mathrm{e}^{-\beta(\epsilon_i - \mu)}\right) = -PV, \tag{7.5}$$

with the upper sign applying to fermions and the lower to bosons. By taking the derivative of Ω with respect to the chemical potential, we obtain the average number of particles

$$\langle N\rangle = -\left(\frac{\partial\Omega}{\partial\mu}\right)_{T,V} = \sum_i \frac{1}{e^{\beta(\epsilon_i-\mu)}\pm 1} \equiv \sum_i \langle n_i\rangle, \tag{7.6}$$

and by differentiating with respect to the temperature, the entropy

$$S = -\left(\frac{\partial\Omega}{\partial T}\right)_{\mu,V} = \mp k_B \sum_i \left[\pm\langle n_i\rangle \ln\langle n_i\rangle + (1\mp\langle n_i\rangle)\ln(1\mp\langle n_i\rangle)\right], \tag{7.7}$$

where $\langle n_i\rangle$ indicates the mean occupation number in the i-th state.

We find again the expression for the Fermi–Dirac and Bose–Einstein distributions obtained previously by the Boltzmann equation (see Eq. (6.6)) and by the most probable distribution method (see Eqs. (D.3) and (D.8) in Appendix D).

It is useful to apply the formulae (7.5) and (7.6) to the case of free particles in a box. We consider then particles contained in a box with side L and volume $V = L^3$. By taking periodic boundary conditions, the quantum numbers are in this case

$$p_x = \frac{2\pi\hbar}{L}n_x, \quad p_y = \frac{2\pi\hbar}{L}n_y, \quad p_z = \frac{2\pi\hbar}{L}n_z \tag{7.8}$$

with $n_x = 0, 1, 2, \ldots$ and similarly n_y and n_z. The energy levels have the standard quadratic form

$$\epsilon(p_x, p_y, p_z) = \frac{1}{2m}\left(p_x^2 + p_y^2 + p_z^2\right) \equiv \epsilon_{\mathbf{p}}. \tag{7.9}$$

In the case of large L, the sum over the quantum numbers can be replaced by an integral over continuous variables. To this end we observe that $\Delta_x = 2\pi\hbar/L$ represents the interval between a discrete value of the quantum number p_x and the successive one. This means that in the infinitesimal interval dp_x, there are dp_x/Δ_x levels. Hence we have the following replacement:

$$\sum_{p_x} \to \int \frac{dp_x}{\Delta_x} = \frac{L}{2\pi\hbar}\int dp_x. \tag{7.10}$$

By reasoning in the same way for the quantum numbers p_y and p_z, we arrive at expressions for the energy and the number of particles:

$$PV = \pm k_B T \frac{sV}{(2\pi\hbar)^3}\int d^3p \ln\left(1 \pm z e^{-\beta\epsilon_{\mathbf{p}}}\right), \tag{7.11}$$

$$U = \frac{sV}{(2\pi\hbar)^3}\int d^3p \frac{\epsilon_{\mathbf{p}} z}{e^{\beta\epsilon_{\mathbf{p}}} \pm z}, \tag{7.12}$$

$$\langle N\rangle = \frac{sV}{(2\pi\hbar)^3}\int d^3p \frac{z}{e^{\beta\epsilon_{\mathbf{p}}} \pm z}, \tag{7.13}$$

where the plus (minus) sign applies to fermions (bosons) and we have introduced the multiplicity of spin s and the fugacity $z = e^{\beta\mu}$.

It is useful to introduce the density of states per unit volume as

$$\nu(\epsilon) = s \int \frac{d^3p}{(2\pi\hbar)^3} \delta\left(\epsilon - \frac{p^2}{2m}\right) = \frac{s(2m)^{3/2}}{4\pi^2\hbar^3}\sqrt{\epsilon}, \tag{7.14}$$

and rewrite the equations (7.11)–(7.13) in the form:

$$PV = \pm k_B T V \frac{s(2m)^{3/2}}{4\pi^2\hbar^3} \int_0^\infty d\epsilon \sqrt{\epsilon} \ln\left(1 \pm z e^{-\beta\epsilon}\right), \tag{7.15}$$

$$U = V \frac{s(2m)^{3/2}}{4\pi^2\hbar^3} \int_0^\infty d\epsilon \frac{\epsilon^{3/2} z}{e^{\beta\epsilon} \pm z} = V \frac{s(2m)^{3/2}(k_B T)^{5/2}}{4\pi^2\hbar^3} I^{\pm}_{3/2}(\ln z), \tag{7.16}$$

$$\langle N \rangle = V \frac{s(2m)^{3/2}}{4\pi^2\hbar^3} \int_0^\infty d\epsilon \frac{\epsilon^{1/2} z}{e^{\beta\epsilon} \pm z} = V \frac{s(2m k_B T)^{3/2}}{4\pi^2\hbar^3} I^{\pm}_{1/2}(\ln z), \tag{7.17}$$

where the Fermi–Dirac and Bose–Einstein integrals $I^{\pm}_{\alpha}$ are defined as

$$I^{\pm}_{\alpha}(x) = \int_0^\infty dy \, \frac{y^\alpha}{e^{y-x} \pm 1}. \tag{7.18}$$

In Appendix E we provide a number of properties of these integrals necessary for the following discussion in this chapter.

The expressions (7.15)–(7.17) will be studied in detail in the next sections. Here we only point out that by integrating by parts the right-hand side of Eq. (7.15) one obtains the relation

$$U = \frac{3}{2} PV = -\frac{3}{2}\Omega, \tag{7.19}$$

which then implies that in the following we can confine ourselves to the study of Eqs. (7.16) and (7.17).

From the above equations, it follows that

$$\frac{2}{3}\frac{U}{N k_B T} = \frac{PV}{N k_B T} = \frac{2}{3}\frac{I^{\pm}_{3/2}(\ln z)}{I^{\pm}_{1/2}(\ln z)}, \tag{7.20}$$

where N stands for $\langle N \rangle$. Equation (7.20) in general states the deviation from the classical limit, which is recovered when $z \ll 1$ and the ratio of the two integrals is equal to $3/2$. This can be easily obtained by noticing that in the classical limit, $z \ll 1$, the integrals $I^{\pm}_{\alpha}(\ln z) \approx z\Gamma(\alpha + 1)$ (see Eqs. (E.3) and (E.6) of Appendix E).

We clarify the physical conditions under which the classical limit is recovered. In the limit $z \ll 1$ the fugacity, via Eq. (7.17), reduces to the classical expression (4.33) and is given by

$$z = \frac{N}{sV}\left(\frac{h^2}{2\pi m k_B T}\right)^{3/2} = \frac{1}{s}\left(\frac{\lambda}{d}\right)^3, \tag{7.21}$$

where $d = (V/N)^{1/3}$ is the mean interparticle distance. Hence the classical limit is recovered when the thermal wave length λ is much smaller than the interparticle distance d. This clearly occurs either when the gas is diluted or at high temperature when λ is small. To estimate when we enter the quantum regime, we use the uncertainty principle. Considering the average value of the energy as $\langle p^2/(2m)\rangle = (3/2)k_B T$, the uncertainty in the location of the particle is $\Delta x \sim h/(2\pi\sqrt{3mk_B T})$. In the limit $d \gg \Delta x$ the particles can be considered as point-like, while for $d \sim \Delta x$ quantum interference appears and we enter the quantum regime.

One then may compute the first quantum corrections to the classical limit of the equation of state by keeping in Eq. (7.20) terms quadratic in z in the expansion of the integrals $I^{\pm}_{3/2}$ and $I^{\pm}_{1/2}$ (see Eqs. (E.3) and (E.6) of Appendix E) to obtain

$$\frac{PV}{Nk_B T} = 1 \pm \frac{1}{2^{5/2}}\left(\frac{\lambda}{d}\right)^3. \tag{7.22}$$

The plus and minus signs signal the different statistical correlations induced by the onset of the two quantum statistics.

7.2 The strongly degenerate Fermi gas

It is instructive to start the analysis of the quantum regime by considering the extreme case of zero temperature. When $T \to 0$, the Fermi distribution function becomes a step function:

$$\lim_{T\to 0} \frac{1}{e^{\beta(\epsilon-\mu)}+1} = \begin{cases} 0, & \epsilon > \mu, \\ 1, & \epsilon < \mu, \\ 1/2, & \epsilon = \mu. \end{cases} \tag{7.23}$$

The form of the distribution function expresses the Pauli exclusion principle according to which, starting from the lowest energy level, all levels are occupied as long as there are particles available to fill them. Each state is occupied by just one particle. The chemical potential at zero temperature is then the energy of the topmost occupied level and this value of the energy is called the Fermi energy:

$$E_F \equiv \mu(T=0). \tag{7.24}$$

The Fermi energy is determined by the number of particles in a given volume, i.e., by the density. To see this for a gas of fermions of spin $1/2$, let us explore the consequences of (7.17) with $s=2$ in the zero temperature limit.

The number of particles becomes

$$N = V\int_0^{E_F} d\epsilon\ \nu(\epsilon) = \frac{2}{3}V\nu(E_F)E_F = V\frac{\sqrt{2}m^{3/2}}{\pi^2\hbar^3}\frac{2}{3}E_F^{3/2} = \frac{V}{3\pi^2}k_F^3. \tag{7.25}$$

We see that the Fermi energy goes like the $2/3$ power of the density, $E_F = \hbar^2 k_F^2/2m = (3\pi^2)^{2/3}(\hbar^2/2m)(N/V)^{2/3}$, where k_F is the Fermi wave vector. This in turn defines the Fermi wave length $\lambda_F = 2\pi/k_F$ of the order of the inter-particle distance.

At zero temperature the energy U of (7.16) may be evaluated in a similar manner:

$$U = V\int_0^{E_F} d\epsilon\ \nu(\epsilon)\,\epsilon = V\frac{\sqrt{2}m^{3/2}}{\pi^2\hbar^3}\frac{2}{5}E_F^{5/2}, \tag{7.26}$$

showing how a Fermi gas, in contrast to a classical gas, has a finite energy at zero temperature. By introducing the Fermi temperature T_F as $E_F \equiv k_B T_F$, we obtain the expressions

$$U = \frac{3}{5} N k_B T_F, \quad PV = \frac{2}{5} N k_B T_F, \tag{7.27}$$

which follow from Eqs. (7.16) and (7.19) by evaluating the integral in the limit $T \to 0$. $T_F \propto n^{2/3}$ is the degeneration temperature below which the quantum effects appear. For a gas of conduction electrons in metals the concentration is typically $n \sim 10^{28}\,\mathrm{m}^{-3}$, the Fermi energy is $E_F \sim 2\,\mathrm{eV}$ and $T_F \sim 2 \times 10^4\,\mathrm{K}$. Even at ambient temperature the gas of conduction electrons is in the strongly degenerate regime $T \ll T_F$. The fermionic isotope three of helium with density of the same order $n \sim 1.6 \times 10^{28}\,\mathrm{m}^{-3}$, but $m \sim 3m_N = 5 \times 10^{-27}\,\mathrm{kg}$, has $E_F \sim 6.7 \times 10^{-23}\,\mathrm{J}$, $T_F \sim 4.9\,\mathrm{K}$ in the Kelvin region and the Fermi velocity $v_F = \hbar k_F/m \sim 163\mathrm{m/s}$.

The isothermal compressibility is then obtained via Eq. (4.35) by evaluating the zero temperature limit of the thermodynamic density of states $\partial n/\partial \mu$:

$$\frac{\partial n}{\partial \mu} = \nu(E_F), \tag{7.28}$$

i.e. in this limit it coincides with the single particle density of states at the Fermi level. The latter also determines the magnetic properties of the strongly degenerate Fermi gas. By introducing a Zeeman term in the Hamiltonian of the Fermi gas of the form

$$H_Z = -\frac{g\mu_B B}{2}\sigma_z, \tag{7.29}$$

where $\mu_B = (e\hbar)/(2mc)$ is the Bohr magneton, g the gyromagnetic factor, B the intensity of the magnetic field and σ_z the standard Pauli matrix, the electron spectrum splits into two branches, $\epsilon_{\pm\mathbf{p}} = p^2/2m \mp \Delta/2$ with $\Delta = g\mu_B B$ the Zeeman splitting. The equation for the number of particles now becomes

$$N = N_+ + N_- \equiv \frac{V}{2}\left(\int_0^{E_F+\Delta/2} d\epsilon\ \nu(\epsilon) + \int_0^{E_F-\Delta/2} d\epsilon\ \nu(\epsilon)\right). \tag{7.30}$$

The magnetization per unit volume of the Fermi gas is given by

$$m(B) = \frac{g\mu_B}{2}\frac{N_+ - N_-}{V}. \tag{7.31}$$

By considering the derivative of the magnetization per unit volume in the zero magnetic field limit, one obtains the Pauli susceptibility

$$\chi = \lim_{B\to 0} \frac{\partial}{\partial B} m(B) = \left(\frac{g\mu_B}{2}\right)^2 \nu(E_F). \tag{7.32}$$

When the temperature is non-zero but small, i.e., when $T \ll T_F$, the Fermi gas is said to be in the degenerate regime and it is possible to analytically evaluate the temperature dependence of the energy and the number of particles. This is achieved by the so-called Sommerfeld expansion and corresponds to the asymptotic value of the Fermi–Dirac integral $I_\alpha^+(x)$ for large positive x. By using Eqs. (E.23) and (E.24) of Appendix E we arrive at the following expressions for the energy and the number of particles:

$$U = V\frac{\sqrt{2}m^{3/2}}{\pi^2\hbar^3}\frac{2}{5}\mu^{5/2}\left(1+\frac{15}{24}\pi^2\frac{(k_B T)^2}{\mu^2}\right), \tag{7.33}$$

$$N = V\frac{\sqrt{2}m^{3/2}}{\pi^2\hbar^3}\frac{2}{3}\mu^{3/2}\left(1+\frac{3}{24}\pi^2\frac{(k_B T)^2}{\mu^2}\right). \tag{7.34}$$

These equations can be solved for U and μ. By inverting the second equation with respect to μ and keeping terms up to quadratic order in T/T_F we get

$$\mu(T, V, N) = E_F\left(1-\frac{\pi^2}{12}\frac{T^2}{T_F^2}\right), \tag{7.35}$$

$$U(T, V, N) = \frac{3}{5}NE_F\left(1+\frac{5\pi^2}{12}\frac{T^2}{T_F^2}\right). \tag{7.36}$$

The specific heat per unit volume, by using (7.36), is

$$c_V = \frac{\pi^2}{3}\nu(E_F)k_B^2 T. \tag{7.37}$$

Notice that c_V can also be derived from Eq. (7.19) and Eq. (7.33) as

$$c_V = -\frac{T}{V}\left(\frac{\partial^2\Omega}{\partial T^2}\right)_{\mu,V}.$$

In the same way the pressure reads

$$P = \frac{N}{V}\left(\frac{2}{5}k_B T_F + \frac{\pi^2}{6}\frac{T}{T_F}k_B T\right). \tag{7.38}$$

The temperature dependent terms in the expression for c_V and P have a clear physical meaning. The ratio $k_B T/E_F$ represents the fraction of energy levels whose occupation is smeared around the Fermi energy by the temperature. Hence the quantity $Nk_B T/E_F$ is an effective number of "active" particles.

We conclude this section by remarking that, as we will see in Chapter 12 with the Landau theory of the Fermi liquid, these equations will be renormalized to include the effects of the interaction between the particles.

7.3 The degenerate Bose gas

Let us consider a Bose gas of N non-interacting bosons with spin 0 in a box of volume $V = L^3$ with energy levels given by Eq. (7.9). The most striking phenomenon of a degenerate Bose gas is the Bose–Einstein condensation (BEC). To understand its origin, we begin by remembering that, in contrast to fermions, bosons may occupy the quantum states with arbitrary occupation numbers and tend to cluster as specified while discussing the most probable distribution in Appendix D. At zero temperature, all the bosons will stay in the lowest single particle energy level corresponding to zero momentum, so that the occupation number for this state coincides with the total number of particles, while all other occupation numbers vanish:

$$n_{\mathbf{p}=\mathbf{0}} = N, \quad n_{\mathbf{p}\neq\mathbf{0}} = 0 \quad \text{for all } \mathbf{p} \neq \mathbf{0}. \tag{7.39}$$

Equation (7.39) states that the ground state is macroscopically occupied in the sense that the value of its occupation number is of the same order as the number of particles. Actually, in this case of non-interacting particles, the two coincide. The Bose–Einstein condensation is a phenomenon such that this macroscopic occupation remains even at finite temperature, i.e., $n_{\mathbf{p}=\mathbf{0}}(T) \sim N$, below a critical value. The word condensation is used to mean that the particles accumulate in the $\mathbf{p} = \mathbf{0}$ state, which corresponds to an ordering in momentum space, in contrast to the ordinary condensation which occurs in real space. This ordering, furthermore, has the behavior typical of a phase transition with a critical temperature T_c, contrary to the Fermi gas for which the approach to "ordered" ground state occurs continuously. Let us now examine in detail the mechanism of the Bose–Einstein condensation, which is relevant for the discussion of the properties of the cold alkali atoms with T_c in the micro-nanokelvin region and of liquid ^{4}He when it becomes superfluid for $T < T_c = 2.19$ K at its own vapor pressure.

We have stated that the classical regime for a Bose gas is obtained when $z \ll 1$, i.e. from Eq. (7.21) the thermal wave length is much smaller than the average interparticle distance. This is clearly so for a mole of ^{4}He at 0^0C and 1 atmosphere, for which $V = 2.241 \times 10^4$ cm^3, $d/\lambda = 63.8$. A refrigeration down to 2–3 K brings z to the order of 1, i.e. to strong degeneration and condensation. Of course, under these physical conditions, as we shall see, it is not possible to avoid considering the interactions.

We now discuss the condensation transition for the Bose gas. From Eqs. (7.16) and (7.17) via (7.20) and (E.3) of Appendix E, the pressure and the number of particles read

$$P = k_B T \lambda^{-3} \mathrm{Li}_{5/2}(z), \tag{7.40}$$

$$n = \frac{\langle N \rangle}{V} \equiv \frac{1}{v} = \lambda^{-3} \mathrm{Li}_{3/2}(z), \tag{7.41}$$

the polylogarithm $\mathrm{Li}_s(z)$ being defined in Eq. (E.4) of Appendix E. Since N is fixed, a temperature decrease, in Eq. (7.41), in the factor $\lambda^{-3} \sim (k_B T)^{3/2}$ in front of the integral must be compensated by an increase of the fugacity $z(T)$ in $\mathrm{Li}_{3/2}(z)$. Since the chemical potential is always non-positive, $\mu \leq 0$ or equivalently $z \leq 1$, this means that the decrease of the factor $(k_B T)^{3/2}$ can be compensated by an increase of z as long as z remains less than unity. The temperature at which z attains the unity value is called the Bose–Einstein condensation temperature, T_c, and is determined by the previous equation with $z = 1$ in $\mathrm{Li}_{3/2}(z)$:

$$\lambda_c^3 = v\zeta_R(3/2), \quad k_B T_c = \frac{2\pi}{\zeta_R(3/2)^{2/3}} \frac{\hbar^2}{m} n^{2/3}, \tag{7.42}$$

where we used $\mathrm{Li}_{3/2}(1) = \zeta_R(3/2)$ (see Eqs. (E.4) and (E.9)) and $\lambda_c = h/(2\pi m k_B T_c)^{1/2}$, where λ_c is the critical thermal wave length corresponding to the critical temperature T_c. For all $T > T_c$ Eq. (7.41) determines z as function of T. A plot of the solution is shown in Fig. 7.1. We notice that the critical temperature scales as the 2/3 power of the density.

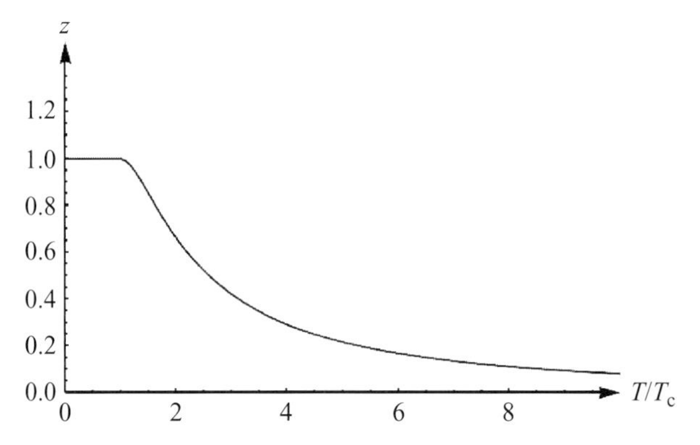

Fig. 7.1 The fugacity as function of the temperature measured in units of T_c.

Grossly speaking we can physically interpret the degeneracy condition given by Eq. (7.42) as the temperature at which the number of thermally accessible states is of the order of the total number of particles. The number of thermally accessible states is obtained by integrating up to $k_B T_c$ the density of states $\nu(\epsilon)$, Eq. (7.14). At the critical temperature, by using the equation of state (7.20) or from Eq. (7.40), we obtain $(2/3)(U/Nk_B T) = (PV/Nk_B T) = \zeta_R(5/2)/\zeta_R(3/2) \sim 0.5134$, about half of the classical value.

A further decrease of the temperature, according to Eq. (7.42), leads inevitably to a decrease in the number of particles. This paradoxical result can be interpreted by observing that the integral of Eq. (7.41) is only taking into account the occupation numbers of those states with non-zero energy, due to the fact that the zero energy state has a vanishing density of states $\sim \sqrt{\epsilon}$. Before taking the continuum limit for the momentum quantum numbers, one should have written

$$N = \sum_{\mathbf{p}} \frac{z}{e^{\beta\epsilon(\mathbf{p})} - z} = \frac{z}{1-z} + \sum_{\mathbf{p}\neq 0} \frac{z}{e^{\beta\epsilon(\mathbf{p})} - z}. \tag{7.43}$$

The second term on the right-hand side corresponds to the integral of Eq. (7.41), whereas the first is the occupation number $n_{\mathbf{p}=\mathbf{0}}$ of the single-particle ground state. At $T > T_c$ this occupation remains of the order of unity and is negligible compared to the total number of particles N. When $T < T_c$ the decrease seen in Eq. (7.42) corresponds to a decrease in the occupation numbers for the particles outside the ground state. At the same time this decrease is compensated by an increase of particles in the ground state. By making use of both Eqs. (7.42) and (7.43) we find an explicit expression for the macroscopic occupation of the lowest level:

$$n_{\mathbf{p}=\mathbf{0}}(T) \equiv N_0(T) = N\left(1 - \left(\frac{T}{T_c}\right)^{3/2}\right), \quad T \leq T_c. \tag{7.44}$$

For the phenomenon of the condensation in momentum space, we now determine the behavior of the isotherms, along with the coexistence of the two phases, in analogy with the ordinary condensation in real space. Equations (7.40) and (7.41), by the elimination of the fugacity z, determine the isotherms of the Bose gas. At fixed temperature a decrease

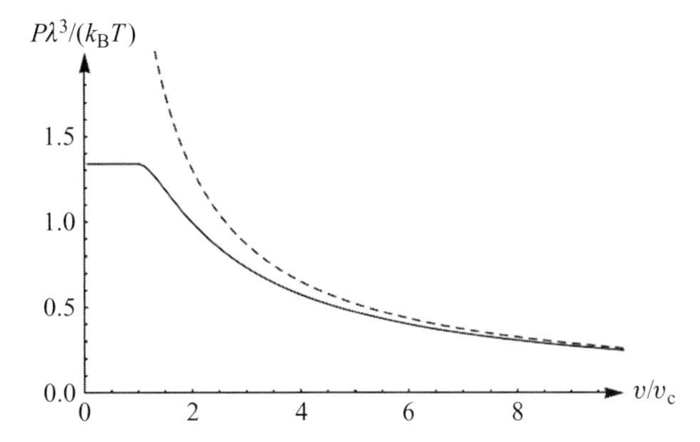

Fig. 7.2 The isotherm of a Bose gas: the pressure in units of $k_B T \lambda^{-3}$ is plotted as a function of the specific volume measured in units of v_c. The flat region is the coexistence between the normal part of the phase at specific volume v and the condensate phase at zero specific voume. The dotted line shows the classical isotherm.

of the volume is compensated by an increase of the fugacity. Let us indicate with v_c the critical volume per particle at which the fugacity attains unity at fixed temperature:

$$v_c = \frac{\lambda^3}{\zeta_R(3/2)}. \tag{7.45}$$

A further decrease of the specific volume below v_c, since it cannot be compensated anymore by an increase of the fugacity, implies a transfer of particles from above the condensate to the condensate. For volumes $v > v_c$, since the function $\mathrm{Li}_{5/2}(z)$ in Eq. (7.40), and $\mathrm{Li}_{3/2}(z)$ in Eq. (7.41) are increasing functions of z, the pressure is a decreasing function of the volume, as required by thermodynamic stability. For volumes $v < v_c$, the pressure remains constant at the value given by Eq. (7.40) with $z = 1$. The behavior of the generic isotherm at fixed temperature T is shown in Fig. 7.2. The flat part of the isotherm in Fig. 7.2 corresponds to the coexistence between the gas phase consisting of the particles outside and inside the condensate. We see that, in contrast to the ordinary liquid–gas transition, there is no further increase of the isotherm after the flat region. This has a counterpart in the T–P diagram, where the coexistence line is obtained by setting $z = 1$ in the expression for the pressure (7.40). There is no allowed room in the phase diagram for pressures greater than the value along the coexistence line because the pure condensed phase in a gas can only exist at zero temperature and pressure.

The entropy of the Bose gas may be obtained via the thermodynamic relation

$$S = -\left(\frac{\partial \Omega}{\partial T}\right)_{V,\mu} = V\left(\frac{\partial P}{\partial T}\right)_{V,\mu}, \tag{7.46}$$

which leads to

$$S(T, V, \mu) = V\left(\frac{5}{2}k_B\lambda^{-3}\mathrm{Li}_{5/2}(z) - k_B\lambda^{-3}\mathrm{Li}_{3/2}(z)\ln z\right), \tag{7.47}$$

where we used the recurrence property of the polylogarithms (E.5) and the fact that

$$\frac{1}{z}\frac{\mathrm{d}z}{\mathrm{d}T} = -k_{\mathrm{B}}\beta \ln z.$$

At fixed volume, for temperatures $T \leq T_{\mathrm{c}}$, the entropy, by setting $z = 1$, reduces to

$$S = \frac{5}{2}k_{\mathrm{B}}\frac{V}{\lambda^3}\zeta_{\mathrm{R}}(5/2) = \frac{5}{2}k_{\mathrm{B}}\frac{\zeta_{\mathrm{R}}(5/2)}{\zeta_{\mathrm{R}}(3/2)}(N - N_0(T)), \tag{7.48}$$

where we used the fact that Eq. (7.41) gives only the number of the particles out of the condensate for $T \leq T_{\mathrm{c}}$. Equation (7.48) then shows that only the particles out of the condensate contribute to the entropy.

By taking the derivative of the pressure, Eq. (7.40), with respect to the temperature along the coexistence line we get

$$\frac{\partial P}{\partial T} = \frac{5}{2}k_{\mathrm{B}}\frac{\zeta_{\mathrm{R}}(5/2)}{\lambda^3} = \frac{5}{2}k_{\mathrm{B}}\frac{\zeta_{\mathrm{R}}(5/2)}{\zeta_{\mathrm{R}}(3/2)}\frac{1}{v}, \tag{7.49}$$

where we used again Eq. (7.41) at $z = 1$. We notice that this last equation is nothing but the Clausius–Clapeyron equation of the condensation transition. On the right-hand side of Eq. (7.49) there appears the ratio of the specific entropy to the specific volume of the phase describing the particles outside the condensate. The particles in the condensate have zero specific entropy and zero specific volume, at least as long as one neglects the interaction.

Finally, we may obtain the specific heat at constant volume. At $T < T_{\mathrm{c}}$, from Eq. (7.48) we get

$$c_V = \langle N\rangle k_{\mathrm{B}}\frac{15}{4}\frac{\zeta_{\mathrm{R}}(5/2)}{\zeta_{\mathrm{R}}(3/2)}\left(\frac{T}{T_{\mathrm{c}}}\right)^{3/2} = \frac{5}{2}\frac{U}{T}. \tag{7.50}$$

For $T > T_{\mathrm{c}}$, we derive the expression for the entropy in Eq. (7.47) with respect to the temperature by imposing that the density of particles remains constant. The derivatives must be conditioned by Eq. (7.41) at fixed density. This condition, by differentiating with respect to T and using the recurrence relation (E.5), gives

$$\frac{T}{z}\frac{\mathrm{d}z}{\mathrm{d}T} = -\frac{3}{2}\frac{\mathrm{Li}_{3/2}(z)}{\mathrm{Li}_{1/2}(z)}. \tag{7.51}$$

According to Eq. (E.16) of Appendix E, $\mathrm{Li}_{1/2}(z)$ diverges when $z \to 1$ as $(-\ln z)^{-1/2}$.

From Eq. (7.47), after using the recurrence relation (E.5) and Eq. (7.51), we get

$$c_V = \langle N\rangle k_{\mathrm{B}}\left(\frac{15}{4}\frac{\mathrm{Li}_{5/2}(z)}{\mathrm{Li}_{3/2}(z)} - \frac{9}{4}\frac{\mathrm{Li}_{3/2}(z)}{\mathrm{Li}_{1/2}(z)}\right) = \frac{5}{2}\frac{U}{T} + \frac{3}{2}\langle N\rangle k_{\mathrm{B}}T\frac{\mathrm{d}\ln z}{\mathrm{d}z}, \tag{7.52}$$

where z must be understood as a function of T and v via Eq. (7.41). From the definition (E.4) of the polylogarithm, we see that at high temperature when $z \ll 1$, $\mathrm{Li}_s(z) \sim z$ so that the expression (7.52) reduces to the classical value $(3/2)\langle N\rangle k_{\mathrm{B}}$. At low temperature the specific heat goes to zero with power $3/2$. At the transition temperature the specific heat is continuous and gets the value

$$(15/4)(\zeta_{\mathrm{R}}(5/2)/\zeta_{\mathrm{R}}(3/2))\langle N\rangle k_{\mathrm{B}} \approx 1.9257\langle N\rangle k_{\mathrm{B}},$$

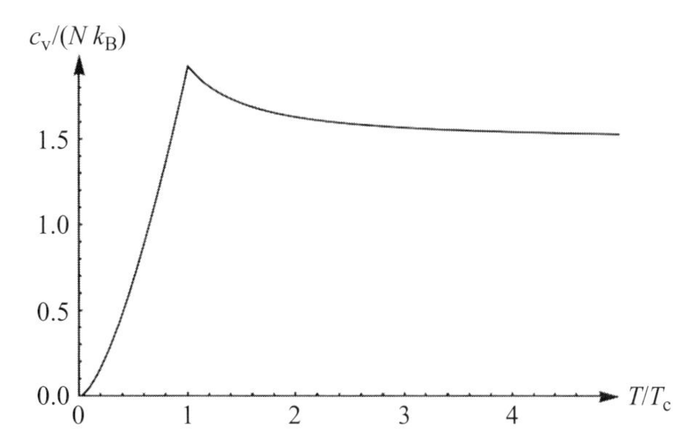

Fig. 7.3 The specific heat per particle of a Bose gas in units of k_B as function of the temperature measured in units of T_c.

which is larger than the classical value. The second term on the right-hand side of Eq. (7.52) vanishes at $T = T_c$ due to the divergence of $\mathrm{Li}_{1/2}(z)$ and remains zero for $T < T_c$ since $\ln z \equiv 0$, while the first one reduces to the same value as Eq. (7.50). A plot of the specific heat is shown in Fig. 7.3. We see that at the critical point the specific heat has a cusp (see Problem 7.5). This is the singularity appearing in the non-interacting case. In liquid ^{4}He, the stronger logarithmic singularity seen experimentally at the superfluid transition is a further indication of the importance of the interactions among the particles for explaining the superfluid phenomenon.

7.4 The Bose and Fermi gases in an harmonic potential

In the last two sections we have considered the homogeneous degenerate Fermi and Bose gases. Their ground states are the Fermi sea and a complete condensation of all N particles in the lowest single-particle state respectively. In the first case the ground state is continuously reached by lowering the temperature through the degeneration temperature $T_F \sim n^{2/3}$, whereas in the Bose case there is a transition in the thermodynamic limit to a Bose–Einstein condensed phase at the crical temperature $T_c \sim n^{2/3}$. The quantum degeneration of a Fermi gas and the Bose–Einstein condensation have been directly observed for ultracold alkali atoms confined in a finite region of space.

A single alkali atom is made of $Z + A$ fermions, Z being the atomic number, A the mass number, depending on the number of neutrons present in the isotope. In the alkali atoms of the first column of the periodic table, one electron occupies the first state with quantum numbers n and s outside the last closed shell and the atomic number Z is always odd. A collection of alkali atoms with odd A obeys therefore Bose–Einstein statistics, whereas a collection of isotope with even A obeys Fermi statistics. The Bose–Einstein condensation was first realized in 1995 for alkali atoms cooled down in the region of microkelvin and later to a few tens of nanokelvin with magnetic or optical confinement and a sample lifetime of tens of seconds: (i) for a number density of 2.5×10^{12} cm^{-3} of rubidium (^{87}Rb) vapor (Anderson et al. (1995)); (ii) for 5×10^5 sodium atoms (^{23}Na) at densities exceeding 10^{14} cm^{-3} (Davis et al. (1995)); and (iii) for vapors of lithium (^{7}Li) with total trapped atom number up to 10^5 (Bradley et al. (1995)).

Quantum degeneracy for an ultracold confined Fermi gas was first observed for potassium atoms (^{40}K) in 1999 (DeMarco and Jin (1999)) and later for lithium (^{6}Li) gas (Schreck et al. (2001); Truscott et al. (2001)). The full theoretical analysis including the effects of the interaction will be given in Chapter 15.

It is then relevant to treat as a starting point the Bose and Fermi gases of N particles confined in a finite region by an harmonic potential:

$$V_{\text{ext}}(\mathbf{r}) = \frac{m}{2}\left(\omega_x^2 x^2 + \omega_y^2 y^2 + \omega_z^2 z^2\right). \tag{7.53}$$

The single particle states are now the harmonic oscillator states with eigenvalues

$$\epsilon(n_x, n_y, n_z) = \left(n_x + \frac{1}{2}\right)\hbar\omega_x + \left(n_y + \frac{1}{2}\right)\hbar\omega_y + \left(n_z + \frac{1}{2}\right)\hbar\omega_z, \tag{7.54}$$

where $n_x = 0, 1, 2, \ldots$ and similarly for n_y and n_z.

The corresponding Bose and Fermi occupation numbers and the total number of particles are again specified by Eq. (7.6), with the single particle energy given by $\epsilon(n_x, n_y, n_z)$ of Eq. (7.54).

For large N we can again switch to the continuum by introducing the single particle density of states $\tilde{\nu}(\epsilon - \epsilon_0)$ for the energy shifted by the zero point energy (see Problem 7.3):

$$\tilde{\nu}(\epsilon - \epsilon_0) = \frac{(\epsilon - \epsilon_0)^2}{2(\hbar\overline{\omega})^3}, \quad \overline{\omega} = (\omega_x\omega_y\omega_z)^{1/3}. \tag{7.55}$$

7.4.1 The Bose gas in an harmonic potential

The Bose–Einstein condensation phenomenon in the presence of the confining potential (de Groot et al. (1950)) manifests as a macroscopic occupation of the lowest single particle energy level $\epsilon_0 \equiv \epsilon(0, 0, 0)$ with eigenstate

$$g_0(\mathbf{r}) = \left(\frac{m\overline{\omega}}{\pi\hbar}\right)^{3/4} \exp\left[-\frac{m}{2\hbar}\left(\omega_x x^2 + \omega_y y^2 + \omega_z z^2\right)\right]. \tag{7.56}$$

When the number N of confined particles is large, the Bose–Einstein condensation has the behavior typical of a phase transition, however finite size effects round the transition.

In the free-particle case, the condensation in the lowest $\mathbf{k} = 0$ state was homogeneous over the volume of the box. In contrast, at zero kelvin, in the presence of the confining potential, all the bosons occupy the state (7.56), all the other states with eigenvalues with $n_{x,y,z} > 0$ are empty and the particle density distribution is given by $n(\mathbf{r}) = N|g_0(\mathbf{r})|^2$. A relevant new feature appears in the trap: the extension of the condensate in real space is now given by the oscillator length

$$a_{ho} = \left(\frac{\hbar}{m\overline{\omega}}\right)^{1/2}. \tag{7.57}$$

At finite temperature, the particles in the excited states have a much wider distribution superimposed onto the narrow peak of the condensed particles.

As for the free particles in a box, lowering the temperature at fixed N increases the value of the chemical potential. When T is such that the chemical potential $\mu(T = T_c)$ reaches the lowest energy value $\mu_c = \epsilon_0$, the lowest state is macroscopically occupied in the sense that the value of its occupation number is of the same order as the number of particles N. For $T < T_c$, analogously to Eq. (7.43), $\mu_c = \epsilon_0$ and the lowest state can be separated from the discrete sum for N:

$$N = N_0(T) + \sum_{n_x,n_y,n_z} \frac{1}{\exp(\beta[\epsilon(n_x, n_y, n_z) - \epsilon_0]) - 1}. \tag{7.58}$$

In the large N limit

$$N = N_0(T) + \int_0^\infty d\epsilon \frac{\tilde{\nu}(\epsilon)}{\exp(\beta\epsilon) - 1} = N_0(T) + \frac{(k_B T)^3}{2(\hbar\overline{\omega})^3} I_2^-(0), \tag{7.59}$$

where the integral $I_2^-(0) = \Gamma(3)\zeta_R(3)$ (see Eq. (E.7)) is defined in (7.18). For $T > T_c$, the average number of particles reads

$$N = \int_0^\infty d\epsilon \frac{\tilde{\nu}(\epsilon)}{\exp(\beta(\epsilon - (\mu - \epsilon_0)) - 1} = \frac{(k_B T)^3}{2(\hbar\overline{\omega})^3} I_2^-(\beta(\mu - \epsilon_0)). \tag{7.60}$$

Equation (7.59) with $N_0 = 0$ for $T = T_c$ gives

$$k_B T_c = \hbar\overline{\omega} \left(\frac{N}{\zeta_R(3)}\right)^{1/3} = \frac{\hbar^2}{m a_{ho}^2} \left(\frac{N}{\zeta_R(3)}\right)^{1/3}, \tag{7.61}$$

which, also in this case, essentially represents the degeneracy condition that the number of thermally accessible states is of the order of the number of particles N.

For $T < T_c$, the condensate as a function of T is obtained from Eq. (7.59) and Eq. (7.61):

$$N_0 = N \left[1 - \left(\frac{T}{T_c}\right)^3\right], \tag{7.62}$$

to be compared with Eq. (7.44).

Equation (7.61) for T_c has to be contrasted with Eq. (7.42) for the homogeneous case. Here T_c is fixed by the frequency of the trapping potential, is proportional to $N^{1/3}$ rather than to $n^{2/3}$ and the thermodynamic limit can only be realized by letting $\overline{\omega} \to 0$ and $N^{1/3} \to \infty$ while keeping their product constant.

For $d \leq 3$, $\tilde{\nu}(\epsilon) \approx \epsilon^{d-1}$. This behavior has to be contrasted with the homogeneous case: the single particle density of states per unit volume, Eq. (7.14), when evaluated in d dimensions gives $\nu(\epsilon) \approx \epsilon^{(d-2)/2}$. Due to this different behavior of the density of states, the integral of Eq. (7.59) is well defined also in $d = 2$ and the condensation can occur at $T < T_c$ with T_c given by

$$k_B T_c^{(2d)} = \hbar\overline{\omega}^{(2d)} \left(\frac{N}{\zeta_R(2)}\right)^{1/2}. \tag{7.63}$$

In $d = 1$, T_c would vanish logarithmically in the proper thermodynamic limit.

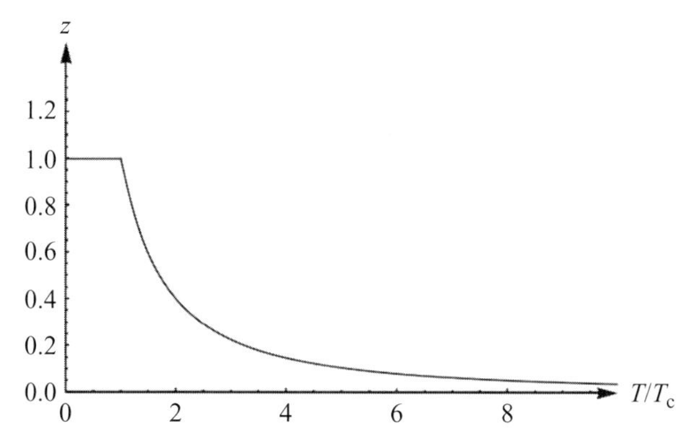

Fig. 7.4 The fugacity of the Bose gas in an harmonic potential in $d = 3$.

Similarly, the energy $U(T)$ and the other thermodynamic quantities can be calculated. For $T < T_c$, taking into account Eqs. (7.55) and (7.59), one obtains in $d = 3$

$$U(T) = \frac{1}{2(\hbar\overline{\omega})^3}\int_0^\infty d\epsilon\, \frac{\epsilon^3}{\exp(\beta\epsilon)-1} = \frac{(k_B T)^4}{2(\hbar\overline{\omega})^3} I_3^-(0),$$

from which the specific heat reads

$$c = Nk_B \frac{4\Gamma(4)\zeta_R(4)}{\Gamma(3)\zeta_R(3)}\left(\frac{T}{T_c}\right)^3. \tag{7.64}$$

For $T > T_c$, we must first find the relation between the fugacity $z = \exp(\beta(\mu - \epsilon_0))$ and the temperature. By using the expression of the particle number (7.60) and of the critical temperature (7.61), we get

$$\frac{T}{T_c} = \left(\frac{\zeta_R(3)}{\mathrm{Li}_3(z)}\right)^{1/3}, \tag{7.65}$$

and in Fig. 7.4 we show the behavior of the fugacity as a function of temperature. Then we evaluate the energy as

$$U(T) = \frac{1}{2(\hbar\overline{\omega})^3}\int_0^\infty d\epsilon\, \frac{\epsilon^3 z}{\exp(\beta\epsilon)-z} = \frac{(k_B T)^4}{2(\hbar\overline{\omega})^3} I_3^-(\ln z) = \frac{\Gamma(4)\mathrm{Li}_4(z)(k_B T)^4}{2(\hbar\overline{\omega})^3},$$

where we have used again the expression of I_3^- in terms of the polylogarithm. After differentiating with respect to the temperature, with the same procedure we have used in the homogeneous case, after some algebra one finds the specific heat

$$c = Nk_B\left[12\frac{\mathrm{Li}_4(z)}{\mathrm{Li}_3(z)} - 9\frac{\mathrm{Li}_3(z)}{\mathrm{Li}_2(z)}\right], \tag{7.66}$$

where the dependence of the temperature is via the fugacity which is implicitly defined by Eq. (7.65). One observes that in the classical limit of $z \ll 1$, when $\mathrm{Li}_s(z) \approx z$, one recovers the classical result for a set of harmonic oscillators $c = 3Nk_B$. At the transition point, $T = T_c$ (or equivalently $z = 1$), the first term of the right-hand side of Eq. (7.66) coincides with the value from $T \to T_c^-$. Hence the second term in the right-hand side of Eq. (7.66)

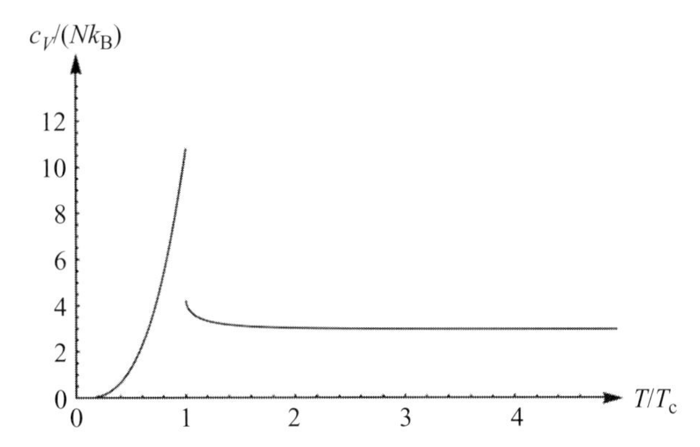

Fig. 7.5 The behavior of the specific heat of the Bose gas in an harmonic potential in $d = 3$.

yields the discontinuity of the specify heat:

$$\Delta c = -9\frac{\zeta_R(3)}{\zeta_R(2)}Nk_B \approx -6.577\ Nk_B. \tag{7.67}$$

7.4.2 The Fermi gas in an harmonic potential

As for the homogeneous case, also the ideal Fermi gas confined in the harmonic potential (7.53) evolves continuously, as $T \to 0$, without any transition, towards the ground state with the full occupation of the single-particle harmonic eigenstates up to the Fermi energy. Again, for a large average value of N, we can switch to the continuum and using the density of states (7.55) $\tilde{\nu}(\epsilon)$ we have

$$N(\mu, T) = s\int_0^\infty \mathrm{d}\epsilon\ \frac{\tilde{\nu}(\epsilon)}{\mathrm{e}^{\beta(\epsilon-(\mu-\epsilon_0))}+1}, \quad U(\mu, T) = \epsilon_0 N(\mu, T) + s\int_0^\infty \mathrm{d}\epsilon\ \frac{\tilde{\nu}(\epsilon)\epsilon}{\mathrm{e}^{\beta(\epsilon-(\mu-\epsilon_0))}+1}. \tag{7.68}$$

At $T = 0$

$$N = \frac{s}{6}\left(\frac{E_F}{\hbar\overline{\omega}}\right)^3, \quad \mu(T=0) \equiv E_F + \epsilon_0 = \hbar\overline{\omega}(6N/s)^{1/3} + \epsilon_0. \tag{7.69}$$

For large N, the Fermi energy is much larger than the level spacing $\hbar\overline{\omega}$, Eq. (7.69) must be compared with Eq. (7.25) for the homogeneous gas for which the Fermi energy is proportional to the power 2/3 of the number density ($E_F \sim n^{2/3}$). The Fermi temperature is now $T_F \sim N^{1/3}$ with the same power of N for the critical temperature Eq. (7.61) of the Bose–Einstein condensation, stating the same degeneracy condition for the two cases.

In the large N limit we can also use the semiclassical approximation with a local Fermi distribution in the phase space $(\mathbf{x}, \mathbf{p})$ with unit cell h^3:

$$f(\mathbf{x}, \mathbf{p}) = \frac{1}{\exp\left\{\beta\left[\frac{p^2}{2m} + V_{\text{ext}}(\mathbf{x}) - (\mu - \epsilon_0)\right]\right\}+1}, \tag{7.70}$$

where $V_{\text{ext}}(x)$ is given by Eq. (7.53). Integration over the momenta $\mathbf{p}$ gives the density profile multiplied by the cube of the Planck constant, $n(\mathbf{x})h^3$, while integration over $\mathbf{x}$ gives the momentum distribution $n(\mathbf{p})h^3$. In the presence of a confining trap $n(\mathbf{p})$ cannot be a step function at $T = 0$ as it obviously is for the homogeneous gas. As $T \to 0$, $f(\mathbf{x}, \mathbf{p})$ is the step function

$$f(\mathbf{x}, \mathbf{p}) = \theta\left(E_{\text{F}} - \frac{p^2}{2m} + V_{\text{ext}}\right).$$

The broadening at $T = 0$ of the momentum distribution around the Fermi momentum $p_{\text{F}} = \sqrt{2mE_{\text{F}}} = (m\hbar\overline{\omega})^{1/2}(48N)^{1/6}$ can be obtained by integrating the step function over x, y and z with the condition $(m/2)(\omega_x^2 x^2 + \omega_y^2 y^2 + \omega_z^2 z^2) < E_{\text{F}} - p^2/2m$. By changing the variables

$$\overline{x} = m^{1/2}\omega_x x, \quad \overline{y} = m^{1/2}\omega_y y, \quad \overline{z} = m^{1/2}\omega_z z,$$

we have

$$\begin{aligned}
n(\mathbf{p}) &= \frac{s}{h^3\overline{\omega}^3 m^{3/2}} \int_{\overline{x}^2+\overline{y}^2+\overline{z}^2<2E_{\text{F}}-p^2/m} \mathrm{d}\overline{x}\mathrm{d}\overline{y}\mathrm{d}\overline{z} \\
&= \frac{1}{6\pi^2}\frac{s}{\hbar^3\overline{\omega}^3}\frac{p_{\text{F}}^3}{m^3}\left(1 - \frac{p^2}{p_{\text{F}}^2}\right)^{3/2} \\
&= \frac{1}{6\pi^2}\frac{1}{p_{\text{F}}^6}(48N)p_{\text{F}}^3\left(1 - \frac{p^2}{p_{\text{F}}^2}\right)^{3/2} \\
&= \frac{8}{\pi^2}\frac{N}{p_{\text{F}}^3}\left(1 - \frac{p^2}{p_{\text{F}}^2}\right)^{3/2}.
\end{aligned} \tag{7.71}$$

The smeared power law behavior substitutes the step function of the homogeneous Fermi sea.

At finite temperature, evolution of Eq. (7.68) for $N(\mu, T)$ and $U(\mu, T)$ via the polylogarithmic functions gives strong deviation from the classical value $(3/2)k_{\text{B}}T$. This quantum degeneration has been tested experimentally in ^{40}K (DeMarco et al. (2001)).

7.5 The photon gas

According to classical electromagnetic theory, the radiation field transfers energy to the electrical charges by accelerating them. Conversely, when the electrical charges are accelerated they radiate, giving up energy to the radiation field. In the description of the interaction between matter and radiation, we define the emissive power of a body, e, as the amount of energy that it radiates per unit surface in unit time and the absorptive power, a, as the fraction of radiant energy absorbed by the body. In general, both e and a are functions of the frequency of the radiation and of the temperature at which the body is kept, as well as of the specific nature of the body. In 1859, Kirchhoff observed that, although e and a may vary when considering different bodies, the ratio e/a becomes a universal function of frequency and temperature when matter and radiation are in equilibrium at temperature T. To appreciate the Kirchhoff argument, let us indicate with $e_0(\nu, T)$ the spectral density of the uniform radiation energy flux per unit solid angle

$$e_0(\nu, T) = \frac{c}{4\pi} u(\nu, T), \tag{7.72}$$

where c is the speed of light and $u(\nu, T)$ the spectral radiation energy density. Let us consider now the amount of radiation energy falling on a body due to an elementary radiation volume. If the direction along which the radiation impinges on the body makes an angle θ with the normal to the body surface, the amount of energy absorbed by the body per unit time and unit surface in the direction θ is $e_0(\nu, T)a(\nu, T)\cos(\theta)$. In thermal equilibrium, the same amount of energy is radiated back from the body, i.e., the emissive power of the body is

$$e(\nu, T) = e_0(\nu, T)a(\nu, T)\cos(\theta). \tag{7.73}$$

If different bodies are in equilibrium and have the same radiation field $e_0(\nu, T)$, then the ratio e/a must be the same for all bodies.

A body for which $a(\nu, T) = 1$, at all frequencies and temperatures, is by definition a black body. Experimentally, a black body can be realized by opening a small hole in a cavity. The radiation entering the cavity will undergo a large number of reflections on the walls of the cavity before having the chance to exit again through the hole. Hence, with great probability all the radiation entering the cavity is absorbed. As a result the radiation emitted from the cavity, to a good approximation, will be given by the universal function $e_0(\nu, T)$. In a real experiment, what is measured is the emissive power of the emitting black body integrated over half of the solid angle:

$$I = \int_0^{\pi/2} d\theta \sin(\theta) \int_0^{2\pi} d\phi \; e_0(\nu, T)\cos(\theta) = \frac{c}{4} u(\nu, T), \tag{7.74}$$

which is a direct measure of the radiation density $u(\nu, T)$.

According to the quantum theory of radiation, the latter is made of photons, relativistic particles with zero rest mass:

$$\epsilon(p_x, p_y, p_z) = c\sqrt{p_x^2 + p_y^2 + p_z^2} \tag{7.75}$$

and spin 1 with two possible polarization states, parallel or antiparallel to the momentum **p**. This fact is a consequence of Lorentz invariance and of massless nature of the photons (Wigner (1957)).[1]

[1] In general, for a massive particle the angle between the spin and velocity is not a Lorentz invariant quantity. For instance, for a massive particle it is always possible to find a reference frame where the particle is at rest. Clearly, in this case, there is no angle at all between spin and velocity, since there is no velocity. In such a frame, the spin degeneracy gives the number of independent components of the wave function that transform among themselves under the group of rotations in three dimensional space. We may say that the existence of one state of polarization implies the existence of the others. For a vector wave function, corresponding to spin 1 particles, this number is three, corresponding to the spin projections $0, \pm 1$.

Wigner has shown that the statement *spin parallel to the velocity* is invariant by Lorentz transformations for zero rest-mass particles. Since a massless particle is never at rest, the group of rotations is restricted to those

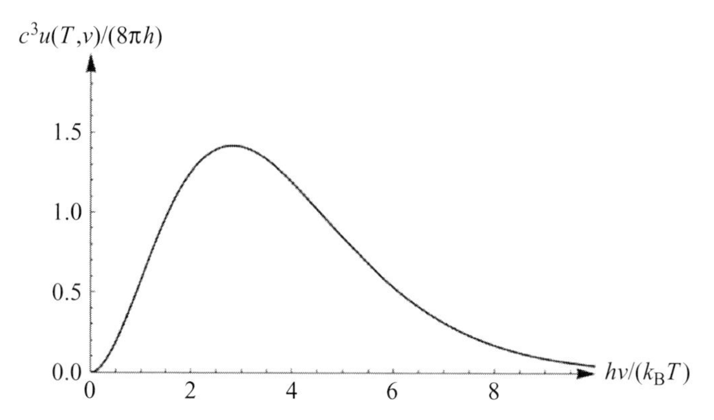

Fig. 7.6 The Planck spectral energy density of Eq. (7.78).

In interaction with matter, photons are absorbed and emitted in such a way that their number is not conserved. The chemical potential, therefore, vanishes as it happens on the coexistence line of the Bose–Einstein condensation. Finally, the momentum is related to the frequency by the well known relations

$$\epsilon = cp = h\nu. \tag{7.76}$$

Let us now consider a gas of photons in thermal equilibrium with matter at temperature T in a box of volume V. According to Bose statistics the total energy of the photon gas reads

$$U(T, V) = 2\frac{V}{h^3}\int d^3p \frac{cp}{e^{\beta cp} - 1}. \tag{7.77}$$

By changing the variable of integration from momentum to frequency, we get the spectral energy density

$$u(T, \nu) = \frac{8\pi}{c^3}\frac{h\nu^3}{e^{\beta h\nu} - 1}, \tag{7.78}$$

which is Planck's law (Planck (1900a, b)) and is shown in Fig. 7.6. The spectral energy density has the form predicted by Wien:

$$u(\nu, T) = \nu^3 f\left(\frac{\nu}{T}\right), \tag{7.79}$$

which implies the existence of a maximum, whose frequency scales with the temperature $\nu_{max} = 2.822k_BT/h$. This latter result is known as the displacement law.

around the momentum direction. The only conserved quantity is the projection of the angular momentum on the momentum axis. Hence the existence of one state of polarization does not imply the existence of the others. Furthermore, due to the extra symmetry by reflection about a plane passing through the momentum axis, the two states with opposite helicity (projection on the momentum axis) are degenerate.

The total energy density, i.e. the spectral energy density integrated over all frequencies, reads

$$\begin{aligned} u(T) &= \int_0^\infty d\nu u(\nu, T) \\ &= \frac{8\pi h}{c^3}\int_0^\infty d\nu \frac{\nu^3}{e^{\beta h\nu}-1} \\ &= \frac{8\pi(k_B T)^4}{(hc)^3}\int_0^\infty dy \frac{y^3}{e^y-1} \\ &= \frac{8\pi(k_B T)^4}{(hc)^3} I_3^-(0) \\ &= \frac{8\pi^5(k_B T)^4}{15(hc)^3}, \end{aligned} \tag{7.80}$$

where we used $I_3^-(0) = \Gamma(4)\zeta_R(4)$ from Eq. (E.7) of Appendix E. By using Eq. (7.80) in the expression for total emissive power of Eq. (7.74), we get

$$I = \sigma T^4, \quad \sigma = \frac{2\pi^5 k_B^4}{15c^2h^3}, \tag{7.81}$$

which is the Stefan–Boltzmann law. The relativistic dispersion changes the equation of state (7.19), which now reads

$$PV = -2k_B TV \int \frac{d^3p}{h^3} \ln\left(1 - e^{-\beta cp}\right) = \frac{U(T,V)}{3}, \tag{7.82}$$

after making an integration by parts. Equation (7.82) is nothing but the radiation pressure $P = u(T)/3$, a well known result from classical electromagnetic theory. By collecting together Eq. (7.82) with (7.80) and recalling (7.74) we obtain

$$P = \frac{4\sigma}{3c}T^4. \tag{7.83}$$

7.6 The phonon gas

In a solid, the thermal agitation makes atoms move around their equilibrium positions. To a good approximation, their motion, for small deviations around equilibrium, can be described in terms of N coupled oscillators, where N is the total number of atoms. The equipartition theorem of classical statistical mechanics, then, tells us that the energy of such an assembly of oscillators is

$$U = 3Nk_B T, \tag{7.84}$$

which yields the classical expression for the specific heat of a solid

$$c_V = 3Nk_B = 3R. \tag{7.85}$$

This result is in agreement with the experimental law of Dulong and Petit, which is valid at high temperatures. At low temperatures the experimental results show that the specific heat goes to zero. On the theoretical side, we notice that in the limit of zero temperature a quantum system will be in its ground state and the entropy, which is proportional to the logarithm of the degeneracy of the ground state, will vanish according to the Nernst principle. Due to the relation

$$c_V = T\left(\frac{\partial S}{\partial T}\right)_V = \frac{\partial S}{\partial \ln T}, \tag{7.86}$$

the specific heat must also tend to zero in the limit of zero kelvin. This is apparent from the fact that in the incremental ratio $\partial S/\partial \ln T$, the denominator goes to $-\infty$ while the numerator remains finite and actually vanishes.

Einstein (1907) pointed out that, at low temperatures, one must use, for the oscillators of the solid, quantum statistics in analogy with the statistical treatment of the electromagnetic radiation at equilibrium. He then proposed a simplified model, in which all the oscillators have the same frequency ν so that one must replace Eq. (7.84) with

$$U = 3N\frac{h\nu}{e^{\beta h\nu} - 1}, \tag{7.87}$$

from which

$$c_V = 3Nk_B\frac{h^2\nu^2}{k_B^2T^2}\frac{e^{\beta h\nu}}{(e^{\beta h\nu} - 1)^2}. \tag{7.88}$$

At high temperatures the above expression reduces to the classical limit (7.85), whereas it goes to zero exponentially when $T \to 0$. Such behavior, although it captures the important aspect of a vanishing specific heat in the zero kelvin limit, shows discrepancies with the experimental data, where the vanishing is rather a power law. Debye (1912) put forwards a modified version of the Einstein model by observing that the assumption of equal frequencies for the oscillators is a too drastic simplification.

According to the theory of normal modes, in terms of which the motion of N coupled harmonic oscillators can be described, there are as many eigenfrequencies as the number of the oscillators. In analogy with the concept of the photon which is associated with the eigenmodes of the electromagnetic radiation in a cavity, the normal modes of a set of N harmonic oscillators can be quantized. Such quantized modes are called phonons and obey Bose–Einstein statistics. As for the photons, the number of phonons is not conserved, which requires the vanishing of the chemical potential.

To proceed quantitatively one must know the spectrum of the normal modes, which in general depends on the specific crystalline structure of a given solid. Debye, however, argued that when confining to the oscillations with large wave length, the spectrum of the normal modes must include the oscillations corresponding to the sound waves, whose dispersion relation is known. By indicating with c the speed of the sound in a solid, we have

$$c = \lambda\nu, \quad \text{or } \epsilon = cp, \tag{7.89}$$

where $k = 2\pi/\lambda = p/\hbar$ and $\epsilon = h\nu$. The energy then becomes

$$U = 3V \int \frac{d^3p}{h^3} \frac{cp}{e^{\beta cp} - 1}. \tag{7.90}$$

The factor of 3 in front of the integral is due to the fact that a sound wave has three possible states of polarization, one longitudinal, in contrast to photons, and two transversal. Wigner's argument, reported in footnote 1, does not apply to the sound velocity. As for the photons, the phonon spectrum of Eq. (7.89) implies that within a volume V there are infinitely many phonon modes. This, however, cannot be due to the finite number $3N$ of the degrees of freedom of the atoms in the solid. Debye then imposed that the number of the phonon modes is fixed to $3N$ by requiring that the linear spectrum (7.89) is valid up to a maximum value of the momentum $p_{\max}$ given by

$$3N = 3V \int_{|\mathbf{p}|<p_{\max}} \frac{d^3p}{h^3}, \Rightarrow p_{\max} = h\left(\frac{3N}{4\pi V}\right)^{1/3}. \tag{7.91}$$

The energy corresponding to this value $p_{\max}$ defines the so-called Debye temperature Θ given by

$$\Theta = \frac{cp_{\max}}{k_B}. \tag{7.92}$$

The expression (7.90) for $U(T, V)$ becomes then

$$\begin{aligned} U &= 3V\frac{4\pi}{h^3}\int_0^{p_{\max}} dp\, p^2 \frac{cp}{e^{\beta cp} - 1} \\ &= 3V\frac{4\pi}{h^3}\frac{(k_BT)^4}{c^3}\int_0^{\Theta/T} dy \frac{y^3}{e^y - 1}. \end{aligned} \tag{7.93}$$

When $T \gg \Theta$, the integral over the variable y is confined to small values and we can expand the exponential in the denominator so that the integral is easily performed:

$$U = \frac{4\pi V k_B^4 \Theta^3 T}{(hc)^3} = 3Nk_BT, \quad T \gg \Theta, \tag{7.94}$$

and we recover the classical value.

For $T \ll \Theta$, we can approximate $\Theta/T \sim \infty$ in the upper limit of the integral (7.93), that becomes $I_3^-(0)$. By using Eq. (E.7), $U(T, V)$ reads

$$U = \frac{4\pi^5 k_B^4 T^4}{5h^3c^3}, \tag{7.95}$$

from which the specific heat per unit volume is

$$c_V = \frac{16\pi^5 k_B^4 T^3}{5h^3c^3}. \tag{7.96}$$

We then see that the specific heat due to the lattice in a solid has a typical temperature dependence $\sim T^3$ in contrast to the linear dependence found for the Fermi gas. At low temperatures then a metal must show the specific heat dependence $aT + bT^3$. Any deviation from this behavior will then be the signal of different low energy excitations of the system under consideration.

7.7 Problems

7.1 Show that, for a gas of N fermions with spin one half and energy dispersion $\epsilon(p_x, p_y, p_z) = cp$,[2] where p is the absolute value of the momentum in d space dimensions, the zero-kelvin energy U has the form

$$U = \frac{d}{d+1} N\epsilon_{\mathrm{F}}. \tag{7.97}$$

7.2 Derive the density of states, the Fermi energy and the local density of a Fermi gas in a constant gravitational field enclosed in a semi-infinite cylinder of basis area A.

7.3 Derive the density of states for particles in the harmonic potential (7.53).

7.4 Derive the Fermi energy for a gas of N fermions in the harmonic potential (7.53).

7.5 Determine the jump in the temperature derivative of the specific heat at the critical temperature for the Bose–Einstein condensation.

[2] We note that such a relativistic dispersion occurs for the energy bands near the Fermi level of graphene layers (in $d = 2$), which are the basic constituents of graphite and were experimentally isolated in 2004 by A. Geim and K. Novoselov.

8 Mean-field theories and critical phenomena

8.1 The van der Waals equation and the classical critical indices

In this chapter we introduce mean-field theory as the simplest variational method to approximate an interacting N-body system with a non-interacting gas of dressed single particles. As a first example, in this section we provide a microscopic derivation of the van der Waals equation which was phenomenologically introduced in the first chapter. We know already that this equation yields a description of the phase transition between the liquid and vapor phases as well as of the critical point.

We consider N classical particles with mass m interacting via a pair-wise potential

$$H = \sum_{i=1}^{N} \frac{\mathbf{p}_i^2}{2m} + \sum_{1\leq i<j\leq N} \phi(|\mathbf{x}_i - \mathbf{x}_j|), \tag{8.1}$$

where

$$\phi(r) = \phi_{\mathrm{hc}}(r) + \phi_{\mathrm{a}}(r), \tag{8.2}$$

with $\phi_{\mathrm{hc}}(r)$ being a hard-core potential

$$\phi_{\mathrm{hc}} = \begin{cases} \infty, & r < r_0, \\ 0, & r > r_0, \end{cases} \tag{8.3}$$

and $\phi_{\mathrm{a}}(r)$ an attractive potential which decreases at infinity in such a way that

$$-\int \mathrm{d}^3 r \phi_{\mathrm{a}}(r) = \alpha < \infty. \tag{8.4}$$

Notice that the potential used in the proof of the Yang–Lee theorems can be ascribed to this category. The partition function can be written as

$$Z = \frac{1}{N!\lambda^{3N}} \int \mathrm{d}^{3N}x \prod_{1\leq i<j\leq N} S(|\mathbf{x}_i - \mathbf{x}_j|) \exp\left(-\beta \sum_{1\leq i<j\leq N} \phi_{\mathrm{a}}(|\mathbf{x}_i - \mathbf{x}_j|)\right), \tag{8.5}$$

where λ is the thermal wave length defined in Eq. (4.13). In Eq. (8.5), the hard-core potential gives rise to the factor

$$S(|\mathbf{x}_i - \mathbf{x}_j|) = \begin{cases} 1, & |\mathbf{x}_i - \mathbf{x}_j| > r_0, \\ 0, & |\mathbf{x}_i - \mathbf{x}_j| < r_0. \end{cases}$$

By introducing the density function

$$n(\mathbf{r}) = \sum_{i=1}^{N} \delta(\mathbf{r} - \mathbf{x}_i), \tag{8.6}$$

we may rewrite the term due to the attractive part of the potential as[1]

$$\frac{1}{2}\sum_{i,j=1}^{N} \phi_a(|\mathbf{x}_i - \mathbf{x}_j|) = \frac{1}{2}\int d^3r d^3r' n(\mathbf{r})\phi_a(|\mathbf{r} - \mathbf{r}'|)n(\mathbf{r}'). \tag{8.7}$$

After introducing the effective potential defined as

$$\phi_{\text{eff}}(\mathbf{r}) = \frac{1}{2}\int d^3r' \phi_a(|\mathbf{r} - \mathbf{r}'|)n(\mathbf{r}'), \tag{8.8}$$

the right-hand side of Eq. (8.7) has the apparent form of a one-body potential. However, one must remember that $\phi_{\text{eff}}(\mathbf{r})$ depends on the particle density via Eq. (8.8). We are now ready to make the mean-field approximation, which amounts to replacing the density function $n(\mathbf{r}')$ with its average $\langle n(\mathbf{r}')\rangle$. In this case $\phi_{\text{eff}}(\mathbf{r})$ acts as as an external field to be determined self-consistently. For an homogeneous system, since N is fixed, the solution is trivially given by $\langle n(\mathbf{r})\rangle = N/V$. The one-particle approximation of an N-body interacting system is equivalent to saying that there is no correlation between the densities at points $\mathbf{r}$ and $\mathbf{r}'$, i.e.

$$\langle n(\mathbf{r})n(\mathbf{r}')\rangle = \langle n(\mathbf{r})\rangle\langle n(\mathbf{r}')\rangle = \left(\frac{N}{V}\right)^2, \tag{8.9}$$

the last equality being valid for homogeneous systems. Then Eq. (8.8) becomes

$$\phi_{\text{eff}}(\mathbf{r}) \approx \frac{1}{2}\frac{N}{V}\int d^3r' \phi_a(|\mathbf{r} - \mathbf{r}'|) = -\frac{1}{2}\frac{N}{V}\alpha. \tag{8.10}$$

The partition function (8.5), making use of (8.10), transforms into

$$Z = e^{\beta\alpha N^2/2V} Z_{\text{hc}}, \tag{8.11}$$

where with Z_{hc} we have indicated the hard-core partition function

$$Z_{\text{hc}} = \frac{1}{N!\lambda^{3N}}\int d^{3N}x \prod_{1\le i<j\le N} S(|\mathbf{x}_i - \mathbf{x}_j|). \tag{8.12}$$

In contrast with the one-dimensional case, where the above partition function can be evaluated exactly (see Problem 8.1), we have to make approximations. We estimate the space integral by assuming that each particle can access only a fraction of the entire volume, since any two particles cannot be closer than r_0:

$$\int d^{3N}x \prod_{1\le i<j\le N} S(|\mathbf{x}_i - \mathbf{x}_j|) \approx (V - Nv_0)^N. \tag{8.13}$$

[1] In the double sum over i and j we have eliminated the restriction $i < j$, which is needed to avoid double counting. Since the attractive part of the potential may be taken as zero for vanishing interparticle distance, the inclusion of $i = j$ causes no harm.

In the above equation, v_0 has the meaning of the volume size of a single particle $\sim r_0^3$. With the two approximations (8.11) and (8.13), the free energy becomes

$$F(T, V, N) = -k_B T N \left[\frac{3}{2} \ln(2\pi m k_B T) + \frac{1}{2}\alpha\beta \frac{N}{V} + \ln\left(\frac{V}{N} - v_0\right) + 1 \right]. \quad (8.14)$$

By deriving the free energy with respect to the volume, one obtains the pressure

$$P = \frac{k_B T N}{V - N v_0} - \frac{1}{2}\alpha \frac{N^2}{V^2}. \quad (8.15)$$

Equation (8.15) is nothing but the van der Waals equation of Eq. (1.74) with the identification $a = \alpha N^2/2$ and $b = N v_0$. We note that the contribution to the pressure from the hard-core potential has the form expected on pure dimensional analysis. Consider the configuration integral in Eq. (8.12). Since it does not depend on the temperature, the pressure must have the form $P_{hc} = (k_B T N/V) f(V/N v_0)$, with the function f depending only on the intensive adimensional variable $V/N v_0$. In the extreme dilute limit the function f tends to unity and hence must have an expansion in terms of the small parameter $v_0 N/V$. Such an expansion is called the *virial expansion*, to which Problem 8.2 is dedicated.

We then get

$$P_{hc} = \frac{k_B T N}{V}\left(1 + \frac{v_0 N}{V} + \cdots\right) \approx \frac{k_B T N}{V - N v_0}.$$

We want now to study in more detail the behavior of Eq. (8.15) close to the critical point. By measuring the thermodynamic variables in units of the critical values ($V_c = 3b$, $N k_B T_c = 8a/27b$ and $P_c = a/27b^2$), we obtain the van der Waals equation in the form (see Eq. (1.85))

$$\left(P + \frac{3}{V^2}\right)\left(V - \frac{1}{3}\right) = \frac{8}{3}T. \quad (8.16)$$

Notice that the above equation does not depend anymore on the parameters α and v_0 specific for each particular system and has a universal behavior which goes under the name of the law of corresponding states. It is useful to introduce reduced variables, which measure their deviation from the critical value in the following way:

$$P = 1 + p, \quad V = 1 + v, \quad T = 1 + t. \quad (8.17)$$

Equation (8.16) becomes

$$p = -1 - \frac{3}{(1+v)^2} + \frac{8}{3}\frac{1+t}{(2/3 + v)}. \quad (8.18)$$

By expanding up to cubic terms

$$\frac{3}{(1+v)^2} = 3 - 6v + 9v^2 - 12v^3 + \cdots,$$

$$\frac{1}{1 + 3v/2} = 1 - \frac{3}{2}v + \frac{9}{4}v^2 - \frac{27}{8}v^3 + \cdots,$$

we arrive at

$$p - 4t = -6tv - \frac{3}{2}v^3 + 9tv^2. \tag{8.19}$$

We know that, on the coexistence curve, the two phases have the same values of p and t, but different values of v. Since the right-hand side of Eq. (8.19) depends on v, the two sides of the equation must be equal to a constant. Hence the coexistence curve, close to the critical point, is a straight line which must pass through the origin. This fixes the constant to zero and the coexistence curve, in the $t - p$ plane, has the equation

$$t = \frac{1}{4}p. \tag{8.20}$$

By the same argument, along the same coexistence curve, the value of v is determined by the equation

$$6tv + \frac{3}{2}v^3 = 0, \tag{8.21}$$

where the term $9tv^2$ can be neglected as $t \to 0$. Equation (8.21) has solutions

$$v = 0, \tag{8.22}$$

$$v = \pm 2\sqrt{-t}. \tag{8.23}$$

When $t > 0$, only (8.22) is a real solution, indeed, above the critical point, the system is in a single phase. When $t < 0$, all solutions exist. The solution (8.22) is inside the unphysical region of specific volume with negative compressibility, whereas the two solutions (8.23) correspond to the low density gas and the high density liquid, respectively. Indeed, Eq. (8.21), for $t < 0$, can also be seen as the result of the Maxwell construction of equal area discussed in Chapter 1 (equal p for the two phase, $p_{\text{gas}} = p_{\text{liquid}}$ and opposite v, $v_{\text{gas}} = -v_{\text{liquid}}$, when applied to Eq. (8.19)).

From Eq. (8.19) at $t = 0$, the critical isotherm is

$$p = -\frac{3}{2}v^3. \tag{8.24}$$

Various physical quantities near criticality behave as powers of the thermodynamic variables stating the deviation from the critical point. We now define the critical indices specifying the power of this anomalous behavior in the mean-field approximation.

In terms of the standard critical index δ, the critical isotherm reads

$$|v| \sim |p|^{1/\delta}, \tag{8.25}$$

with $\delta = 3$. This value does not agree with the experimental data for which δ is in the range 4.5–4.8 as reported in Chapter 16. Despite this quantitative disagreement, the van der Waals theory captures qualitatively the observed power-law behavior. We will come back later on the origin of such a discrepancy and on the way to develop a better theory.

Besides the critical index δ, it is customary to use β to characterize how the temperature drives the system from and to the critical point along the coexistence curve:

$$|v| \sim |t|^{\beta}, \tag{8.26}$$

which gives $\beta = 1/2$ from Eq. (8.23). Also this value disagrees with the measured one, $\beta \sim 0.321$–0.328. This also specifies how the difference of the specific volumes characterizing each of the two phases vanishes on approaching the critical point along the coexistence curve.

Equation (1.87) told us that the isothermal compressibility κ_T diverges at the critical point. We then introduce the indices γ and γ' to describe this behavior as

$$\kappa_T \sim |t|^{-\gamma,\gamma'}, \tag{8.27}$$

where γ and γ' refer to $t > 0$ along the critical isocore and $t < 0$ along the coexistence curve, respectively. By differentiating both members of Eq. (8.19) with respect to v and using Eqs. (8.22) and (8.23), we get

$$\kappa_T \sim -\left(\frac{\partial v}{\partial p}\right)_t = \left(6t + \frac{9}{2}v^2\right)^{-1} \Rightarrow \begin{array}{ll} (6t)^{-1}, & t = 0^+, \\ (-12t)^{-1}, & t = 0^-, \end{array} \tag{8.28}$$

which yields $\gamma = \gamma' = 1$, whereas their experimental value is 1.23–1.28. In both cases, κ_T is positive according to the thermodynamic stability condition. Finally, we calculate the specific heat, distinguishing $t > 0$ and $t < 0$. For $t > 0$, we can approach the critical point along the critical isocore by using the definition

$$c_V = -\frac{T}{V}\left(\frac{\partial^2 F}{\partial T^2}\right)_V. \tag{8.29}$$

This coincides with the specific heat of a gas, $c_V = (3/2)Nk_B/V$.

For $t < 0$, one has to evaluate the additional contribution of the thermodynamic potential along the coexistence curve with the temperature dependent specific volume v given by Eq. (8.23). In order to do this we introduce the Gibbs free energy as the Legendre transformed potential of F with respect to V:

$$G(T, P) = F(T, V(P, T)) + PV(P, T), \tag{8.30}$$

where $V(P, T)$ is obtained by inverting the expression of the pressure (8.15). By using reduced variables and neglecting, for the time being, in Eq. (8.14) the volume-independent term $Nk_BT(\ln\lambda^3 - 1)$ that gives the specific heat of the gas, one gets

$$\frac{G(T, P)}{P_cV_c} = -\left[\frac{3}{V(P, T)} + \frac{8T}{3}\ln\left(V(P, T) - \frac{1}{3}\right)\right] + PV(P, T). \tag{8.31}$$

Close to the critical point, we expand the Gibbs free energy (8.31), using (8.17) and leaving the volume v as a free parameter, to get

$$\frac{G(t, p)}{P_cV_c} = v(p - 4t) + 3tv^2 + \frac{3}{8}v^4, \tag{8.32}$$

where we neglected v-independent terms and a term $3tv^3$ (which behaves $\sim v^5$ after using $t \sim v^2$ *a posteriori*). We see that along the coexistence curve (8.20), the Gibbs free energy becomes an even function of v. The solutions for v can be obtained by requiring G to be a minimum. The extremum condition for G reproduces, as expected, the van der Waals equation (8.19) close to the critical point. As previously mentioned, for $t < 0$, the solution (8.22) is a local maximum, whereas the two symmetric solutions (8.23) are local minima.

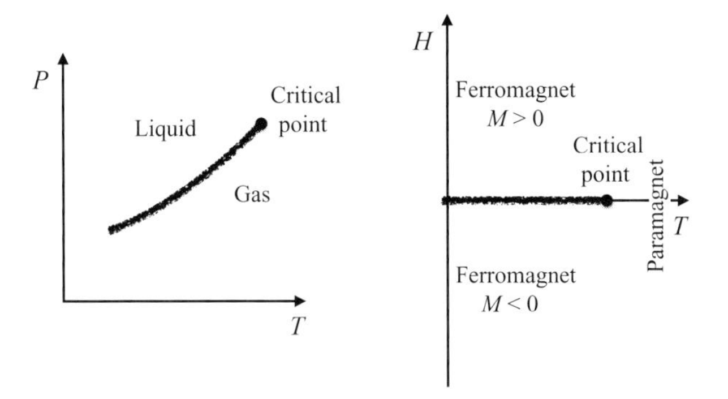

Fig. 8.1 Correspondence between the critical point for the liquid–gas transition and the Curie point for the ferromagnetic–paramagnetic transition.

By using the solutions (8.23), the Gibbs free energy, along the coexistence curve (8.20), reads, with $t = (T - T_c)/T_c$,

$$G(t, p(t)) = -P_c V_c 6t^2 = -6\frac{P_c V_c}{T_c^2}(T - T_c)^2 = -\frac{9Nk_B}{4T_c}(T - T_c)^2, \tag{8.33}$$

which gives a contribution to the specific heat

$$c = -T\frac{\partial^2 G}{\partial T^2} = \frac{9}{2}Nk_B. \tag{8.34}$$

Hence, at the critical point, in going from $t > 0$, along the critical isocore, to $t < 0$, along the coexistence curve, the specific heat has a positive jump $\Delta c = (9/2)Nk_B$.

Finally, the critical indices for the specific heat are defined as

$$c \sim |t|^{-\alpha,\alpha'}$$

with α and α' referring to $t > 0$ and $t < 0$, respectively. Due to the finite discontinuity obtained for the specific heat at the critical temperature, one concludes that $\alpha = \alpha' = 0$, in contrast to the experimental weak singularity with $\alpha = 0.125$.

8.2 The paramagnetic–ferromagnetic transition

In order to discuss the paramagnetic–ferromagnetic transition near the critical Curie point below which the spontaneous magnetization $M_0(T)$ arises, we stress the analogy with the liquid–gas transition near the critical point by comparing the coexistence curve in the T–P plane with the coexistence of the positive and negative spontaneous magnetizations in the T–H plane below the critical temperature (see Fig. 8.1). Also, the isotherms in the P–V plane have a counterpart in the isotherms in the $(-H)$–M_0 plane. In both cases the transitions are characterized by how the jumps in the spontaneous magnetization and in the density along the coexistence curve vanish on approaching the critical point. A paramagnetic system has stable magnetic dipoles that can be aligned along the direction of an applied constant magnetic field. This gives rise to a positive magnetization linear in the field,

$$\mathbf{M} = \chi \mathbf{H}, \tag{8.35}$$

and the total magnetic induction reads

$$\mathbf{B} = \mathbf{H} + 4\pi \mathbf{M} = (1 + 4\pi\chi)\mathbf{H} \equiv \mu \mathbf{H}, \tag{8.36}$$

where μ is the magnetic permeability and χ the (positive for thermodynamic stability in the paramagnetic phase) magnetic susceptibility.

The energy of N independent magnetic dipoles can be written as

$$E = -\sum_{i=1}^{N} \mathbf{H} \cdot \mathbf{m}_i, \tag{8.37}$$

where $\mathbf{m}_i$ is a vector representing the i-th magnetic dipole. The magnetic dipole originates from the total angular momentum of an atom (with both orbital and spin components):

$$\mathbf{m}_i = g_{\mathrm{L}} \mu_{\mathrm{B}} \mathbf{J}_i, \tag{8.38}$$

with the total magnetization $M = N\langle m_z \rangle = N g_{\mathrm{L}} \mu_{\mathrm{B}} \langle J_z \rangle$, where g_{L} and μ_{B} are the Landé gyromagnetic factor and the Bohr magneton. $\mathbf{J}_i$ is a dimensionless angular momentum, z is the direction of the field and the average is taken over a canonical distribution.

The canonical partition function for N independent magnetic dipoles is, the magnetic field being in the z direction,

$$Z = \mathcal{Z}^N = \left(\sum_{J_z=-J}^{J} \mathrm{e}^{2\beta h J_z} \right)^N, \tag{8.39}$$

where $h = g_{\mathrm{L}} \mu_{\mathrm{B}} H/2$ and J the maximum value of the angular momentum. To evaluate the sum in (8.39), let us assume the case of integer J, $J_z = -J, -J+1, \ldots, J$:

$$\begin{aligned} \mathcal{Z} &= \sum_{J_z=0}^{J} \mathrm{e}^{-2\beta h J_z} + \sum_{J_z=0}^{J} \mathrm{e}^{2\beta h J_z} - 1 \\ &= \frac{1 - \mathrm{e}^{-2\beta h (J+1)}}{1 - \mathrm{e}^{-2\beta h}} + \frac{1 - \mathrm{e}^{2\beta h (J+1)}}{1 - \mathrm{e}^{2\beta h}} - 1 \\ &= \frac{\sinh(2\beta h (J + 1/2))}{\sinh(\beta h)}. \end{aligned} \tag{8.40}$$

For half-integer J, $J_z = \pm(2n+1)/2$, with $n = 0, 1, 2, \ldots$, so that

$$\begin{aligned} \mathcal{Z} &= \mathrm{e}^{-\beta h} \sum_{n=0}^{J-1/2} \mathrm{e}^{-2\beta h n} + \mathrm{e}^{\beta h} \sum_{n=0}^{J-1/2} \mathrm{e}^{2\beta h n} \\ &= \mathrm{e}^{-\beta h} \frac{1 - \mathrm{e}^{-2\beta h (J+1/2)}}{1 - \mathrm{e}^{-2\beta h}} + \mathrm{e}^{\beta h} \frac{1 - \mathrm{e}^{2\beta h (J+1/2)}}{1 - \mathrm{e}^{2\beta h}} \\ &= \frac{\sinh(2\beta h (J + 1/2))}{\sinh(\beta h)}. \end{aligned} \tag{8.41}$$

The statistical average of J_z, equal for all atoms, reads

$$\langle J_z \rangle = \frac{1}{\mathcal{Z}} \sum_{J_z=-J}^{J} J_z e^{2\beta h J_z} = \frac{\partial}{\partial(2\beta h)} \ln \mathcal{Z} = B_J(2\beta h), \tag{8.42}$$

where we have introduced the *Brillouin* function, $B_J(2h)$,

$$B_J(2\beta h) = \left(J + \frac{1}{2}\right) \coth\left(2\beta h \left(J + \frac{1}{2}\right)\right) - \frac{1}{2} \coth(\beta h). \tag{8.43}$$

$\langle J_z \rangle$ goes to zero when the external field is sent to zero and no spontaneous magnetization occurs. By recalling the Taylor expansion of the hyperbolic cotangent

$$\coth(x) = \frac{1}{x} + \frac{x}{3} + \cdots,$$

the magnetic susceptibility reads

$$\chi = N\beta(g_L \mu_B)^2 \frac{J(J+1)}{3}, \tag{8.44}$$

which is the Curie expression. Clearly, with independent dipoles, one can only get a paramagnetic behavior. If a ferromagnetic phase develops, there is a magnetization M which stays finite even when the external magnetic field goes to zero. Ferromagnetism needs interaction among the dipoles' moments. In the case of a fluid, to obtain a critical point the equation of state of a gas had to be modified by the interaction to obtain non-linear terms in the volume. Their introduction allows for the two different values of the specific volume of the liquid and of the gas at the same pressure and temperature. In the same way one has to introduce the interaction between dipoles to obtain different values of the magnetization at the same magnetic field ($H = 0$) and temperature. Such an interaction may be an exchange interaction as proposed by Heisenberg. Then, we may take into account the interaction among the magnetic dipoles on average by adding the magnetization to the external field:

$$H \to H + aM = H + a \sum_{i=1}^{N} \langle J_{z,i} \rangle = H + aN\langle J_z \rangle, \tag{8.45}$$

where we have assumed that the average value of the magnetic dipole is the same for all the dipoles and a is a constant that depends on the interaction among the dipoles. The statistical average then must satisfy Eq. (8.43) with the replacement (8.45):

$$\langle J_z \rangle = B_J(2\beta h + a\beta g_L \mu_B N \langle J_z \rangle). \tag{8.46}$$

This is the Curie–Weiss equation and the internal field $g_L \mu_B N \langle J_z \rangle$ due to the average internal interaction is called the Weiss field. This is an equation that determines the value of $\langle J_z \rangle$ as a function of the external field H and the temperature T.

For the important case of half-integer angular momentum, $J = 1/2$, the Curie–Weiss equation, analogous to the van der Waals equation, takes the form

$$\langle J_z \rangle = \frac{1}{2} \tanh\left[\beta h + (a\beta g_L \mu_B N)\langle J_z \rangle/2\right]. \tag{8.47}$$

At zero external field

$$\langle J_z \rangle = \frac{1}{2} \tanh \left[(a\beta g_{\rm L} \mu_{\rm B} N) \langle J_z \rangle / 2 \right]. \tag{8.48}$$

As for the van der Waals equation, we may seek the solution in graphical form (see Fig. 8.2). It is obvious that there is always a solution with $\langle J_z \rangle = 0$. This corresponds to a paramagnetic solution. We see that the emergence of a finite solution corresponds to the possibility of having more than one solution. At sufficiently high temperatures this can never happen. Then the paramagnetic solution is the high temperature phase. At sufficiently low temperatures, instead, there are three solutions. One solution has $\langle J_z \rangle = 0$, while the other two have opposite sign with $\langle J_z \rangle \neq 0$. The last two solutions correspond to ferromagnetic phases with opposite sign for the spontaneous magnetization. There must then be a critical temperature at which the multiple solution starts to develop. This occurs at $T_{\rm c}$ given by

$$k_{\rm B} T_{\rm c} = \frac{1}{4} a g_{\rm L} \mu_{\rm B} N. \tag{8.49}$$

If we take the derivative with respect to H on both sides of Eq. (8.47), we get the magnetic susceptibility

$$\left. \frac{\partial \langle J_z \rangle}{\partial (\beta g_{\rm L} \mu_{\rm B} H)} \right|_{H=0} = \frac{1}{2} \cosh^{-2} \left[(a\beta g_{\rm L} \mu_{\rm B} N) \langle J_z \rangle / 2 \right] (1 - T_{\rm c}/T(1 - 4\langle J_z \rangle^2))^{-1}. \tag{8.50}$$

As in the van der Waals case, the susceptibility becomes infinite at the critical point $T = T_{\rm c}$ and $H = 0$.

8.3 The Curie–Weiss theory and the classical critical indices for the Ising model

The simplest model to introduce an interaction J_{ij} between dipoles in the presence of an external field h_i is the so-called Ising model on a lattice that schematises uniaxial ferromagnets with a preferred orientation

$$H = H_0 + H_{\rm ext} = -\frac{1}{2} \sum_{ij} J_{ij} \sigma_i \sigma_j - \sum_i h_i \sigma_i, \tag{8.51}$$

where the spin variables can only assume two values $\sigma_i = \pm 1$ at the lattice site $\mathbf{x}_i$. In order to reach a better understanding of the correspondence between the paramagnetic–ferromagnetic and gas–liquid transitions, we establish, in Appendix G, the connection between the above Ising and the gas-lattice description of a fluid.

The discussion of Appendix G shows that with a proper correspondence between the variables (chemical potential or pressure vs. magnetic field, density vs. magnetization per spin, see Eqs. (G.9), (G.12), (G.13), the gas-lattice and the Ising model have the same behavior (Lee and Yang (1952)). As a consequence a ferromagnet with a preferred axis, for which the Ising model gives a good schematization, will share the same critical behavior as the gas–liquid transition.

We now discuss the mean-field approximation for the Ising model; the equivalent approximation for the gas-lattice model reproduces in the low-density limit ($nv_0 \to 0$) the van der Waals equation.

The mean-field approximation essentially consists of the following replacement in Eq. (8.51):

$$\sigma_i \sigma_j \simeq \sigma_i \langle \sigma_j \rangle + \langle \sigma_i \rangle \sigma_j - \langle \sigma_i \rangle \langle \sigma_j \rangle, \tag{8.52}$$

where $\langle \sigma_i \rangle = m_i \equiv m(\mathbf{x}_i)$ indicates the ensemble statistical average. Since the approximation done in Eq. (8.52) is made in a symmetric way with respect to the spins, the last term, which is a constant, is necessary to avoid a double counting of the interaction. By taking the ensemble average of Eq. (8.52), we get

$$\langle \sigma_i \sigma_j \rangle \simeq \langle \sigma_i \rangle \langle \sigma_j \rangle, \tag{8.53}$$

which shows how the mean-field approximation amounts to neglecting spatial correlations among the spins, as in the mean-field theory for a fluid (see Eq. (8.9)).

By virtue of the approximation (8.52), the Hamiltonian (8.51) becomes formally like that for a system of non-interacting dipoles

$$H_{\mathrm{MF}} = -\sum_i \tilde{h}_i \sigma_i + \frac{1}{2} \sum_{ij} J_{ij} m_i m_j, \tag{8.54}$$

where the effective field is $\tilde{h}_i = h_i + \sum_j J_{ij} m_j$. Its partition function can be evaluated easily:

$$Z = \sum_{\{\sigma\}} \mathrm{e}^{\beta H_{\mathrm{MF}}} = \mathrm{e}^{-\frac{1}{2}\beta \sum_{ij} J_{ij} m_i m_j} \prod_i 2\cosh(\beta \tilde{h}_i). \tag{8.55}$$

The cost is the introduction of the parameter $\langle \sigma_i \rangle = m_i$, the magnetization per spin, which has to be evaluated self-consistently. The mean-field approximation depends parametrically on the quantity $\langle \sigma_i \rangle \equiv m_i$.

From Eq. (8.55) the free energy is

$$F = -k_{\mathrm{B}} T \sum_i \ln 2\cosh[\beta \tilde{h}_i] + \frac{1}{2} \sum_{ij} J_{ij} m_i m_j. \tag{8.56}$$

The free energy depends parametrically on m_i, which until now has not been fixed. We choose m_i by requiring that F be a minimum. This leads immediately to

$$m_i = \tanh\left[\beta\left(h_i + \sum_j J_{ij} m_j\right)\right], \tag{8.57}$$

which is the Curie–Weiss equation seen in (8.47). An alternative way to get it is to consider self-consistently the ensemble average of σ_i with the mean-field Hamiltonian (8.54)

$$\langle \sigma_i \rangle = \frac{\sum_{\sigma_i = \pm 1} \sigma_i \mathrm{e}^{\beta \tilde{h}_i \sigma_i}}{\sum_{\sigma_i = \pm 1} \mathrm{e}^{\beta \tilde{h}_i \sigma_i}} = \tanh\left[\beta\left(h_i + \sum_j J_{ij} m_j\right)\right]. \tag{8.58}$$

If we take $h_i = h$ at each lattice site the system becomes translationally invariant and $m_i = m$. Then $\tilde{h}_i = h + m\tilde{J}_0$. In the case that the interaction is restricted to the z nearest neighbor sites $\tilde{J}_0 = zJ$, with J the bond interaction (not to be confused with the maximum value of the angular momentum introduced in the previous section), then the self-consistency equation becomes

$$m = \tanh[\beta(h + \tilde{J}_0 m)]. \tag{8.59}$$

Before analyzing the implications of the mean-field results, one may ask when mean-field theory is correct. The basic assumption of the absence of fluctuations becomes sound by increasing the number of spins interacting with a given spin. This is obtained either by increasing the dimensionality of the space, as we shall see, or by extending the range of the interaction. For instance, we consider the simplest spin model with long–range forces, where each spin σ_i at site i is in interaction with any other spin at site j via a constant coupling $J_{ij} = J/N$. The division by N maintains the extensivity of the Hamiltonian, given by

$$H = -\frac{J}{2N}\left(\sum_{i=1}^{N}\sigma_i\right)^2 - h\sum_{i=1}^{N}\sigma_i, \tag{8.60}$$

where h is the constant external field. We show that the exact solution of the model Hamiltonian (8.60) coincides with the solution in the mean-field approximation for the Ising Hamiltonian (8.51).

The partition function reads

$$Z(N,T,h) = \sum_{\{\sigma\}} \exp\left[\beta\frac{J}{2N}\left(\sum_{i=1}^{N}\sigma_i\right)^2 + \beta h\sum_{i=1}^{N}\sigma_i\right]. \tag{8.61}$$

In the above equation the sum runs over all the spin configurations. The exponent in Eq. (8.61) is quadratic in $\sum_{i=1}^{N}\sigma_i$. We can linearize this expression at the expense of introducing an integral over an auxiliary variable x. By exploiting the Gaussian integral identity

$$e^{b^2/2} = \int_{-\infty}^{\infty}\frac{dx}{\sqrt{2\pi}}\exp\left[-\frac{1}{2}x^2 + bx\right], \tag{8.62}$$

with $b = \sqrt{\beta(J/N)}\sum_{i=1}^{N}\sigma_i$ we have

$$Z(N,T,h) = \sum_{\{\sigma\}}\int_{-\infty}^{\infty}\frac{dx}{\sqrt{2\pi}}\exp\left[-\frac{1}{2}x^2 + \left(\sqrt{\beta(J/N)}x + \beta h\right)\sum_{i=1}^{N}\sigma_i\right]. \tag{8.63}$$

In the partition function Z, $\sum_{i=1}^{N}\sigma_i$ appears now linearly and the sum over all the spin configurations can be easily done:

$$Z(N,T,h) = \int_{-\infty}^{\infty}\frac{dx}{\sqrt{2\pi}}e^{-x^2/2}\left[2\cosh\left(\sqrt{\beta(J/N)}x + \beta h\right)\right]^N. \tag{8.64}$$

Changing the auxiliary variable x into $\eta = x/\sqrt{\beta J N}$, the partition function reduces to

$$Z(N, T, h) = \left(\frac{\beta J N}{2\pi}\right)^{1/2} 2^N I(N), \tag{8.65}$$

where

$$I(N) = \int_{-\infty}^{\infty} d\eta \mathrm{e}^{N\phi(\eta)}, \ \phi(\eta) = -\frac{\beta J \eta^2}{2} + \ln\cosh(\beta J\eta + \beta h). \tag{8.66}$$

In the limit $N \to \infty$, the integral $I(N)$ becomes the integrand evaluated at the maximum of $\phi(\eta)$,

$$I(N) = \mathrm{e}^{N\phi(\bar{\eta})}, \quad \bar{\eta} = \max_{-\infty<\eta<\infty}\phi(\eta) \tag{8.67}$$

and

$$Z(N, T, h) = \left(\frac{\beta J N}{2\pi}\right)^{1/2} 2^N \exp\left[N\phi(\bar{\eta})\right]. \tag{8.68}$$

The value $\bar{\eta}$ of the auxiliary variable η, at the maximum of $\phi(\eta)$, is the solution of the following equation:

$$\bar{\eta} = \tanh(\beta J\bar{\eta} + \beta h). \tag{8.69}$$

Equation (8.69), which specifies the condition for the exact solution of model (8.60), coincides with Eq. (8.59) for the mean-field approximate solution of the Ising model. Indeed, the value $\bar{\eta}$ of the auxiliary variable η coincides with the magnetization per spin in this model, as it follows from its definition and Eqs. (8.68) and (8.69):

$$m = \frac{1}{N}\sum_{i=1}^{N}\langle\sigma_i\rangle = k_{\mathrm{B}}T\frac{\partial\ln Z}{\partial h} = \bar{\eta}. \tag{8.70}$$

This is a case in which the magnetization per spin given by the minimum of the thermodynamic potential coincides exactly with its average and there are no fluctuations. In Problem 8.3 we solve the same model (8.60) by combinatorial methods. We now derive the consequences of the mean-field equations. To analyze the possibility of a spontaneous magnetization, we need to find solutions $m \neq 0$ at $h = 0$. This can be studied graphically and it is shown in Fig. 8.2. To this end we define $\alpha = \beta\tilde{J}_0$, the coefficient of m in the argument of the hyperbolic tangent. The solutions of the Curie–Weiss equation correspond to the intersections of the hyperbolic tangent with the bisector. It is easy to see that the types of possible solutions are controlled by the value of α. For $\alpha < 1$, when $h = 0$, only the $m = 0$ solution is possible and this corresponds to the paramagnetic phase. In contrast, when $\alpha > 1$ three solutions are possible: one with $m = 0$ and two symmetric ones with $m \neq 0$. When $\alpha = 1$, the three solutions coincide and we go from one regime of solution to the other. This defines the critical point and the critical temperature is given by

$$T_{\mathrm{c}} = \frac{\tilde{J}_0}{k_{\mathrm{B}}}. \tag{8.71}$$

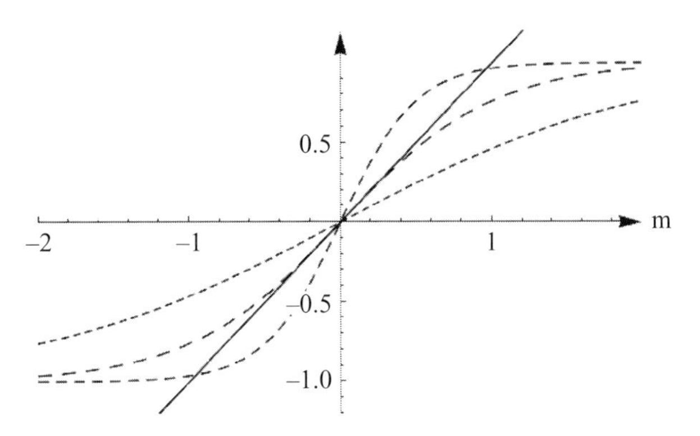

Fig. 8.2 In the figure $\tanh(\alpha m)$, together with the straight line, is plotted for $\alpha = 0.5$ (short-dashed), $\alpha = 1$ (medium-dashed) and $\alpha = 2$ (long-dashed), as examples of the behavior above, at, and below the critical temperature.

In order to decide which solution to choose for $\alpha > 1$ (or $T < T_c$), we make a perturbative expansion about the critical point by exploiting the smallness of m. The Taylor expansion of the hyperbolic tangent reads

$$\tanh x \simeq x - \frac{1}{3}x^3 + \frac{2}{15}x^5 + \cdots, \tag{8.72}$$

which, when inserted into the Curie–Weiss Eq. (8.57), leads to

$$m = \beta h + \frac{\beta}{\beta_c} m - \frac{1}{3}\left(\frac{\beta}{\beta_c}\right)^3 m^3 + \cdots. \tag{8.73}$$

By introducing dimensionless variables

$$t = \frac{T - T_c}{T_c}, \qquad \kappa = \frac{h}{k_B T_c}, \tag{8.74}$$

we finally obtain

$$tm + \frac{1}{3}m^3 = \kappa. \tag{8.75}$$

The last equation is the analog to Eq. (8.19) of the van der Waals theory of the liquid–gas transition. The field κ plays the same role as the combination $p - 4t$. On the coexistence curve, both sides of Eq. (8.75) must be equal to a constant independent of the two intensive variables t and κ and such a constant must be equal to zero. Hence in the t–κ phase diagram the critical point is at the origin and the coexistence line is given by

$$\kappa = 0. \tag{8.76}$$

The solutions for m are then determined by

$$tm_0 + \frac{1}{3}m_0^3 = 0, \tag{8.77}$$

and are

$$m_0 = 0, \tag{8.78}$$

$$m_0 = \pm\sqrt{-3t}. \tag{8.79}$$

For $t < 0$, $m_0 = 0$ corresponds to a local maximum of the free energy (see Eq. (8.84) below) and the last two solutions are two degenerate minima. They correspond to the two opposite values of the spontaneous magnetization m_0 of the coexisting ferromagnetic phases. The critical isotherm ($t = 0$) reads

$$m^3 = 3\kappa, \tag{8.80}$$

which shows that an infinitesimal magnetic field selects one of the two degenerate up or down solutions.

By defining the critical indices δ and β as in the liquid–gas transition:

$$m \sim |\kappa|^{1/\delta}, \quad m \sim |t|^{\beta}, \tag{8.81}$$

we obtain easily $\beta = 1/2$ and $\delta = 3$.

The magnetic susceptibility plays the role of the isothermal compressibility. By taking the derivative with respect to h at $h = 0$ and for $|t| \to 0$ of Eq. (8.59),

$$\chi = \frac{\beta \tilde{J}_0 \chi + \beta}{\cosh^2(\beta \tilde{J}_0 m_0)} = \frac{\beta(1 - m_0^2)}{1 - (1 - m_0^2)\beta \tilde{J}_0} = \frac{\beta}{t + m_0^2} \sim |t|^{-\gamma, -\gamma'} \tag{8.82}$$

with $\gamma = \gamma' = 1$, as in the liquid–gas transition. In the first equality we have used the relation $1/\cosh^2(\beta \tilde{J}_0 m_0) = 1 - m_0^2$ as can be seen from Eq. (8.59) at $h = 0$.

To complete our discussion of the behavior close to the critical point, we examine the specific heat. To this end we consider the free energy of Eq. (8.56) and expand it for small values of m. We obtain

$$F(m) = -Nk_\mathrm{B}T \ln 2 + \left[\frac{T_\mathrm{c}}{2T}(T - T_\mathrm{c})m^2 + \frac{1}{12}\frac{T_\mathrm{c}^4}{T^3}m^4 - \frac{T_\mathrm{c}}{T}hm\right], \tag{8.83}$$

which in terms of reduced variables at $h = 0$ and close to T_c becomes

$$F = -Nk_\mathrm{B}T \ln 2 + Nk_\mathrm{B}T_\mathrm{c}\left(\frac{1}{2}tm^2 + \frac{1}{12}m^4\right). \tag{8.84}$$

The singular dependence is from t and from $m(t)$. The latter differs from zero only in the ferromagnetic phase. By using for $t < 0$, $m^2 = -3t$, one gets

$$c_V = \frac{3}{2}Nk_\mathrm{B}. \tag{8.85}$$

Hence in going from $t > 0$ to $t < 0$, the specific heat has a positive jump $\Delta C_V = (3/2)Nk_\mathrm{B}$. With this type of singularity one associates for the critical indices the value $\alpha = \alpha' = 0$.

Then all the critical indices coincide in the van der Waals and Curie–Weiss theories when the appropriate correspondence among the variables is considered, e.g., $p \to -h$ and $v \to m$. Such a coincidence is due to the mechanism responsible for the formation of the three solutions below the critical point. At a mathematical level this is due to the combined effects of the linear and cubic terms in the equation of state. This can be made even more suggestive by looking at the free energy (8.84). Apart from the term $-Nk_\mathrm{B}T \ln 2$ due to the entropic contribution of the N spins, the free energy close to the critical point depends parametrically on m. When $t > 0$, the only extremum as a function of m is located at the origin and is clearly a minimum. In contrast, when $t < 0$, there are three points satisfying the extremum conditions. The one located at the origin becomes a local maximum, whereas two minima appear symmetrically located with respect to the origin. We will see that this last observation is the starting point of the Landau theory of second order phase transitions.

A comment common to all mean-field theories with relation to correlations and fluctuations is now needed. We first introduce the classical linear response theory and its relation with the correlation and the fluctuations, which will be discussed in general for the quantum case in Chapter 10.

8.4 Classical linear response theory and fluctuations

In the classical context of the Ising Hamiltonian (8.51) there is no problem of non-commutation between the unperturbed Hamiltonian H_0 and the perturbing term $H_{\rm ext}$ or in the ordering of the terms in the response and correlation functions, which instead matter for the quantum case, as we shall see in Chapter 10. For example, the magnetization correlation function now reads

$$f_{\rm m}(\mathbf{x}_i, \mathbf{x}_j) = \langle(\sigma_i - \langle\sigma_i\rangle)(\sigma_j - \langle\sigma_j\rangle)\rangle \equiv \langle\sigma_i\sigma_j\rangle - \langle\sigma_i\rangle\langle\sigma_j\rangle. \tag{8.86}$$

In the linear response we have to calculate, at first order in the perturbation, the difference δm_i between the expectation value $\langle\sigma_i\rangle$ in the presence of δh_i and $\langle\sigma_i\rangle_0$ in the presence of the unperturbed Hamiltonian only:

$$\delta m_i = \frac{\sum_{\{\sigma\}} \sigma_i \mathrm{e}^{-\beta(H_0+H_{\rm ext})}}{\sum_{\{\sigma\}} \mathrm{e}^{-\beta(H_0+H_{\rm ext})}} - \frac{\sum_{\{\sigma\}} \sigma_i \mathrm{e}^{-\beta H_0}}{\sum_{\{\sigma\}} \mathrm{e}^{-\beta H_0}}, \tag{8.87}$$

where $\sum_{\{\sigma\}}$ is the sum extended over all the spin configurations. The above expression can be simply expanded to first order in $H_{\rm ext}$ since the two terms in the Hamiltonian commute. To first order, δm_i reads

$$\delta m_i = -\beta \frac{\sum_{\{\sigma\}} \sigma_i \mathrm{e}^{-\beta H_0} H_{\rm ext}}{\sum_{\{\sigma\}} \mathrm{e}^{-\beta H_0}} + \beta \frac{\sum_{\{\sigma\}} \sigma_i \mathrm{e}^{-\beta H_0}}{\sum_{\{\sigma\}} \mathrm{e}^{-\beta H_0}} \frac{\sum_{\{\sigma\}} \mathrm{e}^{-\beta H_0} H_{\rm ext}}{\sum_{\{\sigma\}} \mathrm{e}^{-\beta H_0}}.$$

By replacing $H_{\rm ext} = -\sum_i \sigma_i \delta h_i$ in the above equation, one has

$$\delta m_i = \beta \sum_j \left[\langle\sigma_i\sigma_j\rangle_0 - \langle\sigma_i\rangle_0\langle\sigma_j\rangle_0\right] \delta h_j, \tag{8.88}$$

where $f_{\rm m}(\mathbf{x}_i, \mathbf{x}_j)$ is evaluated in zero external field and, for translational invariant systems, depends only on the difference $\mathbf{x}_i - \mathbf{x}_j$. In the Fourier space Eq. (8.88) reads

$$\chi(\mathbf{q}) = \delta m(\mathbf{q})/\delta h(\mathbf{q}) = \beta f_{\rm m}(\mathbf{q}) = \beta\langle|\delta\sigma_{\mathbf{q}}|^2\rangle_0. \tag{8.89}$$

This is the classical limit of the quantum fluctuation–dissipation theorem, to be discussed later. In the above we have used the Fourier transforms according to the definition of

Eq. (1.61), where the volume has been replaced by the number of sites N:

$$\delta\sigma_{\mathbf{q}} = \frac{1}{\sqrt{N}}\sum_i e^{-i\mathbf{q}\cdot\mathbf{x}_i}(\sigma_i - \langle\sigma_i\rangle), \quad (\sigma_i - \langle\sigma_i\rangle) = \frac{1}{\sqrt{N}}\sum_{\mathbf{q}} e^{i\mathbf{q}\cdot\mathbf{x}_i}\delta\sigma_{\mathbf{q}}, \tag{8.90}$$

$$f_{\mathrm{m}}(\mathbf{q}) = \sum_i e^{-i\mathbf{q}\cdot(\mathbf{x}_i-\mathbf{x}_j)} f_{\mathrm{m}}(\mathbf{x}_i - \mathbf{x}_j), \quad f_{\mathrm{m}}(\mathbf{x}_i - \mathbf{x}_j) = \frac{1}{N}\sum_{\mathbf{q}} e^{i\mathbf{q}\cdot(\mathbf{x}_i-\mathbf{x}_j)} f_{\mathrm{m}}(\mathbf{q}). \tag{8.91}$$

When δh_i is uniform over the lattice, the susceptibility χ corresponds to the limit $\mathbf{q} \to 0$ of $\beta f_{\mathrm{m}}(\mathbf{q})$, where $f_{\mathrm{m}}(\mathbf{q}) = \langle|\delta\sigma_{\mathbf{q}}|^2\rangle$ coincides with the $\mathbf{q}$-component of the mean square fluctuations of the magnetization.

In both the van der Waals and Curie–Weiss theories, the effective internal field has been derived on the hypothesis of the absence of correlations and fluctuations (see Eq. (8.53)). Whenever the system is far away from criticality, this is a reasonable assumption and these theories are well representative of the experimental results of the equation of state. Correlations in these theories only appear as the first variation with respect to the mean-field equations in the presence of a small external field. For the deviations from the Curie–Weiss theory, for instance, we can start from the mean-field self-consistency equation (8.58) and consider a small variation to the local external field δh_i:

$$\begin{aligned}\delta m_i &= \frac{\beta\delta h_i + \beta\sum_j J_{ij}\delta m_j}{\cosh^2\left(\beta\left(h_i + \sum_j J_{ij} m_j\right)\right)} \\ &= \left(1 - m_i^2\right)\left(\beta\delta h_i + \beta\sum_j J_{ij}\delta m_j\right).\end{aligned} \tag{8.92}$$

For $h_i = 0$ we can use the solution (8.78) and (8.79) for $T < T_{\mathrm{c}}$ and $T > T_{\mathrm{c}}$, respectively.

Equation (8.92) can be easily solved by exploiting the Fourier transform of the convolution there appearing. Let us consider $T > T_{\mathrm{c}}$ for the sake of simplicity. The response function $\chi(\mathbf{q}) = \delta m(\mathbf{q})/\delta(\mathbf{q})$ reads

$$\chi(\mathbf{q}) = \frac{\beta}{1 - \beta J(\mathbf{q})}, \tag{8.93}$$

where $J(\mathbf{q})$ is the Fourier transform of J_{ij} with respect to $i - j$ according to the definition given above.

By taking the small $\mathbf{q}$-expansion of $J(\mathbf{q}) = \tilde{J}_0 - \alpha q^2, \alpha > 0$, the response function $\chi(\mathbf{q})$ can be expressed in terms of the susceptibility χ (8.82):

$$\chi(\mathbf{q}) = \frac{1}{\chi^{-1} + \alpha q^2}, \tag{8.94}$$

which according to the classical linear response Eq. (8.89) is also the correlation multiplied by β. Equation (8.94) has the Ornstein–Zernicke form Eq. (1.64) valid for the fluctuations $\langle|\delta n(\mathbf{q})|^2\rangle$ of the van der Waals approximation. In both cases, the response, and hence the fluctuation, diverges in the limit $\mathbf{q} \to 0$ as T approaches the critical temperature. Note this inconsistency: we started from a theory in the absence of correlations and fluctuations, but a small perturbation makes them diverge. We will see the consequences of this divergence of the fluctuations in Chapter 13 of the Landau theory of second order phase transitions.

8.5 Problems

8.1 Evaluate the hard-core partition function Eq. (8.12) in one space dimension.

8.2 Derive the first correction to the pressure in the virial expansion for a gas of hard spheres with Hamiltonian (8.1).

8.3 Show by combinatorial methods that for the spin model Hamiltonian (8.60) where each spin is interacting with any other spin the mean field solution becomes exact.

9 Second quantization and the Hartree–Fock approximation

9.1 Occupation numbers representation

In many-body theory the representation of the states and of the observables in terms of creation and annihilation operators has the enormous advantage of taking into account simply and directly the combinatorial aspects of the statistics obeyed by Bose and Fermi particles, i.e. the symmetry and antisymmetry of the wave function.

Given a system of N identical particles, we build a complete set of wave functions $\Phi_N(\mathbf{x}_1, \ldots, \mathbf{x}_N)$ as a product of N single-particle wave functions $g_\kappa(\mathbf{x}) = \langle \mathbf{x}|\kappa\rangle$, properly symmetrized for bosons or antisymmetrized for fermions. In general κ labels the single-particle states, while $\mathbf{x}$ is the space coordinate. The symmetrized basic state is obtained by applying the symmetrization operator Λ:

$$|\kappa_1, \kappa_2, \ldots, \kappa_N\rangle = c\Lambda|\kappa_1\rangle|\kappa_2\rangle \cdots |\kappa_N\rangle, \tag{9.1}$$

with c a suitable normalization factor. The symmetrization operator is defined by

$$\Lambda = \frac{1}{N!}\sum_p \lambda_p U_p, \tag{9.2}$$

where U_p is the permutation operator and $\lambda_p = 1$ for bosonic symmetric states. For fermionic antisymmetric states $\lambda_p = 1$ if the permutation is made of an even number of transpositions, whereas $\lambda_p = -1$ when the number of transpositions is odd.

Since $\lambda_p^2 = 1$ it follows that

$$U_p\Lambda = \frac{1}{N!}\lambda_p \sum_q \lambda_p\lambda_q U_p U_q = \lambda_p\Lambda, \tag{9.3}$$

because, for fixed p, $U_p U_q$ runs over all the permutations by varying q. Hence

$$\Lambda^2 = \frac{1}{N!}\sum_p \lambda_p^2 \Lambda = \Lambda. \tag{9.4}$$

Λ is therefore a projection operator which projects the generic ket in subspaces with the required symmetry. Furthermore, the above properties allow one to symmetrize either the ket or the bra. Hence the basic wave function is

$$\Phi_N(\mathbf{x}_1, \ldots, \mathbf{x}_N) = \langle \mathbf{x}_1, \ldots, \mathbf{x}_N|\kappa_1, \kappa_2, \ldots, \kappa_N\rangle, \tag{9.5}$$

where symmetrization of the bra is not necessary.

Once the wave function $\Phi_N(\mathbf{x}_1, \ldots, \mathbf{x}_N)$ is symmetrized, the particles lose their identity and the state can be simply defined by specifying the occupation numbers n_{κ_i} of each single-particle state labelled by the quantum index κ_i:

$$|\kappa_1, \kappa_2, \ldots, \kappa_N\rangle = |n_{\kappa_1}, \ldots, n_{\kappa_i}, \ldots; N\rangle, \quad \sum_{\kappa_i} n_{\kappa_i} = N. \tag{9.6}$$

In terms of the occupation numbers, by exploiting Eq. (9.4) from the normalization condition of the wave function, the factor c^2 reads $c^2 = N!/(n_{k_1}! \ldots n_{k_i}! \ldots)$ (see Problem 9.1). If the system includes an interaction term in the Hamiltonian, its generic state is expressed as a superposition of the above states for free-particle systems.

9.2 Creation and annihilation operators

In order to write in a simple way the set of basic states in the occupation number representation and find the operators whose eigenvalues are the occupation numbers themselves, we introduce the creation $a^\dagger_{\kappa_i}$ and annihilation a_{κ_i} operators, which connect states with n_{κ_i} differing by one.

We start from the bosonic case and define the annihilation operator a_{κ_i} as

$$a_{\kappa_i}|n_{\kappa_1}, \ldots, n_{\kappa_i}, \ldots; N\rangle = \sqrt{n_{\kappa_i}}|n_{\kappa_1}, \ldots, n_{\kappa_i} - 1, \ldots; N - 1\rangle. \tag{9.7}$$

The creation operator is the adjoint $a^\dagger_{\kappa_i}$ and is defined via the matrix elements

$$\begin{aligned}
&\langle n'_{\kappa_1}, \ldots, n'_{\kappa_i}, \ldots; N|a_{\kappa_i}|n_{\kappa_1}, \ldots, n_{\kappa_i}, \ldots; N + 1\rangle \\
&= \sqrt{n_{\kappa_i}}\delta_{n_{\kappa_1}, n'_{\kappa_1}} \cdots \delta_{n_{\kappa_i}-1, n'_{\kappa_i}} \cdots \\
&= \langle n_{\kappa_1}, \ldots, n_{\kappa_i}, \ldots; N + 1|a^\dagger_{\kappa_i}|n'_{\kappa_1}, \ldots, n'_{\kappa_i}, \ldots; N\rangle.
\end{aligned} \tag{9.8}$$

Hence $a^\dagger_{\kappa_i}$ creates a particle in the state $|\kappa_i\rangle$:

$$a^\dagger_{\kappa_i}|n_{\kappa_1}, \ldots, n_{\kappa_i}, \ldots; N\rangle = \sqrt{n_{\kappa_i} + 1}|n_{\kappa_1}, \ldots, n_{\kappa_i} + 1, \ldots; N + 1\rangle. \tag{9.9}$$

From the above definition it follows that the creation and annihilation operators satisfy the commutation rules (see Problem 9.2)

$$\left[a_{\kappa_i}, a_{\kappa_j}\right] = \left[a^\dagger_{\kappa_i}, a^\dagger_{\kappa_j}\right] = 0, \quad \left[a_{\kappa_i}, a^\dagger_{\kappa_j}\right] = \delta_{\kappa_i, \kappa_j}. \tag{9.10}$$

The product operator $\hat{n}_{\kappa_i} = a^\dagger_{\kappa_i} a_{\kappa_i}$ has as eigenvectors the basis vectors $|n_{\kappa_1}, \ldots, n_{\kappa_i}, \ldots; N\rangle$ with eigenvalues given by the occupation numbers n_{κ_i}:

$$a^\dagger_{\kappa_i} a_{\kappa_i}|n_{\kappa_1}, \ldots, n_{\kappa_i}, \ldots; N\rangle = n_{\kappa_i}|n_{\kappa_1}, \ldots, n_{\kappa_i}, \ldots; N\rangle. \tag{9.11}$$

The operator for the total number of particles is

$$\hat{N} = \sum_{\kappa_i} \hat{n}_{\kappa_i} = \sum_{\kappa_i} a^\dagger_{\kappa_i} a_{\kappa_i}. \tag{9.12}$$

We now define the vacuum state $|0\rangle$ such that $a^\dagger_{\kappa_i} a_{\kappa_i}|0\rangle = 0$, $\forall\ \kappa_i$. We can then produce all the basis vectors of an N-particle system by acting with the creation operators on the vacuum state (see Problem 9.3):

$$|n_{\kappa_1}, \ldots, n_{\kappa_i}, \ldots; N\rangle = \frac{1}{\sqrt{n_{\kappa_1}! n_{\kappa_2}! \ldots n_{\kappa_i}! \ldots}} (a^\dagger_{\kappa_1})^{n_{\kappa_1}} (a^\dagger_{\kappa_2})^{n_{\kappa_2}} \cdots (a^\dagger_{\kappa_i})^{n_{\kappa_i}} \cdots |0\rangle, \quad (9.13)$$

with the constraint $\sum_{\kappa_i} n_{\kappa_i} = N$. Due to the commutation rules for the creation operators these states are symmetric for any exchange of the quantum numbers and each single-particle state can be occupied up to N, as it should for bosons.

For fermions it is impossible to add a particle to an occupied single-particle state and the operators $a^\dagger_{\kappa_i} a_{\kappa_i}$ must have eigenvalues 0 and 1. The basis vectors must be antisymmetric and the commutation rules (9.10) must become anticommutation rules. This is achieved by changing the definition of annihilation and creation operators as

$$a_{\kappa_i} |n_{\kappa_1}, \ldots, n_{\kappa_i}, \ldots; N\rangle = (-1)^{P_i} n_{\kappa_i} |n_{\kappa_1}, \ldots, n_{\kappa_i} - 1, \ldots; N-1\rangle, \quad (9.14)$$

and

$$a^\dagger_{\kappa_i} |n_{\kappa_1}, \ldots, n_{\kappa_i}, \ldots; N\rangle = (-1)^{P_i} (1 - n_{\kappa_i}) |n_{\kappa_1}, \ldots, n_{\kappa_i} + 1, \ldots; N+1\rangle, \quad (9.15)$$

with

$$P_i = \sum_{j=1}^{i-1} n_{k_j}, \quad (9.16)$$

which determines the parity of the permutation to bring the operators to act at the right place. The order of the occupation numbers of each single-particle state is arbitrary, but once it is fixed, it must be maintained.

Given the definition (9.14) and (9.15), we have the anticommutation rules (see Problem 9.4)

$$\left\{a_{\kappa_i}, a^\dagger_{\kappa_j}\right\} = \delta_{\kappa_i, k_j}, \quad \{a_{\kappa_i}, a_{k_j}\} = \left\{a^\dagger_{\kappa_i}, a^\dagger_{k_j}\right\} = 0 \quad (9.17)$$

from which it follows that $\hat{n}^2_{\kappa_i} = \hat{n}_{\kappa_i}$. The basis vectors are still given by (9.13).

9.3 Field operators

To give the basis states in the real space representation we introduce the field operators

$$\hat{\psi}(\mathbf{x}) = \sum_{\kappa_i} g_{\kappa_i}(\mathbf{x}) a_{\kappa_i} = \sum_{\kappa_i} \langle \mathbf{x} | \kappa_i \rangle a_{\kappa_i}, \quad (9.18)$$

$$\hat{\psi}^\dagger(\mathbf{x}) = \sum_{\kappa_i} g^*_{\kappa_i}(\mathbf{x}) a^\dagger_{\kappa_i} = \sum_{\kappa_i} \langle \kappa_i | \mathbf{x} \rangle a^\dagger_{\kappa_i}. \quad (9.19)$$

These operators satisfy the commutation rules for bosons (see Problem 9.5)

$$[\hat{\psi}(\mathbf{x}), \hat{\psi}^\dagger(\mathbf{x}')] = \delta(\mathbf{x} - \mathbf{x}'), \quad [\hat{\psi}(\mathbf{x}), \hat{\psi}(\mathbf{x}')] = [\hat{\psi}^\dagger(\mathbf{x}), \hat{\psi}^\dagger(\mathbf{x}')] = 0 \quad (9.20)$$

and the anticommutation rules for fermions

$$\{\hat{\psi}(\mathbf{x}), \hat{\psi}^\dagger(\mathbf{x}')\} = \delta(\mathbf{x} - \mathbf{x}'), \quad \{\hat{\psi}(\mathbf{x}), \hat{\psi}(\mathbf{x}')\} = \{\hat{\psi}^\dagger(\mathbf{x}), \hat{\psi}^\dagger(\mathbf{x}')\} = 0. \quad (9.21)$$

We can then write the Hamiltonian of an N-particle system in the presence of a two-body interaction as

$$H = \int d\mathbf{x} \int d\mathbf{x}' \left[\hat{\psi}^\dagger(\mathbf{x})\epsilon(\mathbf{x}, \mathbf{x}')\hat{\psi}(\mathbf{x}') + \frac{1}{2}\, \hat{\psi}^\dagger(\mathbf{x})\hat{\psi}^\dagger(\mathbf{x}')V(\mathbf{x}, \mathbf{x}')\hat{\psi}(\mathbf{x}')\hat{\psi}(\mathbf{x}) \right], \tag{9.37}$$

where $\epsilon(\mathbf{x}, \mathbf{x}')$ includes the kinetic energy and the chemical potential μ:

$$\epsilon(\mathbf{x}, \mathbf{x}') = \delta(\mathbf{x} - \mathbf{x}') \left(-\frac{\hbar^2}{2m} \nabla^2_{\mathbf{x}} - \mu \right). \tag{9.38}$$

For non-interacting particles in a box of volume V, the wave function is a product of a plane wave labelled by the wave vector $\mathbf{k}$ and a spinor χ_s:

$$g_\kappa(\mathbf{x}) = \frac{1}{\sqrt{V}} e^{i\mathbf{k}\cdot\mathbf{x}} \chi_s, \tag{9.39}$$

where the collective quantum index $\kappa = (\mathbf{k}, s)$ includes the spin quantum number as well. For a translational invariant interaction $V(\mathbf{x}, \mathbf{x}') = V(\mathbf{x} - \mathbf{x}')$ we can write the Hamiltonian (9.37) in the $\mathbf{k}$-representation by using the expansions (9.18) and (9.19) of the field operators with the single-particle wave function (9.39) as

$$H = \sum_{\mathbf{k},s} \epsilon_{\mathbf{k}} a^\dagger_{\mathbf{k},s} a_{\mathbf{k},s} + \frac{1}{2V} \sum_{\mathbf{k},\mathbf{k}',\mathbf{q},s,s'} V(\mathbf{q}) a^\dagger_{\mathbf{k}+\mathbf{q},s} a^\dagger_{\mathbf{k}'-\mathbf{q},s'} a_{\mathbf{k}',s'} a_{\mathbf{k},s}. \tag{9.40}$$

In the above we have introduced the Fourier transforms of the kinetic energy and two-body interaction:

$$\epsilon_{\mathbf{k},\mathbf{k}'} = \frac{1}{V} \int d\mathbf{x} \int d\mathbf{x}'\, e^{-i\mathbf{x}\cdot\mathbf{k}} \epsilon(\mathbf{x} - \mathbf{x}') e^{i\mathbf{x}'\cdot\mathbf{k}'} = \delta_{\mathbf{k},\mathbf{k}'} \left(\frac{\hbar^2 k^2}{2m} - \mu \right) \equiv \delta_{\mathbf{k},\mathbf{k}'} \epsilon_{\mathbf{k}}, \tag{9.41}$$

$$V(\mathbf{q}) = \int d(\mathbf{x} - \mathbf{x}')\, e^{-i(\mathbf{x}-\mathbf{x}')\cdot\mathbf{q}} V(\mathbf{x} - \mathbf{x}'). \tag{9.42}$$

9.6 Equation of motion for field operators

The field operators in the Heisenberg represention obey the standard equation of motion

$$i\hbar \frac{\partial}{\partial t} \hat{\psi}(\mathbf{x}, t) = [\hat{\psi}(\mathbf{x}, t), H], \quad i\hbar \frac{\partial}{\partial t} \hat{\psi}^\dagger(\mathbf{x}, t) = [\hat{\psi}^\dagger(\mathbf{x}, t), H]. \tag{9.43}$$

As a direct consequence of the equation of motion (9.43) one can derive the expression for the number current operator $\hat{\mathbf{j}}(\mathbf{x})$ associated will the density operator $\hat{n}(\mathbf{x})$ of Eq. (9.29). Let us evaluate

$$\begin{aligned} i\hbar \frac{\partial}{\partial t} \hat{n}(\mathbf{x}) &= i\hbar \frac{\partial}{\partial t} \hat{\psi}^\dagger(\mathbf{x}, t)\hat{\psi}(\mathbf{x}, t) \\ &= [\hat{\psi}^\dagger(\mathbf{x}, t), H]\hat{\psi}(\mathbf{x}, t) + \hat{\psi}^\dagger(\mathbf{x}, t)[\hat{\psi}(\mathbf{x}, t), H]. \end{aligned}$$

We notice that in the Hamiltonian (9.37) only the kinetic term contributes to the commutator since the interaction commutes with the density operator. Furthermore, in order to evaluate the commutators, we recall that for any three operators $\hat{A}$, $\hat{B}$ and $\hat{C}$ the following equality holds:

$$[\hat{A}, \hat{B}\hat{C}] = [\hat{A}, \hat{B}]_{\pm}\hat{C} \mp \hat{B}[\hat{A}, \hat{C}]_{\pm}, \tag{9.44}$$

where the upper and lower signs apply to Fermi and Bose statistics and, as subscripts, correspond to the anticommutator or commutator, respectively. Hence

$$[\hat{\psi}^{\dagger}(\mathbf{x}), H] = -\frac{\hbar^2}{2m}\int \mathrm{d}\mathbf{x}'\big[\hat{\psi}^{\dagger}(\mathbf{x}), \hat{\psi}^{\dagger}(\mathbf{x}')\nabla^2_{\mathbf{x}'}\hat{\psi}(\mathbf{x}')\big] = \frac{\hbar^2}{2m}\nabla^2_{\mathbf{x}}\hat{\psi}^{\dagger}(\mathbf{x}).$$

Similarly,

$$[\hat{\psi}(\mathbf{x}), H] = -\frac{\hbar^2}{2m}\nabla^2_{\mathbf{x}}\hat{\psi}(\mathbf{x}).$$

As a consequence

$$\mathrm{i}\hbar\frac{\partial}{\partial t}\hat{n}(\mathbf{x}) = -\frac{\hbar^2}{2m}\nabla_{\mathbf{x}}(\hat{\psi}^{\dagger}(\mathbf{x},t)\nabla_{\mathbf{x}}\hat{\psi}(\mathbf{x},t) - (\nabla_{\mathbf{x}}\hat{\psi}^{\dagger}(\mathbf{x},t))\hat{\psi}(\mathbf{x},t)).$$

The above equation is nothing but the continuity equation, which finally leads to the expression for the number current:

$$\hat{\mathbf{j}}(\mathbf{x}) = -\frac{\mathrm{i}\hbar}{2m}(\hat{\psi}^{\dagger}(\mathbf{x},t)\nabla_{\mathbf{x}}\hat{\psi}(\mathbf{x},t) - (\nabla_{\mathbf{x}}\hat{\psi}^{\dagger}(\mathbf{x},t))\hat{\psi}(\mathbf{x},t)). \tag{9.45}$$

In momentum space it reads

$$\hat{\mathbf{j}}_{\mathbf{q}} = \sum_{\mathbf{k}} a^{\dagger}_{\mathbf{k}-\mathbf{q}/2}a_{\mathbf{k}+\mathbf{q}/2}\frac{\hbar\mathbf{k}}{m}. \tag{9.46}$$

Finally, we give the expression of the charge and current density in the presence of an external electromagnetic field $A^{\mu} = (\phi, \mathbf{A})$ for particles of charge e. For the charge density we simply multiply the number density by e, whereas for the charge current we make the standard replacement $-\mathrm{i}\hbar\nabla_{\mathbf{x}} \to -\mathrm{i}\hbar\nabla_{\mathbf{x}} - (e/c)\mathbf{A}(\mathbf{x})$:

$$\hat{\rho}(\mathbf{x}) = e\hat{\psi}^{\dagger}(\mathbf{x})\hat{\psi}(\mathbf{x}), \tag{9.47}$$

$$\hat{\mathbf{J}}(\mathbf{x}) = e\hat{\mathbf{j}}(\mathbf{x}) - \frac{e^2}{mc}\mathbf{A}(\mathbf{x})\hat{\psi}^{\dagger}(\mathbf{x})\hat{\psi}(\mathbf{x}). \tag{9.48}$$

The second term in Eq. (9.48) is referred to as the diamagnetic contribution. Notice that the charge e can be either positive or negative.

By recalling Eqs. (9.18) and (9.19), we also have the equation of motion for the creation and annihilation operators:

$$\mathrm{i}\hbar\frac{\partial}{\partial t}a_{\kappa_i}(t) = [a_{\kappa_i}(t), H], \quad \mathrm{i}\hbar\frac{\partial}{\partial t}a^{\dagger}_{\kappa_i}(t) = \big[a^{\dagger}_{\kappa_i}(t), H\big]. \tag{9.49}$$

It is instructive and useful to solve the above equations in the case of quantum gases where the Hamiltonian in second quantized form gets the simple expression

$$H = \sum_{\kappa_j}\epsilon_{\kappa_j}a^{\dagger}_{\kappa_j}a_{\kappa_j}, \tag{9.50}$$

where ϵ_{κ_j} are single-particle energy levels. By inserting (9.50) into (9.49), after solving a simple differential equation, we easily get

$$a_{\kappa_i}(t) = \mathrm{e}^{-\mathrm{i}\epsilon_{\kappa_i} t/\hbar} a_{\kappa_i} \equiv \mathrm{e}^{\mathrm{i}Ht/\hbar} a_{\kappa_i} \mathrm{e}^{-\mathrm{i}Ht/\hbar}, \quad a^{\dagger}_{\kappa_i}(t) = \mathrm{e}^{\mathrm{i}\epsilon_{\kappa_i} t/\hbar} a^{\dagger}_{\kappa_i} \equiv \mathrm{e}^{\mathrm{i}Ht/\hbar} a^{\dagger}_{\kappa_i} \mathrm{e}^{-\mathrm{i}Ht/\hbar}. \tag{9.51}$$

The usefulness of these relations is quickly shown by deriving the statistical distribution for the quantum gases. According to Eq. (6.32), the grand canonical average can be performed by a trace operation on a complete set of eigenstates of the Hamiltonian with the Boltzmann factor. Let us evaluate the average of $a^{\dagger}_{\kappa_i} a_{\kappa_j}$:

$$\begin{aligned} \mathrm{Tr}\left(\mathrm{e}^{\beta(\Omega-H)} a^{\dagger}_{\kappa_i} a_{\kappa_j}\right) &= \mathrm{Tr}\left(a_{\kappa_j} \mathrm{e}^{\beta(\Omega-H)} a^{\dagger}_{\kappa_i}\right) \\ &= \mathrm{Tr}\left(\mathrm{e}^{\beta(\Omega-H)} \mathrm{e}^{\beta H} a_{\kappa_j} \mathrm{e}^{-\beta H} a^{\dagger}_{\kappa_i}\right) \\ &= \mathrm{Tr}\big(\mathrm{e}^{\beta(\Omega-H)} \mathrm{e}^{-\beta\epsilon_{\kappa_j}} a_{\kappa_j} a^{\dagger}_{\kappa_i}\big), \end{aligned}$$

where in the first step we used the cyclic property of the trace, in the second $1 = \mathrm{e}^{-\beta H}\mathrm{e}^{\beta H}$ and in the third the relations (9.51) with the replacement $\mathrm{i}t/\hbar = \beta$. The annihilation operator can then further moved to the right by using the commutation relations (9.10) and (9.17) for bosons and fermions, respectively, to yield

$$\begin{aligned} \mathrm{Tr}\big(\mathrm{e}^{\beta(\Omega-H)} a^{\dagger}_{\kappa_i} a_{\kappa_j}\big) &= \mathrm{Tr}\big(\mathrm{e}^{\beta(\Omega-H)} \mathrm{e}^{-\beta\epsilon_{\kappa_j}} a_{\kappa_j} a^{\dagger}_{\kappa_i}\big) \\ &= \delta_{\kappa_i,\kappa_j} \mathrm{e}^{-\beta\epsilon_{\kappa_j}} \mp \mathrm{e}^{-\beta\epsilon_{\kappa_j}} \mathrm{Tr}\big(\mathrm{e}^{\beta(\Omega-H)} a^{\dagger}_{\kappa_i} a_{\kappa_j}\big) \\ &= \frac{\delta_{\kappa_i,\kappa_j}}{\mathrm{e}^{\beta\epsilon_{\kappa_j}} \pm 1}, \end{aligned} \tag{9.52}$$

which are the Fermi–Dirac and Bose–Einstein distribution functions.

9.7 Reduced density matrices

Given the density matrix ρ_N for a system of N particles, we define the p-particle reduced density matrix as

$$\begin{aligned} h_p(\mathbf{x}_1, \ldots, \mathbf{x}_p; \mathbf{x}'_1, \ldots, \mathbf{x}'_p) &= \frac{N!}{(N-p)!} \int \mathrm{d}\mathbf{x}_{p+1} \ldots \mathrm{d}\mathbf{x}_N \\ &\times \rho_N(\mathbf{x}_1, \ldots, \mathbf{x}_p, \mathbf{x}_{p+1}, \ldots, \mathbf{x}_N; \mathbf{x}'_1, \ldots, \mathbf{x}'_p, \mathbf{x}_{p+1}, \ldots, \mathbf{x}_N), \end{aligned} \tag{9.53}$$

where we changed the normalization in such a way that tr $\hat{h}_N = N!$, where the symbol tr stands for the integration of the diagonal density matrix, *viz.*

$$\mathrm{tr}\, h_p \equiv \int \mathrm{d}\mathbf{x}_1, \ldots, \mathrm{d}\mathbf{x}_p\, h_p(\mathbf{x}_1, \ldots, \mathbf{x}_p; \mathbf{x}_1, \ldots, \mathbf{x}_p).$$

The symbol tr must not be confused with the symbol Tr, with which we have indicated the quantum and/or statistical average. The operators

$$\hat{h}_1(\mathbf{x}, \mathbf{x}') = \hat{\psi}^{\dagger}(\mathbf{x}')\hat{\psi}(\mathbf{x}), \tag{9.54}$$

$$\hat{h}_2(\mathbf{x}_1, \mathbf{x}_2; \mathbf{x}'_2, \mathbf{x}'_1) = \hat{\psi}^{\dagger}(\mathbf{x}'_2)\hat{\psi}^{\dagger}(\mathbf{x}'_1)\hat{\psi}(\mathbf{x}_1)\hat{\psi}(\mathbf{x}_2) \tag{9.55}$$

are the second quantization representation of the one-particle and two-particle reduced density matrices. For example, let us consider the one-particle reduced density matrix in a pure state $|\chi\rangle$ with N particles:

$$\begin{aligned} h_1(\mathbf{x}, \mathbf{x}') &= \langle\chi|\hat{h}_1(\mathbf{x}, \mathbf{x}')|\chi\rangle \\ &= \langle\chi|\hat{\psi}^\dagger(\mathbf{x}')\hat{\psi}(\mathbf{x})|\chi\rangle. \end{aligned}$$

By inserting the resolution of the identity for $N-1$-particle states,

$$\begin{aligned} 1 &= \int \mathrm{d}\mathbf{x}_1, \ldots, \mathrm{d}\mathbf{x}_{N-1}\ \Lambda|\mathbf{x}_1, \ldots, \mathbf{x}_{N-1}\rangle\langle\mathbf{x}_1, \ldots, \mathbf{x}_{N-1}|\Lambda \\ &= \frac{1}{(N-1)!}\prod_{j=1}^{N-1}\hat{\psi}^\dagger(\mathbf{x}_j)|0\rangle\langle 0|\left(\prod_{k=1}^{N-1}\hat{\psi}^\dagger(\mathbf{x}_k)\right)^\dagger, \end{aligned}$$

we obtain

$$\begin{aligned} &\langle\chi|\hat{\psi}^\dagger(\mathbf{x}')\hat{\psi}(\mathbf{x})|\chi\rangle \\ &= \frac{1}{(N-1)!}\int \mathrm{d}\mathbf{x}_1\cdots\mathrm{d}\mathbf{x}_{N-1}\ \langle\chi|\hat{\psi}^\dagger(\mathbf{x}')\prod_{j=1}^{N-1}\hat{\psi}^\dagger(\mathbf{x}_j)|0\rangle\langle 0|\left(\prod_{k=1}^{N-1}\hat{\psi}^\dagger(\mathbf{x}_k)\right)^\dagger\hat{\psi}(\mathbf{x})|\chi\rangle \\ &= \frac{N!}{(N-1)!}\int \mathrm{d}\mathbf{x}_1\cdots\mathrm{d}\mathbf{x}_{N-1}\langle\chi|\Lambda|\mathbf{x}', \mathbf{x}_1, \ldots, \mathbf{x}_{N-1}\rangle\langle\mathbf{x}, \mathbf{x}_1, \ldots, \mathbf{x}_{N-1}|\Lambda|\chi\rangle \\ &= N\int \mathrm{d}\mathbf{x}_1\cdots\mathrm{d}\mathbf{x}_{N-1}\ \chi^*(\mathbf{x}', \mathbf{x}_1, \ldots, \mathbf{x}_{N-1})\chi(\mathbf{x}, \mathbf{x}_1, \ldots, \mathbf{x}_{N-1}) \\ &= \frac{N!}{(N-1)!}\int \mathrm{d}\mathbf{x}_1\cdots\mathrm{d}\mathbf{x}_{N-1}\ \rho_N(\mathbf{x}', \mathbf{x}_1, \ldots, \mathbf{x}_{N-1}; \mathbf{x}, \mathbf{x}_1, \ldots, \mathbf{x}_{N-1}), \end{aligned} \tag{9.56}$$

with $\mathrm{tr}\,\hat{h}_1 = N$. Recalling that in this case $\hat{\rho} = |\chi\rangle\langle\chi|$,

$$h_1(\mathbf{x}', \mathbf{x}) = \mathrm{Tr}\,\hat{h}_1\hat{\rho} \tag{9.57}$$

and analogously

$$h_2(\mathbf{x}_1, \mathbf{x}_2; \mathbf{x}_2', \mathbf{x}_1') = \mathrm{Tr}\,\hat{h}_2\hat{\rho}. \tag{9.58}$$

This last definition can be extended to the case of a statistical ensemble where $\hat{\rho} = \sum_m |m\rangle w_m\langle m|$ as in Eq. (6.27). $h_1(\mathbf{x}, \mathbf{x}')$ for instance reads

$$\begin{aligned} h_1(\mathbf{x}, \mathbf{x}') &= N\sum_m w_m\int \mathrm{d}\mathbf{x}_2\mathrm{d}\mathbf{x}_3\cdots\mathrm{d}\mathbf{x}_N\,\psi_m^*(\mathbf{x}', \mathbf{x}_2, \mathbf{x}_3, \ldots, \mathbf{x}_N)\psi_m(\mathbf{x}, \mathbf{x}_2, \mathbf{x}_3, \ldots, \mathbf{x}_N) \\ &\equiv \mathrm{Tr}\big(\mathrm{e}^{\beta(\Omega-H)}\hat{\psi}^\dagger(\mathbf{x}')\hat{\psi}(\mathbf{x})\big), \end{aligned} \tag{9.59}$$

where the average is both quantum and statistical.

For a spinless Bose gas there exists the obvious relation between the two reduced density matrices:

$$h_2(\mathbf{x}_1, \mathbf{x}_2; \mathbf{x}_2', \mathbf{x}_1') = h_1(\mathbf{x}_1, \mathbf{x}_1')h_1(\mathbf{x}_2, \mathbf{x}_2') + h_1(\mathbf{x}_1, \mathbf{x}_2')h_1(\mathbf{x}_2, \mathbf{x}_1') \tag{9.60}$$

and

$$h_1(\mathbf{x}, \mathbf{x}') = \sum_\kappa g_\kappa^*(\mathbf{x}') g_\kappa(\mathbf{x}) n_\kappa, \quad n_\kappa = \mathrm{Tr}\big(\mathrm{e}^{\beta(\Omega - H)} a_\kappa^\dagger a_\kappa\big), \tag{9.61}$$

where the trace is given by (9.52) for the appropriate statistics.

For a spinless Fermi gas, as for a fully polarized gas, the two-particle reduced density matrix $h_2(\mathbf{x}_1, \mathbf{x}_2; \mathbf{x}_2', \mathbf{x}_1')$ also factorizes in a way similar to Eq. (9.60):

$$h_2(\mathbf{x}_1, \mathbf{x}_2; \mathbf{x}_2', \mathbf{x}_1') = h_1(\mathbf{x}_1, \mathbf{x}_1') h_1(\mathbf{x}_2, \mathbf{x}_2') - h_1(\mathbf{x}_1, \mathbf{x}_2') h_1(\mathbf{x}_2, \mathbf{x}_1'), \tag{9.62}$$

showing that a correlation is induced by the exclusion principle, since

$$\lim_{|\mathbf{x}-\mathbf{x}'| \to 0} h_2(\mathbf{x}, \mathbf{x}'; \mathbf{x}', \mathbf{x}) = 0.$$

For a spin $1/2$ Fermi gas it is convenient to keep the explicit trace of the spin quantum number. According to Eqs. (9.18) and (9.19) we may write the field operator as a sum of the field operators for the two spin components $\hat{\psi} = \hat{\psi}_\uparrow + \hat{\psi}_\downarrow$. As a consequence the two-particle density matrix depends explicitly on four spin variables and we obtain

$$h_{2,\alpha\beta\gamma\delta}(\mathbf{x}, \mathbf{x}'; \mathbf{x}', \mathbf{x}) = h_{1,\alpha\delta}(\mathbf{x}, \mathbf{x}) h_{1,\beta\gamma}(\mathbf{x}', \mathbf{x}') - h_{1,\alpha\gamma}(\mathbf{x}, \mathbf{x}') h_{1,\beta\delta}(\mathbf{x}', \mathbf{x}) \tag{9.63}$$

$$= \frac{1}{V^2} \sum_{\mathbf{k},\mathbf{k}'} n_{\mathbf{k},\alpha} n_{\mathbf{k}',\beta} \big(\delta_{\alpha\delta}\delta_{\beta\gamma} - \mathrm{e}^{\mathrm{i}(\mathbf{k}-\mathbf{k}')\cdot(\mathbf{x}-\mathbf{x}')} \delta_{\alpha\gamma}\delta_{\beta\delta}\big). \tag{9.64}$$

By using the well known Fierz identity for the Pauli matrices

$$\delta_{\alpha\gamma}\delta_{\beta\delta} = \frac{1}{2}(\delta_{\alpha\delta}\delta_{\beta\gamma} + \boldsymbol{\sigma}_{\alpha\delta} \cdot \boldsymbol{\sigma}_{\beta\gamma}),$$

we may rewrite (9.64) in terms of the invariants under rotations in spin space:

$$h_{2,\alpha\beta\gamma\delta}(\mathbf{x}, \mathbf{x}'; \mathbf{x}', \mathbf{x})$$
$$= \frac{1}{V^2} \sum_{\mathbf{k},\mathbf{k}'} n_{\mathbf{k},\alpha} n_{\mathbf{k}',\beta} \left(\delta_{\alpha\delta}\delta_{\beta\gamma} \left(1 - \frac{1}{2}\mathrm{e}^{\mathrm{i}(\mathbf{k}-\mathbf{k}')\cdot(\mathbf{x}-\mathbf{x}')}\right) - \frac{1}{2}\mathrm{e}^{\mathrm{i}(\mathbf{k}-\mathbf{k}')\cdot(\mathbf{x}-\mathbf{x}')} \boldsymbol{\sigma}_{\alpha\delta} \cdot \boldsymbol{\sigma}_{\beta\gamma} \right). \tag{9.65}$$

For a paramagnetic Fermi gas at $T = 0$ with the density $n = N/V$ given by Eq. (7.25), we get for $\mathbf{x} = \mathbf{x}'$

$$h_{2,\alpha\beta\gamma\delta}(\mathbf{x}, \mathbf{x}; \mathbf{x}, \mathbf{x}) = \frac{n^2}{8}(\delta_{\alpha\delta}\delta_{\beta\gamma} - \boldsymbol{\sigma}_{\alpha\delta} \cdot \boldsymbol{\sigma}_{\beta\gamma}). \tag{9.66}$$

9.8 Hartree–Fock method for a Fermi system

We now consider the Hartree–Fock method, the quantum counterpart of the classical mean-field approximation, for an interacting Fermi system with the Hamiltonian (9.37). In the Hartree–Fock method the ground state energy is approximated by the average of H with the best trial state $|\phi_\mathrm{T}\rangle$ antisymmetrized product of single-particle states $|\kappa\rangle$.

For a Fermi gas at $T = 0$, the average occupation number $\langle a_\kappa^\dagger a_\kappa \rangle$ becomes the step function assuming the values 1 and 0 over the occupied and unoccupied single-particle states, respectively, and the one-particle reduced density matrix satisfies the constraint

$$\hat{h}_1^2 = \hat{h}_1, \quad \text{or} \quad \int d\mathbf{x}'' \, h_1(\mathbf{x}, \mathbf{x}'')h_1(\mathbf{x}'', \mathbf{x}') = h_1(\mathbf{x}, \mathbf{x}'). \tag{9.67}$$

The best trial state will be determined by minimizing $W_0 = \langle \phi_T | H | \phi_T \rangle$ with the constraint (9.67).

The Hartree–Fock approximation can be developed in analogy with the classical mean-field theory, i.e., by factorizing the two-particle correlation function into products of single-particle correlation functions as for the spin correlation function in Eq. (8.52). In the present case we use therefore the relation (9.64) between h_2 and h_1 valid for a gas.

The linearization procedure of the two-body interaction in the Hamiltonian (9.37) is given by

$$\begin{aligned}\hat{h}_{2,\alpha\beta\gamma\delta}(\mathbf{x}, \mathbf{x}'; \mathbf{x}', \mathbf{x}) \to\; & h_{1,\alpha\delta}(\mathbf{x}, \mathbf{x})\hat{h}_{1,\beta\gamma}(\mathbf{x}', \mathbf{x}') + \hat{h}_{1,\alpha\delta}(\mathbf{x}, \mathbf{x})h_{1,\beta\gamma}(\mathbf{x}', \mathbf{x}') \\ & - h_{1,\alpha\gamma}(\mathbf{x}, \mathbf{x}')\hat{h}_{1,\beta\delta}(\mathbf{x}', \mathbf{x}) - \hat{h}_{1,\alpha\gamma}(\mathbf{x}, \mathbf{x}')h_{1,\beta\delta}(\mathbf{x}', \mathbf{x}) \\ & - [h_{1,\alpha\delta}(\mathbf{x}, \mathbf{x})h_{1,\beta\gamma}(\mathbf{x}', \mathbf{x}') - h_{1,\alpha\gamma}(\mathbf{x}, \mathbf{x}')h_{1,\beta\delta}(\mathbf{x}', \mathbf{x})],\end{aligned} \tag{9.68}$$

where the expectation values are taken with respect to the state $|\phi_T\rangle$. The last term in Eq. (9.68) is introduced to restore the validity of (9.64). A generalization of this linearization procedure will be used to present the microscopic BCS theory for superconductivity in Chapter 15.

By adopting the convention according to which a summation is implied when a spinor index appears twice, the linearized Hamiltonian then reads

$$H_{\text{lin}} = \int d\mathbf{x} \int d\mathbf{x}' \left(\hat{\psi}_\alpha^\dagger(\mathbf{x}) \nu_{\alpha\beta}(\mathbf{x}, \mathbf{x}') \hat{\psi}_\beta(\mathbf{x}') - \frac{1}{2} U_{\alpha\beta}(\mathbf{x}, \mathbf{x}') h_{1,\beta\alpha}(\mathbf{x}', \mathbf{x}) \right), \tag{9.69}$$

where we have introduced the single-particle energy

$$\nu_{\alpha\beta}(\mathbf{x}, \mathbf{x}') = \epsilon(\mathbf{x}, \mathbf{x}')\delta_{\alpha\beta} + U_{\alpha\beta}(\mathbf{x}, \mathbf{x}'), \tag{9.70}$$

with the self-consistent potential $U_{\alpha\beta}(\mathbf{x}, \mathbf{x}')$ given by

$$U_{\alpha\beta}(\mathbf{x}, \mathbf{x}') = \delta(\mathbf{x} - \mathbf{x}')\delta_{\alpha\beta} \int d\mathbf{x}'' V(\mathbf{x}, \mathbf{x}'')h_{1,\gamma\gamma}(\mathbf{x}'', \mathbf{x}'') - V(\mathbf{x}, \mathbf{x}')h_{1,\alpha\beta}(\mathbf{x}, \mathbf{x}'). \tag{9.71}$$

The expectation value of H_{lin} is

$$W_0 = \frac{1}{2} \int d\mathbf{x} \int d\mathbf{x}' [\epsilon(\mathbf{x}, \mathbf{x}') + \nu(\mathbf{x}, \mathbf{x}')]_{\alpha\beta} h_{1,\beta\alpha}(\mathbf{x}', \mathbf{x}) \equiv \frac{1}{2} \text{tr} \left[(\epsilon + \nu) h_1 \right], \tag{9.72}$$

where the symbol tr together with the products $(\epsilon + \nu)h_1$ must be understood in a matrix sense with respect to the *indices* $\mathbf{x}, \alpha$ and $\mathbf{x}', \beta$. We have to minimize W_0 with respect to the single-particle density matrix with the constraint (9.67) by introducing a Lagrangian multiplier λ:

$$\delta \left[W_0 + \text{tr}\lambda \left(h_1 - h_1^2 \right) \right] = 0. \tag{9.73}$$

The corresponding equation is

$$\nu = -\lambda + \lambda h_1 + h_1 \lambda. \tag{9.74}$$

If we multiply both sides of this equation by h_1 we have

$$[\nu, h_1] = 0. \tag{9.75}$$

The last equation tells us that the effective Hamiltonian and the one-particle reduced density matrix have to be diagonalized by the same complete set of single-particle states and that at $T = 0$ the matrix element of the single-particle energy between an occupied and an empty state must vanish, which is the standard form of the self-consistent Hartree–Fock equation.

By recalling Eq. (6.23) for the density matrix, we notice that (9.75) is nothing but the time-independent limit of the self-consistent equation for the reduced one-particle density matrix. Hence, the time-dependent self-consistent Hartree–Fock method requires solving the equation

$$\mathrm{i}\hbar \frac{\partial h_1}{\partial t} = [\nu, h_1]. \tag{9.76}$$

For particles in a box and with a translationally invariant interaction, by assuming a paramagnetic ground state, the effective single-particle energy ν is diagonalized in the $\mathbf{k}$-representation with the diagonal value $\nu_{\mathbf{k}}$ given by a Fourier transform as in Eq. (9.41). In the $\mathbf{k}$-representation, by solving Eq. (9.74) with respect to h_1, we get

$$h_{1\mathbf{k},\alpha\beta} = \delta_{\alpha\beta} \frac{1}{2} \left(1 + \frac{\nu_{\mathbf{k}}}{\lambda_{\mathbf{k}}}\right). \tag{9.77}$$

The value of $\lambda_{\mathbf{k}}$ is determined by using the constraint (9.67), which implies $\lambda_{\mathbf{k}} = \pm\nu_{\mathbf{k}}$. The plus (minus) sign applies for $|\mathbf{k}| < (>) k_{\mathrm{F}}$, showing that the Fermi surface in the Hartree–Fock approximation keeps its volume unchanged.

At $T \neq 0$ the same equation (9.72) is valid provided we replace h_1, which is the average of $\hat{h}_1$ over the ground state, with the average over a statistical ensemble defined via the linearized Hamiltonian (9.69). To determine h_1 we must minimize the grand potential $\Omega = W_0 - TS$, where the entropy is given by (see Eq. (7.7))

$$S = -k_{\mathrm{B}} \mathrm{tr}\left[h_1 \ln h_1 + (1 - h_1) \ln(1 - h_1)\right]. \tag{9.78}$$

In the place of (9.74), after making the variation of Ω, we have

$$\nu + k_{\mathrm{B}} T \ln\left(\frac{h_1}{1 - h_1}\right) = 0, \tag{9.79}$$

which implies for h_1 the standard Fermi function form in terms of the dressed $\nu_{\mathbf{k}}$ single-particle energies.

9.9 Problems

9.1 Determine the normalization factor c introduced in Eq. (9.1).

9.2 Prove the commutation relations (9.10).

9.3 Show that the state with $n_{\kappa_i} = n$ and $n_{\kappa_j} = 0$ for all $j \neq i$ is

$$|n\rangle = \frac{(a^\dagger_{\kappa_i})^n}{\sqrt{n!}}|0\rangle.$$

9.4 Prove the anticommutation relations (9.17).

9.5 Derive the relations (9.20).

10 Linear response and the fluctuation–dissipation theorem in quantum systems: equilibrium and small deviations

10.1 Linear response

In most of the experimental setups, the measurement of an observable, $\hat{\mathcal{O}}_A$, implies the observation of how its value changes upon the switching on of an external perturbation. Such a change depends both on the perturbation itself and on the properties of the system to which the observable refers. For sufficiently weak perturbations, the system response will be a linear function of the perturbation strength. In this section we develop a general scheme and show how the linear response of a system to an external perturbation provides a great deal of information on the unperturbed system's spectral and statistical properties, in particular on its fluctuations.

We consider the Hamiltonian made of two parts. The first part, H_0, describes the system in the absence of the perturbation and the second part, H_{ext}, the perturbation due to an external field. Quite generally, the perturbation involves the coupling of an external field, U_B, with an observable, $\hat{\mathcal{O}}_B$, of the system, such as the magnetic field coupled to the magnetization. In general, the observable $\hat{\mathcal{O}}_B$ can be different from the observable $\hat{\mathcal{O}}_A$, which we wish to measure in the presence of the perturbation U_B. The indices A and B specify the observables. We will consider several examples.

The Hamiltonian describing the coupling with the external field then reads

$$H_{\mathrm{ext}}(t) = \int \mathrm{d}\mathbf{x}' U_B(\mathbf{x}', t)\hat{\mathcal{O}}_B(\mathbf{x}'), \tag{10.1}$$

where $U_B(\mathbf{x}', t)$ specifies the external field coupling with the observable $\hat{\mathcal{O}}_B(\mathbf{x}')$. The external field $U_B(\mathbf{x}', t)$ may depend on time, and so does the associated Hamiltonian. Here we consider the quantum case. The classical case, already discussed in Chapter 8 for the Ising model, can be obtained by taking the limit $\hbar \to 0$.

At each time, the average (quantum and statistical) of an observable is expressed in terms of the density matrix as in Eq. (6.13). The density matrix obeys Eq. (6.23), which to linear order in the perturbation reads

$$i\hbar\partial_t\hat{\rho}^{(1)}(t) = \left[H_0, \hat{\rho}^{(1)}(t)\right] + \left[H_{\mathrm{ext}}(t), \hat{\rho}^{(0)}\right], \tag{10.2}$$

where $\hat{\rho}^{(0)}$ is the density matrix in the absence of the perturbation. The solution of Eq. (10.2), with the boundary condition $H_{\text{ext}} = 0$ for $t \leq t_0$, reads

$$\hat{\rho}^{(1)}(t) = -\frac{\mathrm{i}}{\hbar} \int_{t_0}^{t} \mathrm{d}t' \, \mathrm{e}^{-\mathrm{i}H_0(t-t')/\hbar} \left[H_{\text{ext}}(t'), \hat{\rho}^{(0)} \right] \mathrm{e}^{\mathrm{i}H_0(t-t')/\hbar}, \tag{10.3}$$

which can be verified by differentiation. To linear order in the perturbation the average of the observable, at time t, is

$$\delta\mathcal{O}_A(\mathbf{x}, t) \equiv \langle \hat{\mathcal{O}}_A(\mathbf{x}) \rangle_t^{(1)} = \mathrm{Tr}\big(\hat{\rho}^{(1)}(t)\hat{\mathcal{O}}_A(\mathbf{x})\big), \tag{10.4}$$

which becomes, after using the expression (10.3) for $\hat{\rho}^{(1)}(t)$,

$$\delta\mathcal{O}_A(\mathbf{x}, t) = -\frac{\mathrm{i}}{\hbar} \int_{t_0}^{t} \mathrm{d}t' \int \mathrm{d}\mathbf{x}' \mathrm{Tr}\big(\hat{\rho}^{(0)}[\hat{\mathcal{O}}_A(\mathbf{x}, t), \hat{\mathcal{O}}_B(\mathbf{x}', t')]\big) U_B(\mathbf{x}', t'), \tag{10.5}$$

where we have introduced the Heisenberg representation of the operators with respect to the unperturbed Hamiltonian H_0:

$$\hat{\mathcal{O}}_A(\mathbf{x}, t) = \mathrm{e}^{\mathrm{i}H_0 t/\hbar} \hat{\mathcal{O}}_A(\mathbf{x}) \mathrm{e}^{-\mathrm{i}H_0 t/\hbar}, \quad \hat{\mathcal{O}}_B(\mathbf{x}', t') = \mathrm{e}^{\mathrm{i}H_0 t'/\hbar} \hat{\mathcal{O}}_B(\mathbf{x}') \mathrm{e}^{-\mathrm{i}H_0 t'/\hbar}. \tag{10.6}$$

In deriving Eq. (10.5) we have used the fact that at equilibrium $\hat{\rho}^{(0)}$ can be replaced by its time-independent ensemble form. The variation of the average value of the observable $\hat{\mathcal{O}}_A(\mathbf{x})$ at time t, Eq. (10.5), can be written as

$$\delta\mathcal{O}_A(\mathbf{x}, t) = \int_{-\infty}^{\infty} \mathrm{d}t' \int \mathrm{d}\mathbf{x}' R^{AB}(\mathbf{x}, \mathbf{x}'; t - t') U_B(\mathbf{x}', t'), \tag{10.7}$$

where, according to Eq. (10.5), we have introduced the response function $R^{AB}(\mathbf{x}, \mathbf{x}'; t)$ defined as

$$\begin{aligned} R^{AB}(\mathbf{x}, \mathbf{x}'; t) &= -\frac{\mathrm{i}}{\hbar} \theta(t) \mathrm{Tr}\big(\hat{\rho}^{(0)}[\hat{\mathcal{O}}_A(\mathbf{x}, t), \hat{\mathcal{O}}_B(\mathbf{x}', 0)]\big) \\ &= -\frac{\mathrm{i}}{\hbar} \theta(t) \mathrm{Tr}\left(\frac{\mathrm{e}^{-\beta H_0}}{Z} [\hat{\mathcal{O}}_A(\mathbf{x}, t), \hat{\mathcal{O}}_B(\mathbf{x}', 0)] \right), \end{aligned} \tag{10.8}$$

which depends on the properties of the unperturbed system only. In the above $\theta(t)$ is the step function and we use the equilibrium density matrix of the canonical ensemble $\hat{\rho}^{(0)} \Rightarrow \mathrm{e}^{-\beta H_0}/Z = \mathrm{e}^{\beta(F - H_0)}$ for the unperturbed system. Notice that, since H_0 is time-independent, the response function depends only on the time difference of the arguments of the two Heisenberg operators (10.6). $R^{AB}(\mathbf{x}, \mathbf{x}'; t)$ must be a bounded function of t for the system to react smoothly to any external perturbation. The physical interpretation of (10.7) is that the variation of an observable at a given space–time point is the integral over all space–time points $(\mathbf{x}', t')$ of the perturbation of an external field transmitted via the response function connecting the pairs of space–time points $(\mathbf{x}, t)$ and $(\mathbf{x}', t')$.

We introduce the Fourier transform, with respect to both space and time, of the response function $\tilde{R}^{AB}(\mathbf{q}, \omega)$ defined by

$$R^{AB}(\mathbf{x}, \mathbf{x}'; t - t') = \int_{-\infty}^{\infty} \frac{\mathrm{d}\omega}{2\pi} \sum_{\mathbf{q}, \mathbf{q}'} \tilde{R}^{AB}_{\mathbf{q}, \mathbf{q}'}(\omega) \, \mathrm{e}^{-\mathrm{i}\omega(t - t')} \, \mathrm{e}^{\mathrm{i}(\mathbf{q}\cdot\mathbf{x} + \mathbf{q}'\cdot\mathbf{x}')}. \tag{10.9}$$

If translational invariance in space is present in the unperturbed system, the response function depends only on the difference of the space arguments, i.e. $R^{AB}(\mathbf{x},\mathbf{x}';t-t') = R^{AB}(\mathbf{x}-\mathbf{x}',t-t')$ and $\tilde{R}^{AB}_{\mathbf{q},\mathbf{q}'}(\omega) = \tilde{R}^{AB}_{\mathbf{q}}(\omega)\delta_{\mathbf{q}+\mathbf{q}',0}$, where the response in momentum space can be written as

$$\tilde{R}^{AB}_{\mathbf{q}}(\omega) = -\int_{-\infty}^{\infty} \mathrm{d}t\, \mathrm{e}^{\mathrm{i}\omega t} \frac{\mathrm{i}}{\hbar}\theta(t)\mathrm{Tr}\big(\hat{\rho}^{(0)}[\hat{\mathcal{O}}_A(\mathbf{q},t), \hat{\mathcal{O}}_B(-\mathbf{q},0)]\big), \tag{10.10}$$

since only the components $\mathbf{q}$ and $-\mathbf{q}$ of the operators $\hat{\mathcal{O}}_A(\mathbf{x})$ $\hat{\mathcal{O}}_B(\mathbf{x}')$ are connected in the response function.

By Fourier transforming Eq. (10.7) we obtain

$$\delta\mathcal{O}_A(\mathbf{q},\omega) = \tilde{R}^{AB}_{\mathbf{q}}(\omega)\tilde{U}_B(\mathbf{q},\omega), \tag{10.11}$$

where $\tilde{U}_B(\mathbf{q},\omega)$ is the Fourier transform of the external potential.

For a static perturbation $U_B(\mathbf{x}',t') = U_B(\mathbf{x}')$, Eq. (10.7) contains only the $\omega =$ 0-component of the response function and, in momentum space, reads

$$\delta\mathcal{O}_A(\mathbf{q},0) = \tilde{R}^{AB}_{\mathbf{q}}(0)\tilde{U}_B(\mathbf{q}). \tag{10.12}$$

Morover, if the perturbing field is also uniform, $U_B(\mathbf{x}') = U_B$, the static response corresponds to the generalized susceptibility and is given by taking also the zero-momentum limit

$$\chi_{AB} = \frac{\partial\mathcal{O}_A}{\partial U_B} = \lim_{\mathbf{q}\to 0}\lim_{\omega\to 0} \tilde{R}^{AB}_{\mathbf{q}}(\omega), \tag{10.13}$$

where the order of limits must be maintained.

We now show a few remarkable properties of the response function. By performing the trace over a set of eigenstates $|m\rangle$ of H_0 and by introducing the resolution of the identity $1 = \sum_l |l\rangle\langle l|$ between the two observables, we get

$$\begin{aligned} R^{AB}_{\mathbf{q}}(t) &= -\frac{\mathrm{i}}{\hbar}\theta(t)\sum_{l,m} \mathrm{e}^{\beta F}\mathrm{e}^{-\beta E_m}\left(A_{ml}(\mathbf{q})B^*_{ml}(-\mathbf{q})\mathrm{e}^{-\mathrm{i}\omega_{lm}t} - A^*_{ml}(\mathbf{q})B_{ml}(-\mathbf{q})\mathrm{e}^{\mathrm{i}\omega_{lm}t}\right) \\ &= -\frac{\mathrm{i}}{\hbar}\theta(t)\sum_{l,m} \mathrm{e}^{\beta F}\mathrm{e}^{-\beta E_m} A_{ml}(\mathbf{q})B^*_{ml}(-\mathbf{q})(1-\mathrm{e}^{-\beta\hbar\omega_{lm}})\mathrm{e}^{-\mathrm{i}\omega_{lm}t}, \end{aligned} \tag{10.14}$$

where $\omega_{lm} = (E_l - E_m)/\hbar$, E_l being the eigenvalues of H_0 in the set of eigenstates $|l\rangle$ and

$$A_{ml}(\mathbf{q}) = \langle m|\hat{\mathcal{O}}_A(\mathbf{q})|l\rangle, \quad B^*_{ml}(-\mathbf{q}) = \langle l|\hat{\mathcal{O}}_B(-\mathbf{q})|m\rangle. \tag{10.15}$$

In the second line of Eq. (10.14), we have interchanged the indices m and l and used the fact that $A^*_{lm}(\mathbf{q})B_{lm}(-\mathbf{q}) = A_{ml}(\mathbf{q})B^*_{ml}(-\mathbf{q})$ since the observables are Hermitian operators and we have assumed space inversion symmetry.

By means of the formula

$$-\mathrm{i}\int_0^{\infty} \mathrm{d}t\, \mathrm{e}^{\mathrm{i}(\omega+\mathrm{i}\eta-\omega_{lm})t} = \frac{1}{\omega+\mathrm{i}\eta-\omega_{lm}}, \quad \eta = 0^+, \tag{10.16}$$

we obtain the Fourier transform with respect to time of the response function as

$$\tilde{R}^{AB}_{\mathbf{q}}(\omega) = \sum_{l,m} e^{\beta F} e^{-\beta E_m} A_{ml}(\mathbf{q}) B^*_{ml}(-\mathbf{q}) \frac{1 - e^{-\beta\hbar\omega_{lm}}}{\hbar\omega - \hbar\omega_{lm} + i\eta}. \tag{10.17}$$

This shows immediately that $\tilde{R}^{AB}_{\mathbf{q}}(\omega)$ is an analytical function in the upper complex plane of the variable ω and obeys the Kramers–Kronig relation

$$\tilde{R}^{AB}_{\mathbf{q}}(\omega) = \frac{1}{\pi}\int_{-\infty}^{\infty} d\omega' \, \frac{\mathrm{Im}\tilde{R}^{AB}_{\mathbf{q}}(\omega')}{\omega' - \omega - i\eta}, \tag{10.18}$$

where by exploiting the property

$$\frac{1}{x - i\eta} = \mathcal{P}\left(\frac{1}{x}\right) + i\pi\delta(x)$$

the imaginary part is

$$\mathrm{Im}\tilde{R}^{AB}_{\mathbf{q}}(\omega) = -\frac{\pi}{\hbar}(1 - e^{-\beta\hbar\omega}) \sum_{l,m} e^{\beta F} e^{-\beta E_m} A_{lm}(\mathbf{q}) B^*_{ml}(-\mathbf{q}) \delta(\omega - \omega_{lm}). \tag{10.19}$$

From the expression in Eq. (10.19) and taking into account (10.15), we deduce that

$$\mathrm{Im}\tilde{R}^{AB}_{-\mathbf{q}}(-\omega) = -\mathrm{Im}\tilde{R}^{AB}_{\mathbf{q}}(\omega). \tag{10.20}$$

A common example is, for instance, given by considering the density operator (9.29) as the observable, $\hat{\mathcal{O}}_A = \hat{n}(\mathbf{x})$, and the perturbation due to a scalar field $U(\mathbf{x}', t)$, which also couples to the density, i.e. $\hat{\mathcal{O}}_B = \hat{n}(\mathbf{x}')$,

$$H_{\mathrm{ext}} = \int d\mathbf{x}' \, U(\mathbf{x}', t)\hat{n}(\mathbf{x}'). \tag{10.21}$$

If the external field is static and homogeneous, $U(\mathbf{x}, t) = U = -\delta\mu$, it is the chemical potential μ that couples to the density and from Eq. (10.13), the linear response reads

$$\chi_{nn} = \frac{\partial n}{\partial U} = \lim_{\mathbf{q}\to 0}\lim_{\omega\to 0} \tilde{R}^{nn}_{\mathbf{q}}(\omega) = -\frac{\partial n}{\partial \mu}, \tag{10.22}$$

which in this case corresponds to the thermodynamic density of states, related to the isothermal compressibility via Eq. (4.35).

In general the response to a static and homogeneous field corresponds to the susceptibility of the system, e.g. the spin susceptibility for a magnetic field or the specific heat for a temperature variation.

The linear response of a generic observable $\hat{\mathcal{O}}_A$ to an external perturbation coupled to an observable $\hat{\mathcal{O}}_B$ has been expressed in terms of the response function, R^{AB}, which is a property of the system at equilibrium in the absence of the external perturbation. The correlation function of the same pair of observables, f^{AB}, to be defined in Section 10.3, is also a property of the system at equilibrium. It is natural to ask whether the response and the correlation functions are related. This is the content of the so-called fluctuation–dissipation theorem, which we will derive in Section 10.3. We first consider explicitly the important example of a response to an electromagnetic field.

10.2 The response to an electromagnetic field and gauge invariance

The linear response to an external electromagnetic field is one of the most important applications of the concepts developed in the previous paragraph. In fact, many experimental investigation techniques are based on how the physical system responds to an electromagnetic field.

To derive the interaction Hamiltonian, we begin by noticing that the interaction with a scalar electric potential $\phi(\mathbf{x}, t)$ has the form (10.21)

$$H_\phi = \int d\mathbf{x}\phi(\mathbf{x}, t)\hat{\rho}(\mathbf{x}), \tag{10.23}$$

where the charge density operator is defined in terms of the density operator $\hat{\rho}(\mathbf{x}) = -e\hat{n}(\mathbf{x})$ for particles with negative charge, having in mind electrons in a metal. We take the unit charge, $e > 0$, positive throughout the book.

The interaction with a vector potential is given by the minimal coupling (see Problem 10.1)

$$H_\mathbf{A} = -\frac{1}{c}\int d\mathbf{x}\mathbf{A}(\mathbf{x}, t) \cdot \hat{\mathbf{J}}(\mathbf{x}), \tag{10.24}$$

where c is the speed of light and the charge current density operator reads

$$\hat{\mathbf{J}}(\mathbf{x}) = -\frac{e}{m}\sum_{i=1}^{N} \delta(\mathbf{x} - \mathbf{x}_i)\hat{\mathbf{p}}_i. \tag{10.25}$$

This last result could have also been obtained by requiring relativistic Lorentz covariance. Indeed, we may combine both terms (10.23) and (10.24) and describe the coupling with an external electromagnetic field with the Hamiltonian[1]

$$H_{\phi-\mathbf{A}} = \frac{1}{c}\int d\mathbf{x}A^\mu(\mathbf{x})J_\mu(\mathbf{x}) \tag{10.26}$$

where the Greek superscript runs over time ($\mu = 0$) and space indices ($\mu = 1, ..., d$) and we adopt the standard convention that when an upper and lower index are equal, a sum over their values is understood. Space indices will be later on indicated by Latin letters. As is standard, lower indices have space components with a minus sign, e.g., $J_\mu = (c\hat{\rho}, -\hat{\mathbf{J}})$. Our goal is to study the system response to an external electromagnetic field within the linear response. By the compact notation $x = (t, \mathbf{x})$, the linear response is given by

$$J^\mu(x) = \frac{1}{c}\int dx' \; R^{\mu\nu}(x, x')A_\nu(x'), \tag{10.27}$$

where the response kernel $R^{\mu\nu}(x, x')$,

$$R^{\mu\nu}(x, x') = -\frac{i}{\hbar}\theta(t - t')\langle[J^\mu(x), J^\nu(x')]\rangle, \tag{10.28}$$

[1] We adopt the relativistic notation: upper and lower indices indicate contravariant and covariant vectors, respectively, i.e. $A^\mu = (\phi, \mathbf{A})$ and $A_\mu = (\phi, -\mathbf{A})$.

is a four-current response function such as that defined in Eq. (10.8) and includes both the density–density and current–current response functions. If the unperturbed system is translationally invariant and has a time-independent Hamiltonian, we can use Fourier transforms with respect to both $\mathbf{x} - \mathbf{x}'$ and $t - t'$:

$$R^{\mu\nu}(\mathbf{x} - \mathbf{x}', t - t') = \int_{-\infty}^{\infty} \frac{d\omega}{2\pi} \sum_{\mathbf{q}} e^{i\mathbf{q}\cdot(\mathbf{x}-\mathbf{x}')-i\omega(t-t')} \tilde{R}^{\mu\nu}_{\mathbf{q}}(\omega). \tag{10.29}$$

The sum over the momenta is left unspecified for the time being. It depends on the choice of boundary conditions. In the limit of an infinite system, the sum gets replaced by an integral over all space in the standard way (see Eq. (7.10)). In Fourier space, Eq. (10.27) becomes local:

$$J^{\mu}(q) = \frac{1}{c} \tilde{R}^{\mu\nu}(q) A_{\nu}(q), \tag{10.30}$$

where $q = (\omega, \mathbf{q})$. Physical observables are now readily obtained. For instance, the DC electrical conductivity, by making the choice of a time-dependent vector gauge, $\mathbf{E} = -\partial_t \mathbf{A}(t)/c$, reads

$$\sigma_{ij} = -\lim_{\omega \to 0} \lim_{|\mathbf{q}| \to 0} \frac{\tilde{R}^{ij}_{\mathbf{q}}(\omega)}{i\omega}. \tag{10.31}$$

This is the famous Kubo formula for the electrical conductivity. Notice the order of limits in Eq. (10.31). This corresponds to considering a uniform electric field, infinitely slowly varying in time. Of course the result cannot depend on the gauge choice. In fact gauge invariance poses some constraints on the elements of the response function tensor $\tilde{R}^{\mu\nu}_{\mathbf{q}}(\omega)$.

Charge conservation is expressed by the continuity equation

$$\partial_t \rho + \nabla \cdot \mathbf{J} = 0, \tag{10.32}$$

while gauge invariance requires that the physics, i.e. the electric field, the magnetic field, and the response, are unchanged by the replacement

$$A^{\mu}(x) \to A^{\mu}(x) - \partial^{\mu} f(x) \tag{10.33}$$

with f an arbitrary function and $\partial^{\mu} = (c^{-1}\partial_t, -\nabla)$. Equations (10.32) and (10.33) imply

$$q_{\mu} \tilde{R}^{\mu\nu} = 0, \tag{10.34}$$

$$\tilde{R}^{\mu\nu} q_{\nu} = 0. \tag{10.35}$$

More explicitly, one has the following relations connecting the various response functions:

$$\omega \tilde{R}^{00} = q^j \tilde{R}^{j0}, \tag{10.36}$$

$$\omega \tilde{R}^{0i} = q^j \tilde{R}^{ji}, \tag{10.37}$$

$$\omega \tilde{R}^{00} = q^j \tilde{R}^{0j}, \tag{10.38}$$

$$\omega \tilde{R}^{i0} = q^j \tilde{R}^{ij}, \tag{10.39}$$

from which $\tilde{R}^{0i} = \tilde{R}^{i0}$, $\tilde{R}^{ij} = \tilde{R}^{ji}$ and

$$\omega^2 \tilde{R}^{00} = q^i \tilde{R}^{ij} q^j. \tag{10.40}$$

The tensor structure of the spatial part of the response function $\tilde{R}^{ij}_{\mathbf{q}}(\omega)$ depends only on the momentum $\mathbf{q}$. We may introduce longitudinal and transverse components of $\tilde{R}^{ij}(\mathbf{q},\omega)$ as

$$\tilde{R}^{ij}_{\mathbf{q}}(\omega) \equiv \tilde{R}_{\mathrm{L}}(|\mathbf{q}|,\omega)\hat{q}^i\hat{q}^j + \tilde{R}_{\mathrm{T}}(|\mathbf{q}|,\omega)(\delta_{ij} - \hat{q}^i\hat{q}^j), \tag{10.41}$$

where $\hat{q}^i = \mathbf{q}^i/|\mathbf{q}|$. Hence Eq. (10.40) can be expressed as

$$\tilde{R}_{\mathrm{L}}(|\mathbf{q}|,\omega) = \frac{\omega^2}{|\mathbf{q}|^2}\tilde{R}^{00}_{\mathbf{q}}(\omega). \tag{10.42}$$

In an isotropic system, the tensor response function must be diagonal, $\tilde{R}^{ij}_{\mathbf{q}}(\omega) = \tilde{R}(\mathbf{q},\omega)\delta_{ij} = \tilde{R}_{\mathrm{L}}(\mathbf{q},\omega)\delta_{ij}$. As a result of this, taking into account the definition (10.31) and the relation (10.42), we can express the electrical conductivity as the dynamic limit, i.e. sending the momentum $\mathbf{q}$ to zero first and then the frequency, of the density–density response function

$$\sigma = \mathrm{i} \lim_{\omega\to 0}\lim_{|\mathbf{q}|\to 0} \frac{\omega}{|\mathbf{q}|^2}\tilde{R}^{00}_{\mathbf{q}}(\omega). \tag{10.43}$$

By recalling Eq. (10.22), the thermodynamic density of states is given instead by the static limit of the same density–density response function

$$\left(\frac{\partial n}{\partial \mu}\right) = -\frac{1}{e^2}\lim_{|\mathbf{q}|\to 0}\lim_{\omega\to 0}\tilde{R}^{00}_{\mathbf{q}}(\omega). \tag{10.44}$$

The factor $1/e^2$ appears since we are considering here charge density instead of the number density and the minus sign is due to the fact that a shift in the chemical potential is equivalent to negative shift in the potential.

10.3 The principle of detailed balance and the fluctuation–dissipation theorem

Preliminarily, we introduce the correlation function at equilibrium for two observables $\hat{O}_A$ and $\hat{O}_B$ which for non-commuting operators must be defined in terms of the anticommutator. By assuming translational invariance, we define the correlation function at a given momentum $\mathbf{q}$ as

$$f^{AB}_{\mathbf{q}}(t) = \mathrm{Tr}\left(\frac{\mathrm{e}^{-\beta H_0}}{Z}\frac{1}{2}\{\hat{O}_A(\mathbf{q},t),\hat{O}_B(-\mathbf{q},0)\}\right), \tag{10.45}$$

where the curly brackets stand for the anticommutator. We will understand in the following that the average value has been subtracted from each single operator. The above definition of the correlation function for commuting operators reduces to the classical form introduced in Section 8.4.

By proceeding as in the case of the response function by performing the trace over a set of eigenstates, we get

$$\begin{aligned} f_{\mathbf{q}}^{AB}(t) &= \frac{1}{2}\sum_{mn} \mathrm{e}^{\beta F}\mathrm{e}^{-\beta E_m}\left(A_{mn}(\mathbf{q})B_{mn}^*(-\mathbf{q})\mathrm{e}^{-\mathrm{i}\omega_{nm}t} + A_{mn}^*(\mathbf{q})B_{mn}(-\mathbf{q})\mathrm{e}^{\mathrm{i}\omega_{nm}t}\right) \\ &= \frac{1}{2}\sum_{mn} \mathrm{e}^{\beta F}\mathrm{e}^{-\beta E_m} A_{mn}(\mathbf{q})B_{mn}^*(-\mathbf{q})\mathrm{e}^{-\mathrm{i}\omega_{nm}t}(1+\mathrm{e}^{\beta\hbar\omega_{nm}}). \end{aligned} \tag{10.46}$$

We notice that, in Eq. (10.46), besides the sign change between the two terms in comparison with the corresponding expression (10.14) for the response function, here also the function $\theta(t)$ is missing: the correlation extends to all times, whereas the response starts at the perturbing time. This difference reflects itself in the resulting expression of the Fourier transform with respect to time:

$$f_{\mathbf{q}}^{AB}(t) = \int_{-\infty}^{\infty} \frac{\mathrm{d}\omega}{2\pi}\, \tilde{f}_{\mathbf{q}}^{AB}(\omega)\, \mathrm{e}^{-\mathrm{i}\omega t}. \tag{10.47}$$

Using Eq. (10.46), the Fourier transform $\tilde{f}_{\mathbf{q}}^{AB}(\omega)$, also known as the power spectrum of fluctuations, reads

$$\tilde{f}_{\mathbf{q}}^{AB}(\omega) = (1+\mathrm{e}^{-\beta\hbar\omega})\pi \sum_{mn} \mathrm{e}^{\beta F}\mathrm{e}^{-\beta E_m} A_{mn}(\mathbf{q})B_{mn}^*(-\mathbf{q})\delta(\omega-\omega_{nm}). \tag{10.48}$$

In general we may decompose the anticommutator into two terms

$$\begin{aligned} \tilde{f}_{\mathbf{q}}^{AB}(\omega) &= \mathrm{Tr}\left(\frac{\mathrm{e}^{-\beta H_0}}{Z}\frac{1}{2}\{\hat{O}_A(\mathbf{q},\omega), \hat{O}_B(-\mathbf{q},-\omega)\}\right) \\ &\equiv \frac{1}{2}\left(\bar{f}_{\mathbf{q}}^{AB}(\omega) + \bar{f}_{-\mathbf{q}}^{BA}(-\omega)\right). \end{aligned} \tag{10.49}$$

By taking into account Eq. (10.48) and the matrix elements (10.15), we get

$$\bar{f}_{-\mathbf{q}}^{BA}(-\omega) = \mathrm{e}^{-\beta\hbar\omega}\, \bar{f}_{\mathbf{q}}^{AB}(\omega). \tag{10.50}$$

This relation expresses the ratio between the probabilities for emitting and absorbing energy ω with a fluctuation of momentum $\mathbf{q}$. The two processes are the time reversals of one another and the relation (10.50) is also known as the detailed balance, stemming from the property of microscopic reversibility.[2] Both emission and absorption processes are described as fluctuations with respect to the same equilibrium state at temperature T. We will see in Section 10.5 that such a detailed balance relation can be generalized to situations involving non-equilibrium transitions between different equilibrium states.

For $\hat{O}^A = \hat{O}^B$, $\tilde{f}_{\mathbf{q}}^{AA}(\omega)$ is nothing but the quantum expression for the mean square of the dynamical fluctuations of $\hat{O}_A$ for each wave vector $\mathbf{q}$.

[2] We note that in Chapter 2 we encountered a different form of the detailed balance, when relating the scattering probability for the two-particle collisions.

Direct comparison of (10.48) with (10.19) yields the fluctuation–dissipation theorem (Callen and Welton (1951)):

$$\tilde{f}_{\mathbf{q}}^{AB}(\omega) = -\hbar \coth\left(\frac{\beta\hbar\omega}{2}\right) \mathrm{Im}\tilde{R}_{\mathbf{q}}^{AB}(\omega). \tag{10.51}$$

The fluctuation–dissipation theorem is valid in the same form also in the $\mathbf{x}$, $\mathbf{x}'$ representation and connects $f(\mathbf{x}, \mathbf{x}'; \omega)$ to $\mathrm{Im}R(\mathbf{x}, \mathbf{x}'; \omega)$.

By antifourier transforming Eq. (10.51) one obtains

$$f_{\mathbf{q}}^{AA}(t) = -\hbar \int_{-\infty}^{\infty} \frac{d\omega}{2\pi} e^{-i\omega t} \coth\left(\frac{\beta\hbar\omega}{2}\right) \mathrm{Im}R_{\mathbf{q}}^{AA}(\omega). \tag{10.52}$$

At equal times, $t = 0$, for $\hat{O}^A = \hat{O}^B$, we have

$$f_{\mathbf{q}}^{AA}(0) = -\hbar \int_{-\infty}^{\infty} \frac{d\omega}{2\pi} \coth\left(\frac{\beta\hbar\omega}{2}\right) \mathrm{Im}R_{\mathbf{q}}^{AA}(\omega). \tag{10.53}$$

Equation (10.53) establishes a connection between the mean square fluctuations of $\hat{O}^A$ at each wave vector $\mathbf{q}$ and the integral over the frequency of the imaginary part of the corresponding response function weighted by the $\coth(\beta\hbar\omega/2)$.

In the limit $\hbar \to 0$ or at high temperature and low frequencies, we obtain the classical limit of the fluctuation–dissipation theorem:

$$\tilde{f}_{\mathbf{q}}^{AB}(\omega) = -\frac{2k_B T}{\omega} \mathrm{Im}\tilde{R}_{\mathbf{q}}^{AB}(\omega). \tag{10.54}$$

Finally, by integrating the above relation over the frequency, taking into account (10.18) and (10.47), we obtain

$$f_{\mathbf{q}}^{AB}(0) = -k_B T \tilde{R}_{\mathbf{q}}^{AB}(0). \tag{10.55}$$

This last equation is the classical static version of Eq. (10.53). For $\hat{O}^A = \hat{O}^B$, Eq. (10.55) establishes a direct relation between the amplitude of the fluctuations of $\hat{O}^A$ at a given wave vector $\mathbf{q}$ and the zero-frequency response $\tilde{R}_{\mathbf{q}}^{AA}(0)$, which represents the response to an external static field coupled to $\hat{O}^A$. Equation (10.55) for the spin σ ($\hat{O}^A = \hat{O}^B = \sigma$) coincides with Eq. (8.89) for the spin susceptibility.

In the example of the density operator $\hat{O}_A = \hat{O}_B = \hat{n}$, the $\mathbf{q} = 0$-limit of Eq. (10.55), considering (10.22), reproduces the expression for the mean square of the density fluctuations given by Eqs. (1.59) and (4.39).

The power spectrum of the density fluctuations (10.48) satisfies a useful sum rule, the f-sum rule, which applies to a many-body system of N identical particles of mass m interacting via a momentum independent potential.

We consider the power spectrum defined by Eqs. (10.49) and (10.48):

$$\bar{f}_{\mathbf{q}}^{nn}(\omega) = \pi \sum_{mn} \frac{e^{-\beta E_m}}{Z} \langle m|\hat{n}(\mathbf{q})|n\rangle \langle n|\hat{n}(-\mathbf{q})|m\rangle \delta(\omega - \omega_{nm}), \tag{10.56}$$

where the density operator in $\mathbf{q}$-space is $\hat{n}(\mathbf{q}) = \sum_{i=1}^{N} e^{i\mathbf{q}\cdot\mathbf{x}_i}$ in first quantized form and $\hat{n}(\mathbf{q}) = \sum_{\mathbf{k}} a_{\mathbf{k}}^{\dagger} a_{\mathbf{k}+\mathbf{q}}$ in second quantized form.

A simple lengthy calculation, by inserting the resolution of the identy $\sum_n |n\rangle\langle n| = 1$, shows that the average value of the double commutator reads

$$\langle[\hat{n}(-\mathbf{q}, t), [H, \hat{n}(\mathbf{q}, t)]]\rangle = 2\sum_{mn} \frac{e^{-\beta E_m}}{Z} |\langle m|\hat{n}(\mathbf{q})|n\rangle|^2 \hbar\omega_{nm}, \tag{10.57}$$

where the time-dependent density operator is $\hat{n}(\mathbf{q}, t) = e^{iHt/\hbar}\hat{n}(\mathbf{q})e^{-iHt/\hbar}$. Notice that we have again assumed space inversion symmetry.

The time evolution of the operators in the double commutator can be taken into account by the time evolution of the double commutator. The latter drops out when taking the trace in the quantum statistical average over an equilibrium ensemble as for any time-dependent operator.

For the system under consideration, the double commutator can be directly computed either in first or in second quantization (see Problem 10.2):

$$[\hat{n}(-\mathbf{q}), [H, \hat{n}(\mathbf{q})]] = \frac{\hbar^2 q^2}{m}\hat{N}, \tag{10.58}$$

where $\hat{N}$ is the total number operator.

Finally, by combining Eq. (10.58) with Eqs. (10.57) and (10.56) we get the f-sum rule

$$\int_{-\infty}^{\infty} \frac{d\omega}{\pi}\, \hbar\omega\, \bar{f}_{\mathbf{q}}^{nn}(\omega) = \frac{\hbar^2 q^2}{2m} N. \tag{10.59}$$

10.4 Onsager's symmetry relations

The principle of microscopic reversibility implies that the correlation function must be invariant under the time reversal operation, provided that both the Hamiltonian and the observables $\hat{O}_A(\mathbf{x})$ and $\hat{O}_B(\mathbf{x}')$ do not change sign by sending $t \to -t$. In Fourier frequency space, the time inversion corresponds to the sign change $\omega \to -\omega$. The microscopic reversibility implies for the correlation function (10.49)

$$\tilde{f}^{BA}(\mathbf{x}', \mathbf{x}; \omega) = \tilde{f}^{AB}(\mathbf{x}, \mathbf{x}'; -\omega) = \tilde{f}^{AB}(\mathbf{x}, \mathbf{x}'; \omega), \tag{10.60}$$

where the first equality stems from the definition (i.e. the interchange $\mathbf{x} \to \mathbf{x}'$ and $A \to B$ corresponds to a change of sign in the frequency) and the second corresponds to invariance under time reversal.

Via the fluctuation–dissipation relation this implies a symmetry for the response function as well:

$$\mathrm{Im}\tilde{R}^{AB}(\mathbf{x}, \mathbf{x}'; \omega) = \mathrm{Im}\tilde{R}^{BA}(\mathbf{x}', \mathbf{x}; \omega). \tag{10.61}$$

Owing to the Kramers–Kronig relation of Eq. (10.18), we obtain in general

$$\tilde{R}^{AB}(\mathbf{x}, \mathbf{x}'; \omega) = \tilde{R}^{BA}(\mathbf{x}', \mathbf{x}; \omega), \tag{10.62}$$

which are known as Onsager's relations (Onsager (1931a, b)).

In the presence of a time-reversal breaking term in the Hamiltonian, as in the case of an external magnetic field, the above relation must be modified, since the Hamiltonian remains invariant under time reversal if simultaneously we reverse the sign of the magnetic field. The relation (10.60) then becomes

$$\tilde{f}^{BA}(\mathbf{x}', \mathbf{x}; \omega, \mathcal{H}) = \tilde{f}^{AB}(\mathbf{x}, \mathbf{x}'; \omega, -\mathcal{H}). \tag{10.63}$$

The relations (10.63), in turn, imply for the response functions the symmetry relations

$$\tilde{R}^{BA}(\mathbf{x}', \mathbf{x}; \omega, \mathcal{H}) = \tilde{R}^{AB}(\mathbf{x}, \mathbf{x}'; \omega, -\mathcal{H}). \tag{10.64}$$

Finally, we must envisage the case corresponding to observables which are not time reversal invariant. Let us indicate with $\varepsilon_A = \pm 1$ the parity of the observable $\mathcal{O}_A$ under time reversal. The general Onsager's relations become

$$\tilde{R}^{BA}(\mathbf{x}', \mathbf{x}; \omega, \mathcal{H}) = \varepsilon_A \varepsilon_B \tilde{R}^{AB}(\mathbf{x}, \mathbf{x}'; \omega, -\mathcal{H}). \tag{10.65}$$

10.5 Generalized fluctuations theorem and minimum availability function

As anticipated, the detailed balance relation (10.50) connecting time reversed fluctuations processes can be generalized to situations in which the system, starting from thermal equilibrium at some given temperature T, performs a non-equilibrium transformation under the action of an external time-dependent field. To follow the convention established in the literature, we denote by λ_t the parameter which specifies at a given instant of time, t, the action of the external field and with $H(\lambda_t)$ the corresponding perturbed Hamiltonian.[3] λ is usually coupled linearly to an observable $\hat{\mathcal{O}}$ of the system and the total time-dependent Hamiltonian is

$$H(\lambda_t) = H_0 - \lambda_t \hat{\mathcal{O}}, \tag{10.66}$$

which for simplicity is assumed to be invariant under time reversal. The function λ is referred to as the force protocol applied to the system and it is understood that it lasts for a finite amount of time τ. At the end of the process the system may be out of equilibrium. $H(\lambda_0)$ and $H(\lambda_\tau)$ are the Hamiltonian at the beginning and at the end of the driving process. The limit of infinite τ, with low variation in time of the function λ, corresponds to an adiabatic variation of the Hamiltonian parameters and the transformation becomes a quasi-equilibrium one, during which, at each instant of time, the system is in equilibrium determined by the instantaneous Hamiltonian. The opposite limit with vanishing τ corresponds to a sudden variation of the Hamiltonian parameters.

For non-autonomous systems, i.e. for systems driven by a time-dependent Hamiltonian as specified above, there exists a microreversibility principle for the evolution of states as

[3] For a recent review on this topics see Esposito et al. (2009) and Campisi et al. (2011). The symbol λ without the suffix indicates the entire function and not just the value at a given time.

for autonomous systems in the following sense. For each realization of the protocol we may define a reverse protocol $\tilde{\lambda}$ which retraces backward the time evolution of the forward protocol $\tilde{\lambda}_t = \lambda_{\tau-t}$. Let us consider, for instance, a quantum system intially found in a pure state $|i^{\lambda_0}\rangle$. The present analysis can be naturally rephrased to include the conceptually simpler classical case. During the action of the forward protocol λ, the system evolves to a final state $|f^{\lambda_\tau}\rangle$:

$$|f^{\lambda_\tau}\rangle = \hat{U}_{\tau,0}(\lambda)|i^{\lambda_0}\rangle, \tag{10.67}$$

$\hat{U}_{\tau,0}(\lambda)$ being the time evolution operator in the presence of the protocol λ. It obeys the equation

$$i\hbar\partial_t \hat{U}_{t,0}(\lambda) = H(\lambda_t)\hat{U}_{t,0}(\lambda). \tag{10.68}$$

We may then consider the time-reversed state $\mathcal{T}|f^{\lambda_\tau}\rangle$ of the finale state, $\mathcal{T}$ being the time reversal operator. We define the reverse evolution operator, $\hat{U}_{t,0}(\tilde{\lambda})$, in the presence of the reverse driving $\tilde{\lambda}$, as the operator which, acting on the time reverse of the final state, evolves into the time reverse of the initial state:

$$\mathcal{T}|i^{\lambda_0}\rangle = \hat{U}_{\tau,0}(\tilde{\lambda})\mathcal{T}|f^{\lambda_\tau}\rangle \quad \text{or} \quad |i^{\lambda_0}\rangle = \mathcal{T}^\dagger\hat{U}_{\tau,0}(\tilde{\lambda})\mathcal{T}|f^{\lambda_\tau}\rangle. \tag{10.69}$$

By comparing with Eq. (10.67) one obtains the condition expressing the microreversibility:

$$\mathcal{T}^\dagger\hat{U}_{\tau,0}(\tilde{\lambda})\mathcal{T} = \hat{U}^\dagger_{\tau,0}(\lambda). \tag{10.70}$$

After recalling that $\tilde{\lambda}_t = \lambda_{\tau-t}$, the microreversibility can be formulated through a relation between the probability $p_{i\mapsto f}$ of a state $|i^{\lambda_0}\rangle$ evolving in the time τ into the state $|f^{\lambda_\tau}\rangle$ and $p_{f\mapsto i}$ for the reverse process:

$$p_{i\mapsto f} \equiv |\langle f^{\lambda_\tau}|\hat{U}_{\tau,0}(\lambda)|i^{\lambda_0}\rangle|^2 \to p_{f\mapsto i}. \tag{10.71}$$

Similarly, the microreversibility is valid for the evolution of a classical system in the Γ phase space, where, given a trajectory solution of the equations of motion, its time-reverse is also a solution.

The non-equilibrium transformation of the system under the action of the protocol (10.66) can be classically characterized by the the variation of the energy of the system during the process and by the heat and work exchanged with the environment. Since each starting state on which the protocol is applied is distributed according to an ensemble, the resulting process will also be distributed. In particular, in the classical case the trajectories of the representative point in the phase space are to be distributed and so will be the work defined along each of these trajectories as (see Problem 10.3):

$$W = -\int_0^\tau dt\, \dot{\lambda}_t \mathcal{O}_t = H(\lambda_\tau) - H(\lambda_0). \tag{10.72}$$

The definition of W as the energy of the system measured at the end and at the beginning of the protocol process can be extended to quantum systems with the due proviso that the time of the energy measurement may be as long as the preassigned precision requires.

In general, starting from a system in thermal equilibrium, the work, W, performed during such a non-equilibrium transformation is a fluctuating quantity in the sense that if

11 Brownian motion and transport in disordered systems

11.1 Einstein's theory of Brownian motion

According to Einstein (1905) a colloid, a fine suspension of particles or grains in a liquid, is not qualitatively different from a solution. A solution is made of molecules of a substance mixed with those of another substance, in a fine suspension; instead, the grains can be made of many molecules and may have dimensions ranging from the microscopic to macroscopic size.

As a consequence of this analogy, a colloid must show the phenomenon of osmotic pressure, i.e. if we indicate with n the number of grains per unit volume, they act as a gas of diluted particles and the pressure exerted by the grains of the solute follows the law

$$P = nk_{\mathrm{B}}T. \tag{11.1}$$

Still using the analogy with solutions, one may expect the grains of the solute be in thermal motion as the molecules of the solvent. But given the big difference in mass between solvent and solute, the motion of the grains of the latter is much slower than that of the former. The molecules of the solvent, being thermally agitated, move randomly in all directions and in so doing collide with the grains. The exchanged momentum in each collision is randomly oriented. As a result the grains, after a transient time, will move with a pattern typical of Brownian motion. Einstein showed that such a motion can be described in terms of a diffusive process.

Let us assume that the colloidal solution is in the field of gravity and indicate with M the mass of the grains. In statistical equilibrium the force of gravity acting on the grains must compensate the difference in osmotic pressure at two different levels, in complete analogy with the barometric formula valid for a perfect gas (see Eq. (N.1) of the answer to Problem 4.1). Let us take the force of gravity directed towards the negative x-axis and let $n(x)$ be the density of grains at level x. The total mass of the grains contained in a volume element $\mathrm{d}x\,\mathrm{d}y\,\mathrm{d}z$ is $Mn(x)\mathrm{d}x\,\mathrm{d}y\,\mathrm{d}z$. The gravity acting on this total mass is then compensated by the difference in osmotic pressure at levels x and $x + \mathrm{d}x$. This leads to a stationary condition

$$Mgn(x)\mathrm{d}x\,\mathrm{d}y\,\mathrm{d}z = -\frac{\partial P}{\partial x}\mathrm{d}x\,\mathrm{d}y\,\mathrm{d}z,$$

which by using Eq. (11.1) leads to

$$Mg\,n(x) + k_{\mathrm{B}}T\frac{\partial n(x)}{\partial x} = 0. \tag{11.2}$$

Following Einstein, the statistical equilibrium described by Eq. (11.2) may be thought, from the point of view of the motions of the grains, as the compensation of two motions. On the one side, the grains fall in the field of gravity in a viscous medium (the solvent). In general a particle moving with velocity $\mathbf{v}$ in a viscous medium is subject to a force

$$\mathbf{f}_\mathrm{v} = -\lambda \mathbf{v} + \mathbf{X}(t), \tag{11.3}$$

where the first term on the right-hand side is the viscous damping (due to the solvent in this case) and the second one is a stochastic force (*complementary* in the words of Langevin) describing the collisions between the grains and the molecules of the solvent, which are subjected to a disordered thermal motion. The stochastic force, due to the independent action of a great number of molecules, is isotropic with zero average, uncorrelated over times longer than the characteristic time τ. If τ is much shorter than any time of measure, the stochastic force can be taken to be Gaussian and delta correlated in time:

$$\overline{X_i(t)} = 0, \quad \overline{X_i(t)X_j(t')} = \delta_{ij} C \delta(t - t'), \tag{11.4}$$

where C is a constant to be determined. λ depends on the geometry of the moving body and on the viscosity η of the liquid. If we assume the grains to be spheres with radii r_0, λ is given by the Stokes formula $\lambda = 6\pi r_0 \eta$.

The viscosity of the medium is the macroscopic manifestation of the collisions that, at microscopic level, the molecules of the solvent exert on the grains as a reaction of the motion of the grains due to the external (gravitational) field. The grains eventually will move on average with constant velocity

$$\langle v_0 \rangle = \frac{Mg}{\lambda}, \tag{11.5}$$

since averaging over time $\langle X(t) \rangle = 0$. Combining Eq. (11.2) and Eq. (11.5) we obtain the equilibrium condition

$$\langle v_0 \rangle n(x) + \frac{k_\mathrm{B} T}{\lambda} \frac{\partial n(x)}{\partial x} = 0. \tag{11.6}$$

This last equation states the compensation between the downwards drift current $\langle v_0 \rangle n(x)$ and the upwards diffusion current $v_{\mathrm{diff}} = -(k_\mathrm{B} T/\lambda)\partial_x n(x) = -D\partial_x n(x)$ with no net flow of matter on average. By definition D is the diffusion coefficient

$$D = \frac{k_\mathrm{B} T}{\lambda} = \frac{M\overline{v^2}}{d\lambda}. \tag{11.7}$$

The last relation follows from the equipartition theorem, $M\overline{v^2}/2 = (d/2)k_\mathrm{B} T$, d being the space dimensionality. Relation (11.7) is known as the Einstein relation and is far more general than the physical situation considered here. Its conceptual and historical importance is due to the fact that it allows, by observing the Brownian motion of the grains with a microscope, the measure of the Boltzmann constant k_B or the Avogadro number N_A. This measurement was indeed carried out by Jean Baptiste Perrin in 1913.[1]

[1] The experiment and its interpretation are described in the famous book by Perrin (1913).

To make the connection with a diffusion process explicit, Einstein used a kinetic equation-type argument. He argued that there is a time scale τ (which in the Drude formula, to be discussed later in this chapter, coincides with the scattering time) such that the motions of the grains occurring in successive time intervals τ are independent events. Then a kinetic equation is written for the function $n(\mathbf{x}, t)$, which describes the number of grains in the volume element $\mathrm{d}\mathbf{x}$ at time t. To simplify our discussion, let us confine ourselves for the time being to the x-axis coordinate in the absence of external forces. The equation for $n(x, t)$ reads

$$n(x, t + \tau)\mathrm{d}x = \mathrm{d}x \int_{-\infty}^{\infty} \mathrm{d}\Delta n(x - \Delta, t)\phi(\Delta), \tag{11.8}$$

where the function $\phi(\Delta)$ expresses the probability that a grain will stochastically move by the distance Δ along the x-axis in the time interval τ and satisfies the properties

$$\int_{-\infty}^{\infty} \mathrm{d}\Delta\phi(\Delta) = 1, \quad \phi(\Delta) = \phi(-\Delta).$$

When considering experimental observation times that are large compared to τ and noticing that $\phi(\Delta)$ mostly differs from zero for small values of Δ, one can Taylor expand both sides of Eq. (11.8) in t and x, obtaining the diffusion equation (the first derivative averages out to zero)

$$\frac{\partial n}{\partial t} = D\frac{\partial^2 n}{\partial^2 x}, \tag{11.9}$$

where the diffusion coefficient is expressed in terms of the function $\phi(\Delta)$ as

$$D = \frac{1}{\tau}\int_{-\infty}^{\infty} \mathrm{d}\Delta\phi(\Delta)\frac{\Delta^2}{2}. \tag{11.10}$$

In three dimensions the diffusion equation (11.9) becomes

$$\frac{\partial n}{\partial t} = D\nabla^2 n, \tag{11.11}$$

with

$$D = \frac{1}{\tau}\int_{-\infty}^{\infty} \mathrm{d}\Delta\phi(\Delta)\frac{\Delta^2}{6} \equiv \frac{1}{\tau}\frac{1}{6}\overline{\Delta^2}. \tag{11.12}$$

As a final remark we consider that the solution of the diffusion equation (11.9) reads

$$n(x, t) = n_0\frac{\mathrm{e}^{-x^2/(4Dt)}}{\sqrt{4\pi Dt}}, \tag{11.13}$$

where $n(x, 0) = n_0\delta(x)$ represents an initial density concentrated at the origin. The mean square displacement along the x-axis is clearly

$$\overline{x^2} = \frac{1}{n_0}\int_{-\infty}^{\infty} \mathrm{d}x\, x^2 n(x, t) = 2Dt, \tag{11.14}$$

which is linear in t rather than t^2 as in ballistic motion.

In the presence of an external force $\mathbf{F} = -\nabla U$ produced by a potential U and a limit velocity $\mathbf{v}_0 = \mathbf{F}/\lambda$, the probability of a motion Δ depends on the position $\mathbf{x}$ as well. If $\mathbf{F}$ is directed along the negative x-axis as for the gravity force Eq. (11.8) becomes

$$n(x, t + \tau)\mathrm{d}x = \mathrm{d}x \int_{-\infty}^{\infty} \mathrm{d}\Delta n(x - \Delta, t)\phi(\Delta, x - \Delta).$$

By making a Taylor expansion as that of Eq. (11.8) one obtains a linear gradient term given by

$$-\frac{\partial}{\partial x}\left(n(x,t)\frac{1}{\tau}\int_{-\infty}^{\infty} \mathrm{d}\Delta\ \Delta\ \phi(\Delta, x)\right) = -\frac{\partial}{\partial x}\left(n(x,t)\frac{\overline{\Delta(x)}}{\tau}\right) = -\frac{\partial}{\partial x}\left(n(x,t)\frac{F(x)}{\lambda}\right),$$

having identified $\overline{\Delta(x)}/\tau$ with the unidirectional limit velocity $F(x)/\lambda$ (see e.g. (11.5)). By returning to three dimensions, one needs to add a term $-\nabla \cdot (n\mathbf{F}/\lambda)\tau$ and the diffusive equation (11.11) becomes the Fokker–Planck equation (Fokker (1914); Planck (1917))

$$\frac{\partial n}{\partial t} = -\nabla \cdot \left(\frac{\mathbf{F}}{\lambda} n - D\nabla n\right), \tag{11.15}$$

which has the form of a continuity equation. The related current is made of two terms, one due to the external force and the second one due to the diffusion of the Brownian motion. At equilibrium the two terms must compensate each other as in Eq. (11.6) for the unidirectional force.

The compatibility of Eqs. (11.12) and (11.7) requires that $\overline{\Delta^2} = 2\overline{v^2}\tau^2$ with $\tau = M/\lambda$. Indeed, as we shall see in the next section, the quantity M/λ is the characteristic time scale which describes the exponential approach to the asymptotic velocity v_0 given in Eq. (11.5). Then, we see that observable macroscopic quantities such as the viscosity η or the time scale τ can be obtained in terms of the second moment of the function $\phi(\Delta)$, which must be computed by a microscopic theory.

11.2 Langevin's theory of Brownian motion

The Langevin treatment of Brownian motion (Langevin (1908)) elucidates the origin of the transition from ballistic to diffusive motion. The idea is to use a stochastic equation of motion in $d = 1$ and in the absence of external forces for the time being:

$$M\frac{\mathrm{d}^2 x}{\mathrm{d}t^2} = -\lambda\frac{\mathrm{d}x}{\mathrm{d}t} + X, \tag{11.16}$$

where, as before, the first term on the right-hand side is the viscous damping due to the Stokes law, while the second one is the delta-correlated stochastic force. A formal integration of (11.16) gives

$$v(t) = v_0 \mathrm{e}^{-t/\tau} + \frac{1}{M}\int_0^t \mathrm{d}s\ \mathrm{e}^{-(t-s)/\tau}\ X(s), \quad \tau = M/\lambda$$

with the average over the fluctuations of the stochastic force

$$\overline{v^2(t)} = v_0^2 e^{-2t/\tau} + \frac{C}{2M\lambda}(1 - e^{-2t/\tau}).$$

In thermal equilibrium $\overline{v^2(t)}$ (when t tends to infinity) must satisfy the equipartition theorem and this fixes the constant in Eq. (11.4), $C = 2\lambda k_B T$.

After multiplying Eq. (11.16) by x, one obtains

$$Mx\frac{d^2x}{dt^2} = -\lambda x \frac{dx}{dt} + xX. \tag{11.17}$$

By observing that

$$x\frac{dx}{dt} = \frac{1}{2}\frac{dx^2}{dt}, \quad x\frac{d^2x}{dt^2} = \frac{1}{2}\frac{d^2x^2}{dt^2} - \left(\frac{dx}{dt}\right)^2,$$

Eq. (11.17) becomes

$$\frac{d^2x^2}{dt^2} = 2\left(\frac{dx}{dt}\right)^2 - \frac{\lambda}{M}\frac{dx^2}{dt} + \frac{2}{M}xX. \tag{11.18}$$

As a result the averaged equation of motion reads

$$\frac{d^2\overline{x^2}}{dt^2} = 2\frac{k_B T}{M} - \frac{\lambda}{M}\frac{d\overline{x^2}}{dt} \tag{11.19}$$

and describes the evolution with time of the average value of the square of the displacement. By setting

$$y = \frac{d\overline{x^2}}{dt},$$

one obtains the first order differential equation

$$\frac{dy}{dt} = -\frac{\lambda}{M}y + 2\frac{k_B T}{M}, \tag{11.20}$$

which has the solution

$$y = y_0 e^{-(\lambda/M)t} + 2\frac{k_B T}{\lambda}, \tag{11.21}$$

with y_0 a suitable integration constant.

As a result the mean square displacement reads

$$\overline{x^2}(t) = -y_0\frac{M}{\lambda}e^{-(\lambda/M)t} + 2\frac{k_B T}{\lambda}t. \tag{11.22}$$

One then sees that after a rapid exponential transient with typical time $\tau = M/\lambda$, the mean square displacement increases linearly with time with diffusion constant $D = k_B T/\lambda$ and agrees with Eq. (11.14) obtained from the solution of the diffusion equation.

In general the diffusion coefficient is given in terms of the velocity correlations. Remember that, in d dimensions,

$$\mathbf{r}(t) = \int_0^t dt' \mathbf{v}(t')$$

and

$$\overline{\mathbf{r}^2(t)} = \int_0^t \mathrm{d}t_1 \int_0^t \mathrm{d}t_2 \overline{\mathbf{v}(t_1) \cdot \mathbf{v}(t_2)},$$

then the diffusion coefficient is

$$\begin{aligned} D &= \frac{1}{2d} \lim_{t\to\infty} \frac{\mathrm{d}}{\mathrm{d}t}\left(\overline{\mathbf{r}^2(t)}\right) \\ &= \lim_{t\to\infty} \frac{1}{d} \int_0^t \mathrm{d}t' \, \langle \mathbf{v}(t) \cdot \mathbf{v}(t') \rangle \\ &= \frac{1}{d}\overline{v^2}\tau. \end{aligned} \tag{11.23}$$

In the last relation, we have used the property that, if the distribution of the forces is Gaussian, the velocity distribution also is Gaussian with a spread $\overline{v^2}$ and depends only on the time difference with width τ.

11.3 The Johnson–Nyquist noise

The conduction electrons in a metal may experience, at microscopic scales, a fluctuating electric field due to the thermal motion of the positively charged ions. This gives rise to Johnson–Nyquist noise (Johnson (1928); Nyquist (1928)), which can be described as the Brownian motion of colloidal particles. Let us describe the conductor as a series of a resistance R and inductance L closed with a fluctuating electromotive force $\mathcal{E}$. The circuit equation is then

$$L\frac{\mathrm{d}^2 q}{\mathrm{d}t^2} = -R\frac{\mathrm{d}q}{\mathrm{d}t} + \mathcal{E}. \tag{11.24}$$

This has the same form as the Langevin equation (11.16), where the charge q plays the role of the displacement x with the correspondences $M \to L$, $\lambda \to R$ and $X \to \mathcal{E}$.

At a formal level one may go through the same steps as for the Langevin equation. It is, however, useful to indicate the physical mechanism at the basis of the equilibrium between the charges in the conductor and the surrounding medium. Let us suppose that at certain instant of time there is a fluctuation from equilibrium of the current in the inductance. As a result there will be a magnetic energy

$$\frac{L}{2}\left(\frac{\mathrm{d}q}{\mathrm{d}t}\right)^2,$$

which eventually is dissipated in the resistance as Joule heating. In this way the circuit is giving energy to the surrounding medium as the grains in the colloidal solution give away kinetic energy to the molecules of the solvent during the collisions. The heat received by the positively charged ions corresponds to an increased thermal energy, which gives rise to a fluctuation in the electric field felt in the circuit. Then this electric field transfers work to the circuit building up the magnetic energy in the inductance.

The discussion of the previous section has been carried out mainly following historical arguments, which provide several physical insights. It is, however, useful to cast the results obtained into the more general framework of the fluctuation–dissipation theorem derived in the previous chapter. This clarifies on what general assumptions the theory of Brownian motion and electrical noise is based.

In the case of the electrical circuit described by Eq. (11.24) the observable is the electrical current flowing in the circuit and we want to determine its response to an applied voltage $\phi(t)$. The Kirchhoff laws applied to the circuit yield an equation for the current similar to (11.24), where the fluctuating electromotive force is now replaced by the electric voltage $\phi(t)$:

$$L\frac{\mathrm{d}I(t)}{\mathrm{d}t} = -RI(t) + \phi(t).$$

We must solve the above equation to linear order in $\phi(t)$. By a Fourier transform we easily obtain

$$I(\omega) = \frac{G}{1 - \mathrm{i}\omega\tau}\phi(\omega),$$

where we have introduced the conductance $G = 1/R$ and the characteristic time $\tau = L/R$. Since the voltage $\phi(t)$ couples with the charge $q(t)$, the above relation yields the current–charge response function $\tilde{R}^{Iq}(\omega)$ given by

$$\tilde{R}^{Iq}(\omega) = \frac{G}{1 - \mathrm{i}\omega\tau}. \tag{11.25}$$

In order to use the fluctuation–dissipation theorem, we introduce, according to the notation of Chapter 10, the current–charge correlation function $f^{Iq}(t)$ and its Fourier transform $\tilde{f}^{Iq}(\omega)$ defined by

$$f^{Iq}(t) = \langle I(t)q(0)\rangle, \quad \tilde{f}^{Iq}(\omega) = \int_{-\infty}^{\infty} \mathrm{d}t\, \mathrm{e}^{\mathrm{i}\omega t}\, f^{Iq}(t). \tag{11.26}$$

In the above the angular brackets indicate the statistical average. The classical limit (10.55) of the fluctuation–dissipation theorem gives

$$\langle I(t)q(t)\rangle = k_{\mathrm{B}}T\tilde{R}^{Iq}(\omega = 0) = \frac{k_{\mathrm{B}}T}{R}. \tag{11.27}$$

With respect to Eq. (10.55) there is a minus sign of difference. This is due to the fact that the coupling Hamiltonian for the charge in the circuit is $H_{\mathrm{ext}} = -q\phi(t)$, which has a relative minus sign with respect to Eq. (10.1). By observing that

$$\frac{\mathrm{d}}{\mathrm{d}t}\langle(q(t))^2\rangle = 2\langle I(t)q(t)\rangle = 2k_{\mathrm{B}}TG, \tag{11.28}$$

one obtains again the diffusive character of the charge fluctuations $\langle(q(t))^2\rangle = 2Dt$ with $D = k_{\mathrm{B}}TG$ reproducing the expression (11.7) for the diffusion coefficient since $G = 1/R \to \lambda^{-1}$.

The power spectrum of the fluctuations can be obtained by the spectral form of the fluctuation–dissipation theorem (10.54). By noticing, via $I(\omega) = -\mathrm{i}\omega q(\omega)$, that

$$\mathrm{Im}\tilde{R}^{II}(\omega) = \omega\mathrm{Re}\tilde{R}^{Iq}(\omega)$$

and, after using (11.25) into (10.54) (again with a sign change), we get

$$\tilde{f}^{II}(\omega) = \frac{2k_{\mathrm{B}}T}{R}\frac{1}{1+\omega^2\tau^2}. \tag{11.29}$$

In the laboratory, Johnson–Nyquist noise is usually studied by expressing the power spectrum in terms of the cosine Fourier transform of $f^{II}(s)$, defined by

$$\tilde{P}(\nu) = 4\int_0^{\infty} \mathrm{d}s\cos(2\pi\nu s)\, f^{II}(s) = 2\tilde{f}^{II}(2\pi\nu). \tag{11.30}$$

We then get the well-known Johnson–Nyquist formula

$$\tilde{P}(\nu) = \frac{4k_{\mathrm{B}}T}{R}\frac{1}{1+(2\pi\nu\tau)^2} \approx \frac{4k_{\mathrm{B}}T}{R}, \tag{11.31}$$

with the last equality holding in the low frequency regime $\nu\tau \ll 1$ valid in most used circuit devices.

As a final remark, we stress that the procedure here used is very general, since it relates any given expression for the response function to the fluctuations of the physical observable which is responding to an external perturbation. The same strategy can be used to analyze the fluctuations, say, of the density close to the critical point. Indeed, by using the van der Waals equation in Chapter 1, we were able to derive the static response, i.e. the compressibility. Then, the clever use of the Boltzmann principle made by Einstein allowed us to estimate the fluctuations. Such a connection is generally provided by the fluctuation–dissipation theorem. The analysis of the fluctuations in the Landau theory of the phase transitions will make use of the same connection, as will be done in Chapter 13.

11.4 Transport description in the disordered Fermi gas

The theory of Brownian motion developed in the previous sections can be applied to the description of the electrical conduction in metals. The conventional theory of electrical transport in metals is due to Drude (1900). Although formulated a quarter of a century before the advent of quantum mechanics and hence based on classical mechanics, some of its main conclusions, such as the Einstein relation (11.7), remain valid in the quantum mechanical treatment when the parameters of the Brownian motion are properly interpreted. Modifications to the Drude description nevertheless arise due to quantum interference effects in much the same spirit as wave optics does for geometrical optics. This occurs for strong disorder and will be discussed in Chapter 21.

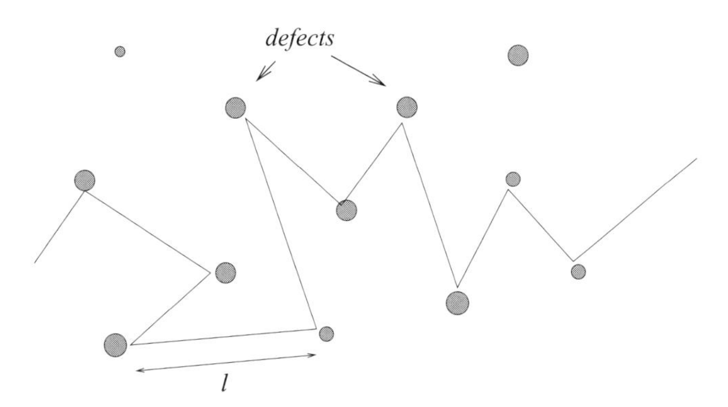

Fig. 11.1 A pictorial representation of the semiclassical theory of transport.

11.4.1 The Drude model of electrical conduction

In his original formulation, Drude suggested that electrons, under the action of an externally applied electric field, are accelerated according to the Newton equation of motion until they collide with the fixed positive ions after a time τ. The distance between two successive collisions determines the mean free path l.[2] The motion is then the result of independent aleatory events with characteristic time τ. Due to this sequence of independent scattering events, under the action of an external electric field, the electrons yield a steady current, which can be described by the Ohm law

$$\mathbf{j} = \sigma \mathbf{E}, \tag{11.32}$$

where the electrical conductivity σ, is given by the Drude formula

$$\sigma = \frac{e^2 n \tau}{m}, \tag{11.33}$$

where m is the electron mass, $e > 0$ the unit charge (we take as $-e$ the electron charge throughout the book), n the electron density.

After the birth of quantum mechanics, Sommerfeld reformulated the Drude theory to comply with the Fermi statistics of electrons, providing the correct relation between τ and l via the Fermi velocity v_{F}. More importantly, with the work of Bloch, it was realized that the relaxation of electron momentum and the finite value of the conductivity (which is infinite in a perfect crystal) is due to imperfections of the ion lattice, i.e. to *disorder*.

A pictorial description of Drude's model of electrical conduction is shown in Fig. 11.1. In this limit of independent electrons, their motion is diffusive and can be treated as the Brownian motion of one particle in a fluid with viscous forces.

Ohm's law plays the role of the Stokes formula of the previous discussion of the colloidal solution. The viscosity of the electron gas is described as the electrical resistance. It is then easy to show that a perfect correspondence can be made between the two situations by the

[2] Notice that in Chapter 2 we discussed the mean free path due to collisions among the molecules of a perfect gas. Here we consider the collisions between the gas of electrons and the fixed impurity ions, which act as random centers of force.

substitution $\lambda \to m/\tau$, so that the diffusion coefficient reads

$$D = \frac{k_B T \tau}{m} = \frac{\overline{v^2}\tau}{d} \to \frac{v_F^2 \tau}{d}. \tag{11.34}$$

As we now see, this is indeed the correct expression for the diffusion coefficient provided the thermal velocity is replaced by the Fermi velocity, or that classical mechanics is replaced by quantum mechanics. The Einstein argument and the Einstein relation remain unchanged.

Let us consider the situation in which an external potential $U(\mathbf{x}, t)$ yields a perturbation of the density and indicate with $\delta n(\mathbf{x}, t)$ the variation of the local density with respect to the equilibrium value n as in Eq. (10.21). By using the Fokker–Planck equation (11.15), we obtain the equation for $\delta n(\mathbf{x}, t)$

$$(\partial_t - D\nabla^2)\delta n(\mathbf{x}, t) = \frac{n}{\lambda}\nabla^2 U(\mathbf{x}, t). \tag{11.35}$$

The electrical current which satisfies the above continuity equation is

$$\mathbf{j}(\mathbf{r}, t) = e\nabla U \frac{n}{\lambda} + eD\nabla\delta n(\mathbf{r}, t) \tag{11.36}$$

and $\nabla U = e\mathbf{E}$, $\mathbf{E}$ being the electric field.

By Fourier transforming Eq. (11.35) with respect to space and time we obtain the density–density response function $\tilde{R}^{nn}_{\mathbf{q}}(\omega)$ introduced in Section 10.1:

$$\tilde{R}^{nn}_{\mathbf{q}}(\omega) \equiv \frac{\delta n(\mathbf{q}, \omega)}{\delta U(\mathbf{q}, \omega)} = -\frac{n}{\lambda}\frac{q^2}{Dq^2 - \mathrm{i}\omega}. \tag{11.37}$$

According to Eq. (10.22) the static limit ($\mathbf{q} \to 0$, at $\omega = 0$) of $\tilde{R}^{nn}_{\mathbf{q}}(\omega)$ reads

$$\lim_{\mathbf{q}\to 0} \tilde{R}^{nn}_{\mathbf{q}}(0) \equiv -\frac{\partial n}{\partial \mu} = -\frac{n}{\lambda D}. \tag{11.38}$$

Equation (11.38) can also be checked by using the fluctuation–dissipation theorem in the form (10.55). This is left as an exercise for the reader in Problem 11.1.

The density–density response function now assumes the form

$$\tilde{R}^{nn}_{\mathbf{q}}(\omega) = -\frac{\partial n}{\partial \mu}\frac{Dq^2}{Dq^2 - \mathrm{i}\omega}. \tag{11.39}$$

The electrical conductivity σ is then obtained by using Eq. (10.43) with the identification $\mathrm{e}^2 \tilde{R}^{nn}_{\mathbf{q}}(\omega) = \tilde{R}^{00}_{\mathbf{q}}(\omega)$:

$$\sigma = e^2 D \frac{\partial n}{\partial \mu}. \tag{11.40}$$

The last expression is a generalized Einstein relation for the electrical conductivity. By observing that $\lambda = m/\tau$ the Drude formula for the electrical conductivity (11.33) follows.

11.4.2 The kinetic equation for a disordered Fermi gas

The Drude formula, and in general the transport coefficients, can also be directly derived by using the kinetic Boltzmann equation, where the collision mechanism can be explicitly specified in the collision integral. As an important paradigmatic example we derive now the transport coefficients for fermions in the semiclassical approximation assuming that the only source of scattering is from static impurities. However, the method is rather general and can also be applied to the classical perfect gas, as shown in Problem 11.3.

The Boltzmann equation Eq. (6.3), for fermions in the presence of impurity scattering only, reads

$$(\partial_t + \mathbf{v}\cdot\nabla_{\mathbf{r}} - e\mathbf{E}\cdot\nabla_{\mathbf{p}})f_{\mathbf{p}} = -\sum_{\mathbf{p}'} W_{\mathbf{pp}'}(f_{\mathbf{p}}(1-f_{\mathbf{p}'}) - f_{\mathbf{p}'}(1-f_{\mathbf{p}})), \tag{11.41}$$

where $f_{\mathbf{p}} \equiv f(\mathbf{r},\mathbf{p},t)$ is the distribution function and $\mathbf{E}$ the electric field.

We assume that the scattering from impurities is incoherent and, therefore, we first consider the scattering from a single impurity located at the origin of the coordinates and then we multiply the scattering rate by the number of impurities N_{i}. The scattering potential is assumed to have a delta-like form

$$V(\mathbf{r}) = v_0\delta(\mathbf{r}), \tag{11.42}$$

v_0 being a constant. The presence of the impurity breaks the translational invariance, so that there is a probability $W^{(\mathrm{i})}_{\mathbf{p},\mathbf{p}'}$ for transitions from a state $\langle \mathbf{r}|\mathbf{p}\rangle = \mathrm{e}^{\mathrm{i}\mathbf{p}\cdot\mathbf{r}/\hbar}/\sqrt{V}$ of momentum $\mathbf{p}$ to a state with momentm $\mathbf{p}'$. $W^{(\mathrm{i})}_{\mathbf{p},\mathbf{p}'}$ refers to the scattering probability for a single impurity. By limiting ourselves to the Fermi Golden Rule approximation we obtain

$$\begin{aligned} W^{(\mathrm{i})}_{\mathbf{pp}'} &= \frac{2\pi}{\hbar}\delta(\mathbf{p}^2/2m - \mathbf{p}'^2/2m)\left|\frac{1}{V}\int \mathrm{d}^3 r\, \mathrm{e}^{\mathrm{i}(\mathbf{p}-\mathbf{p}')\cdot\mathbf{r}/\hbar} V(\mathbf{r})\right|^2 \\ &= \frac{2\pi v_0^2}{\hbar V^2}\delta(\mathbf{p}^2/2m - \mathbf{p}'^2/2m). \end{aligned} \tag{11.43}$$

The scattering rate is connected to the scattering probability by integrating over all possible final states with momentum $\mathbf{p}'$ and summing over all impurities:

$$\frac{1}{\tau} = \sum_{\mathbf{p}'} W_{\mathbf{pp}'} = N_{\mathrm{i}}\sum_{\mathbf{p}'} W^{(\mathrm{i})}_{\mathbf{pp}'} = N_{\mathrm{i}} V \int \frac{\mathrm{d}^3 p'}{(2\pi\hbar)^3} W^{(\mathrm{i})}_{\mathbf{pp}'}. \tag{11.44}$$

The explicit evaluation of the integral gives

$$\frac{1}{\tau} = \frac{2\pi}{\hbar}\frac{\nu(\epsilon_{\mathbf{p}})}{s} n_i v_0^2, \tag{11.45}$$

where we have used the single-particle density of states of Eq. (7.14) and the number of impurities per unit volume $n_{\mathrm{i}} = N_{\mathrm{i}}/V$. Equation (11.41) then becomes

$$(\partial_t + \mathbf{v}\cdot\nabla_{\mathbf{r}} - e\mathbf{E}\cdot\nabla_{\mathbf{p}})f_{\mathbf{p}} = -\frac{1}{\tau}(f_{\mathbf{p}} - \langle f_{\mathbf{p}'}\rangle), \tag{11.46}$$

where the angular brackets indicate the integration over the solid angle of the directions of $\mathbf{p}'$:

$$\langle \ldots \rangle = \int \frac{\mathrm{d}\Omega_{\hat{\mathbf{p}}'}}{4\pi}\ldots .$$

Strictly speaking the collision time defined by Eq. (11.45) depends on the initial momentum through the density of states $\nu(\epsilon_{\mathbf{p}})$ and the *in*-term of the collision integral cannot be written simply in terms of τ. However, since in a degenerate Fermi gas only the states around the Fermi surface are relevant, we may keep the density of states evaluated at the Fermi energy, neglecting its energy and momentum dependence and obtaining Eq. (11.46).

In the presence of a static and uniform electric field, we seek a solution in the form $f_{\mathbf{p}} = f_0(\mathbf{p}) + \delta f_{\mathbf{p}}$, where $f_0(\mathbf{p})$ is the equilibrium Fermi function in the absence of the electric field. The linearized Boltzmann equation reads

$$-e\mathbf{E} \cdot \mathbf{v}_{\mathbf{p}} \frac{\partial f_0(\mathbf{p})}{\partial \epsilon} = -\frac{1}{\tau}(\delta f_{\mathbf{p}} - \langle \delta f_{\mathbf{p}'} \rangle), \tag{11.47}$$

where $\mathbf{v}_{\mathbf{p}} = \mathbf{p}/m$. Alternatively we could have assumed local equilibrium for $f_{\mathbf{p}}$ given by the Fermi distribution with a weak spatial gradient in the electrochemical potential:

$$\mu(\mathbf{r}) = \mu - e\phi(\mathbf{r}) = \mu + e\mathbf{E} \cdot \mathbf{r}. \tag{11.48}$$

By multiplying both sides of Eq. (11.47) by $\mathbf{p}/m$ and integrating over the momentum, taking into account the spin degeneracy ($s = 2$) for a strongly degenerate Fermi gas with $\partial f_0(\mathbf{p})/\partial \epsilon_{\mathbf{p}} = -\delta(\epsilon_{\mathrm{F}} - \epsilon_{\mathbf{p}})$, we obtain the electrical current in the form

$$\begin{aligned}
\mathbf{j} &= -2e \int \frac{\mathrm{d}^3 p}{(2\pi\hbar)^3} \frac{\mathbf{p}}{m} \delta f_{\mathbf{p}} \\
&= -2e^2\tau \int \frac{\mathrm{d}^3 p}{(2\pi\hbar)^3} \frac{\mathbf{p}}{m} \mathbf{E} \cdot \mathbf{v}_{\mathbf{p}} \frac{\partial f_0(\mathbf{p})}{\partial \epsilon_{\mathbf{p}}} \\
&= e^2 \nu_{\mathrm{F}} \frac{v_{\mathrm{F}}^2 \tau}{d} \mathbf{E} \\
&= e^2 \nu_{\mathrm{F}} D \mathbf{E} && (11.49) \\
&= \sigma \mathbf{E}. && (11.50)
\end{aligned}$$

The above equation is valid in d dimensions. Notice that the second term in the right-hand side of Eq. (11.47), since it does not depend on the direction of $\mathbf{p}$, drops out due to the angle average when is multiplied by $\mathbf{p}$. By observing that for the Fermi gas the single-particle density of states coincides with the thermodynamic density of state $\partial n/\partial \mu$, we obtain the Einstein relation (11.40).

In Eq. (11.49), we have explicitly obtained the microscopic expression of the diffusion coefficient for a Fermi gas. It is instructive to see directly that the combination $v_{\mathrm{F}}^2 \tau/d$ is indeed the diffusion coefficient by looking at the coefficient of the gradient of the density in the expression of the current. To appreciate this point it is worthwhile showing a general technique for obtaining the diffusion equation from the kinetic equation. After linearizing with respect to the electric field, the Boltzmann equation (11.46) can be written as

$$(1 + \tau\partial_t + \tau\mathbf{v}_{\mathbf{p}} \cdot \nabla_{\mathbf{r}}) f_{\mathbf{p}} = \langle f_{\mathbf{p}'} \rangle + e\tau\mathbf{E} \cdot \mathbf{v}_{\mathbf{p}} \frac{\partial f_0(\mathbf{p})}{\partial \epsilon_{\mathbf{p}}}.$$

We may, formally, invert the differential operator acting on the left-hand side of the above equation, obtaining

$$f_{\mathbf{p}} = (1 + \tau\partial_t + \tau\mathbf{v}_{\mathbf{p}} \cdot \nabla_{\mathbf{r}})^{-1} \left(\langle f_{\mathbf{p}'} \rangle + \tau e\mathbf{E} \cdot \mathbf{v}_{\mathbf{p}} \frac{\partial f_0(\mathbf{p})}{\partial \epsilon_{\mathbf{p}}} \right).$$

The diffusive approximation arises when the time and space scales over which the distribution function varies are large compared to τ and $l = v_{\mathrm{F}}\tau$. In this case the derivative terms in the differential operator are small compared to unity and one obtains an equation for $f_{\mathbf{p}}$ in terms of $\langle f_{\mathbf{p}'} \rangle$ and the electric field term:

$$f_{\mathbf{p}} = (1 - \tau\partial_t - \tau\mathbf{v}_{\mathbf{p}} \cdot \nabla_{\mathbf{r}} + \cdots) \left(\langle f_{\mathbf{p}'} \rangle + \tau e\mathbf{E} \cdot \mathbf{v}_{\mathbf{p}} \frac{\partial f_0(\mathbf{p})}{\partial \epsilon_{\mathbf{p}}} \right).$$

If now we multiply both sides of the above equation by $-e\mathbf{v}_{\mathbf{p}}$ and integrate over the momentum, taking into account the spin degeneracy, we get

$$\begin{aligned} \mathbf{j}(\mathbf{r}, t) &= (-2e) \int \frac{\mathrm{d}^3 p}{(2\pi\hbar)^3} \mathbf{v}_{\mathbf{p}} \left(-\tau\mathbf{v}_{\mathbf{p}} \cdot \nabla_{\mathbf{r}} \langle f_{\mathbf{p}'} \rangle + e\tau\mathbf{E} \cdot \mathbf{v}_{\mathbf{p}} \frac{\partial f_0(\mathbf{p})}{\partial \epsilon_{\mathbf{p}}} \right) \\ &= e\frac{v_{\mathrm{F}}^2\tau}{d} \nabla_{\mathbf{r}} n(\mathbf{r}, t) + e^2 \frac{n\tau}{m}\mathbf{E}, \end{aligned} \tag{11.51}$$

which reproduces the Fokker–Planck equation (11.36). In obtaining the first term on the right-hand side of Eq. (11.51) we have used the fact that the non-equilibrium, position dependent part of $\langle f_{\mathbf{p}'} \rangle$ differs from zero for values of the momentum near the Fermi surface. This justifies the evaluation of the velocity at the Fermi surface.

The thermal transport can be studied in a similar fashion by assuming again local equilibrium with a weak temperature gradient $\nabla_{\mathbf{r}} T(\mathbf{r})$. In this case the Boltzmann equation becomes

$$-\frac{\epsilon_{\mathbf{p}} - \mu}{T} \frac{\partial f_0(\mathbf{p})}{\partial \epsilon_{\mathbf{p}}} \mathbf{v}_{\mathbf{p}} \cdot \nabla_{\mathbf{r}} T = -\frac{1}{\tau}(\delta f_{\mathbf{p}} - \langle \delta f_{\mathbf{p}'} \rangle). \tag{11.52}$$

The energy current can be split into two contributions:

$$\begin{aligned} \mathbf{J}_{\mathrm{Q}} &= 2 \int \frac{\mathrm{d}^3 p}{(2\pi\hbar)^3} \epsilon_{\mathbf{p}} \frac{\mathbf{p}}{m} \delta f_{\mathbf{p}} \\ &= -\frac{\mu}{e}\mathbf{j}(\mathbf{r}, t) + 2 \int \frac{\mathrm{d}^3 p}{(2\pi\hbar)^3} (\epsilon_{\mathbf{p}} - \mu) \frac{\mathbf{p}}{m} \delta f_{\mathbf{p}}. \end{aligned} \tag{11.53}$$

The thermal conductivity is defined as the thermal current response to a temperature gradient in the absence of electrical current, $\mathbf{j}(\mathbf{r}, t) = 0$,

$$\mathbf{J}_{\mathrm{Q}} = -\kappa \nabla_{\mathbf{r}} T(\mathbf{r}). \tag{11.54}$$

To obtain it, we multiply both sides of Eq. (11.52) by $(\epsilon_{\mathbf{p}} - \mu)\mathbf{p}/m$ and integrate over the momentum. The contribution at $T = 0$ vanishes and the first correction in the Sommerfeld expansion must be evaluated. Taking into account the spin degeneracy we get

$$\begin{aligned}\mathbf{J}_{\mathrm{Q}} &= 2\int \frac{\mathrm{d}^3p}{(2\pi\hbar)^3}\frac{(\epsilon(\mathbf{p})-\mu)^2}{T}\frac{\mathbf{p}}{m}\frac{\partial f_0(\mathbf{p})}{\partial \epsilon_{\mathbf{p}}}\tau \mathbf{v}_{\mathbf{p}}\cdot\nabla_{\mathbf{r}}T(\mathbf{r})\\ &= -\frac{\pi^2 k_{\mathrm{B}}^2}{3}T\nu_{\mathrm{F}}D\nabla_{\mathbf{r}}T(\mathbf{r})\\ &= -c_V D\nabla_{\mathbf{r}}T(\mathbf{r})\\ &= -\kappa\nabla_{\mathbf{r}}T(\mathbf{r}),\end{aligned} \tag{11.55}$$

where D is given by Eq. (11.34), c_V comes from Eq. (7.37) and the integral over the momentum has been transformed into an integral over the particle energy, thus showing the single-particle density of states ν_{F}. The resulting energy integral can be dealt with by the methods of Appendix E as in Eq. (E.22). We see that an Einstein relation is valid also for thermal transport, connecting the equilibrium thermodynamic response to a change of temperature to the thermal conductivity via the diffusion coefficient $\kappa = c_V D$. We also note that the ratio of the thermal to the electrical conductivity depends only on the temperature:

$$\frac{\kappa}{\sigma} = \frac{\pi^2 k_{\mathrm{B}}^2}{3e^2}T, \tag{11.56}$$

a result known as the Wiedemann–Franz law, experimentally established in 1853. The constant $\pi^2 k_{\mathrm{B}}^2/(3e^2)$ is called the Lorenz number.

Finally, we note that spin transport may be studied in a similar way by introducing a weakly space dependent magnetic field via

$$\mu(\mathbf{r}) = \mu - \frac{1}{2}g\mu_{\mathrm{B}}B^a(\mathbf{r})\sigma^a, \tag{11.57}$$

where σ^a are the set of Pauli matrices with $a = x, y, z$. In writing the Boltzmann equation, we have to introduce a distribution function with a matrix structure in spin space:

$$f_{\mathbf{p}} = f_{\mathbf{p}}^0\sigma^0 + f_{\mathbf{p}}^a\sigma^a, \tag{11.58}$$

$f_{\mathbf{p}}^0$ being the charge distribution function considered previously and $f_{\mathbf{p}}^a$ means the distribution function for particles with spin polarization along the a axis. As an example, consider the case of a spin quantization axis along the x axis. We can distinguish the distribution functions for up and down spin particles. The sum and difference of these distribution functions give the charge $f_{\mathbf{p}}^0$ and spin $f_{\mathbf{p}}^z$ distribution functions, respectively. The Boltzmann equation for the a-th spin component then becomes

$$\frac{g\mu_{\mathrm{B}}}{2}\nabla_{\mathbf{r}}B^a(\mathbf{r})\cdot\mathbf{v}_{\mathbf{p}}\frac{\partial f_0(\mathbf{p})}{\partial\epsilon} = -\frac{1}{\tau}\left(\delta f_{\mathbf{p}}^a - \langle\delta f_{\mathbf{p}'}^a\rangle\right). \tag{11.59}$$

By going through the same steps as for the electrical transport, we find the spin current as

$$\begin{aligned}
\mathbf{j}^a &= -\frac{g\mu_B}{2}\int\frac{dp^3}{(2\pi\hbar)^3}\frac{\mathbf{p}}{m}\delta f^a_{\mathbf{p}} \\
&= -\frac{(g\mu_B)^2}{4}\nu_F D\nabla_{\mathbf{r}}B^a(\mathbf{r}) \\
&= -\chi D\nabla_{\mathbf{r}}B^a(\mathbf{r}) \\
&= \sigma_s(-\nabla_{\mathbf{r}}B^a(\mathbf{r})),
\end{aligned} \tag{11.60}$$

where we have introduced the spin conductivity $\sigma_s = \chi D$ satisfying another Einstein relation involving the Pauli spin susceptibility χ given by Eq. (7.32).

In each of the above cases the Einstein relation involves the static response to the homogeneous corresponding field, the chemical potential for the charge density, the temperature for the energy density and the magnetic field for the spin magnetization, as is clear in the context of the linear response theory.

11.5 Problems

11.1 Show by a direct calculation that the fluctuation–dissipation theorem in the form (10.55) is satisfied by the static limit ($\mathbf{q} \to 0, \omega = 0$) (11.38) of the response function $\tilde{R}^{nn}_{\mathbf{q}}(\omega)$.

11.2 Derive the scattering probability and the scattering time by a single impurity with potential (11.42) in the Born approximation.

11.3 Derive the electrical conductivity for a classical gas of negatively charged particles interacting with fixed random impurities.

12 Fermi liquids

12.1 Temperature range of application and quasiparticles

The fermionic isotope of helium (^{3}He) with three nucleons and total spin 1/2, together with its bosonic counterpart (^{4}He) with four nucleons and total spin zero, is the only element which does not solidify under its own vapor pressure and remains liquid down to absolute zero temperature, unless a pressure of around 30 bar is applied (25 bar for ^{4}He and 34 bar for ^{3}He). This is due to the combined effect of the high zero-point energy for these systems with light atomic mass (London (1959); Atkins (1959)) and by the weakness of the van der Waals dipolar forces since the He atoms are structureless spherical particles. At normal pressure the liquid is formed at $T = 3.2$ K and $T = 4.2$ K for ^{3}He and ^{4}He, respectively. For a liquid, as for a gas, quantum effects strongly manifest themselves when, at the degenerate temperature, the thermal wave length $\lambda = h/(2\pi m k_B T)^{1/2}$ is of the order of the interatomic distance $d = n^{-1/3}$, $n = N/V$ (see Eq. (7.21)). For a bosonic gas this leads directly to the Bose–Einstein condensation critical temperature of Eq. (7.42) with a related thermodynamic phase transition. Similarly, liquid ^{4}He undergoes the superfluid transition at a critical temperature $T_\lambda = 2.14$ K of the same order of magnitude. For the fermionic gas the degenerate temperature coincides with the Fermi temperature $k_B T_F = (3\pi^2)^{2/3}(\hbar^2/2m)n^{2/3}$ which can be obtained from Eq. (7.25). At the same density, the Fermi temperature for a fermionic gas and the condensation temperature for a bosonic gas coincide up to a factor of order unity. For the ^{3}He atoms T_F has been evaluated in Chapter 7 to be few Kelvin. For liquid ^{3}He, however, the superfluid critical temperature, in the millikelvin region, is much lower then T_F. In the interval of temperature between T_F and the temperature of the onset for superfluidity, ^{3}He behaves as a *normal* Fermi liquid. In the same way, the conduction electrons in metals with well screened Coulomb long range interaction are considered as a normal Fermi liquid in the interval of temperatures $T \ll T_F \sim 10^4$ K and the critical temperature for the onset of superconductivity. By definition, for a system to be a normal Fermi liquid, its state must evolve continuously from the non-interacting state by adiabatically switching on the interaction and the energy of the particles of the gas must be in one-to-one correspondence with the dressed one-particle low-lying excitations of the interacting system. These are the so-called quasiparticles with the same spin and charge (when present) as the original particles. Perturbation theory for the interaction can be used in this case.

The formation of the superfluid or superconducting phase of fermions, as we shall see in the Chapter 15, must involve the formation of bound pairs to form pseudo-bosons which undergo a Bose condensation. The super-ground state cannot evolve continuously and perturbatively from the non-interacting ground state by an adiabatic switching of the interaction.

We give an heuristic evaluation of the critical temperature of the superfluid (or superconducting) state of fermions to substantiate theoretically the above statement about the difference between $T_{\rm F}$ and $T_{\rm c}$. A fermionic superfluid system, whatever the mechanism and the symmetry of the paired state, is characterized, as we shall see, by a gap Δ in the spectrum of the excitations whose magnitude, independently of its symmetry, must be of the order of the energy involved in the transition, $\Delta \sim k_{\rm B}T_{\rm c}$. On the other hand, there are two characteristic quantities in the systems, i.e. the Fermi velocity $v_{\rm F} = \hbar k_{\rm F}/m$ ($k_{\rm F}$ being the Fermi wave vector) and the extension of the pair bound state ξ_0. The gap Δ must be expressed in terms of these two quantities. By a dimensional argument we have $k_{\rm B}T_{\rm c} \sim \Delta \sim \hbar\xi_0^{-1}v_{\rm F} = E_{\rm F}/k_{\rm F}\xi_0$. The super-transition temperature $T_{\rm c}$ is therefore reduced by a factor $k_{\rm F}\xi_0$ with respect to the Fermi temperature of the corresponding gas, i.e. by the ratio between the extension of the pairs and the average distance between the particles. The interval of temperatures $T_{\rm F}/k_{\rm F}\xi_0 < T < T_{\rm F}$ is the region of application of the Fermi-liquid theory. In neutral liquid ^{3}He, $T_{\rm F} \approx 1$ K, ξ_0 is of the order of 600 Å and therefore $T_{\rm c}$ is in the millikelvin region. For an ordinary metal, $T_{\rm F} \sim 10^4$ K, the pairs are strongly overlapping with typical size $\xi_0 \sim 10^3$ Å, so that the onset of superconductivity is around 10 K. The high-temperature superconductors are bad metals with fewer charge carriers than ordinary metals and with low Fermi energy ($T_{\rm F} \sim 10^3$ K), the pairs are boson-like particles with typical $\xi_0 \sim 10$ Å and therefore $T_{\rm c} \sim 10$–100 K. The normal (non-superconducting) phase of these systems, however, as will be shown in Chapter 15, is not a normal Fermi liquid. Besides the mechanism of formation of high-temperature superconductivity, a major (up to now) still open problem of condensed matter theory is how to characterize the normal phase of these systems. By the way, one should notice that in ^{4}He $\xi_0 n^{1/3} \sim 1$ and indeed the superfluid critical temperature is of the order of the Bose–Einstein condensation temperature.

We now present the Landau theory of a normal Fermi liquid (Landau (1956, 1957, 1958); Nozières (1964); Nozières and Pines (1966); Baym and Pethick (1978, 1991)) that finds its application in the interval of temperatures between the Fermi temperature $T_{\rm F}$ and the critical temperature $T_{\rm c}$ of superfluidity in neutral liquid ^{3}He (Wheatley (1966)) and of superconductivity in the charged electron gas of metals (apart from anisotropy effects introduced by the lattice).

In both cases, neutral liquid ^{3}He and the charged electron liquid in metals, Landau theory describes the behavior of the normal (non-super) phase and also explains why the independent particle approximation works so well for thermodynamic and transport (Drude–Sommerfeld theory) properties of electrons in metals despite their strong interaction.

Neglecting for the moment the possibility of superfluid (superconducting) transition, the low-temperature behavior of a fermionic liquid is determined by the low-energy excitations from the ground state. These excitations with a well defined relation between their energy and momentum $\mathbf{p}$ (from which they derive the name of quasiparticles) are put in one-to-one correspondence with the free particles[1] excited from the Fermi sea. The quasiparticles are well defined as long as their energy $\epsilon(\mathbf{p})$ is much larger than their indeterminacy due to their lifetime τ, i.e. $\epsilon(\mathbf{p}) \gg \hbar/\tau$. In the opposite case, the quasiparticle concept itself loses its validity. Since the quasiparticles are fermions, their distribution is the Fermi distribution. The scattering is peaked around the Fermi surface and leads to

$$\tau = \frac{\hbar E_{\mathrm{F}}}{a(\epsilon - E_{\mathrm{F}})^2 + b(k_{\mathrm{B}}T)^2}, \tag{12.1}$$

where a and b are coefficients of the order of unity. At $\epsilon \sim \epsilon_{\mathrm{F}}$ the fraction of particles contributing to the scattering is of the order $k_{\mathrm{B}}T/E_{\mathrm{F}}$ and the number of states available is $k_{\mathrm{B}}T$, thus leading to Eq. (12.1). (See Problem 12.1.) At $\epsilon \neq E_{\mathrm{F}}$, also the term $(\epsilon - E_{\mathrm{F}})^2$ has to be included in the scattering rate. The Fermi liquid theory works very well as long as we are at $T \ll T_{\mathrm{F}}$ and energies close to the Fermi surface, i.e. it is applicable for $T \lesssim 0.1$ K in ^{3}He and even at room temperature for metals since $T_{\mathrm{F}} \sim 10^4$ K.

In the first approximation therefore the system of interacting particles is transformed into a gas of fermionic quasiparticles. The energy $\hat{\epsilon}(\mathbf{p}, \boldsymbol{\sigma})$ of each quasiparticle, however, depends via the interaction on the momentum distribution $\hat{n}(\mathbf{p}, \boldsymbol{\sigma})$ and becomes a functional of $\hat{n}(\mathbf{p}, \boldsymbol{\sigma})$. Here $\boldsymbol{\sigma}$ is the vector of Pauli matrices related to the spin operator by $\mathbf{s} = (1/2)\boldsymbol{\sigma}$ and both the quasiparticle energy $\hat{\epsilon}$ and the momentum distribution $\hat{n}$ are matrices in spin space denoted by the hat symbol. To be explicit $\hat{\epsilon} = \epsilon^0\sigma^0 + \epsilon^a\sigma^a$, $\hat{n} = n^0\sigma^0 + n^a\sigma^a$, σ^a $(a = x, y, z)$ is the set of Pauli matrices with σ^0 as the identity and satisfies the properties

$$\{\sigma^a, \sigma^b\} = 2\delta_{ab}, \quad [\sigma^a, \sigma^b] = 2\mathrm{i}\varepsilon_{abc}\sigma^c, \quad \mathrm{Tr}\sigma^a = 0,$$

ε_{abc} being the completely antisymmetric Ricci tensor.

The quasiparticle energy can therefore be obtained as the functional derivative with respect to the distribution function $\hat{n}(\mathbf{p}, \boldsymbol{\sigma})$ of the total energy E of the liquid:

$$\hat{\epsilon}(\mathbf{p}, \boldsymbol{\sigma}) = \frac{\delta E}{\delta \hat{n}(\mathbf{p}, \boldsymbol{\sigma})}. \tag{12.2}$$

When we perturb the system and produce a change $\delta\hat{n}(\mathbf{p}, \boldsymbol{\sigma})$ in the momentum distribution, the quasiparticle energy changes because of the rearrangement of all the others as

$$\delta\hat{\epsilon}(\mathbf{p}, \boldsymbol{\sigma}) = \mathrm{Tr}_{\sigma'} \int \frac{\mathrm{d}^3 p'}{(2\pi\hbar)^3} f(\mathbf{p}, \mathbf{p}'; \boldsymbol{\sigma}, \boldsymbol{\sigma}') \delta\hat{n}(\mathbf{p}', \boldsymbol{\sigma}'). \tag{12.3}$$

Notice that the quasiparticle phenomenological interaction function $f(\mathbf{p}, \mathbf{p}'; \boldsymbol{\sigma}, \boldsymbol{\sigma}')$ has the dimension of an energy times a volume. In the microscopic treatment of the interacting Fermi gas of Chapter 19, the quasiparticle interaction can be obtained from the dynamic limit of the scattering amplitude (see Eq. (19.27)). Since $\hat{\epsilon}(\mathbf{p}, \boldsymbol{\sigma})$ is a functional of the distribution $\hat{n}(\mathbf{p}, \boldsymbol{\sigma})$ (like in the Hartree–Fock approximation of Chapter 9), the total energy E is not given by the sum over all momenta and spin of $\hat{\epsilon}(\mathbf{p}, \boldsymbol{\sigma})\hat{n}(\mathbf{p}, \boldsymbol{\sigma})$. In the Landau

[1] We notice here the difference between the quasiparticle for degenerate Fermi systems and degenerate Bose systems as liquid ^{4}He below the superfluid temperature. In this last case there is no one-to-one correspondence between the particles and the quasiparticles, the low-lying excitations are phonon-like and there is no perturbative formulation of the problem.

theory only variations of the total energy can be considered:

$$\delta E = \mathrm{Tr}_\sigma \int \frac{V\mathrm{d}^3 p}{(2\pi\hbar)^3} \hat{\epsilon}^0(\mathbf{p},\sigma)\delta\hat{n}(\mathbf{p},\sigma) + \frac{1}{2}\mathrm{Tr}_{\sigma\sigma'} \int \frac{V\mathrm{d}^3 p}{(2\pi\hbar)^3} \int \frac{\mathrm{d}^3 p'}{(2\pi\hbar)^3} f(\mathbf{p},\mathbf{p}';\sigma,\sigma')\delta\hat{n}(\mathbf{p},\sigma)\delta\hat{n}(\mathbf{p}',\sigma'), \tag{12.4}$$

where $\hat{\epsilon}^0(\mathbf{p},\sigma)$ is the quasiparticle energy in isolation and the $f(\mathbf{p},\mathbf{p}';\sigma,\sigma')$ is the interaction function given by

$$\frac{1}{V} f(\mathbf{p},\mathbf{p}';\sigma,\sigma') = \frac{\delta^2 E}{\delta\hat{n}(\mathbf{p},\sigma)\delta\hat{n}(\mathbf{p}',\sigma')}. \tag{12.5}$$

In the isotropic case and in the absence of spin–orbit coupling, the interaction function can be decomposed in terms of the invariants under rotations in the spin space, i.e. the identity matrix $\mathbf{I}$ and the product $\boldsymbol{\sigma}\cdot\boldsymbol{\sigma}'$:

$$f(\mathbf{p},\mathbf{p}';\sigma,\sigma') = F(\mathbf{p},\mathbf{p}')\mathbf{I} + G(\mathbf{p},\mathbf{p}')\sigma\cdot\sigma'. \tag{12.6}$$

In the absence of an external magnetic field or spin polarization, $\hat{\epsilon}(\mathbf{p},\sigma)$ does not depend on the spin and therefore can be replaced by $\epsilon(\mathbf{p})$, similarly $\hat{n}(\mathbf{p},\sigma)$ by $n(\mathbf{p})$. Therefore in all the equilibrium properties, except for the spin susceptibility, $f(\mathbf{p},\mathbf{p}';\sigma,\sigma')$ will be replaced by $F(\mathbf{p},\mathbf{p}')$ and the trace over the spin variable gives only a multiplicity factor 2 for spin-1/2 particles. When the spin quantization axis is directed along the z-axis, we may write

$$f(\mathbf{p},\mathbf{p}';\uparrow,\uparrow) = f(\mathbf{p},\mathbf{p}';\downarrow,\downarrow) = F(\mathbf{p},\mathbf{p}') + G(\mathbf{p},\mathbf{p}'),$$
$$f(\mathbf{p},\mathbf{p}';\uparrow,\downarrow) = f(\mathbf{p},\mathbf{p}';\downarrow,\uparrow) = F(\mathbf{p},\mathbf{p}') - G(\mathbf{p},\mathbf{p}').$$

12.2 Equilibrium properties

The subtle feature of Fermi-liquid theory is that, even though ^{3}He and the electron gas in metals are strongly interacting systems, they can be described by an independent particle model due to the strong suppression of scattering by the exclusion principle for $T \ll T_\mathrm{F}$ and energies close to the Fermi surface. We are then dealing with a gas of quasiparticles with a Fermi gas distribution $n(\mathbf{p})$ and entropy S given by Eqs. (7.6) and (7.7), respectively. However, the energy of each quasiparticle depends functionally on the distribution of all the others and the expressions for all the observables, even though similar to those of the gas, have to involve the function $f(\mathbf{p},\mathbf{p}';\sigma,\sigma')$ with $\mathbf{p}$ and $\mathbf{p}'$ evaluated at the Fermi surface. The interaction is then effectively taken into account by changing the coefficients of the thermodynamic expressions valid for the gas.

12.2.1 Effective mass and specific heat

The starting point is the definition of the effective mass m^* of the quasiparticles in terms of the group velocity at the Fermi surface as an obvious generalization of the free particle mass:

$$\nabla_{\mathbf{p}}\epsilon(\mathbf{p})|_{p=p_{\mathrm{F}}} = \frac{\mathbf{p}_F}{m^*} \equiv \mathbf{v}_{\mathrm{F}}^{\mathrm{L}}. \tag{12.7}$$

The superscript L stands for Landau and m^* is different from m and effectively takes into account the backflow of the liquid around the moving particle. It can be quite different from m as for instance in the so-called heavy fermions (see Hewson (1997)) (e.g. UPt_3, $CeIn_3$, ...).

The total momentum $\mathbf{P}$ of the system must be equal to the mass flow: in terms of quasiparticles we can write

$$\mathbf{P} = 2V\int \frac{\mathrm{d}^3 p}{(2\pi\hbar)^3}\mathbf{p}n(\mathbf{p}) = 2mV\int \frac{\mathrm{d}^3 p}{(2\pi\hbar)^3}\nabla_{\mathbf{p}}\epsilon(\mathbf{p})n(\mathbf{p}), \tag{12.8}$$

where m is the bare mass and $\nabla_{\mathbf{p}}\epsilon(\mathbf{p})$ is the group velocity of the quasiparticles. The second equality in Eq. (12.8) follows from the observation that the mass flow must be equal to the total number current of the liquid times the mass of a single particle. The functional derivative with respect to $n(\mathbf{p})$ of the identity (12.8) gives

$$\begin{aligned}\mathbf{p} &= m\nabla_{\mathbf{p}}\epsilon(\mathbf{p}) + 2m\int \frac{\mathrm{d}^3 p'}{(2\pi\hbar)^3}\nabla_{\mathbf{p}'}F(\mathbf{p},\mathbf{p}')n(\mathbf{p}')\\ &= m\nabla_{\mathbf{p}}\epsilon(\mathbf{p}) - 2m\int \frac{\mathrm{d}^3 p'}{(2\pi\hbar)^3}F(\mathbf{p},\mathbf{p}')\nabla_{\mathbf{p}'}n(\mathbf{p}'),\end{aligned} \tag{12.9}$$

where in obtaining the last line we have integrated by parts. As $T \to 0$, the one-to-one correspondence between the quasiparticle and the particle states implies that

$$\nabla_{\mathbf{p}'}n(\mathbf{p}') = -\hat{\mathbf{p}}'\delta(p_{\mathrm{F}} - p') = \nabla_{\mathbf{p}'}\epsilon(\mathbf{p})\frac{\partial n(\mathbf{p}')}{\epsilon(\mathbf{p}')}, \tag{12.10}$$

so that, evaluating the previous equation at $p = p_{\mathrm{F}}$ and taking the scalar product with $\hat{\mathbf{p}}$ of both sides, we obtain

$$\frac{1}{m} = \frac{1}{m^*} + \frac{\nu_{\mathrm{F}}}{m}\int \frac{\mathrm{d}\Omega}{4\pi}F(\mathbf{p}_{\mathrm{F}},\mathbf{p}'_{\mathrm{F}})\cos(\theta), \tag{12.11}$$

where $\mathrm{d}\Omega$ is the element of the solid angle at the Fermi surface, θ is the angle between $\mathbf{p}_{\mathrm{F}}$ and $\mathbf{p}'_{\mathrm{F}}$ and ν_{F} is the single particle density of states of the free gas defined in Eq. (7.14) evaluated at the Fermi surface.

We notice that F, when evaluated at $p = p' = p_{\mathrm{F}}$, depends on the angle θ only and the integral on the right-hand side of the above equation corresponds to selecting the first harmonic in a Legendre polynomial expansion of the function F. We then introduce the expansion of F in terms of Legendre polynomials $P_l(\cos\theta)$

$$F(\mathbf{p}_{\mathrm{F}},\mathbf{p}'_{\mathrm{F}}) \equiv F(\theta) = \frac{m}{m^*\nu_{\mathrm{F}}}\sum_{l=0}^{\infty}F_l P_l(\cos\theta), \tag{12.12}$$

where the normalization factor has been chosen to make the expansion coefficients dimensionless. From the orthogonality relations of the Legendre polynomials

$$\frac{2l+1}{2}\int_{-1}^{1} \mathrm{d}\cos\theta \; P_l(\cos\theta)P_{l'}(\cos\theta) = \delta_{ll'}, \tag{12.13}$$

we obtain for the coeffcients F_l

$$F_l = \frac{m^*\nu_F}{m}\frac{2l+1}{2}\int_{-1}^{1} \mathrm{d}\cos\theta \; P_l(\cos\theta)F(\theta). \tag{12.14}$$

Similarly for the function G:

$$G(\mathbf{p}_F, \mathbf{p}'_F) \equiv G(\theta) = \frac{m}{m^*\nu_F}\sum_{l=0}^{\infty} G_l P_l(\cos\theta). \tag{12.15}$$

In the following we will express all the thermodynamical physical quantities in terms of the first four coefficients F_0, G_0, F_1 and G_1 corresponding to the coefficients of the expansion for the first two polynomials $P_0(\cos\theta) = 1$ and $P_1(\cos\theta) = \cos\theta$.

The integral term in Eq. (12.11) amounts to $F_1/(3m^*)$ and therefore

$$\frac{m^*}{m} = 1 + \frac{F_1}{3}. \tag{12.16}$$

Since the number of particles is equal to the number of quasiparticles and p_F proportional to $n^{1/3}$ is invariant, the Fermi velocity is reduced as $v_F^L = v_F m/m^*$. The procedure to evaluate the quasiparticle specific heat c_V^L goes through exactly as for the free gas expression c_V of Eq. (7.37). We have to notice, however, that in the present case the effective mass m^* appears in the quasiparticle density of states at the Fermi surface ν_F^L. We then obtain

$$\nu_F^L = \frac{m^*}{m}\nu_F, \quad c_V^L = \frac{m^*}{m}c_V. \tag{12.17}$$

12.2.2 Compressibility and sound velocity

We can deal with the sound propagation in the quasi-classical regime for fermions, which is defined by Eq. (6.2), and take the classical expression for the sound velocity in a liquid with density $\rho = mn$ as the hydrodynamic response to a variation of the pressure P:

$$s^2 = \frac{1}{m}\frac{\partial P}{\partial n}. \tag{12.18}$$

Recall that the isothermal compressibility is given by Eq. (4.35):

$$\kappa_T = -(1/v)(\partial v/\partial P)_T = (1/n^2)(\partial n/\partial \mu)_T,$$

where v is the specific volume. Then, the sound velocity s^2 can be expressed in terms of the compressibility κ_T:

$$s^2 = \frac{1}{mn\kappa_T} = \frac{n}{m}\left(\frac{\partial \mu}{\partial n}\right)_{T,V}. \tag{12.19}$$

As $T \to 0$, for a degenerate Fermi gas $\mu = E_F \sim n^{2/3}$ and therefore

$$s^2 = \frac{2}{3}\frac{E_F}{m} = \frac{1}{3}v_F^2. \tag{12.20}$$

For a degenerate Fermi gas at the concentration of liquid ^{3}He, from the Fermi energy evaluated in Chapter 7, $E_F = 6.7 \times 10^{-23}$J and $m = 5.01 \times 10^{-27}$Kg we obtain $v_F = 163$m/s and $s = 94$m/s.

For a degenerate Fermi liquid the number of quasiparticles is equal to the number of particles but we have to make explicit the dependence of the energy of the quasiparticles on the interaction function $F(\mathbf{p}, \mathbf{p}')$. In this case

$$n\left(\frac{\partial \mu}{\partial n}\right)_{T,V} = n\left(\frac{\partial \mu}{\partial p_F}\right)_{T,V}\left(\frac{\partial p_F}{\partial n}\right)_{T,V} = \frac{1}{3}p_F\left(\frac{\partial \mu}{\partial p_F}\right)_{T,V}. \tag{12.21}$$

The derivative of μ with respect to p_F introduces the function F since a variation of p_F induces also a variation of the distribution function:

$$\begin{aligned}\frac{\partial \mu}{\partial p_F} &= \frac{\partial \epsilon_F}{\partial p_F} + 2\int \frac{d^3 p'}{(2\pi\hbar)^3} F(\mathbf{p}_F, \mathbf{p}')\frac{\partial n(\mathbf{p}')}{\partial p_F} \\ &= \frac{p_F}{m^*} + \frac{p_F \nu_F}{m}\int \frac{d\Omega}{4\pi}F(\theta) \\ &= \frac{p_F}{m^*}(1 + F_0).\end{aligned} \tag{12.22}$$

Then from Eq. (12.16) and Eq. (12.19), for the sound velocity we obtain

$$(s^L)^2 = \frac{1}{3}\frac{p_F^2}{m}\frac{1+F_0}{m^*} = s^2\frac{1+F_0}{1+F_1/3}. \tag{12.23}$$

The Landau expressions for the isothermal compressibility κ_T and the thermodynamic density of states $\partial n/\partial \mu$ are

$$\kappa_T^L = \kappa_T\frac{1+F_1/3}{1+F_0}, \quad \left(\frac{\partial n}{\partial \mu}\right)^L = \left(\frac{\partial n}{\partial \mu}\right)\frac{1+F_1/3}{1+F_0} = \frac{\nu_F^L}{1+F_0}, \tag{12.24}$$

where $\kappa_T = \nu_F/n^2$ is the free gas compressibility. A Fermi liquid is stiffer than a gas and its thermodynamic density of states is different from the single-particle density of states.

12.2.3 Spin susceptibility

In order to calculate the spin susceptibility, we have to reintroduce the spin dependence explicitly. For small magnetic fields, the variation of the energy $\hat{\epsilon}(\mathbf{p}, \boldsymbol{\sigma})$ is linear in the field. However, in the variation of the quasiparticle energy we shall have a term due to the change of the distribution induced by the field itself:

$$\delta\hat{\epsilon}(\mathbf{p}, \boldsymbol{\sigma}) = -\gamma_0 \mathbf{H}\cdot\boldsymbol{\sigma} + \mathrm{Tr}_{\sigma'}\int \frac{d^3 p'}{(2\pi\hbar)^3} f(\mathbf{p}, \mathbf{p}'; \boldsymbol{\sigma}, \boldsymbol{\sigma}')\delta\hat{n}(\mathbf{p}', \boldsymbol{\sigma}') \tag{12.25}$$

$$= -\gamma(\mathbf{p})\mathbf{H}\cdot\boldsymbol{\sigma}, \tag{12.26}$$

where γ_0 is the magnetic moment associated with the particle and $\gamma(\mathbf{p})$ is the full magnetic moment of the quasiparticle. Notice that

$$\frac{\delta\hat{n}(\mathbf{p}',\boldsymbol{\sigma}')}{\delta\hat{\epsilon}(\mathbf{p}',\boldsymbol{\sigma}')}|_{\mathbf{H}=0}\delta\hat{\epsilon}(\mathbf{p}',\boldsymbol{\sigma}') = -\frac{\partial n(\mathbf{p}')}{\partial\epsilon(\mathbf{p}')}\gamma(\mathbf{p}')\mathbf{H}\cdot\boldsymbol{\sigma}'$$

and in the strongly degenerate limit

$$\frac{\partial n(\mathbf{p}')}{\partial\epsilon(\mathbf{p}')} = -\frac{m^*}{p_F}\delta(p'-p_{\mathrm{F}}).$$

By using the above equations in Eq. (12.25) and observing that

$$\mathrm{Tr}_{\sigma'}(\mathbf{H}\cdot\boldsymbol{\sigma}') = 0, \quad \mathrm{Tr}_{\sigma'}(\boldsymbol{\sigma}\cdot\boldsymbol{\sigma}'\boldsymbol{\sigma}'\cdot\mathbf{H}) = 2\boldsymbol{\sigma}\cdot\mathbf{H},$$

one obtains

$$\gamma(\mathbf{p}_{\mathrm{F}}) = \gamma_0 - 2m^* p_{\mathrm{F}}\gamma(\mathbf{p}_F)\int\frac{\mathrm{d}\Omega}{(2\pi\hbar)^3}G(\theta) = \frac{\gamma_0}{1+G_0}. \tag{12.27}$$

Hence the spin susceptibility is given by

$$\chi^{\mathrm{L}} = \chi\frac{m^*}{m}\frac{1}{1+G_0}, \tag{12.28}$$

where the normalization of G_0 is according to Eq. (12.15) and $\chi = \gamma_0^2\nu_{\mathrm{F}}$ is the spin susceptibility of the Fermi gas with $\gamma_0 = g\mu_{\mathrm{B}}/2$ in the notation of Eq. (7.32). Note that the compressibility and the spin susceptibility become infinite for $F_0 = -1$ and $G_0 = -1$, respectively. In the first case, the system would be unstable for density variations with the related phase separation. In the second case, the system would undergo a ferromagnetic ordering. In general, the stability criterion of a Fermi liquid is that the expansion parameters must satisfy $F_l > -(2l+1)$ and $G_l > -(2l+1)$ (Pomeranchuk instabilities). This criterion follows from the requirement that $E - \mu N$ be an absolute minimum with respect to a variation of the distribution $\hat{n}(\mathbf{p},\boldsymbol{\sigma})$ (Baym and Pethick (1978)).

For liquid ^{3}He the expansion parameters can be determined experimentally by measuring the specific heat, the sound velocity, or the compressibility and the spin susceptibility. From the measured values (Wheatley (1966), Abel et al. (1966)) of the specific heat, the Fermi velocity $v_{\mathrm{F}} \sim 53.8\,\mathrm{m/s}$ and the sound velocity $s \sim 187\,\mathrm{m/s}$ at ~ 0.3 atm, one obtains large and positive values of $F_0 \sim 10.77$ and $F_1 \sim 6.25$. In units of the free spin susceptibility, the measured value of χ is around 10. This leads to $G_0 \sim -0.76$. Liquid ^{3}He is a polarized homogeneous liquid near to a ferromagnetic instability.

12.3 Transport properties

12.3.1 Zero sound

We start by noticing that the equilibrium properties of liquid ^{3}He test the qualitative validity of the Landau theory: all the quantities have the same behavior as those of a gas, i.e. the specific heat is linear in temperature, the compressibility and spin susceptibility are finite. As for ^{4}He with the second sound in the Landau theory of the two fluids, which we shall encounter in Chapter 15, a great success of the Landau normal Fermi-liquid theory was decided by the prediction of a new phenomenon, the zero sound, a collective excitation with the interaction as the restoring force, which has been observed in 1966 (Abel et al. (1966)).

Ordinary sound, being a compressional wave, produces a local modulation of the density and, in turn, an isotropic deformation (swelling and shrinking) of the Fermi surface. Zero sound instead produces a volume-preserving distortion of the Fermi surface.

The hydrodynamical compressional wave of the first sound must have a wave length much larger than the interatomic distance. In the quantum Fermi liquid the characteristic time of propagation, inverse of the frequency of the wave ω^{-1}, must be much longer than the quasiparticle lifetime $\tau \sim \hbar/\epsilon(\mathbf{p})$, i.e. $\omega\tau \ll 1$. On the other hand, as $T \to 0$, $\tau \to \infty$ and there is no time to locally thermalize the system: the ordinary sound is strongly damped.

For the zero sound instead, $\omega\tau \gg 1$, there is no thermalization and the velocity in the gas turns out to be equal to v_F, instead of $v_F/\sqrt{3}$ as for the ordinary sound.

Discussing the variation with respect to local oscillations, the simplest approach is to introduce a distribution of quasiparticles as a function of position and time $n(\mathbf{p}, \mathbf{r}, t)$, potentially in conflict with the Heisenberg principle. On the one hand, as done in Chapter 6, zero sound can be discussed in the semiclassical approximation because the De Broglie wave length $\lambda_F = \hbar/p_F$ is of the order of the interatomic distance and can be easily considered as point-like with respect to the distance over which the distribution function $n(\mathbf{p}, \mathbf{r}, t)$ varies.[2] At finite temperature, on the other hand, the spread Δp of the momentum in the distribution is of the order $\Delta p \sim k_B T/v_F$. A corresponding variation of the position in the distribution is $\Delta x \sim \hbar/\Delta p \sim \lambda_F(T_F/T)$. As long as $T \ll T_F$, $\Delta x \gg \lambda_F$ and there is no inconsistency with the Heisenberg principle.

We can then use the Boltzmann equation (6.3) for $n(\mathbf{p}, \mathbf{r}, t)$, where now the collision integral provides the variation of the quasiparticle number per unit volume in phase space due to the collisions:

$$\frac{\partial n}{\partial t} + \nabla_{\mathbf{p}}\epsilon(\mathbf{p}, \mathbf{r}, t) \cdot \nabla_{\mathbf{r}} n(\mathbf{p}, \mathbf{r}, t) - \nabla_{\mathbf{r}}\epsilon(\mathbf{p}, \mathbf{r}, t) \cdot \nabla_{\mathbf{p}} n(\mathbf{p}, \mathbf{r}, t) = I(n)$$

$$\frac{\partial n}{\partial t} - \{\epsilon(\mathbf{p}, \mathbf{r}, t), n(\mathbf{p}, \mathbf{r}, t)\}_{PB} = I(n), \qquad (12.29)$$

where $\{\ldots, \ldots\}_{PB}$ are the Poisson brackets and the collisionless evolution is governed by the classical equations of motion with the (space and time dependent) quasiparticle energy $\epsilon(\mathbf{p}, \mathbf{r}, t)$ as the Hamiltonian:

$$\frac{d\mathbf{r}}{dt} \equiv \mathbf{v_p} = \nabla_{\mathbf{p}}\epsilon(\mathbf{p}, \mathbf{r}, t), \qquad \frac{d\mathbf{p}}{dt} = -\nabla_{\mathbf{r}}\epsilon(\mathbf{p}, \mathbf{r}, t). \qquad (12.30)$$

[2] In the opposite case when no semiclassical approximation is valid, we should introduce the Wigner distribution or use the quantum response functions of Chapter 10 and the Poisson brackets should be replaced by $\hbar^{-1}$ times the commutator.

To find the conservation law related to the total quasiparticle number, we integrate Eq. (12.29) over the momentum. We observe that

$$\nabla_{\mathbf{p}}\epsilon(\mathbf{p},\mathbf{r},t)\cdot\nabla_{\mathbf{r}}n(\mathbf{p},\mathbf{r},t)=\nabla_{\mathbf{r}}\cdot n(\mathbf{p},\mathbf{r},t)(\nabla_{\mathbf{p}}\epsilon(\mathbf{p},\mathbf{r},t))-n(\mathbf{p},\mathbf{r},t)\nabla_{\mathbf{p}}\cdot(\nabla_{\mathbf{r}}\epsilon(\mathbf{p},\mathbf{r},t)).$$

The second term on the right-hand side of the above equation combines with the third on the right-hand side of Eq. (12.29) to yield a total derivative with respect to the momentum, which hence vanishes upon integration since the distribution function vanishes at infinity. By integrating over the momentum, the collision integral of Eq. (6.3) vanishes due to the exact compensation of scattering *in* and *out* terms, meaning the local conservation of the quasiparticle number. As a result the momentum-integrated kinetic equation yields the continuity equation for the quasiparticle number in the form

$$\frac{\partial n(\mathbf{r},t)}{\partial t}+\nabla\cdot\mathbf{j}(\mathbf{r},t)=0, \tag{12.31}$$

where the number density reads

$$n(\mathbf{r},t)=2\int\frac{\mathrm{d}^3p}{(2\pi\hbar)^3}n(\mathbf{p},\mathbf{r},t)$$

and the number current density

$$\mathbf{j}(\mathbf{r},t)=2\int\frac{\mathrm{d}^3p}{(2\pi\hbar)^3}\nabla_{\mathbf{p}}\epsilon(\mathbf{p},\mathbf{r},t)\,n(\mathbf{p},\mathbf{r},t).$$

Similarly, the conservation related to the energy and momentum may be obtained by integration of Eq. (12.29) over the momentum after multiplication by the quasiparticle energy $\epsilon(\mathbf{p},\mathbf{r},t)$ and momentum $\mathbf{p}$, respectively.

For small deviations from equilibrium, with $n(\mathbf{p},\mathbf{r},t)=n_0(\mathbf{p})+\delta n(\mathbf{p},\mathbf{r},t)$ and $\epsilon(\mathbf{p},\mathbf{r},t)=\epsilon_0(\mathbf{p})+\delta\epsilon(\mathbf{p},\mathbf{r},t)$, we can linearize Eq. (12.29):

$$\frac{\partial\delta n}{\partial t}+\mathbf{v}_{\mathbf{p}}\cdot\nabla_{\mathbf{r}}\delta n-\nabla_{\mathbf{r}}\delta\epsilon\cdot\nabla_{\mathbf{p}}n_0=I(n). \tag{12.32}$$

$\mathbf{v}_{\mathbf{p}}$ is the equilibrium value and

$$\nabla_{\mathbf{p}}n_0(\mathbf{p})=\frac{\partial n_0(\mathbf{p})}{\partial\epsilon(\mathbf{p})}\nabla_{\mathbf{p}}\epsilon(\mathbf{p}).$$

Equation (12.32) reads

$$\frac{\partial\delta n}{\partial t}+\mathbf{v}_{\mathbf{p}}\cdot\nabla_{\mathbf{r}}\left(\delta n-\frac{\partial n_0(\mathbf{p})}{\partial\epsilon(\mathbf{p})}\delta\epsilon\right)=I(n). \tag{12.33}$$

As $T\to 0$, we reach the collisionless regime $I(n)\sim\delta n/\tau=(1/\omega\tau)\omega\delta n\ll\omega\delta n$ and we set $I(n)=0$. In the absence of an external field, the third term in Eq. (12.33) is present only in the Fermi liquid due to the dependence of the quasiparticle energy on the space-dependent quasiparticle distribution (see Eqs. (12.3)–(12.6)):

$$\nabla_{\mathbf{r}}\delta\epsilon(\mathbf{p},\mathbf{r},t)=2\int\frac{\mathrm{d}^3p'}{(2\pi\hbar)^3}F(\mathbf{p},\mathbf{p}')\nabla_{\mathbf{r}}\delta n(\mathbf{p}',\mathbf{r},t). \tag{12.34}$$

In the presence of an external field $U(\mathbf{r},t)$ coupled with the local density $n(\mathbf{r},t)$, a term $\nabla_{\mathbf{r}}U(\mathbf{r},t)$ should be added on the right-hand side of Eq. (12.34). In this context it is worth noticing that the gradient of the chemical potential appears as an effective field. By recalling Eq. (12.10), we observe that a non-trivial solution only develops at the Fermi surface. We then look for a solution in the form of a wave:

$$\delta n(\mathbf{p}, \mathbf{r}, t) = \delta(p - p_{\mathrm{F}}) u(\theta, \phi) \mathrm{e}^{\mathrm{i}(\mathbf{r}\cdot\mathbf{k} - \omega t)}, \tag{12.35}$$

where θ and ϕ designate the direction of $\hat{\mathbf{p}}$ with respect to the direction of $\mathbf{k}$.

Naming $s_0 = \omega/k$ the zero sound velocity and α the angle between $\mathbf{p}_{\mathrm{F}}$ and $\mathbf{p}'_{\mathrm{F}}$ and inserting the form Eq. (12.35) into Eq. (12.33) one obtains

$$\left(\frac{s_0}{v_F} - \cos(\theta)\right) u(\theta, \phi) = \frac{2 p_F m^* \cos(\theta)}{(2\pi\hbar)^3} \int \mathrm{d}\Omega' F(\alpha) u(\theta', \phi'). \tag{12.36}$$

The solution of Eq. (12.36) depends on the form of $F(\alpha)$ and is usually very involved. We can study the simple case where the expansion in Eq. (12.12) of F involves F_0 only:

$$\left(\frac{s_0}{v_F} - \cos(\theta)\right) u(\theta, \phi) = \frac{\cos(\theta) F_0}{4\pi} \int \mathrm{d}\Omega' u(\theta', \phi'). \tag{12.37}$$

From Eq. (12.37) we get

$$u(\theta, \phi) = \mathrm{const} \frac{\cos(\theta)}{s_0/v_{\mathrm{F}} - \cos(\theta)} \tag{12.38}$$

and the Fermi surface is elongated along the $\mathbf{k}$ direction with the zero sound velocity determined by

$$\frac{1}{F_0} = \frac{1}{4\pi} \int \mathrm{d}\Omega \frac{\cos(\theta)}{s_0/v_{\mathrm{F}} - \cos(\theta)}. \tag{12.39}$$

After integrating the right-hand side of Eq. (12.39), we obtain

$$\frac{1}{F_0} = \frac{s_0}{2v_{\mathrm{F}}} \ln \frac{s_0/v_{\mathrm{F}} + 1}{s_0/v_{\mathrm{F}} - 1} - 1. \tag{12.40}$$

For an attractive force $F_0 < 0$ there is no real solution for s_0 and the zero sound mode is damped. When $F_0 \to 0$

$$\frac{s_0}{v_{\mathrm{F}}} \sim 1 + \frac{2}{\mathrm{e}^2} \mathrm{e}^{-2/F_0} \to 1. \tag{12.41}$$

The zero sound velocity coincides with the Fermi velocity and there is no thermalization.

In liquid ^{3}He the zero sound has been observed with a value $s_0/v_{\mathrm{F}} = 3.60 \pm 0.01$ (Abel et al. (1966)). With the value $F_0 = 10.77$ given in Section 12.2.3, from Eq. (12.40) one obtains $s_0/v_{\mathrm{F}} = 2.05$. A better agreement with the experimental value is obtained with the inclusion of F_1 even though there is no control on the exclusion of further terms in the expansion. One could also study the spin wave propagation by including the function $G(\alpha)$. Since $G_0 < 0$, spin wave modes are damped in liquid ^{3}He.

12.3.2 Transport coefficients

To evaluate the transport coefficients and their temperature dependence, we must solve the kinetic equation for the quasiparticle distribution function in the presence of the external fields specific for each transport phenomenon considered. One could, for instance, assume the system to be in local equilibrium with a local coarse grained distribution of quasiparticles, characterized by a local chemical potential $\mu(\mathbf{r}, t)$, temperature $T(\mathbf{r}, t)$ and magnetic field $\mathbf{H}(\mathbf{r}, t)$, whose spatial gradient are responsible for charge (mass), thermal and spin transport, respectively. We must also specify how the collision integral acts in each case. This leads, generally, to involved expressions for the transport scattering time, which are different for the charge, heat and spin diffusion. Furthermore, the results may depend on the approximation considered.

To be simple and avoid these complications here, we do not consider scattering among the quasiparticles and only take into account scattering by impurities. Indeed, recall that the scattering time among quasiparticles near the Fermi surface goes to infinity as $(\hbar/k_B T)T_F/T$, when $T \to 0$. As long as T is sufficiently low so that the scattering time among the quasiparticles is longer than the scattering time τ from impurities, the system is in the collisionless regime as far as scattering among the quasiparticles is concerned. The scattering integral is determined by τ due to the impurities, with mean free path of the quasiparticles given by $l = v_F\tau$. With these limitations, we can evaluate the transport coefficients in normal Fermi liquids by referring to the correspondence established in Chapter 11 between the semiclassical resistive motion of the electrons and the viscous motion of colloidal particles in the Brownian diffusion. In this section, we first present a few heuristic arguments to obtain the expressions of the kinetic coefficients of a Fermi liquid and then provide a better justification on the basis of the Boltzmann equation for quasiparticles introduced in the previous subsection.

Let us start with the case of charge transport. The local coarse grained density $n(\mathbf{r}, t)$ satisfies the Fokker–Planck equation (11.15), which by means of Eqs. (11.35) and (11.38) leads to the charge-conserving density–density response function (11.39):

$$\frac{\delta n(\mathbf{q}, \omega)}{\delta \mu(\mathbf{q}, \omega)} = \frac{\partial n}{\partial \mu} \frac{Dq^2}{Dq^2 - \mathrm{i}\omega}, \tag{12.42}$$

where $n(\mathbf{q}, \omega)$, $\mu(\mathbf{q}, \omega)$ are the Fourier transforms of $n(\mathbf{r}, t)$ and $\mu(\mathbf{r}, t)$, respectively, and to the generalized Einstein relation for the electrical conductivity Eq. (11.40):

$$\sigma = e^2 D \frac{\partial n}{\partial \mu}, \quad D = \frac{n}{m}\left(\frac{\partial n}{\partial \mu}\right)^{-1} \tau = \frac{nl}{p_F}\left(\frac{\partial n}{\partial \mu}\right)^{-1}. \tag{12.43}$$

The expression for D in Eq. (12.43) can be readily adapted to the Fermi liquid. The mean free path l depends on the distribution of the impurities only, n and p_F are invariant, whereas $\frac{\partial n}{\partial \mu}$ varies according to Eq. (12.24), hence the Landau diffusion coefficient for charge D^L reads

$$D^L = D\frac{1 + F_0}{1 + F_1/3}. \tag{12.44}$$

Corrections at finite T, due to the explicit evaluation of the effect of the collisions among the quasiparticles, would modify l and therefore D. The electrical conductivity σ^L, via

Eq. (12.43), reads

$$\sigma^{\mathrm{L}} = e^2 \left(\frac{\partial n}{\partial \mu}\right)^{\mathrm{L}} D^{\mathrm{L}} = e^2 \frac{\partial n}{\partial \mu} D. \tag{12.45}$$

For the thermal transport, by assuming the validity of the Wiedemann–Franz law (11.56), we may conclude that also the thermal conductivity is unchanged with respect to the Fermi gas value, i.e. $\kappa^{\mathrm{L}} = \kappa$. Since for the Fermi gas, according to Eq. (11.55), $\kappa = c_V D_{\mathrm{Q}}$ with $D_{\mathrm{Q}} = D$, we deduce how the kinetic coefficient $D_{\mathrm{Q}}^{\mathrm{L}}$ must be renormalized by using the specific heat renormalization of Eq. (12.17):

$$\kappa^{\mathrm{L}} = c_V^{\mathrm{L}} D_{\mathrm{Q}}^{\mathrm{L}}, \quad D_{\mathrm{Q}}^{\mathrm{L}} = \frac{\pi^2 k_{\mathrm{B}}^2}{3e^2} \sigma T \left(c_V^{\mathrm{L}}\right)^{-1} = \frac{D}{1 + F_1/3} = \frac{D^{\mathrm{L}}}{1 + F_0}. \tag{12.46}$$

An argument in the same spirit of the generalized Einstein relation (12.45) for the electrical conductivity can be used to express the spin conductivity in terms of the spin susceptibility and the spin diffusion coefficient. Within the local equilibrium assumption we are adopting here, spin transport arises as a consequence of a magnetic field gradient so that spin-up (spin-down) quasiparticles will move towards regions of higher (lower) Zeeman energy. The spin flow will be characterized by the spin diffusion coefficient $D_{\mathrm{spin}}^{\mathrm{L}}$, which must contain the spin flow velocity. The question arises whether the spin flow velocity is the same as the charge (mass) flow. When we derived the effective mass expression (see Eq. (12.16)), we imposed the condition (see Eq. (12.9)) that the quasiparticle mass flow is the same as that of the liquid. This led to the condition that the operator $\mathbf{p}/m$ must be equal to the operator $(1 + F_1/3)\mathbf{p}/m^*$. For spin flow there is no such compelling condition to apply. Hence, all we can say is that the quasiparticle spin flow will be associated with the operator $(1 + G_1/3)\mathbf{p}/m^* = ((1 + G_1/3)/(1 + F_1/3))\mathbf{p}/m$. This leads then to a spin conductivity of the form

$$\begin{aligned} \sigma_{\mathrm{spin}}^{\mathrm{L}} &= \chi^{\mathrm{L}} D_{\mathrm{spin}}^{\mathrm{L}} = \frac{1 + G_1/3}{1 + F_1/3} \chi D, \\ D_{\mathrm{spin}}^{\mathrm{L}} &= \frac{(1 + G_0)(1 + G_1/3)}{(1 + F_1/3)^2} D = D^{\mathrm{L}} \frac{(1 + G_0)(1 + G_1/3)}{(1 + F_0)(1 + F_1/3)}. \end{aligned} \tag{12.47}$$

If one took into account the modifications introduced by the explicit evaluation of the collision integral for each of the transport phenomena considered above, different temperature corrections to the diffusion constant D would be present.

After the above heuristic considerations, let us evaluate the kinetic coefficients by using the linearized Boltzmann equation for the quasiparticles (12.33). By integrating over the momentum $\mathbf{p}$ we obtain the continuity equation (12.31) with the density

$$\delta n(\mathbf{r}, t) = 2 \int \frac{d^3 p}{(2\pi\hbar)^3} \delta n(\mathbf{p}, \mathbf{r}, t) \tag{12.48}$$

and the current

$$\mathbf{j}(\mathbf{r}, t) = 2 \int \frac{d^3 p}{(2\pi\hbar)^3} \mathbf{v}_{\mathbf{p}} \left(\delta n(\mathbf{p}, \mathbf{r}, t) - \frac{\partial n_0(\mathbf{p})}{\partial \epsilon_0(\mathbf{p})} 2 \int \frac{d^3 p'}{(2\pi\hbar)^3} F(\mathbf{p}, \mathbf{p}') \delta n(\mathbf{p}', \mathbf{r}, t) \right). \tag{12.49}$$

We interchange the order of integration over $\mathbf{p}$ and $\mathbf{p}'$ in the second term on the right-hand side of the above equation and note that

$$2\int \frac{d^3p}{(2\pi\hbar)^3}\frac{\partial n_0(\mathbf{p})}{\partial\epsilon_0(\mathbf{p})}\mathbf{v}_{\mathbf{p}}F(\mathbf{p},\mathbf{p}') = \frac{F_1}{3}\mathbf{v}_{\mathbf{p}'}.$$

Then, after taking into account $\mathbf{v}_{\mathbf{p}}(1+F_1/3) = \mathbf{p}_F/m$ from Eq. (12.9), we get the simple result

$$\mathbf{j}(\mathbf{r},t) = 2\int \frac{d^3p}{(2\pi\hbar)^3}\mathbf{v}_{\mathbf{p}}\delta n(\mathbf{p},\mathbf{r},t)\left(1+\frac{F_1}{3}\right) = 2\int \frac{d^3p}{(2\pi\hbar)^3}\frac{\mathbf{p}}{m}\delta n(\mathbf{p},\mathbf{r},t), \qquad (12.50)$$

expressing the fact that the total current of the quasiparticles coincides with that of the original particles. The factor $1+F_1/3$ compensates for the effective mass in the expression of the quasiparticle velocity.

To obtain the explicit expression for the kinetic coefficient relating the charge (12.48) and current (12.49) densities, we employ the same collision integral in the presence of static impurities used for the Fermi gas in Section 11.4:

$$I(n) = -\frac{1}{\tau}(\delta n(\mathbf{p},\mathbf{r},t) - \langle\delta n(\mathbf{p}',\mathbf{r},t)\rangle), \qquad (12.51)$$

where for brevity $\langle\ldots\rangle$ indicates integration over the solid angle of $\mathbf{p}'$, whose absolute value is fixed at the Fermi surface. We then insert the collision integral (12.51) into the linearized Boltzmann equation (12.33), multiply both sides by the free particle velocity $\mathbf{p}/m$, integrate over the momentum and sum over the spin degeneracy. By interchanging again the order of integration over $\mathbf{p}$ and $\mathbf{p}'$ in the quasiparticle interaction term and neglecting the time derivative compared to the collision integral term (justified in the low frequency or diffusive regime $\omega\tau \ll 1$), we get

$$-\frac{2}{\tau}\int \frac{d^3p}{(2\pi\hbar)^3}\frac{\mathbf{p}}{m}\delta n(\mathbf{p},\mathbf{r},t)$$
$$= \frac{\partial}{\partial r_i}2\int \frac{d^3p}{(2\pi\hbar)^3}\delta n(\mathbf{p},\mathbf{r},t)\left(v^i_{\mathbf{p}}\frac{\mathbf{p}}{m} - 2\int \frac{d^3p'}{(2\pi\hbar)^3}v^i_{\mathbf{p}'}\frac{\mathbf{p}'}{m}F(\mathbf{p},\mathbf{p}')\frac{\partial n_0(\mathbf{p}')}{\partial\epsilon_0(\mathbf{p}')}\right).$$

Carrying out the integral over the momentum $\mathbf{p}'$ we get

$$\mathbf{j}(\mathbf{r},t) = -\tau\frac{\partial}{\partial r_i}\int \frac{d^3p}{(2\pi\hbar)^3}\delta n(\mathbf{p},\mathbf{r},t)v^i_{\mathbf{p}}\frac{\mathbf{p}}{m}(1+F_0). \qquad (12.52)$$

In the presence of a weak static gradient of the chemical potential

$$\nabla_{\mathbf{r}}\delta n(\mathbf{p},\mathbf{r},t) = -\frac{\partial n_0(\mathbf{p})}{\partial\epsilon_0(\mathbf{p})}\nabla_{\mathbf{r}}\mu(\mathbf{r}), \qquad (12.53)$$

the integral over the momentum in Eq. (12.52) comes mostly from the region around the Fermi surface and, considering the expressions (12.48) and (12.50) for the density and the current, we obtain the expression for the diffusive current in the form

$$\mathbf{j}(\mathbf{r},t) = -D\frac{1+F_0}{1+F_1/3}\nabla_{\mathbf{r}}\delta n(\mathbf{r},t) = -D^{\mathrm{L}}\nabla_{\mathbf{r}}\delta n(\mathbf{r},t), \qquad (12.54)$$

with $D = v_F^2\tau/d$, thus confirming the heuristic derivation of Eq. (12.44).

To obtain the kinetic coefficients for thermal transport, we first determine the expression for the energy current $\mathbf{J}_Q(\mathbf{r},t)$. By multiplying the linearized Boltzmann equation (12.33) by the quasiparticle energy $\epsilon(\mathbf{p})$, we get

$$\mathbf{J}_Q(\mathbf{r},t) = 2\int \frac{d^3p}{(2\pi\hbar)^3}\mathbf{v}_\mathbf{p}\epsilon(\mathbf{p})\left(\delta n(\mathbf{p},\mathbf{r},t) - \frac{\partial n_0(\mathbf{p})}{\partial \epsilon_0(\mathbf{p})}2\int \frac{d^3p'}{(2\pi\hbar)^3}F(\mathbf{p},\mathbf{p}')\delta n(\mathbf{p}',\mathbf{r},t)\right), \tag{12.55}$$

which has the same form as Eq. (12.49) for the number current. Hence, in a similar way, the presence of the Landau function F in the integral over $\mathbf{p}'$ produces a factor $1 + F_1/3$, which compensates the effective mass in the quasiparticle velocity. As a result we obtain

$$\mathbf{J}_Q(\mathbf{r},t) = 2\int \frac{d^3p}{(2\pi\hbar)^3}\epsilon(\mathbf{p})\frac{\mathbf{p}}{m}\delta n(\mathbf{p},\mathbf{r},t) \tag{12.56}$$

$$= \mu\mathbf{j}(\mathbf{r},t) + 2\int \frac{d^3p}{(2\pi\hbar)^3}(\epsilon(\mathbf{p})-\mu)\frac{\mathbf{p}}{m}\delta n(\mathbf{p},\mathbf{r},t). \tag{12.57}$$

In Eq. (12.57), we have split the energy current in two terms. By imposing the condition of no charge (mass) flow, only the second term must be evaluated. To get an explicit expression, the distribution function must be obtained from the solution of the linearized Boltzmann equation (12.33). By taking into account that the temperature gradient is a static perturbation in analogy with Eq. (12.53), the Boltzmann equation to be solved becomes

$$\begin{aligned} &\mathbf{v}_\mathbf{p}\cdot\nabla_\mathbf{r}T\left(\frac{\partial n_0(\mathbf{p})}{\partial T} - \frac{\partial n_0(\mathbf{p})}{\partial \epsilon_0(\mathbf{p})}2\int \frac{d^3p'}{(2\pi\hbar)^3}F(\mathbf{p},\mathbf{p}')\frac{\partial n_0(\mathbf{p}')}{\partial T}\right)\\ &= -\frac{1}{\tau}(\delta n(\mathbf{p},\mathbf{r},t) - \langle\delta n(\mathbf{p}',\mathbf{r},t)\rangle). \end{aligned}\tag{12.58}$$

After multiplying both sides by $(\epsilon(\mathbf{p})-\mu)\mathbf{p}/m$, integrating over the momentum and summing over the spin degeneracy, we get

$$\begin{aligned} \mathbf{J}_Q(\mathbf{r},t) &= 2\int \frac{d^3p}{(2\pi\hbar)^3}\frac{(\epsilon(\mathbf{p})-\mu)}{T}\frac{\mathbf{p}}{m}\frac{\partial n_0(\mathbf{p})}{\partial \epsilon_\mathbf{p}}\tau\mathbf{v}_\mathbf{p}\cdot\nabla_\mathbf{r}T\\ &\quad\times\left[(\epsilon(\mathbf{p})-\mu) - 2\int \frac{d^3p'}{(2\pi\hbar)^3}(\epsilon(\mathbf{p}')-\mu)F(\mathbf{p},\mathbf{p}')\frac{\partial n_0(\mathbf{p}')}{\partial \epsilon_0(\mathbf{p}')}\right]\\ &= 2\int \frac{d^3p}{(2\pi\hbar)^3}\frac{(\epsilon(\mathbf{p})-\mu)^2}{T}\frac{\mathbf{p}}{m}\frac{\partial n_0(\mathbf{p})}{\partial \epsilon_\mathbf{p}}\tau\mathbf{v}_\mathbf{p}\cdot\nabla_\mathbf{r}T\\ &= -\frac{\pi^2k_B^2}{3}T\nu_F^L D\frac{1}{1+F_1/3}\nabla_\mathbf{r}T\\ &= -c_V^L D_Q^L\nabla_\mathbf{r}T, \end{aligned}\tag{12.59}$$

where the integral over the momentum has been transformed into an integral over the quasiparticle energy, thus showing the quasiparticle density of states ν_F^L. The second term in the square brackets of the first line of Eq. (12.59) vanishes because each integration over the momentum contains a factor $(\epsilon(\mathbf{p})-\mu)$, which is odd with respect to the distance in energy from the Fermi surface. The resulting energy integral has the form (11.55) previously encountered when dealing with the Fermi gas. We see that the effective mass entering the

The order parameter in this case is the average magnetization and is defined by

$$m = \frac{1}{N}\sum_i m_i = -\frac{1}{N}\frac{\partial F}{\partial h}. \tag{13.3}$$

According to (13.3), m depends on h. We define the spontaneous symmetry breaking as

$$\lim_{h\to 0} m(h) \neq 0. \tag{13.4}$$

Equation (13.3) defines m as the conjugate field of h. We then introduce a thermodynamic potential $\tilde{F}(T, m)$ by a Legendre transformation as

$$\tilde{F}(T, m) = F(T, h(m)) + Nhm, \tag{13.5}$$

where $h(m)$ is determined via Eq. (13.3). As a consequence

$$\frac{1}{N}\frac{\partial \tilde{F}}{\partial m} = h. \tag{13.6}$$

When the external field is not uniform, via Eq. (13.1), we may also define $\tilde{F}$ as a function of the set $\{m_i\}$:

$$\tilde{F}(T, \{m_i\}) = F(T, \{h_j(\{m_i\})\}) + \sum_j h_j m_j, \qquad \frac{\partial \tilde{F}}{\partial m_i} = h_i. \tag{13.7}$$

Following Landau, in general for a scalar order parameter φ as for the Ising model, when the system is homogeneous we introduce the free energy per unit volume, $f(T, \varphi)$, in terms of the order parameter. Near the critical point ($T = T_c$, $h = 0$) the order parameter is small and $f(T, \varphi)$ is expanded in terms of the invariants of the symmetry group characterizing the invariance of the original Hamiltonian. In the present case the invariants are the even powers of the order parameter:

$$f(T, \varphi) = a\varphi^2 + b\varphi^4 + \cdots, \tag{13.8}$$

with a and b coefficients which depend on the specific nature of the system and are continuous functions of thermodynamic parameters like the temperature. In other words, the free energy $f(T, \varphi)$ is invariant with respect to the symmetry transformation $\varphi \to -\varphi$, which is the only possible one for a real scalar order parameter, as in an axial ferromagnet. Given the expression (13.8), the value of φ at zero external field is self-consistently determined by minimizing f:

$$2a\varphi_0 + 4b\varphi_0^3 = 0. \tag{13.9}$$

As in previous formulations of the mean-field approximation, the fluctuations are neglected. For vanishing external field, there are three solutions:

$$\varphi_0 = 0, \quad \varphi_0 = \pm\sqrt{a/2b}. \tag{13.10}$$

The first must be stable at $T > T_c$ and the last two at $T < T_c$. The conditions for having a minimum of f is $2a + 12b\phi_0^2 > 0$, i.e. $2a > 0$ at $T > T_c$ and $-4a > 0$ at $T < T_c$. Landau assumed $a = a'(T - T_c) \equiv a'T_c t$, $a' > 0$. At $T = T_c$, any value of φ different from zero must increase the free energy. b must be positive at $T = T_c$ and therefore is positive in the

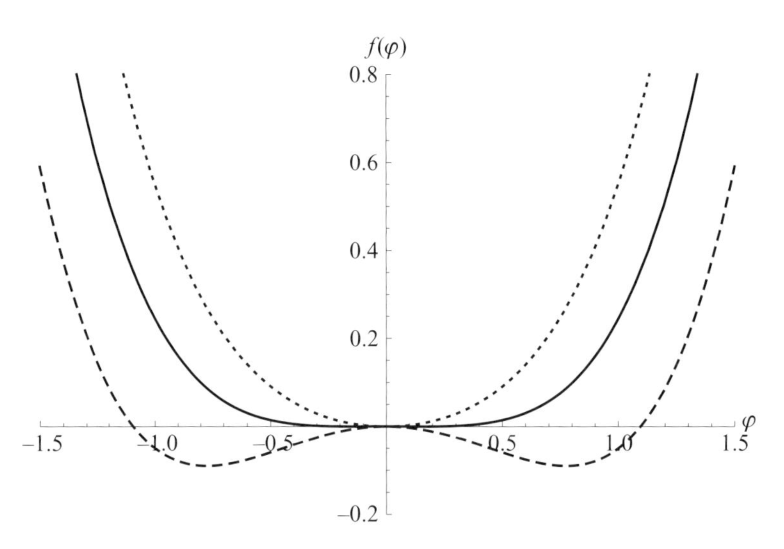

Fig. 13.1 Thermodynamic potential for $a > 0$ (dotted line), $a = 0$ (full line) and $a < 0$ (dashed line).

neighbourhood of T_c. It can be taken as a positive constant. When $T < T_c$, the vanishing solution is a relative maximum of f. The spontaneous symmetry breaking occurs when one chooses one of the two equivalent solutions with $\varphi_0 \neq 0$. Hence, by choosing the solution corresponding to an ordered phase we lower the symmetry. The symmetry we are considering is an example of a discrete symmetry since there is a finite (two in the present case) number of equivalent minima (see Fig. 13.1). Before considering other types of symmetry, let us see the predictions of the Landau theory for the critical indices. From (13.10) and recalling the definition of the critical indices (8.81) it follows that $\beta = 1/2$. In the presence of the external field Eq. (13.9) becomes

$$2a\bar{\varphi} + 4b\bar{\varphi}^3 = h. \tag{13.11}$$

The critical isotherm $a = 0$ gives $\delta = 3$.

By deriving (13.11) with respect to h one gets the susceptibility χ:

$$\chi \equiv \frac{\partial \varphi}{\partial h} = \frac{1}{2a + 12b\bar{\varphi}^2}, \tag{13.12}$$

which, as $h \to 0$ and $\bar{\varphi} \to \varphi_0$ reads

$$\chi = \frac{1}{2a}, \quad \text{for } T > T_c,$$

$$\chi = -\frac{1}{4a}, \quad \text{for } T < T_c.$$

In both cases $\gamma = \gamma' = 1$ and the susceptibility is positive, in agreement with the thermodynamic stability condition.

Finally, by using (13.10) into (13.8),

$$f(T, \varphi_0) = a\varphi_0^2 + b\varphi_0^4 = -\frac{a^2}{4b}, \tag{13.13}$$

which yields for the specific heat discontinuity per unit volume at T_c

$$\Delta c_V = c_V(T \to T_c^-) - c_V(T \to T_c^+) = \lim_{T \to T_c} \left(-T \frac{\partial^2 f}{\partial T^2} \right) = T_c \frac{a'^2}{2b}$$

and we get the indices $\alpha = \alpha' = 0$. By comparing Eq. (8.84) for the expansion of the free energy per spin in the mean field approximation for the Ising model with Eq. (13.8) for the Landau free energy per unit volume one has $a' = (N/V)k_B/2$, $b = (N/V)k_B T_c/12$, so that

$$\Delta c_V = \frac{3}{2} k_B \frac{N}{V}, \tag{13.14}$$

as previously found.

13.2 Fluctuations and Ginzburg criterion for the validity of mean-field theories

Up to now we discussed the thermodynamic quantities related to a variation of the temperature and the external field conjugate to the order parameter, which control the approach to the critical point. We want to introduce now the possibility of local fluctuations. In the discrete Ising model case the latter are described by the correlation functions

$$f_m(x_i - x_j) = \langle (\sigma_i - \langle \sigma_i \rangle)(\sigma_j - \langle \sigma_j \rangle) \rangle = -\frac{1}{\beta} \frac{\partial^2 F}{\partial h_i \partial h_j}. \tag{13.15}$$

The last equality follows by a further derivative of Eq. (13.1) with respect to the local magnetic field. The correlation function Eq. (13.15) is related to the response function by the linear response theory

$$\delta m_i = \beta \sum_j f_m(x_i - x_j) \delta h_j, \tag{13.16}$$

which is the simplest classical form of the fluctuation–dissipation theorem (Eq. (8.88)) for the discrete case.

By letting the lattice spacing go to zero, the site index becomes a continuous variable $\mathbf{x}$. The external field is then a function of the position $h(\mathbf{x})$, where $\mathbf{x}$ denotes a point in a d-dimensional space. In the same way, the scalar order parameter becomes a function of position, $m(\mathbf{x}) \to \varphi(\mathbf{x})$. The free energy becomes then a functional of $h(\mathbf{x})$ and $\varphi(\mathbf{x})$, and its correlation function $f_\varphi(\mathbf{x}, \mathbf{x}')$ is introduced via functional derivatives. It is now natural to generalize Eq. (13.1) for m_i and Eq. (13.15) for f_m and the Legendre transformation (13.7) to the continuum for a generic scalar order parameter $\varphi(\mathbf{x})$ and its conjugate field $h(\mathbf{x})$:

$$\varphi(\mathbf{x}) = -\frac{\delta F}{\delta h(\mathbf{x})}, \quad f_\varphi(\mathbf{x}, \mathbf{x}') = \frac{1}{\beta} \frac{\delta \varphi(\mathbf{x})}{\delta h(\mathbf{x}')} = -\frac{1}{\beta} \frac{\delta^2 F}{\delta h(\mathbf{x}) \delta h(\mathbf{x}')}, \tag{13.17}$$

$$\tilde{F}(T, \varphi) = \int \mathrm{d}\mathbf{x}\, h(\mathbf{x}) \varphi(\mathbf{x}) + F(T, h), \tag{13.18}$$

which generalizes Eq. (13.5).

To allow local fluctutations with slow space variation of the order parameter, the Landau functional and its functional derivatives read

$$\tilde{F} = \int d^d x[a\varphi^2(\mathbf{x}) + b\varphi^4(\mathbf{x}) + c|\nabla\varphi(\mathbf{x})|^2], \tag{13.19}$$

$$\frac{\delta \tilde{F}}{\delta\varphi(\mathbf{x})} = 2a\varphi(\mathbf{x}) + 4b\varphi^3(\mathbf{x}) - 2c\nabla^2\varphi(\mathbf{x}) = h(\mathbf{x}). \tag{13.20}$$

The homogeneous solution must be stable and the coefficient c must be positive since any dishomegeneity must increase the free energy. Within this approximation, recalling that $\delta h(\mathbf{x})/\delta h(\mathbf{x}') = \delta(\mathbf{x} - \mathbf{x}')$, the differential equation for $f_\varphi(\mathbf{x}, \mathbf{x}')$ reads

$$\left(2a + 12b\bar{\varphi}^2 - 2c\nabla_x^2\right) f_\varphi(\mathbf{x}, \mathbf{x}') = k_B T\delta(\mathbf{x} - \mathbf{x}'), \tag{13.21}$$

where $\bar{\varphi}$ is the solution of the Eq. (13.11) for the case of a uniform field h. By Fourier transforming Eq. (13.21) we get

$$f_\varphi(\mathbf{k}) = \frac{k_B T}{2a + 12b\bar{\varphi}^2 + 2ck^2} \equiv \frac{k_B T}{2c} \frac{1}{\xi^{-2}(T) + k^2} = \langle|\delta\varphi(\mathbf{k})|^2\rangle, \tag{13.22}$$

where we have introduced the correlation length $\xi(T)$ defined by

$$\xi^2(T) = \frac{c}{a + 6b\bar{\varphi}^2}. \tag{13.23}$$

At $h = 0$, $\bar{\varphi}^2 = \varphi_0^2$ and

$$\xi^2(T) = \frac{c}{a}, \quad T > T_c,$$
$$\xi^2(T) = \frac{c}{-2a}, \quad T < T_c. \tag{13.24}$$

For each wave vector $\mathbf{k}$ the correlation function defines the mean square root of the fluctuations as the first correction to the mean-field approximation. The analog of (13.22) for the gas-lattice model is Eq. (1.64).

From the linear response expression for the continuum (10.7), after anti-Fourier transforming,

$$\delta\varphi(\mathbf{x}) = \beta \int d\mathbf{x}' f_\varphi(\mathbf{x}, \mathbf{x}')\delta h(\mathbf{x}'), \tag{13.25}$$

it follows that the zero-field limit of the $k = 0$-component of the Fourier transform is the static susceptibility

$$\chi = \beta \int d\mathbf{x}' f_\varphi(\mathbf{x}, \mathbf{x}') = \beta f_\varphi(\mathbf{k} = 0), \tag{13.26}$$

which coincides with the result of the fluctuation–dissipation theorem, Eq. (10.55). By combining Eq. (13.26) with the approximated expression of the correlation function (13.22), at zero external field, the correlation length can be expressed in terms of the static susceptibility:

$$\chi = \frac{1}{2c}\xi^2(T)|_{h=0}. \tag{13.27}$$

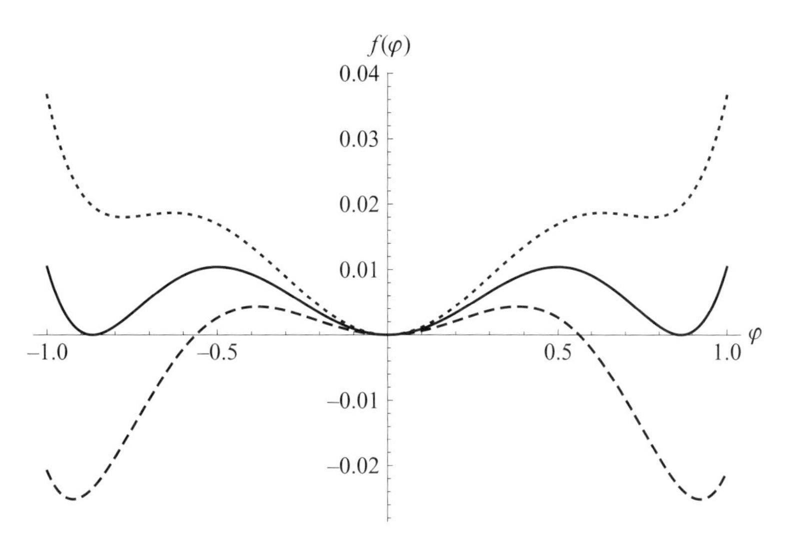

Fig. 13.2 A plot of the thermodynamic potential f for three different values of a: $a = 0.24, a = 3/16$ and $a = 0.125$ having chosen $b = -d_0 = -1$. The full line shows the coincidence of the three minima as imposed by Eq. (13.36).

From (13.27) and from the critical behavior (13.12) of the static susceptibility we obtain that, in the present approximation,

$$\xi(T) \sim |t|^{-\nu,\nu'}, \quad 2\nu = \gamma = \gamma' = 2\nu' = 1. \tag{13.28}$$

From the divergence of the susceptibility and of the correlation function $f_\varphi(k)$ as $k \to 0$ and $T \to T_c$ at zero external field h, it is clear that the effect of the fluctuations becomes more and more important on approaching the critical point, in contradiction with the initial assumption of the absence of fluctuations. The Landau theory will maintain its validity as long as the order parameter is larger than its fluctuations (Ginzburg criterion):

$$\varphi_0^2 \gg \int_{|\mathbf{k}|<\xi^{-1}(T)} \mathrm{d}^d k \frac{k_B T_c}{2c} \frac{1}{\xi^{-2}(T) + k^2}. \tag{13.29}$$

For $T < T_c$, in mean-field approximation, φ_0^2 has the same temperature behavior as $\xi^{-2}(T)$ (Eq. (13.24)) and the integral of the fluctuations is proportional to $\xi^{2-d}(T)$. The Ginzburg criterion becomes $\xi^{4-d}(T) \ll$ const depending on a', b and c. This states that the mean-field approximation is valid for $d > 4$ up to the critical point. For $d < 4$, instead, the mean-field approximation breaks down in the so-called critical region when T is sufficiently close to T_c. The extension of the critical region is not universal and depends on the parameters a', b and c specific to each particular system.

13.3 First order transition and tricritical point

In the standard treatment of the Landau theory of the second order phase transitions the coefficient a is temperature dependent and can be both positive and negative, whereas

b is assumed to be constant, $b > 0$, for thermodynamic stability. $a(T_c) = 0$ determines the critical point. In the case that a depends on two variables, for instance T and P as for the superfluid transition line of ^{4}He (see the phase diagram in Fig. 15.1 of Chapter 15), $a(T, P) = 0$ determines a critical line of second order. It is also possible to treat the first order phase transition with a discontinuity of the order parameter along a coexistence line provided this discontinuity is not too large, i.e. the system is not too far from a second order transition. In the Landau approach a first order transition with a discontinuity in the order parameter from zero to a finite value occurs when, together with a, b also depends on the thermodynamic variables and b assumes negative values. Then, in order to maintain thermodynamic stability, one needs to keep terms up to sixth order in φ in the expansion of the thermodynamic potential per unit volume:

$$f = \frac{1}{2}a\varphi^2 + \frac{1}{4}b\varphi^4 + \frac{1}{6}d_0\varphi^6, \tag{13.30}$$

where $d_0 > 0$. We have slightly changed the coefficients in the expansion with respect to Eq. (13.8) for convenience. The equation determining the value of φ is obtained by minimizing f. The extremum condition reads

$$\frac{\partial f}{\partial \varphi} = \varphi(a + b\varphi^2 + d_0\varphi^4) = h. \tag{13.31}$$

At $h = 0$, Eq. (13.31) has the trivial solution $\varphi = 0$ with zero energy $f = 0$ and the solutions $\varphi = \pm\varphi_0$ satisfying

$$a + b\varphi_0^2 + d_0\varphi_0^4 = 0, \quad \varphi_0^2 = \frac{-b + \sqrt{b^2 - 4ad_0}}{2d_0}. \tag{13.32}$$

All three solutions are local minima. The solution with $\varphi \neq 0$ for the ordered phase only exists when $a < b^2/(4d_0)$ in order that φ_0^2 be real. The energy for the disordered solution $\varphi = 0$ is clearly zero, whereas the energy of the ordered solution φ_0 reads

$$\begin{aligned} f_0 &= \frac{1}{2}a\varphi_0^2 + \frac{1}{4}b\varphi_0^4 + \frac{1}{6}d_0\varphi_0^6 \\ &= \frac{1}{2}a\varphi_0^2 + \frac{1}{4}b\varphi_0^4 - \frac{1}{6}b\varphi_0^4 + \frac{1}{6}\left(b\varphi_0^4 + d_0\varphi_0^6\right) \\ &= \frac{1}{2}a\varphi_0^2 + \left(\frac{1}{4} - \frac{1}{6}\right)b\varphi_0^4 - \frac{1}{6}a\varphi_0^2 \\ &= \frac{1}{3}a\varphi_0^2 + \frac{1}{12}b\varphi_0^4. \end{aligned} \tag{13.33}$$

In order that the solution φ_0 be a global minimum, its energy must be negative relative to the disordered solution at $\varphi = 0$. This starts when

$$a = -\frac{b\varphi_0^2}{4}, \tag{13.34}$$

i.e. in the sector $a > 0$ and $b < 0$ of the a–b plane.

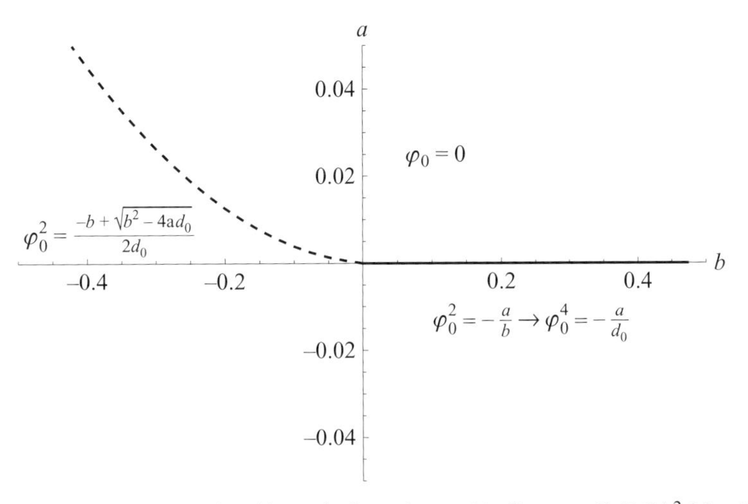

Fig. 13.3 Phase diagram in the a–b plane. The dotted line is the first order transition line $a = (3/16)b^2/d_0$, whereas the full line is the second order transition line $a = 0, b > 0$. The origin is the tricritical point. In the sector $a < 0$, $b > 0$ of the a–b plane, the solution φ_0 crosses from $\varphi_0^2 = -a/b$ near $a = 0$ to $\varphi_0^4 = -a/d_0$ near $b = 0$ when $b < d_0\varphi_0^2$.

For the onset, the three solutions $f(\varphi_0) = 0$ coincide. By inserting Eq. (13.32) into the above equation one gets

$$a = -\frac{b}{4}\frac{-b + \sqrt{b^2 - 4ad_0}}{2d_0}. \tag{13.35}$$

By solving with respect to a one obtains the relation for the coexistence line:

$$a = \frac{3}{16}\frac{b^2}{d_0}. \tag{13.36}$$

This last equation, for $b < 0$ and $a > 0$, describes the first order transition line in the b–a plane with the finite positive jump $\varphi_0^2 = 3b/4d_0$ of the order parameter squared from the disordered to the ordered phase. At the origin, $b = 0$ and $a = 0$, such a line joins the line of second order critical points $b > 0$ and $a = 0$ (see Fig. 13.3). The origin is called the tricritical point.

The reason (Griffiths (1970)) for such a name arises from the fact that for $h \neq 0$, the first order transition no longer exists, but in the space b–a–h three lines of second order critical points depart from the origin. One of the three lines is the standard one ($a = 0$, $h = 0$, $b > 0$) mentioned above. We refer to Problem 13.1 for a full discussion.

Finally, we may comment on the values of the critical indices for the tricritical point. In mean field and therefore in Landau theory the critical indices are determined by the powers in the expansion of the thermodynamic potential. The values of the classical indices for the tricritical point will consequently be different from those of the standard critical point. By repeating the same arguments as for the critical point, the tricritical indices are obtained. From Eq. (13.32), starting from $b = 0$ and assuming again the deviation of a linear in $T - T_{\mathrm{TCP}}$, or approaching the tricritical point along the coexistence line

Eq. (13.34), $\beta = 1/4$. Equation (13.31) along the tricritical isotherm $a = b = 0$ gives $\varphi \sim h^{1/5}$, i.e. $\delta = 5$. By deriving Eq. (13.31) with respect to h in the limit $h = 0$ and $b = 0$ the susceptibility $\chi \propto 1/|a|$ and $\gamma = 1$. The specific heat can be calculated from Eqs. (13.32) and (13.33) either approaching the tricritical point with $a \to 0$ starting at $b = 0$ or along a line $a = kb$ to obtain a divergence with $\alpha = 1/2$. After the introduction in Eq. (13.30) of a local order parameter $\varphi(\mathbf{x})$ and of a surface energy term $c|\nabla\varphi(\mathbf{x})|^2$, one obtains the analog of Eq. (13.22) for the correlation function with the index $\nu = 1/2$ for the correlation length ξ. The same condition Eq. (13.29) for the validity of mean field theory now leads to $d = 3$ instead of $d = 4$ for the separating dimensionality.

We understand better how to switch from a second order transition line with $b > 0$ to a first order line with $b' < 0$ if we couple the order parameter φ of the second order transition with a non-critical field ψ. Two terms are then added to the Landau expansion Eq. (13.30) of the thermodynamic potential with two new constants e and r:

$$e\psi|\varphi|^2 + \frac{r}{2}\psi^2. \tag{13.37}$$

The first term indicates that the presence of the non-ordering field introduces a cost in energy for the formation of the ordered phase. The second one represents the energy associated with the presence of the non-critical field ψ. By minimizing with respect to ψ we obtain the relation between ψ and φ:

$$\psi = -\frac{e}{r}|\varphi|^2, \quad b' = b - 2\frac{e^2}{r}. \tag{13.38}$$

The expression for b' given above is obtained by substituting back in the themodynamic potential, which then amounts to renormalizing b in b'. $b' = 0$ together with $a = 0$ characterizes the tricritical point.

The classical example of a tricritical point is given by the mixture of the two isotopes ^{3}He and ^{4}He of liquid helium, below the superfluid critical temperature $T_c = 2.17$ K (see Chapter 15). The experiments (Graf et al. (1967)) show that below a given molar fraction $x_{\mathrm{TCP}} = 0.67$ of ^{3}He and for $T > T_{\mathrm{TCP}} = 0.87$ K a superfluid second order transition is present along a critical λ line $T_\lambda(x)$ which is a decreasing function of x. The system at higher x and lower T undergoes a first order phase separation transition of the mixture into a superfluid ^{4}He-rich phase and a normal ^{3}He-rich phase. (See Fig. 13.4 for the phase diagram in the T–x plane.) φ in this case is the two-component order parameter of the superfluid phase transition (see Chapter 15) of the ^{4}He component and ψ is the molar fraction x of ^{3}He which tends to reduce T_λ. The phase diagram of Fig. 13.4 can be understood by considering that for $T > T_{\mathrm{TCP}}$ the presence of ^{3}He with molar fraction x competes with the superfluid ordering of ^{4}He and makes $T_\lambda(x)$ decreasing with increasing x. At fixed temperature $T < T_{\mathrm{TCP}}$ but near it, the phase separation gives two values $x_\pm$ of the molar fraction of ^{3}He, where x_{TCP} is the average value of the ^{3}He molar fraction present in the two phase-separated components. With the solution x_-, the ^{4}He-rich component has a molar fraction $x_{^4\mathrm{He}} = 1 - x_- > 1 - x_{\mathrm{TCP}}$ with a $T_\lambda > T$. This phase-separated component is then well inside the superfluid region with a well developed order parameter. The opposite is true for x_+ and this coexisting phase is in the normal phase. The two coexisting phases tend to coincide on approaching the tricritical point where $x_- = x_+ = x_{\mathrm{TCP}}$, $T_\lambda(x_-) = T_\lambda(x_{\mathrm{TCP}})$

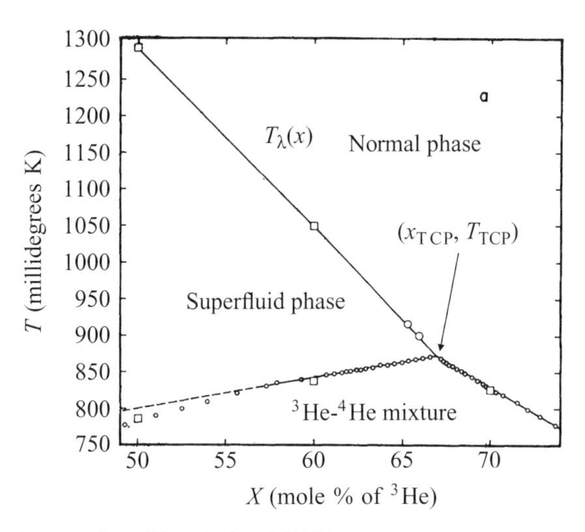

Fig. 13.4 ^{3}He-^{4}He mixture phase diagram adapted from Graf et al. (1967).

and $\varphi = 0$. A complete mean-field account of the ^{4}He–^{3}He mixture near to the tricritical point is given with an effective three component spin model on a lattice. The $s_i = \pm 1$ components represent that the lattice site i is occupied by a ^{4}He atom, the $s_i = 0$ component represents the ^{3}He occupation (given by $1 - \langle s_i^2 \rangle$) and the superfluid transition is replaced by a magnetic transition with order parameter $\varphi = \langle s_j \rangle$ (Blume et al. (1971)).

13.4 Continuous symmetries and the lower critical dimensionality

Before proceeding with the corrections to the Landau theory, we want to examine the consequences of using a vector order parameter instead of the scalar order parameter considered so far. This will allow us to comment on continuous symmetry breaking. For the sake of simplicity, we take a two-dimensional order parameter, $\boldsymbol{\varphi} = (\varphi_1, \varphi_2)$. The Landau functional (13.19) and its variation (13.21) become

$$F_{\mathrm{L}}(T, \boldsymbol{\varphi}) = \int \mathrm{d}\mathbf{x}(a\boldsymbol{\varphi}(\mathbf{x}) \cdot \boldsymbol{\varphi}(\mathbf{x}) + b(\boldsymbol{\varphi}(\mathbf{x}) \cdot \boldsymbol{\varphi}(\mathbf{x}))^2 + c|\nabla \boldsymbol{\varphi}(\mathbf{x})|^2), \tag{13.39}$$

$$2a\boldsymbol{\varphi}(\mathbf{x}) + 4b(\boldsymbol{\varphi}(\mathbf{x}) \cdot \boldsymbol{\varphi}(\mathbf{x}))\boldsymbol{\varphi}(\mathbf{x}) - 2c\nabla^2 \boldsymbol{\varphi}(\mathbf{x}) = \mathbf{h}(\mathbf{x}). \tag{13.40}$$

The free energy, being a function of the scalar product of the order parameter with itself, is manifestly rotationally invariant. The equation determining the order parameter becomes a vector equation.

As in the scalar case, we begin by considering the homegeneous solution $\bar{\boldsymbol{\varphi}}(\mathbf{h})$. In the absence of the external field we introduce polar coordinates in the (φ_1, φ_2) plane:

$$\varphi_1 = \varphi_0 \cos\theta, \quad \varphi_2 = \varphi_0 \sin\theta,$$

Eq. (13.40) becomes

$$2a\varphi_0 + 4b\varphi_0^3 = 0, \tag{13.41}$$

and all the critical indices for the thermodynamic quantities remain equal to those for the scalar order parameter, contrary to the experimental results which, for instance, distinguish between the isotropic, planar and axial ferromagnet.

In contrast to the case of a scalar order parameter with two minima, we have now an infinity of equivalent minima $0 \leq \theta \leq 2\pi$ (the so-called Mexican hat shape for the thermodynamic potential), and passing from one minimum to another by varying θ from 0 to 2π does not cost energy. This is typical of a continuous symmetry. To show how this happens in the present situation, let us choose one particular minimum, say the one corresponding to $\theta = 0$, which amounts to the presence of a vanishingly small homogeneous external field h_1 in the direction $\theta = 0$. Hence the chosen solution of Eq. (13.40) reads

$$2a\bar{\varphi} + 4b\bar{\varphi}^3 = h_1, \quad \varphi_0 = \lim_{h_1 \to 0} \bar{\varphi}(h_1). \tag{13.42}$$

By allowing the fluctuations $\delta\varphi_1$ and $\delta\varphi_2$ with respect to the solution (13.42), the deviation of the Landau free energy density up to quadratic terms is

$$\Delta f_{\rm L} = \delta\varphi_1^2\left(a + 6b\varphi_0^2\right) + \delta\varphi_2^2\left(a + 2b\varphi_0^2\right) + h_1(\varphi_0)\delta\varphi_1. \tag{13.43}$$

In the limit of vanishing h_1 we see that the transverse fluctuations, $\delta\varphi_2$, do not cost energy, while the quadratic longitudinal fluctuations, $\delta\varphi_1$, have an energy cost which, in the ordered phase, is $-2a\delta\varphi_1^2$, twice the value in the symmetrical high temperature phase.

The functional derivatives of Eq. (13.40) with respect to $h_1(\mathbf{x}')$ and $h_2(\mathbf{x}')$ evaluated at $h_1(\mathbf{x}') = h_1$ and $h_2(\mathbf{x}') = 0$ yield the equations for the longitudinal (parallel to the order parameter) and transverse correlation functions $\delta\varphi_1(\mathbf{x})/\delta h_1(\mathbf{x}')$ and $\delta\varphi_2(\mathbf{x})/\delta h_2(\mathbf{x}')$, respectively:

$$2(a + 6b\bar{\varphi}^2 - c\nabla^2) f_{\varphi_\parallel}(\mathbf{x} - \mathbf{x}') = k_{\rm B} T_{\rm c} \delta(\mathbf{x} - \mathbf{x}'), \tag{13.44}$$

$$2(a + 2b\bar{\varphi}^2 - c\nabla^2) f_{\varphi_\perp}(\mathbf{x} - \mathbf{x}') = k_{\rm B} T_{\rm c} \delta(\mathbf{x} - \mathbf{x}'), \tag{13.45}$$

where we have used $\delta h(\mathbf{x})/\delta h(\mathbf{x}') = \delta(\mathbf{x} - \mathbf{x}')$.

For an n-component order parameter, we would have $n - 1$ transverse equivalent correlation functions. The longitudinal correlation function has exactly the same behavior as the one for a scalar order parameter and the Ginzburg criterion follows in the same way. The transverse correlation function instead shows problems when the limit of zero external field is taken at low dimensionality. After substituting in Eq. (13.45) $2(a + 2b\bar{\varphi}^2)$ with $h_1/\bar{\varphi}$, the Fourier transform of $f_{\varphi_\perp}(\mathbf{x} - \mathbf{x}')$ reads

$$f_{\varphi_\perp}(k) = \frac{k_{\rm B} T_{\rm c}}{h_1/\bar{\varphi} + 2ck^2}, \tag{13.46}$$

which is infinite at any T when k and h_1 go to zero and $\bar{\varphi} \to \varphi_0 \neq 0$. This implies that the anti-Fourier transform of $f_{\varphi_\perp}(k)$ is proportional to $(h_1/\bar{\varphi})^{(d-2)/2}$. For $d \leq 2$, the limit $h_1 \to 0$ requires that $\bar{\varphi} \to 0$, otherwise one would get an infinite value for the expectation value of the square of the local order parameter. We have therefore a *lower* critical dimensionality

$d = 2$ for the existence of a transition at finite critical temperature characterized by an order parameter. This is a phenomenological derivation of the so-called Mermin and Wagner (1966) theorem, extended to the occurrence of Bose–Einstein condensation by Hohenberg (1967).

For a scalar order parameter, the lower critical dimensionality is $d = 1$. Consider for example the Ising model for a chain of N spins $\sigma_i = \pm 1$ with nearest-neighbor interaction $J > 0$:

$$H = -J \sum_{i=1}^{N-1} \sigma_i \sigma_{i+1}. \tag{13.47}$$

The complete alignment of the N spins in one direction gives an internal energy U and an entropy S given by

$$U = -J(N-1), \quad S = 0. \tag{13.48}$$

We can disorder the system by introducing n cuts and for each cut we turn all the spins to the right. The free energy variation due to n cuts is

$$\Delta F = \Delta U - T\Delta S = 2nJ - k_B T \ln \frac{(N-1)!}{(N-1-n)!n!}. \tag{13.49}$$

The entropy variation ΔS is obtained by calculating the number of possible configurations of the N spins with n cuts in the $N - 1$ bonds. At each temperature the stability condition is given by minimization of ΔF with respect to n, which at large N and n via the Stirling formula yields

$$\frac{\bar{n}/N}{1 - \bar{n}/N} = e^{-2\beta J}. \tag{13.50}$$

For each finite temperature the number of cuts is a finite fraction of the total number of bonds and the number of spins up is equal to the number of spins down with vanishing spontaneous magnetization. No transition at a finite temperature with a finite order parameter is possible. The argument remains valid as long as the interaction is short ranged.

13.5 Problem

13.1 Determine the curves of the second order critical points at $h \neq 0$.

14 The Landau–Wilson model for critical phenomena

14.1 The Gaussian transformation

In this chapter, we introduce the functional formulation of statistical mechanics and find the approximation of the partition function corresponding to Landau theory. We then proceed to develop its corrections. To be simple we start from the Ising model and then extend the result to what is now called the Landau–Wilson model.

We generalize the Gaussian transformation (8.62) which has been used to calculate the partition function in the thermodynamic limit of the long-range spin model (8.60), where each spin interacts with every other spin.

The Gaussian integral

$$\int_{-\infty}^{\infty} \mathrm{d}x \mathrm{e}^{-ax^2+bx} = \left(\frac{\pi}{a}\right)^{1/2} \mathrm{e}^{\frac{b^2}{4a}} \tag{14.1}$$

is extended to a symmetric positive definite $N \times N$ matrix, $\mathbf{A}$, which admits N real eigenvalues λ_i and is diagonalised by an orthogonal matrix $\mathbf{U}$. Let $\mathbf{x}$ be an N-component column vector and $\mathbf{x}^{\mathrm{T}}$ its transpose. We then have

$$\mathbf{x}^{\mathrm{T}}\mathbf{A}\mathbf{x} = \mathbf{x}'^{\mathrm{T}}\tilde{\mathbf{A}}\mathbf{x}',$$

where $\tilde{\mathbf{A}}$ is the diagonal form and $\mathbf{x}' = \mathbf{U}\mathbf{x}$ is the transformed vector. By using (14.1) we obtain

$$\int \mathrm{d}\mathbf{x} \mathrm{e}^{-\mathbf{x}^{\mathrm{T}}\mathbf{A}\mathbf{x}+\mathbf{b}^{\mathrm{T}}\mathbf{x}} = \frac{\pi^{N/2}}{(\det\mathbf{A})^{1/2}} \mathrm{e}^{\frac{1}{4}\mathbf{b}^{\mathrm{T}}\mathbf{A}^{-1}\mathbf{b}}. \tag{14.2}$$

We apply the above results to the Ising Hamiltonian (8.51).

We define the interaction matrix $\mathbf{K} = \beta\mathbf{J}$ and the vector $\boldsymbol{\sigma} = (\sigma_1, ..., \sigma_N)$ for the N spins. The partition function reads

$$Z = \sum_{\{\sigma\}} \mathrm{e}^{\sum_i \beta h_i \sigma_i} \mathrm{e}^{\frac{1}{2}\boldsymbol{\sigma}^{\mathrm{T}}\mathbf{K}\boldsymbol{\sigma}}.$$

With the transformation (14.2), the term in the exponent quadratic in $\boldsymbol{\sigma}$ can be eliminated from Z at the cost of introducing the integrals on the auxiliary variables ψ_l:

$$\begin{aligned} Z &= \sum_{\{\sigma\}} \left(\frac{2}{\pi}\right)^{N/2} (\det\mathbf{K}^{-1})^{1/2} \int \delta\psi \, \mathrm{e}^{-\frac{1}{2}\boldsymbol{\psi}^{\mathrm{T}}\mathbf{K}^{-1}\boldsymbol{\psi}} \mathrm{e}^{\sum_i(\beta h_i+\psi_i)\sigma_i} \\ &= \left(\frac{2}{\pi}\right)^{N/2} (\det\mathbf{K}^{-1})^{1/2} \int \delta\psi \, \mathrm{e}^{-\frac{1}{2}\boldsymbol{\psi}^{\mathrm{T}}\mathbf{K}^{-1}\boldsymbol{\psi}} \Pi_{i=1}^{N} 2\cosh(\beta h_i + \psi_i), \end{aligned} \tag{14.3}$$

where $\delta\psi = \prod_l \mathrm{d}\psi_l$. The summation over the configurations of the spins has been easily performed on an equivalent set of independent spins. Hence, apart from a normalization factor,

$$Z = \int \delta\psi \mathrm{e}^{-S(\mathbf{h},\boldsymbol{\psi})} \tag{14.4}$$

with

$$S(\mathbf{h},\boldsymbol{\psi}) = \frac{1}{2}\sum_{ij}\psi_i(\mathbf{K}^{-1})_{ij}\psi_j - \sum_i \ln 2\cosh(\beta h_i + \psi_i). \tag{14.5}$$

It is useful at this point to make the linear transformation

$$\hat{\varphi}_i = \sum_j (\mathbf{K}^{-1})_{ij}\psi_j \tag{14.6}$$

so that

$$Z = \int \delta\hat{\varphi}\mathrm{e}^{-S(\mathbf{h},\hat{\boldsymbol{\varphi}})} \tag{14.7}$$

with

$$S(\mathbf{h},\hat{\boldsymbol{\varphi}}) = \frac{1}{2}\sum_{ij}\hat{\varphi}_i K_{ij}\hat{\varphi}_j - \sum_i \ln\cosh\left(\beta h_i + \sum_j K_{ij}\hat{\varphi}_j\right) - N\ln 2. \tag{14.8}$$

The last term represents the adimensional free energy of free spins. Equations (14.7) and (14.8) give an integral representation of the partition function in the sense that each configuration of the set of values $\hat{\boldsymbol{\varphi}}$ is weighted by the exponential of the effective action S.

The magnetization per spin is

$$m_i = \langle\sigma_i\rangle = \frac{1}{\beta}\frac{\partial \ln Z}{\partial h_i} = \left\langle \sum_j (\mathbf{K}^{-1})_{ij}\psi_j \right\rangle = \langle\hat{\varphi}_i\rangle \equiv \varphi_i. \tag{14.9}$$

The identification of the magnetization per spin with the average of the field $\hat{\boldsymbol{\varphi}}$ has been obtained by considering the derivative of the logarithm of Eq. (14.3) with respect to h_i as equivalent to the derivative with respect to ψ_i of the linear term in the exponent and finally integrating by parts. Further derivatives identify higher order correlation functions, as for instance

$$f_m(\mathbf{x}_i,\mathbf{x}_j) = \frac{1}{\beta}\frac{\partial\langle\sigma_i\rangle}{\partial h_j} = \frac{1}{\beta}\frac{\partial\langle\hat{\varphi}_i\rangle}{\partial h_j} = f_\varphi(\mathbf{x}_i,\mathbf{x}_j) \tag{14.10}$$

and the two formulations are completly equivalent.

From the free energy

$$F(T,\mathbf{h}) = -k_{\mathrm{B}}T\ln\int\delta\hat{\varphi}\ \mathrm{e}^{-S(\mathbf{h},\hat{\boldsymbol{\varphi}})} \tag{14.11}$$

one obtains the Legendre transformed potential

$$\tilde{F}(T,\boldsymbol{\varphi}) = F(T,\mathbf{h}) + \sum_i h_i\varphi_i. \tag{14.12}$$

We notice that from Eq. (14.9) and the second line of Eq. (14.3)

$$\langle\hat{\varphi}_i\rangle = \left\langle \tanh\left(\beta h_i + \sum_j K_{ij}\hat{\varphi}_j\right)\right\rangle. \tag{14.13}$$

Mean-field theory is reproduced when the average of the hyperbolic tangent can be replaced by the hyperbolic tangent of the averaged field, i.e., when the fluctuations of the order parameter can be neglected and the value of the latter is obtained by minimizing S.

The action (14.8) is a non-polynomial function of $\hat{\boldsymbol{\varphi}}$. Its expansion is

$$S(\mathbf{h}, \hat{\boldsymbol{\varphi}}) = \frac{1}{2}\sum_{ij}\hat{\varphi}_i\left(K_{ij} - \sum_l K_{il}K_{jl}\right)\hat{\varphi}_j + \frac{1}{12}\sum_i\left(\sum_j K_{ij}\hat{\varphi}_j\right)^4 - \beta\sum_{ij} h_i K_{ij}\hat{\varphi}_j + \cdots, \tag{14.14}$$

where only the first term of the expansion in h_i has been taken. Powers in $\hat{\varphi}_i$ larger than four have been omitted since they are irrelevant near criticality for reasons which will be clarified later.

By introducing the Fourier transforms

$$\hat{\varphi}_i = \frac{1}{\sqrt{N}}\sum_{\mathbf{k}} \mathrm{e}^{-\mathrm{i}\mathbf{k}\cdot\mathbf{r}_i}\,\hat{\varphi}_{\mathbf{k}},$$
$$K_{ij} = \frac{1}{N}\sum_{\mathbf{k}} \mathrm{e}^{-\mathrm{i}\mathbf{k}\cdot(\mathbf{r}_i-\mathbf{r}_j)} K(k),$$

the quadratic term of the action (14.14) is

$$S_2 = \frac{1}{2}\sum_{\mathbf{k}} K(k)(1 - K(-k))\hat{\varphi}_{\mathbf{k}}\hat{\varphi}_{-\mathbf{k}}. \tag{14.15}$$

The singular critical behavior manifests itself in the $k = 0$ limit. In approaching criticality we can take therefore the expansion of $K(k)$ up to the quadratic term $K(k) = K(0)(1 - \rho^2k^2)$ so that S reads

$$\begin{aligned} S = &\frac{1}{2}\sum_{\mathbf{k}}[K(0)(1 - K(0)) - K(0)\rho^2k^2(1 - 2K(0))]\hat{\varphi}_{\mathbf{k}}\hat{\varphi}_{-\mathbf{k}} \\ &+ \frac{K(0)^4}{12N}\sum_{\mathbf{k}_1,\mathbf{k}_2,\mathbf{k}_3,\mathbf{k}_4}\delta_{\mathbf{k}_1+\mathbf{k}_2+\mathbf{k}_3+\mathbf{k}_4,0}\,\hat{\varphi}_{\mathbf{k}_1}\hat{\varphi}_{\mathbf{k}_2}\hat{\varphi}_{\mathbf{k}_3}\hat{\varphi}_{\mathbf{k}_4} + \cdots - \beta K(0)\sum_{\mathbf{k}} h_{\mathbf{k}}\hat{\varphi}_{-\mathbf{k}}. \end{aligned} \tag{14.16}$$

The Landau form is easily obtained by recalling that in the mean-field approximation $K(0) = T_c^0/T$.

In order to make contact with Landau theory we have to generalize the form (14.7)–(14.8) to the continuum. To this end we keep in mind that

$$\sum_i \to \frac{N}{V}\int \mathrm{d}^d x, \quad \hat{\varphi}_i \to \left(\frac{V}{N}\right)^{1/2}\hat{\varphi}(\mathbf{x}), \quad K_{ij} \to \left(\frac{V}{N}\right)K(\mathbf{x}, \mathbf{x}'), \tag{14.17}$$

so that

$$\frac{1}{2}\sum_{ij}\hat{\varphi}_i K_{ij}\hat{\varphi}_j \rightarrow \frac{1}{2}\int \mathrm{d}^d x\, \mathrm{d}^d x' \hat{\varphi}(\mathbf{x}) K(\mathbf{x},\mathbf{x}')\hat{\varphi}(\mathbf{x}') \tag{14.18}$$

and

$$\sum_i \ln\cosh\left(\beta h_i + \sum_j K_{ij}\hat{\varphi}_j\right) \tag{14.19}$$
$$\rightarrow \frac{N}{V}\int \mathrm{d}^d x\ \ln\cosh\left[\left(\frac{V}{N}\right)^{1/2}\left(\beta h(\mathbf{x}) + \int \mathrm{d}^d x' K(\mathbf{x},\mathbf{x}')\hat{\varphi}(\mathbf{x}')\right)\right],$$

having used

$$h_i \rightarrow \left(\frac{V}{N}\right)^{1/2} h(\mathbf{x}) \tag{14.20}$$

and the product of the integrals on $\hat{\varphi}_i$ goes into the functional integral $\mathcal{D}\hat{\varphi}$. Accordingly the partition function is

$$Z\,[h(\mathbf{x})] = \int \mathcal{D}\hat{\varphi}\ \mathrm{e}^{-S[h(\mathbf{x}),\hat{\varphi}(\mathbf{x})]} \tag{14.21}$$

and the action becomes a functional of the field and, close to the critical point ($K(0) \approx 1$) and for small $h(\mathbf{x})$, has the Landau form

$$k_{\mathrm{B}} T S\,[h(\mathbf{x}),\hat{\varphi}(\mathbf{x})] = \int \mathrm{d}^d x (a\hat{\varphi}^2(\mathbf{x}) + b\hat{\varphi}^4(\mathbf{x}) + c|\nabla\hat{\varphi}(\mathbf{x})|^2 - h(x)\hat{\varphi}(\mathbf{x})), \tag{14.22}$$

where $a = K(0)(1-K(0))/2\beta \propto T - T_{\mathrm{c}}^0$.

In analogy with Eq. (14.12) the Legendre transform reads

$$\tilde{F}(T,\varphi) = F(T, h\{\varphi\}) + \int \mathrm{d}^d x h(\mathbf{x})\varphi(\mathbf{x}), \tag{14.23}$$

where $\tilde{F}(T,\varphi)$ is a functional of $\varphi = \langle\hat{\varphi}\rangle$ and h is also a functional of φ.

For an n-component field, the action can be generalized by replacing $\hat{\varphi}^2(\mathbf{x}) = \sum_{i=1}^n \hat{\varphi}_i^2(\mathbf{x})$, $\hat{\varphi}^4(\mathbf{x}) = (\sum_{i=1}^n \hat{\varphi}_i^2(\mathbf{x}))^2$ and $\mathbf{h}(\mathbf{x})\cdot\hat{\boldsymbol{\varphi}}(\mathbf{x}) = \sum_{i=1}^n h_i(\mathbf{x})\hat{\varphi}_i(\mathbf{x})$.

14.2 The Gaussian approximation

The mean-field approximation and Landau theory are obtained by imposing a non-fluctuating order parameter and taking the maximum of the integrand of the partition function instead of the full functional integration:

$$Z(\mathbf{h}) = \int \mathcal{D}\hat{\varphi}\ \mathrm{e}^{-S[\mathbf{h}(\mathbf{x}),\hat{\boldsymbol{\varphi}}(\mathbf{x})]} \sim Z(\bar{\boldsymbol{\varphi}}(\mathbf{h}),\mathbf{h}) = \mathrm{e}^{-S(\bar{\boldsymbol{\varphi}}(\mathbf{h}),\mathbf{h})}, \tag{14.24}$$

where $\bar{\boldsymbol{\varphi}}$ obeys the equation (13.11) or (13.40) for the scalar and the n-component field, respectively.

The first correction to Landau theory is the Gaussian approximation, which consists of retaining in the action only quadratic terms in the deviations of the order parameter $\Delta\hat{\boldsymbol{\varphi}}$ from $\hat{\boldsymbol{\varphi}}$:

$$Z = \int \mathcal{D}\hat{\varphi} \mathrm{e}^{-S} \approx \mathrm{e}^{-S(\bar{\varphi}(\mathbf{h}),\mathbf{h})} \int \mathcal{D}\Delta\hat{\varphi}\, \mathrm{e}^{-\Delta S(\bar{\varphi},\Delta\hat{\varphi})}, \tag{14.25}$$

$$\hat{\boldsymbol{\varphi}}(\mathbf{k}) = \frac{1}{V^{1/2}} \int_V \mathrm{d}^d x\, \mathrm{e}^{-\mathrm{i}\mathbf{k}\cdot\mathbf{x}} \Delta\hat{\boldsymbol{\varphi}}(\mathbf{x}), \tag{14.26}$$

with the corresponding correction to the thermodynamic potential:

$$\Delta F_{\text{fluct}} = -k_{\mathrm{B}} T \ln \int \mathcal{D}\hat{\varphi}\, \mathrm{e}^{-\Delta S(\bar{\varphi},\Delta\hat{\varphi})}. \tag{14.27}$$

Above the critical temperature in the limit $\mathbf{h} \to 0$

$$k_{\mathrm{B}} T \Delta S = \frac{1}{2} \sum_{i,\mathbf{k}} (a + ck^2) |\hat{\varphi}_i(\mathbf{k})|^2. \tag{14.28}$$

Below T_{c} we have to be careful in doing the limit at zero field and we get

$$k_{\mathrm{B}} T \Delta S = \frac{1}{2} \sum_{\mathbf{k}} \left[(a + 6b\bar{\varphi}^2 + ck^2) |\hat{\varphi}_1(k)|^2 + \sum_{i=2}^{n} (a + 2b\bar{\varphi}^2 + ck^2) |\hat{\varphi}_i(k)|^2 \right], \tag{14.29}$$

$$\Delta F_{\text{fluct}} = -k_{\mathrm{B}} T \frac{n}{2} \sum_{\mathbf{k}} \ln \left(\frac{2\pi / k_{\mathrm{B}} T}{(a + ck^2)} \right), \quad T > T_{\mathrm{c}}, \tag{14.30}$$

$$\Delta F_{\text{fluct}} = -k_{\mathrm{B}} T \frac{1}{2} \sum_{\mathbf{k}} \left[\ln \left(\frac{2\pi / k_{\mathrm{B}} T}{(-2a + ck^2)} \right) + (n-1) \ln \left(\frac{2\pi / k_{\mathrm{B}} T}{ck^2} \right) \right], \quad T < T_{\mathrm{c}}. \tag{14.31}$$

The fluctuation induced term of the specific heat which is singular as T goes to T_{c} reads

$$\Delta C_{\text{fluct}} = -T \frac{\partial^2 \Delta F}{\partial T^2} \sim T_{\mathrm{c}}^2 \int \mathrm{d}^d k \frac{2(a')^2}{(-2a + ck^2)^2} \sim \frac{1}{\xi^{d-4}(T)}. \tag{14.32}$$

This contribution has to be compared with the constant mean-field value $T_{\mathrm{c}} \frac{a'^2}{2b}$ and the Ginzburg criterion follows again.

The calculation of the correlation function in $\mathbf{k}$ space reduces to moments of the Gaussian distribution

$$f_{\varphi}(k) = \frac{\int \mathcal{D}\hat{\varphi}\, |\hat{\boldsymbol{\varphi}}(\mathbf{k})|^2 \mathrm{e}^{-\Delta S}}{\int \mathcal{D}\hat{\varphi}\, \mathrm{e}^{-\Delta S}} \tag{14.33}$$

and yields the same expression (13.22) derived from the small corrections to the self-consistency equation for the order parameter in Chapter 13. When dealing with the n-component model, the transverse correlation functions in the limit $\mathbf{h} \to 0$ yield the Mermin and Wagner (1966) theorem as from Eq. (13.29).

14.3 Hydrodynamics for the isotropic ferromagnet

The isotropic ferromagnet corresponds to $n = 3$ for the number of components of the order parameter. The transverse massless modes correspond physically to the spin wave excitations. We follow a treatment due to Halperin and Hohenberg (1969a). Our starting point is the expression for the transverse (phase) fluctuations of a three-dimensional ferromagnet, whose spontaneous magnetization per unit volume m_0 is along the z-axis. Then we may write the free energy density variation with respect to the equilibrium value $f_0(m_z = m_0)$

$$\Delta f(m_x, m_y) = \frac{1}{2}\frac{\rho_s}{m_0^2}[|\nabla m_x|^2 + |\nabla m_y|^2], \tag{14.34}$$

where the stiffness constant ρ_s has been introduced to describe the ability of the system to perform transverse fluctuations. The stiffness constant ρ_s is related, at the level of the mean-field approximation, to the constant c of Landau theory as $\rho_s = 2c\varphi_0^2$. This is easily understood in the case of a complex order parameter, $\varphi(\mathbf{x}) = \varphi_0 e^{i\theta(\mathbf{x})}$, if we allow only phase fluctuations $|\nabla\varphi(\mathbf{x})|^2 = \varphi_0^2(\nabla\theta(\mathbf{x}))^2$. Notice that the inclusion of fluctuations modifies the value of the stiffness constant. The argument can be directly generalized to the transverse fluctuations of a generic order parameter with a number of components $n \geqslant 2$.

Let us now introduce along the x-axis a weak magnetic field of the form $h_x = h\cos(\mathbf{k}\cdot\mathbf{x})$ with h small. By requiring Δf to be a minumum one gets

$$-\frac{\rho_s}{m_0^2}\nabla^2 m_x = h\cos(\mathbf{k}\cdot\mathbf{x}), \tag{14.35}$$

$$-\frac{\rho_s}{m_0^2}\nabla^2 m_y = 0. \tag{14.36}$$

The perturbed solution reads

$$m_x = \frac{m_0^2}{\rho_s k^2} h\cos(\mathbf{k}\cdot\mathbf{x}), \tag{14.37}$$

$$m_y = 0. \tag{14.38}$$

Let us now consider small deviations from equilibrium. Since the magnetization is a conserved quantity we may write

$$\frac{\partial m_x}{\partial t} = -\nabla\cdot\mathbf{J}^x, \tag{14.39}$$

$$\frac{\partial m_y}{\partial t} = -\nabla\cdot\mathbf{J}^y - hm_0\cos(\mathbf{k}\cdot\mathbf{x}), \tag{14.40}$$

where $\mathbf{J}^x$ and $\mathbf{J}^y$ are magnetization currents and the last term in Eq. (14.40) is due to the precession $\mathbf{h}\times\mathbf{m}$ due to the field h_x. For small deviations from equilibrium the currents $\mathbf{J}^x$ and $\mathbf{J}^y$ can be expressed as

$$\mathbf{J}^x = A\nabla m_x + B\nabla m_y, \tag{14.41}$$

$$\mathbf{J}^y = C\nabla m_x + D\nabla m_y, \tag{14.42}$$

and by symmetry we require $A = D$ and $B = -C$. In equilibrium in the presence of the small magnetic field, the compatibility of Eqs. (14.39) and (14.40) with Eqs. (14.35) and (14.36) gives $A = 0$ and $B = -\rho_s/m_0$. Then, in the absence of the small magnetic field, the equations of motion for the spin densities read

$$\frac{\partial m_x}{\partial t} = \frac{\rho_s}{m_0} \nabla^2 m_y, \tag{14.43}$$

$$\frac{\partial m_y}{\partial t} = -\frac{\rho_s}{m_0} \nabla^2 m_x, \tag{14.44}$$

which describe spin waves excitations with dispersion relation $\omega = c_s k^2$ and velocity $c_s = \rho_s/m_0$.

15 Superfluidity and superconductivity

15.1 Introduction

This chapter will deal with the phenomena of superfluidity and superconductivity, emphasizing those concepts which are common to the understanding of both. For this reason we will proceed in parallel for the two phenomena by discussing the main experimental facts, the emergence of the phenomenological theories and finally their microscopic description.

Some systems under given conditions have the property of sustaining a flow without friction, i.e. with no viscosity. This phenomenon, named superfluidity, is present for transport velocities and temperatures below critical values specific to each system.

The liquid helium with isotope number 4 (^{4}He) becomes superfluid below the critical temperature $T_c = 2.17$ K. It was established first in Leiden (in 1924 and 1932) by some anomalies in the thermodynamic properties and later (1936–38) by transport experiments (Kapitza (1938)). The superfluidity of liquid helium with isotope number 3 (^{3}He) occurs at much lower temperatures in the millikelvin region ($T_c = 2.7$ mK near the liquid–solid coexistence curve). For this reason it was discovered much later, in 1971, by Osheroff et al. (1972a) at Cornell, when the improvement in the required technology had been achieved.

Such different values of the critical temperatures for the two isotopes must be related to their different statistics and not to their different atomic masses. ^{4}He has nuclear spin zero and therefore obeys Bose–Einstein statistics, whereas the isotope ^{3}He has semi-integer nuclear spin and obeys Fermi–Dirac statistics.

The absence of friction in liquid helium corresponds to the absence of dissipation of the electrical current of the electrons in many metals at temperatures below critical values typically of the order of $T_c \approx 1$–10 K. The phenomenon named superconductivity was discovered by Kamerling-Onnes (1911), for mercury with $T_c = 4$ K. Higher values of T_c were first obtained for alloys and then in families of copper oxides which, normally insulators, become, when suitably doped, superconductors at much higher temperatures ($T_c \approx 30$–150 K) as initially discovered by Bednorz and Müller (Bednorz et al. (1987)). These superconductors show anomalies in the behavior of both the normal and superconducting phases with respect to the previous ordinary superconductors.

Superconductivity is by now a common property of many different compounds whereas superfluidity is present in the two isotopes of the liquid helium only, besides the presence of the Bose–Einstein condensation (BEC) for gases of alkaline atoms since 1995 (Anderson et al. (1995); Davis et al. (1995); Bradley et al. (1995)).

A few basic concepts like ordering and its consequences will be the guiding ideas of this chapter. References to specific systems will only be made to select general principles. Cold atoms and ^{3}He will therefore be referred to very briefly and high temperatures superconductors will be considered merely to point out the present difficulties in fully understanding them.

15.2 The Landau criterion

Dealing with superfluidity and superconductivity, we begin with a criterion of flow stability due to Landau (1941), who stated the conditions under which a superfluid flow can be obtained.

We consider a moving fluid in a capillary at $T = 0$ so that the fluid is in its ground state. In the presence of stable superfluid flow, there is no interaction between the fluid helium and the walls of the capillary, otherwise the viscosity would manifest itself with a loss of kinetic energy of the fluid and the appearance of inner dissipative motions. Landau assumed that the fluid can interact with an external body (the wall) only through the creation or annihilation of elementary excitations with well defined dispersion law of energy $\epsilon(\mathbf{p})$ versus momentum $\mathbf{p}$. In a frame moving with the fluid at velocity $\mathbf{v}$, the total energy and momentum are those of the elementary excitations. Instead, in a frame at rest with respect to the capillary, in the presence of a single excitation $\epsilon(\mathbf{p})$, the total energy E and momentum $\mathbf{P}$ are, according to the Galilean transformations,

$$E = \epsilon(\mathbf{p}) + \mathbf{p} \cdot \mathbf{v} + \frac{1}{2} M v^2, \tag{15.1}$$

and

$$\mathbf{P} = \mathbf{p} + M\mathbf{v}, \tag{15.2}$$

where M is the total mass of the fluid. The condition for the creation of excitations and the appearance of dissipation, i.e. of viscosity, requires that the energy (15.1) is lower than the energy in the absence of dissipation, $Mv^2/2$, i.e. it must be that $\Delta E = \epsilon(\mathbf{p}) + \mathbf{p} \cdot \mathbf{v} < 0$. Clearly, the most favorable circumstance is when the momentum $\mathbf{p}$ of the excitation is opposite to the velocity $\mathbf{v}$. The above condition implies the existence of a critical velocity, v_c, determined by the energy dispersion relation of the excitations:

$$v > \left. \frac{\epsilon(\mathbf{p})}{\mathbf{p}} \right|_{\min} \equiv v_c. \tag{15.3}$$

This condition must apply to any superflow. An immediate consequence of the Landau criterion is that a Bose gas cannot be a superfluid. In fact the free-like spectrum of a Bose gas has $v_c = 0$. The superfluid transition of ^{4}He cannot be simply modelled as a Bose gas even though the Bose–Einstein condensation is relevant, as we shall see, for the interpretation of its phenomenology.

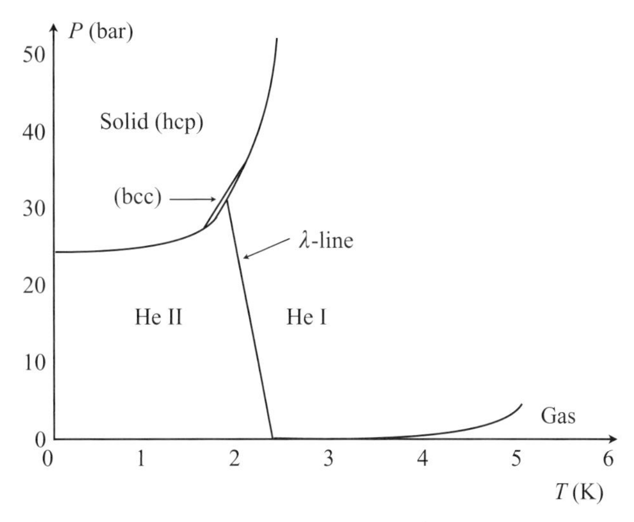

Fig. 15.1 The phase diagram of ^{4}He in the pressure vs. temperature plane. When ^{4}He is in the liquid state, there are two distinct phases indicated as He I and He II. In the phase He I, ^{4}He behaves as a normal liquid, whereas in phase He II it is superfluid. The solid phase has either hexagonal close packed (hcp) or body centered cubic (bcc) symmetry.

15.3 Phenomenology of superfluid ^{4}He

15.3.1 Experimental facts

A few specific properties of superfluids and superconductors are selected to achieve a theoretical characterization of these phases. We start by introducing the phenomenology of superfluid ^{4}He, since the correspondence between its properties and their theoretical interpretation is most direct.

Helium is the element with the simplest structure after hydrogen. The attractive tail of the van der Waals forces between the atoms with diameter $d = 2.6$ Å is extremely weak. This, together with the small value of the atomic mass and the high zero point energy, allows this system and its isotope ^{3}He to remain liquid for pressures below a finite value (25 Bars for ^{4}He and 35 Bars for ^{3}He) down to zero kelvin with a very low boiling point (4.2 K for ^{4}He and 3.2 K for ^{3}He) and no triple point between the solid, liquid and gas phases, see Fig. 15.1.

As early as in 1911 Kamerling-Onnes found that at about 2.2 K when the liquid is cooled, it expands instead of contracting. Later in 1924 Kamerling-Onnes and Boks (see London (1959)) found that the density has a sharp maximum at that temperature with a discontinuity in the slope. The low temperature phase, named He II in 1928 (see London (1959)), has a configuration with larger mean distance between the atoms with respect to the normal liquid (He I) above T_c. Specific heat measurements (Keesom and Clusius (1932)) showed a singularity at T_c ($T_c = T_\lambda(P)$) with the characteristic form of the Greek letter λ after which the transition has been named. Later measurements determined this singularity to be logarithmic or even a finite cusp (see Table 16.3 in Chapter 16). We now know that this behavior characterizes the λ-transition as a disorder–order transition of second order with $n = 2$ components of the order parameter (see also Chapter 16). This ordering transition at the λ-line is confirmed by the shape of the melting-pressure curve, which below 1 K is practically horizontal (see Fig. 15.1) and, from the Clausius–Clapeyron equation (1.72), we can deduce that the entropy of the liquid is the same as the entropy of the solid. Between $T = 0$ K and $T = 1$ K the liquid becomes as ordered as the solid. As shown by X-ray experiments, its order cannot be reached in real space, where actually the configurations become less dense with an average distance between the atoms $\overline{d} = 3.6$ Å.

Experiments on transport properties clarified the phenomenon of superfluidity underlying the λ-transition. The thermal conductivity in liquid He II becomes infinite with decreasing temperature gradient (1936). The anomaly in heat transport leads also to the strange phenomenon called the fountain effect.[1] The most enlightening experiment was performed in 1938 by measuring an enormous viscosity drop through the λ-point ($\eta < 10^{-11}$ poises, to be compared with the typical viscosity of a normal fluid $\eta \sim 10^{-2}$–10) by the capillary flow method by Kapitza (1938) and simultaneously by Allen and Misener (1938), confirming an early observation of Keesom and van den Ende (1930). However, measurements with the rotating disk method gave a continuously decreasing function of η with values of the order of 10^{-5} poises (Keesom and MacWood (1938)).

The enormous difference in the results of η by the two capillary flow and rotating disk methods, together with other observations like the fountain effect, led Tisza (1938) to propose the two-fluid model.

15.3.2 The two-fluid model

Liquid helium below the λ-line is assumed to be formed of two interpenetrating fluids, one with normal properties, the other flowing without dissipation and carrying no entropy and therefore highly ordered. In the capillary flow method, provided the capillary is sufficiently thin, only the superfluid component is allowed to move with zero viscosity, whereas with the rotating disk method the normal component contribution is captured with finite viscosity. The flow is generally characterized by the mass density $\rho = mn$ (m being the mass of the particles) and momentum current $\mathbf{J}$ which are now the sum of the corresponding quantities for the two components:

$$\rho = \rho_n(T) + \rho_s(T) \equiv \rho_n + \rho_s, \tag{15.4}$$

$$\mathbf{J} = \mathbf{J}_n + \mathbf{J}_s = \rho_n \mathbf{v}_n + \rho_s \mathbf{v}_s. \tag{15.5}$$

The superfluid density ρ_s is equal to the total density at $T = 0$ and for temperatures above 1 K decreases rapidly to zero at $T = T_c$. Vice versa for ρ_n which is measured by rotating a pile of closely packed disks in He II. The superfluid component slides on it, whereas the normal component is dragged along. The corresponding variation of the inertia can be measured

[1] The fountain effect is a particular manifestation of the thermomechanical effect according to which a gradient of pressure induces a gradient in the temperature and vice versa. Its explanation will be given within the two-fluid model.

and as a result a rapid increase of ρ_n for $1.3 < T < 2.17$ K is obtained (Andronikashvili (1946)), as shown in Fig. 15.2. According to the London (1938) early suggestion, the superfluid component has been associated with the condensate of a Bose gas, which has zero entropy, all the entropy being carried by the particles out of the condensate, as shown in Eq. (7.48). Even if the thermodynamic behavior of various physical quantities like the specific heat, the superfluid $\rho_s(T)$ and normal density $\rho_n(T)$, is not well reproduced, the critical temperature for a Bose gas (Eq. (7.42)) with the density of liquid helium (27.6 m^3/mole) would be $T_c = 3.15$ K, not too far from the real value $T_c = 2.17$ K considering the crudeness of the Bose gas model.

The two-fluid model by itself and the related quantum hydrodynamics properly describe the most striking phenomena of liquid He II. On the basis of the two Eqs. (15.4) and (15.5), one may derive an hydrodynamic description, where besides the standard sound excitation, one obtains the so-called second sound excitations. The latter are entropy density waves, as opposed to particle density waves, arising from the fact that a temperature gradient creates regions with different relative amounts of normal and superfluid components which can oscillate out of phase. The two-fluid hydrodynamics is briefly recalled in Appendix I. Here we content ourselves with a qualitative explanation of the thermomechanical effect.

Let us imagine two vessels filled with He II connected by a thin capillary. A sufficiently thin capillary increases the viscosity and makes the flow of the normal component almost impossible. The superfluid component, because it has zero viscosity, can flow in the capillary, which then acts as a selective membrane. Let us assume that initially the two vessels are in equilibrium so that the relative amounts of normal and superfluid components are the same in both vessels. If now we heat up one of the two vessels, there will be an increase of its normal component and a decrease of the superfluid one. To restore equilibrium there will be a flow of the superfluid component from the other vessel so that the amount of fluid in the first vessel will increase and, eventually, the fluid itself will spill out of the vessel giving rise to the fountain effect.

The initial shortcomings of the two-fluid model, of being strictly linked to the Bose–Einstein condensation of a gas, have been overcome by Landau (1941, 1947).

The interaction between the helium atoms, besides their statistics, has to be taken into account. The interacting helium atoms must have a particular ground state showing condensation. The excitations from the ground state must satisfy the Landau criterion (15.3). The spectrum of excitations of ^{4}He has been obtained in 1961 (Henshaw and Woods (1961)) by measuring the energy loss of diffused neutrons of incident momentum $\mathbf{p}$.[2] The low energy excitations are phonons with $\epsilon(\mathbf{p}) = c|\mathbf{p}| \equiv cp$, c being the sound velocity corresponding to compressional waves of the fluid. On increasing the momentum, the spectrum reaches a maximum and then a minimum, named the roton minimum Δ_0, at momentum $p_0 \sim 2\pi\hbar/\bar{d}$, where $\bar{d}$ is of the order of the mean distance among the atoms. Around the minimum the roton spectrum can be described by $\epsilon_r(\mathbf{p}) = \Delta_0 + (p - p_0)^2/2\mu_r$, where the empirical values are $\Delta_0/k_B = 8.68$ K and $\mu_r = 0.16\, m_{He}$. This spectrum clearly satisfies the Landau criterion with $v_c \sim 60$ m/s. This value of v_c is, however, much higher

[2] For the connection between the spectrum of excitation and the measured cross section of scattered neutrons see the discussion in the next subsection and in particular Eq. (15.17).

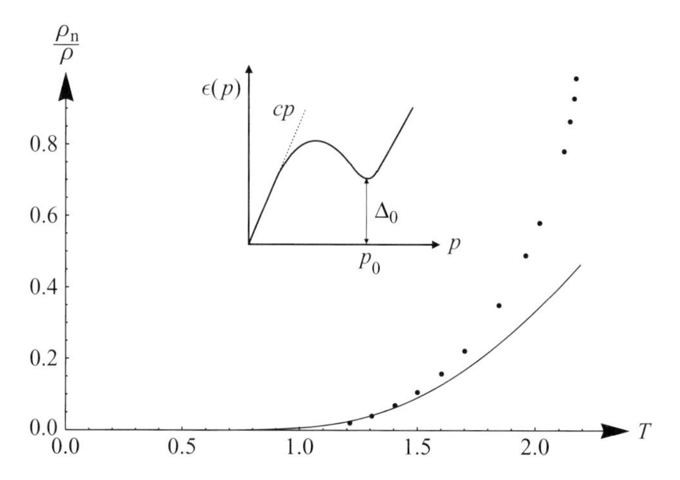

Fig. 15.2 The normal density $\rho_n(T)$ normalized to the value of the total density as a function of temperature using Eq. (15.11) (full line) with values $\Delta_0 = 8.68$ K, $p_0/\hbar = 1.91$ Å^{-1}, $\mu_r = 0.16\,\mathrm{m_{He}}$ measured at 1.1 K by inelastic neutron scattering (see Brooks and Donnelly (1977) for extensive discussions on these values). The dots are the measured values obtained by Dash and Taylor (1957) in a rotating disks experiment using the technique of Andronikashvili. The inset shows a schematic plot of the spectrum of excitations as obtained by neutron scattering, showing the roton minimum at momentum p_0, after the initial linear behavior with slope given by the speed of sound.

than the experimental values of the order of 1 cm/s. Other excitations related to the vorticity must be considered, which, as we shall see, account for the experimental values of v_c. The spectrum of excitations was phenomenologically foreseen by Landau (1947) by analyzing the thermodynamic properties. In particular, at low temperature the specific heat behaves as $c_V \sim T^3$ (not $c_V \sim T^{3/2}$ as for a Bose gas in Eq. (7.50)) followed by an exponential behavior at higher temperatures, $c_V \sim \exp(-\beta\Delta_0)$. The T^3 term is well represented by the presence of phonons (Eq. (7.96)) as the lowest excitations of the spectrum excited at low temperature. The appearance of a Boltzmann exponential behavior in the specific heat at higher temperature is related to the roton gap Δ_0 (see the inset in Fig. 15.2). The roton contribution to the low-temperature specific heat is evaluated by describing it as a classical gas with zero chemical potential since $k_B T \ll \Delta_0$ and there is no conservation of the number of rotons. The free energy of a classical gas of N_r rotons is given by

$$F_r = -k_B T \ln \frac{1}{N_r!}\left(\frac{V}{(2\pi\hbar)^3}\int d\mathbf{p}\; e^{-\beta\epsilon_r(\mathbf{p})}\right)^{N_r}. \tag{15.6}$$

By requiring that F is a minimum with respect to N_r, as shown in Problem 15.1, we have

$$F_r = -k_B T N_r \equiv -k_B T \frac{V}{(2\pi)^3}\int d\mathbf{p}\; e^{-\beta\epsilon_r(\mathbf{p})} = -k_B T\left(\frac{2Vp_0^2 e^{-\Delta_0/k_B T}(\mu_r k_B T)^{1/2}}{(2\pi)^{3/2}\hbar^3}\right). \tag{15.7}$$

The specific heat then follows:

$$c_{V,\mathrm{r}} = k_\mathrm{B} N_\mathrm{r} \left(\frac{3}{4} + \frac{\Delta_0}{k_\mathrm{B} T} + \left(\frac{\Delta_0}{k_\mathrm{B} T} \right)^2 \right). \tag{15.8}$$

Moreover, Landau overcame the two-fluid paradox of having two interpenetrating fluids flowing without friction with respect to each other. He related the normal component to the excitations (their inertia well reproduces $\rho_\mathrm{n}(T)$) and the substrate of these excitations to the superfluid component, which at $T = 0$ coincides with the total density, no excitations being present. The temperature dependence of the normal density component is obtained by considering that, when the fluid flows with velocity $\mathbf{v}$, the associated momentum density reads

$$\mathbf{P} = \frac{1}{(2\pi\hbar)^3} \int \mathrm{d}\mathbf{p}\, \mathbf{p}\, n(\epsilon(\mathbf{p}) - \mathbf{p} \cdot \mathbf{v}), \tag{15.9}$$

where $n(\epsilon(\mathbf{p}) - \mathbf{p} \cdot \mathbf{v})$ represents the number of excitations of momentum $\mathbf{p}$ and energy $\epsilon(\mathbf{p})$, if $n(\epsilon(\mathbf{p}))$ is the same number of excitations when the fluid is at rest. This result can be obtained by Galilean transformations by considering a frame where the fluid is at rest. With respect to the frame where the fluid is in motion the extra energy of an excitation is $\epsilon(\mathbf{p}) - \mathbf{p} \cdot \mathbf{v}$. An equivalent way to arrive at this result is to consider how the distribution function changes when a gas of quasiparticles as a whole is performing a motion with velocity $\mathbf{v}$.[3]

At small values of $\mathbf{v}$, by expanding $n(\epsilon(\mathbf{p}) - \mathbf{p} \cdot \mathbf{v})$ with respect to $\mathbf{v}$, one has $\mathbf{P} = \rho_\mathrm{n} \mathbf{v}$ with

$$\rho_\mathrm{n} = -\frac{1}{3(2\pi\hbar)^3} \int \mathrm{d}\mathbf{p}\, p^2 \frac{\partial n(\epsilon(\mathbf{p}))}{\partial \epsilon(\mathbf{p})}, \tag{15.10}$$

where the factor $1/3$ comes from the angular average. One then obtains the phonon and roton contributions, respectively,

$$\rho_{\mathrm{n,ph}}(T) = \frac{2\pi^2 k_\mathrm{B}^4 T^4}{45\hbar^3 c^5}, \qquad \rho_{\mathrm{n,r}}(T) = \frac{2 p_0^4 \mathrm{e}^{-\Delta_0/k_\mathrm{B} T} \mu_\mathrm{r}^{1/2}}{3(2\pi)^{3/2} (k_\mathrm{B} T)^{1/2} \hbar^3}. \tag{15.11}$$

The phonon contribution is responsible for the slow decrease of ρ_s at low temperature, whereas the roton part determines the rapid vanishing of ρ_s upon increasing the temperature above 1 K. In Fig. 15.2 we plot $\rho_\mathrm{n}(T)/\rho$ as obtained from Eq. (15.11) together with the experimental values. The agreement up to $T = 1.5$ K shows that, within this range of temperature, phonon and roton excitations behave as a gas of bosonic independent particles. For $T > 1.5$ K, the excitations can no longer be considered as independent entities, even though the neutron scattering experiments still probe the single-excitation spectrum up to temperature ~ 4 K.[4]

[3] This can be analyzed by adding to the constraints (3.11) the further constraint of total momentum conservation $\mathbf{P} = \sum_i \mathbf{p}_i n_i$ with the vector Lagrange multiplier $-\beta \mathbf{v}$.

[4] The experiments by Henshaw and Woods (1961) show that the energy and width of the neutron scattering peak decrease and increase rapidly, respectively, on getting close to T_λ. However, above T_λ, their temperature dependence weakens again.

15.3.3 The Feynman theory

Landau never showed that condensation is present in the ground state, nor gave any indication of the type of order appearing at the λ-transition.

Feynman (1953b, a, 1954); Feynman and Cohen (1956) reconsidered the Landau arguments given above and derived from first principles the basic aspect of superfluidity, i.e. the *scarcity* of available low-energy excitated states, limited to those few states with excitations compatible with the Landau criterion.

The ground state function $\phi_0(\mathbf{x}_1, \ldots, \mathbf{x}_N)$ is a real positive totally symmetric function of the positions $\{\mathbf{x}_i\}$ of the N particles, as shown in the answer to Problem 15.2. Its amplitude must be large for probable configurations with well separated and evenly spaced atoms with averaged distance $\overline{d}$ ($\overline{d} = 3.6\,\text{Å} \gg d = 2.7\,\text{Å}$, d being the diameter of the helium atoms). The amplitude is small for improbable configurations when there is clumping. Feynman then hypothesized for ϕ_0

$$\phi_0(\mathbf{x}_1, \ldots, \mathbf{x}_N) = \prod_{ij} F(\mathbf{x}_{ij}), \tag{15.12}$$

a product of all pairs of particles of a function $F(\mathbf{x}_{ij})$, which depends only on the relative distance $\mathbf{x}_{ij}$ of the atoms of each pair and vanishes for $|\mathbf{x}_{ij}| \leq d$. Feynman then showed that the only low-energy excitations are phonons, corresponding to states orthogonal with respect to the ground state ϕ_0 with low variation of the density over large distances and small wave vectors. The Bose statistics makes the phonons the only low-lying excited states because variation in the density cannot be accomplished by permuting atoms starting from an homogeneous configuration. All other states either are equivalent to the ground state by permutations or involve movements of atoms on distances less than the average atomic distance $\overline{d}$, i.e. are rotons separated from the ground state by an energy at least $\Delta_0 \sim \hbar^2/(2m\overline{d}^2)$, corresponding to a wave vector $k_0 \sim 2\pi/\overline{d}$.

In the Feynman theory, to each wave vector $\mathbf{k}$ corresponds a given excitation energy $\epsilon(\hbar\mathbf{k})$ as in the Landau scheme. We will impose this condition on the dynamical structure factor which relates the measurement of the spectrum to the properties of the system. The spectrum is measured via the neutron diffusion by the atoms of the material. The potential responsible for neutron diffusion can be written as $V(\mathbf{x}) = v_0 \sum_{\mathbf{R}} \delta(\mathbf{x} - \mathbf{R})$, where the constant v_0 measures the strength of the neutron–atom interaction and $\mathbf{R}$ runs over the positions of the atoms of the material. Notice that, apart from the constant v_0, the potential coincides with the density operator for the Bose liquid. Formally, the problem is similar to the diffusion of electrons by impurities discussed in the answer to Problem 11.2. In the Born approximation, the probability of a neutron exchanging an energy $\hbar\omega$ and momentum $\hbar\mathbf{q} = \hbar(\mathbf{k} - \mathbf{k}')$ with the liquid can be obtained by the Fermi Golden Rule as in Eq. (11.43), where the delta function for the conservation of energy now reads $\delta(\hbar^2 k'^2/2m + E_n - \hbar^2 k^2/2m - E_0)$. Here $\mathbf{k}$ and $\mathbf{k}'$ are the initial and final momenta of the neutron.[5] E_0 and E_n are the energies

[5] For simplicity we neglect here the spin of the neutron and the magnetic interaction with the atomic magnetic moments.

of the initial and final states $|0\rangle$ and $|n\rangle$ of the system. We must also replace the scattering potential in Eq. (11.43) with the expectation value $\langle n|V(\mathbf{x})|0\rangle$, i.e. the matrix elements of the density operator. Finally, after summing over all final states $|\mathbf{k}', n\rangle$ of the neutron and the system, the scattering cross section turns out proportional to the power spectrum of the density fluctuations $\overline{f}^{nn}_{\mathbf{q}}(\omega)$,[6] defined in Eq. (10.56) with the limitation of performing the average with respect to the ground state $|0\rangle$. This corresponds to set $m = 0$ in Eq. (10.56). The power spectrum of the density fluctuations is also known as the structure factor $S(\mathbf{q}, \omega) = \overline{f}^{nn}_{\mathbf{q}}(\omega)/\pi$ of the liquid. The static structure factor $S(\mathbf{q}) = N^{-1}\langle 0|\hat{n}(\mathbf{q})\hat{n}(-\mathbf{q})|0\rangle$ is obtained via the integral

$$S(\mathbf{q}) = \frac{1}{N}\int_0^\infty \mathrm{d}\omega S(\mathbf{q}, \omega) = \frac{1}{N}\sum_n |(n_{\mathbf{q}})_{n0}|^2. \tag{15.13}$$

$(N/V)S(\mathbf{q})$ is the $\mathbf{q}$-representation of the two-particle reduced density matrix $h_2(\mathbf{x}, \mathbf{x}'; \mathbf{x}', \mathbf{x})$ of Eq. (9.55).

The dynamical structure factor for any system of N particles interacting via a two-body potential, which is independent of the momentum, satisfies the f-sum rule (10.59). The last one now reads

$$\hbar\int_0^\infty \mathrm{d}\omega\, \omega \bar{f}^{nn}_{\mathbf{q}}(\omega) = \pi\hbar\sum_n \omega_{0n}|\langle 0|\hat{n}(\mathbf{q})|n\rangle|^2 = N\frac{\hbar^2 q^2}{2m}. \tag{15.14}$$

The specific quantum nature of low-temperature liquid helium is represented by assuming that for each $\mathbf{q}$ there is only one frequency of excitation $\omega(\mathbf{q})$ of the modulated state $\hat{n}(\mathbf{q})|0\rangle$, $n_{\mathbf{q}}\phi_0 = \sum_{i=1}^N \mathrm{e}^{\mathrm{i}\mathbf{q}\cdot\mathbf{x}_i}\phi_0$, which exhausts the sum rule. Then

$$S(\mathbf{q}, \omega) = \sum_n |(n_{\mathbf{q}})_{n0}|^2\delta(\omega - \omega_{n0}) = NS(\mathbf{q})\delta(\omega - \omega(\mathbf{q})) \tag{15.15}$$

and the sum rule (15.14) implies

$$\hbar\int_0^\infty \mathrm{d}\omega\, \omega S(\mathbf{q}, \omega) = \hbar NS(\mathbf{q})\omega(\mathbf{q}) = N\frac{\hbar^2 q^2}{2m}. \tag{15.16}$$

The excitation energy $\epsilon(\hbar\mathbf{q}) = \hbar\omega(\mathbf{q})$ of the state $n_{\mathbf{q}}\phi_0$ is then[7]

$$\epsilon(\hbar\mathbf{q}) = \frac{\hbar^2 q^2}{2mS(\mathbf{q})}. \tag{15.17}$$

The experimental $S(\mathbf{q})$ starts linear in $\mathbf{q}$, giving rise to the phonons and has a maximum at $q \sim 2\pi/\bar{d}$ where the roton minimum is. The roton part of the spectrum requires, however, a more refined treatment (Feynman and Cohen (1956)) to be treated quantitatively.

[6] The suffix n in $\bar{f}^{nn}$ for the density is not to be confused with the quantum number of the state.

[7] This expression was derived as an inequality by Feynman via a variational method using a trial wave function $F(\mathbf{q})\phi_0$ for the excited states in the presence of excitations with definite momentum $\mathbf{q}$. The solution of his variational equation is $F(\mathbf{q}) = n_{\mathbf{q}}$.

15.4 Phenomenology of superconductors

In this section we will briefly recall those experimental aspects of ordinary superconductors which we believe are important to the discussion of the basic concepts of the theory. We will also point out some phenomenological interpretations of these experimental facts.

Many metals and alloys below their superconductive transition temperature T_c have a stable flux of charge carriers also in the absence of an external driving field, i.e. the resistivity approaches zero at the transition. The possibility of having persistent currents is not true for every magnitude of the current. There exists a critical value of the current above which the superconductivity disappears. Related to this critical current there is a critical magnetic field which goes to zero at T_c as

$$H_c(T) = H_0(1 - (T/T_c)^2), \tag{15.18}$$

with H_0 of the order of 10^2 G. The slope of the separation curve $H_c(T)$ at T_c is finite and negative. For ordinary superconductors T_c is of the order of few Kelvin and usually has a crystal ionic mass dependence $T_c \sim M^{-1/2}$ (isotope effect) (Serin et al. (1950)), which gives a strong indication for the microscopic interpretation of the mechanism leading to superconductivity.

Infinite conductivity implies that inside a superconductor the electric field $\mathbf{E} = 0$ and, owing to the Maxwell equation

$$\nabla \times \mathbf{E} = -\frac{1}{c}\frac{\partial \mathbf{B}}{\partial t}, \tag{15.19}$$

the magnetic induction $\mathbf{B}$ is constant. As a result, if a magnetic field is applied after reducing the temperature below T_c, $\mathbf{B}$ remains zero inside the sample. Vice versa, if the temperature were lowered below T_c after applying the magnetic field, it would remain different from zero. Then the constancy of $\mathbf{B}$ would imply a dependence of the final superconductive state on the history of how this state has been achieved, i.e. if the external magnetic field is applied before or after reducing the temperature below T_c. This would make questionable the very concept of superconductivity as a *true* equilibrium thermodynamic state. Superconductors, however, exhibit the Meissner–Ochsenfeld effect discovered in 1933 (Meissner and Ochsenfeld (1933)): the magnetic field is expelled from the bulk of a superconductive sample, thus always imposing $\mathbf{B} = 0$ in the final superconductive state. The thermodynamic description applies to the superconductive phase. Actually $\mathbf{B}$ and $\mathbf{E}$ are vanishing everywhere except in a thin layer of depth λ at the surface. The penetration depth λ is temperature dependent and goes to infinity at the critical temperature when the system becomes normal:

$$\lambda = \lambda_0[1 - (T/T_c)^4]^{-1/2}, \quad \lambda \sim (\lambda_0/2)[1 - T/T_c]^{-1/2} \quad \text{for } T \to T_c. \tag{15.20}$$

Its zero-temperature value λ_0 is of the order of a few hundred Ångstroms.

The phenomenological conditions for superconductivity are therefore the conductivity $\sigma \to \infty$ and $\mathbf{B} = 0$ of a perfect conductor and of a perfect diamagnet up to the critical field (superconductors of the first type or type I).

In a temperature–magnetic field phase diagram, along the coexistence curve between the normal and the superconducting phase, the Clausius–Clapeyron equation (1.72) reads

$$\frac{\mathrm{d}H_c(T)}{\mathrm{d}T} = \frac{s_n - s_s}{m_s - m_n}, \tag{15.21}$$

where s_n, s_s, m_n, m_s are the entropies and magnetizations per unit volume for normal and superconducting state respectively. For the normal state we can put $m_n \sim 0$ and because of the Meissner–Ochsensfeld effect, $m_s = -(H/4\pi)$; the entropy difference between the normal and the superconducting state is then given by

$$s_n - s_s = -\frac{1}{4\pi} H_c \frac{\mathrm{d}H_c}{\mathrm{d}T}. \tag{15.22}$$

$s_n - s_s$ is positive for any $T < T_c$ ($\mathrm{d}H_c/\mathrm{d}T < 0$), except at T_c where H_c vanishes and the transition is of second order. The superconducting phase is more ordered, although X-ray experiments show no structural modifications of the conduction electrons. Actually, for $T < T_c$, the free energy for the superconducting phase F_s must be less than F_n for the normal phase. Their difference $F_n - F_s$ in zero external magnetic field is equivalent to the magnetic cost for the expulsion of the critical field at that temperature:

$$F_n - F_s = \frac{H_c^2}{8\pi}. \tag{15.23}$$

Indeed, the free energy difference in the presence of the critical field $H_c(T)$ must vanish and with the previous positions $m_n \sim 0$ and $m_s = -H/4\pi$, can be written as

$$F_n(T, H_c) - F_s(T, H_c) \sim F_n(T, 0) - F_s(T, 0) - \frac{H_c^2}{8\pi}, \tag{15.24}$$

from which Eq. (15.23) follows.

The specific heat of a superconductor has two peculiarities. (i) It is discontinuous at the transition from the normal to the superconducting phase (see e.g. Phillips (1959)). (ii) At low temperatures it goes to zero exponentially, in contrast to the linear-T behavior of ordinary metals (see e.g. Corak et al. (1956)).

The jump Δc in the specific heat is accounted for by taking the derivative with respect to the temperature of the entropy difference of Eq. (15.22) and is given by $\Delta c = (T_c/4\pi)(\mathrm{d}H_c/\mathrm{d}T)^2|_{T=T_c} > 0$. We recall that the discontinuity in the specific heat is a common feature of the second order phase transitions when treated in the mean-field approximation, as shown in Eq. (13.14) of Chapter 13.

The low temperature exponential decrease of the specific heat gives evidence for the existence of a temperature dependent gap $\Delta(T)$ in the excitation spectrum, which would also in this case be in agreement with the Landau criterion for a super-flow. A gap in the spectrum has also been shown in several other experiments (see e.g. Corak et al. (1956); Biondi et al. (1957a, b); Glover and Tinkham (1957)) by infrared transmission and reflection measurements and by tunneling. Near T_c, $\Delta(T)$ behaves as

$$\Delta \sim (1 - T/T_c)^{1/2}. \tag{15.25}$$

As discussed in Chapter 13, the second order phase transitions are characterized by the existence of an order parameter φ specifying the amount of ordering in the low temperature phase, which goes to zero at the critical point. Within the mean-field approximation the order parameter goes to zero as the square root of the deviation from the critical temperature, thus suggesting that Δ is related to the order parameter of the superconductive transition, which seems to be well described in the mean-field approximation.

The phenomenological theory proposed by the London brothers (London and London (1935)) rationalizes the problems arising from the above listed experimental facts.

The first observation concerns the consequences we can deduce for a system which has infinite conductivity. If we call $\mathbf{v}_s$ the velocity of the charge carriers in the absence of dissipation, the equation of motion can be written as

$$m^* \frac{d\mathbf{v}_s}{dt} = -e^* \mathbf{E},$$

where e^* and m^* are the effective charge and mass, respectively. We have assumed here that the carriers have negative charge, having in mind that in metals the electrons are responsible for the electrical transport. Combining this equation with the Maxwell equation (15.19) we get

$$\frac{d}{dt}\left(\nabla \times \mathbf{J}_s + \frac{n_s e^{*2}}{m^* c}\mathbf{B}\right) = 0. \tag{15.26}$$

Here we have introduced the supercurrent $\mathbf{J}_s = -e^* n_s \mathbf{v}_s$, n_s being the carrier density. Inside a superconductor, moreover, $\mathbf{B} = 0$. The London brothers showed that this last condition is obtained by introducing an additional equation in analogy with the skin effect[8] for the penetration of an a.c. electric field in a conductor

$$\nabla \times \mathbf{J}_s + \frac{n_s e^{*2}}{m^* c}\mathbf{B} = 0. \tag{15.27}$$

In fact, substituting now in Eq. (15.27), the expression for the current given by the Maxwell equation $\mathbf{J} = (c/4\pi)\nabla \times \mathbf{B}$, we obtain

$$\nabla^2 \mathbf{B} = \lambda^{-2}\mathbf{B}, \quad \lambda^2 = \frac{m^* c^2}{4\pi n_s e^{*2}}. \tag{15.28}$$

It is easy to show (see Problem 15.3) that for a bulk superconductor occupying the half space $x > 0$, and in the presence of an external field parallel to the boundary surface with the vacuum, Eq. (15.28) implies for the magnetic induction $\mathbf{B}$ exponentially decreasing solutions

$$\mathbf{B}(x) = \mathbf{B}(0)e^{-x/\lambda},$$

and the Meissner effect follows. Here the length λ is called the London penetration depth and gives a measure of n_s. From the temperature dependence of λ near T_c given by

[8] We recall that in a conductor the Maxwell equations for an a.c. electric field reduce to

$$\nabla^2 \mathbf{E}(\mathbf{x}, \omega) = -\frac{\omega^2}{c^2}\left(1 + \frac{4\pi i\sigma}{\omega}\right)\mathbf{E}(\mathbf{x}, \omega),$$

where σ is the Drude conductivity of Eq. (11.33). As a result, the electric field decays exponentially in space, so that the electric current is concentrated in a thin layer close to the surface of the conductor.

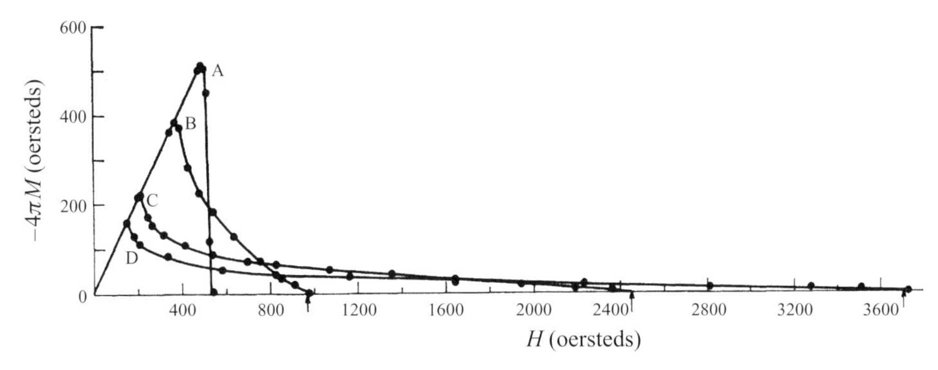

Fig. 15.3 Magnetization curves for annealed polycrystalline lead and lead–indium alloys as functions of magnetic field at 4.2 K. Curve A is the magnetization curve for a type I material (lead). Curves B, C and D are magnetization curves for type II materials (lead with 2.08%, 8.23%, 20.4% indium, respectively). The area under the four curves is the same. (After Livingston (1963).)

Eq. (15.20), we deduce now that n_s should behave as the square of the gap Δ, i.e. as the square of the order parameter. This consideration will be important in building up a theory of superconductivity.

A local relation between the current and the field has been implicitly assumed in the London equation. In fact if we choose the London gauge for the vector potential $\nabla \cdot \mathbf{A} = 0$, $A_\nu = 0$ (A_ν being the component of $\mathbf{A}$ perpendicular to the boundary surface) and consider an isolated simply connected superconductor ($\nabla \cdot \mathbf{J} = 0$), Eq. (15.27) implies

$$\mathbf{J}_s = -\frac{c}{4\pi\lambda^2}\mathbf{A}. \tag{15.29}$$

In general, as was first indicated by Pippard (1955) and successively shown by the microscopic theory, the current depends on the vector potential through a non-local relation

$$\mathbf{J}_s(\mathbf{x}) = \int d\mathbf{x}' \; K(\mathbf{x}, \mathbf{x}')\mathbf{A}(\mathbf{x}').$$

The typical distance over which $\mathbf{A}$ and $\mathbf{J}$ vary is given by the London penetration depth λ, while the kernel K varies over a characteristic distance ξ_0, whose expression can be derived by dimensional analysis. The characteristic energy related to the superconducting transition is $k_B T_c$ with an associated characteristic time $\hbar / k_B T_c$. The latter combined with the Fermi velocity v_F yields a characteristic distance of the order of $\xi_0 \sim \hbar v_F / k_B T_c$. The introduction of the coherence distance ξ_0 allows for the distinction between superconductors of the first type and second type, when ξ_0 is compared with the zero kelvin penetration depth λ_0. The superconductors for which $\xi_0 > \lambda_0$ are named Pippard or type I superconductors, while those for which $\xi_0 < \lambda_0$ with the local relation (15.29) are named London or type II superconductors.

For reasons which will be made clear later, a complete diamagnetism is present for type I superconductors only. In Fig. 15.3 the different behavior of the magnetization for type II superconductors is shown in an alloy of lead and indium. The penetration of the field starts at the critical field $H_{c1}(T)$ which is less than the thermodynamic critical field $H_c(T)$ and is completed at an upper critical field $H_{c2}(T)$, larger than $H_c(T)$.

The London equation (15.29) has a very deep meaning. We recall that in the presence of a static vector potential the momentum of a particle is expressed (see Problem 10.1) as $\mathbf{p} = m\mathbf{v} - \frac{e^*}{c}\mathbf{A}$. The carriers responsible for superconductivity are in a state of zero momentum when the London equation is satisfied:

$$\mathbf{p} = -\frac{e^*}{c}\left(\frac{m^*c}{n_s e^{*2}}\mathbf{J}_s + \mathbf{A}\right) = 0. \tag{15.30}$$

In ordinary metals in the absence of a scalar field, the current $\mathbf{J}$ instead vanishes and $\mathbf{p}$ varies with $\mathbf{A}$ from point to point.

The above phenomenological analysis shows that the transition from the normal to the superconducting phase is an ordering transition. Equation (15.30) recalls the Bose–Einstein condensation and this ordering seems to occur in momentum space. How this can be achieved for fermions is discussed in the next section.

15.5 The condensation criterion and the order parameter

As we have seen in the previous sections, two requirements have to be satisfied by a theory of superfluidity and superconductivity.

1. The Landau criterion, which is related to the stability of the flow and gives a condition to be imposed on the excitation spectrum of the system from its ground state.
2. The superfluid phase is highly ordered, and the ordering occurs in momentum space.

Each condition by itself is not sufficient to describe the *super* phase. The first does not say what kind of transition the system undergoes at the critical temperature. The second one does not ensure superfluidity, as for example in the case of the spinless Bose gas. In an interacting system the condensation must, however, affect the excitations from the ground state where the order is, in any case, maximum. There is no explicit connection between the two conditions; however, every explicit model of superconductivity or superfluidity satisfying one of the two criteria satisfies the other.

A general criterion for condensation has been formulated first by Penrose and Onsager (1956) for bosons and later in general by Yang (1962). We discuss now a system of bosons which is relevant for studying the superfluidity of ^{4}He. We shall show afterwards which modifications are to be made for a system of fermions. In both cases the order parameter will be derived and the condensation criterion itself will give us hints on how a microscopic theory can be formulated.

15.5.1 The boson case

In the case of a spinless Bose gas we speak of a BEC if in the thermodynamic limit a finite fraction of the total number of particles are in the zero momentum state (see Eq. (7.44)).

Such a definition of condensation is strictly related to the existence of single-particle levels. These are in general not well defined for interacting systems. In order to avoid this concept we use the reduced density matrices introduced in Eq. (9.53) and generalize the condensation criterion.

We recall that for the Bose gas, the one-particle reduced density matrix h_1 is diagonal in momentum space and its elements are the occupation numbers $n_{\mathbf{k}}$. Here the zero momentum state plays a special role and h_1 can be accordingly written (see Eq. (7.44))

$$h_1(\mathbf{x}, \mathbf{x}') = \frac{1}{V}\sum_{\mathbf{k}} \mathrm{e}^{-\mathrm{i}\mathbf{k}\cdot(\mathbf{x}-\mathbf{x}')} n_{\mathbf{k}} = \frac{N_0(T)}{V} + \frac{1}{V}\sum_{\mathbf{k}\neq 0} \mathrm{e}^{-\mathrm{i}\mathbf{k}\cdot(\mathbf{x}-\mathbf{x}')} n_{\mathbf{k}}, \tag{15.31}$$

$N_0(T) = \alpha(T)N$, $0 < \alpha \leq 1$, V being the volume of the system and

$$N = N_0(T) + \sum_{\mathbf{k}\neq \mathbf{0}} n_{\mathbf{k}}.$$

The Riemann–Lebesgue theorem on the Fourier transforms states that, if $f(x)$ is an integrable function on the real axis, then

$$\lim_{x\to\pm\infty} f(x) = \lim_{x\to\pm\infty} \int_{-\infty}^{\infty} \mathrm{d}t\, \mathrm{e}^{\mathrm{i}xt} \hat{f}(t) = 0.$$

Therefore in the limit of an infinite system where the sum over $\mathbf{k}$ is replaced by a continuous integration in Eq. (15.31),

$$\begin{aligned} \lim_{|\mathbf{x}-\mathbf{x}'|\to\infty} h_1(\mathbf{x}, \mathbf{x}') &= \frac{N_0(T)}{V} = \frac{\alpha N}{V} = N_0(T)|\phi_0|^2, \quad T < T_{\mathrm{c}}, \\ \lim_{|\mathbf{x}-\mathbf{x}'|\to\infty} h_1(\mathbf{x}, \mathbf{x}') &= 0, \quad T > T_{\mathrm{c}}. \end{aligned} \tag{15.32}$$

where $\phi_0 = 1/\sqrt{V}$ is the $\mathbf{k} = \mathbf{0}$ single-particle state with macroscopic occupation at $T < T_{\mathrm{c}}$. The order in momentum space given by the BEC has therefore a counterpart in configuration space. This is the appearance of a correlation in the off-diagonal matrix elements of the one-particle reduced density matrix named as off-diagonal long range order (ODLRO).

This formulation of the BEC is generalizable to interacting systems, since h_1 and its eigenstates are well defined in principle also in the presence of interactions.

By definition a system shows condensation when the largest eigenvalue n_σ of $h_1(\mathbf{x}, \mathbf{x}')$ is a finite fraction of the total number of particles. Once the eigenvalue equation

$$\int \mathrm{d}\mathbf{x}' h_1(\mathbf{x}, \mathbf{x}')\phi_i(\mathbf{x}') = n_i \phi_i(\mathbf{x}) \tag{15.33}$$

is solved, h_1 can be written in terms of its eigenvalues and eigenfunctions:

$$h_1(\mathbf{x}, \mathbf{x}') = \sum_i n_i \phi_i^*(\mathbf{x}')\phi_i(\mathbf{x}), \tag{15.34}$$

which generalizes Eq. (15.31). Condensation is then specified by

$$\lim_{N,V\to\infty}\frac{n_\sigma}{N}=\alpha,\quad 0<\alpha\le 1, \tag{15.35}$$

$$\lim_{N,V\to\infty}\frac{n_j}{N}=0, \tag{15.36}$$

where $j\neq\sigma$ and $N/V=$ const. It is possible to show in general that the above definition of condensation, as for the Bose gas, is equivalent to the appearance of ODLRO:

$$\lim_{|\mathbf{x}-\mathbf{x}'|\to\infty} h_1(\mathbf{x},\mathbf{x}')=\alpha N f^*(\mathbf{x}')f(\mathbf{x}),\qquad \int_V \mathrm{d}\mathbf{x}|f(\mathbf{x})|^2=1, \tag{15.37}$$

with $\alpha N=n_\sigma$ and $f=\phi_\sigma$. Condensation as defined by Eqs. (15.35) and (15.36) leads to ODLRO when we consider the decomposition (15.34) of h_1 in terms of its eigenvalues and eigenfunctions. Vice versa, if Eq. (15.37) is satisfied with the normalization condition

$$|f(\mathbf{x})|\le\frac{c}{\sqrt{V}}, \tag{15.38}$$

we show that in the thermodynamic limit $f(\mathbf{x})$ tends to the eigenfunction of h_1 corresponding to the maximum eigenvalue n_σ:

$$f(\mathbf{x})\to\phi_\sigma(\mathbf{x}),\qquad \alpha N\to n_\sigma. \tag{15.39}$$

To show this, let us consider the following functional of any normalized trial wave function $\phi_\mathrm{t}=\sum_i c_i\phi_i(\mathbf{x})$, $\sum_i|c_i|^2=1$:

$$F(\phi_\mathrm{t})=\int\mathrm{d}\mathbf{x}\int\mathrm{d}\mathbf{x}'\phi_\mathrm{t}(\mathbf{x})h_1(\mathbf{x},\mathbf{x}')\phi_\mathrm{t}^*(\mathbf{x}')=\sum_i|c_i|^2 n_i\le n_\sigma,$$

where the equality holds only if $\phi_\mathrm{t}=\phi_\sigma$. If we choose

$$\phi_\mathrm{t}(\mathbf{x})=f(\mathbf{x})$$

then

$$F(f)=\alpha N+I,$$

$$I=\int\mathrm{d}\mathbf{x}\int\mathrm{d}\mathbf{x}'f(\mathbf{x})\tilde{h}_1(\mathbf{x},\mathbf{x}')f^*(\mathbf{x}'),$$

$$\tilde{h}_1(\mathbf{x},\mathbf{x}')=h_1(\mathbf{x},\mathbf{x}')-\alpha N f(\mathbf{x})f^*(\mathbf{x}').$$

In the thermodynamic limit I was shown by Penrose and Onsager to be negligible when compared with αN. In fact

$$I\le\int\mathrm{d}\mathbf{x}\int\mathrm{d}\mathbf{x}'|f(\mathbf{x})||\tilde{h}_1(\mathbf{x},\mathbf{x}')||f^*(\mathbf{x}')|\le c^2\int\mathrm{d}\mathbf{r}|\tilde{h}_1(\mathbf{r})|,$$

where we have used the inequality (15.38) and the translational invariance: $\tilde{h}_1(\mathbf{x},\mathbf{x}')=\tilde{h}_1(\mathbf{r})$ is a function of the relative variable $\mathbf{x}-\mathbf{x}'=\mathbf{r}$ only. Because of Eq. (15.37), for any $\epsilon>0$ there exists $R>0$ such that

$$|\tilde{h}_1(\mathbf{r})|<\frac{\epsilon}{2c^2}\ \text{ if }\ |\mathbf{r}|>R,$$

$$\frac{c^2\tilde{h}_\mathrm{M}V_R}{V}<\frac{\epsilon}{2},$$

where $\tilde{h}_M$ is the maximum of $\tilde{h}_1$ in the limited domain V_R and for large enough V. Then

$$I \leq c^2 \int d\mathbf{r}|\tilde{h}_1(\mathbf{r})| \leq \frac{\epsilon}{2} \int_{r>R} d\mathbf{r} + c^2 \tilde{h}_M \int_{r<R} d\mathbf{r} \leq \epsilon V.$$

This is clearly enough to show that

$$\lim_{V,N\to\infty, N/V=\text{const}} F(f) = \alpha N \leq n_\sigma \tag{15.40}$$

and there exists a macroscopic eigenvalue of h_1. Moreover, any trial wave function ϕ_t satisfying the normalization condition Eq. (15.38) can be obtained as a superposition of f with a function g orthogonal to it:

$$\phi_t = af + bg, \quad a^2 + b^2 = 1, \quad \int d\mathbf{x} f^*(\mathbf{x}) g(\mathbf{x}) = 0.$$

The value (15.40) of F corresponds to the maximum of the functional $F(\phi_t) = a^2 \alpha N$ in this class of functions ϕ_t. The identification (15.39) follows. The argument assumes that the eigenfunction itself belongs to the specified class of wave functions.

The normalized expression of the eigenfunction corresponding to the eigenvalue n_σ obtained in the off-diagonal long range limit,

$$\psi(\mathbf{x}) = (\alpha N)^{1/2} \phi_\sigma(\mathbf{x}), \tag{15.41}$$

can be taken as the order parameter which goes to zero at the transition temperature, and characterizes the type of ordering occurring in superfluids, i.e. condensation in the effective single-particle state, obtained as an eigenfunction of the one-particle reduced density matrix.

The function $\psi(\mathbf{x})$ can be considered therefore as a kind of weighted wave function for the condensate. It has a semiclassical character, i.e. its modulus squared coincides with a density rather than a probability, because of the macroscopic occupancy of such a state.

It is interesting to consider the case of a limited number of identical interacting bosons confined in a finite region of space by an external potential.

Due to the limited size of the system, we can only give a less precise definition of BEC. BEC exists if an eigenvalue of h_1, n_σ, is of the order of the total number of particles. The corresponding eigenfunction $\phi_\sigma(\mathbf{x})$ is the condensate wave function and, similarly, the "order parameter" is $\psi(\mathbf{x}) = (n_\sigma)^{1/2} \phi_\sigma(\mathbf{x})$, though we have to remember that, strictly speaking, there is no phase transition.

15.5.2 The fermion case

In superconductors the electrons, being fermions, cannot condense into a single-particle state, because of the Pauli principle. The simplest possibility is that the two-particle reduced density matrix has an eigenvalue $n_\sigma = \alpha N$. Taking into account the translational invariance we can introduce the variables

$$\mathbf{r} = \mathbf{x}_1 - \mathbf{x}_2, \quad \mathbf{r}' = \mathbf{x}'_1 - \mathbf{x}'_2, \tag{15.42}$$

$$\mathbf{R} = \mathbf{x}_1 - \mathbf{x}'_1 + \mathbf{x}_2 - \mathbf{x}'_2, \tag{15.43}$$

as h_2 depends on only three independent vectors. For simplicity we do not change the notation when we consider such a dependence. The associated eigenvalue problem is

$$\int d\mathbf{x}_1' \int d\mathbf{x}_2' h_2(\mathbf{x}_1, \mathbf{x}_2; \mathbf{x}_2', \mathbf{x}_1',)\phi_i(\mathbf{x}_1', \mathbf{x}_2') = \lambda_i \phi_i(\mathbf{x}_1, \mathbf{x}_2), \tag{15.44}$$

$$\int d\mathbf{x}_1' \int d\mathbf{x}_2' |\phi_i(\mathbf{x}_1', \mathbf{x}_2')|^2 = 1. \tag{15.45}$$

On the assumption that there exists a two-particle eigenfunction $\phi_\sigma(\mathbf{x}_1, \mathbf{x}_2)$ with the corresponding macroscopic eigenvalue $\lambda_\sigma = \alpha N$, $0 < \alpha < 1$, it is then possible to show the asymptotic ODLRO property

$$\lim_{\mathbf{R}\to\infty, |\mathbf{r}|, |\mathbf{r}'| < \xi_0} h_2(\mathbf{r}, \mathbf{r}', \mathbf{R}) = \alpha N \phi_\sigma^*(\mathbf{r}) \phi_\sigma(\mathbf{r}'). \tag{15.46}$$

To show that, vice versa, ODLRO

$$\lim_{\mathbf{R}\to\infty, |\mathbf{r}|, |\mathbf{r}'| < \xi_0} h_2(\mathbf{r}, \mathbf{r}', \mathbf{R}) = \alpha N f^*(\mathbf{r}) f(\mathbf{r}') \tag{15.47}$$

implies macroscopic occupancy of a pair state requires some care. In the case of Bose systems, in order to show that the asymptotic factorization property of the one-particle reduced density matrix is equivalent to condensation, it has been enough to assume that the function f in which h_1 factorizes is normalised as the eigenfunctions of h_1 itself, i.e. the function f has to be chosen in the same class as the eigenfunctions of h_1. For fermions, in order to repeat in a straightforward way the same argument as for bosons, the function f in which the two-particle reduced density matrix asymptotically factorizes must be in the reduced class of normalizable functions f such that $f(\mathbf{x}_1 - \mathbf{x}_2) = 0$ if $|\mathbf{x}_1 - \mathbf{x}_2|$ is greater than an arbitrary but finite distance ξ_0. Physically this means that f varies in the restricted class of functions which describe two-particle bound states, whose size must be of the order of ξ_0. The distance ξ_0 between $\mathbf{x}_1$ and $\mathbf{x}_2$ beyond which $\phi_\sigma \to 0$ corresponds to the coherence length. We can include, as Yang did, in this class exponentially decreasing functions. When the above restrictive normalization condition is not strictly satisfied, a detailed knowledge on how $h_2 \to 0$ for $R \to \infty$ with arbitrary $\mathbf{r}$ and $\mathbf{r}'$ is necessary to show the existence of an eigenvalue of h_2 of order N. This is, for instance, the case of the BCS (Bardeen et al. (1957)) theory[9] which had to be dealt with explicitly because $f(\mathbf{r})$ has a tail like r^{-2} for large r. In the previous section we have seen that a phenomenological explanation of the superconductivity implies condensation. If the pair eigenfunction $\phi_\sigma(\mathbf{x}_1, \mathbf{x}_2)$ did not go to zero for large separation between $\mathbf{x}_1$ and $\mathbf{x}_2$, it would factorize into the product of two single-particle states. $h_2(\mathbf{x}_1, \mathbf{x}_2; \mathbf{x}_1', \mathbf{x}_2')$ would then be approximated by products of h_1s at different points, as for a Fermi gas. Eigenvalues of the order of N for h_2 cannot be obtained in this case, because the eigenvalues of h_1 are all infinitesimal with respect to N as N goes to infinity.

We have achieved the result that fermions may condense provided pair formation occurs. Therefore any microscopic theory of superconductivity showing condensation must be based on electron pair formation, as indeed the BCS theory does. The order parameter is

[9] The BCS theory of superconductivity will be presented later in this Chapter in Section 15.9.

with the chemical potential μ determined by

$$\int \mathrm{d}^d x |\psi(\mathbf{x})|^2 = n.$$

For a vortex line ($n = 1$) discussed in Section 15.6.1, in cylindrical coordinates $\varphi = \theta$ and r, Eq. (15.73) is satisfied with $\nabla^2\theta(\mathbf{x}) = 0$ and $|\psi(\mathbf{x})|$ varying along r orthogonally to $\hat{\varphi}$. Equation (15.72) with $V_{\text{ext}} = 0$ is

$$-\frac{\hbar^2}{2m}\frac{1}{r}\frac{\partial}{\partial r}\left(r\frac{\partial|\psi(\mathbf{x})|}{\partial r}\right) + \left[\frac{\hbar^2}{2m}\frac{1}{r^2} - \mu + g|\psi(\mathbf{x})|^2\right]|\psi(\mathbf{x})| = 0. \quad (15.74)$$

Far from the centre of the vortex line, Eq. (15.74) has a constant solution $|\psi(\mathbf{x})|^2 \approx \rho_s/m = \rho/m$ and $\mu = g\rho/m$. $|\psi(\mathbf{x})|^2$ decreases to zero on approaching the center of the line at $r = 0$ where the velocity is singular. The variation to zero of $m|\psi|^2 = \rho_s$ occurs on the characteristic length $\xi_0 = \hbar/(2mng)^{1/2} = \hbar/(8\pi n a_s)^{1/2}$, called the healing length. Substituting this solution for the density instead of the constant ρ in the integral (15.65) for the energy, the density vanishes when the velocity becomes singular and the cutoff a is not necessary anymore. Whenever $|\psi(\mathbf{x})|^2$ has to vanish at a given boundary, its recovery to the constant value $\rho/m = n$ occurs on a distance of the order of the healing length ξ_0.

The derivation of the Gross–Pitaevskii equation (15.70) can be easily done starting from the Hamiltonian (9.37), in second quantization, with the above contact two-body potential (15.69), which may include the external potential term $\int \mathrm{d}\mathbf{x}\hat{\psi}^\dagger(\mathbf{x}, t)V_{\text{ext}}(\mathbf{x}, t)\hat{\psi}(\mathbf{x}, t)$. The Heisenberg Eq. (9.43) for the field reads

$$i\hbar\frac{\partial\hat{\psi}(\mathbf{x}, t)}{\partial t} = \left[-\frac{\hbar^2}{2m}\nabla^2 - \mu + V_{\text{ext}}(\mathbf{x}, t) + g|\hat{\psi}(\mathbf{x}, t)|^2\right]\hat{\psi}(\mathbf{x}, t). \quad (15.75)$$

If we assume that the system is in a condensed state for which ODLRO (15.37) in $h_1(\mathbf{x}, \mathbf{x}'; t) = \langle\hat{\psi}^\dagger(\mathbf{x}', t)\hat{\psi}(\mathbf{x}, t)\rangle$ exists, then in this state $\langle\hat{\psi}(\mathbf{x}, t)\rangle$ coincides with the order parameter $\psi(\mathbf{x}, t)$. Assuming furthermore that all correlations can be neglected so that $\langle|\hat{\psi}(\mathbf{x}, t)|^2\hat{\psi}(\mathbf{x}, t)\rangle = |\psi(\mathbf{x}, t)|^2\psi(\mathbf{x}, t)$, Eq. (15.75), when averaged over this state, becomes the Gross–Pitaevskii Eq. (15.70). For a finite system this mean-field approach is still valid provided the number of particles is sufficiently high that in the dilute limit we can neglect the fluctuations.

In the time-independent case, the ground state $\phi_N^0(\mathbf{x}_1, \mathbf{x}_2, \ldots, \mathbf{x}_N)$ corresponding to the above approximation of having all the particles in a single-particle state coincides with the simple symmetrized product of the same single-particle state $\phi_\sigma(\mathbf{x}) = \phi_0(\mathbf{x})$, in which condensation occurs:

$$\phi_N^0(\mathbf{x}_1, \mathbf{x}_2, \ldots, \mathbf{x}_N) = \prod_{i=1}^N \phi_0(\mathbf{x}_i). \quad (15.76)$$

If we evaluate the Hartree–Fock energy with the trial state (15.76)

$$\left\langle\phi_N^0(\mathbf{x}_1, \mathbf{x}_2, \ldots, \mathbf{x}_N)|H|\phi_N^0(\mathbf{x}_1, \mathbf{x}_2, \ldots, \mathbf{x}_N)\right\rangle \quad (15.77)$$

and then minimize the result with respect to the single-particle function $\phi_0(\mathbf{x})$, we obtain the time-independent Gross–Pitaevskii equation (as shown in Problem 15.5). Even though the wave function (15.76) does not contain any correlation (Leggett (2001)), in most cases it gives good results for studying BEC in alkali metals. Perturbation from the ground state condensate can be obtained by adding to the stationary order parameter a small perturbation

$$\psi(\mathbf{x},t) = \psi(\mathbf{x})e^{-i\mu t/\hbar} + \delta\psi(\mathbf{x},t),$$

then linearizing the Gross–Pitaevskii equation and its complex conjugate. This procedure is equivalent to the Bogolubov approximation for evaluating the excitations from a condensed ground state, which will be discussed in Section 15.8.

15.6.3 Bose–Einstein condensation for alkali gases

The Bose–Einstein condensation of a limited number of identical bosons confined in a finite region of space by an external potential has been discussed in Section 7.4.1 and characterized by ODLRO at the end of Section 15.5.1.

We do not consider the very refined experimental technique to trap the system for the time of the experiment to such low temperatures. For this we refer to the many existing reviews (Chu (1998); Cohen-Tannoudji (1998); Phillips (1998); Ketterle et al. (1999)). We only point out here that the harmonic trapping potential Eq. (7.53)

$$V_{\text{ext}}(\mathbf{r}) = \frac{m}{2}\left(\omega_x^2 x^2 + \omega_y^2 y^2 + \omega_z^2 z^2\right) \tag{15.78}$$

can be spherically ($\omega_x = \omega_y = \omega_z$) or axially ($\omega_x = \omega_y = \omega_\perp$, ω_z) symmetric. In the last configuration, when $\omega_\perp > \omega_z$ the trap is disk shaped, whereas for $\omega_\perp < \omega_z$, it is cigar shaped as for an ^{87}Rb gas.

For comprehensive presentation of the Bose vapors in general see (Dalfovo et al. (1999); Tino et al. (1999); Leggett (2001); Pethick and Smith (2008); Pitaevskii and Stringari (2003)).

As we have seen previously, liquid ^{4}He is strongly interacting and locally there is no sensitivity to boundary conditions. The alkali vapors are very diluted, allow for quantitative perturbative calculations and are strongly sensitive to the confining boundary conditions.

In the presence of an harmonic potential the position dependent "order parameter" $\psi(\mathbf{x},t) = |\psi(\mathbf{x},t)|\exp(i\theta(\mathbf{x},t))$ is defined as the normalized eigenfuntion of the one-particle reduced density matrix h_1 with eigenvalue n_σ of order N. n_σ coincides now with the occupation N_0 of the lowest single-particle eigenstate Eq. (7.56). With good approximation in the dilute limit, $\psi(\mathbf{x},t) = |\psi(\mathbf{x},t)|\exp(i\theta(\mathbf{x},t))$, in the presence of interparticle interaction, satisfies the Gross–Pitaevkii equation (15.70).

Most results discussed in the present chapter for an infinite system can be extended to this confined case. In particular the phase of the order parameter is the superfluid velocity field related to the fraction of condensed particles according to Eqs. (15.60) and (15.61):

$$v_s(\mathbf{x},t) = \frac{\hbar}{m}\nabla\theta(\mathbf{x},t). \tag{15.79}$$

In the absence of interparticle interaction, as already indicated in Section 7.4.1, the appearance of a narrow peak (of width a_{ho}, Eq. (7.57), typically of the order of 1 μm in the experiments) centered at zero velocity on top of a broad thermal velocity distribution is a signature of Bose–Einstein condensation. The fraction of atoms described by the same wave function and belonging to the narrow peak was indeed observed to increase abruptly by decreasing the temperature, e.g. for ^{87}Rb vapor by expanding the atomic cloud (see Fig. 2 in Anderson et al. (1995)). In first approximation the critical temperature for the appearance of the condensate can be evaluated with Eq. (7.61) valid for a gas with large N. The factor $N^{1/3}$ makes $k_B T_c \gg \hbar\omega_{ho}$. The typical frequency of the trapping potential used in the experiments is of the order of tens to hundreds of Hertz and with $N \approx 10^5$, $T_c = (\hbar\omega_{ho})(N/\zeta_R(3))^{1/3} \approx 210$ nK. Of course, finite size and interparticle interaction effects have to be considered. Finite sizes, besides rounding the transition, reduce the condensate fraction and the reduced critical temperature with corrections of the order of $N^{-1/3}$, already negligible for $N = 10^3$–10^4. The interaction effects are dealt with the Gross–Pitaevskii equation introduced in Subsection 15.6.2 which, for the static situation in the ground state, is given by Eq. (15.72) with $\nabla\theta = 0$:

$$\left[-\frac{\hbar^2}{2m}\nabla^2 - \mu + V_{\text{ext}}(\mathbf{x}) + g|\psi(\mathbf{x})|^2\right]|\psi(\mathbf{x})| = 0, \tag{15.80}$$

with the coupling g given in terms of the s-wave scattering length as by Eq. (15.69), with $a_s > 0$ (repulsive interaction) except when otherwise stated.

Several length scales are intrinsic to Eq. (15.80), the s-wave scattering length (Eq. (15.69)), the healing length, $\xi_0 = \hbar(8\pi n a_s)^{-1/2}$, the harmonic oscillator length a_{ho} and the average distance $d = n^{-1/3} = (a_{ho}^3/N)^{1/3}$. Under realistic experimental conditions in alkali gases, a hierarchy of these lengths can be established (Leggett (2001)):

$$a_s \approx 2 - 5 \text{ nm} < d \approx 20 \text{ nm} < \xi_0 \approx 40 \text{ nm} < a_{ho} = \left(\frac{\hbar}{m\overline{\omega}}\right)^{1/2} \approx 10^3 \text{ nm}. \tag{15.81}$$

The above inequalities set up a hierarchy between the corresponding energies. In particular we can see that even in the dilute limit $(a_s/d)^3 \ll 1$, the effect of the interparticle interaction can be relevant. Indeed, the interaction energy E_{int} can be much larger than the kinetic energy E_{kin} as in most experimental conditions:

$$E_{\text{int}} \approx Ngn \approx g\frac{N^2}{a_{ho}^3} = N^2\frac{\hbar^2}{m a_{ho}^2}\frac{a_s}{a_{ho}} \approx E_{\text{kin}} N\frac{a_s}{a_{ho}}. \tag{15.82}$$

In the above derivation, we have considered the N particles confined in the volume a_{ho}^3, Eq. (15.69) for the coupling g and the kinetic energy $E_{\text{kin}} \approx N(\hbar)^2/m a_{ho}^2$. The presence of the condensate strengthens the effect of the interaction. When the factor specifying the relevance of the interaction is such that $N a_s/a_{ho} \gg 1$, the kinetic energy term can be neglected in Eq. (15.80) and the density profile is determined by the confining potential (Thomas–Fermi approximation):

$$n(\mathbf{x}) = |\psi(\mathbf{x})|^2 = \frac{1}{g}\left(\mu - V_{\text{ext}}(\mathbf{x})\right), \tag{15.83}$$

as long $\mu > V_{\text{ext}}(\mathbf{x})$, and zero otherwise. Actually the vanishing of $n(\mathbf{x})$ is not abrupt, but it is smeared over the healing length ξ_0 where the Laplacian term becomes relevant. It is also clear that the repulsive interparticle potential extends the condensate to a length

b beyond the harmonic oscillator length a_{ho}. Consider for simplicity the isotropic case, $\omega_x = \omega_y = \omega_z$,

$$\mu = \frac{m\overline{\omega}^2 b^2}{2}. \tag{15.84}$$

On the other hand, by integrating Eq. (15.83) with $V_{\rm ext}$ given by Eq. (7.53) on a volume $4\pi b^3/3$ to give the total number of particles, we obtain b as a function of N:

$$b = \left(\frac{15gN}{4\pi m\overline{\omega}^2}\right)^{1/5} = a_{ho}\left(15\frac{a_{\rm s}}{a_{ho}}N\right)^{1/5}. \tag{15.85}$$

For the last equality we have used the expression (15.69) for the coupling g. The interparticle interaction extends the region of the condensate even in the dilute limit via the factor $15N$. With the typical values given above, Eq. (15.81), and $N \approx 10^5$, $b \approx 3 - 4\ a_{ho}$. The peak density at $\mathbf{x} = 0$ is instead depressed by the interparticle interaction.

The homogeneous condensate is always unstable against attractive interaction ($a_{\rm s} < 0$). Condensation has been instead observed in ^{7}Li with negative s-wave scattering (Bradley et al. (1995)). In a trap the zero-point kinetic energy can stabilize the system provided it is, on average, larger than the interaction energy. Roughly speaking, from Eq. (15.82) this condition is realized for $N < N_{\rm c}$ with $N_{\rm c}$ given by $N_{\rm c}|a_{\rm s}|/a_{ho} \approx 1$.

The extension to finite temperature of the Gross–Pitaevskii equation (Dalfovo et al. (1999)) can be carried out by means of the temperature-dependent Hartree–Fock method. We consider here only how the critical temperature is reduced by the repulsive interparticle interaction with respect to the value for the non-interacting gas in an harmonic trap. At finite temperature $T \gtrsim T_{\rm c}$, the interaction energy per particle is of the order of $(1/2)gn(T)$, where $n(T)$ is the particle density evaluated over the extension of the classical oscillator $L^3(T) = (2k_{\rm B}T/m\bar{\omega}^2)^{3/2}$. Grossly speaking, the relative variation of $T_{\rm c}$ of the gas (7.53) due to the interaction can be evaluated as

$$\frac{k_{\rm B}\Delta T_{\rm c}}{k_{\rm B}T_{\rm c}} = gN\left(\frac{m\bar{\omega}^2}{2k_{\rm B}T_{\rm c}}\right)^{3/2}\frac{1}{k_{\rm B}T_{\rm c}} \sim \frac{a_{\rm s}}{a_{ho}}N^{1/6}, \tag{15.86}$$

where we omitted all numerical factors. In deriving the last form of Eq. (15.86) we have used Eq. (7.61) for $T_{\rm c}$ of the gas, Eq. (7.57) for the oscillator length a_{ho} and Eq. (15.69) for g. Typical configurations of BEC in experiments lead to corrections of a few percent with a negative correction for the repulsive interaction opposite to the homogeneous case (Ferrier-Barbut et al. (2014)).

15.6.4 Superconductors

Following Section 15.6, a sound assumption is that the superconducting state is well approximated by a coherent macroscopic occupation of the pair state at fixed phase rather than at fixed N, giving rise to a spontaneous symmetry breaking. For superfluid helium (^{4}He) the above conclusion was reached by using the one-particle reduced density matrix h_1. Until now we have considered mainly global phase transformation, i.e. the phase θ is independent of space and time. As is well known from electrodynamics, local gauge invariance is obtained with the introduction of the scalar and vector potentials V and $\mathbf{A}$ and their transformations with a space- and time-dependent scalar function f:

$$\mathbf{A} \to \mathbf{A} + \nabla f(\mathbf{x}, t), \tag{15.87}$$

$$V \to V - \frac{\partial f(\mathbf{x}, t)}{c\partial t}. \tag{15.88}$$

Since the order parameter ψ, obtained from the factorization of the two-particle reduced density matrix h_2, corresponds to a weighted "single particle wave function" describing the distribution of the center of the mass of the bound pair, under local gauge transformation it transforms according to

$$\psi \to \mathrm{e}^{-\mathrm{i}(e^*/\hbar c) f(\mathbf{x},t)} \psi, \tag{15.89}$$

e^* being the effective electric charge of the charge carriers ($e^* = 2e$ for the present case).

Now we want to show that our interpretation of the order parameter as a semiclassical wave function is sufficient to explain all the super-effects. We start from the expression of the Schrödinger equation for a "particle" in the presence of a scalar and vector potential:

$$\mathrm{i}\hbar \frac{\partial \psi}{\partial t} = \frac{1}{2m^*} \left(\frac{\hbar}{\mathrm{i}} \nabla + \frac{e^*}{c} \mathbf{A} \right)^2 \psi - e^* V \psi, \tag{15.90}$$

where the effective mass $m^* = 2m$, having in mind pair formation, and m being the mass of the electron. Using the general expression for the current in quantum mechanics, we get the supercurrent associated to $\psi = \psi_0 \mathrm{e}^{\mathrm{i}\theta(\mathbf{x})}$:

$$\mathbf{J}_\mathrm{s} = -e^* |\psi_0|^2 \frac{\hbar}{m^*} \left(\nabla\theta + \frac{e^*}{c\hbar} \mathbf{A} \right). \tag{15.91}$$

Whenever $\mathbf{A}$ is equal to zero, the superfluid current is given by

$$\mathbf{J}_\mathrm{s} = -\frac{e^* |\psi_0|^2 \hbar}{m^*} \nabla\theta. \tag{15.92}$$

In the presence of a vector potential we choose the gauge $\nabla \cdot \mathbf{A} = 0$. When the system is isolated, $\nabla \cdot \mathbf{J}_\mathrm{s} = 0$ gives

$$\nabla^2 \theta = 0, \tag{15.93}$$

which is equivalent to $\theta =$ const. The London relation Eq. (15.29) between the current and the vector potential then follows from Eq. (15.91), provided $|\psi|^2 = n_\mathrm{s}$, as hypothesized for the normalization of the order parameter in the previous section.

The quantization of the flux (Deaver and Fairbank (1961); Byers and Yang (1961)) and the Josephson effect (Josephson (1964); Shapiro et al. (1964); Jaklevic et al. (1964)) are also easily derived from the expression for the current.

Let us consider a superconducting ring with transverse dimension much larger than the penetration depth below its transition temperature. We consider a closed contour C well inside the ring. Along the contour the current must be zero because the field cannot penetrate into the bulk. Hence performing the contour integration we obtain

$$\int_\mathrm{c} \mathbf{J}_\mathrm{s} \cdot \mathbf{dl} = -\int_\mathrm{c} e^* |\psi_0|^2 \frac{\hbar}{m^*} \left(\nabla\theta + \frac{e^*}{c\hbar} \mathbf{A} \right) \cdot \mathbf{dl} = -e^* |\psi_0|^2 \frac{\hbar}{m^*} \left(\Delta\theta + \frac{e^*}{c\hbar} \Phi(B) \right), \tag{15.94}$$

where $\Phi(B)$ is the magnetic flux trapped inside the ring. Since the wave function must be single-valued, $\Delta\theta = 2\pi n$, the flux is quantized:

$$\Phi(B) = n\frac{ch}{e^*}, \tag{15.95}$$

n being an integer.

The experiments gave $e^* = 2e$ (Deaver and Fairbank (1961)) according to the basic assumption of pair condensation.

We now briefly discuss the Josephson effect. It may be useful to make an analogy with ferromagnets. In a ferromagnet below its Curie point the order parameter is the spontaneous magnetization M_1 which breaks the invariance under rotations, setting a preferred orientation in space. Another ferromagnet with a magnetization M_2 and oriented at an angle θ with respect to the previous one, when sufficiently near, gives rise to an interaction energy $U(\theta) \sim M_1 M_2 \cos(\theta)$. In the case of two superconductors we may apply the same concept. The complex order parameter may also be viewed as a two-component vector. In the phase of broken symmetry, a choice of phase for ψ corresponds to a preferred orientation in the complex plane. Then it is natural to write the interaction energy of two superconductors, separated by a insulating junction, as a periodic function of the phase difference $\theta_1 - \theta_2$ of the order parameter, exactly as for ferromagnets. In order to convince ourselves of the presence of such an energy term in the Josephson junction we add the following considerations. If we let the thickness d of the insulating layer become very large ($d \gg \xi_0$), the system is made of two independent isolated superconductors, each one with a constant phase uncorrelated to the other one. On the other hand, when the thickness d goes to zero, the two superconductors will tend to behave as a single one with an overall constant phase and the phase difference between the two must go to zero. There is therefore an interaction energy $U(\theta_1 - \theta_2)$ which goes to zero as $d \gg \xi_0$ and becomes important when $d \ll \xi_0$. U must be an even periodic function of θ, $U = C\cos\theta$. When the two superconductors are separated by a small distance d (usually d is of the order 10–50 Å) much less than the penetration length λ and the coherence distance ξ_0 over which the pair wave function extends, we may approximate the phase difference as $\theta_1 - \theta_2 = \nabla\theta \cdot \mathbf{d}$. From gauge invariance, we know that only the combination of the phase gradient and the vector potential is physically meaningful. Hence, using the general fact that the derivative with respect to the vector potential gives the current, we may write

$$J_s \sim \frac{\mathrm{d}U(\theta)}{\mathrm{d}\nabla\theta} = J_c \sin(\theta), \tag{15.96}$$

from which we see that a supercurrent flows across the junction, i.e. pairs can tunnel from one side to the other of the junction without a d.c. voltage. The maximum Josephson current J_c can be determined only by microscopic calculations.

The presence of an electromagnetic field modifies the phase of the pair wave function according to Eq. (15.90). A scalar potential V across the junction gives rise to an a.c. Josephson current with frequency (see Eq. (15.54))

$$\dot{\theta} = \frac{e^* V}{\hbar} = -\frac{\Delta\mu}{\hbar}, \tag{15.97}$$

where μ here is the electrochemical potential and $\Delta\mu = -e^* V$ its difference across the junction. This current oscillates very rapidly due to the appearance of the Planck constant and therefore averages to zero.

Combining two Josephson junctions in a circuit, it is possible to study interference effects on the outcoming current as a function of the magnetic flux through the loop in units of the quantum $\hbar c/e^*$.

The experimental observation of the quantization of flux (Deaver and Fairbank (1961)), the d.c. Josephson current and the interference effects in Josephson junctions (Jaklevic et al. (1964)) are the most direct evidence of the appearance of quantum phenomena on macroscopic level via the coherent macroscopic occupation of a single pair state due to the ODLRO of h_2.

Extension of Eq. (15.97) in superfluid liquid helium (^{4}He), the formula (15.54) analogous to Eq. (15.97) has led to the observation of the phase slippage phenomenon (Richards and Anderson (1965); Khorana and Chandrasekhar (1967)). When liquid HeII is put with different levels of filling in two vessels linked to each other through a small orifice, Eq. (15.97) reads

$$2\pi \frac{\mathrm{d}n}{\mathrm{d}t} = \Delta\mu = mgz, \tag{15.98}$$

where $\mathrm{d}n/\mathrm{d}t$ is the rate of vorticity (each vortex introduces a change of 2π in the phase) through the orifice and z is the level difference in the two vessels. If we can sustain the constant flux of vorticity through the orifice, then the difference z remains constant. This has been achieved by putting a quartz transducer of the right frequency in front of the aperture.

15.7 The Landau–Ginzburg equations

In the discussion of the Josephson effect, we made the analogy between a superconductor and a ferromagnet below its Curie point. In both cases spontaneous symmetry breaking occurs in the sense that the phase of ψ or the direction of the spontaneous magnetization is fixed among all the possible degenerate configurations.

In the case of magnetic transitions, an infinitesimal external magnetic field is necessary in order to fix the direction of the magnetization among all the equivalent possible degenerate orientations. In the language of thermodynamics, the magnetic field and the magnetization are conjugate variables. All physical properties of a ferromagnet can be described once we have introduced the free energy $F(T, h)$ as a functional of the temperature T and the magnetic field $h(\mathbf{x})$. The magnetization per unit volume $m(\mathbf{x})$, i.e. the order parameter, is defined as the functional derivative of the free energy with respect to the magnetic field h, as illustrated in Eq. (13.17). The spontaneous magnetization, i.e. the order parameter, is obtained in the limit of zero external field. We consider the thermodynamic potential as a function of the order parameter by performing a Legendre transformation from $F(T, h)$ to $\tilde{F}(T, m)$, as indicated in Eq. (13.18) of Chapter 13. Going back to superconductivity, it is possible to associate with the order parameter an external conjugate field, even though its physical interpretation is no more as direct as in the magnetic phenomena. Accordingly we introduce the free energy for a superconducting sample as a functional of the temperature T and of a complex external field η, and define the order parameter ψ as the functional derivative of the free energy with respect to this field η in the same way as was done in Eq. (13.17). It is again convenient to perform a Legendre transformation by the introduction of the term

$$\int d\mathbf{x}\,(\eta(\mathbf{x})\psi^*(\mathbf{x}) + \eta^*(\mathbf{x})\psi(\mathbf{x}))$$

and consider the thermodynamic potential as a functional of the order parameter ψ instead of the external field η.

By recalling that the main assumption of the Landau theory, which for superconductors goes under the name of Landau–Ginzburg, is that the thermodynamic potential per unit volume near the transition point can be expanded in a power series of the order parameter and of the gradient of the order parameter, and we get

$$F_s(T,\psi) = F_n + \frac{1}{V}\int d\mathbf{x}\left(a|\psi(\mathbf{x})|^2 + \frac{b}{2}|\psi(\mathbf{x})|^4 + c|\nabla\psi(\mathbf{x})|^2 + \cdots\right), \quad (15.99)$$

where we have dropped the tilde from the transformed thermodynamic potential and the subscripts s and n stand for superconducting and normal phase, respectively. $|\psi(\mathbf{x})|^2$ and $|\nabla\psi(\mathbf{x})|^2$ can only appear in the expansion, since F cannot depend on the phase of the order parameter. The coefficients a, b and c must be regular and smooth functions of T, as discussed in Chapter 13 (see, in particular the discussion after Eq. (13.10) leading to $a = a'(T - T_c)$, b and c constant). Notice that, for later convenience, we have inserted a factor $1/2$ in front of the coefficient b with respect to Eq. (13.10).

In order to preserve local gauge invariance the kinetic term in the expansion for the free energy in the presence of an external magnetic field becomes

$$c|\nabla\psi(\mathbf{x})|^2 \to c\left|\left(\nabla + i\frac{e^*}{\hbar c}\mathbf{A}\right)\psi\right|^2. \quad (15.100)$$

Further, to have the complete expression for the thermodynamic potential in the presence of a magnetic field we add a term due to the energy density of the field $H^2/8\pi$. To avoid confusion with the coefficient c introduced in Eq. (15.99), from now on we put the light velocity equal to unity.

By varying the thermodynamic potential with respect to ψ^* and $\mathbf{A}$ with the gauge choice $\nabla \cdot \mathbf{A} = 0$ one obtains the Landau–Ginzburg equations for the order parameter and the current:

$$a\psi + b|\psi|^2\psi - c\left(\nabla + i\frac{e^*}{\hbar}\mathbf{A}\right)^2\psi = \eta, \quad (15.101)$$

$$\mathbf{J} = i\frac{e^*c}{\hbar}\left(\psi^*\left(\nabla + i\frac{e^*}{\hbar}\mathbf{A}\right)\psi - \text{c.c.}\right), \quad (15.102)$$

where c.c. indicates complex conjugate.

Variation of the free energy with respect to ψ leads to the complex conjugate of Eq. (15.101). Equation (15.102) is the usual expression relating the vector potential and the current density. We note in addition that Eq. (15.102) is equivalent to Eq. (15.91) for the supercurrent provided we identify the parameter $c = \hbar^2/2m^* = \hbar^2/4m$. With a uniform η and in the absence of an external electromagnetic field Eq. (15.101) reduces to

$$(a + b|\bar{\psi}|^2)\bar{\psi} = \eta, \tag{15.103}$$

which is nothing but Eq. (13.11). In the limit $\eta \to 0$, $\bar{\psi} \to \psi_0$ and the order parameter has two homogeneous solutions

$$\psi_0 \equiv 0, \quad |\psi_0|^2 = -\frac{a}{b}. \tag{15.104}$$

Substituting the second of the above solutions in Eq. (15.99), we obtain the difference between the thermodynamic potentials for the normal and the superconductive systems:

$$F_{\mathrm{n}}(T) - F_{\mathrm{s}}(T) = \frac{a'^2}{2b}(T - T_{\mathrm{c}})^2 = \frac{H_{\mathrm{c}}^2(T)}{8\pi}. \tag{15.105}$$

The last equality has been obtained by means of Eq. (15.23). The temperature dependence of the critical field near the critical temperature given by Eq. (15.18) shows that the superconductive transition is well represented as a second order phase transition in the Landau approximation. The specific heat jump we discussed in Section 15.4 now reads

$$\Delta c = \frac{a'^2 T_{\mathrm{c}}}{b} = \frac{T_{\mathrm{c}}}{4\pi}\left(\frac{\partial H_{\mathrm{c}}(T)}{\partial T}\right)^2_{T=T_{\mathrm{c}}}. \tag{15.106}$$

The Landau–Ginzburg parameters a', b are then related to physical quantities such as the critical field and the specific heat jump and can be obtained by fitting the experiments.

Up to now we considered homogeneous solutions only. If we allow for small space variation of the order parameter, either due to fluctuations or to the presence of an external field, two lengths characterize the Landau–Ginzburg theory, the penetration depth λ and the correlation length $\xi(T)$. As was already stressed, the London relation Eq. (15.29) between the current and the vector potential has a proportionality factor containing the penetration depth λ given by (velocity of light $c = 1$)

$$\lambda^2 = \frac{m^*}{4\pi e^{*2}}\frac{1}{|\psi_0|^2} = \frac{m^*}{4\pi e^{*2}}\frac{b}{a'|T - T_{\mathrm{c}}|}, \tag{15.107}$$

provided $|\psi|^2$ can be approximated by $|\psi_0|^2$ in the range of variation of $\mathbf{A}$. Otherwise the electromagnetic penetration is no longer exponential and the rise of ψ to the value ψ_0 is no longer abrupt.

We discuss now the non-homogeneous case, allowing for fluctuations of the order parameter. Using again the analogy with ferromagnets, we write the complex order parameter in terms of its real and imaginary parts, which behave as the components of a two dimensional vector in the complex plane, $\psi = \psi_1 + \mathrm{i}\psi_2$. From Eq. (15.101) and its complex conjugate it is easy to derive the equations for ψ_1 and ψ_2 at $\mathbf{A} = 0$:

$$a\psi_1 + b\left(\psi_1^2 + \psi_2^2\right)\psi_1 - c\nabla^2\psi_1 = \eta_1, \tag{15.108}$$

$$a\psi_2 + b\left(\psi_1^2 + \psi_2^2\right)\psi_2 - c\nabla^2\psi_2 = \eta_2, \tag{15.109}$$

having set $\eta_1 = (\eta + \eta^*)/2$ and $\eta_2 = (\eta - \eta^*)/2\mathrm{i}$. Equation (15.104) determines the homogeneous value of the order parameter below T_c but for a phase factor. We now choose the homogeneous solution by taking $\eta_2 \equiv 0$, $\eta_1 \to 0$, i.e. fixing the gauge with the real homogeneous solution of Eqs. (15.108) and (15.109): $\psi_1 = \psi_0$, $\psi_2 = 0$. On the assumption of small variations of the order parameter with respect to the homogeneous value $\bar{\psi}$ in the presence of the external field η_1, we set $\psi_1 = \bar{\psi} + \delta\psi_1$, $\psi_2 = \delta\psi_2$ and the linearized form of Eqs. (15.108) and (15.109) reads

$$(a + 3b\bar{\psi}^2)\delta\psi_1 - c\nabla^2\delta\psi_1 = \delta\eta_1, \tag{15.110}$$

$$(a + b\bar{\psi}^2)\delta\psi_2 - c\nabla^2\delta\psi_2 = \delta\eta_2. \tag{15.111}$$

Within our gauge choice the variations of ψ_2 correspond to the variations of the phase of the order parameter. In fact, by writing $\psi = |\psi|\mathrm{e}^{\mathrm{i}\theta} \sim (|\psi_0| + \delta|\psi|)(1 + \mathrm{i}\theta) \sim |\psi_0| + \delta|\psi| + \mathrm{i}\theta|\psi_0|$, we see that Eq. (15.111) for ψ_2 when $\delta\eta_2 = 0$ is equivalent to Eq. (15.93) for the phase of the order parameter. Due to gauge invariance all these rotations lead to equivalent configurations.

According to the theory of linear response as applied in Chapter 13 when discussing the fluctuations of the order parameter in the Landau theory, we introduce the correlation function $f_{\psi,\parallel}$ and $f_{\psi,\perp}$ associated with the "longitudinal" and "transverse" fluctuations of ψ as functional derivatives of ψ_1 and ψ_2 with respect to η_1 and η_2:

$$f_{\psi,\parallel}(\mathbf{x}, \mathbf{x}') = k_B T \frac{\delta\psi_1(\mathbf{x})}{\delta\eta_1(\mathbf{x}')}, \tag{15.112}$$

$$f_{\psi,\perp}(\mathbf{x}, \mathbf{x}') = k_B T \frac{\delta\psi_2(\mathbf{x})}{\delta\eta_2(\mathbf{x}')}. \tag{15.113}$$

Considering the mathematical property

$$\frac{\delta\eta_{1,2}(\mathbf{x})}{\delta\eta_{1,2}(\mathbf{x}')} = \delta(\mathbf{x} - \mathbf{x}'), \tag{15.114}$$

from Eqs. (15.110) and (15.111) we obtain the equations for the correlation functions $f_{\psi,\parallel}$ and $f_{\psi,\perp}$:

$$(-2a - c\nabla^2)f_{\psi,\parallel} = k_B T\delta(\mathbf{x} - \mathbf{x}'), \tag{15.115}$$

$$((\eta_1/\bar{\psi}) - c\nabla^2)f_{\psi,\perp} = k_B T\delta(\mathbf{x} - \mathbf{x}'), \tag{15.116}$$

with $\bar{\psi}(\eta_1) \to \psi_0$ when $\eta_1 \to 0$. Because of translational invariance we can solve these equations by standard methods using Fourier transforms. In fact, if we define

$$f_{\psi,\parallel}(\mathbf{k}) = \int \mathrm{d}(\mathbf{x} - \mathbf{x}') f_{\psi,\parallel}(\mathbf{x}, \mathbf{x}')\mathrm{e}^{-\mathrm{i}\mathbf{k}\cdot(\mathbf{x}-\mathbf{x}')}, \tag{15.117}$$

$$f_{\psi,\perp}(\mathbf{k}) = \int \mathrm{d}(\mathbf{x} - \mathbf{x}') f_{\psi,\perp}(\mathbf{x}, \mathbf{x}')\mathrm{e}^{-\mathrm{i}\mathbf{k}\cdot(\mathbf{x}-\mathbf{x}')}, \tag{15.118}$$

for $T < T_c$, we get from Eqs. (15.115) and (15.116)

$$f_{\psi,\parallel}(\mathbf{k}) = k_B T \frac{1}{-2a + ck^2} = \frac{k_B T}{c} \frac{1}{\xi^{-2}(T) + k^2}, \tag{15.119}$$

$$f_{\psi,\perp}(\mathbf{k}) = k_B T \frac{1}{(\eta_1/\bar{\psi}) + ck^2}, \tag{15.120}$$

where we have introduced the temperature-dependent correlation length $\xi(T)$ for the fluctuations of the amplitude of the order parameter:

$$\xi^{-2}(T) = \frac{-2a}{c} = \frac{2a'}{c}|T - T_c| \equiv 2\xi_{0,LG}^{-2} \frac{|T - T_c|}{T_c}, \quad T < T_c,$$

$$\xi^{-2}(T) = \frac{a}{c} = \frac{a'}{c}|T - T_c| \equiv \xi_{0,LG}^{-2} \frac{|T - T_c|}{T_c}, \quad T > T_c, \tag{15.121}$$

where $\chi^{-1}/c = \xi^{-2}(T)$, $\xi_{0,LG} = (c/a'T_c)^{1/2}$ being the zero-temperature correlation length. The value of $\xi(T)$ gives the characteristic distance over which the order parameter varies. From Eq. (15.121), we see that the transverse fluctuations are massless, i.e. global ($\mathbf{k} = \mathbf{0}$) rotations of the order parameter do not cost energy.

The ratio K of the two characteristic lengths λ and $\xi(T)$ is the Landau–Ginzburg parameter which characterizes the classification of superconductors of the first and second kind introduced in Section 15.4. This quantity can be directly connected to experimentally observable quantities:

$$K = \frac{\lambda(T)}{\xi(T)} = \frac{2^{1/2} e^*}{\hbar} \lambda^2(T) H_c(T), \tag{15.122}$$

as can be easily derived by using Eqs. (15.105) and (15.121) and the expression for λ given by Eq. (15.107).

As discussed by Abrikosov (1957), when $K > \sqrt{2}$ the disadvantage of the vanishing order parameter in a region of dimension $\xi(T)$ is less than the cost of expelling the magnetic field in a region of dimension λ. The magnetic field can in this case penetrate into filaments, forming a triangular lattice (Abrikosov lattice) in planes perpendicular to the field. Each filament extends on a region of size λ and has a core of size $\xi(T)$ where the order parameter decreases to zero. Vice versa, when $K < \sqrt{2}$ the full diamagnetic expulsion of the field occurs according to the Meissner effect.

According to the Ginzburg criterion of Chapter 13, the Landau theory is valid as long as

$$\frac{\langle|\delta\psi|^2\rangle}{\psi_0^2} = \frac{T_c}{2(\Delta C/k_B)\xi_{0,LG}^2 (2\pi)^d (T - T_c)} \int_0^{\xi^{-1}(T)} d^d k \frac{1}{k^2 + \xi^{-2}(T)} < 1, \tag{15.123}$$

which means that the contribution of fluctuations is small with respect to the homogeneous solution ψ_0. The evaluation of the ratio Eq. (15.123) gives in $d = 3$

$$\frac{|T - T_c|}{T_c} \gg \frac{|T_0 - T_c|}{T_c} \sim \frac{1}{(\Delta C/k_B)^2} \frac{1}{\xi_{0,LG}^6}, \tag{15.124}$$

from which we see that Landau–Ginzburg theory is valid for temperatures which differ from the critical point more than a certain value $|T_0 - T_c|$ which depends on the parameters of the system. In particular we stress the sixth power dependence on the zero temperature correlation length $\xi_{0,\mathrm{LG}}$.

In ordinary superconductors $\xi_{0,\mathrm{LG}}$ is quite large, of the order of 10^4 Å, and $\Delta c = 10^4$ erg/cm K. With these values $|T_0 - T_\mathrm{c}| \sim 10^{-15} T_\mathrm{c}$ is outside experimental possibilities of observing any deviations from the Landau–Ginzburg theory.

In two dimensions, however, the effect of fluctuations is enhanced. This is apparent from Eq. (15.123) for the magnitude of the fluctuations of the order parameter. Effective two dimensional systems are films whose thickness d is less than the correlation length ξ ($d < \xi$). In this case the inverse sixth power dependence on ξ is changed to

$$\frac{|T_0 - T_\mathrm{c}|}{T_\mathrm{c}} \sim \frac{1}{(\Delta C/k_\mathrm{B})} \frac{1}{\xi_{0,\mathrm{LG}}^2 d}. \tag{15.125}$$

In going from three dimensions to two effective dimensions we make the the substitution

$$\frac{1}{V}\sum_{\mathbf{k}} \cdots = \frac{1}{(2\pi)^3}\int \mathrm{d}^3 k \ldots, \rightarrow \frac{1}{V}\sum_{\mathbf{k}} \cdots = \frac{1}{d}\frac{1}{(2\pi)^2}\int \mathrm{d}^2 k \ldots,$$

where d takes care of the reduced effective dimension of the region of integration in Eq. (15.123). A further increase of the effect of the fluctuations is due to the presence of impurities in the superconductive samples. In systems where there is a strong concentration of impurities (dirty superconductors), $\xi_{0,\mathrm{LG}}$ is usually much smaller ($\xi \sim 10^2$ Å) than in pure samples (clean superconductors) ($\xi_{0,\mathrm{LG}} \sim 10^4$ Å). To qualitatively understand the origin of the change of the value of $\xi_{0,\mathrm{LG}}$ we observe that in clean systems the electronic motion is ballistic whereas in dirty samples the electrons diffuse to a distance $R(t)$ in a time t according to the diffusion law $\langle R^2(t)\rangle = Dt$, where $D = v_\mathrm{F} l/3$ is the diffusion coefficient and l the mean free path. The energy involved in the correlation of the electrons leading to superconductivity is of the order of $k_\mathrm{B} T_\mathrm{c}$, which is also proportional to the gap Δ_0 of the excitation spectrum. In clean systems the electrons at the Fermi surface are correlated and hence can form a pair over a typical distance $\xi_\mathrm{c} \sim \xi_0 \sim \hbar v_\mathrm{F}/k_\mathrm{B} T_\mathrm{c}$. We will see later from the microscopic theory of superconductivity that the zero temperature correlation length $\xi_{0,\mathrm{LG}}$ and the coherence distance ξ_0 are proportional to each other. In dirty superconductors electrons at the Fermi surface are correlated over times of the order of $\tau \sim \hbar/k_\mathrm{B} T_\mathrm{c}$ and the zero temperature coherence distance is estimated by the distance over which a wave packet may diffuse, $\xi_\mathrm{d} \sim (D\hbar/k_\mathrm{B} T_\mathrm{c})^{1/2} = (\hbar v_\mathrm{F} l/3k_\mathrm{B} T_\mathrm{c})^{1/2} \sim (\xi_\mathrm{c} l)^{1/2}$. The subscripts c and d stand for clean and dirty, respectively. The mean free path is typically of the order of 10^2 Å.

Various effects due to fluctuations have been considered (see the book by Larkin and Varlamov (2005) for a comphensive discussion), in particular Glover (1967) used thin films of amorphous bismuth in order to take best advantage of the two previous effects. He found that when approaching T_c from the high temperature side as a sign of superconductive fluctuations and as a precursor effect of superconductivity itself, the conductivity σ diverges with a characteristic temperature behavior. This phenomenon, named paraconductivity, is presented in detail in Appendix K.

In the new high T_c superconductors the value of ξ as measured by critical field and fluctuation conductivity is of the order of 10 Å, making plausible the possibility of observing deviations from mean-field behavior and fluctuation effects also in three dimensional clean samples.

We conclude this section by showing how the present approach can be generalized to time-dependent phenomena. For instance, in order to investigate transport phenomena like paraconductivity, we must know how the fluctuating order parameter relaxes towards its equilibrium value.

According to the general principles of thermodynamics, at equilibrium the free energy must be a minimum so that the functional derivative with respect to the order parameter, e.g. the conjugate field, must vanish. When the system is out of equilibrium, for small deviations from equilibrium values of the order parameter, the restoring force is proportional to the conjugate field. Hence the rate of variation of the order parameter reads

$$\frac{\partial \psi}{\partial t} = -\frac{1}{\gamma}\frac{\delta F(\psi)}{\delta \psi^*}, \tag{15.126}$$

where the quantity $1/\gamma$ is a kinetic coefficient.

The equation for the order parameter we have written in the case of non-homogeneous fluctuations now reads

$$a\psi + b|\psi|^2\psi - c\nabla^2\psi = -\gamma\frac{\partial \psi}{\partial t}. \tag{15.127}$$

Above T_c, small uniform deviations from the equilibrium value $\psi_0 = 0$ lead to an exponential relaxation towards equilibrium with the characteristic Landau–Ginzburg relaxation time

$$\tau_{LG}^{-1} = \frac{a}{\gamma} = \frac{a'}{\gamma}(T - T_c). \tag{15.128}$$

Below T_c, small deviations from the equilibrium value $|\psi_0|^2 = -a/b$ are obtained by linearizing Eq. (15.127) around the value ψ_0 and using the gauge invariance Eq. (15.88):

$$(-2a - c\nabla^2)|\psi| = -\gamma\frac{\partial |\psi|}{\partial t}, \tag{15.129}$$

$$\gamma\left(\dot{\theta} - \frac{e^*}{\hbar}V\right) = c\nabla^2\theta, \tag{15.130}$$

and the characteristic Landau–Ginzburg relaxation time is given by

$$\tau_{LG}^{-1} = -\frac{a}{\gamma} = \frac{a'}{\gamma}(T_c - T). \tag{15.131}$$

When the Josephson relation Eq. (15.97) is satisfied, Eq. (15.130) reproduces Eq. (15.93).

In both cases, above and below T_c, the relaxation of the order parameter tends to infinity with the susceptibility, giving rise to the so-called critical slowing down of the fluctuations: the order parameter relaxes infinitely slowly towards the equilibrium value. For non-uniform fluctuations, the relaxation time becomes wave vector dependent:

$$\tau_{\mathbf{k}} = \frac{1}{\tau_{LG}^{-1} + (c/\gamma)k^2}. \tag{15.132}$$

By considering the extra current carried out by superconductive fluctuations above the critical temperature and taking into account that such fluctuations relax according to Eq. (15.127), one can explain the diverging behavior of the correction to the electrical conductivity as a precursor of the superconducting transition. This explanation was provided by Aslamazov and Larkin (1968) and its technical details can be found in Appendix K.

15.8 The Bogoliubov model for superfluidity

The first microscopic approximate solution to the problem of the low-lying excitations from the ground state of an interacting Bose gas was achieved in the limit of low density and weak coupling by Bogoliubov (1947), within a generalized Hartree–Fock approximation. Let us consider the Hamiltonian of Eq. (9.40) in Fourier space, which we report here for convenience:

$$H = \sum_{\mathbf{p}} \epsilon_{\mathbf{p}} a^{\dagger}_{\mathbf{p}} a_{\mathbf{p}} + \frac{1}{2V} \sum_{\mathbf{p}_1, \mathbf{p}_2, \mathbf{q}} V(\mathbf{q}) a^{\dagger}_{\mathbf{p}_1+\mathbf{q}} a^{\dagger}_{\mathbf{p}_2} a_{\mathbf{p}_2+\mathbf{q}} a_{\mathbf{p}_1}. \tag{15.133}$$

For a free Bose gas, we have seen that at $T = 0$ all N particles are in the state with zero momentum. When the interaction is present but not too strong, we may expect that some particles are depleted from the lowest state and

$$a^{\dagger}_0 a_0 = N_0, \quad \sum_{\mathbf{p} \neq \mathbf{0}} a^{\dagger}_{\mathbf{p}} a_{\mathbf{p}} = N - N_0, \tag{15.134}$$

the main assumption for the Bogoliubov solution being that for very dilute systems $N_0 \simeq N$ and $N - N_0 \ll N_0$. In the thermodynamic limit, the requirement that the occupation of the zero-momentum state is of the order of N implies that unity in the commutation rules for bosons can be neglected and we then can treat $a_0 = \sqrt{N_0}$ and $a^{\dagger}_0 = \sqrt{N_0}$ as commuting c-numbers. In the Hamiltonian we separate the contributions corresponding to the zero-momentum state from the other states. This can be done by assigning zero momentum to one or more of the creation and annihilation operators in the Hamiltonian. The quadratic term in the creation and annihilation operators trivially separates into two terms corresponding to zero and non-zero momentum states, respectively. In the interaction term, in principle, we may assign zero momentum to one, two, three and four operators. There is clearly only one term where all operators have zero momentum. We will see in a few moments that such a term gives the leading contribution to the ground state energy. Assigning zero momentum to only one creation or annihilation operator is forbidden by momentum conservation. Next, we may assign zero momentum to two operators. This can be done in four ways by selecting one creation and one annihilation operator, whereas there are two more ways in which one selects either two creation or two annihilation operators. Finally, there are four ways to assign zero momentum to only one operator. We then gather in H_0 all the terms up to those that are quadratic in $a_{\mathbf{p}}$ and $a^{\dagger}_{\mathbf{p}}$ with $\mathbf{p} \neq \mathbf{0}$, while we collect in H_{int} all the remaining terms.

Hence we get the linearized Hamiltonian H_0 in the presence of a chemical potential:

$$H_0 = \frac{V(\mathbf{0})N_0^2}{2V} - \mu N_0 + \sum_{\mathbf{p}\neq\mathbf{0}} \nu_{\mathbf{p}} a_{\mathbf{p}}^\dagger a_{\mathbf{p}} + \frac{1}{2}\sum_{\mathbf{p}\neq\mathbf{0}} \mu_{\mathbf{p}}\left(a_{\mathbf{p}}^\dagger a_{-\mathbf{p}}^\dagger + a_{\mathbf{p}} a_{-\mathbf{p}}\right), \tag{15.135}$$

where

$$\nu_{\mathbf{p}} = \frac{p^2}{2m} - \mu + \frac{N_0}{V}\left(V(\mathbf{0}) + V(\mathbf{p})\right), \quad \mu_{\mathbf{p}} = \frac{N_0}{V}V(\mathbf{p}). \tag{15.136}$$

$\nu_{\mathbf{p}}$ corresponds to the standard Hartree–Fock terms, whereas $\mu_{\mathbf{p}}$ (not to be confused with the chemical potential μ) is anomalous and corresponds to the excitation (adsorption) from (onto) the condensate allowed by momentum conservation.

The interacting Hamiltonian H_{int} reads

$$H_{\text{int}} = \frac{\sqrt{N_0}}{2V}\sum_{\mathbf{p},\mathbf{q}\neq\mathbf{0}} 2V(\mathbf{q})\left(a_{\mathbf{q}}^\dagger a_{\mathbf{p}}^\dagger a_{\mathbf{p}+\mathbf{q}} + a_{\mathbf{p}+\mathbf{q}}^\dagger a_{\mathbf{p}} a_{\mathbf{q}}\right) + \frac{1}{2V}\sum_{\mathbf{p}_1,\mathbf{p}_2,\mathbf{q}\neq\mathbf{0}} V(\mathbf{q}) a_{\mathbf{p}_1+\mathbf{q}}^\dagger a_{\mathbf{p}_2}^\dagger a_{\mathbf{p}_2+\mathbf{q}} a_{\mathbf{p}_1}. \tag{15.137}$$

Let us now concentrate on H_0 at the level of the first approximation. N_0 is the parameter introduced by the linearization procedure and has to be determined variationally. The diagonalization of H_0 is more involved than in the simple Hartree–Fock case. We have to introduce new bosonic operators $\gamma_{\mathbf{p}}$ and $\gamma_{\mathbf{p}}^\dagger$ for quasiparticles, a linear combination of the particle operators $a_{\mathbf{p}}$ and $a_{-\mathbf{p}}^\dagger$. $\gamma_{\mathbf{p}}$ of course gives zero when applied to the ground state where no excitations are present:

$$a_{\mathbf{p}} = u_{\mathbf{p}}\gamma_{\mathbf{p}} + v_{\mathbf{p}}\gamma_{-\mathbf{p}}^\dagger, \quad \gamma_{\mathbf{p}} = u_{\mathbf{p}} a_{\mathbf{p}} - v_{\mathbf{p}} a_{-\mathbf{p}}^\dagger, \tag{15.138}$$

with the canonicity condition required by imposing the bosonic commutation relations

$$u_{\mathbf{p}}^2 - v_{\mathbf{p}}^2 = 1. \tag{15.139}$$

H_0 is then transformed into a constant term W_0, a diagonal term in $\gamma_{\mathbf{p}}^\dagger\gamma_{\mathbf{p}}$ and an off-diagonal term with two creation and two annihilation quasiparticle operators:

$$\begin{aligned} H_0 = W_0 &+ \sum_{\mathbf{p}\neq\mathbf{0}} \left[\nu_{\mathbf{p}}\left(v_{\mathbf{p}}^2 + u_{\mathbf{p}}^2\right) + 2\mu_{\mathbf{p}} v_{\mathbf{p}} u_{\mathbf{p}}\right]\gamma_{\mathbf{p}}^\dagger\gamma_{\mathbf{p}} \\ &+ \sum_{\mathbf{p}\neq\mathbf{0}} \left(\nu_{\mathbf{p}} v_{\mathbf{p}} u_{\mathbf{p}} + \frac{1}{2}\mu_{\mathbf{p}}\left(v_{\mathbf{p}}^2 + u_{\mathbf{p}}^2\right)\right)\left(\gamma_{\mathbf{p}}^\dagger\gamma_{-\mathbf{p}}^\dagger + \gamma_{\mathbf{p}}\gamma_{-\mathbf{p}}\right), \end{aligned} \tag{15.140}$$

with the ground state energy W_0

$$W_0 = \frac{V(\mathbf{0})N_0^2}{2V} - \mu N_0 + \sum_{\mathbf{p}\neq\mathbf{0}}\left(\nu_{\mathbf{p}} v_{\mathbf{p}}^2 + \mu_{\mathbf{p}} v_{\mathbf{p}} u_{\mathbf{p}}\right). \tag{15.141}$$

The first term in Eq. (15.141) for W_0 corresponds to the energy of an homogeneous condensate in the simplified Gross–Pitaevskii–Hartree type ground state, Eq. (15.76).

At the lowest level of approximation, the chemical potential μ is determined in terms of the parameter N_0 by minimization of $W_0 \simeq V(\mathbf{0})N_0^2/2V - \mu N_0$, with respect to N_0 itself:

$$\mu = \frac{V(\mathbf{0})N_0}{V}. \tag{15.142}$$

The chemical potential (15.142) coincides with the one given by the stationary Gross–Pitaevskii equation (15.71) in the absence of V_{ext} and constant solution $|\psi(\mathbf{x})|^2 = \rho/m = N_0/V$.

The diagonalization condition for H_0 is then obtained by the vanishing of the third term in Eq. (15.140):

$$2\nu_{\mathbf{p}} v_{\mathbf{p}} u_{\mathbf{p}} + \mu_{\mathbf{p}}\left(v_{\mathbf{p}}^2 + u_{\mathbf{p}}^2\right) = 0. \tag{15.143}$$

The above equation has to be solved together with the canonicity condition (15.139). The solution is

$$u_{\mathbf{p}}^2\left(v_{\mathbf{p}}^2\right) = \pm\frac{1}{2} + \frac{\nu_{\mathbf{p}}}{2E_{\mathbf{p}}}, \quad u_{\mathbf{p}} v_{\mathbf{p}} = -\frac{1}{2}\frac{\mu_{\mathbf{p}}}{E_{\mathbf{p}}}, \quad E_{\mathbf{p}} = \sqrt{\nu_{\mathbf{p}}^2 - \mu_{\mathbf{p}}^2}. \tag{15.144}$$

H_0 in terms of quasiparticle operators is then

$$H_0 = W_0 + \sum_{\mathbf{p}\neq 0} E_{\mathbf{p}} \gamma_{\mathbf{p}}^{\dagger} \gamma_{\mathbf{p}}. \tag{15.145}$$

The quasiparticle energy $E_{\mathbf{p}}$ reads

$$E_{\mathbf{p}} = \sqrt{(p^2/2m)(p^2/2m + 2nV(\mathbf{p}))}, \tag{15.146}$$

where $N_0/V \sim N/V = n$.

We can now draw some physical conclusions. First, we notice that at low values of momentum, the spectrum of excitations becomes linear, describing the phonon modes associated with the sound propagation with speed given by

$$c = \sqrt{\frac{nV(\mathbf{0})}{m}} = \sqrt{\frac{\rho V(\mathbf{0})}{m^2}}, \tag{15.147}$$

$\rho = nm$ is the matter density. $E_{\mathbf{p}}$ evolves into the spectrum of the free particles at high $\mathbf{p}$. The leading contribution to the ground state energy in Eq. (15.141) gives the pressure

$$P = -\left(\frac{\partial W_0}{\partial V}\right)_{\mu} = \frac{1}{2}V(\mathbf{0})n^2 = \frac{V(\mathbf{0})\rho^2}{2m^2}. \tag{15.148}$$

Then the square of the sound speed is related to the derivative of the pressure with respect to the mass density, as required by standard hydrodynamics.

Corrections (Gavoret and Nozières (1964)) to this simple first approximation by considering the cubic and quartic terms in H_{int} can be obtained in perturbation theory, which however is plagued by infrared singularities. The latter, however, must cancel at any order in perturbation theory, since we are dealing with a stable liquid phase and all physical quantities are regular functions. This gives rise to a subtle field-theoretic problem (Castellani et al. (1997)), which we do not discuss further here.

15.9 The microscopic theory of superconductivity

In Section 15.5 we established the basis for a microscopic theory of superconductivity. The latter must reproduce the fundamental requirement of ODLRO in the two-particle reduced density matrix and therefore the electrons have to be bound in pairs. In Chapter 9 we discussed how the Hartree–Fock approximation arises from the extension to an interacting system of the factorization, valid for a gas, of the two-particle density matrix in terms of the one-particle density matrix (see Eq. (9.68)). The factorization of Eq. (9.68) does not show ODLRO for fermions since h_1 cannot have macroscopic eigenvalues. According to its definition (15.46), ODLRO is present if we can construct a ground state $|s\rangle$ which yields a generalized Hartree–Fock approximation such that

$$h_{2,\sigma_1\sigma_2\tau_2\tau_1}(\mathbf{x}_1,\mathbf{x}_2;\mathbf{x}_2',\mathbf{x}_1') = h_{1,\sigma_1\tau_1}(\mathbf{x}_1,\mathbf{x}_1')h_{1,\sigma_2\tau_2}(\mathbf{x}_2,\mathbf{x}_2') - h_{1,\sigma_1\tau_2}(\mathbf{x}_1,\mathbf{x}_2')h_{1,\sigma_2\tau_1}(\mathbf{x}_2,\mathbf{x}_1') + \chi^*_{\tau_1\tau_2}(\mathbf{x}_1',\mathbf{x}_2')\chi_{\sigma_1\sigma_2}(\mathbf{x}_1,\mathbf{x}_2), \tag{15.149}$$

where the lower indices indicate the spin degrees of freedom and $\chi_{\sigma\sigma'}$ is related to the pair eigenfunction. Bardeen, Cooper and Schrieffer accomplished this task (Bardeen et al. (1957)) and produced the so-called BCS theory. We shall now study their microscopic theory following the generalized Hartree–Fock approach in the form introduced by Valatin (1962), starting from the generic Hamiltonian given by Eq. (9.40) with two-body interaction and spin 1/2:

$$H = \sum_{\mathbf{k},\sigma} \epsilon_{\mathbf{k}} a^\dagger_{\mathbf{k},\sigma} a_{\mathbf{k},\sigma} + \frac{1}{2V} \sum_{\mathbf{k},\mathbf{k}',\mathbf{q},\sigma,\sigma'} V(\mathbf{q}) a^\dagger_{\mathbf{k}+\mathbf{q},\sigma} a^\dagger_{\mathbf{k}'-\mathbf{q},\sigma'} a_{\mathbf{k}',\sigma'} a_{\mathbf{k},\sigma}. \tag{15.150}$$

From now on we consider all quantities per unit volume and set $V = 1$ to avoid confusion with the potential $V(\mathbf{q})$.

In order for the superconducting phase to be stable with respect to the normal phase, the state $|s\rangle$ we are looking for must lead to a lower energy than that of the ground state for the normal system $|n\rangle$, which in the Hartree–Fock approach can be approximated by a Slater determinant of Bloch functions:

$$|n\rangle = \prod_{0<|\mathbf{k}|<k_F} a^\dagger_{\mathbf{k}\uparrow} a^\dagger_{-\mathbf{k}\downarrow} |0\rangle. \tag{15.151}$$

Because of the isotope effect the interaction leading to superconductivity must be due to phonons. The effective electron–electron interaction via phonons is weak and attractive i.e. negative in a region around the Fermi surface, and it has not been included in the Bloch states which approximate the normal electron states. Because of this interaction, pairs of electrons will be scattered from states $\mathbf{k}_1$ and $\mathbf{k}_2$ to states $\mathbf{k}_1'$ and $\mathbf{k}_2'$ with momentum conservation. Even though the potential is negative, the matrix elements of the potential between different spin configurations can be either positive or negative, depending on the number of transpositions of indices we have to make on the ordered state to calculate these matrix elements. Statistically, therefore, even with an attractive potential, we would have a zero mean energy gain with respect to the normal case. We have therefore to choose a subset of configurations such as to allow the potential matrix elements to remain negative. If the states are occupied by pairs of electrons in the sense that if one is occupied another related to it must be also occupied, then the number of transpositions is always even and we have fulfilled our task. These pairs must have the same momentum so that the number of allowed collisions is minimum and the energy gain is maximum. Of course, if there is no net flow, this momentum must be zero. We have reached again the point discussed in Section 15.4, i.e. the superconducting state must be formed by bound pairs of electrons with opposite momentum $\mathbf{k}$ and spin σ. The ground state $|s\rangle$ is therefore a coherent mixing of Bloch states occupied in pairs.

A normalized state satisfying all the requirements discussed above is

$$|s\rangle = \prod_{\mathbf{k}} \left((1-h_{\mathbf{k}})^{1/2} + h_{\mathbf{k}}^{1/2} a^{\dagger}_{\mathbf{k}\uparrow} a^{\dagger}_{-\mathbf{k}\downarrow}\right)|0\rangle. \tag{15.152}$$

In the general case the best choice of single particle states $g_\kappa(\mathbf{x})$ has to be determined by the variational principle. Of course, in the absence of an external field and for homogeneous systems, the single particle states are momentum states (see Eq. (9.39)), $g_\kappa(\mathbf{x}) = g_{\mathbf{k}}(\mathbf{x})\chi_\sigma = e^{i\mathbf{k}\cdot\mathbf{x}}\chi_\sigma$ with $\kappa = (\mathbf{k}, \sigma)$, $\sigma = \uparrow, \downarrow$ being the spin projection. Space inversion and time reversal symmetry require $h_{-\mathbf{k}} = h_{\mathbf{k}}$ and $h^*_{-\mathbf{k}} = h_{\mathbf{k}}$, respectively. Hence $h_{\mathbf{k}}$ must be an even real function of $\mathbf{k}$.

The state $|s\rangle$ does not conserve the total number of particles and allows for the factorization (15.149) of h_2. There is in fact a probability different from zero for any number of pairs. The conservation on average of the number of particles N introduces a constraint on the coefficients $h_{\mathbf{k}}$ which coincides with the $\mathbf{k}$-representation of h_1:

$$\begin{aligned} N &= 2\sum_{\mathbf{k}} h_{\mathbf{k}}, \\ h_{1\mathbf{k}\sigma,\mathbf{k}'\sigma'} &= \langle s|a^{\dagger}_{\mathbf{k}'\sigma'} a_{\mathbf{k}\sigma}|s\rangle = h_{\mathbf{k}}\delta_{\mathbf{k},\mathbf{k}'}\delta_{\sigma\sigma'}. \end{aligned} \tag{15.153}$$

Together with $h_{\mathbf{k}}$, we have to define the new quantity $\chi_{\mathbf{k}}$ which coincides with the $\mathbf{k}$-representation of the pair wave function $\chi(\mathbf{x}, \mathbf{x}')$. To see this, we observe that

$$\begin{aligned} \chi_{\mathbf{k}\sigma,\mathbf{k}'\sigma'} &= \langle s|a_{\mathbf{k}'\sigma'} a_{\mathbf{k}\sigma}|s\rangle \\ &= (h_{\mathbf{k}})^{1/2}(1-h_{\mathbf{k}})^{1/2}\delta_{\mathbf{k},-\mathbf{k}'}\mathbf{I}_{\sigma\sigma'} \\ &\equiv \mathbf{I}_{\sigma\sigma'}\chi_{\mathbf{k}}\delta_{\mathbf{k},-\mathbf{k}'}, \end{aligned} \tag{15.154}$$

where the antisymmetric matrix $\mathbf{I}_{\sigma\sigma'} = \sigma\delta_{\sigma,-\sigma'}$ reflects the fermion anticommutation rules. In real space we have

$$\begin{aligned} \chi_{\sigma\sigma'}(\mathbf{x}, \mathbf{x}') &= \langle s|\hat{\psi}_{\sigma'}(\mathbf{x}')\hat{\psi}_\sigma(\mathbf{x})|s\rangle \\ &= \mathbf{I}_{\sigma\sigma'}\sum_{\mathbf{k},\mathbf{k}'} g_{\mathbf{k}}(\mathbf{x})g_{\mathbf{k}'}(\mathbf{x}')\chi_{\mathbf{k}\sigma,\mathbf{k}'\sigma'} \\ &= \mathbf{I}_{\sigma\sigma'}\sum_{\mathbf{k}} \chi_{\mathbf{k}} g_{\mathbf{k}}(\mathbf{x})g_{-\mathbf{k}}(\mathbf{x}') \\ &\equiv \mathbf{I}_{\sigma\sigma'}\chi(\mathbf{x}, \mathbf{x}'). \end{aligned} \tag{15.155}$$

To make the treatment similar to that used in the case of Hartree–Fock theory, we introduce a generalized single-particle reduced density matrix

$$K(\mathbf{x}, \mathbf{x}') = \langle \Psi^\dagger(\mathbf{x}')\Psi(\mathbf{x})\rangle, \tag{15.156}$$

where we have introduced the four-component spinor

$$\mathbf{\Psi}(\mathbf{x}) = \begin{pmatrix} \hat{\psi}_\uparrow(\mathbf{x}) \\ \hat{\psi}_\downarrow^\dagger(\mathbf{x}) \\ \hat{\psi}_\downarrow(\mathbf{x}) \\ \hat{\psi}_\uparrow^\dagger(\mathbf{x}) \end{pmatrix}. \tag{15.157}$$

The four-by-four generalized density matrix K factorizes into two blocks:

$$K(\mathbf{x}, \mathbf{x}') = \begin{pmatrix} \mathbf{K}_\uparrow(\mathbf{x}, \mathbf{x}') & 0 \\ 0 & \mathbf{K}_\downarrow(\mathbf{x}, \mathbf{x}') \end{pmatrix}, \quad \mathbf{K}_\sigma(\mathbf{x}, \mathbf{x}') = \begin{pmatrix} h_{1,\sigma\sigma}(\mathbf{x}, \mathbf{x}') & \chi_{\sigma\overline{\sigma}}(\mathbf{x}, \mathbf{x}') \\ \chi^*_{\sigma\overline{\sigma}}(\mathbf{x}, \mathbf{x}') & 1 - h_{1,\overline{\sigma\sigma}}(\mathbf{x}, \mathbf{x}') \end{pmatrix}, \tag{15.158}$$

where $\overline{\sigma} = -\sigma$. By using Eqs. (15.153)–(15.155), we can represent the generalized density matrix in $\mathbf{k}$-space as

$$\mathbf{K}_\sigma(\mathbf{x}, \mathbf{x}') = \sum_{\mathbf{k}} g_{\mathbf{k}}(\mathbf{x})g^*_{\mathbf{k}}(\mathbf{x}')\, \mathbf{K}_\sigma(\mathbf{k}), \quad \mathbf{K}_\sigma(\mathbf{k}) = \begin{pmatrix} h_{\mathbf{k}} & \mathbf{I}_{\sigma\overline{\sigma}}\chi_{\mathbf{k}} \\ \mathbf{I}_{\sigma\overline{\sigma}}\chi^*_{-\mathbf{k}} & 1 - h_{-\mathbf{k}} \end{pmatrix}. \tag{15.159}$$

In deriving (15.159), we have used the fact that $g_{-\mathbf{k}}(\mathbf{x}) = g^*_{\mathbf{k}}(\mathbf{x})$, which reflects the time reversal invariance of the Hamiltonian. We then have the new constraint in place of Eq. (9.67):

$$K^2 = K \quad \text{or} \quad \int d\mathbf{x}'' K(\mathbf{x}, \mathbf{x}'')K(\mathbf{x}'', \mathbf{x}') = K(\mathbf{x}, \mathbf{x}'). \tag{15.160}$$

Because of the pair wave function, we have, in addition to the self-consistent potential U of Eq. (9.71), the pair potential (again not to be confused with the chemical potential)

$$\mu(\mathbf{x}, \mathbf{x}') = V(\mathbf{x} - \mathbf{x}')\chi(\mathbf{x}, \mathbf{x}'), \tag{15.161}$$

or in the $\mathbf{k}$-space representation

$$\mu_{\mathbf{k},\mathbf{k}'} = \int d\mathbf{x} \int d\mathbf{x}' g^*_{\mathbf{k}}(\mathbf{x})\mu(\mathbf{x}, \mathbf{x}')g^*_{\mathbf{k}'}(\mathbf{x}') = \mu_{\mathbf{k}}\delta_{\mathbf{k},-\mathbf{k}'}, \tag{15.162}$$

where

$$\mu_{\mathbf{k}} = \sum_{\mathbf{k}'} V(\mathbf{k} - \mathbf{k}')\chi_{\mathbf{k}'}, \tag{15.163}$$

with $V(\mathbf{q})$ defined in Eq. (9.42). The linearized Hamiltonian reads

$$\begin{aligned} H_{\text{lin}} &= \frac{1}{2}\int d\mathbf{x} \int d\mathbf{x}' (\mathbf{\Psi}^\dagger(\mathbf{x})M(\mathbf{x}, \mathbf{x}')\mathbf{\Psi}(\mathbf{x}') + \nu(\mathbf{x}, \mathbf{x}')) \\ &\quad - \frac{1}{2}\sum_\sigma \int d\mathbf{x} \int d\mathbf{x}' [U(\mathbf{x}, \mathbf{x}')h_{1,\sigma\sigma}(\mathbf{x}', \mathbf{x}) + V(\mathbf{x}, \mathbf{x}')|\chi_{\sigma\overline{\sigma}}(\mathbf{x}, \mathbf{x}')|^2], \end{aligned} \tag{15.164}$$

with the matrix M defined by

$$M = \begin{pmatrix} \nu & \mu & 0 & 0 \\ \mu^* & -\nu & 0 & 0 \\ 0 & 0 & \nu & -\mu \\ 0 & 0 & -\mu^* & -\nu \end{pmatrix}. \tag{15.165}$$

The last term in round brackets in the first line of Eq. (15.164) is obtained from the fermion anticommutation rules by observing that

$$\hat{\psi}_\sigma^\dagger \hat{\psi}_\sigma = \frac{1}{2}\left(\hat{\psi}_\sigma^\dagger \hat{\psi}_\sigma - \hat{\psi}_\sigma \hat{\psi}_\sigma^\dagger + 1\right).$$

The ground state energy is then given by

$$\begin{aligned} W_0 &= \langle s|H_{\rm lin}|s\rangle \\ &= \frac{1}{2}\sum_\sigma \int d\mathbf{x} \int d\mathbf{x}'[\epsilon(\mathbf{x},\mathbf{x}') + \nu(\mathbf{x},\mathbf{x}')]h_{1,\sigma\sigma}(\mathbf{x}',\mathbf{x}) \\ &\quad + \frac{1}{2}\sum_\sigma \int d\mathbf{x} \int d\mathbf{x}'\mu^*(\mathbf{x},\mathbf{x}')\chi_{\sigma\bar{\sigma}}(\mathbf{x}',\mathbf{x}) \\ &= \frac{1}{2}\sum_{\mathbf{k},\sigma} \left[(\epsilon_\mathbf{k} + \nu_\mathbf{k})h_\mathbf{k} + \mu_\mathbf{k}\chi_\mathbf{k}\right], \end{aligned} \tag{15.166}$$

or in matrix notation

$$W_0 = \frac{1}{2}{\rm tr}[\epsilon h_1 + \nu h_1 + \mu^* \chi]. \tag{15.167}$$

Variation of W_0 with respect to h_1 and χ under the constraint (15.160) in the **k**-space representation, where $\chi_\mathbf{k}$ can be taken real,

$$h_\mathbf{k}^2 + \chi_\mathbf{k}^2 = h_\mathbf{k}, \tag{15.168}$$

leads to

$$\nu_\mathbf{k}\delta h_\mathbf{k} + \mu_\mathbf{k}\delta\chi_\mathbf{k} + E_\mathbf{k}(2\chi_\mathbf{k}\delta\chi_\mathbf{k} - (1 - 2h_\mathbf{k})\delta h_\mathbf{k}) = 0, \tag{15.169}$$

where $E_\mathbf{k}$ is a Lagrange multiplier. The solutions of the above equations are obtained by setting to zero the coefficients of $\delta h_\mathbf{k}$ and $\delta\chi_\mathbf{k}$ and read

$$h_\mathbf{k} = \frac{1}{2}\left(1 - \frac{\nu_\mathbf{k}}{E_\mathbf{k}}\right), \quad \chi_\mathbf{k} = -\frac{1}{2}\frac{\mu_\mathbf{k}}{E_\mathbf{k}}. \tag{15.170}$$

Using the constraint (15.168) leads to the expression for the Lagrange multiplier $E_\mathbf{k}$, which, as we shall see later on, has the meaning of a quasiparticle energy excitation as in the Hartree–Fock case:

$$E_\mathbf{k} = \sqrt{\nu_\mathbf{k}^2 + \mu_\mathbf{k}^2}. \tag{15.171}$$

The linearized Hamiltonian (15.164) in the **k**-space representation reads

$$H_{\rm lin} = \sum_{\mathbf{k},\sigma} \nu_\mathbf{k} a_{\mathbf{k},\sigma}^\dagger a_{\mathbf{k},\sigma} + \frac{1}{2}\sum_{\mathbf{k}\sigma} \sigma\mu_\mathbf{k}\left(a_{\mathbf{k},\sigma}^\dagger a_{-\mathbf{k},\bar{\sigma}}^\dagger + a_{-\mathbf{k},\bar{\sigma}} a_{\mathbf{k},\sigma}\right) + {\rm const.} \tag{15.172}$$

This expression is not diagonal and we have therefore to introduce a canonical transformation in order to diagonalize it. The factor $\sigma = \pm 1$ refers to the upper (lower) block of the matrix M and shows that two distinct transformations are necessary for diagonalizing the two blocks. New operators $\xi_{\mathbf{k},\sigma}$ and $\xi^\dagger_{\mathbf{k},\sigma}$ must be introduced satisfying the same anticommutation rules as $a_{\mathbf{k},\sigma}$ and $a^\dagger_{\mathbf{k},\sigma}$. The Bogoliubov–Valatin transformation reads

$$\xi_{\mathbf{k},\sigma} = (1 - h_{\mathbf{k}})^{1/2} a_{\mathbf{k},\sigma} - \sigma h_{\mathbf{k}}^{1/2} a^\dagger_{-\mathbf{k},-\sigma}, \tag{15.173}$$

$$\xi^\dagger_{\mathbf{k},\sigma} = (1 - h_{\mathbf{k}})^{1/2} a^\dagger_{\mathbf{k},\sigma} - \sigma h_{\mathbf{k}}^{1/2} a_{-\mathbf{k},-\sigma}. \tag{15.174}$$

The ground state $|s\rangle$ given by Eq. (15.152) is the vacuum state for the operators $\xi_{\mathbf{k},\sigma}$ and $\xi^\dagger_{\mathbf{k},\sigma}$, i.e. $\xi_{\mathbf{k},\sigma}|s\rangle = 0$ for any $\mathbf{k}$. Provided that $h_{\mathbf{k}}$ and $\chi_{\mathbf{k}}$ are given by Eq. (15.170), after the transformation (15.173), (15.174), the Hamiltonian (15.172) becomes

$$H_{\text{lin}} = W_0 + \sum_{\mathbf{k},\sigma} E_{\mathbf{k}} \xi^\dagger_{\mathbf{k},\sigma} \xi_{\mathbf{k},\sigma}, \tag{15.175}$$

where $E_{\mathbf{k}}$, given by Eq. (15.171), is the new quasiparticle energy.

To see explicitly the consequences of the theory developed so far, let us take for simplicity a model in which the Hartree–Fock self-consistent potential U is zero and the potential V is the electron–electron interaction due to phonon exchange. As a result, the potential $V(\mathbf{q}, \omega)$ is a retarded interaction and depends on both transferred momentum $\mathbf{q}$ and frequency ω. This interaction has typically the form (see Abrikosov et al. (1963))

$$V(\mathbf{q}, \omega) = g \frac{\omega_{\mathbf{q}}^2}{\omega^2 - \omega_{\mathbf{q}}^2}, \tag{15.176}$$

$\omega_{\mathbf{q}} \sim \omega_{\text{D}}$ being the phonon dispersion relation with the Debye frequency defined in Eq. (7.92) and g the electron–phonon coupling constant. Conservation of momentum and energy require $\mathbf{k} - \mathbf{k}' = \mathbf{q}$ and $\epsilon_{\mathbf{k}} - \epsilon_{\mathbf{k}'} = \omega$. The so-called BCS model amounts to assuming $V(\mathbf{q}, \omega)$ constant and negative in a region $2\omega_{\text{D}}$ around the Fermi surface:

$$V_{\mathbf{k},\uparrow,-\mathbf{k},\downarrow;\mathbf{k}',\uparrow,-\mathbf{k}',\downarrow} = -g\theta(\omega_{\text{D}} - \epsilon_{\mathbf{k}})\theta(\omega_{\text{D}} - \epsilon_{\mathbf{k}'}). \tag{15.177}$$

As a result of the above position, the expression (15.163) for the pairing potential becomes

$$\mu_{\mathbf{k}} = -g \sum_{\mathbf{k}'} \theta(\omega_{\text{D}} - |\epsilon_{\mathbf{k}}|)\theta(\omega_{\text{D}} - |\epsilon_{\mathbf{k}'}|)\chi_{\mathbf{k}'}. \tag{15.178}$$

By using Eq. (15.170) which expresses $\chi_{\mathbf{k}}$ in terms of $\mu_{\mathbf{k}}$ one obtains the self-consistency equation of the BCS mean-field theory of superconductivity:

$$\mu_{\mathbf{k}} = g \sum_{\mathbf{k}'} \theta(\omega_{\text{D}} - |\epsilon_{\mathbf{k}}|)\theta(\omega_{\text{D}} - |\epsilon_{\mathbf{k}'}|)\frac{\mu_{\mathbf{k}'}}{2E_{\mathbf{k}'}}. \tag{15.179}$$

Due to the factorization of the interaction potential with respect to $\mathbf{k}$ and $\mathbf{k}'$, the pairing potential is constant and non-vanishing only for $|\epsilon_{\mathbf{k}}| < \omega_{\text{D}}$, so that, by writing $\mu_{\mathbf{k}} = \Delta\theta(\omega_{\text{D}} - |\epsilon_{\mathbf{k}}|)$, we have finally

$$\Delta = g\Delta \sum_{\mathbf{k}'} \theta(\omega_{\text{D}} - |\epsilon_{\mathbf{k}'}|)\frac{1}{2E_{\mathbf{k}'}}. \tag{15.180}$$

$\omega_{\rm D}$ is much smaller than the Fermi energy and the density of states in the shell $2\omega_{\rm D}$ around the Fermi surface can be considered constant. The equation for Δ then becomes

$$\frac{1}{N_0 g} = \int_0^{\omega_{\rm D}} {\rm d}\epsilon \frac{1}{\sqrt{\epsilon^2 + \Delta^2}} = sh^{-1}\left(\frac{\omega_{\rm D}}{\Delta}\right), \tag{15.181}$$

where $N_0 = \nu(E_{\rm F})/2$ is the single-particle density of states at the Fermi energy per spin. For most superconductors $N_0 g \ll 1$ and we get

$$\Delta = 2\omega_{\rm D} {\rm e}^{-1/N_0 g}. \tag{15.182}$$

Solution (15.182) is clearly non-perturbative. This is the reason for the difficulty in explaining superconductivity, i.e. superconductive behavior cannot be obtained by any perturbative expansion from the normal state. In fact we had to imagine a completely new state.

We have been guided by many considerations, among which is the factorization property (15.149) for h_2 in order to have condensation. Bardeen, Cooper and Schrieffer have calculated, at $T = 0$, within their model, the behavior of the function $\phi(r) = 1/(\alpha N)^{1/2}\chi(r)$ appearing in Eq. (15.46). By using Eq. (15.170) for $\chi_{\mathbf{k}}$, we have, with $r = |\mathbf{x}_1 - \mathbf{x}_2|$,

$$\begin{aligned}\chi(\mathbf{r}) &= \int \frac{{\rm d}^3 k}{(2\pi)^3} {\rm e}^{{\rm i}\mathbf{k}\cdot\mathbf{r}} \chi_{\mathbf{k}} \\ &= -\frac{\Delta}{2}\int \frac{{\rm d}^3 k}{(2\pi)^3} {\rm e}^{{\rm i}\mathbf{k}\cdot\mathbf{r}} \frac{\theta(\omega_{\rm D} - |\epsilon_{\mathbf{k}}|)}{\sqrt{\epsilon_{\mathbf{k}}^2 + \Delta^2}} \\ &= -\frac{N_0\Delta}{2}\int_{-\omega_{\rm D}}^{\omega_{\rm D}} {\rm d}\epsilon \frac{1}{\sqrt{\epsilon^2 + \Delta^2}} \frac{\sin(k(\epsilon)r)}{k(\epsilon)r},\end{aligned}$$

where in the last step we changed variable and replaced the density of states by its value at the Fermi energy $E_{\rm F}$. By expressing the modulus of the momentum in terms of the energy, we use $k(\epsilon) = \hbar^{-1}\sqrt{2m(E_{\rm F} + \epsilon)} \approx k_{\rm F} + \epsilon/(\hbar v_{\rm F})$. As a result, by keeping only leading terms in ϵ the integral can be written as

$$\begin{aligned}\chi(r) &= -N_0\Delta \frac{\sin(k_{\rm F} r)}{k_{\rm F} r}\int_0^{\omega_{\rm D}} {\rm d}\epsilon \frac{\cos\left(\frac{\epsilon r}{\hbar v_{\rm F}}\right)}{\sqrt{\epsilon^2 + \Delta^2}} \\ &= -N_0\Delta \frac{\sin(k_{\rm F} r)}{k_{\rm F} r}\int_0^{\omega_{\rm D}/\Delta} {\rm d}x \frac{\cos\left(\frac{\Delta r x}{\hbar v_{\rm F}}\right)}{\sqrt{1 + x^2}} \\ &= -N_0\Delta \frac{\sin(k_{\rm F} r)}{k_{\rm F} r}\left\{\int_0^{\infty} {\rm d}x \frac{\cos\left(\frac{r x}{\pi\xi_0}\right)}{\sqrt{1 + x^2}} - \int_{\omega_{\rm D}/\Delta}^{\infty} {\rm d}x \frac{\cos\left(\frac{r x}{\pi\xi_0}\right)}{\sqrt{1 + x^2}}\right\},\end{aligned} \tag{15.183}$$

where we have introduced the coherence distance $\xi_0 = \hbar v_{\rm F}/\pi\Delta$, which measures the size of the pair. Apart from the conventional factor of π, the expression of ξ_0 has a very clear physical meaning. Pairs form with electrons belonging to an energy shell around the Fermi surface of amplitude 2Δ such that the associated momentum uncertainty $\Delta/v_{\rm F}$ implies the spatial extent of the wave packets $\hbar/(\Delta/v_{\rm F})$.

As a result, the BCS pair wave function turns out to be

$$\lim_{r\to\infty} \phi^{\rm BCS}(r) = C\left[\frac{1}{r}K_0\left(\frac{r}{\pi\xi_0}\right) + \frac{\Delta\pi\xi_0}{\omega_{\rm D}r^2}\sin\left(\frac{\omega_{\rm D}r}{\Delta\pi\xi_0}\right)\right]. \tag{15.184}$$

K_0 is the modified Bessel function with asymptotic exponential decrease and originates from the first integral, whereas the second term, proportional to $1/r^2$ and originating from the second integral, is usually neglected because of the smallness of the ratio $\Delta/\omega_{\rm D}$. However, we stated that ODLRO automatically leads to condensation for the restricted class of functions sensibly different from zero in a finite region of r. Hence, even in BCS theory, in principle, in order to show that the state $|s\rangle$ given in Eq. (15.152) leads to condensation we have to solve explicitly the eigenvalue equation for the two-partice reduced density matrix h_2 given by Eq. (15.149), considering the BCS solution (15.170) for h_1 and χ, and show that there is an eigenvalue of the order of N. This is done in Problem 15.6.

At $T = 0$, by neglecting the Hartree–Fock terms in $\nu_{\mathbf{k}}$, the ground state energy W_0 is given by Eq. (15.166) with $\nu_{\mathbf{k}} = \epsilon_{\mathbf{k}}$. It is instructive to evaluate W_0 explicitly in this simple case. After inserting the expressions (15.170) in Eq. (15.166), we get

$$\begin{aligned}
W_0 &= \sum_{\mathbf{k},\sigma}\left[\epsilon_{\mathbf{k}}\frac{1}{2}\left(1-\frac{\epsilon_{\mathbf{k}}}{E_{\mathbf{k}}}\right) - \frac{1}{4}\frac{\Delta^2}{E_{\mathbf{k}}}\right] \\
&= \sum_{\mathbf{k},\sigma}\left[\epsilon_{\mathbf{k}}\frac{1}{2}\left(1-\frac{\epsilon_{\mathbf{k}}}{|\epsilon_{\mathbf{k}}|}+\frac{\epsilon_{\mathbf{k}}}{|\epsilon_{\mathbf{k}}|}-\frac{\epsilon_{\mathbf{k}}}{E_{\mathbf{k}}}\right) - \frac{1}{4}\frac{\Delta^2}{E_{\mathbf{k}}}\right] \\
&= W_{0,\rm FG} + \sum_{\mathbf{k},\sigma}\left[\epsilon_{\mathbf{k}}\frac{1}{2}\left(\frac{\epsilon_{\mathbf{k}}}{|\epsilon_{\mathbf{k}}|}-\frac{\epsilon_{\mathbf{k}}}{E_{\mathbf{k}}}\right) - \frac{1}{4}\frac{\Delta^2}{E_{\mathbf{k}}}\right],
\end{aligned}$$

where $W_{0,\rm FG}$ is the ground state energy of the normal state, i.e. of the Fermi gas. The first term in square brackets represents the extra contribution of the kinetic energy term in the BCS ground state, whereas the second term is due to the interaction. The latter, by transforming the sum over $\mathbf{k}$ to an integral over the energy $\epsilon_{\mathbf{k}}$ and using the gap equation (15.180), is easily written as

$$-\frac{\Delta^2}{g}.$$

The first term, after passing from the sum to the integral, reads

$$\begin{aligned}
\sum_{\mathbf{k},\sigma}\epsilon_{\mathbf{k}}\frac{1}{2}\left(\frac{\epsilon_{\mathbf{k}}}{|\epsilon_{\mathbf{k}}|}-\frac{\epsilon_{\mathbf{k}}}{E_{\mathbf{k}}}\right) &= 2N_0\int_0^{\omega_{\rm D}} d\epsilon_{\mathbf{k}}\left(\epsilon_{\mathbf{k}} - \frac{\epsilon_{\mathbf{k}}^2}{\sqrt{\Delta^2+\epsilon_{\mathbf{k}}^2}}\right) \\
&= 2N_0\left(\frac{\omega_{\rm D}^2}{2} - \frac{\omega_{\rm D}}{2}\sqrt{\omega_{\rm D}^2+\Delta^2} + \frac{\Delta^2}{2}\ln\left(\frac{\omega_{\rm D}+\sqrt{\omega_{\rm D}^2+\Delta^2}}{\Delta}\right)\right) \\
&\approx 2N_0\left(-\frac{\Delta^2}{4} + \frac{\Delta^2}{2}\ln\frac{2\omega_{\rm D}}{\Delta}\right) \\
&= -N_0\frac{\Delta^2}{2} + \frac{\Delta^2}{g},
\end{aligned}$$

where in the last term we used Eq. (15.182) for the gap in order to eliminate ω_D. This term exactly compensates the contribution coming from the interacting Hamiltonian so that the first term represents the energy gain in the superconducting state, whose ground state energy finally reads

$$W_0 = W_{0,\mathrm{FG}} - N_0 \frac{\Delta^2}{2}. \tag{15.185}$$

By expressing the density of states in terms of density and Fermi energy $N_0 \sim n/E_F$, we see that the energy gain is of order $n(\Delta/E_F)\Delta$, which means that a fraction Δ/E_F of the electrons gain an energy Δ. This is in agreement with the fact that, according to Eq. (15.170), $h_{\mathbf{k}}$ differs from its Fermi gas value only in a region of width Δ around the Fermi surface.

In the remainder of this section we shall briefly sketch how to extend the BCS theory at finite temperature referring for a detailed derivation to the original paper (Bardeen et al. (1957)).

The ground state $|s\rangle$ is the vacuum state for the quasiparticles obtained by the Bogoliubov–Valatin transformations defined in Eqs. (15.173) and (15.174) with the Hamiltonian given by Eq. (15.175). At $T = 0$ no quasiparticles are present so that the quasiparticle number operator applied to the BCS ground state is zero. At $T \neq 0$ the entropy of a Fermi gas of quasiparticles is known in terms of their occupation number $f_{\mathbf{k}}$ (see Eq. (7.7)). So the free energy is $F = \langle H_{\mathrm{lin}} \rangle - TS$:

$$F = \overline{W}_0 + \sum_{\mathbf{k},\sigma} [\overline{E}_{\mathbf{k}} f_{\mathbf{k}} + k_B T (f_{\mathbf{k}} \ln f_{\mathbf{k}} + (1 - f_{\mathbf{k}}) \ln(1 - f_{\mathbf{k}}))],$$

where $f_{\mathbf{k}} = \langle \xi^\dagger_{\mathbf{k},\sigma} \xi_{\mathbf{k},\sigma} \rangle$ and the average is a quantum statistical average. Furthermore, we shall indicate whenever necessary the quantities at $T \neq 0$ with a bar. Minimization of F with respect to the distribution function $f_{\mathbf{k}}$ leads to the Fermi statistics of the quasiparticles:

$$f_{\mathbf{k}} = \langle \xi^\dagger_{\mathbf{k},\sigma} \xi_{\mathbf{k},\sigma} \rangle = (e^{\beta \overline{E}_{\mathbf{k}}} + 1)^{-1}. \tag{15.186}$$

From the transformations (15.173) and (15.174), it is easy to show that

$$\begin{aligned}
\langle a^\dagger_{\mathbf{k},\sigma} a_{\mathbf{k},\sigma} \rangle &= \langle ((1 - h_{\mathbf{k}})^{1/2} \xi^\dagger_{\mathbf{k},\sigma} + \sigma h_{\mathbf{k}}^{1/2} \xi_{-\mathbf{k},-\sigma})((1 - h_{\mathbf{k}})^{1/2} \xi_{\mathbf{k},\sigma} + \sigma h_{\mathbf{k}}^{1/2} \xi^\dagger_{-\mathbf{k},-\sigma}) \rangle \\
&= (1 - h_{\mathbf{k}}) \langle \xi^\dagger_{\mathbf{k},\sigma} \xi_{\mathbf{k},\sigma} \rangle + h_{\mathbf{k}} \langle \xi_{-\mathbf{k},-\sigma} \xi^\dagger_{-\mathbf{k},-\sigma} \rangle \\
&= (1 - h_{\mathbf{k}}) f_{\mathbf{k}} + h_{\mathbf{k}} (1 - f_{\mathbf{k}})
\end{aligned}$$

and hence

$$\begin{aligned}
1 - 2\langle a^\dagger_{\mathbf{k},\sigma} a_{\mathbf{k},\sigma} \rangle &\equiv (1 - 2\overline{h}_{\mathbf{k}}) \\
&= 1 - 2\,[(1 - h_{\mathbf{k}}) f_{\mathbf{k}} + h_{\mathbf{k}} (1 - f_{\mathbf{k}})] \\
&= (1 - 2h_{\mathbf{k}})(1 - 2f_{\mathbf{k}}) \\
&= (1 - 2h_{\mathbf{k}}) \tanh\left(\frac{1}{2}\beta \overline{E}_{\mathbf{k}}\right)
\end{aligned} \tag{15.187}$$

and similarly

$$\overline{\chi}_{\mathbf{k}} = \chi_{\mathbf{k}}(1 - 2f_{\mathbf{k}}) = \chi_{\mathbf{k}} \tanh\left(\frac{1}{2}\beta \overline{E}_{\mathbf{k}}\right).$$

Further minimization of the free energy with respect to $h_{\mathbf{k}}$, $\chi_{\mathbf{k}}$ leads to the same expression (15.169) with barred quantities together with the constraint Eq. (15.160). The new quasiparticle energies $\overline{E}_{\mathbf{k}}$ have the same expression as Eq. (15.171) with the self-consistent potentials $\overline{\nu}$ and $\overline{\mu}$. $\overline{\nu}$ and $\overline{\mu}$ are obtained from the definition (9.70) and (15.161) of ν and μ with the temperature-dependent $\overline{h}$ and $\overline{\chi}$ appearing in place of h and χ. Under the hypothesis of Eq. (15.177) and assuming the Hartree–Fock potential $U = 0$, the gap equation reads

$$\frac{1}{N_0 g} = \int_0^{\omega_D} d\epsilon_{\mathbf{k}} \frac{1}{\sqrt{\epsilon_{\mathbf{k}}^2 + \overline{\Delta}^2}} \tanh\left(\frac{1}{2}\beta\sqrt{\epsilon_{\mathbf{k}}^2 + \overline{\Delta}^2}\right). \tag{15.188}$$

At the critical temperature the gap energy vanishes:

$$\begin{aligned}\frac{1}{N_0 g} &= \int_0^{\omega_D} d\epsilon_{\mathbf{k}} \frac{1}{\epsilon_{\mathbf{k}}} \tanh\left(\frac{1}{2}\beta_c \epsilon_{\mathbf{k}}\right) \\ &= \ln\frac{\beta_c\omega_D}{2} \tanh\left(\frac{\beta_c\omega_D}{2}\right) - \int_0^{\beta_c\omega_D/2} dx \frac{\ln x}{\cosh^2(x)}.\end{aligned} \tag{15.189}$$

For $\beta_c\omega_D \gg 1$, one can approximate the hyperbolic tangent with unity and send to infinity the upper limit of the second integral with the result[11]

$$k_B T_c = 1.13\omega_D e^{-1/N_0 g} \tag{15.190}$$

and the isotope effect follows via ω_D which is proportional to $M^{-1/2}$, M being the crystal ion mass of the host material.[12] The explicit behavior of $\overline{\Delta}$ as a function of the temperature near T_c can be obtained by expanding the gap equation for small values of Δ (see Problem 15.7) and reads

$$\overline{\Delta} = \sqrt{\frac{8}{7\zeta_R(3)}}\pi k_B T_c \sqrt{\frac{T_c - T}{T_c}}, \tag{15.191}$$

which, but for a normalization factor, coincides with the Landau–Ginzburg result.

It must be noticed that the coherence distance $\xi_0 = \hbar v_F/\pi\Delta$ introduced in (15.183), as is shown in Appendix K, apart from a numerical factor of order unity, coincides with the zero-temperature correlation length introduced in Eq. (15.121), when considering the fluctuations of the order parameter. Coming back to the microscopic expressions of Appendix K and keeping in mind the expression for $\xi(T)$ in (15.121), we obtain

$$\xi(T) = \xi_0 \frac{\pi}{e^{\gamma_E}} \sqrt{\frac{7\zeta_R(3)}{49} \frac{T_c}{|T - T_c|}} \approx 0.37\xi_0 \sqrt{\frac{T_c}{|T - T_c|}}. \tag{15.192}$$

[11] The second integral, when the upper limit goes to infinity, is equal to $-\ln(4e^{\gamma_E}/\pi) \equiv I$, $\gamma_E \approx 0.5772$ being the Euler constant.

[12] This is the typical dependence of an harmonic oscillator. This is not surprising since, as discussed in Section 7.6, phonons arise from the quantization of the normal modes of a system of coupled oscillators.

A specific heat jump as given by Eq. (15.106) is also obtained. This latter result can be easily obtained by using the entropy of a gas of quasiparticles with energy $\overline{E}_{\mathbf{k}}$:

$$S = -2k_{\mathrm{B}} \sum_{\mathbf{k}} \left(f_{\mathbf{k}} \ln f_{\mathbf{k}} + (1 - f_{\mathbf{k}}) \ln(1 - f_{\mathbf{k}})\right). \tag{15.193}$$

The specific heat in the superconducting phase is then given by

$$\begin{aligned} c_{\mathrm{s}} &= T \frac{\partial S}{\partial T} \\ &= 2k_{\mathrm{B}} \sum_{\mathbf{k}} \frac{e^{\beta \overline{E}_{\mathbf{k}}}}{(1 + e^{\beta \overline{E}_{\mathbf{k}}})^2} \beta^2 \left(\overline{E}_{\mathbf{k}}^2 + \frac{\beta}{2} \frac{\partial \overline{E}_{\mathbf{k}}^2}{\partial \beta} \right). \end{aligned} \tag{15.194}$$

In the limit of low temperature we recover the decreasing exponential term $c_{\mathrm{s}} \sim e^{-\beta \Delta(0)}$ as $T \to 0$. The integral over the energy $\epsilon_{\mathbf{k}}$ can be extendend to infinity because of the fast convergence of the integral. We get

$$\begin{aligned} c_{\mathrm{s}} &\approx 2k_{\mathrm{B}} N_0 \beta^2 \Delta^2 \int_{-\infty}^{\infty} d\epsilon_{\mathbf{k}} e^{-\beta \sqrt{\Delta^2 + \epsilon_{\mathbf{k}}^2}} \\ &\approx 2k_{\mathrm{B}} N_0 \beta^2 \Delta^2 e^{-\beta \Delta} \int_{-\infty}^{\infty} d\epsilon_{\mathbf{k}} e^{-\beta \epsilon_{\mathbf{k}}^2 / 2\Delta} \\ &= 2k_{\mathrm{B}} N_0 \sqrt{2\pi} \Delta \left(\frac{\Delta}{k_{\mathrm{B}} T} \right)^{3/2} e^{-\Delta / k_{\mathrm{B}} T}. \end{aligned}$$

Near T_{c}, in the difference between c_{s} and the specific heat in the normal phase the only term which survives reads

$$\begin{aligned} c_{\mathrm{s}} - c_{\mathrm{n}} &\to_{T \to T_{\mathrm{c}}} k_{\mathrm{B}} \beta^3 N_0 \int_{-\infty}^{\infty} d\epsilon_{\mathbf{k}} \frac{e^{\beta \overline{E}_{\mathbf{k}}}}{(1 + e^{\beta \overline{E}_{\mathbf{k}}})^2} \frac{\partial \Delta^2(T)}{\partial \beta} |_{T = T_{\mathrm{c}}} \\ &\approx k_{\mathrm{B}} \beta^3 N_0 \int_{-\infty}^{\infty} d\epsilon_{\mathbf{k}} \frac{e^{\beta \overline{\epsilon}_{\mathbf{k}}}}{(1 + e^{\beta \overline{\epsilon}_{\mathbf{k}}})^2} \frac{\partial \Delta^2(T)}{\partial \beta} |_{T = T_{\mathrm{c}}} \\ &= \frac{8}{7 \zeta_{\mathrm{R}}(3)} \pi^2 N_0 k_{\mathrm{B}}^2 T_{\mathrm{c}}, \end{aligned} \tag{15.195}$$

where in the last step we use the result (15.191). The result (15.195) provides a finite jump in the specific heat at T_{c} in agreement with the behavior of a second order transition in the mean-field approximation.

The results obtained so far make clear that we can establish a direct connection between the phenomenological Landau–Ginzburg theory of Section 15.7 and the generalized Hartree–Fock type of approach. One can show that the Landau–Ginzburg equation (15.101) for the order parameter can be derived from the microscopic theory on the assumption of being near the critical temperature where the magnitude of the order parameter is small. This is shown in Appendix K. Indeed, it is possible to show that the self-consistency BCS equation for the gap (15.188), for T close to T_{c} reproduces the phenomenological Landau–Ginzburg equation and provides an explicit microscopic expression for the Landau coefficients reported in Appendix K. The ratio $a'^2 T_{\mathrm{c}}/b$ expressing the specific heat jump in the Landau–Ginzburg theory (see Eq. (15.106)) agrees with the BCS value in

Eq. (15.195). Notice that the derivation of the kinetic coefficient γ of Eq. (15.126) requires the generalization of the time-dependent Hartree–Fock equation (9.76) in the presence of anomalous pairing. In terms of the generalized single-particle density matrix, the time-dependent self-consistent Hartree–Fock equations in matrix form become

$$i\hbar \frac{\partial K}{\partial t} = [M, K] . \qquad (15.196)$$

This last equation is the starting point of derivation of the Landau–Ginzburg equations performed in Appendix K.

15.10 Crossover between BCS and Bose–Einstein condensation

For a gas of fermions of mass m interacting via an attractive potential $V(\mathbf{x})$, great attention has been recently devoted to the smooth crossover between the two limits of superfluidity: the BCS limit with Cooper pairs in the spin singlet and $l = 0$ s-state extending over a distance ξ_0 and the BEC limit of almost point-like Bose particles made of strongly bound pairs (Leggett (1980); Nozières and Schmitt-Rink (1985)). This problem, which seemed to be merely academic, is now experimentally accessible in ultracold atomic Fermi gas in a trap. (For general reviews see Ketterle and Zwierlein (2008); Giorgini et al. (2008); Bloch et al. (2008).) The interaction among atoms is controlled via the Feshbach resonance (Chin et al. (2010)), which can span continously between the two limits. It is beyond the scope of this book to illustrate the extraordinary technical achievements and the theoretical improvements to reach this goal, for which we refer to the reviews mentioned above.

We only briefly consider the crossover following Nozières and Schmitt-Rink, which has been extended to the system in a trap (see e.g. Perali et al. (2004)).

The BCS weak coupling limit is valid for $n\xi_0^3 \gg 1$, i.e. when the extension of the pair wave function ξ_0 is much larger than the average interparticle distance k_F^{-1}. The homogeneous system is characterised by a Fermi surface with an occupation $h_{\mathbf{k}}$, Eq. (15.170) smeared in a region of the size of the gap Δ around the Fermi energy. The two characteristic scales ξ_0 and k_F^{-1} are both present. Then, according to the argument mentioned at the beginning of Chapter 12, the critical temperature (15.190) can be cast in the form

$$T_c = \frac{1.13}{\pi} \frac{T_F}{k_F \xi_0} \ll T_F . \qquad (15.197)$$

In the opposite limit of strong coupling, bound pairs do not overlap. The internal structure of the pair wave function has no role and the Fermi surface loses its meaning. We do not have to worry about the exclusion principle, at least for dilute systems, and the factor $k_F\xi_0$ must disappear from T_c. The critical temperature is the one of a gas of $N/2$ point-like bosons of mass $M = 2m$:

$$T_c = \frac{2\pi}{\zeta_R(3/2)^{2/3}} \frac{\hbar^2}{2m} \left(\frac{n}{2}\right)^{2/3} = \pi \left(\frac{\sqrt{2}}{3\pi^2 \zeta_R(3/2)} \right)^{2/3} T_F = 0.21 T_F , \qquad (15.198)$$

where we have used Eq. (7.42) for T_c and Eq. (7.25) for E_F. For a Fermi gas in a trap in the BEC limit, Eq. (7.42) for T_c and Eq. (7.25) for E_F have to be replaced by Eq. (7.61) and Eq. (7.69) respectively, leading to

$$\frac{T_c}{T_F} = \frac{1}{2^{1/3}(6\zeta_R(3))^{1/3}} = 0.41. \tag{15.199}$$

We now discuss how this crossover can be realized within the BCS formalism. For simplicity we consider $\nu_{\mathbf{k}} = \epsilon_{\mathbf{k}}$, with no Hartree–Fock terms in the self-consistency equation (15.169), whose formal solutions are Eq. (15.170). By eliminating $E_{\mathbf{k}}$, the solutions (15.170) and Eq. (15.163) for $\mu_{\mathbf{k}}$ can be combined as

$$\left(\frac{\hbar^2 k^2}{m} - 2\mu\right)\chi_{\mathbf{k}} = -(1 - 2h_{\mathbf{k}})\sum_{\mathbf{k}'} V(\mathbf{k},\mathbf{k}')\chi_{\mathbf{k}'}, \tag{15.200}$$

where the chemical potential μ has to be self-consistently determined by the conservation (15.153) of the number N of particles in terms of $h_{\mathbf{k}}$.

In the dilute strong coupling limit of tightly bound pairs $h_{\mathbf{k}} \approx \chi_{\mathbf{k}}^2$, i.e. half of the occupation of fermions coincides with the pair occupation. Hence, the constraint Eq. (15.168) implies $h_{\mathbf{k}} \ll 1$. Since all the Fermi particles are bound in isolated pairs, at leading order, Eq. (15.200) reduces to the Schrödinger equation for a single bound pair:

$$\left(\frac{\hbar^2 k^2}{m} - 2\mu\right)\chi_{\mathbf{k}} = -\sum_{\mathbf{k}'} V(\mathbf{k},\mathbf{k}')\chi_{\mathbf{k}'}, \tag{15.201}$$

with $2\mu \equiv -\epsilon_0$ as the eigenvalue. Notice that a bound state requires $2\mu < 0$ as for a Bose gas, so that ϵ_0 represents the energy necessary to break the pair.

In this case the Fourier component (Eq. (N.54) of the answer to Problem 15.6) of the eigenfunction of the two-particle reduced density matrix h_2 is $\phi(\mathbf{k}) = \chi_{\mathbf{k}}/(N/2)^{1/2}$. The bound pairs are described by the composite bosons

$$b_{\mathbf{q}} = \sum_{\mathbf{k}} \phi(\mathbf{k}) a^\dagger_{\mathbf{k}+\mathbf{q}/2,\uparrow} a^\dagger_{-\mathbf{k}+\mathbf{q}/2,\downarrow}. \tag{15.202}$$

The wave function $\phi(\mathbf{k})$ extends in real space over a distance $\sim \epsilon_0^{-1/2}$. Hence, for sufficiently strong interaction when the size of the pair is much smaller than the interparticle distance, such that the pairs do not overlap, i.e. $(\hbar^2/m)^{3/2} n \epsilon_0^{-3/2} \ll 1$, the system can be regarded as a gas of independent bosons. In the ground state of this gas, all the composite bosons are coherently in the same $\mathbf{q} = 0$ state with creation operator $b_0^\dagger = \sum_{\mathbf{k}} \phi(\mathbf{k}) a^\dagger_{\mathbf{k}\uparrow} a^\dagger_{-\mathbf{k}\downarrow}$:

$$|\mathrm{BEC}\rangle = \exp\left((N/2)^{1/2} b_0^\dagger\right)|0\rangle. \tag{15.203}$$

By writing the exponential of the sum over $\mathbf{k}$ as the product over $\mathbf{k}$ of the exponential and observing that $(a^\dagger_{\mathbf{k}\uparrow} a^\dagger_{-\mathbf{k}\downarrow})^2$ vanishes because $a^\dagger_{\mathbf{k}\uparrow}$ and $a^\dagger_{-\mathbf{k}\downarrow}$ are fermion creation operators, the state $|\mathrm{BEC}\rangle$ can be written in the BCS form with $(1 - h_{\mathbf{k}})^{1/2} \sim 1$ and $h_{\mathbf{k}}^{1/2} \sim \chi_{\mathbf{k}}$.

Either by increasing the density or by decreasing the bounding interaction, the pairs start to overlap and the effect of the fermion exclusion principle reveals itself, and we cannot write the ground state in the form Eq. (15.203). In the weak coupling limit the BCS state

(15.152) is recovered with the full constraint (15.168). Hence, the mean-field BCS ground state describes limits of both weak and strong coupling and may be used as an approximate interpolation scheme in order to study the evolution from one limit to the other.

The interpolation between the above limits can be made explicit by solving Eq. (15.200) for an attractive separable potential:

$$V(\mathbf{k},\mathbf{k}') = -g\frac{1}{\sqrt{1+k^2/k_0^2}}\frac{1}{\sqrt{1+(k')^2/k_0^2}}. \tag{15.204}$$

The equations for the chemical potential (15.153) fixing the number of particles and the equation for the gap become (see Problem 15.8 for details)

$$n = \int \frac{\mathrm{d}^3k}{(2\pi)^3}\left(1-\frac{\epsilon_{\mathbf{k}}}{E_{\mathbf{k}}}\right), \tag{15.205}$$

$$\frac{1}{g} = \int \frac{\mathrm{d}^3k}{(2\pi)^3}\frac{1}{2E_{\mathbf{k}}}\frac{1}{1+k^2/k_0^2}, \tag{15.206}$$

where $\epsilon_{\mathbf{k}} = \hbar^2k^2/2m - \mu$ and $E_{\mathbf{k}} = \sqrt{\epsilon_{\mathbf{k}}^2 + \mu_{\mathbf{k}}^2}$ with $\mu_{\mathbf{k}} = \Delta/\sqrt{1+k^2/k_0^2}$.

Similarly, Eq. (15.201) for a single bound pairs reads (see Problem 15.8)

$$\frac{1}{g} = \frac{m}{\hbar^2k_0^2}\sum_{\mathbf{k}}\frac{1}{\left(\frac{k^2}{k_0^2}+\frac{m\epsilon_0}{\hbar^2k_0^2}\right)\left(1+\frac{k^2}{k_0^2}\right)}. \tag{15.207}$$

We note that the separable potential (15.204) has an s-wave bound state with binding energy $\epsilon_0 = -2\mu > 0$ ($\mu < 0$ since the bound state eigenvalue is 2μ), which can be determined by transforming the sum into an integral over momentum. The integral is elementary and the solution of Eq. (15.207) is

$$\epsilon_0 = \frac{\hbar^2k_0^2}{m}\left(\frac{g}{g_{\mathrm{c}}}-1\right)^2. \tag{15.208}$$

The derivation of the last equation shows that, contrary to the Cooper problem of Appendix J, there exists a minimum potential strength $g_{\mathrm{c}} = 4\pi\hbar^2/(mk_0)$ for the existence of a bound state in isolation, corresponding to the solution of Eq. (15.207) with $\epsilon_0 = 0$. As we shall see, the evolution from the weak to the strong coupling limit is smooth in the mean-field approximation with g_{c} setting the scale over which the crossover occurs.

The wave function of the bound state is obtained by anti-Fourier transforming $\chi_{\mathbf{k}}$ given by Eq. (15.200):

$$\chi_{\mathbf{k}} = -\frac{\text{constant}}{\left(\frac{\hbar^2k^2}{m}+\epsilon_0\right)\sqrt{1+\frac{k^2}{k_0^2}}},$$

which shows that the extension of the bound state is proportional to $\epsilon_0^{-1/2}$ for g near g_{c} and to k_0^{-1} for larger values of g. Hence, the above discussion is valid in the strong coupling provided we are in the dilute limit $nk_0^{-3} \ll 1$, which for a system in a trap is the one relevant for the experiments. When $nk_0^{-3} \gtrsim 1$, the pairs overlap for large values of g and one cannot regard the system as made of free point-like bosons anymore.

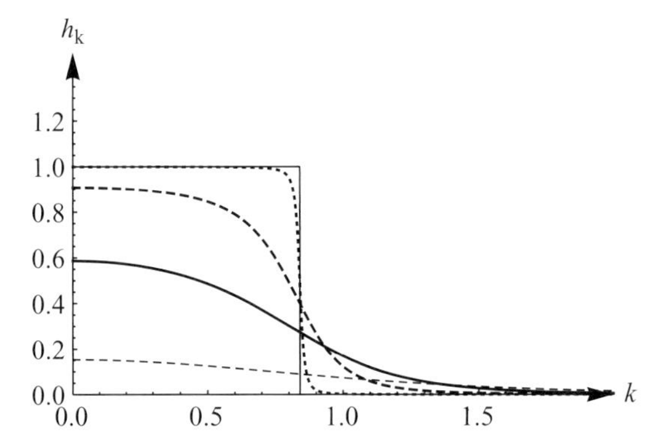

Fig. 15.4 The ground state occupation $h_{\mathbf{k}}$ at fixed density $n = 0.02$ in units of k_0^3 as a function of k in units of k_0 for four values of g_c/g: $g_c/g = 3.06$ (thin line), $g_c/g = 1.7$ (dotted line), $g_c/g = 0.94$ (dashed line), $g_c/g = 0.63$ (thick line) and $g_c/g = 0.3$ (thin dashed line).

In the dilute limit it is then possible to study both the BCS limit and the BEC of bound pairs as well as the crossover between the two by solving the coupled Eqs. (15.205) and (15.206). Notice that the convergence of the integral in Eq. (15.206) is now ensured by the factor containing k_0. By measuring momenta in units of k_0, energies in units of $\hbar^2 k_0^2/2m$ and the density in units of k_0^{-3}, one sees that the above equations depend only on the dimensionless parameter g/g_c.

In the dilute limit it is possible to follow the smooth evolution of the occupation $h_{\mathbf{k}}$ from the almost step-like function for small coupling $g/g_c \ll 1$ to the smeared function in k-space for $g/g_c \gg 1$, reflecting the point-like form in real space of the pair wave function. In Fig. 15.4 this evolution is presented by solving the above equations numerically for several values of g/g_c. In particular, even for $g_c/g = 0.94$, i.e. close to the critical value of the coupling for the problem of the bound state, the behavior of $h_{\mathbf{k}}$ shows no anomalous behavior.

At finite temperature the problem is more involved. In particular we sketch the evolution of the critical temperature and of the temperature for breaking the pairs as depicted in Fig. 15.5. The two asymptotic limits are clear. In the weak coupling $g/g_c \ll 1$ the standard BCS critical temperature Eq. (15.197) is recovered. The vanishing of the order parameter is determined by the presence of single-particle Fermi excitations, the particle–hole excitations created by breaking the loose BCS pairs, and the pair-breaking temperature T_{pair} coincides, at finite temperature, with T_c. Upon increasing the coupling, by confining to the finite-temperature BCS equations, T_{pair} and T_c separate from one another since at large coupling T_{pair} must asymptotically coincide with the pair-breaking temperature of unbinding a single tight pair, $T_{\text{single pair}} \sim \epsilon_0/k_B$. In the strong coupling limit $g/g_c \gg 1$, T_c, given by Eq. (15.198), is determined by the onset of a macroscopic coherent occupation of the lowest single-particle state of the bosonic tightly bound $N/2$ pairs. T_c is much larger than than the BCS value (15.197) but is much lower than the pair-breaking temperature $T_{\text{single pair}}$, which in this limit according to Eq. (15.208) grows as the square of the coupling g.

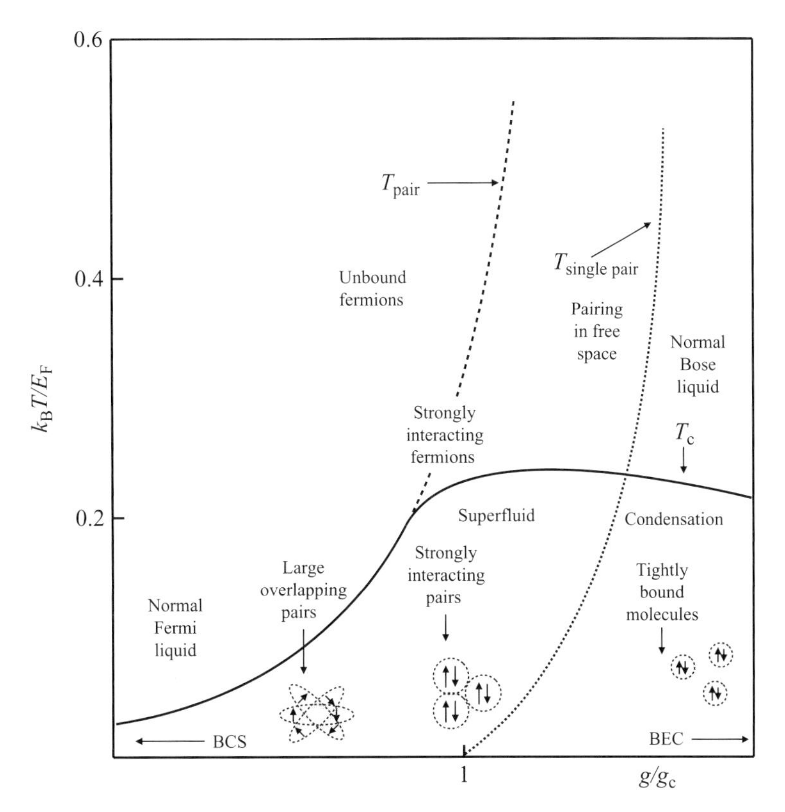

Fig. 15.5 Qualitative phase diagram of the BCS to BEC crossover as a function of the dimensionless parameter g/g_c and the temperature T in units of E_F. The full line gives (qualitatively) the evolution of the critical temperature between the BCS and BEC limits. The pictures show the evolution from the BCS limit with large, overlapping Cooper pairs to the BEC limit with tightly bound molecules. In the intermediate region there are strongly interacting pairs whose size is comparable to the mean interparticle distance $1/k_F$. The dotted and dashed lines indicate the pair-breaking temperatures $T_{single\ pair}$ and T_{pair} at which fermions unbound in free space and in the BCS ground state, respectively.

However, by decreasing g, the pairs start to overlap and interact as hard core bosons. At finite temperature they can be excited from the condensate without being broken. T_c is determined by the thermal excitation of these bosonic collective modes (15.202) at finite $\mathbf{q}$. By considering the effect of these pair fluctuations (not included in the BCS approach at finite temperature) on the thermodynamic potential and on the chemical potential, T_c, the onset of the pair condensation, is determined by the singularity in the pair response function at $\mathbf{q} = 0$ and $\omega = 0$, the susceptibility of the present transition. At the simplest approximation, the pair fluctuations are included (see e.g. Perali et al. (2004)) as the Gaussian corrections around the mean-field BCS solution. The numerical solution of the self-consistent coupled equations for the critical temperature and the chemical potential at fixed density gives, for a dilute interacting Fermi gas, the qualitative phase diagram of Fig. 15.5. The maximum in T_c around $g = g_c$ has been attributed to the residual repulsive interaction effects among the

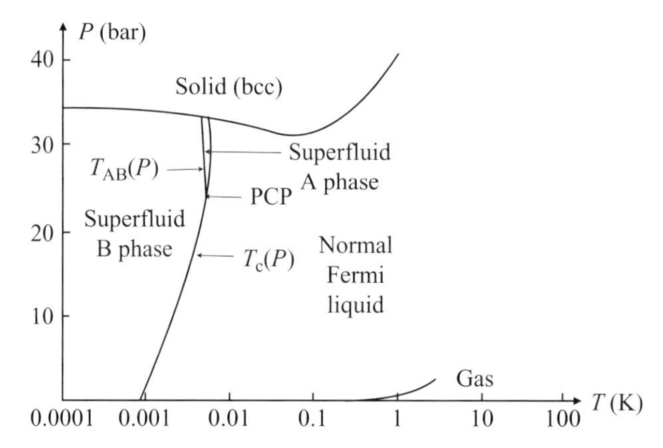

Fig. 15.6 Phase diagram of ^{3}He with the two superfluid phases A and B in the absence of a magnetic field. The dark full line $T_c(P)$ marks the critical temperature for the onset of superfluidity at various pressures. T_{AB} is the first order separation line between the two superfluid phases. PCP is the polycritical point.

bound pairs. In any case a controlled quantitative evaluation of the curve $T_c(g/g_c)$ remains a major theoretical challenge.

15.11 Superfluid ^{3}He

Soon after the BCS theory was developed, it became evident that the pairing phenomenon could occur in other degenerate Fermi systems, in particular in liquid ^{3}He (Pitaevskii (1959); Emery and Sessler (1960); Brueckner et al. (1960); Anderson and Morel (1961)). In this case, however, there are no lattice phonons to act as mediators of the required effective attraction among fermions, which has therefore to be provided by the interparticle interaction. Estimates of the critical temperature were wildly ranging from 10^{-1} to 10^{-2} Kelvin, discouraging the experimentalists who could not find any sign of it in this range of temperature. The superfluid phases of ^{3}He were instead discovered in the millikelvin region while studying the magnetic properties of the solid along the melting curve (Osheroff et al. (1972a)). During compressional cooling along this curve, an anomaly was found at around $T \approx 2.6$ mK and 34.36 bar, named the A point, followed by another one, point B, at lower temperature, around $T \approx 2$ mK in the pressurization curve as a function of time, with uncertainty on temperatures due to the difficulty in defining the absolute temperature scale. These anomalies were initially attributed to nuclear spin ordering in the solid. Soon after, NMR experiments (Osheroff et al. (1972b)) attributed these anomalies to the liquid phase as a signature of the onset of two different superfluid phases (as shown in the phase diagram of Fig. 15.6) (for a full account of this discovery see e.g. Wheatley (1975)). Phase B extends below the melting curve down to the lowest temperature and pressure. Phase A occupies a triangular region between the A and B points on the melting curve and a polycritical point at around 21 bar on the critical $T_c(P)$ line for the onset of superfluidity, which ends at the A point on the melting curve. The polycritical point is the crossing point between $T_c(P)$ and the transition line $T_{AB}(P)$ between the A and B phases. More surprising is the effect of an external field on the phase diagram. In the presence of an external magnetic field, T_c is split into two values, with their difference proportional to the magnetic field. A new phase named A_1 appears as a wedge between the normal and the A and B phases. The transition line T_{AB} between the A and B phases does not intersect the line $T_c(P)$ at the polycritical point anymore, i.e. the phase A extends to lower pressures by increasing the magnetic field (see Fig. 1 of Wheatley (1975)).

The superfluidity of the two new phases has been established, for instance, as for ^{4}He, via vibrating-wire experiments. The inferred viscosity drops by three orders of magnitude in the A phase and, according to the two-fluid model, the normal density decreases by lowering the temperature in the A phase (Alvesalo et al. (1973)). The ratio between the superfluid and the normal density is linearly vanishing as the temperature approaches the critical temperature, as required by the mean-field theory.

Comprehensive reviews on superfluid ^{3}He exist both on experiments (Wheatley (1975)) and on theory (Leggett (1975)) and more recently (Vollhardt (1998); Lee and Leggett (2011)).

We discuss here only the main properties suitable for identifying the nature of the superfluid transition and to reveal the new paired states of the superfluid phases occurring in ^{3}He with respect to the s-wave electron pairing of ordinary superconductors. We do not consider many interesting effects like the spin dynamics, orientational effects and, above all, the results of combining the Landau quasiparticle Fermi-liquid theory of Chapter 12 with superfluidity as developed mainly by Leggett (1965, 1975), which give rise to molecular field corrections in the thermodynamic quantities.

Specific heat measurements give clear indications on the nature of the single particle excitation spectrum: (i) for normal ^{3}He with the linear-in-T behavior (as seen in Chapter 12), (ii) in superfluid ^{4}He with the low temperature phononic and higher temperature rotonic, Eq. (15.8), contributions, and (iii) in ordinary superconductors with the exponential behavior signalling an energy gap, Eq. (15.194). Moreover, a λ-like singularity at the superfluid transition suggested a non-mean-field two-component order parameter for ^{4}He, whereas a discontinuity at T_c for ordinary superconductors supported a mean-field second order phase transition from the normal metal. For superfluid ^{3}He, it turned out to be very difficult to measure heat capacity at these low temperatures due to the small heat differences involved and the uncertainty in the absolute temperature scale at these low temperatures. Nevertheless, a discontinuity in the specific heat at T_c according to a mean-field second order phase transition was established along $T_c(P)$ from the melting pressure down to 24 bar (see Fig. 15.7(a) and (b), respectively). The $T_{AB}(P)$ line instead seems to be first order, as for a transition between two states with two developed order parameters in different symmetries, which cannot evolve smoothly into one other.

The magnetic field effects provide more specific information on pairing. The susceptibility in the A phase is constant and nearly equal to the value of the normal liquid χ_N and the A phase must be in an equal-spin-pairing state. In the B phase the nuclear magnetic susceptibility χ_B is reduced with respect to the normal phase value, decreases with the

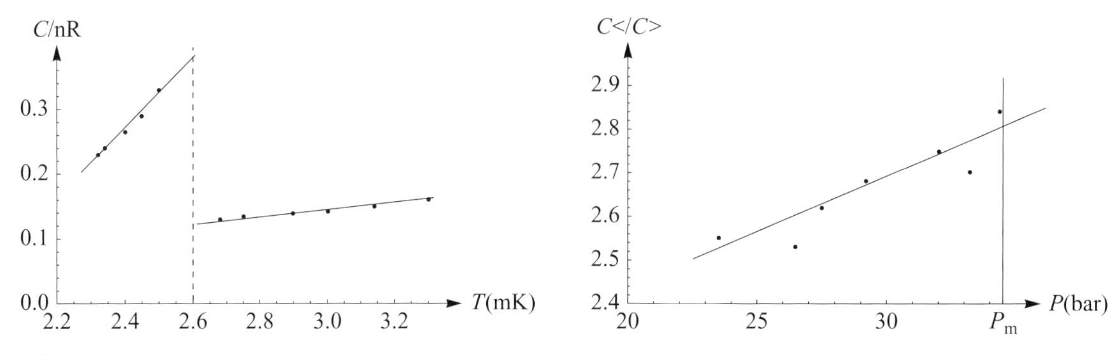

Fig. 15.7 (a) Molar specific heat relative to the gas constant R for liquid ^{3}He at pressure 33.4 bar near the second order transition. (b) Ratio $C_</C_>$ of the specific heat of liquid ^{3}He just below to just above the second-order transition as a function of pressure with P_m the melting pressure. From Figs. 5 and 6, respectively, in Wheatley (1975). (After Webb et al. (1973).)

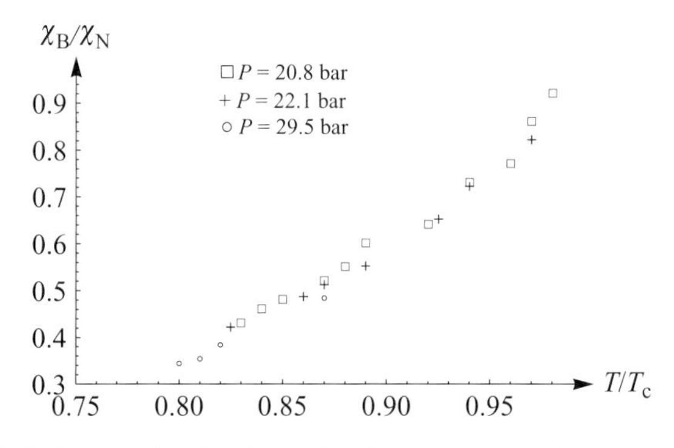

Fig. 15.8 Spin susceptibility of ^{3}He B-phase as a function of the reduced temperature T/T_c at various pressures in a field of 378 Gauss (after Paulson et al. (1974)). The measures at higher pressures stop at lower fractions of $T_c(P)$ since T_{AB} becomes significantly different from $T_c(P)$.

temperature and extrapolates to a finite value as $T \to 0$, in contrast with the results of a singlet pairing. This is shown for instance by the data of χ_B/χ_N in a field of 378 G reported at various pressures ($13.45 \leq P \leq 29.5$ bar) in Fig. 15.8 as function of the reduced temperature $T/T_c(P)$ (Paulson et al. (1974)). The data fall on an almost unique function of T/T_c and the reduced susceptibility decreases to a few percent for points far below the melting pressure, when the B phase approaches the second order transition line from the normal phase $T_c(P)$ (Wheatley (1975)).

In ordinary superconductors, the attraction mediated by lattice phonons leads to isotropic pair states with spin singlet and s-wave angular momentum $\psi = \psi_0(|\uparrow\downarrow\rangle - |\downarrow\uparrow\rangle)$. In ^{3}He, as mentioned above, a pair instability must be produced by an attraction between the fermionic particles, actually between quasiparticles, since the normal phase is a Landau

Fermi-liquid phase. At these low temperatures (millikelvin region) the quasiparticles are stable entities (their lifetime, Eq. (12.1), is becoming infinitely long). The derivation of the effective interaction between quasiparticles from interparticle interaction is rather involved, as is evident from the spread of the theoretical values of the evaluated critical temperature, which were obtained in the sixties by the various authors. We can qualitatively assume (Leggett (1975)) that some main features of the interatomic potential are maintained in the quasiparticles' effective potential.[13] In particular, the interatomic potential of ^{3}He is strongly repulsive at short distances ($r < 2.5$ Å) thus impeding the isotropic s-wave pairing. The potential becomes weakly attractive at a distance $r_0 \sim 3$ Å. To optimize the benefit of the attraction, the relative distance of the partners of the pairs must be of the order of r_0 i.e. $k_{\mathrm{F}}r_0 \sim 1$ and finite relative angular momentum l. The effective interaction V can be decomposed in terms of Legendre polynomials $P_l(\cos(\theta))$,

$$V(|\mathbf{k}-\mathbf{k}'|) = V(k, k', \cos(\theta)) = \sum_l (2l+1)V_l(k, k')P_l(\cos(\theta)), \qquad (15.209)$$

where θ is the angle between $\mathbf{k}$ and $\mathbf{k}'$ and the coefficients V_l are given by

$$V_l(k, k') = \int \frac{\mathrm{d}\Omega}{4\pi} V(|\mathbf{k}-\mathbf{k}'|)P_l(\cos(\theta)). \qquad (15.210)$$

l is even for a spin singlet and odd for a spin triplet. One should determine which V_l is most attractive out of $l = 1, 2, 3$ and then evaluate the critical temperature. V_l can be taken, instead, as a phenomenological parameter to be derived from the critical temperature rather than vice versa. In general, for $l > 0$, the gap energy has nodes at points or lines on the Fermi surface, which may lead to a power law behavior of the low temperature specific heat rather than exponential as for $l = 0$. The pair wave function is anisotropic and vanishes at zero relative distance as $(kr)^l$ (see Problem 15.9) and the effect of the repulsion is reduced. The above mentioned experimental indication of equal spin pairing for the A phase is in favor of an $l = 1$ spin triplet pairing, odd in the orbital variables as required by the Pauli principle. Indeed, liquid ^{3}He, as we have seen in Chapter 12, has a strong Stoner enhancement in the spin susceptibility and spin fluctuations tend to favor the $l = 1$ with respect to the $l = 2$ pairing, because the latter is a spin-singlet. For $l = 1$ the pair wave function is in general given by a linear superposition of spin ($S = 1$) and orbital angular ($l = 1$) momentum states:

$$\psi = \psi_{l,l_z}^{S_z=1}|\uparrow\uparrow\rangle + \psi_{l,l_z}^{S_z=0}(|\uparrow\downarrow\rangle + |\downarrow\uparrow\rangle) + \psi_{l,l_z}^{S_z=-1}|\downarrow\downarrow\rangle. \qquad (15.211)$$

Altogether nine ($(2l+1)(2S+1)$) complex parameters would in principle be required to describe this intrinsically anisotropic pair state. The broken symmetry in the superfluid transition of ^{3}He is more complex than for an isotropic structureless order parameter, as for the s-wave superconductors, where the gap $\mu_{\mathbf{k}}$ does not depend on the direction of

[13] One could alternatively obtain information on the quasiparticles effective interaction from the Fermi liquid response functions as developed in Chapter 12.

k. Besides the gauge invariance, rotational symmetry in the spin and orbital space can be involved.

At the time of its discovery, the expectations for a superfluid phase in ^{3}He with $l = 1$ p-state can be summarized in two proposals having the total angular momentum $S + l = 0$: (i) the Anderson–Morel state (Anderson and Morel (1961)) with the property of the A phase of equal spin state pairing with total spin $S = 1$, spin projection $S_z = \pm 1$ and equal angular momentum $l = 1$ with projection $l_z = \mp 1$; and (ii) the Balian–Werthamer state (Balian and Werthamer (1963)) with $l = 1$ and all three components of the spin triplet, with the properties of the B phase.

In the equal spin pairing of Anderson and Morel, the up and down spin particles are paired independently and the BCS ground state Eq. (15.152) can now be generalized as

$$\begin{aligned}\Psi &= \Psi_\uparrow \Psi_\downarrow \\ &\equiv \prod_{\mathbf{k}} \left((1 - h_{\mathbf{k}\uparrow})^{1/2} + (h_{\mathbf{k}\uparrow})^{1/2} a^\dagger_{\mathbf{k}\uparrow} a^\dagger_{-\mathbf{k}\uparrow}\right)\left((1 - h_{\mathbf{k}\downarrow})^{1/2} + (h_{\mathbf{k}\downarrow})^{1/2} a^\dagger_{\mathbf{k}\downarrow} a^\dagger_{-\mathbf{k}\downarrow}\right)|0\rangle,\end{aligned} \tag{15.212}$$

with $\sqrt{h_{\mathbf{k}\uparrow}} = -\sqrt{h_{-\mathbf{k}\uparrow}}$. Two independent gap equations for $\mu_{\mathbf{k}\uparrow}$ and $\mu_{\mathbf{k}\downarrow}$ generalize Eq. (15.188) to include the $l = 1$ projected odd part of the interaction potential (15.209) given by $V(\mathbf{k}, \mathbf{k}') = -3g\hat{\mathbf{k}} \cdot \hat{\mathbf{k}}'\theta(\omega_\mathrm{D} - \epsilon_\mathbf{k})\theta(\omega_\mathrm{D} - \epsilon_{\mathbf{k}'})$, where $\hat{\mathbf{k}}$ is a unit vector along the direction of **k**. By making the equal spin pairing ansatz with an anisotropic gap whose symmetry corresponds to the spherical harmonics with $l_z = -1$ for the spin up and $l_z = 1$ for the spin down:[14]

$$\mu_{\mathbf{k}\uparrow} = \Delta\theta(\omega_\mathrm{D} - \epsilon_\mathbf{k})\sqrt{3/2}(\hat{k}_x - \mathrm{i}\hat{k}_y), \tag{15.213}$$

$$\mu_{\mathbf{k}\downarrow} = \Delta\theta(\omega_\mathrm{D} - \epsilon_\mathbf{k})\sqrt{3/2}(\hat{k}_x + \mathrm{i}\hat{k}_y), \tag{15.214}$$

one obtains (see Problem 15.10 for details)

$$\Delta = 2\omega_\mathrm{D} \mathrm{e}^{-1/gN_0} \times 0.94. \tag{15.215}$$

Balian and Werthamer (1963) showed that within the generalized weak coupling BCS approach any equal spin p-state would be unstable with respect to the formation of a ground state with pairs of relative angular momentum l opposite to the total spin S, i.e. zero total angular momentum J of each pair as a special case of all possible relative orientation of the two vectors. More specifically, they arranged the pair state of Eq. (15.211) for the $l = 1$ case in a two-by-two symmetric matrix proportional to a unitary matrix:

$$\hat{\mu}_\mathbf{k} = \begin{pmatrix} \mu_\mathbf{k}^{\uparrow\uparrow} & \mu_\mathbf{k}^{\uparrow\downarrow} \\ \mu_\mathbf{k}^{\downarrow\uparrow} & \mu_\mathbf{k}^{\downarrow\downarrow} \end{pmatrix}, \quad \hat{\mu}_\mathbf{k}\hat{\mu}_\mathbf{k}^\dagger = |\mu_\mathbf{k}|^2\hat{\mathbf{1}}. \tag{15.216}$$

The four entries of the above matrix correspond to the coefficients of the four two-spin states of Eq. (15.211). The unitarity condition ensures that the square of the matrix behaves

[14] The factor $\sqrt{3/2}$ has been introduced so that the solid angle average of the square modulus of the gap gives Δ^2 as in the isotropic case.

as a scalar and properly defines the quasiparticle excitation energies $E_{\mathbf{k}}^2 = \epsilon_{\mathbf{k}}^2 + |\mu_{\mathbf{k}}|^2$. The entries of the matrix $\hat{\mu}_{\mathbf{k}}$ must be written as linear combinations of spherical harmonics with $l = 1$ in the present case. According to this description the equal spin pairing state of Eqs. (15.213) and (15.214) reads

$$\hat{\mu}_{\mathbf{k}}^{\mathrm{AM}} = \Delta\sqrt{\frac{3}{2}}\begin{pmatrix} \hat{k}_x - \mathrm{i}\hat{k}_y & 0 \\ 0 & -\hat{k}_x - \mathrm{i}\hat{k}_y \end{pmatrix}. \tag{15.217}$$

In the BW state the symmetry of the state with $S_z = 0$ and $l_z = 0$ (k_z component in the $\mu_k^{\sigma\bar{\sigma}}$) has to be added:

$$\hat{\mu}_{\mathbf{k}}^{\mathrm{BW}} = \Delta\begin{pmatrix} \hat{k}_x - \mathrm{i}\hat{k}_y & \hat{k}_z \\ \hat{k}_z & -\hat{k}_x - \mathrm{i}\hat{k}_y \end{pmatrix}. \tag{15.218}$$

This state has lower energy (Balian and Werthamer (1963)). In such a state all the three spin substates have equal weight in the pairing state of Eq. (15.211). The energy gap (see the unitarity condition in Eq. (15.216), $\hat{\mu}_{\mathbf{k}}\hat{\mu}_{\mathbf{k}}^{\dagger} = \Delta^2$) and the quasiparticle energy become isotropic. In this case, with the same interaction potential used for the equal spin pairing state, it is not difficult to convince ourselves that the gap equation has the same form as in the standard s-wave BCS case with

$$\Delta = 2\omega_{\mathrm{D}}\mathrm{e}^{-1/gN_0}. \tag{15.219}$$

By using Eq. (15.185) for the ground state energy one has

$$W_0^{\mathrm{AM}} - W_0^{\mathrm{BW}} = \frac{N_0}{2}(2\omega_{\mathrm{D}}\mathrm{e}^{-1/gN_0})^2(1-(0.94)^2), \tag{15.220}$$

showing that the state (15.218) has indeed lower energy. Actually it was shown (Balian and Werthamer (1963)) that this state is an absolute minimum, although it is degenerate with respect to other solutions. We do not discuss the issue further here and refer to the already mentioned paper by Balian and Werthamer. We note that, as a consequence of the isotropy of the gap, all the spin related properties are also isotropic; the paramagnetic spin susceptibility is reduced with respect to the normal state value and decreases to a finite value as T decreases. As we have seen, these are the properties related to the B phase (Wheatley (1975)).

The A phase, when related to the Anderson and Morel state with no $S_z = 0$ component of the triplet, has the same temperature-independent susceptibility as the normal phase. The external magnetic field does not hinder the pairing between equal spin states and, vice versa, the formed pairs do not hinder the field polarization. On the other hand, the presence of an external magnetic field splits the population of the two spin components linearly in the field, the up-spin Fermi surface extends at the expense of the down-spin Fermi surface. The critical temperature is then also split as in the experiments. In the $\mathrm{A_I}$ phase, stable in the presence of an external magnetic field, only the $S_z = 1$ component survives. So far so good, however, how can the A phase be stable in the Anderson–Morel state with respect to the Balian–Werthamer state? It was shown (Anderson and Brinkman (1973)) that a feedback effect of spin fluctuations can stabilize the Anderson–Morel state. The first-order transition from the A phase to the B phase occurs with a coherent mixing of the three components of the spin triplet and a subsequent reduction of the spin susceptibility due to the presence of the $S_z = 0$ component. In the A phase, the presence of the two components $|\uparrow\uparrow\rangle$ and $|\downarrow\downarrow\rangle$ gives an intrinsic anisotropy to the spin part of the order parameter with fixed direction with respect to the angular momentum direction of all pairs. Contrary to the Balian–Werthamer state, the gap turns out to be anisotropic along the Fermi surface with two nodes and a power law behavior of the specific heat. The degeneracy of all the states with different **l** and **S** orientation may be lifted by additional forces such as dipolar forces between the nuclear spins, which leads to parallel orientation between **l** and **S**. We will, however, not proceed further in discussing more specific properties of superfluid ^{3}He.

15.12 High temperature superconductivity: a brief presentation

The discovery of superconductive materials with high critical temperatures has been the chimera of physicists for decades. Until 1986 the record value was $T_c = 23.2$ K for an alloy of composition Nb_3Ge. In that year, J. G. Bednorz and K. A. Müller found that a copper oxide compound, made of Ba-La-Cu-O, becomes superconductive in the 30 K range. The superconducting compound was later identified by them as La_2CuO_4 doped with Ba. When the insulating La_2CuO_4 compound is doped with barium, Ba^{+2} ions substitute the La^{+3} ions, providing charge carriers (holes[15]). The critical temperature for the onset of superconductivity is doping dependent and reaches its maximum at a specific doping named the optimal doping. There was soon a rush in the discovery of different families of copper oxides, the so-called cuprates, with increasing maximum critical temperature well above the nitrogen boiling point ($T = 77$ K), far more accessible than the helium required to cool materials below the critical temperatures of traditional superconductors. To mention a few of them: the lanthanum family LSCO ($La_{2-x}Sr_xCuO_4$), where the insulating La_2CuO_4 compound is doped with strontium, in 1987 (Bednorz et al. (1987)) with $0 \le x \le 0.3$ and $T_c = 42$ K at $x = 0.15$ (besides the already mentioned barium, the dopant can also be calcium); the YBCO family ($YBa_2Cu_3O_{6+x}$) in 1987 (Wu et al. (1987); Hor et al. (1987)) with $0.6 \le x \le 1$ and $T_c = 92$ K at $x \approx 0.95$; the BSCCO family with maximum $T_c =$ 110 K in $Bi_{2-x}Sr_xCa_2Cu_3O_{10+\delta}$. Bismuth can be replaced with thallium or mercury (Gao et al. (1994)) with the highest critical temperature $T_c = 135$ K for the $m = 3$ family $HgBa_2Ca_{m-1}Cu_mO_{2m+2+\delta}$. The record is $T_c = 164$ K under pressure (31 GPa). Several other different compounds have been added to the high T_c-superconductors after the cuprates. Besides MgB_2 and carbon-based compounds (the prototypical case being C_{60} doped with potassium (Hebard et al. (1991))), we mention the iron arsenide based compounds found to be superconductive in 2008 in fluorine-doped LaFeAsO (Kamihara et al. (2008)). These

[15] We recall that, in the band theory of solids, since a completely filled band does not carry any current, the current due to a partially filled band can be described as due to fictitious particles with positive charge residing in those band levels which are unoccupied by the electrons. These fictitious particles are called holes.

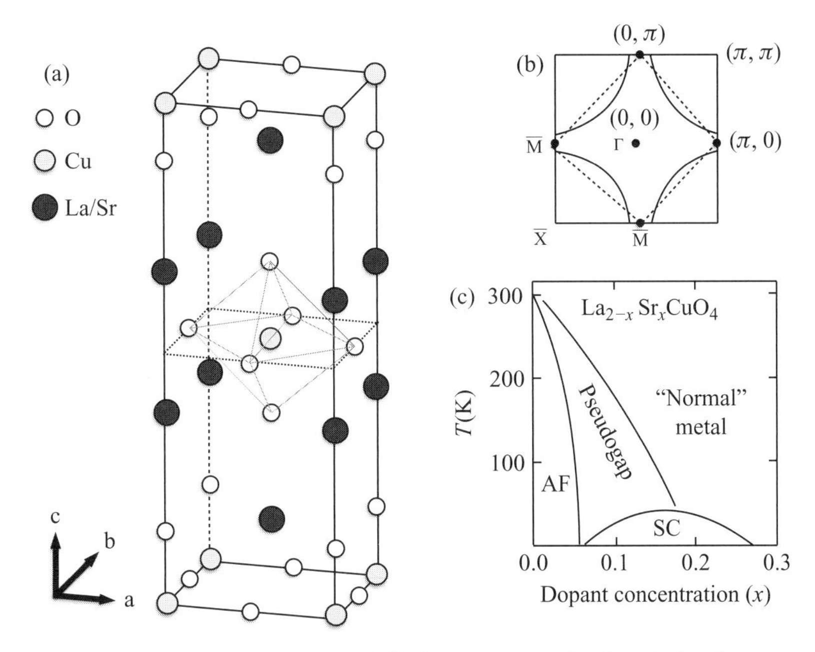

Fig. 15.9 (a) The structure of the $La_{2-x}Sr_xCuO_4$ compound. (b) The open Fermi surface for underdoped $La_{2-x}Sr_xCuO_4$. (c) Temperature versus doping schematic phase diagram of $La_{2-x}Sr_xCuO_4$. It includes the antiferromagnetic (AF), the pseudogap, the superconducting (SC) and metallic phases.

compounds are made of layers of iron and a pnictogen like arsenic or phosphorus.[16] They have the second highest critical temperature to date, up to around 55 K in $PrO_{1-x}F_xFeAs$ (Ishida et al. (2009)). For these systems, with multi-band complex electronic structure, the pairing mechanism seems to be of a magnetic nature and the pair interaction to be mediated by the spin fluctuations produced by the superimposed or nearby magnetic order (see e.g. the review by Kordyuk (2012)). In the following we shall only refer to the cuprates to point out the basic problems opened by their phenomenology in this rapidly evolving field.

All cuprate compounds have a layered structure characterized (see, e.g. the LSCO structure in Fig. 15.9) by weakly coupled CO_2 copper–oxygen planes of Cu-atoms each being surrounded by four oxygens. The interstitial insulating layers, besides oxygen which lies at the vertex of a pyramid having the in-plane oxygens as corners (it is then denoted apical oxygen not always present in other materials), are usually made of rare earth elements like lanthanum, yttrium, barium, etc. depending on the various families of cuprates, thus reducing the coupling among the CuO_2 planes. The relevant electronic structure is therefore quasi-two-dimensional. When the chemical composition is such that the CuO_2 planes contain one hole per unit cell, consisting of one Cu and two O atoms, these materials are antiferromagnetic insulators, despite the fact that they have an odd number of electrons per unit cell, which should, according to band theory, make them metals. For instance, in the

[16] A pnictogen is a member of group 5A of the periodic table including nitrogen, phosphorus, arsenic, antimony and bismuth.

undoped parent compound La_2CuO_4, the Cu atoms have the level $3d^9$ occupied with one hole per Cu site and so does the unit CuO_2 cell in the plane. In contrast to band theory, the system is a Mott insulator (i.e. an insulator due to strong interaction among the charge carriers) with antiferromagnetic order, thus suggesting the presence of strong correlation among the holes, i.e. the insulating property is due to the strong local repulsion among the electrons (holes) and not to the filling of the bands. The undoped system, in first approximation, can therefore be schematized as a planar lattice with one hole per Cu site described in terms of the Hubbard model introduced in 1963 (Hubbard (1963)) just to represent electron correlation effects:

$$H = -t \sum_{\langle ij \rangle, s} a^{\dagger}_{i,s} a_{j,s} + \text{h.c.} + U \sum_{i} a^{\dagger}_{i,\uparrow} a_{i,\uparrow} a^{\dagger}_{i,\downarrow} a_{i,\downarrow}. \tag{15.221}$$

The first term is the kinetic energy which favors the charge mobility via the hopping t from nearest neighbor sites, while the local interaction term U, representative of the local Coulomb repulsion, makes unfavorable the double occupation of a single site with holes of opposite spin, thus hindering their motion. When U prevails the system becomes a (Mott) insulator, the holes tend to localize and their spins are antiferromagnetically ordered.[17] Chemical doping, e.g. the substitution of the trivalent La^{+3} with the x divalent Sr^{+2} (or the excess of oxygen in other cuprates, e.g. YBCO in which the hole-doping level is, however, different from x) introduces x additional holes in the CuO_2 planes. By increasing the doping percentage x, the antiferromagnetic order is rapidly destroyed and then the insulator transforms into a metal, albeit a bad metal (due to the paucity of charge carriers). This bad metal, nevertheless, becomes a superconductor when the temperature is lowered below the critical temperature $T_c(x)$ (see Fig. 15.9), which has a dome shaped form with the maximum T_c at an optimal doping.

The doping can also add x electrons to the CuO_2 planes as in the $A_{2-x}Ce_xCuO_4$ (A =Nd, La, Sm, Pr) compound. In this case the critical temperature is usually lower and the antiferromagnetic order is more robust and persists up to the doping at which superconductivity is already set in (see Fig. 15.10).

In the metal phase, a strong anisotropy is measured in the electrical resistivity which, in the CO_2 planes, is orders of magnitude smaller than the transversal resistivity. (See, e.g., Fig. 15.11 for single crystals of LSCO (Nakamura and Uchida (1993)); the transport mainly occurs in the CuO_2 planes.) The metal in question must be schematized as a doped quasi-two-dimensional Mott insulator. One has to face the still open problem of strongly correlated quasi-two-dimensional (electron) holes and its consequences. We shall give in the following only few hints on this open problem.

In ordinary superconductors, it is usually assumed that superconductivity sets in as an instability of the metallic phase, well described in terms of a normal Fermi liquid with a well defined Fermi surface. (See Appendix J for the Cooper instability of the Fermi sea.) In the

[17] The origin of the antiferromagnetic order can be physically understood by considering that two nearest neighbor sites occupied with particles with opposite spin may go to a virtual state where one site remains empty and the other is doubly occupied. Because of the Pauli principle, double occupancy is only possible for particles of opposite spin.

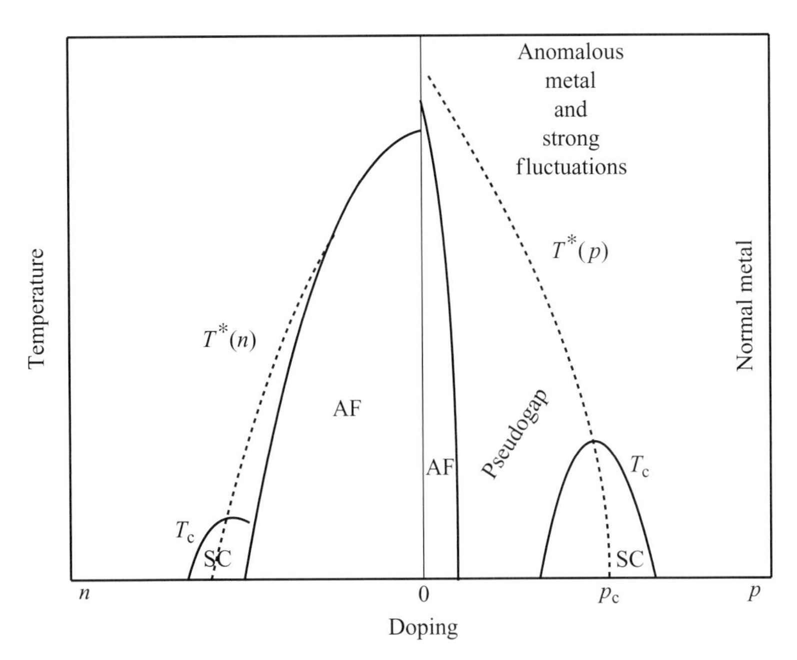

Fig. 15.10 Schematic phase diagram of high T_c superconductors in arbitrary units: temperature versus hole p (electron n) doping. SC (AF) stands for the superconducting (antifarromagnetic) phase limited by the critical (full) line. The dotted lines mark the onset T of the pseudogap regions. p_c indicates the possible "hidden" quantum critical point.

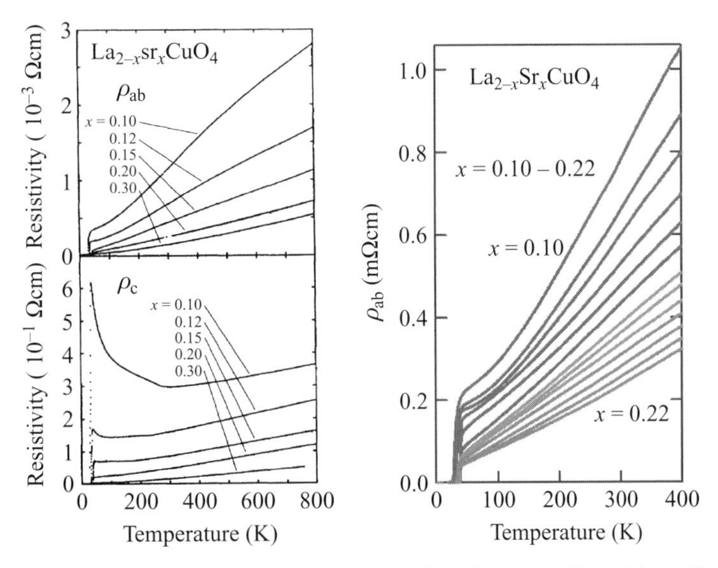

Fig. 15.11 Temperature dependence of the in-plane (upper left and right panels) and transverse (lower left panel) resistivity at various doping compositions in the metallic phase of LSCO left panel from Nakamura and Uchida (1993). The right panel is adopted with permission from Ando et al. (2004). Copyright by the American Physical Society.

cuprates the metallic phase is not in line with the predictions of normal Fermi-liquid theory. The highly doped material is well described as a Fermi liquid. By lowering the doping the in-plane resistivity starts to deviate from the Fermi liquid T^2-behavior. Nearby optimal doping ($x \sim 0.16$–0.18 for LSCO) above T_c, the resistivity is linear in T up to a very high temperature (see e.g. Fig. 15.11 and Ando et al. (2004) for various cuprate families). This behavior is clearly in conflict with the T^2 behavior of the quasiparticle scattering in the normal Fermi liquid (see Chapter 12) and cannot be ascribed to phonon scattering which is linear only for temperatures well above the high Debye temperature. A pseudogap (Timusk and Statt (1999); Timusk (2003)) opens in the electronic structure around the Fermi surface for a doping level less than the optimal doping and temperatures below a crossover temperature $T^*(x)$, well above $T_c(x)$. The pseudogap opening is accompanied by a decreasing density of states. For $T > T^*(x)$ the in-plane resistivity is still linear and a downward deviation (see Fig. 15.11) of the in-plane resistivity from linearity in T (Ito et al. (1993); Ando et al. (2004)) is usually identified as marking the T^* at a given x (see Fig. 15.10). In the pseudogap region, due to the opening of the gap, the branches of the Fermi surface (see Fig. 15.9(b)) shrink into Fermi arcs, which become shorter upon reducing doping starting from the M-points. Does a new state of matter appear below $T^*(x)$ in this quasi-two-dimensional strongly correlated electronic system?

Coming to the superconducting phase, the other valuable description of ordinary superconductors, the BCS theory, is also questioned. In traditional superconductors, the necessary attraction for the formation of electron pairs is mediated by phonons, the elastic excitations of the crystal lattice, as indicated by the isotope effect, whereby the critical temperature for the onset of superconductivity depends on the isotopic mass of the lattice ion and therefore on the lattice excitations. The metallic and the superconductive phases have two well separated energy scales. The electronic energy scale in metals, the Fermi energy E_F, is much greater than the Debye frequency ω_D related to pair formation. Moreover, the correlation length $\xi_{0,LG}$ extends over several lattice sites and according to the Ginzburg criterion the order parameter has negligible fluctuations. All this makes the mean-field BCS theory applicable. In the new superconductors T_c is higher and T_F is lower due to the paucity of charge carriers. The isotope effect is mainly present in the pseudogap onset temperature $T^*(x)$ rather than in T_c (see Raffa et al. (1998); Williams et al. (2000); Rubio Temprano et al. (2000)). The order parameter is still formed by spin singlet pair states, as in BCS theory, with a momentum structure of the pair wavefunction of the same symmetry as the $d_{x^2-y^2}$ wave function, with nodes (see Fig. 15.12) along the diagonal of the Brillouin zone (Γ-X direction), rather than s-wave symmetry of ordinary superconductors (see e.g. the review by Tsuei and Kirtley (2000)). A structure factor $w = \cos(k_x) - \cos(k_y)$ appears in the gap equation for a pure two-dimensional lattice and, below T_c, the Fermi surface reduces to four points along the diagonals. The correlation length $\xi_{0,LG}$ extends over a few lattice sites with strong phase fluctuations. The pair phenomenon is intermediate between the two limiting cases of BEC with local tightly bound pairs and BCS superconductivity. In the first case, the pair formation temperature can be very different from the pair condensation (see Fig. 15.5); in the second case they both coincide with the superconductivity critical temperature. No marked energy scale separation between the normal and the superconductive

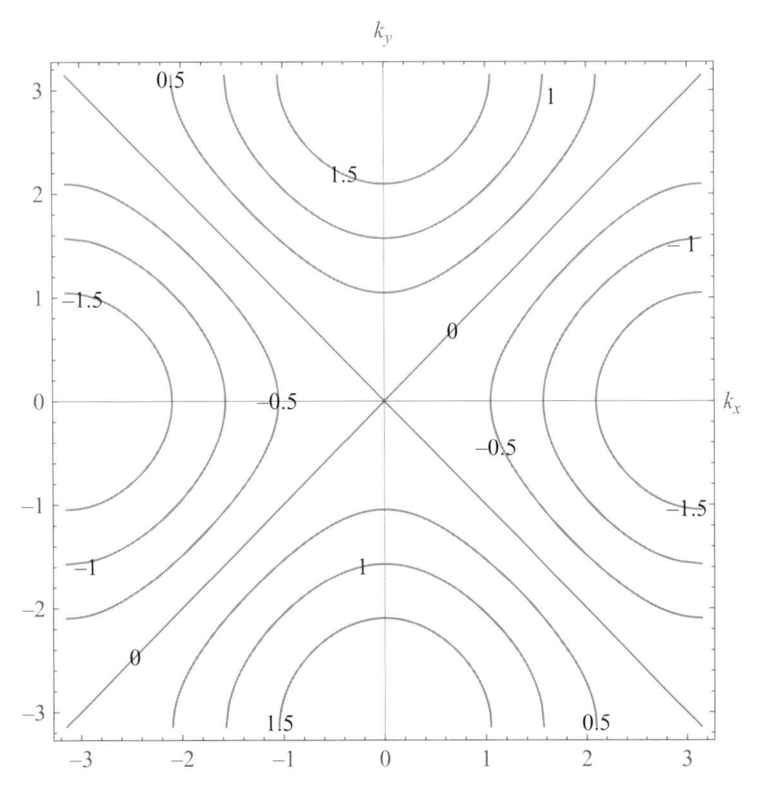

Fig. 15.12 Contour plot of the d-wave symmetry of the pairing function $w = \cos(k_x) - \cos(k_y)$.

states, together with strong almost local pairing with d-wave symmetry, makes mean-field theory questionable and is in favor of a non-phononic pairing mechanism of electronic nature, although the fermion–phonon interaction must be involved, due to its effect on T^*.

An international debate thus ensued among theoretical physicists, whether a strongly correlated hole (or electron) system, with an occupation of approximately one hole per lattice site of a quasi-two-dimensional system can become a superconductor; how repulsion could generate the attraction necessary for pairing; what the new paired state will be, and by what mechanisms it might be formed. At the same time, the other, perhaps even more stimulating, theoretical problem presented itself: understanding the anomalous metallic phase to which the superconductivity was linked and how it evolves from the Fermi liquid valid at high doping. Both problems, despite an intense research effort which has already lasted more than a quarter of century, still remain unsolved and no commonly accepted comprehensive paradigm has yet appeared.

At the moment of writing this book, two distinct lines of thought are debated. On the one hand pairing is mostly due to instantaneous interactions arising from the short-range magnetic correlations as a by-product of strong electron–electron correlations ("Mottness" physics, see e.g. Anderson (2007); Lee et al. (2006) and references therein). In this framework the pseudogap results from the nearly local uncondensed Cooper pairs. The other point of view is instead based on the idea that an elusive form of (local and dynamical) order is competing with the Fermi liquid and superconducting states and T^* would be representative of the temperature for the onset of this order, which also could not be fully developed as a true phase transition (for reviews see, e.g., Vojta (2009); Seibold et al. (2012) and references therein).

Along the first line it has been hypothesized (see e.g. Randeria and Trivedi (1998)) that, because of correlation in quantum systems, the pseudogap is due to formation, at T^*, of tight pairs in singlet states with no coherence. T^* would coincide with T_{pair} of Fig. 15.5 and superconductivity would in this case appear at a lower temperature when these quasi-bosonic preformed pairs condense. In this quasi-bosonic limit, the loss of the Fermi liquid description and of the concomitant Fermi surface in the pseudogap allows for a bound pair formation temperature different from the condensation of pairs.

Alternatively, the pseudogap state may appear at $T^*(x)$ in relation to an ordered or quasi-ordered state of varied nature. A tendency towards states with modulated spin and charge densities is a common feature of cuprates and may originate from the tendency of strongly correlated electron systems to phase separate in charge-rich and charge-poor domains. Strong correlation hinders the mobility of charge carriers and reduces the kinetic energy and its homogenizing effect: the system becomes more prone to phase separate. Unless the counter-ions are mobile, the macroscopic electronic phase segregation is prohibited by the Coulomb energy cost. The system compromises by self-organizing in charge (spin) modulated states. In particular, most frequently unidirectional stripe-like ordering (and sometimes its nematic precursor) appears in underdoped systems as a consequence of the expulsion of holes from the antiferromagnetic background of the undoped system (Zaanen and Gunnarsson (1989); Machida (1989); Emery and Kivelson (1993); Löw et al. (1994)). The onset of unidirectional order develops also around optimal doping as a charge density wave from an instability of the homogeneous correlated Fermi liquid along a critical line $T_{\text{CDW}}(x)$ of the order of T^* (Castellani et al. (1995); Andergassen et al. (2001)) and eventually evolves into well formed stripes by underdoping. A true thermodynamic transition can be smeared out either by quenched disorder or by the low dimensionality. Another proposal of an ordered state (Varma (1999); Simon and Varma (2002) and references therein) which does not break translational invariance but time reversal symmetry, is the loop-current order on lattice plaquettes of the CuO_2 planes. An ordering of nematic type has also been obtained as a consequence of an instability of the symmetric Fermi surface of the type discussed in Section 12.2.3 (Yamase et al. (2005)). Starting from stripes (first observed as a static order at $x = 1/8$ in $La_{2-x}Ba_xCuO_4$ and $La_{2-x-y}Nd_ySr_xCuO_4$ (Tranquada et al. (1995, 1996, 1997)) and then at various dopings as incommensurate spin modulation in a long list of experiments), all these orderings have been partially observed in some doping regions, in some experiments and in some families (Tranquada (2012); Abbamonte et al. (2005); Chang et al. (2012); Ghiringhelli et al. (2012); Comin et al. (2014); da Silva Neto et al. (2014); Wu et al. (2011); Fauqué et al. (2006); Kaminski et al. (2002), to mention a few of them). To extend the study of the unusual normal state and uncover the eventual order underlying superconductivity, a strong magnetic field has been often applied to destroy superconductivity. The $T^*(x)$ line extends in this case down to zero kelvin at a doping around optimal doping, uncovering in this way a quantum criticality (i.e. criticality around zero kelvin) normally hidden by the superconducting dome, as was suggested theoretically (Varma (1993); Castellani et al. (1995, 1997, 1998); Tallon and Loram (2001)). The strong fluctuations associated with the onset of this order could mediate a retarded effective interaction with the double result of spoiling the Fermi liquid and producing strong pairing while entering the pseudogap region below $T^*(x)$. Magnetic fluctuations, which persist above the disruption, by doping, of the long range antiferromagnetic order, have also been considered as good candidates to produce the pseudogap phase as a modified Fermi liquid (for a review see Abanov et al. (2003)) and as mediators of incommensurate charge density wave order (Wang and Chubukov (2014)). To further complicate the picture, in the presence of a high magnetic field, quantum oscillations, proper to a Fermi liquid, have been observed in electronic responses of $YBa_2Cu_3O_{6+x}$ (LeBoeuf et al. (2007); Sebastian et al. (2008)). Unfortunately, because each family is accessible by different experimental techniques, it is then hard to obtain a coherent set of experiments for samples of the same cuprate family. A unified scenario of high temperature superconductors remains a challenge.

15.13 Problems

15.1 Derive (15.7) and (15.8).

15.2 Show that the ground-state wave function of N bosons interacting via a pair potential is real, positive and totally symmetric.

15.3 Find the space dependence in a superconductor residing in the $x > 0$ half space.

15.4 Estimate the fraction of the condensate (15.58).

15.5 Consider the Hamiltonian for N interacting spinless bosons:

$$H = \int d\mathbf{x}\hat{\psi}^\dagger(\mathbf{x})\left[-\frac{\hbar^2}{2m}\nabla^2 + V_{\text{ext}}(\mathbf{x})\right]\hat{\psi}(\mathbf{x}) + \frac{1}{2}\int d\mathbf{x}\int d\mathbf{x}'\hat{\psi}^\dagger(\mathbf{x})\hat{\psi}^\dagger(\mathbf{x}')V(\mathbf{x}-\mathbf{x}')\hat{\psi}(\mathbf{x})\hat{\psi}(\mathbf{x}')$$

with the contact effective interaction (15.69).

- Calculate the Hartree–Fock energy with the trial N-body wave function (15.76).
- Minimize it with respect to the single-particle wave function $\phi_0(\mathbf{x})$ to obtain the Gross–Pitaevskii equation.

15.6 Show that the BCS theory yields an eigenvalue of order N for the the two-particle reduced density matrix.

15.7 Determine the temperature dependence of the gap $\Delta(T)$ close to T_c by expanding the gap equation (15.188) and making use of the identity

$$\tanh\left(\frac{x}{2}\right) = \sum_{l=0}^{\infty} \frac{4x}{x^2 + (2l+1)^2\pi^2}.$$

15.8 Derive the form of the BCS gap equation (15.200) for the separable potential of Eq. (15.204).

15.9 Derive the radial function behaviour at small relative distance for two interacting particles of the previous problem.

15.10 Derive the result (15.215) for the anisotropic pairing specified by Eqs. (15.213–15.214).

16 Scaling theory

16.1 The scaling laws

Mean-field theory shows that, close to the critical point, various physical quantities have singular power-law behaviors, whose exponents are integers or ratios of integers as a direct consequence of the expansion in powers of the order parameter of the thermodynamic potential and/or of the equation of state.[1] In Table 16.1 we report the synoptic view of the definition of the critical exponents and their values in mean-field theory, discussed in Chapters 8 and 13.

The *melting* of a phase is generally due to the increase of thermal fluctuations as the temperature T is increased. Near the critical points, according to the Ginzburg criterion, indeed, one expects strong effects of the fluctuations of the order parameter for $d < 4$ and sizeable deviations from the mean-field results. This is confirmed by exact and numerical solutions, whenever they are available (see Table 16.2), and by experiments (see Table 16.3) since the above physical quantities, although showing a power law behavior, have critical indices different from those evaluated in mean-field theory.

The measured indices and those obtained by exact or numerical calculations depend on the dimensionality of space, d, and the number of components, n, of the order parameter. From a general point of view, the role of d and n was already emphasized by the absence of a phase transition at finite temperature for $d = 1$ when $n = 1$ and $d \leq 2$ for $n \geq 2$. In this respect the mean-field theories are too universal and their predictions do not depend on n and d. All the systems, near criticality, can, instead, be grouped in different universality classes, insofar the critical indices are the same for completly different systems, provided that d, n and any other underlying symmetry are the same. For instance, axial ferromagnets with one-component order parameter have critical indices different from those of a planar or isotropic ferromagnet, but compatible with those of the liquid–gas transition or binary mixture, again with a scalar order parameter. Moreover, numerical calculations (Jasnow and Wortis (1968)) on the spin-infinity anisotropic Heisenberg model showed that the critical indices assume only three sets of values, corresponding to the symmetry of the Ising ($n = 1$), X-Y ($n = 2$) and Heisenberg model ($n = 3$) and do not change by varying the degree of anisotropy. Therefore, once the proper choice of the *relevant* variables is made, for example

[1] For a summary of the phenomenology of critical phenomena and the development of scaling theory (Patashinskij and Pokrovskij (1966); Kadanoff (1966)), see the classical review by Kadanoff et al. (1967), the articles by Fisher (1967) and Widom (1974) and the books by Patashinskij and Pokrovskij (1979) and Stanley (1987).

Table 16.1 The critical exponents of the power law behavior of the physical observables as $t = T - T_c$ and $h \to 0$. Here α, γ, ν and α', γ', ν' refer to the approach to the critical point from above ($t > 0$) and from below ($t < 0$), respectively.

Observable	Symbol	Power law
Order parameter	$\varphi_0(t, h=0) \sim \|t\|^{\beta}, \beta > 0$	$\beta = 1/2$
Specific heat	$c(t, h=0) \sim \|t\|^{-\alpha,\alpha'}$	$\alpha = \alpha' = 0$
Susceptibility	$\chi(t, h=0) \sim \|t\|^{-\gamma,\gamma'} \gamma, \gamma' > 0$	$\gamma = \gamma' = 1$
Critical isotherm	$\bar{\varphi}(t=0, h) \sim \|h\|^{1/\delta}, \delta > 1$	$\delta = 3$
Correlation length	$\xi(t, h=0) \sim \|t\|^{-\nu,\nu'}, \nu, \nu' > 0$	$\nu = \nu' = 1/2$
Correlation function	$f_{\varphi}(k, t=0, h=0) \sim k^{-(2-\eta)}, \eta \geq 0$	$\eta = 0$

Table 16.2 Theoretical critical indices for the exact solution of the two-dimensional Ising model, the numerical solution of the three dimensional Ising model, the anisotropic (X-Y) and isotropic (Heisenberg) models as reported in [1]Jasnow and Wortis (1968); [2]Ferrenberg and Landau (1991); [3]Holm and Janke (1993).

	Ising 2D $n=1$ $d=2$	Ising 3D[2] $n=1$ $d=3$	X-Y[1] $n=2$ $d=3$	Heisenberg[3] $n=3$ $d=3$
β	0.125	0.0326 ± 0.004	0.35 ± 0.007	0.362
α	0(ln)	0.119 ± 0.006	-0.022 ± 0.009	-0.08 ± 0.04
γ	1.75	1.2470 ± 0.003	1.32 ± 0.01	1.389
δ		4.8 ± 0.05	–	4.65 ± 0.29
ν	1	0.6289 ± 0.002	0.67 ± 0.01	0.704
η	0.25	0.024 ± 0.007	–	0.07 ± 0.06

Table 16.3 Experimental critical indices for the liquid–gas transition, the transition in uniaxial systems, the superfluid transition in ^{4}He, and the ferromagnetic transition in Fe. Data are from [1]Patashinskij and Pokrovskij (1979), [2]Lipa et al. (1996), [3]Goldner and Ahlers (1992) (second sound measurement).

	Liquid–gas[1] $n=1$ $d=3$	Uniaxial magnets[1] $n=1$ $d=2$	He $n=2$ $d=3$	Isotropic magnets[1] $n=3$ $d=3$
β	0.34 ± 0.01	0.33–0.35	–	0.38–0.39
α	$0.11 \pm 0.01/0.12 \pm 0.01$	0.08–0.1	-0.001–-0.0026[2]	-0.09–-0.12
γ	1.22 ± 0.02	1.15 ± 0.02	–	1.34–1.40
δ	4.6 ± 0.02	–		4.42–4.46
ν	0.63 ± 0.01	–	0.6705 ± 0.0002[3]	0.7 ± 0.02
η	0.05 ± 0.01	–		/

in the simplest case the deviation from the critical temperature, t, and the field, h, conjugate to the order parameter; the other parameters specifying the Hamiltonian (e.g. the anisotropy parameter) are *irrelevant* for determining the critical behavior as long as they do not assume a value which changes the symmetry of the system.[2] Different physical systems with the same symmetry of the order parameter show the same critical behavior once the proper correspondence between their relevant variables or their physical observables is made, e.g. compressibility for the gas–liquid and susceptibility for the axial ferromagnets. We can now take these observations as the universality principle concerning second order phase transitions.

Let us consider a physical system S whose relevant physical quantities are t and h, while $\{\zeta_i\}$ represents the set of variables, specifying the details of the Hamiltonian, considered to be irrelevant, such as for instance the parameters controlling the degree of anisotropy in a Heisenberg model. S' is a system whose parameters ζ_i' are different from ζ_i. According to the universality principle, if ζ_i and ζ_i' do not assume values making the symmetry of S and S' different, the critical behavior of the two systems, and therefore the dependence of the free energy on the relevant quantities, must be the same for both systems. At the same time the order parameter, the correlation function and all the other physical quantities are, at most, changed by a multiplying function depending only on the irrelevant parameters, i.e.

$$F_S(t, h; \{\zeta_i\}) = F_{S'}(t', h'; \{\zeta_i'\}), \tag{16.1}$$

$$\varphi(t, h; \{\zeta_i\}) = \lambda_\varphi(\zeta_i, \zeta_i')\varphi(t', h'; \{\zeta_i'\}), \tag{16.2}$$

$$f_\varphi(t, h; \{\zeta_i\}) = \lambda_\varphi'(\zeta_i, \zeta_i') f_\varphi(t', h'; \{\zeta_i'\}). \tag{16.3}$$

Notice that F should be invariant near criticality. The two systems S and S' must be critical simultaneously, therefore $t = 0$ and $h = 0$ imply $t' = 0$ and $h' = 0$. In addition to the universality principle, we have to consider two further ingredients to build up a phenomenological theory of the critical phenomena: the correlation distance approaching infinity at criticality and an asymptotic power law behavior of the physical quantities. By implementing these two further conditions on the universality relations (16.1), (16.2), (16.3) we build up the so-called scaling theory.

Since $\xi \to \infty$, one of the irrelevant parameters ζ_i is the unit of length l_0. For instance, for a spin model on a lattice, l_0 is the lattice spacing. In this case, the universality does not connect different physical systems when a suitable correspondence of relevant variables is established, but the relations (16.1)–(16.3) connect two pictures of the same system obtained with a different spatial resolution. Let s be the scaling factor for changing the unit length from $\zeta = l_0$ to $\zeta' = sl_0$ when going from the higher resolution to the lower one. Since at the critical point the correlation length diverges, the resolution scale does not matter anymore, in other words, the system must become scale invariant. The basic idea to obtain scaling is that the long-range correlations of the fluctuations of the order parameter are responsible for the singular critical behavior and the correlation length ξ is the only relevant length, i.e., the details of the system on short scales less than ξ are irrelevant for criticality.

[2] A less vague definition of relevant and irrelevant variable will naturally emerge in the development of the theory.

Let us now see how to translate these conjectures into more mathematical terms.

When going to a lower resolution, a measure of length L and a volume V change into L' and V':

$$L' = s^{-1}L, \; V' = s^{-d}V. \tag{16.4}$$

In the following we consider a system of units with length as the only fundamental unit. This implies that every quantity must have physical dimensions expressed in terms of powers of the unit length. We are studying the anomalous critical behavior of the observables for small deviations of T from its critical value T_c. $k_B T$, when it is multiplying the logarithm of the partition function, for each system can therefore be taken as a finite fixed value $\beta = \beta_c$. F and βF (or H and βH) have equivalent zero dimension as far as criticality is concerned.

The free energy per unit volume, f, according to Eqs. (16.1) and (16.4), scales then as

$$f(t, h, \zeta) = s^{-d} f(t'(s), h'(s), \zeta'(s)). \tag{16.5}$$

Concerning the rescaling of the relevant variables t and h, the simplest choice is that the rescaled parameters are proportional to the original ones as in Eq. (16.2) for the order parameter with $\lambda_\varphi(l_0, l_0')$ depending only on the ratio $s = l_0'/l_0$:

$$t'(s) = \lambda_t(s)t, \quad h'(s) = \lambda_h(s)h. \tag{16.6}$$

To obtain power laws for the physical quantities, one naturally assumes

$$t'(s) = s^{D_t}t, \quad h'(s) = s^{D_h}h, \tag{16.7}$$

where we have introduced D_t and D_h as the scaling dimensions of the corresponding quantities in terms of the inverse length. Equation (16.5) for the free energy per unit volume becomes

$$f(t, h) = s^{-d} f(s^{D_t}t, s^{D_h}h). \tag{16.8}$$

Analogously for the order parameter and the correlation function:

$$\varphi(t, h) = s^{-D_\varphi}\varphi(s^{D_t}t, s^{D_h}h), \tag{16.9}$$

$$f_\varphi(r, t, h) = s^{-2D_\varphi} f_\varphi(s^{-1}r, s^{D_t}t, s^{D_h}h). \tag{16.10}$$

Here, D_φ is the scaling dimension of the order parameter and $2D_\varphi$ is the corresponding dimension for the two-point correlation function of the order parameter. The physics cannot depend on the fictitious parameter s. We note, therefore, that in the above equations, the dependence on the scaling parameter s of the relevant variables must compensate for the dependence on s due to the overall rescaling of the free energy density, the order parameter and the correlation function.

The relations (16.8)–(16.10) tell us that the corresponding quantities must be homogeneous functions of t^{1/D_t} and h^{1/D_h} with the degree given by the scaling dimension d, D_φ and $2D_\varphi$. Independently from being above or below the critical temperature, the homogeneous functions may get the two alternative forms

$$f(t,h) = |t|^{d/D_t} \bar{f}_t\left(\frac{h}{|t|^{D_h/D_t}}\right) = h^{d/D_h} \bar{f}_h\left(\frac{|t|}{h^{D_t/D_h}}\right), \tag{16.11}$$

$$\varphi(t,h) = |t|^{D_\varphi/D_t} \bar{\varphi}_t\left(\frac{h}{|t|^{D_h/D_t}}\right) = h^{D_\varphi/D_h} \bar{\varphi}_h\left(\frac{|t|}{h^{D_t/D_h}}\right), \tag{16.12}$$

$$\begin{aligned} f_\varphi(r,t,h) &= |t|^{2D_\varphi/D_t} f_1\left(\frac{r}{|t|^{-1/D_t}}, \frac{h}{|t|^{D_h/D_t}}\right) \\ &= r^{-2D_\varphi} f_2\left(\frac{r}{h^{-1/D_h}}, \frac{|t|}{h^{D_t/D_h}}\right). \end{aligned} \tag{16.13}$$

The overall assumption is that all the singular behavior is contained in the power law in front of the homogeneous functions of degree zero, which are considered regular functions of their dimensionless arguments, $|t|/h^{D_t/D_h}$, $h/|t|^{D_h/D_t}$ and $r/|t|^{-1/D_t}$ or $r/h^{-1/D_h}$. We stress that the homogeneity holds close to criticality and at distances much larger than the unit length. Assuming that the long-range part of the correlation function transforms into the critical behavior of its Fourier transform at long wavelength, in **k**-space we have

$$\hat{f}_\varphi(k,t,h) = |t|^{(2D_\varphi-d)/D_t} \hat{f}_1\left(k|t|^{-1/D_t}, \frac{h}{|t|^{D_h/D_t}}\right) = k^{2D_\varphi-d} \hat{f}_2\left(k|t|^{-1/D_t}, \frac{|t|}{h^{D_t/D_h}}\right). \tag{16.14}$$

The change in the scaling dimension $2D_\varphi \to 2D_\varphi - d$ is due to the volume integration implied by the Fourier transform. We now identify the scaling dimensions in terms of critical indices. First of all D_φ and D_h, above, are not independent. In fact, by recalling that the product of any pair of thermodynamically conjugate variables must scale as an energy density, the sum of their scaling dimensions must be equal to the dimensionality of space:

$$D_\varphi + D_h = d. \tag{16.15}$$

From Eq. (16.14), in the limit $h = 0$ and $t \to 0$, we get

$$D_t = \frac{1}{\nu} = D_t^0 + \left(\frac{1}{\nu} - \frac{1}{\nu_{\text{mf}}}\right), \quad D_\varphi = \frac{d-2}{2} + \frac{\eta}{2} = D_\varphi^0 + \frac{\eta}{2}, \tag{16.16}$$

where $D_t^0 = 2$ and $D_\varphi^0 = (d-2)/2$ are the mean-field ($\nu_{\text{mf}} = 1/2$ and $\eta = 0$) scaling dimensions of t and φ. The deviations of the critical indices from the mean-field values are named anomalous dimensions. The first result follows from $k|t|^{-1/D_t}$, must be dimensionless and the only relevant scale is ξ ($\xi \equiv |t|^{-\nu}$). The second result is obtained by identifying the Fourier transform of the correlation function at $t = 0$ and $h = 0$ with its asymptotic behavior $k^{-(2-\eta)}$ ($\hat{f}_\varphi(k,0,0) \sim k^{\eta-2}$), provided that $\hat{f}_2(0,0)$ is finite. The identification of $D_t = 1/\nu$ follows also by noticing that

$$\xi' \sim |t'|^{-\nu} \sim \frac{\xi}{s} \sim \frac{|t|^{-\nu}}{s}. \tag{16.17}$$

This result does not depend on approaching the critical temperature from above or from below and therefore

$$\nu = \nu'. \tag{16.18}$$

The scaling (16.6) now implies that $|t| \sim (\xi^{-1})^{D_t}$ and in general upon a scaling transformation, a generic observable O gets the value O' given by

$$O' = s^{D_O} O, \tag{16.19}$$

where D_O is the scaling dimension associated with O. This is equivalent to expressing the quantities in terms of an inverse correlation length. The scaling dimension allows us to define in mathematical terms the relevance or irrelevance of a physical quantity. A physical quantity O is said to be *relevant* if $d > D_O > 0$. Remember that due to the relation (16.15) between the scaling dimensions of two conjugate variables, for values of the scaling dimensions outside the above range, one of the two scaling dimension would become negative. If $D_O < 0$, O is said to be *irrelevant*. When $D_O = 0$, we say that the quantity is *marginal*. Indeed, by approaching the critical point the rescaling procedure can be iterated to infinity and the variables with negative scaling dimensions disappear. On the contrary, the set of the relevant variables when vanishing determine the critical surface, i.e. the simple critical point $h = 0$ and $T = T_c$ ($t = 0$), in the case we are considering. The occasional presence of marginal variables in a few specific low-dimensional models leads to non-universal critical indices depending on their value.

Besides the homogeneity relations, the other consequence of the scaling is seven relations among the nine critical indices since only D_φ and D_t appear as independent parameters.

1. Specific heat

$$c = -T_c \frac{\partial^2 f}{\partial t^2}|_{t,h=0} \sim |t|^{d/D_t - 2} \sim |t|^{-\alpha} \Rightarrow d\nu = 2 - \alpha, \quad \alpha = \alpha'. \tag{16.20}$$

2. Order parameter

$$\varphi(t, 0) \sim |t|^{D_\varphi/D_t} = |t|^\beta \Rightarrow \beta = \nu \frac{d - 2 + \eta}{2}. \tag{16.21}$$

3. Critical isotherm ($T = T_c$)

$$\varphi(0, h) \sim |h|^{D_\varphi/D_h} \sim |h|^{1/\delta} \Rightarrow \delta = \frac{d - 2 + \eta}{d + 2 - \eta}. \tag{16.22}$$

4. Susceptibility

$$\chi_\varphi = \frac{\partial \varphi}{\partial h}|_{h=0} \sim |t|^{(D_\varphi - D_h)/D_t} \sim |t|^{-\gamma} \Rightarrow \gamma = \nu(2 - \eta), \quad \gamma = \gamma'. \tag{16.23}$$

   The physical meaning of the scaling relation $2 - \eta = \gamma/\nu$ is that the two asymptotic forms of the correlation function (16.14), $f_\varphi(k, 0, 0) \sim k^{-(2-\eta)}$ and $f_\varphi(0, t, 0) \sim |t|^{-\gamma}$, must join smoothly at $k = \xi^{-1}$, both above and below the critical temperature. This connection between the critical behavior above and below the critical point (in Fig. 16.1 region III is connected to region I via the critical region II) implies, in addition to $\nu = \nu'$ of Eq. (16.18), the equalities $\alpha = \alpha'$ and $\gamma = \gamma'$.

All the above scaling relations have been determined on the assumption that the homogeneous functions of degree zero have a finite value when evaluated either at $h = 0$ or $t = 0$, which can only be checked by microscopic calculations.

The seven scaling laws (16.18), (16.20), (16.21), (16.22), (16.23) reduce the number of independent critical indices to two. Typically, ν and η are taken as independent critical indices. Furthermore, the scaling relations can be written in the following way:

$$2 = \alpha + 2\beta + \gamma, \tag{16.24}$$

$$\beta\delta = \beta + \gamma, \tag{16.25}$$

$$\gamma = \nu(2 - \eta), \tag{16.26}$$

$$\alpha = 2 - \nu d. \tag{16.27}$$

We notice that, apart from the last relation arising from the behavior of the specific heat, the space dimension has dropped out of the scaling laws.

If the scaling theory is valid in general, it should apply also to mean-field theory, which has $\nu = 1/2$, $\eta = 0$, $\beta = 1/2$, $\gamma = 1$, $\alpha = 0$ and $\delta = 3$. We see that the first three scaling laws (16.24)–(16.26) are satisfied, while the last one, (16.27), is only satisfied when $d = 4$. This is due to the presence of so-called *dangerous irrelevant* variables. To appreciate this point, let us consider the free energy density of Landau theory. It depends on the coefficients of its expansion (13.8), $a \sim t$ and b. According to the scaling at $h = 0$, f has the general form

$$f(t, 0, b) = |t|^{d/D_t^0} \bar{f}_t\left(0, \frac{b}{|t|^{D_b^0/D_t^0}}\right), \tag{16.28}$$

with $D_t^0 = 2$ and where we added the explicit dependence on the irrelevant variable b. The latter has scaling dimension $D_b^0 = d - 4D_\varphi^0 = 4 - d$ which is negative for $d > 4$, where we used $D_\varphi^0 = (d-2)/2$ ($\eta = 0$) and the fact that the scaling dimensions of b and φ^4 must add to the space dimensionality. The scaling $\alpha = 2 - \nu d$ follows only if the function $\bar{f}_t(0, x)$ is regular when its argument goes to zero. We know, however, from the explicit form of the free energy density (13.13) in Landau theory that $\bar{f}_t(0, x) \sim 1/x$. This brings an additional dependence on $|t|$ so that

$$f \sim |t|^{\nu d + \nu D_b^0} = |t|^{4\nu}, \tag{16.29}$$

which leads finally to the scaling law

$$\alpha = 2 - 4\nu = 0. \tag{16.30}$$

$\alpha = 0$ denotes the specific heat jump at criticality, proper to the mean-field approximation. Hence, although for $d > 4$, b is irrelevant, it must be considered to get the correct scaling behavior of the relevant quantities. Then b is referred to as a dangerous irrelevant variable. We conclude that Landau theory for $d > 4$ is perfectly in agreement with the scaling laws. In general we can combine the two expressions (16.27) and (16.30) for the fourth scaling law in the form

$$\alpha = 2 - \nu \min(4, d). \tag{16.31}$$

As a further example of application of the scaling laws, let us consider the tricritical point discussed in Section 13.3, where one has $\beta = 1/4$, $\delta = 5$ and $\nu = 1/2$. By using the scaling law $\beta\delta = \beta + \gamma$, one gets $\gamma = 1$. With the scaling law $2 = \alpha + 2\beta + \gamma$ one gets $\alpha = 1/2$. Finally, with $\alpha = 2 - \nu d$, one sees that the classical behavior with $\nu = 1/2$ corresponds to $d = 3$. Also the scaling law $\gamma = \nu(2 - \eta)$ is satisfied with $\eta = 0$. Direct calcuations instead of the use of scaling give the same results. The scaling dimension of the coupling d_0 is given by $6 - 2d$, telling us that it becomes irrelevant for $d > 3$. In $d = 3$ there are logarithmic corrections to the above classical power law behaviors.

The phenomenological scaling theory is based on strong assumptions, whose validity must be confirmed by experiment and the exact and numerical solutions of specific models. The first prediction is the homogeneity of the thermodynamical relations. For instance, all the experimental results accumulated in decades for the equation of state in many systems, such as gas–liquid and magnetic transitions, collapse into the universal form (16.12). The second strong prediction is given by the seven relations among the nine critical indices, which are well satisfied by the sets of values given by a few exact solutions of models in low dimensions, such as the Ising 2D model, or by experiments such as reported in Tables 16.2 and 16.3.

An additional scaling law can be added to the seven scaling relations given above. This is the so-called Josephson scaling for the critical index ζ of the superfluid density $\rho_s \sim |t|^\zeta$ or the *stiffness* of a ferromagnet with multi-component ($n \geq 2$) order parameter (Josephson (1966)). The extra contribution to the thermodynamic potential of a uniform superfluid flow with order parameter slowly varying in phase $\varphi(\mathbf{x}) = \varphi_0 \exp(\mathrm{i}\theta(\mathbf{x}))$ is

$$\beta \Delta \tilde{F} = \frac{1}{2} \int \mathrm{d}^d x \; A_\perp(T) |(\nabla\varphi)_\perp|^2, \tag{16.32}$$

where $(\nabla\varphi)_\perp = \mathrm{i}(\nabla\theta(\mathbf{x}))\varphi_0 \exp(\mathrm{i}\theta(\mathbf{x}))$. $A_\perp(T)$ gives the *exact* hydrodynamic form of the kinetic energy when expressed as a functional of the order parameter beyond the Landau approximation. By recalling Eq. (15.60) for the superfluid velocity $\mathbf{v}_s = (\hbar/m)\nabla\theta$ in terms of the phase θ of the order parameter, the extra contribution $\Delta\tilde{F}$ reads

$$\Delta \tilde{F} = \frac{1}{2} \int \mathrm{d}^d x \; \rho_s \mathbf{v}_s^2 = \frac{1}{2\beta} \int \mathrm{d}^d x \; A_\perp(T) \varphi_0^2 (\nabla\theta(\mathbf{x}))^2 = \frac{1}{2\beta} A_\perp(T) \varphi_0^2 \sum_{\mathbf{k}} k^2 \theta_{\mathbf{k}}^2, \tag{16.33}$$

with $\rho_s = (m^2/\hbar^2\beta) A_\perp(T)\varphi_0^2$. The temperature dependence of the coefficient $A_\perp(T)$ represents the difference between the superfluid density ρ_s and the modulus squared of the order parameter. The two quantities, instead, coincide in the Landau mean-field approximation. As a result, we may relate the static response in the low momentum limit to the superfluid density as a generalization, but for a factor $k_B T$, of Eq. (13.46) with $\delta\varphi_{\mathbf{k}} = -\theta_{\mathbf{k}}\varphi_0$ and $\delta\varphi_{\mathbf{k}}^* = -\theta_{\mathbf{k}}^*\varphi_0$:

$$\chi_{\mathbf{k}}^{-1} = \frac{\delta^2 \Delta\tilde{F}}{\delta\varphi_{\mathbf{k}} \delta\varphi_{\mathbf{k}}^*} = \frac{1}{2\beta} A_\perp(T) k^2 = \frac{1}{2} \frac{\hbar^2 \rho_s k^2}{m^2 \varphi_0^2}. \tag{16.34}$$

By anti-Fourier transforming, the response function obtained in Eq. (16.34), in $d = 3$ reads

$$\frac{k_B T \chi(\mathbf{x} - \mathbf{x}')}{\varphi_0^2} = \frac{m^2}{\hbar^2} \frac{k_B T}{\rho_s} \frac{1}{2\pi |\mathbf{x} - \mathbf{x}'|}, \tag{16.35}$$

whereas in $d \leq 2$ it is infrared divergent. Near criticality the order parameter correlation length ξ is the only relevant length. In d dimensions, the anti-Fourier transform of the response function goes like $\sim |\mathbf{x}-\mathbf{x}'|^{-(d-2)}$ and, according to the scaling theory, must be an homogeneous function of $(\xi/|\mathbf{x}-\mathbf{x}'|)^{d-2}$ of degree zero, i.e. the scaling dimension (16.19) of ρ_s is $D_{\rho_s}=d-2$:

$$\rho_s \sim \xi^{-(d-2)}. \tag{16.36}$$

Notice that we have identified $\xi = m^2 k_B T/\hbar^2 \rho_s$. The Josephson scaling relation relation for ζ is then

$$\zeta = \nu(d-2) = 2\beta - \eta\nu. \tag{16.37}$$

Therefore, $A_\perp(T) \sim |t|^{-\eta\nu} = \xi^\eta$. This scaling relation implies that for $d=3$ $\zeta=\nu$ and hence the measurement of ρ_s provides information about ν. Historically, a logarithmic singularity ($\alpha=0$) was reported for the specific heat of helium, implying via (16.27) $\nu = 2/3 = \zeta$ in agreement with the measured value of ζ at the time. More recent experiments reported in Table 16.3 give a negative value of α compatible with the value of ν of Table 16.3.

In $d=2$, $\zeta=0$, whereas in $d=1$, ζ loses any physical meaning. The question arises: does ODLRO exist in $d=2$? As we mention in Chapter 13, the answer, to be discussed in the next section, is the Hohenberg version (Hohenberg (1967)) of the Mermin–Wagner theorem.

We finally point out that a variable can be relevant while its response stays finite. In fact, the condition for a variable to be relevant is less stringent than the requirement for the divergence of its response function. Let X and Y be a pair of conjugate variables with scaling dimensions $D_X + D_Y = d$, the response $\chi_X = \partial X/\partial Y \sim t^{-\gamma_X}$ diverges if $\gamma_X = (D_Y - D_X)/D_t > 0$, that is $D_X < d/2$ or $D_Y > d/2$.

If we take X as the order parameter φ, its response is divergent for any d and n, as can be easily checked by inspecting Tables 16.2 and 16.3. If, instead, Y is equal to t, various possibilities arise. For the classical theories and $d \geq 4$, $D_t = 1/\nu \leq d/2$ and the specific heat stays finite. Going beyond the classical theories and taking the values for the index ν from Tables 16.2 and 16.3, we see that for $n=1$ and $d=3$ the condition $D_t > d/2$ is satisfied and the response, which is the specific heat, in this case diverges. For $n=1$ and $d=2$, $D_t = d/2$ and the specific heat diverges only logarithmically. For $n=2,3$ and $d=3$, $D_t < d/2$ and the specific heat is finite, with a cusp singularity α which is negative for liquid ^{4}He at the λ-point and for ferromagnets at the Curie transition.

A foundation of scaling theory will be given in the next chapter via the introduction of a general transformation, the Renormalization Group transformation, which becomes the scaling transformation asymptotically near the critical point.

16.2 On the existence of ODLRO at low dimension

To answer the question on the existence of ODLRO at low dimension we begin by defining the order parameter–order parameter response function for superfluids according to the general definition of Eq. (10.8) of Chapter 10:

$$\chi(\mathbf{x}, t) = \frac{\mathrm{i}}{\hbar}\theta(t)\mathrm{Tr}(\mathrm{e}^{\beta(\Omega-K)}[\hat{\psi}(\mathbf{x}, t), \hat{\psi}^{\dagger}(0, 0)]), \tag{16.38}$$

apart from a minus sign which is necessary in order to have compatibility with the definition (16.34).[3]

The expectation value of the density operator, when expressed in terms of field operators (9.22), can be connected to the order parameter–order parameter correlation function, obtained from Eq. (10.45) after replacing the observables $\hat{O}_A(\mathbf{x}, t)$ and $\hat{O}_B(\mathbf{x}, t)$ with $\hat{\psi}(\mathbf{x}, t)$ and $\hat{\psi}^{\dagger}(\mathbf{x}, t)$. In Fourier space such a connection gets the form

$$n_{\mathbf{k}} + \frac{1}{2} = f_{\mathbf{k}}^{\hat{\psi}\hat{\psi}^{\dagger}}(t = 0) = \int_{-\infty}^{\infty} \frac{\mathrm{d}\omega}{2\pi} \tilde{f}_{\mathbf{k}}^{\hat{\psi}\hat{\psi}^{\dagger}}(\omega), \tag{16.39}$$

where $f_{\mathbf{k}}^{\hat{\psi}\hat{\psi}^{\dagger}}(t = 0)$ is the equal-time correlation function for the order parameter and $\tilde{f}_{\mathbf{k}}^{\hat{\psi}\hat{\psi}^{\dagger}}(\omega)$ the associated power spectrum (see (10.48)).

Thanks to the fluctuation–dissipation theorem (10.51), the power spectrum is directly related to the imaginary part of the response function:

$$\tilde{f}_{\mathbf{k}}^{\hat{\psi}\hat{\psi}^{\dagger}}(\omega) = \hbar \coth\left(\frac{\beta\hbar\omega}{2}\right) \mathrm{Im}\, \tilde{\chi}_{\mathbf{k}}(\omega). \tag{16.40}$$

By substituting Eq. (16.40) into Eq. (16.39) and using the inequality $x \coth(x) \geq 1$, we get

$$n_{\mathbf{k}} + \frac{1}{2} = \hbar \int_{-\infty}^{\infty} \frac{\mathrm{d}\omega}{2\pi} \mathrm{Im}\tilde{\chi}_{\mathbf{k}}(\omega) \coth\left(\frac{\beta\hbar\omega}{2}\right) \geq \int_{-\infty}^{\infty} \frac{\mathrm{d}\omega}{\pi} \frac{\mathrm{Im}\tilde{\chi}_{\mathbf{k}}(\omega)}{\beta\omega} = \frac{\tilde{\chi}_{\mathbf{k}}(0)}{\beta}, \tag{16.41}$$

where the integral over the imaginary part gives the static limit by using the Kramers–Kronig relation Eq. (10.18):

$$\tilde{\chi}_{\mathbf{k}}(0) = \int_{-\infty}^{\infty} \frac{\mathrm{d}\omega}{\pi} \frac{\mathrm{Im}\tilde{\chi}_{\mathbf{k}}(\omega)}{\omega}. \tag{16.42}$$

Finally, by recalling Eq. (16.34), we obtain

$$n_{\mathbf{k}} + \frac{1}{2} \geq \frac{2}{\beta} \frac{m^2}{\hbar^2} \frac{\varphi_0^2}{\rho_s k^2}. \tag{16.43}$$

Since the integral over momentum cannot be larger than the total number of depleted particles $N - N_0$, Eq. (16.43) shows that this condition is satisfied in $d \leq 2$ provided $\varphi_0^2 = 0$, then forbidding ODLRO.

[3] In Eq. (10.1) the external field enters with a minus sign difference with respect to the external field in the Ising Hamiltonian (8.51) used in the formulation of Landau theory.

16.3 Dynamical scaling

The static critical behavior of several systems has been understood first by analyzing the role of the classical theories and then by establishing the limits of their validity. At the phenomenological level these theories have been replaced by the scaling theory, whose predictions have been extensively tested by comparison with the results of exactly solvable models, numerical analysis and experiments. A similar unified description of critical dynamics is more difficult to reach due to the variety of excitation modes proper to each specific system, which interact with the fluctuations of the order parameter. A review can be found in the paper by Hohenberg and Halperin (1977).

The first prediction of a critical dynamical behavior along the line of classical theories has been given by Van Hove (1954) to determine the critical slowing down, i.e. the divergence at the critical point of the characteristic time τ_ξ for the spin density fluctuations in ferromagnetic systems, as measured by neutron scattering. Above the Curie point of the ferromagnetic transition, the diffusion equation for the magnetization $m(\mathbf{x}, \tau)$ is

$$\frac{\partial m(\mathbf{x}, \tau)}{\partial \tau} = D_s \nabla^2 m(\mathbf{x}, \tau), \tag{16.44}$$

where $D_s = \Lambda/\chi_m$ is the diffusion coefficient, Λ is the transport coefficient and χ_m the magnetic susceptibility. This last connection is the equivalent of (11.38), with the correspondence $n/\lambda \to \Lambda$ and $\partial n/\partial\mu \to \chi_m$. Classically, the transport coefficient Λ is assumed temperature independent or at least non-singular, the susceptibility in mean-field diverges as $\chi_m \sim |t|^{-\gamma} \sim \xi^{\gamma/\nu}$, with the classical value $\gamma/\nu = 2$. D_s, therefore, vanishes at the critical point, while the characteristic relaxation time $\tau_\xi = \xi^2/D_s$ of the model goes to infinity as $\chi_m\xi^2 \sim \xi^4$. In the case of the ferromagnetic transition the order parameter $m(\mathbf{x}, \tau)$ is a conserved quantity. In this case the diffusion equation follows from the hydrodynamics with the flow equations expressing the conservation laws. When the order parameter is not conserved, we can rely on the time-dependent Landau–Ginzburg equation (15.127) and the characteristic relaxation time τ_{LG} (15.128) diverges as the susceptibility, being proportional to ξ^2 instead of ξ^4.

The critical slowing down in the case of superconductors has been experimentally verified (Glover (1967)) by the diverging correction to the conductivity. This is the so-called paraconductivity effect, arising when approaching the critical temperature from above (see Appendix K).

The classical results for the dynamical criticality fail exactly as in the static case when considered too near the critical point where transport coefficients would not be constant.

Recall that in the static case the scaling theory is based on the long distance behavior or small wave vector $\mathbf{k}$ of the correlations in particular of the order parameter, whose correlation function is an homogeneous function of $r = |\mathbf{x}|$ and ξ or k and ξ, $\xi = |t|^{-\nu}$ being the order parameter correlation length. Near the critical point ξ becomes the only relevant length in the description of any system.

Dynamical scaling (Ferrell et al. (1968); Halperin and Hohenberg (1969b)) extends these ideas to include the frequency and the correlation time. Generally, a system is described in terms of conserved thermodynamic quantities whose densities satisfy a set of macroscopic (hydrodynamic) equations valid at long distances and wavelength (small k) and times long with respect to any microscopic time scale in the system (small frequencies). The order parameter may belong to this set of quantities, as in the ferromagnetic transition. If not, as for superconductors and superfluids, its equation must be added. The spectrum of the dynamical correlation function of most observables, even in the hydrodynamical limit, consists of various modes with different dispersions and strengths.

If a single mode $\omega_\xi(k)$ or a single process of characteristic time τ_ξ dominates the behavior of the order parameter dynamical correlation function at small k and ω, in general the order parameter dynamical correlation function at zero external field $f_\varphi(k, \omega, t)$ can be written as

$$f_\varphi(k, \omega, t) = \frac{2\pi}{\omega_\xi(k)} f_\varphi(k, t) g_{k,\xi}\left(\frac{\omega}{\omega_\xi}\right), \quad \int \mathrm{d}x\, g_{k,\xi}(x) = 1. \tag{16.45}$$

The normalization condition on the function $g_{k,\xi}(x)$ in Eq. (16.45) is required by Eq. (10.47) for the correlation function at $\tau = 0$. If the dynamical correlation function is peaked at a specific frequency, this frequency identifies $\omega_\xi(k)$. Alternatively, if the spectral distribution is a Lorentzian, the corresponding width coincides with $\omega_\xi(k)$. We may formulate the *dynamical scaling hypothesis* by assuming that near criticality, the homogeneity condition holds in terms of $k\xi$ and $\omega\tau_\xi$ both for the order parameter dynamical correlation function $f_\varphi(k, \omega, t)$ and for the frequency of the dominating mode:

$$\omega_\xi(k) = k^z \Omega(k\xi) \text{ or } k = \omega^{z'} K(\omega\tau_\xi). \tag{16.46}$$

In Eq. (16.46) z is the dynamical scaling index and the function Ω depends on the product $k\xi$ and not on k and ξ separately. Given the homogeneity of the static correlation function, for the dynamical case, it is enough to implement the condition that $g_{k,\xi}(x) = g_{k\xi}(x)$ i.e. $g_{k\xi}(x)$ depends only on the product $k\xi$.

The two equations (16.45) and (16.46) are the mathematical formulation of the *dynamical scaling hypothesis* and from them we deduce that the scaling index for the dynamical correlation function is equal to that of the static one minus the dynamical scaling index z.

The general strategy for applying the dynamical scaling hypothesis (16.45) and (16.46) consists of connecting the expressions of the dispersion relation of the dominating mode from the ordered and disordered hydrodynamic regions I ($T < T_\mathrm{c}$) and III ($T > T_\mathrm{c}$) where $k\xi \ll 1$ to the critical region II with $k\xi \gg 1$, as shown in Fig. 16.1.

To be specific, we now analyze two distinct situations in which the order parameter is and is not conserved. We follow the hystorical development of dynamical scaling pioneered by Ferrell et al. (1968), which deals with the non-conserved case of the superfluid transition of ^{4}He. For superfluid ^{4}He, the two-fluid hydrodynamics has been derived and extensively studied (Khalatnikov (1965)). The critical mode corresponds to the second sound. The two-fluid hydrodynamics, which we have seen in Chapter 15 and in Appendix I, would give for $f_\varphi(\mathbf{x}, \tau, t)$ the same approximation which has been used in the static case to obtain the parallel Eq. (13.44) and the transverse Eq. (13.45) component of $f_\varphi(\mathbf{x}, t)$ corresponding to the amplitude Eq. (13.22) and phase fluctuations (16.34), respectively, actually for the v_s correlation function.

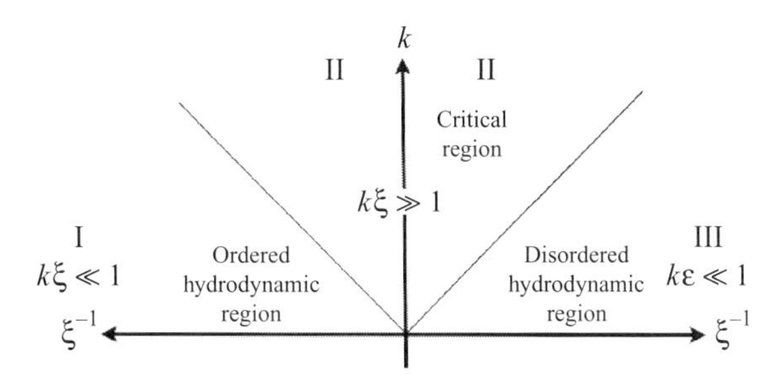

Fig. 16.1 The different regions in the (ξ^{-1}, k) plane. Regions I and III identified by the condition $k\xi \ll 1$ correspond to the ordered ($T < T_c$) and disordered ($T > T_c$) phases where the hydrodynamic description is valid. Region II with $k\xi \gg 1$ is the strongly fluctuating critical region. The inverse correlation length $\xi^{-1} = |t|^{\nu}$ measures the distance from the critical point.

According to the two-fluid hydrodynamics, at large distances $|\mathbf{x}|$ or small wave vectors $|\mathbf{k}|$, $f_{\varphi\perp}(k)$ dominates. Below T_λ, $f_\varphi(\mathbf{k}, \omega, t)$ is dominated by the pole of the second sound mode determined by the condition (see Eq. (I.14) of Appendix I)

$$\omega^2 + i\omega k^2 D_2 - c_2^2 k^2 = 0, \quad c_2^2 = \frac{\rho_s}{\rho_n} \frac{Ts^2}{\rho c_P}, \tag{16.47}$$

where D_2 is the second sound attenuation coefficient and c_2 is the second sound velocity. s is the entropy per unit volume and c_P the specific heat at constant pressure. The simultaneous measure of specific heat and second sound velocity permits a measure of ρ_s. At low k the mode dispersion reads

$$\omega_2(k) = c_2 k \left[1 - \frac{D_2^2 k^2}{8c_2^2} \mp i \frac{D_2 k}{2c_2} \right]. \tag{16.48}$$

According to the dynamic scaling hypothesis, the mode dispersion (16.48), derived in the hydrodynamic region I, must evolve continuously into the expression valid in the critical region II. On the line $k\xi = 1$ separating the regions I and II, we may express the mode dispersion in terms of the correlation length ξ instead of k. The linear-in-k behavior of the mode dispersion (16.48) yields a scaling $\sim \xi^{-1}$, to which we must add the scaling of the second sound velocity. The latter can be obtained from the expression in Eq. (16.47) by using the Josephson scaling Eq. (16.36), $\rho_s \sim \xi^{-(d-2)}$, i.e. is inversely proportional to the correlation length at $d = 3$. Neglecting the weak divergency of c_P, we obtain $c_2 \sim \xi^{-(d-2)/2}$. This yields for the dynamical critical index the value

$$z = d/2. \tag{16.49}$$

When $c_P/c_V \neq 1$ and first and second sounds are coupled, the order parameter correlation function has a component related to a pole of the first sound. For most temperatures up to very near to T_λ, this effect can be neglected in the first approximation, since $c_P/c_V - 1$ is at most 10^{-2} up to $T \sim 10^{-5}$ K from T_λ (Hohenberg and Martin (1965)).

By applying the homogeneity condition in $k\xi$ to the second sound mode solution of Eq. (16.48) at fixed real k, one obtains that $kD_2 \sim c_2 \sim \xi^{-(d-2)/2}$, so that the second sound attenuation coefficient diverges as $\xi^{(4-d)/2}$. If instead we use the homogeneity condition in $\omega\tau_\xi$ on the solution of Eq. (16.48) for real fixed ω we obtain for the second sound the characteristic time $\tau_\xi \sim \xi/c_2 \sim D_2/c_2^2 \sim \xi^{d/2}$.

For $T > T_\lambda$ the frequency dependence of f_φ is not derivable from the hydrodynamics since the order parameter is not a constant of motion and the two-fluid model is valid only below T_λ. However, one can guess that $f_\varphi(k, \omega, t)$ is peaked around $\omega = 0$ with a width $\Gamma(k) \neq 0$ for $k \to 0$ corresponding to a non-propagating mode. Both above and below T_λ, $f_\varphi(k, \omega, t)$ is not directly detectable and all the predictions of scaling theory must be transposed to other observables which in the limit $k \to 0$ couple to the order parameter, the so-called extended scaling theory.

The heat–heat correlation function $f_{\mathrm{Q}}(\mathbf{k}, \omega, t)$ as derived by the two-fluid hydrodynamics in the limit $c_P/c_V - 1 \to 0$ is also dominated by the second sound pole as a pure temperature wave mode for $T < T_\lambda$. Above T_λ, when $k \to 0$, $f_\varphi(\mathbf{k}, \omega, t)$ is dominated by the thermal diffusion mode $\omega_{\mathrm{Q}} = \mathrm{i}D_T k^2$, where D_T is the thermal diffusion coefficient. In the hydrodynamic region $k\xi \ll 1$ below T_λ the propagating second sound mode $\omega_2(k)$ coincides with the thermal wave mode $\omega_{\mathrm{Q}} = c_2 k$. The homogeneity condition implies that the characteristic mode persists in the same homogeneous form in every $k\xi$ = constant line both below and above T_λ with the same power. In particular, the extrapolated form of $f_{\mathrm{Q}}(\mathbf{k}, \omega, t)$ at $k \to 0$ at fixed ξ both above and below T_λ when $k\xi = 1$ match with the form $\xi \to \infty$ at fixed k. Therefore $c_2(k) \sim k^{(d-2)/2}$, $\omega_\xi(k) \sim \omega_{\mathrm{Q}}(k) \sim k^{d/2}$, $k \to 0$ at $\xi = \infty$, i.e. in the so-called critical zone II.

Above T_λ in zone III the thermal diffusion mode must be homogeneous and match with the propagating mode below T_λ, $\omega_{\mathrm{Q}} = \mathrm{i}D_T k^2 = \mathrm{i}k^{d/2}\Omega(k\xi)$. The thermal diffusion constant D_T, then, must diverge at criticality as $D_T \sim \xi^{(4-d)/2}$, i.e. as $\xi^{1/2}$ in $d = 3$. If we use the hydrodynamic relation between D_T and the thermal conductivity κ (analogous to Eq. (11.55)),

$$D_T = \frac{\kappa}{c_P}, \tag{16.50}$$

we obtain that the thermal conductivity in $d = 3$ diverges also as

$$\kappa \sim \xi^{1/2}. \tag{16.51}$$

If, in the matching conditions, we had considered the divergence of c_P^+ and c_P^- as approaching T_λ from above and below, the divergence of the thermal conductivity in $d = 3$ would read

$$\kappa \sim \xi^{1/2} \frac{c_P^+}{(c_P^-)^{1/2}}. \tag{16.52}$$

The prediction of thermal conductivity was originally verified qualitatitevely by Kerrisk and Keller (1967) and quantitatively by Ahlers (1968), including the specific heat behavior with a value of the index $\nu/2 = 0.334 \pm 0.005$ for $10^{-7} < (T - T_\lambda) < 5 \times 10^{-3}$ K.

The ferromagnetic transition provides an example in which the order parameter is conserved. Below the Curie temperature, the dominant mode is that associated with spin waves. From hydrodynamics, as seen in Section 14.3, the mode dispersion reads

$$\omega_{\mathrm{s}}(k) = c_{\mathrm{s}} k^2, \quad c_{\mathrm{s}} = \frac{\rho_{\mathrm{s}}}{m_0}, \tag{16.53}$$

where the velocity of the spin waves is related to the *stiffness* constant ρ_s and to the magnetization density m_0. The *stiffness* constant ρ_s is introduced with the same symbol as in the case of ^{4}He since it is associated in both cases with the phase fluctuations of the order parameter. For ρ_s we can use again the Josephson scaling (16.36) $\rho_s \sim \xi^{-(d-2)}$ together with the scaling of $m_0 \sim \xi^{-\beta/\nu}$ of Eq. (16.21) to get in critical region II

$$\omega_s(\xi) \sim \xi^{-(d+2-\eta)/2}, \tag{16.54}$$

so that $z = 5/2$ at $d = 3$ for $\eta = 0$ in any case ($\eta \sim 0.07$).

Above the Curie temperature, the spin diffusion provides the dominating mode (16.44):

$$\omega_s(k) = D_s k^2. \tag{16.55}$$

Along the line $k\xi = 1$ between the regions II and III, we obtain, taking into account Eq. (16.54),

$$\xi^{-(d+2-\eta)/2} \sim D_s \xi^{-2}, \quad D_s \sim \xi^{-(d-2-\eta)/2}. \tag{16.56}$$

One sees that, in contrast to the expectation of the classical theory of Van Hove, the vanishing of the diffusion coefficient does not compensate for the divergence of the susceptibility, which goes as $\xi^{2-\eta}$ according to Eq. (16.23). This implies that the transport coefficient also has a scaling behavior

$$\Lambda \sim \xi^{-(d-6+\eta)/2}. \tag{16.57}$$

One sees that in three dimensions Λ diverges as $\xi^{3/2}$. Furthermore, $d = 6$ appears as the upper critical dimension at which the classical behavior is recovered with finite Λ at the transition.

16.4 Problem

16.1 Evaluate the long-distance behavior of the one-particle reduced density matrix in the presence of phase fluctuations of the order parameter in order to show the non-existence of ODLRO in $d = 1, 2$.

17 The renormalization group approach

As discussed in the previous chapter, the phenomenological theory of critical phenomena finds a sound basis in the logical sequence of universality, scaling, relevant and irrelevant variables. The difficult problem of a large number of degrees of freedom strongly correlated within the coherence distance, which diverges at criticality, is thereby, in principle, reduced to the determination of the homogeneous form of the relevant response functions and of the evaluation of a few critical indices related to the relevant parameters, whereas the details specifying each system do not matter.

In order to give a microscopic foundation to universality and to the scaling theory, an exact transformation is needed, which changes into a scaling transformation asymptotically near the critical point. This is the renormalization group (RG) approach, recently presented in a huge number of review papers and books. Historically, the first comprehensive overview was given by the sixth volume of the series edited by Domb and Green (1976) and by the classic books by Patashinskij and Pokrovskij (1979) and by Ma (1976).

Along with scaling theory, the field-theoretic renormalization group (RG) approach (Gell-Mann and Low (1954); Bogoliubov and Shirkov (1959); Bonch-Bruevich and Tyablikov (1962)) was introduced in critical phenomena (Di Castro and Jona-Lasinio (1969)) by showing how the RG equations generalize the universality relations in the sense that they relate one model system to another by varying the coupling and suitably rescaling the other variables and the correlation functions. At the same time a detailed analysis of the pertubative diagrammatic structure of the correlation functions together with the use of Ward identities yielded a renormalized theory (Migdal (1969); Polyakov (1968, 1969)). By iterating the RG transformation the coupling disappears from the equations near to the critical point, the homogeneous form of the static order-parameter correlation function is obtained and a microscopic definition of the critical indices is given. The scaling picture then acquires a theoretical basis.

Wilson (Wilson (1971a, b); Wilson and Kogut (1974)) gave a great contribution to the understanding of the physics underlying the RG procedure and to the explicit calculation of the critical indices. He proposed a simple mathematical realization of the other very physical idea of universality due to Kadanoff of grouping together degrees of freedom on the scale s associated with larger and larger cells. Since the correlation length goes to infinity, the degrees of freedom at short distance do not influence the critical behavior. Wilson, then, proposed a procedure of elimination of the short-wavelength fluctuations, with momenta between the upper momentum[1] cutoff Λ and Λ/s, and a subsequent rescaling of

[1] In this chapter we will use momentum and wave vector as equivalent words, since the problem at hand is classical, so that no confusion about the factor $\hbar$ may arise.

the resulting variables. By iteration of this procedure the mechanism of disappearance of the original coupling constants was then clarified and scaling was obtained. At the same time non-classical critical indices were calculated for the first time by various methods. Here we will only recall their evaluation by means of the famous ϵ-expansion ($\epsilon = 4 - d$) (Wilson (1972); Wilson and Fisher (1972)). It turned out that ϵ, assumed to be small, is the control parameter in a perturbative evaluation of the RG transformations starting from the known critical behavior at $d = 4$. In the field-theoretic approach the ϵ-expansion was also formulated soon after (Di Castro (1972); Schroer (1973); Brezin et al. (1973)).

Few theories have been supported by so many successes as the renormalization group approach; nevertheless a certain knowledge of the fundamental symmetries inherent to the problem, in particular of the order parameter, has to be assumed in order to make the proper choice of the basic variables on which to operate the transformation. One must therefore translate a given problem in terms of an effective Hamiltonian expressed through physically significant variables. This effective model should account for a qualitative thermodynamic description of the system at mean-field level. This is then assumed as the zeroth approximation suitable for the application of the RG transformation.

In the next sections we deal with the simplest case of critical phenomena within the Landau–Wilson model of Chapter 14, where the starting mean-field theory is well characterized by the Landau theory. For instance, in the case of an ordinary magnetic critical point, the spontaneous magnetization (the order parameter φ), the deviation from the critical temperature t and the coupling constant are considered as the basic variables.

17.1 General properties of the renormalization group

In the phenomenological scaling theory we have hypothesized a transformation without building it explicitly. For instance, for the Ising model (see Eq. (8.51)) on a lattice with nearest neighbors' interaction, due to the divergence of the correlation length, it was implicitly assumed that, by changing the lattice spacing $r_0 \to l_0' = sl_0$, the coupling $K = \beta J$ goes into a new coupling K' and the order parameter $\varphi(\mathbf{x}_i) = \langle \sigma_i \rangle$ transforms to $\varphi'(\mathbf{x}_j) = \langle \sigma_j' \rangle$, where the block variable, σ_j', associated with the j-cell is defined as the arithmetic average of the original variables σ_i on the lattice site i belonging to the cell j (see Fig. 17.1):

$$\sigma_j' = \frac{1}{s^d} \sum_{i \in j} \sigma_i. \tag{17.1}$$

At the same time the correlation length must satisfy the relation

$$\xi(K') = \frac{1}{s} \xi(K). \tag{17.2}$$

Stretching the scaling argument of the previous chapter, one could think that on the scale $sl_0 \ll \xi$ all the spins σ_i, in the cell j, are aligned either up or down and the cell variable σ_j' would take the values ± 1 only. As a consequence the original Ising model with coupling K

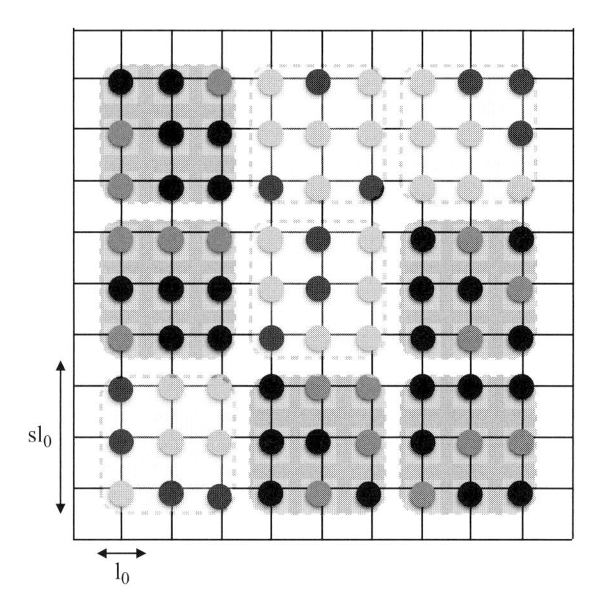

Fig. 17.1 Pictorial representation of a configuration of the two-dimensional Ising lattice. Black and gray dots represent the values $\sigma_i = 1$ and $\sigma_i = -1$ at each lattice site i, respectively. The regions delimited by dashed lines are block cells j containing nine of the original lattice sites with the choice $s = 3$. Dark and light gray regions indicate a positive $\sigma'_j > 0$ and negative $\sigma'_j < 0$ average of the block variables σ'_j, respectively.

would transform into an Ising model with coupling K'. This, however, turns out to be too restrictive. The variables of two adjacent cells would interact via the bonds belonging to the common boundary and therefore K' would be equal to $s^{d-1}K$. The original and scaled systems must become critical simultaneously, $K_c = K'_c = s^{d-1}K_c$, with the only possible values $K_c = 0$ or $K_c = \infty$ ($T_c = \infty$ or $T_c = 0$ as trivially expected), corresponding to complete disorder or complete order, respectively. Criticality ($\xi \to \infty$) must occur at finite temperature and this means that the Ising model cannot be simply transformed into a scaled Ising model. We allow for non-complete alignment of the spin within a cell by substituting the linear relation between K and K' with a more involved one

$$K' = f(K), \tag{17.3}$$

provided that

$$K'_c = f(K_c) = K_c. \tag{17.4}$$

We further assume that the transformation cannot introduce singularities and therefore $f(K)$ can be expanded around K_c:

$$f(K) \approx f(K_c) + \left(\frac{\mathrm{d}f}{\mathrm{d}K}\right)_{K=K_c} (K - K_c) + \cdots . \tag{17.5}$$

Recalling the power law behavior of $\xi \sim |K - K_c|^{-\nu}$, Eq. (17.2) now reads

$$\frac{1}{s} = \frac{|K' - K_c|^{-\nu}}{|K - K_c|^{-\nu}} = \left| \left(\frac{df}{dK} \right)_{K=K_c} \right|^{-\nu}. \tag{17.6}$$

The critical index $\nu = 1/D_t$ is then associated with the linearized transformation

$$\frac{|t'|}{|t|} = s^{D_t} = \left(\frac{df}{dK} \right)_{K=K_c}. \tag{17.7}$$

We can now extend the above considerations for the Ising model to a generic Hamiltonian specified by a set of p parameters $\{\mu_i\}$ and identify the conditions that a proper transformation depending on the scale s must satisfy. To begin with, Eq. (17.3) is replaced by a set of p transformations depending on the scale s and the full set of parameters $\{\mu_i\}$:

$$\mu_i' = f_i^s(\{\mu_l\}), \quad i = 1, \ldots, p. \tag{17.8}$$

This set of transformations identifies what is called a renormalization group (RG) transformation between H and H' acting on a preassigned Hamiltonian space and can be symbolically indicated as

$$H' = R_s(H). \tag{17.9}$$

In principle, the preassigned space must be larger than the subspace of the initial Hamiltonian, since generally new coupling constants are generated by the RG procedure. However, most of the coupling constants (the irrelevant ones) are subsequently washed away by the RG flow, so that these transient variables are only important in determining the approach and the corrections to scaling.

The parameter s is interpreted as the change in the length scale (scaling factor) and two subsequent applications of the RG transformation (17.9) with parameters s and s' must coincide with a single transformation with parameter ss' (the semigroup property):

$$R_{s's}(H) = R_{s'}(R_s(H)). \tag{17.10}$$

In the remaining part of this section the properties of the group transformation will be presented (Cassandro and Gallavotti (1975); Griffiths and Pearce (1978)). The RG transformation must leave the thermodynamic potential invariant:

$$F\left[R_s(H)\right] = F\left[H\right], \tag{17.11}$$

which means that the macroscopic thermodynamic behavior must not depend on the degree of resolution at which the system is described.

As before, the RG transformation must be regular. Let $H + \tau \delta H$ be a Hamiltonian close to H for a small parameter $\tau \ll 1$, i.e., if $\{\mu_i^\tau\}$ are the parameters specifying $H + \tau\delta H$, then $\delta\mu_i \equiv \mu_i^\tau - \mu_i \sim O(\tau)$. We assume then that

$$R_s(H + \tau\delta H) = R_s(H) + \tau L_s(H)\delta H, \tag{17.12}$$

where L_s is a linear operator which depends on H. This amounts to saying that spurious unphysical singularities are not introduced by the transformation. In terms of the parameters of the Hamiltonian, Eq. (17.12) is equivalent to

$$\delta\mu_i' = \sum_l \frac{\partial f_i^s}{\partial \mu_l}\delta\mu_l, \quad i = 1, \ldots, p. \tag{17.13}$$

The RG transformation must be constructed to extract the dominant long-distance behavior of the system near criticality and reach, within the present scheme, the scale invariance once the irrelevant variables are eliminated by iterating the transformations. Associated with a given RG transformation there must be a set of fixed points H^* in the Hamiltonian space, defined by

$$R_s(H^*) = H^*, \quad \mu_i^* = f_i^s(\{\mu_l^*\}). \tag{17.14}$$

As previously discussed, two trivial examples of fixed points must be found when the system is completely ordered ($T = 0$) or completely disordered ($T = \infty$): in both cases an RG transformation should leave the system unchanged. These are the low-temperature and high-temperature fixed points, respectively. Any system out of criticality is attracted by one of them, depending on whether $T < T_c$ or $T > T_c$, respectively. At criticality, instead, the correlation length is infinite and remains infinite after any iteration of the transformation (17.9). The system, by elimination of the short range behavior, should then be attracted by a fixed point with $\xi = \infty$. In parameter space we define a critical surface as the set of points whose Hamiltonians satisfy

$$\lim_{s\to\infty} R_s(H_c) = H^*. \tag{17.15}$$

Each Hamiltonian H_c belonging to the critical surface is, in general, different from H^* and the difference $H_c - H^*$ disappears in the limit $s \to \infty$. The physical interpretation is that $H_c - H^*$ describes the details of the system which become irrelevant in the long-range limit.

In the following we assume that there is at least one non-trivial fixed-point Hamiltonian H^* for an interacting system, specified by a set $\{\mu_i^*\}$, separating the two non-critical regions.

From the composition law Eq. (17.10) and the smoothness postulate Eq. (17.12), we have

$$R_{s'}(R_s(H + \tau\delta H)) = R_{s's}(H + \tau\delta H) = R_{s's}(H) + \tau L_{s's}(H)\delta H$$

or

$$\begin{aligned} R_{s'}(R_s(H + \tau\delta H)) &= R_{s'}(R_s(H) + \tau L_s(H)\delta H) \\ &= R_{s's}(H) + \tau L_{s'}(R_s(H))L_s(H)\delta H \end{aligned}$$

and hence, in general

$$L_{s's}(H) = L_{s'}(R_s(H))L_s(H), \tag{17.16}$$

and only at a fixed point does the composition law for the linearized transformation simplify as

$$L_{s's}(H^*) = L_{s'}(H^*)L_s(H^*). \tag{17.17}$$

Let us assume that $L_s(H)$ is diagonalizable and indicate with $\{\lambda_l^*\}$ its eigenvalues. Equation (17.17) implies

$$\lambda_l^* = s^{D_l}, \quad l = 1, 2, \ldots \tag{17.18}$$

and D_l are the scaling indices (or scaling dimensions), when s is identified with the change of the length scale. They are an intrinsic property of the fixed-point Hamiltonian H^*. The Hamiltonians which flow to the same fixed point under the RG transformations belong to the domain of attraction of H^*, and lead therefore to the same scaling indices. Universality amounts then to saying that all the Hamiltonians within the same domain of attraction share the same critical behavior. We denote by h_l^* the eigenvectors of $L_s(H^*)$ which correspond to the eigenvalues λ_l^*. We can use the eigenvectors to expand with coefficients $\bar{\mu}_i$ any perturbation of the fixed-point Hamiltonian:

$$\tau\delta H = \sum_l \bar{\mu}_l h_l^* \Rightarrow (\tau\delta H)' = L_s(H^*)\tau\delta H = \tau \sum_l s^{D_l}\bar{\mu}_l h_l^*. \tag{17.19}$$

From Eq. (17.19) we read

$$\bar{\mu}_l' = s^{D_l}\bar{\mu}_l. \tag{17.20}$$

The $\{\bar{\mu}_l\}$ are combinations of the original $\{\delta\mu_i\}$ and are named linear scaling fields (Wegner and Riedel (1973); Riedel and Wegner (1974)). If $D_l > 0$ and the corresponding $\bar{\mu}_l \neq 0$, the RG transformations drive the system away from H^*, eventually moving it toward another fixed point. The eigenvectors with positive scaling dimensions define the directions of escape from the critical surface related to a given fixed point and are the relevant fields. On the other hand, the eigenvectors with negative scaling dimensions define the tangent plane to the critical surface at H^*, and correspond to the irrelevant variables. They determine the transient behavior of the system, when approaching criticality.

17.2 The Kadanoff–Wilson transformation

In the previous section we listed the general properties of a RG transformation. Here, we explicitly build up the Wilson transformation. We first introduce the conceptually simpler transformation in real configuration space, which, following the Kadanoff block-variable idea, groups together variables belonging to a cell of size s in units of lattice spacing. Then we introduce the transformation in momentum space as Wilson himself did. Although the main ingredient is again the elimination of degrees of freedom at a short distance, it is of simpler implementation.

Following Ma (1976) (but see also Cassandro and Jona-Lasinio (1978)), let us consider a statistical system defined by a given probability distribution

$$P\left[\{\hat{\varphi}(\mathbf{x}_i)\}\right] = \frac{1}{Z}\mathrm{e}^{-H[\{\hat{\varphi}(\mathbf{x}_i)\}]}, \tag{17.21}$$

where $\hat{\varphi}(\mathbf{x}_i)$ is a continuous random variable defined on the site $\mathbf{x}_i$ of a d-dimensional lattice (as is customary, the temperature has been absorbed in the parameters of the Hamiltonian

H), the order parameter being the thermal average $\langle \hat{\varphi}(\mathbf{x}_i) \rangle$ of $\hat{\varphi}(\mathbf{x}_i)$, as for instance in the Landau–Wilson model. If $2\pi \Lambda^{-1} \equiv l_0$ is the original lattice spacing, we construct a new lattice with lattice spacing $s\Lambda^{-1}$ as follows. If j labels the position of the center of the cell with s^d points i, the block variables are defined as

$$\hat{\varphi}^{(s)}(\mathbf{x}_j) = \frac{1}{s^d} \sum_{i \in j} \hat{\varphi}(\mathbf{x}_i). \tag{17.22}$$

Before we can compare the new probability distribution P_s, associated with the block variables, with P, we have to measure all the physical quantities in units of the new lattice spacing $s\Lambda^{-1}$. In order to obtain a non-trivial asymptotic probability distribution we also need to rescale the field variable while measuring the distances in the new units, i.e. $\mathbf{x}_j \to \mathbf{x}_j/s$ and at the same time

$$\hat{\varphi}^{(s)}(\mathbf{x}_j) \to s^{D_\varphi} \hat{\varphi}^{(s)}(\mathbf{x}_j) \equiv \hat{\varphi}(\mathbf{x}_j/s). \tag{17.23}$$

Notice that we have on purpose used for the normalization index the same symbol D_φ as the scaling index of (16.9), since they will be shown to be equal. The requirement of the introduction of a different normalization of the block variable with respect to the arithmetic average can indeed be argued in general. Let $\{\hat{\varphi}(\mathbf{x}_i)\}$ be a set of random independent variables, identically distributed with zero average and finite variance:

$$\langle \hat{\varphi}(\mathbf{x}_i) \rangle = 0, \ \langle \hat{\varphi}(\mathbf{x}_i) \hat{\varphi}(\mathbf{x}_{i'}) \rangle = \delta_{ii'} \sigma^2.$$

Then, construct a block of N variables $\bar{\varphi}_N = \sum_{i=1,N} \hat{\varphi}(\mathbf{x}_i)$ (with $N = s^d$ in the block variable, which in the limit $s \to \infty$ becomes an extensive variable), whose variance is

$$\left\langle \bar{\varphi}_N^2 \right\rangle = N\sigma^2 \tag{17.24}$$

and the fluctuations satisfy the standard square-root law $\sqrt{\bar{\varphi}_N^2}/N \sim 1/\sqrt{N}$. Since the variance of $\bar{\varphi}_N$ increases linearly with N, the variance of the normalized variable $\bar{\varphi}_N/\sqrt{N}$ is finite, i.e., it has a well defined asymptotic probability distribution and in this non-critical case $D_\varphi = d/2$. Indeed, the distribution for the sum $(1/\sqrt{N}) \sum_{i=1,N} \hat{\varphi}(\mathbf{x}_i)$ of independent variables tends to a Gaussian distribution when $N \to \infty$ according to the central limit theorem (see Appendix A). If the variables are not independent, Eq. (17.24) becomes

$$\left\langle \bar{\varphi}_N^2 \right\rangle = \sum_{ii'} \langle \hat{\varphi}(\mathbf{x}_i) \hat{\varphi}(\mathbf{x}_{i'}) \rangle \equiv \sum_{ii'} f_\varphi(\mathbf{x}_i, \mathbf{x}_{i'}) \sim N \sum_{i=1}^{N} f_\varphi(|\mathbf{x}_i|) \sim N\chi \quad \text{for} \quad N \to \infty, \tag{17.25}$$

where in the last step we made use of translational invariance and of the expression of the susceptibility as the Fourier transform of the correlation function at the wave vector $|\mathbf{q}| = 0$ (see Eq. (13.26)). For a non-critical system χ is finite, the square-root law for the fluctuations is still valid and the finite susceptibility replaces the variance σ^2. This means that, although there are correlations in the system, these are short-ranged, so that one can always find independent subsystems (block variables), and the central limit theorem follows. In this case, the only scaling dimension giving a well-defined asymptotic probability distribution is $D_\varphi = d/2$, as in the case of independent variables. On the other hand, if the system is critical the correlations are long ranged ($f_\varphi(r,0,0) \sim 1/r^{2D_\varphi}$) and

$$\langle \bar{\varphi}_N^2 \rangle \sim s^d \int_0^s \mathrm{d}x \; x^{d-1-2D_\varphi} = s^d s^{d-2D_\varphi}. \tag{17.26}$$

At criticality a divergent χ implies that $d - 2D_\varphi > 0$, i.e., $D_\varphi < d/2$. The square-root law is no longer valid and the variables are strongly correlated. The correct normalization of the block variables still guarantees a well-defined asymptotic probability distribution, which in general will not be Gaussian. The choice of the proper normalization of the block variables is related to the choice of the correct asymptotic behavior of the correlation function of the order parameter. This last procedure was used by Wilson to select the proper fixed point.

The probability distribution for the block variables is

$$P_s\left[\{\hat{\varphi}(\mathbf{x}_l/s)\}\right] = \int \left[\prod_i \mathrm{d}\hat{\varphi}(\mathbf{x}_i)\right] P\left[\{\hat{\varphi}(\mathbf{x}_i)\}\right] \prod_j \delta\left(\hat{\varphi}(\mathbf{x}_j/s) - \frac{s^{D_\varphi}}{s^d}\sum_{i\in j}\hat{\varphi}(\mathbf{x}_i)\right), \tag{17.27}$$

where the delta-function constraints the system to the transformation (17.22) and (17.23). We define a new Hamiltonian H_s such that $P_s = \mathrm{e}^{-H_s - F_s}$, and the transformation associated with the probability distribution corresponds to a transformation R_s acting on $H \in \mathcal{H}$, $H_s = R_s(H)$, which is defined modulo a constant term which is fixed by putting $H_s[\{\hat{\varphi}(\mathbf{x}_j/s) = 0\}] = 0$, and associating the part not containing the field $\hat{\varphi}(\mathbf{x}_j/s)$ with F_s.

Although intuitively simple, the idea of constructing block variables encounters several technical difficulties when it is applied in practice. These difficulties are partially avoided by a technically simpler realization of the physical idea that short-distance degrees of freedom should be eliminated. This is the momentum-space formulation proposed by Wilson. We first introduce the Fourier transform for the random variables $\hat{\varphi}_i$:

$$\hat{\varphi}_{\mathbf{k}} = \frac{1}{\sqrt{N}}\sum_i \mathrm{e}^{-\mathrm{i}\mathbf{k}\cdot\mathbf{x}_i}\hat{\varphi}(\mathbf{x}_i), \quad \hat{\varphi}(\mathbf{x}_i) = \frac{1}{\sqrt{N}}\sum_i \mathrm{e}^{\mathrm{i}\mathbf{k}\cdot\mathbf{x}_i}\hat{\varphi}_{\mathbf{k}}, \tag{17.28}$$

where $\mathbf{x}_i$ is the vector position of the site i and N is the number of lattice sites ($N \to \infty$ in the thermodynamic limit). Then we realize that the short-distance details of a given configuration $\{\hat{\varphi}(\mathbf{x}_i)\}$ are related to the large-momentum components of the Fourier transform $\hat{\varphi}_{\mathbf{k}}$. The elimination of the irrelevant degrees of freedom in real space, for $0 < |\mathbf{x}| < s\Lambda^{-1}$, can be reformulated as:

1. the elimination of the large momentum components of the *fast modes* $\hat{\varphi}_{\mathbf{k}}$ with $\Lambda/s < |\mathbf{k}| < \Lambda$, where the inverse lattice spacing acts as an ultraviolet cutoff for momentum integrals;
2. the rescaling of the length for the *slow modes* $\hat{\varphi}_{\mathbf{q}}$ with $0 < |\mathbf{q}| < \Lambda/s \to 0 < |\mathbf{q}'| = s|\mathbf{q}| < \Lambda$; and
3. the normalization of the block variables $\hat{\varphi}_{\mathbf{q}} \to s^{d/2 - D_\varphi}\hat{\varphi}_{s\mathbf{q}}$, where we have taken into account Eq. (17.28) to find the proper rescaling for the Fourier mode $\hat{\varphi}_{\mathbf{k}}$.

The RG-transformed probability distribution is then

$$\mathrm{e}^{-H_s(\{\hat{\varphi}_{\mathbf{q}'}\})-F_s} = \left(\int \left[\prod_{\Lambda/s<|\mathbf{k}|<\Lambda} \mathrm{d}\hat{\varphi}_{\mathbf{k}}\right] \mathrm{e}^{-H[\{\hat{\varphi}_{\mathbf{k}},\hat{\varphi}_{\mathbf{q}}\}]}\right)_{\mathbf{q}\to\mathbf{q}'=s\mathbf{q},\ \hat{\varphi}_{\mathbf{q}}\to s^{d/2-D_\varphi}\hat{\varphi}_{s\mathbf{q}}} . \tag{17.29}$$

Like Eq. (17.27), Eq. (17.29) induces a transformation R_s on the space H.[2]

We are now ready to implement the general scenario of critical phenomena introduced in the previous sections and use the Wilson transformation (17.29) in momentum space. On the basis of the phenomenological approaches discussed so far, we adopt the simplest form of the coarse-grained Landau–Wilson Hamiltonian,

$$H\left[\hat{\varphi}(\mathbf{x})\right] = \int \mathrm{d}^d x \quad \left(\frac{r_0}{2}\hat{\varphi}^2(\mathbf{x}) + \frac{c}{2}|\nabla\hat{\varphi}(\mathbf{x})|^2 + u\hat{\varphi}^4(\mathbf{x})\right), \tag{17.30}$$

which, as discussed in Section 14.1, represents an n-component spin model in a d-dimensional space with $\hat{\varphi}^2(\mathbf{x}) = \sum_{i=1}^n \hat{\varphi}_i^2(\mathbf{x})$, $\hat{\varphi}^4(\mathbf{x}) = (\sum_{i=1}^n \hat{\varphi}_i^2(\mathbf{x}))^2$. In Eq. (17.30), as is customary, r_0 and u are used instead of a and b. As also noted in Section 14.1, the thermodynamic potential described by the Hamiltonian (17.30) coincides with the Landau form (13.19) if the partition function is evaluated in the saddle-point approximation, i.e. in the absence of fluctuations of the order parameter. In momentum space the Hamiltonian (17.30) reads

$$H\left[\hat{\varphi}_{\mathbf{k}}\right] = \frac{1}{2}\sum_{0<|\mathbf{k}|<\Lambda,i} (r_0 + ck^2)\hat{\varphi}_{i,\mathbf{k}}\hat{\varphi}_{i,-\mathbf{k}} + \frac{u}{L^d}\sum_{\{k_i\},i,j} \delta_{\{\mathbf{k}_i\}}\hat{\varphi}_{i,\mathbf{k}_1}\hat{\varphi}_{i,\mathbf{k}_2}\hat{\varphi}_{j,\mathbf{k}_3}\hat{\varphi}_{j,\mathbf{k}_4}, \tag{17.31}$$

where $\{\mathbf{k}_i\} \equiv \{0 < |\mathbf{k}_1|, |\mathbf{k}_2|, |\mathbf{k}_3|, |\mathbf{k}_4| < \Lambda\}$, $\delta_{\{\mathbf{k}_i\}} \equiv \delta_{\mathbf{k}_1+\mathbf{k}_2+\mathbf{k}_3+\mathbf{k}_4,0}$, $i, j = 1, \ldots, n$, and $L^{d/2}$ replaces $\sqrt{N}$ in Eq. (17.28) for the Fourier transform. In the n-component model the product over the fast modes in the integral in Eq. (17.29) extends to the n components.

One of the requirements of the RG approach is the regularity of the transformation. This means that in general we can expand the transformation with respect to a perturbing term. To this end it is convenient to split the Hamitonian into two parts:

$$H = H_0 + H_1, \tag{17.32}$$

e.g. including in H_0 all the terms quadratic in $\hat{\varphi}$ and leaving as a perturbation in H_1 the quartic term, whose explicit expression can be read off from Eq. (17.31). The perturbative expansion of (17.29) in H_1 reads

$$\mathrm{e}^{-(H_s+F_s)} = \left(\int \prod_{i,\Lambda/s<|\mathbf{k}|<\Lambda} \mathrm{d}\hat{\varphi}_{i,\mathbf{k}}\mathrm{e}^{-H_0}\sum_{m=0}^{\infty}\frac{(-1)^m}{m!}H_1^m\right)_{\mathbf{q}\to\mathbf{q}'=s\mathbf{q},\ \hat{\varphi}_{\mathbf{q}}\to s^{d/2-D_\varphi}\hat{\varphi}_{s\mathbf{q}}} . \tag{17.33}$$

By calling $H_{s,m} + F_{s,m}$ the term of order H_1^m of the transformed Hamiltonian, expanding the exponential in the left-hand side of Eq. (17.33) and equating the two sides order by

[2] Different realizations of the RG transformation must lead to the same asymptotic scaling behavior near a fixed point, even though the approach to it, i.e. the elimination of the irrelevant variables, may be different for different groups.

order, we obtain the so-called cumulant expansion of the transformation:

$$H_{s,0} + F_{s,0} = \int \prod_{i,\Lambda/s<|\mathbf{k}|<\Lambda} \mathrm{d}\hat{\varphi}_{i,\mathbf{k}} \mathrm{e}^{-H_0}, \tag{17.34}$$

$$H_{s,1} + F_{s,1} = \langle H_1 \rangle, \tag{17.35}$$

$$H_{s,2} + F_{s,2} = -\frac{1}{2}\left(\langle H_1^2 \rangle - \langle H_1 \rangle^2\right), \tag{17.36}$$

$$\cdots = \cdots, \tag{17.37}$$

where the average symbol here indicates the integration over the fast variables:

$$\langle \dots \rangle = \frac{\int \prod_{i,\Lambda/s<|\mathbf{k}|<\Lambda} \mathrm{d}\hat{\varphi}_{i,\mathbf{k}} \mathrm{e}^{-H_0} \dots}{\int \prod_{i,\Lambda/s<|\mathbf{k}|<\Lambda} \mathrm{d}\hat{\varphi}_{i,\mathbf{k}} \mathrm{e}^{-H_0}}. \tag{17.38}$$

It is understood that the rescaling, $\mathbf{q} \to \mathbf{q}' = s\mathbf{q}$, $\hat{\varphi}_{\mathbf{q}} \to s^{d/2-D_\varphi}\hat{\varphi}_{s\mathbf{q}} = s^{1-\eta/2}\hat{\varphi}_{s\mathbf{q}}$, must be performed after the integration over the fast variables.

17.3 Derivation of the RG equations

17.3.1 Gaussian approximation

A trivial application of the Wilson transformation (17.29) is the derivation of the Gaussian approximation (see Section 14.1) with the RG approach. Although trivial, this illustrates well how the RG approach works. In this case ($u \equiv 0$ in the Hamiltonian Eqs. (17.30) and (17.31)) a strong simplification occurs because one can exactly separate the fast from the slow modes in the Hamiltonian. The integration over the fast modes leads to

$$F_{s,0} = -\frac{n}{2} \sum_{\Lambda/s<|\mathbf{k}|<\Lambda} \ln\left(\frac{2\pi}{r_0 + ck^2}\right), \tag{17.39}$$

which is the contribution to the free energy (14.30) in the Gaussian approximation from the fast modes and coincides with the full free energy when all the modes are taken into account ($\Lambda/s \to 0$). The untouched Hamiltonian of the slow modes, when rescaled, gives

$$H_{s,0} = \sum_{i,0<|\mathbf{q}'|<\Lambda} \frac{1}{2}\left(r_0' + c'(q')^2\right)|\hat{\varphi}_{i,\mathbf{q}'}|^2, \tag{17.40}$$

with the flow equations for this simple case

$$c' = s^{-\eta}c, \quad r_0' = s^{2-\eta}r_0, \quad u \equiv 0. \tag{17.41}$$

There are three fixed-point solutions of the flow equations:

$$\eta = 2, \quad D_\varphi = \frac{d}{2}, \qquad c^* = 0, \quad r_0^* = r_0 > 0 \quad \text{or} \quad r_0^* = r_0 < 0, \tag{17.42}$$

$$\eta = 0, \quad D_\varphi = \frac{d-2}{2}, \quad c^* = c, \quad r_0^* = 0. \tag{17.43}$$

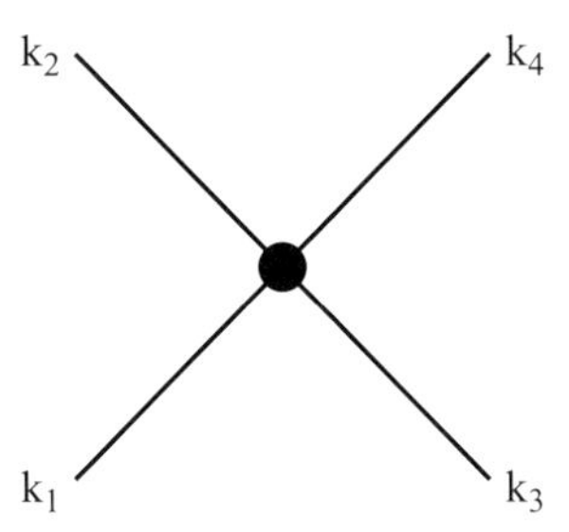

Fig. 17.2 Four-leg interaction vertex due to H_1. The black dot represents the coupling u and the four legs the four fields in H_1.

The first two solutions correspond to a non-critical behavior ($D_\varphi = \frac{d}{2}$) with the two fixed points describing the two uncorrelated cases of $T = \infty$ ($r_0 > 0$) and $T = 0$ ($r_0 < 0$). For the latter, one has to introduce a small stabilizing $u\hat{\varphi}^4$ term.

The $\eta = 0$ solution (critical Gaussian fixed point) represents the critical behavior of the Landau theory with a diverging response and diverging coherence distance with critical index $\nu = 1/2$ according to

$$r_0' = s^{D_{r_0}} r_0, \quad D_{r_0} = \frac{1}{\nu}. \tag{17.44}$$

The first two solutions correspond to an asymptotic distribution Gaussian in the fields, the last one corresponds to a distribution Gaussian in the gradient of the fields. In the last case, neighboring blocks of spins are coupled.

In order to analyze the effect of the quartic term with $u \neq 0$, we have to linearize the group equations around the critical Gaussian fixed point. To this end we use the first order term of the transformation Eq. (17.35), where the average is performed with respect to H^* specified by Eq. (17.43) and the perturbing Hamiltonian $H_1 = H_1^{r_0} + H_1^u$ now includes both the r_0 and u terms of Eq. (17.31). Since $\hat{\varphi}_\mathbf{k}$ has a Gaussian distribution $\exp(-(ck^2/2)\hat{\varphi}_\mathbf{k}\hat{\varphi}_{-\mathbf{k}})$, the average of the product of any number of fields $\hat{\varphi}_\mathbf{k}$ can be evaluated in terms of products of the basic quadratic contraction (average)

$$\langle \hat{\varphi}_{i,\mathbf{k}_1} \hat{\varphi}_{j,\mathbf{k}_2} \rangle = \delta_{ij} \frac{\delta(\mathbf{k}_1 + \mathbf{k}_2)}{ck_1^2}. \tag{17.45}$$

In Eq. (17.35), the average of the r_0-term in H_1 leads to a contribution to F_s due to the integrated fast modes, while the slow modes reproduce the free quadratic term $r_0 s^{2-\eta}$ ($\eta = 0$) of H_s.

To calculate the average of the u-term of H_1, we graphically represent it in Fig. 17.2 as a vertex with four legs where each leg represents a field $\hat{\varphi}_{i,\mathbf{k}}$. The corresponding mathematical expression from Eq. (17.31) reads

$$H_1^u = \frac{u}{L^d} \sum_{ij,\{\mathbf{k}_i\}} \delta_{\{\mathbf{k}_i\}} \hat{\varphi}_{i,\mathbf{k}_1} \hat{\varphi}_{i,\mathbf{k}_2} \hat{\varphi}_{j,\mathbf{k}_3} \hat{\varphi}_{j,\mathbf{k}_4}. \tag{17.46}$$

We assign to each leg an incoming or an outgoing direction depending on the sign of $\mathbf{k}$. The contraction defined in Eq. (17.45) corresponds to a line joining two legs as in Fig. 17.3.

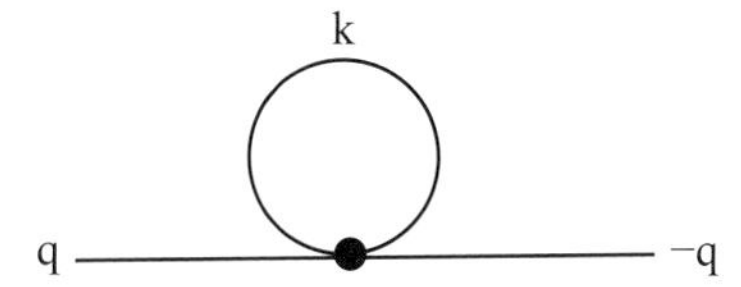

Fig. 17.3 First order correction to the quadratic term r_0. The full closed loop represents the contraction of two legs of the vertex.

When all four momenta of the legs are in the fast modes region, the average produces a contribution to F_s. When all four momenta of the legs are in the slow modes region, the average leaves this term untouched in H_s and rescaling the volume L^d (s^{-d}) and the field $\hat{\varphi}_{i,\mathbf{k}} \to \hat{\varphi}_{i,\mathbf{k}s} s^{1-\eta/2}$ with $\eta = 0$ leads to

$$u' = s^{4-d} u. \tag{17.47}$$

We produce a correction to r_0 proportional to u by contracting two out of the four legs, leaving the remaining two legs untouched. There is a multiplicity factor $M = 2n + 4$, corresponding to the number of equivalent ways of making the contraction (see Fig. 17.3):

$$\begin{aligned}\delta r_0 &= (2n+4)u \sum_{i,0<|\mathbf{q}|<\Lambda/s} \left(\frac{1}{L^d} \sum_{\Lambda/s<|\mathbf{k}|<\Lambda} \frac{1}{ck^2} \right) |\hat{\varphi}_{i,\mathbf{q}}|^2 \\ &= (2n+4)u \frac{K_d}{d-2} \frac{\Lambda^{d-2}}{c} (1 - s^{2-d}) \sum_{i,0<|\mathbf{q}|<\Lambda/s} |\hat{\varphi}_{i,\mathbf{q}}|^2. \end{aligned} \tag{17.48}$$

In Eq. (17.48), the sum over $\mathbf{k}$ is performed in the continuum limit and K_d is the area of a unit hypersphere in d dimensions divided by $(2\pi)^d$. In the multiplicity factor, $2n$ terms come from a contraction of two legs of the type $\langle \hat{\varphi}_{i,\mathbf{k}_1} \hat{\varphi}_{i,\mathbf{k}_2} \rangle$, while the factor of 4 arises with contractions of the type $\langle \hat{\varphi}_{i,\mathbf{k}_1} \hat{\varphi}_{j,\mathbf{k}_2} \rangle$. After rescaling the term of (17.48) and adding it to the contribution due to the r_0-term in H_1, we obtain

$$r_0' = s^2 (r_0 + u B_d (1 - s^{2-d})), \tag{17.49}$$

where B_d is

$$B_d = 2(2n+4) \frac{K_d}{c} \frac{\Lambda^{d-2}}{d-2}. \tag{17.50}$$

By combining (17.47) with (17.49), we obtain the linearized transformation around the Gaussian fixed point as

$$\begin{pmatrix} r_0' \\ u' \end{pmatrix} = L_s(H^*) \begin{pmatrix} r_0 \\ u \end{pmatrix}, \tag{17.51}$$

where

$$L_s(H^*) = \begin{pmatrix} s^2 & (s^2 - s^{4-d}) B_d \\ 0 & s^{4-d} \end{pmatrix}. \tag{17.52}$$

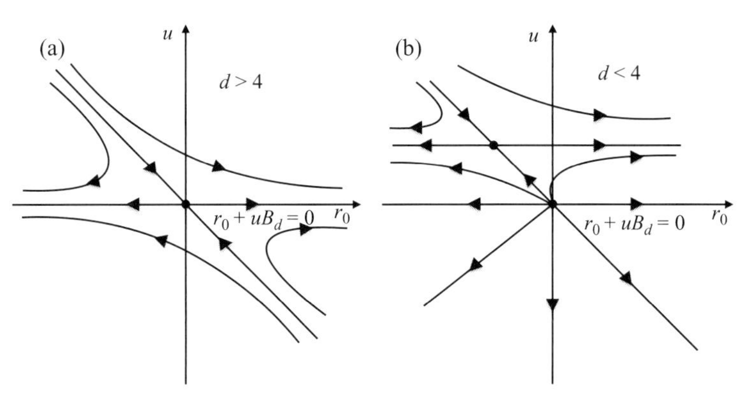

Fig. 17.4 (a) RG trajectories for $d > 4$. The origin ($r_0 = 0$, $u = 0$) is the Gaussian fixed point. The straight line $r_0 + uB_d = 0$ is the critical line, whose points scale to the fixed point by iterating the RG transformation. The physical solutions are for $u \geq 0$. For $d < 4$ the Gaussian fixed point becomes unstable also along the critical line. (b) RG trajectories for $d < 4$. Besides the Gaussian fixed point as in $d > 4$, there appears a non-Gaussian fixed point along the critical line $r_0 + uB_d = 0$. Whereas the Gaussian fixed point is unstable, the new fixed point is stable along the critical line. In all cases for $t > 0$ and $t < 0$ the flow scales to $r_0^* = \infty$ and $-\infty$, respectively.

The two eigenvalues and the corresponding eigenvectors are

$$\lambda_1 = s^2, \quad \lambda_2 = s^{4-d}, \quad h_1^* = \begin{pmatrix} 1 \\ 0 \end{pmatrix}, \quad h_2^* = \begin{pmatrix} -B_d \\ 1 \end{pmatrix}. \tag{17.53}$$

For $d > 4$ the critical Gaussian fixed point is stable with respect to the introduction of the u-term along the critical line corresponding to $t = r_0 + uB_d = 0$. Deviation from criticality is governed by $t' = s^2 t$, t being the scaling field combination of r_0 and u. Given $\eta = 0$ and $\nu = 1/2$ all the critical indices have the classical values and are independent on the number of components n. The correction to the critical temperature introduced by u is instead dependent on n. A picture of the RG flow in the parameter space (r_0, u) is shown in Fig. 17.4.

For $d < 4$, u becomes relevant with respect to the Gaussian critical fixed point. The latter becomes unstable for the introduction of u, since Eq. (17.47) scales away from the fixed point even for $t = 0$.

In order to describe the critical behavior of the standard second-order phase transitions, one has to find a new fixed point.

17.3.2 ϵ-expansion

For space dimensionality $d < 4$ but sufficiently near to four, we may ask if the slow running away of $u \propto s^{(4-d)}$ from the Gaussian fixed point could be compensated by the non-linear terms arising from $(H_1^u + H_1^{r_0})^2$ in Eq. (17.36) to obtain a non-Gaussian fixed point. The new equation for u' in differential form would read

$$\frac{du'}{d \ln s} = \epsilon u' - a u'^2,$$

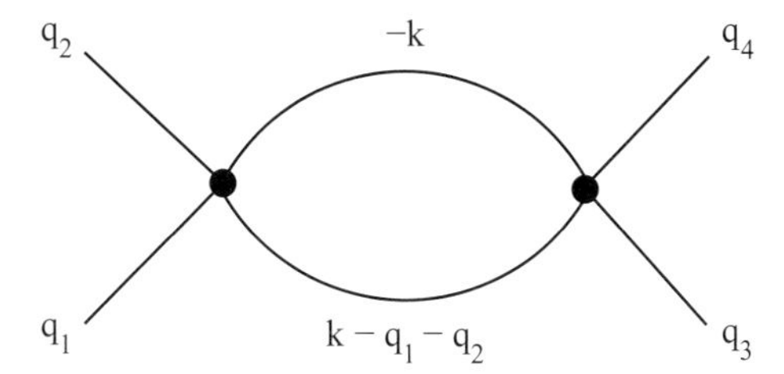

Fig. 17.5 First correction to the four-leg vertex.

where $\epsilon = 4 - d$ measures the deviation with respect to the dimensionality $d = 4$. With a constant $a > 0$ independent of ϵ, a new fixed point $u^* = \epsilon/a$ besides $u^* = 0$ would appear. Linearizing the above equation around the new fixed point, for $\delta u' = u' - u^*$ one obtains $\delta u' = s^{-\epsilon}\delta u$. The new fixed point would be stable with respect to the perturbation δu for $d < 4$. We proceed to the explicit calculation. Let us begin with $(H_1^u)^2$. By considering the term where four legs have a fast momentum, which are averaged out by the contraction (17.45) to form a loop (each line of the loop is a contraction), and the other four legs a slow momentum, a correction is obtained to the quartic term in the Hamiltonian as shown in Fig. 17.5. In mathematical terms

$$
\begin{aligned}
-\frac{1}{2}\langle (H_1^u)^2 \rangle &= -4(n+8)\frac{u^2}{L^d} \sum_{ij,\{0<|\mathbf{q}_i|<\Lambda/s\}} \left(\frac{1}{L^d} \sum_{\Lambda/s<|\mathbf{k}|<\Lambda} \frac{1}{ck^2} \frac{1}{c(k-q_1-q_2)^2} \right) \\
&\quad \times \delta_{\{\mathbf{q}_i\}} \hat{\varphi}_{i,\mathbf{q}_1} \hat{\varphi}_{i,\mathbf{q}_2} \hat{\varphi}_{j,\mathbf{q}_3} \hat{\varphi}_{j,\mathbf{q}_4} \\
&= -4(n+8)u^2 \frac{K_d \Lambda^{d-4}}{c^2} \frac{1-s^{4-d}}{d-4} \frac{1}{L^d} \sum_{ij,\{0<|\mathbf{q}_i|<\Lambda/s\}} \delta_{\{\mathbf{q}_i\}} \hat{\varphi}_{i,\mathbf{q}_1} \hat{\varphi}_{i,\mathbf{q}_2} \hat{\varphi}_{j,\mathbf{q}_3} \hat{\varphi}_{j,\mathbf{q}_4} .
\end{aligned}
\tag{17.54}
$$

Notice that the integral over the fast momenta has been evaluated for large s. In the multiplicity factor $4(n+8)$, the $4n$-term comes from selecting at each vertex a pair of equal components j of the field $\hat{\varphi}_{j,\mathbf{k}}$, whereas 32 comes by selecting pairs with different components in each vertex. The factor $1/2$ of the cumulant expansion is compensated by the fact that, once we have chosen a pair of legs from each vertex for the contraction, there remain two possible ways of doing it. The scaling for u of Eq. (17.47) now becomes

$$
u' = s^{4-d} \left(u - 4(n+8)u^2 \frac{K_d \Lambda^{d-4}}{c^2} \frac{1-s^{4-d}}{d-4} \right). \tag{17.55}
$$

The fixed point condition becomes from (17.55)

$$
u^* = s^{\epsilon} u^* \left(1 - 4(n+8)u^* \frac{K_{4-\epsilon}\Lambda^{-\epsilon}}{c^2} \frac{s^{\epsilon}-1}{\epsilon} \right). \tag{17.56}
$$

When ϵ is small, at first order in ϵ, $s^{\epsilon} = 1 + \epsilon \ln s + \cdots$, the correction term is proportional to $\ln s$ and Eq. (17.56) has, besides the solution $u^* = 0$, the new fixed point

$$
u^* = \frac{c^2}{4(n+8)K_4}\epsilon, \tag{17.57}
$$

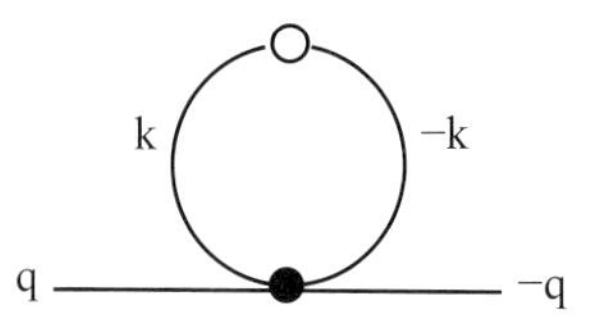

Fig. 17.6 Renormalization of the quadratic term r_0 due to $H_1^u H_1^{r_0}$. The empty dot (r_0) is the two-leg vertex of $H_1^{r_0}$.

which, as we have seen, is stable for $d < 4$. Since the fixed point occurs for a value of u of order ϵ, we may rely on a perturbative evaluation of the RG transformation in powers of u. Moreover, all the integrals appearing in the perturbative expansion must be expanded in powers of ϵ, so that the entire procedure involves a simultaneous expansion in ϵ and u.

From the term $-(1/2)H_1^2$, by contracting only two legs, we generate an interaction with six fields. The last one can also be generated from $(1/3!)H_1^3$ by contracting three pairs of legs. This term can also generate, at this order, an interaction at eight fields by contracting two pairs, and so on. However, as long as we consider $3 < d \leq 4$, all these interactions intervene only in the transient of the transformation, because their scaling dimensions $D_m = d - (m/2)(d - 2 + \eta)$ are negative for $m > 4$.

In order to find the new critical line for $d < 4$, we now consider the contribution to the renormalization of the quadratic terms of the Hamiltonian arising from $2H_1^u H_1^{r_0}$. We have the diagram of Fig. 17.6 where the empty dot represents the two-leg vertex $(1/2)r_0\hat{\varphi}_i\hat{\varphi}_i$ associated with the r_0-term. The corresponding mathematical expression reads

$$-\frac{1}{2}\langle 2H_1^{r_0}H_1^u\rangle = -\frac{r_0}{2}u2(2n+4)\sum_{i,0<|\mathbf{q}|<\Lambda/s}\left(\frac{1}{L^d}\sum_{\Lambda/s<|\mathbf{k}|<\Lambda}\frac{1}{(ck^2)^2}\right)|\hat{\varphi}_{i,\mathbf{q}}|^2$$
$$= -\frac{r_0}{2}uB_d(d-2)\frac{\Lambda^{-2}}{c}\frac{1-s^{4-d}}{d-4}\sum_{i,0<|\mathbf{q}|<\Lambda/s}|\hat{\varphi}_{i,\mathbf{q}}|^2. \quad (17.58)$$

By combining the above term (17.58) with that coming from Eq. (17.48) and adding it to the r_0-term due to $H_1^{r_0}$, we have

$$r_0 + uB_d(1 - s^{2-d}) - uB_d\frac{r_0(d-2)}{c\Lambda^2}\frac{1-s^{4-d}}{d-4}. \quad (17.59)$$

We see that the second term in Eq. (17.59) yields to lowest order in ϵ a correction, which becomes independent of s when s is large, at least for small ϵ. Since u^* is of order ϵ in Eq. (17.57), for the search of the fixed point at order ϵ, the last term would be of order ϵ^2 and can be neglected. The fixed point value at order ϵ for r_0 gives

$$r_0^* = -u^*B_4 = -\frac{(n+2)c\Lambda^2}{2(n+8)}\epsilon, \quad (17.60)$$

where the integral B_d has been evaluated to zero order in ϵ. This suggests that, as for the Gaussian fixed point, the inclusion of u produces a shift in the critical temperature, so that the real parameter controlling the deviation with respect to the critical temperature is

t defined by

$$t = r_0 + uB_4. \tag{17.61}$$

r_0^* is obviously on the critical line $t = 0$. Equation (17.59) becomes, after replacing r_0 with t,

$$t - uB_4 \frac{2}{c\Lambda^2} \ln s\,(t - uB_4) \approx t - utB_4 \frac{2}{c\Lambda^2} \ln s. \tag{17.62}$$

In the above equation we do not consider the term proportional to u^2 even if it is of the order ϵ^2 because, as will be shown later in Eq. (17.69), it is exactly cancelled by considering from the u^2-terms, due to $(H_1^u)^2$, those contributing to the renormalization of the quadratic part of the Hamiltonian H_s. This cancellation confirms that t is the correct deviation with respect to the critical temperature. Then the RG transformation for t reads

$$t' = s^2 t\left(1 - uB_4 \frac{2}{c\Lambda^2} \ln s\right), \tag{17.63}$$

which, as expected, has a fixed point for $t = 0$. By linearising the group equations (17.63) and (17.55) around the new fixed point $t^* = 0$ and u^* given by (17.57), after exponentiating $\ln s$ we get

$$\begin{pmatrix} \delta t' \\ \delta u' \end{pmatrix} = \begin{pmatrix} s^{\lambda_1} & 0 \\ 0 & s^{\lambda_2}. \end{pmatrix} \begin{pmatrix} \delta t \\ \delta u \end{pmatrix}, \tag{17.64}$$

where

$$\lambda_1 = 2\left(1 - \frac{n+2}{2(n+8)}\epsilon\right), \quad \lambda_2 = -\epsilon. \tag{17.65}$$

From this we obtain the index $\nu = \lambda_1^{-1}$ as

$$\nu = \frac{1}{2}\left(1 + \frac{n+2}{2(n+8)}\epsilon\right). \tag{17.66}$$

The $L_s(H^*)$ given by Eq. (17.64) has the proper form of the scaling field representation of Eq. (17.20). It is instructive to re-express the linearized RG transformation in term of the original parameters r_0 and u, by using

$$\begin{pmatrix} \delta t \\ \delta u \end{pmatrix} = \begin{pmatrix} 1 & B_4 \\ 0 & 1 \end{pmatrix} \begin{pmatrix} \delta r_0 \\ \delta u \end{pmatrix}. \tag{17.67}$$

Off-diagonal terms must of course appear in $L_s(H^*)$, but $L_s(H^*)$ must be expressed in terms of its eigenvalues of the linearized transformation only. Indeed,

$$L_s(H^*) = \begin{pmatrix} 1 & B_4 \\ 0 & 1 \end{pmatrix}^{-1} \begin{pmatrix} s^{\lambda_1} & 0 \\ 0 & s^{\lambda_2} \end{pmatrix} \begin{pmatrix} 1 & B_4 \\ 0 & 1 \end{pmatrix} = \begin{pmatrix} s^{\lambda_1} & B_4(s^{\lambda_1} - s^{\lambda_2}) \\ 0 & s^{\lambda_2} \end{pmatrix}, \tag{17.68}$$

which has the same form as the corresponding linearized transformation around the Gaussian fixed point, of course with different exponents. Equations (17.55) and (17.63) are the RG transformations by keeping terms of order u with respect to the leading terms. In this

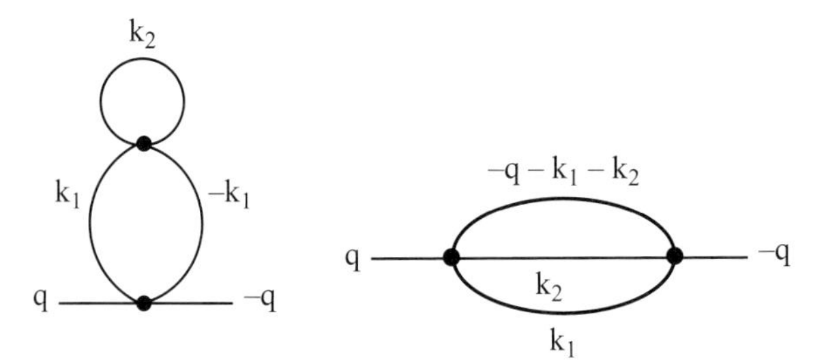

Fig. 17.7 Renormalization of the quadratic term at second order in H_1^2.

respect, the equation for u contains terms up to second order in u, which, however, represent a correction to the leading term of order u.

With the terms considered so far, there are no corrections to the term $(1/2)\sum_{i,\mathbf{k}} ck^2|\hat{\varphi}_{i,\mathbf{k}}|^2$ in the Hamiltonian. Hence to this order, according to Eq. (17.43), the new fixed point also has $\eta = 0$, the value of the Landau theory. Within this approximation the two independent critical indices are ν given by Eq. (17.66) ($\nu = 1/2 + \epsilon/12$ for the universality class $n = 1$) and $\eta = 0$. This is in agreement with the small values of η found experimentally. Corrections to the index η arise when taking into account terms of second order in u and in ϵ. We must then analyze the terms arising from $(H_1^u)^2$ contributing to the renormalization of the quadratic part of the Hamiltonian H_s. There are two such terms, as shown in Fig. 17.7, which have to be evaluated in $d = 4$. The first gives

$$-\frac{1}{2}\left\langle \left(H_1^u\right)^2 \right\rangle_a^{(2)} = -16(n+2)^2u^2 \sum_{i,0<|\mathbf{q}|<\Lambda/s} \left[\sum_{\Lambda/s<|\mathbf{k}_1|<\Lambda} \frac{1}{c^2k_1^4} \sum_{\Lambda/s<|\mathbf{k}_2|<\Lambda} \frac{1}{ck_2^2} \right] \hat{\varphi}_{i,\mathbf{q}}\hat{\varphi}_{i,-\mathbf{q}}$$

$$= -B_4^2 \frac{2}{c\Lambda^2} u^2 \ln s \sum_{i,0<|\mathbf{q}|<\Lambda/s} \hat{\varphi}_{i,\mathbf{q}}\hat{\varphi}_{i,-\mathbf{q}}, \tag{17.69}$$

where the superscript (2) indicates that it contributes to the quadratic part. As anticipated, the term of Eq. (17.69) exactly cancels the term of order u^2 neglected in Eq. (17.60). The other term of order u^2 contributing to the renormalization of the quadratic part is

$$-\frac{1}{2}\left\langle \left(H_1^u\right)^2 \right\rangle_b^{(2)}$$

$$= -32(n+2)u^2 \sum_{i,0<|\mathbf{q}|<\Lambda/s} \left[\sum_{\Lambda/s<|\mathbf{k}_1|,|\mathbf{k}_2|<\Lambda} \frac{1}{c^3} \frac{1}{k_1^2k_2^2(\mathbf{q}+\mathbf{k}_1+\mathbf{k}_2)^2} \right] \hat{\varphi}_{i,\mathbf{q}}\hat{\varphi}_{i,-\mathbf{q}}$$

$$\equiv -\sum_{i,0<|\mathbf{q}|<\Lambda/s} \hat{\varphi}_{i,\mathbf{q}}\Sigma(\mathbf{q})\hat{\varphi}_{i,-\mathbf{q}}, \tag{17.70}$$

where

$$\Sigma(\mathbf{q}) = 32(n+2)u^2 \sum_{\Lambda/s<|\mathbf{k}_1|,|\mathbf{k}_2|<\Lambda} \frac{1}{c^3} \frac{1}{k_1^2k_2^2(\mathbf{q}+\mathbf{k}_1+\mathbf{k}_2)^2}. \tag{17.71}$$

When $\mathbf{q} = \mathbf{0}$, the term $\Sigma(\mathbf{0})$ is proportional to Λ^2 and contributes to the renormalization of the parameter t by adding a correction of order u^2 on the right-hand side of Eq. (17.61).

The contribution to the index η arises when considering the dependence on $\mathbf{q}$. Assuming that there were no restrictions in k space, by Fourier transforming we may write

$$\Sigma(\mathbf{q}) = 32(n+2)u^2 \int \mathrm{d}^d x \mathrm{e}^{\mathrm{i}\mathbf{q}\cdot\mathbf{x}} G(\mathbf{x})^3, \tag{17.72}$$

where we have introduced the real-space contraction (see Problem 17.1)

$$G(\mathbf{x}) = \langle \hat{\varphi}(\mathbf{x})\hat{\varphi}(0)\rangle = \sum_{\mathbf{q}} \frac{\mathrm{e}^{\mathrm{i}\mathbf{q}\cdot\mathbf{x}}}{cq^2} = \frac{\Gamma(d/2-1)}{4\pi^{d/2}c|\mathbf{x}|^{d-2}}. \tag{17.73}$$

We have

$$\Sigma(\mathbf{q}) - \Sigma(\mathbf{0}) = 32(n+2)u^2 \int \mathrm{d}^d x G(\mathbf{x})^3(\mathrm{e}^{\mathrm{i}\mathbf{q}\cdot\mathbf{x}} - 1) \approx \left(\frac{32(n+2)}{2d}u^2 \int \mathrm{d}^d x x^2 G(\mathbf{x})^3\right) q^2. \tag{17.74}$$

We take into account the original restrictions in k space on the Wilson integrals in Eq. (17.70) by limiting the above integral in real space to $\Lambda^{-1} < x < s\Lambda^{-1}$. One then finds at $d = 4$

$$\Sigma(\mathbf{q}) - \Sigma(\mathbf{0}) = cq^2 \frac{16(n+2)u^2 K_4^2}{2c^4} \ln s, \tag{17.75}$$

which leads to the RG transformation for the parameter c:

$$c' = (c + \delta c)s^{-\eta}\left(1 + \frac{16(n+2)u^2 K_4^2}{2c^4} \ln s\right) \tag{17.76}$$

from which, by using the fixed-point value u^* of Eq. (17.57) and removing $\ln s$ by adjusting η, one obtains

$$\eta = \frac{n+2}{2(n+8)^2}\epsilon^2. \tag{17.77}$$

Notice that since the first correction for the index η arises at second order in u, it is sufficient to use the fixed-point value for u with accuracy to first order in ϵ. However, if one wants to use the value for η with corrections up to ϵ^2, one must consistently consider corrections up to the same order to ν. This would imply that we should derive the RG transformations for t (r_0) and u up to second and third order in u, respectively.

We have then accomplished the task of determining the non-trivial fixed point and the associated critical indices to order ϵ. Following Wilson, the renormalization group transformation has been developed by exploiting the physical idea of grouping together degrees of freedom at the small scale. In the next section we see how these results follow from the general structure of the renormalization group in field theory.

17.4 Field-theoretic renormalization group

The field theoretic group approach to critical phenomena finds a powerful technical support for the explicit evaluation of the group equations and of the critical indices in the previous knowledge of the well known perturbation theory in terms of Feynman diagrams. See for instance the book by D. J. Amit and V. Martín-Mayor (Amit and Martín-Mayor (2005)).[3]

In this presentation we will try to be self-consistent and will not assume any previous knowledge of field theory. The most general polynomial Hamiltonian for a continuous field theory can be obtained by generalizing the Landau–Wilson model of Eq. (17.31) with the introduction of a generic interaction term $u_m \int \mathrm{d}\mathbf{x}\hat{\varphi}^m(\mathbf{x})$ ($u_4 = u$). As we have already seen in Section 17.3.2, its dimensions D_m are, for $m > 4$, negative for $3 < d \leq 4$ and the corresponding couplings scale to zero. In studying critical phenomena we can therefore limit our discussion to the coarse-grained Landau–Wilson Hamiltonian (17.30)–(17.31), known in field theory as the $\hat{\varphi}^4$-model, with an ultraviolet cutoff Λ:

$$H = H_0 + H_1, \tag{17.78}$$

$$H_0 = \frac{1}{2}\sum_i \int \mathrm{d}^d x \left(r_0\hat{\varphi}_i^2(\mathbf{x}) + c|\nabla\hat{\varphi}_i(\mathbf{x})|^2\right) = \frac{1}{2}\sum_{|\mathbf{k}|<\Lambda, i} (r_0 + c|\mathbf{k}|^2)|\hat{\varphi}_{i,\mathbf{k}}|^2, \tag{17.79}$$

$$H_1 = \int \mathrm{d}^d x u \left(\sum_i \hat{\varphi}_i^2(\mathbf{x})\right)^2 = \frac{u}{L^d}\sum_{\{\mathbf{k}_i\}, i, j} \delta_{\{\mathbf{k}_i\}}\hat{\varphi}_{i,\mathbf{k}_1}\hat{\varphi}_{i,\mathbf{k}_2}\hat{\varphi}_{j,\mathbf{k}_3}\hat{\varphi}_{j,\mathbf{k}_4}. \tag{17.80}$$

Remember that $\hat{\varphi}^*_{i,\mathbf{k}} = \hat{\varphi}_{i,-\mathbf{k}}$ for a real field. The index i runs over the n real components of the field. The interaction term H_1 of Eq. (17.80), quartic in the field, is the perturbing term in an eventual perturbation theory with the quadratic term H_0 of Eq. (17.79) as the starting "free" Hamiltonian. One would have the same cumulant expansion as in Section 17.2, but now the average Eq. (17.38) would be with no restriction on the fast variables.

For simplicity, let us consider the scalar case with $n = 1$ components of the field $\hat{\varphi}(\mathbf{x})$. The extension to any n is straightforward. In the presence of an external field $h(\mathbf{x})$ the free energy $F(T, \{h(\mathbf{x})\})$ is a functional of $h(\mathbf{x})$.

By extending to higher functional derivatives, the relation in Eq. (13.17) for the order parameter $\varphi(\mathbf{x}) = \langle\hat{\varphi}(\mathbf{x})\rangle = -\delta F/\delta h(\mathbf{x})$ and for its connected correlation function $f_\varphi(\mathbf{x}, \mathbf{x}') = -\beta^{-1}\delta^2 F/\delta h(\mathbf{x})\delta h(\mathbf{x}')$, F is shown to be the generating functional of the connected correlation functions of any order m:

$$f^m_\varphi(\mathbf{x}_1, \ldots, \mathbf{x}_m) = -\frac{\delta^m F}{\delta h(\mathbf{x}_1)\cdots\delta h(\mathbf{x}_m)}. \tag{17.81}$$

The correlation function $f_\varphi(\mathbf{x}, \mathbf{x}')$ in field theory is referred to as the propagator $G(\mathbf{x}, \mathbf{x}')$ of the classical field $\hat{\varphi}(\mathbf{x})$. By means of the Legendre transform (13.18), one can eliminate $h(\mathbf{x})$ in favor of the order parameter $\varphi(\mathbf{x})$ and define the functional $\Gamma(T, \{\varphi(\mathbf{x})\})$:

$$\Gamma(T, \{\varphi(\mathbf{x})\}) = \beta\tilde{F}(T, \{\varphi(\mathbf{x})\}) = \beta F(T, \{h(\mathbf{x})\}) + \beta\int \mathrm{d}^d x h(\mathbf{x})\varphi(\mathbf{x}), \tag{17.82}$$

with

$$\frac{\delta\Gamma}{\delta\varphi(\mathbf{x})} = \beta h(\mathbf{x}).$$

[3] Indeed, most of the actual calculations of the critical indices have been carried out within this context.

Γ is then the generator of the m-point vertex functions defined via the m-functional derivatives

$$\Gamma^{(m)}(\mathbf{x}_1, \ldots, \mathbf{x}_m) = \frac{\delta^m \Gamma}{\delta\varphi(\mathbf{x}_1)\cdots\delta\varphi(\mathbf{x}_m)}. \tag{17.83}$$

Because of translational invariance, the Fourier transform of $\Gamma^{(m)}$ reads

$$\delta(\mathbf{k}_1 + \cdots + \mathbf{k}_m)\Gamma^{(m)}(\mathbf{k}_1, \ldots, \mathbf{k}_m) = \int \mathrm{d}^d x_1 \cdots \mathrm{d}^d x_m \mathrm{e}^{-\mathrm{i}(\mathbf{k}_1\cdot\mathbf{x}_1 + \cdots + \mathbf{k}_m\cdot\mathbf{x}_m)}\Gamma^{(m)}(\mathbf{x}_1, \cdots, \mathbf{x}_m). \tag{17.84}$$

In particular, the second derivative of Γ coincides with the inverse correlation function:

$$\Gamma^{(2)}(\mathbf{x}, \mathbf{x}') = \frac{\delta^2\Gamma}{\delta\varphi(\mathbf{x})\delta\varphi(\mathbf{x}')} = \beta\frac{\delta h(\mathbf{x})}{\delta\varphi(\mathbf{x})} = f_\varphi^{-1}(\mathbf{x}, \mathbf{x}'), \tag{17.85}$$

and for a translational invariant system $\Gamma^{(2)}(\mathbf{k}) = f_\varphi^{-1}(\mathbf{k})$.

The quadratic Hamiltonian H_0 (17.79) is considered as the basic "free" Hamiltonian to which corresponds the Gaussian approximation of Chapter 14. In this approximation $f_\varphi(\mathbf{k}) = \langle|\hat{\varphi}_\mathbf{k}|^2\rangle$ is the average over the Gaussian distribution (see Eq. (14.33)) and coincides with the basic quadratic contraction (17.45) of one pair of legs in the graphic representation discussed in the previous section. Now H_0 of Eq. (17.79) includes both terms quadratic in the field and Eq. (17.45) has to be generalized here to include r_0. The corresponding $\Gamma^{(2)}(\mathbf{k})$ reads

$$\Gamma_0^{(2)}(\mathbf{k}) = k^2 + r_0 \equiv G^{-1}(\mathbf{k}). \tag{17.86}$$

We recall that r_0 coincides with the inverse squared of the correlation length in the Landau approximation and controls the approach to the critical point where it vanishes.

The vertex $\Gamma_0^{(4)}(\mathbf{k}_1, \mathbf{k}_2, \mathbf{k}_3, \mathbf{k}_4)$ coincides with the coupling constant ($\Gamma_0^{(4)} = u$) of four fields represented by four legs departing, in the graphic representation in Fig. 17.2, from a black dot denoting u.

As we have already mentioned, the perturbative structure of the theory in the interaction coupling u takes advantage of the standard procedure based on Feynman diagrams. The latter allow us to represent in a graphic fashion the perturbative contributions to the various physical quantities, once the general rules for calculating the analytical expressions corresponding to the different diagrams have been established. As in Section 17.3.1, continuous lines represent contractions of two legs, i.e. the bare propagators $(\Gamma_0^{(2)}(\mathbf{k}))^{-1} = G(\mathbf{k})$, which join the bare interaction vertices u. Momentum conservation is understood at each vertex. The momentum of each closed continuous line must be integrated with no restriction on k, yielding therefore the explicit dependence on the momentum cutoff Λ. Diagrams contributing to the dressed inverse propagator $\Gamma^{(2)}$ and to the dressed $\Gamma^{(4)}$ (big black dot) are shown in Fig. 17.8 up to third order in u. For an explicit expression of the lowest order diagrams see the answers to Probems 17.2 and 17.5. The first evident effect of introducing a non-zero coupling constant u is a shift of the transition temperature with respect to $r_0 = 0$, as we have already seen in Eq. (17.61) at first order in the ϵ-expansion within the Wilson approach. Hereafter we assume we know the full shift of the critical temperature and include it in

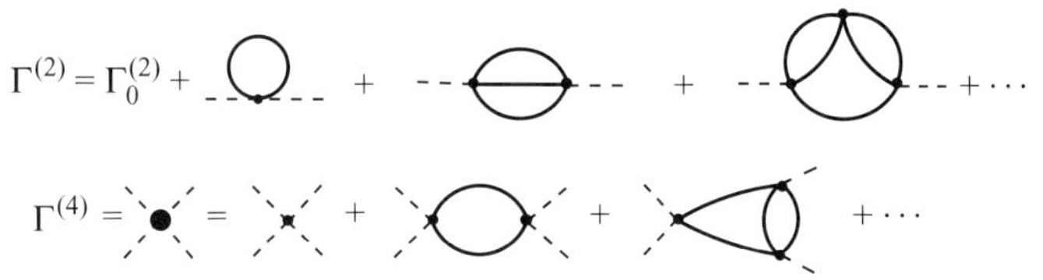

Fig. 17.8 Diagrams contributing up to third order in u to the inverse propagator $\Gamma^{(2)}$ (top). Diagrams contributing to the dressed vertex $\Gamma^{(4)}$ (bottom). The dotted lines represent the field $\hat{\varphi}$ which in the Hamiltonian depart from the bare vertices. The diagrams of $\Gamma^{(2)}$ with full instead of dotted lines give the graphical representation of the propagator $G \equiv f_\varphi$. The small black dot is the bare vertex $\Gamma^{4(0)}$, whereas the big dot indicates the fully dressed one.

the definition of the new mass $t = r_0 + \delta r_0(u, \Lambda)$ in substitution of r_0, as in Eq. (17.61). $\Gamma^{(2)}(\mathbf{k})$ is then expressed in terms of t and named the bare mass in field theory, which vanishes at the true critical temperature. Perturbation theory in terms of the bare coupling constant u is, however, ill-defined in $d < 4$ due to the infrared divergences which arise when the critical point is approached. The dimensionless coupling $\overline{u} = u/t^{(4-d)/2}$ diverges while approaching criticality and makes the bare perturbative expansion meaningless. We need a renormalized theory, where the renormalized coupling constant remains finite in the limit $t \to 0$, i.e., when the square of the inverse coherence length ξ^{-2}, which plays the role of the true mass of the field $\hat{\varphi}$, vanishes. Therefore we are interested in a renormalized theory in which the ultraviolet cutoff Λ is much larger than the mass of the propagator. If we can control, as in a traditional renormalization procedure, the ultraviolet divergences generated by the momentum integrations, we can take the limit $\Lambda \to \infty$, make the cutoff disappear from the RG equations and also control the infrared singular behavior (small $\mathbf{k}$) ruled by the dimensionless variable k/Λ. The resulting scaling theory will also be valid at finite Λ as long as $\Lambda \gg \xi^{-1}$. For our derivation of the infrared renormalized theory we shall be guided by considerations on the invariance properties of the repetitive diagrammatic structure shown in Fig. 17.8, which is a consequence of the general definitions (17.81) and (17.83) of propagators and vertices. At each order in perturbation theory, an additional vertex, two propagator lines and a momentum integration are added, which are present in each additional loop in the diagram. In constructing the RG transformation, the relevant variables which enter in the description of the system are t and h (or its conjugate variable φ). As we have seen previously, another quantity which is needed to describe the approach to the fixed point is the first irrelevant coupling, described by u. We are therefore led to find a transformation of these variables and of the corresponding vertices via the introduction of the standard renormalizations of field theory, i.e. the so-called wave function φ, mass t, and vertex u renormalizations, multiplicatively implemented by three parameters Z_φ, Z_t, Z_u. It is convenient to introduce the dimensionless vertices

$$\gamma_\varphi = \frac{\partial \Gamma^{(2)}}{\partial k^2}, \quad \gamma_t = \frac{\partial \Gamma^{(2)}}{\partial t}, \quad \gamma_u = \frac{\Gamma^{(4)}}{u}, \tag{17.87}$$

where obviously their bare expressions are

$$\gamma_\varphi^0 = \frac{\partial}{\partial k^2}(ck^2 + t) = c, \quad \gamma_t^0 = \gamma_u^0 = 1.$$

$$\gamma_t = \bigcirc = \circ + \ldots + \ldots + \cdots$$

$$\gamma_t^0 = 1$$

Fig. 17.9 Diagrammatic expansion for γ_t (big empty dot). Each small empty dot represents the mass insertion due to the derivative of $\Gamma^{(2)}$ with respect to t.

The diagrammatic representation of γ_t is given in Fig. 17.9. It is obtained by inserting in each propagator line an empty dot $\gamma_t^0 = 1$ to represent the derivative with respect to t of $\Gamma^{(2)}$. Indeed

$$\frac{\partial G^0}{\partial t} = -\frac{1}{(ck^2+t)^2} = -(G^0)^2 \frac{\partial \gamma_t^0}{\partial t}$$

and each derivative breaks a bare propagator line in two and introduces the vertex γ_t^0. The same applies to the full quantities G and γ_t.

Let us introduce for the moment the trivial invariance property under each renormalization separately. Under a *wave-function renormalization* of the variables

$$\varphi \to Z_\varphi^{-1/2}\varphi, \quad u \to Z_\varphi^2 u, \quad t \to Z_\varphi t,$$

the invariance holds provided at the same time

$$\gamma_\varphi \to Z_\varphi \gamma_\varphi.$$

Similarly, under a *mass renormalization*

$$t \to Z_t^{-1} t, \quad \gamma_t \to Z_t \gamma_t.$$

Finally, under a *vertex renormalization*

$$u \to Z_u^{-1} u, \quad \gamma_u \to Z_u \gamma_u.$$

When the three transformations are used simultaneously, the new variables read

$$\bar{\varphi} = Z_\varphi^{-1/2}\varphi, \ \bar{t} = Z_\varphi Z_t^{-1} t, \ \bar{u} = Z_\varphi^2 Z_u^{-1} u. \tag{17.88}$$

If we define the invariant functional in terms of the new variables $\bar{\varphi}, \bar{t}, \bar{u}$ as

$$\bar{\Gamma}(\bar{\varphi}, \bar{t}, \bar{u}) = \Gamma\big(Z_\varphi^{1/2}\bar{\varphi}, Z_\varphi^{-1} Z_t \bar{t}, Z_\varphi^{-2} Z_u \bar{u}\big),$$

it follows that

$$\bar{\Gamma}^{(n)}(\bar{\varphi}, \bar{t}, \bar{u}) = Z_\varphi^{n/2}\Gamma^{(n)}\big(Z_\varphi^{1/2}\bar{\varphi}, Z_\varphi^{-1} Z_t \bar{t}, Z_\varphi^{-2} Z_u \bar{u}\big).$$

$\bar{\Gamma}$ and $\bar{\Gamma}^{(n)}$ are given by the same expansion as Γ and $\Gamma^{(n)}$ provided the correspondence between the old and new variables is made. Similarly, we define

$$\bar{\gamma}_i(k/\Lambda, \bar{\varphi}, \bar{t}, \bar{u}) = Z_i \gamma_i\big(k/\Lambda, Z_\varphi^{1/2}\bar{\varphi}, Z_\varphi^{-1} Z_t \bar{t}, Z_\varphi^{-2} Z_u \bar{u}\big), \quad i = \varphi, t, u. \tag{17.89}$$

The diagrammatic structure is invariant by renormalizing the coupling u and rescaling the field φ, the mass t and the four-point vertex γ_u. A proper choice of these renormalizations will take care of the singularities in perturbation theory. In this way we implement and generalize the universality condition which relates one model system to another by varying the coupling and suitably rescaling the other variables and the correlation functions (vertices and propagators).

Above, we have assumed that each variable is transformed independently from the others and the transformation is not yet well defined. However, we can still use the arbitrariness of the multiplicative factors in the correlation functions to obtain a one-parameter transformation. This can be achieved by imposing suitable normalization conditions on the γ_is at a given normalization point (hereafter n.p.) specified by an auxiliary variable λ with the dimensions of $(\text{length})^{-1}$. In the dimensionless quantities γ_φ, γ_t and γ_u, the variable k/Λ will be replaced by $\bar{k}/\bar{\Lambda}$ with a rescaled momentum and a rescaled cut off $\bar{\Lambda} = \Lambda/\lambda$. This rescaling physically represents the Kadanoff–Wilson block-variable idea of rescaling the length unit $l_0 \sim \Lambda^{-1}$. The normalization condition permits us to express the multiplicative factors Z_i in terms of λ and only one of the variable sets $\bar{\varphi}, \bar{t}, \bar{u}$ or φ, t, u. We can therefore change $\bar{\varphi}, \bar{t}, \bar{u}$ by varying λ at fixed φ, t, u, or vice versa. Since we expect $\bar{\varphi}$ and $\bar{t}$ to be the relevant variables which drive the system towards the critical point, where they must vanish, we can parametrize the transformation in terms of $\bar{u}$ only. This procedure implements the physical idea, discussed in Chapter 16, that, when the first irrelevant coupling u is changed, a proper multiplicative rescaling of the relevant variables yields a physical system which, asymptotically near the critical point, shares the same critical properties with the initial system. Then we make the following choice:

$$\bar{\gamma}_\varphi|_{\text{n.p.}} = 1, \quad \bar{\gamma}_t|_{\text{n.p.}} = 1, \quad \bar{\gamma}_u|_{\text{n.p.}} = 1, \tag{17.90}$$

where the normalization point is chosen as $\text{n.p.} = (k^2 = 0, \bar{\varphi} = 0, \bar{t} = \lambda^2, \text{ for all } \bar{u})$. In this way the transformation is multiplicative and linear in the variables $\bar{\varphi}$ and $\bar{t}$. Different choices of the normalization point would give different parametrizations of the transformation, without changing the final expression for the critical exponents. The asymptotic equivalence among the different versions of the field-theoretic approach and the Wilson approach is discussed in Benettin et al. (1977) and Jona-Lasinio (1973). It is useful to introduce the following dimensionless variables:

$$t_\lambda = \frac{\bar{t}}{\lambda^{D_t^0}}, \quad D_t^0 = 2; \quad \varphi_\lambda = \frac{\bar{\varphi}}{\lambda^{D_\varphi^0}}, \quad D_\varphi^0 = 1 - \epsilon/2; \quad u_\lambda = \frac{\bar{u}}{\lambda^{D_u^0}}, \quad D_u^0 = \epsilon, \tag{17.91}$$

by rescaling them with their bare dimensions. We also define the parameters of the transformation leading from one n.p. λ to another n.p. λ' by iterating the multiplicative transformation with scaling parameter given by $\lambda \to \lambda' = \lambda/s$:

$$Z_{\varphi,\lambda'/\lambda} = \frac{Z_{\varphi,\lambda'}}{Z_{\varphi,\lambda}}, \quad Z_{t,\lambda'/\lambda} = \frac{Z_{t,\lambda'}}{Z_{t,\lambda}}, \quad Z_{u,\lambda'/\lambda} = \frac{Z_{u,\lambda'}}{Z_{u,\lambda}}, \tag{17.92}$$

where the first suffix specifies the renormalization and the second one parametrizes the transformation. The variables change now according to the following multiplicative rules:

$$\varphi_{\lambda'} = \frac{\varphi_\lambda}{(\lambda'/\lambda)^{1-\epsilon/2}} Z_{\varphi,\lambda'/\lambda}^{-1/2},$$
$$t_{\lambda'} = \frac{t_\lambda}{(\lambda'/\lambda)^2} Z_{\varphi,\lambda'/\lambda} Z_{t,\lambda'/\lambda}^{-1},$$
$$u_{\lambda'} = \frac{u_\lambda}{(\lambda'/\lambda)^\epsilon} Z_{\varphi,\lambda'/\lambda}^2 Z_{u,\lambda'/\lambda}^{-1}. \tag{17.93}$$

In this way the correlation functions are multiplicatively related by

$$\bar{\gamma}_\varphi\left(\frac{k^2}{\lambda^2}, t_\lambda, \varphi_\lambda, u_\lambda\right) = Z_{\varphi,\lambda'/\lambda}^{-1} \bar{\gamma}_\varphi\left(\frac{k^2}{\lambda'^2}, t_{\lambda'}, \varphi_{\lambda'}, u_{\lambda'}\right),$$
$$\bar{\gamma}_t\left(\frac{k^2}{\lambda^2}, t_\lambda, \varphi_\lambda, u_\lambda\right) = Z_{t,\lambda'/\lambda}^{-1} \bar{\gamma}_t\left(\frac{k^2}{\lambda'^2}, t_{\lambda'}, \varphi_{\lambda'}, u_{\lambda'}\right),$$
$$\bar{\gamma}_u\left(\frac{k^2}{\lambda^2}, t_\lambda, \varphi_\lambda, u_\lambda\right) = Z_{u,\lambda'/\lambda}^{-1} \bar{\gamma}_u\left(\frac{k^2}{\lambda'^2}, t_{\lambda'}, \varphi_{\lambda'}, u_{\lambda'}\right), \tag{17.94}$$

and the thermodynamic functional is invariant:

$$\bar{\Gamma}\left(t_\lambda, \varphi_\lambda, u_\lambda\right) = \bar{\Gamma}\left(t_{\lambda'}, \varphi_{\lambda'}, u_{\lambda'}\right). \tag{17.95}$$

In Eqs. (17.94) and (17.95) we have omitted the dependence on the ultraviolet cutoff Λ/λ because it will disappear in the renormalized theory. The normalization point (17.90) at λ' with $t_{\lambda'} = 1$, $k^2/\lambda'^2 = 0$, $\varphi_{\lambda'} = 0$ and t_λ given by Eq. (17.93) determines the Zs in terms of the correlation functions:

$$\bar{\gamma}_\varphi\left(0, \left(\frac{\lambda'}{\lambda}\right)^2 Z_{\varphi,\lambda'/\lambda}^{-1} Z_{t,\lambda'/\lambda}, 0, u_\lambda\right) = Z_{\varphi,\lambda'/\lambda}^{-1},$$
$$\bar{\gamma}_t\left(0, \left(\frac{\lambda'}{\lambda}\right)^2 Z_{\varphi,\lambda'/\lambda}^{-1} Z_{t,\lambda'/\lambda}, 0, u_\lambda\right) = Z_{t,\lambda'/\lambda}^{-1},$$
$$\bar{\gamma}_u\left(0, \left(\frac{\lambda'}{\lambda}\right)^2 Z_{\varphi,\lambda'/\lambda}^{-1} Z_{t,\lambda'/\lambda}, 0, u_\lambda\right) = Z_{u,\lambda'/\lambda}^{-1}. \tag{17.96}$$

Equations (17.96) must be solved self-consistently. If these equations can be inverted, as can be shown order by order in perturbation theory, they implicitly define the renormalization parameters as functions of λ'/λ and u_λ only:

$$Z_{i,\lambda'/\lambda} = Z_i(\lambda'/\lambda, u_\lambda), \quad Z_i(\lambda'/\lambda, u_\lambda)_{\lambda'=\lambda} = 1, \quad i = \varphi, t, u. \tag{17.97}$$

To illustrate the inversion procedure we compute the correlation functions γ_φ, γ_t and γ_u in perturbation theory. Actually, having in mind the ϵ-expansion, most of the integrals are calculated in $d = 4$ recalling the results from Section 17.3.2. Here, however, the integrals over $\mathbf{k}$ are in the interval $|\mathbf{k}| < \Lambda$ and the infrared singularities at $|\mathbf{k}| = 0$ are cut off by t as long as it is finite.

Let us begin with the evaluation of γ_u, which up to terms of order u is obtained from the first two diagrams for the expansion of $\Gamma^{(4)}$ shown in Fig. 17.8. See Problem 17.5 for details. Recall that γ_u is obtained from $\Gamma^{(4)}$ by dividing by u according to the definition given in

Eq. (17.87). The evalutation of these diagrams involves the same integral encounterd for the evaluation of $\langle (H_1^u)^2 \rangle$ performed in Eq. (17.54) within the first step of the Wilson approach for the RG transformation. We then have (from now on we put $c = 1$) in $d = 4$

$$\gamma_u(k_i, t, u, \Lambda) = 1 + \frac{4(n+8)}{2} K_4 u \left[\ln \frac{t}{\Lambda^2} + \cdots \right], \tag{17.98}$$

where the appearance of t is due to the use of the bare inverse propagator $\Gamma_0^{(2)}(\mathbf{k}) = t + k^2$ obtained from Eq. (17.86) after the shift of the mass term.

For the evaluation of γ_t we need the first two diagrams of Fig. 17.9 (see Problem 17.3). The evaluation of the second one involves the same integral encountered in considering the term $\langle H_1^{r_0} H_1^u \rangle$ of Eq. (17.58). We then have

$$\gamma_t(k_i, t, u, \Lambda) = 1 + \frac{(2n+4)}{2} K_4 u \left[\ln \frac{t}{\Lambda^2} + \cdots \right]. \tag{17.99}$$

Finally, to obtain γ_φ we need to find the correction to the k^2-coefficient in the inverse $\Gamma^{(2)}$ propagator. To first order in u there is no such a correction, since the second diagram in the expansion for $\Gamma^{(2)}$ in Fig. 17.8 only yields a shift in the value of the mass, as we have already remarked. A correction to the k^2-coefficient and to the relation between t and ξ^{-2} arises to second order in u by considering the third diagram of the expansion of $\Gamma^{(2)}$ in Fig. 17.8 (see Problem 17.4). Again the evaluation of the diagram parallels the calculation of the corresponding term in the Wilson scheme, $\langle (H_1^u)^2 \rangle_b^{(2)}$ discussed in Eq. (17.70). We then have

$$\gamma_\varphi(0, t, u, \Lambda) = 1 - \frac{4(2n+4)}{2} K_4^2 u^2 \left[\ln \frac{t}{\Lambda^2} + \cdots \right], \tag{17.100}$$

where the dots indicate the terms in higher order in u. The infrared singularities as $\Lambda \to \infty$, $k \to 0$ (logarithmic in $d = 4$, stronger in $d < 4$) render the direct calculation of the physical quantities like the correlation functions and the vertices impossible as long as the renormalization group equations do not produce the scaling behavior via the fixed point condition and, as e.g. in $d = 4 - \epsilon$, the logarithmic singularities are exponentiated to powers. We stress that Eqs. (17.98), (17.99) and (17.100) are used to calculate the transformation, not γ_u, γ_t and γ_φ. By using the perturbative form of the correlation functions (17.98)–(17.100), imposing the normalization condition of Eq. (17.96), we indeed obtain

$$Z_{u,\lambda'/\lambda}^{-1} = 1 + 4(n+8) K_4 u \left[\ln \frac{\lambda'}{\lambda} - \frac{1}{2} \ln \frac{Z_{\varphi,\lambda'/\lambda}}{Z_{t,\lambda'/\lambda}} \right], \tag{17.101}$$

$$Z_{t,\lambda'/\lambda}^{-1}(\lambda, u, \Lambda) = 1 + (2n+4) K_4 u \left[\ln \frac{\lambda'}{\lambda} - \frac{1}{2} \ln \frac{Z_{\varphi,\lambda'/\lambda}}{Z_{t,\lambda'/\lambda}} \right], \tag{17.102}$$

$$Z_{\varphi,\lambda'/\lambda}^{-1}(\lambda, u, \Lambda) = 1 - 4(2n+4) K_4^2 u^2 \left[\ln \frac{\lambda'}{\lambda} - \frac{1}{2} \ln \frac{Z_{\varphi,\lambda'/\lambda}}{Z_{t,\lambda'/\lambda}} \right]. \tag{17.103}$$

The equations determining the Zs can be solved by considering that we only need to keep terms to first order in u for Eqs. (17.101) and (17.102) and second order in u for (17.103). Then, the Zs under the logarithm sign can be set to unity (the zero order approximation) and we get finally

$$Z_{u,\lambda'/\lambda}(u_\lambda) = 1 - 4(n+8)K_4 u \ln\frac{\lambda'}{\lambda}, \tag{17.104}$$

$$Z_{t,\lambda'/\lambda}(u_\lambda) = 1 - (2n+4)K_4 u \ln\frac{\lambda'}{\lambda}, \tag{17.105}$$

$$Z_{\varphi,\lambda'/\lambda}(u_\lambda) = 1 + 4(2n+4)K_4^2 u^2 \ln\frac{\lambda'}{\lambda}. \tag{17.106}$$

We remark that, whereas the Zs relating the bare quantities to those renormalized at the point λ still depend on the ultaviolet cutoff Λ, the latter drops out when considering the transformation from the normalization point λ to λ'. We also see that, as required from Eqs. (17.97) the Zs defining a transformation connecting two normalization points depend only on the coupling constant at the initial scale and the rescaling factor λ'/λ for the momenta. Such a ratio is consistent with the Wilson scaling factor as $s = \lambda/\lambda'$. The infrared regime is studied as $\lambda' \to 0$, i.e. $s \to \infty$.

After this illustrative digression, we come back to the general framework of the field theoretic RG transformation and to Eqs. (17.93)–(17.95), which represent its functional form. We begin by showing that these equations can be cast in a differential form by formally differentiating both sides with respect to the scaling parameter λ'/λ and setting $\lambda' = \lambda$ after that. For the running variables, transforming according to (17.93), we obtain

$$\lambda\frac{d\varphi_\lambda}{d\lambda} = -\varphi_\lambda \beta_\varphi(u_\lambda), \tag{17.107}$$

$$\lambda\frac{dt_\lambda}{d\lambda} = -t_\lambda \beta_t(u_\lambda), \tag{17.108}$$

$$\lambda\frac{du_\lambda}{d\lambda} = -u_\lambda \beta_u(u_\lambda), \tag{17.109}$$

where we have introduced the Gell-Mann–Low β-functions (Gell-Mann and Low (1954))

$$\beta_\varphi(u_\lambda) = D_\varphi^0 + \frac{1}{2}\left(\frac{\partial Z_\varphi}{\partial \lambda'/\lambda}\right)_{\lambda'=\lambda}, \tag{17.110}$$

$$\beta_t(u_\lambda) = D_t^0 - \left(\frac{\partial Z_\varphi}{\partial \lambda'/\lambda}\right)_{\lambda'=\lambda} + \left(\frac{\partial Z_t}{\partial \lambda'/\lambda}\right)_{\lambda'=\lambda}, \tag{17.111}$$

$$\beta_u(u_\lambda) = D_u^0 - 2\left(\frac{\partial Z_\varphi}{\partial \lambda'/\lambda}\right)_{\lambda'=\lambda} + \left(\frac{\partial Z_u}{\partial \lambda'/\lambda}\right)_{\lambda'=\lambda}. \tag{17.112}$$

Similarly, from Eq. (17.94) we get

$$\left[\lambda\frac{\partial}{\partial\lambda} - \varphi_\lambda\beta_\varphi(u_\lambda)\frac{\partial}{\partial\varphi_\lambda} - t_\lambda\beta_t(u_\lambda)\frac{\partial}{\partial t_\lambda} - u_\lambda\beta_u(u_\lambda)\frac{\partial}{\partial u_\lambda} - \left(\lambda\frac{\partial \ln Z_{i,\lambda'/\lambda}}{\partial\lambda'}\right)_{\lambda'=\lambda}\right]\bar{\gamma}_i = 0, \tag{17.113}$$

where $\bar{\gamma}_i = \bar{\gamma}_i\left(\frac{k^2}{\lambda^2}, t_\lambda, \varphi_\lambda, u_\lambda\right)$ and $i = \varphi, t, u$. An analogous equation is valid for the free energy per unit volume $\bar{\gamma}$:

$$\left[-\varphi_\lambda\beta_\varphi(u_\lambda)\frac{\partial}{\partial\varphi_\lambda} - t_\lambda\beta_t(u_\lambda)\frac{\partial}{\partial t_\lambda} - u_\lambda\beta_u(u_\lambda)\frac{\partial}{\partial u_\lambda} + d\right]\bar{\gamma} = 0. \tag{17.114}$$

Actually, Eqs. (17.113) for $\bar{\gamma}_i$ can be obtained from the last one by taking the functional derivatives with respect to φ and t. $\lambda\partial_{\lambda'} \ln Z_{i,\lambda'\lambda}$ in the equation for $\bar{\gamma}_i$ are expressed in terms of the bare dimensions D_i^0 and the functions $\beta_i(u_\lambda)$ via Eqs. (17.110)–(17.112).

The fixed point of the transformation is obtained by the condition

$$\frac{\mathrm{d}\varphi_\lambda}{\mathrm{d}\lambda} = \frac{\mathrm{d}t_\lambda}{\mathrm{d}\lambda} = \frac{\mathrm{d}u_\lambda}{\mathrm{d}\lambda} = 0.$$

On the critical surface $t = 0$ and $\varphi = 0$, the fixed point of the transformation is given by u^*, solution of the following equation:

$$u^* \beta_u(u^*) = 0. \tag{17.115}$$

This equation yields the trivial Gaussian fixed point $u^* = 0$ and the non-trivial fixed point $u^* = u^*(\epsilon)$ with the critical scaling dimensions of φ and t given by $D_\varphi = \beta_\varphi(u^*_\lambda)$ and $D_t = \beta_t(u^*_\lambda)$, as solutions of the linearized equations (17.107) and (17.108) around $u_\lambda = u^*$. The critical index for u,

$$D_u = \left(\frac{\partial(u_\lambda \beta_u)}{\partial u_\lambda} \right)_{u_\lambda = u^*},$$

specifies how $\Delta u = u_\lambda - u^*$ approaches the fixed point and gives the first correction to the asymptotic behavior.

At the fixed point, the second, third and fourth terms in the square brackets of Eq. (17.113) vanish and the last one gives the anomalous part of the scaling dimension of the correlation functions from Eqs. (17.110) and (17.111) at u^*: $d - 2 - 2\beta_\varphi(u^*_\lambda) = -\eta$ for $\overline{\gamma}_\varphi$, $d - 2\beta_\varphi(u^*_\lambda) - \beta_t(u^*_\lambda) = 2 - 1/\nu - \eta$ for $\overline{\gamma}_t$ and the scaling indices $d - 4\beta_\varphi(u^*_\lambda)$ for $\overline{\gamma}_u$ and d for the free energy per unit volume. The corresponding scaling laws are then achieved.

The general consequence of scaling theory, i.e. the homogeneity of the correlation functions and of the thermodynamic potential, is achieved by considering that Eqs. (17.114) and (17.113) near the fixed point reduce to homogeneity relations for the corresponding quantities in terms of the variables k/λ, $t_\lambda^{1/\beta_t(u^*)}$ and $\varphi_\lambda^{1/\beta_\varphi(u^*)}$.

To understand the meaning of the renormalization notice that the fixed point condition (17.115) implies the scaling index $\epsilon - 2\eta$ for the four-point vertex $\overline{\gamma}_u \propto Z_u^{-1}$ and $-\eta$ for $\overline{\gamma}_\varphi \propto Z_\varphi^{-1}$. In the renormalized coupling $u_{\lambda'}$, given by Eq. (17.93), their asymptotic behaviors compensate each other for the anomalous part and what is left compensates the bare dimension ϵ of u. The renormalized coupling has in this way a well defined asymptotic limit and corresponds to the building block of the diagrammatic representation of the perturbation theory that at each order adds two propagator lines, a vertex and a momentum integration. This explains how the renormalization takes care of the infrared singularities plaguing the critical phenomena at $d < 4$. The problem of the divergence of the bare dimensionless coupling mentioned previously has been overcome in this way with the use of renormalized couplings and fields.

As we have seen in Section 17.3.2 in the Wilson approach, the existence of a stable non-trivial fixed point $u^* \neq 0$ is verified within the ϵ-expansion, where by using the explicit expressions for the Zs from Eqs. (17.104)–(17.106) we have to order u

$$\beta_\varphi(u_\lambda) = 1 - \frac{\epsilon}{2} + \frac{1}{2}(8n+16)K_4^2u^2, \quad (17.116)$$

$$\beta_t(u_\lambda) = 2 - (2n+4)K_4u, \quad (17.117)$$

$$\beta_u(u_\lambda) = \epsilon - (4n+32)K_4u. \quad (17.118)$$

The vanishing of Eq. (17.118) is equivalent to the fixed point condition of Eq. (17.56) of the Wilson scheme and yields $u^* = \epsilon/((4n+32)K_4)$, as given in Eq. (17.57). By using this fixed-point value u^* in Eqs. (17.116) and (17.117) we get the critical indices with $D_u = -\epsilon$

$$D_\varphi = 1 - \frac{\epsilon}{2} + \frac{1}{2}\frac{n+2}{2(n+8)^2}\epsilon^2, \quad (17.119)$$

$$D_t = 2 - \frac{n+2}{2(n+8)}\epsilon. \quad (17.120)$$

The first equation, by recalling the definition of the anomalous dimension η given in Eq. (16.16), yields $\eta = \epsilon^2(n+2)/(2(n+8)^2)$ as obtained in Eq. (17.77). The second equation reproduces, instead, the value of the eigenvalue λ_1 of Eq. (17.65) and yields $\nu = (1/2)(1+(n+2)/(2(n+8))\epsilon)$.

17.5 Problems

17.1 Evaluate the bare propagator $G(\mathbf{x})$ at the critical point $t = 0$.

17.2 Write the explicit expression of $\Gamma^{(2)}$ up to second order in u.

17.3 By using the result of Problem 17.2, evaluate the vertex function γ_t to first order in u for $d = 4$.

17.4 Evaluate the vertex function γ_φ.

17.5 Evaluate the vertex function γ_u.

18 Thermal Green functions

In Chapter 9 we applied the second quantization to the Hartree–Fock method. In this chapter we provide the basic aspects of the thermal Green function method, which is convenient for systematic perturbative expansion beyond the mean-field approximation. Several problems in quantum statistical mechanics can be elegantly and compactly described in terms of the so-called Matsubara Green functions (Matsubara (1955); Abrikosov et al. (1963)). Hence, if on one hand, the density matrix description, often used in the previous chapters, has a more direct physical interpretation, on the other hand the Green function method is far more powerful and takes advantage of a handy diagrammatic perturbation expansion. In the following chapters, we rely on the Green function for a more complete treatment of some many-body problems. To be self contained we introduce here this technical chapter with a short presentation of the by now standard Green function method. For a comprehensive treatment we refer the reader to the many books available and in particular to Abrikosov et al. (1963).

18.1 The Matsubara Green function

The main aim of quantum statistical mechanics is the evaluation of the partition function (see Eqs. (6.31) and (6.32)), e.g. in the grand canonical ensemble

$$\mathcal{Z} = \mathrm{Tr}\left[\mathrm{e}^{-\beta(H-\mu N)}\right]. \tag{18.1}$$

In the case of quantum gases one can evaluate $\mathcal{Z}$ exactly as we did in Chapter 7. In general, the evaluation of the partition function is a difficult task and it is useful to use perturbation theory to obtain approximate solutions. To set up a perturbation theory, one begins by splitting the Hamiltonian as $H = H_0 + H_\mathrm{i}$, where H_0 represents the "easy" part, which can be solved exactly. In the second step, one could naturally think of expanding the exponential of the interacting part H_i in order to write the partition function as a perturbative series. However, such a procedure requires extreme care due to the non-commutative nature of operators in quantum mechanics. The way of solving this problem is via the "*time*" *ordering* of operators in each term of the perturbative expansion. It can be shown, as we shall see in the present chapter, that the quantum statistical average of any time-ordered sequence of operators with respect to the unperturbed Hamiltonian H_0 can be expressed in terms of elementary quantum statistical averages of two field operators. Such elementary averages are called Green functions and in this section we introduce the basic definitions concerning them.

We begin by defining the single-particle Matsubara Green function $\mathcal{G}$ as an *"imaginary" "time"-ordered product* of *time-dependent* field operators

$$\mathcal{G}(\mathbf{x}, \tau; \mathbf{x}', \tau') = -\langle \mathrm{T}_\tau \hat{\psi}(\mathbf{x}, \tau)\bar{\hat{\psi}}(\mathbf{x}', \tau')\rangle, \tag{18.2}$$

where the fictitious imaginary time is $i\tau$ with τ which varies in the interval $(0, \beta)$, $\beta = 1/k_\mathrm{B}T$. The connection between the imaginary time and the temperature is suggested from the relation between β and H in Eq. (18.1) for $\mathcal{Z}$, which is analogous to the relation between it and H in the evolution operator in quantum mechanics. This relation will be further clarified in the following discussion. For the sake of simplicity we omit spin indices and introduce them again wherever necessary. The time dependence of the field operators is of the "Heisenberg form"

$$\hat{\psi}(\mathbf{x}, \tau) = \mathrm{e}^{K\tau}\hat{\psi}(\mathbf{x})\mathrm{e}^{-K\tau}, \quad \bar{\hat{\psi}}(\mathbf{x}, \tau) = \mathrm{e}^{K\tau}\hat{\psi}^\dagger(\mathbf{x})\mathrm{e}^{-K\tau}, \tag{18.3}$$

where $K = H - \mu N$ is the appropriate operator for the grand canonical formalism. Notice that $\bar{\hat{\psi}}(\mathbf{x}, \tau)$ is not the Hermitian conjugate of $\hat{\psi}(\mathbf{x}, \tau)$, contrary to what happens to standard *real-time* Heisenberg operators. The T_τ-product orders operators by placing them according their decreasing "time" arguments. In the case of the definition of (18.2), we have

$$\mathrm{T}_\tau\hat{\psi}(\mathbf{x}, \tau)\bar{\hat{\psi}}(\mathbf{x}', \tau') = \theta(\tau - \tau')\hat{\psi}(\mathbf{x}, \tau)\bar{\hat{\psi}}(\mathbf{x}', \tau') \mp \theta(\tau' - \tau)\bar{\hat{\psi}}(\mathbf{x}', \tau')\hat{\psi}(\mathbf{x}, \tau), \tag{18.4}$$

where the upper and lower signs apply to Fermi and Bose statistics, respectively. Finally, the angular brackets in Eq. (18.2) represent the quantum statistical average

$$\langle \ldots \rangle = \mathrm{Tr}[\mathrm{e}^{-\beta K + \beta\Omega} \ldots], \tag{18.5}$$

the grand canonical potential Ω in terms of the variables T, V and μ providing the normalization. The trace is over all the states in the system at a fixed N and then the sum over all possible numbers of particles.

Once we know the Green function, the one-particle reduced density matrix of Eq. (9.54) and the total number of particles N can be easily obtained in the equal-"time" limit as

$$h_1(\mathbf{x}, \mathbf{x}') = \pm\mathcal{G}(\mathbf{x}, 0; \mathbf{x}', 0^+), \tag{18.6}$$

$$N = \pm \int \mathrm{d}\mathbf{x}\, \mathcal{G}(\mathbf{x}, 0; \mathbf{x}, 0^+), \tag{18.7}$$

where the symbol 0^+ indicates that the second "time" argument is infinitesimally larger than the first. This procedure ensures that we get the correct order of the field operators.

Similarly to the m-particle reduced density matrix, the m-particle Matsubara Green functions can be introduced as a "time"-ordered product of m fields $\hat{\psi}$ and m fields $\bar{\hat{\psi}}$:

$$\mathcal{G}^{(2m)}(x_1, \ldots, x_m; x_1', \ldots, x_m') = (-1)^m \langle \mathrm{T}_\tau \hat{\psi}(x_1) \cdots \hat{\psi}(x_m)\bar{\hat{\psi}}(x_1') \cdots \bar{\hat{\psi}}(x_m')\rangle, \tag{18.8}$$

where $x_i \equiv (\mathbf{x}_i, \tau_i)$, $i = 1, \ldots, m$ and $x_i' \equiv (\mathbf{x}_i', \tau_i')$, $i = 1, \ldots, m$. For instance, the two-particle Green function reads

$$\mathcal{G}^{(4)}(x_1, x_2, x_3, x_4) = \langle \mathrm{T}_\tau \hat{\psi}(x_1)\hat{\psi}(x_2)\bar{\hat{\psi}}(x_3)\bar{\hat{\psi}}(x_4)\rangle. \tag{18.9}$$

By exploiting the cyclic property of the trace, it is not hard to see that the Green function depends on the difference of the "time" arguments only. Then we set $\tau' = 0$ in the following discussion. For positive "times" $\tau > 0$ we have

$$\mathcal{G}(\mathbf{x}, \tau; \mathbf{x}', 0) = -\mathrm{Tr}[\mathrm{e}^{-\beta K+\beta\Omega}\mathrm{e}^{K\tau}\hat{\psi}(\mathbf{x})\mathrm{e}^{-K\tau}\hat{\psi}^{\dagger}(\mathbf{x}')],$$

while for negative ones $\tau < 0$ and using the cyclic property of the trace,

$$\begin{aligned}\mathcal{G}(\mathbf{x}, \tau; \mathbf{x}', 0) &= \pm\mathrm{Tr}[\mathrm{e}^{-\beta K+\beta\Omega}\hat{\psi}^{\dagger}(\mathbf{x}')\mathrm{e}^{K\tau}\hat{\psi}(\mathbf{x})\mathrm{e}^{-K\tau}] \\ &= \pm\mathrm{Tr}[\mathrm{e}^{K\tau}\hat{\psi}(\mathbf{x})\mathrm{e}^{-K\tau}\mathrm{e}^{-\beta K+\beta\Omega}\hat{\psi}^{\dagger}(\mathbf{x}')] \\ &= \pm\mathrm{Tr}[\mathrm{e}^{-\beta K+\beta\Omega}\mathrm{e}^{K(\tau+\beta)}\hat{\psi}(\mathbf{x})\mathrm{e}^{-K(\tau+\beta)}\hat{\psi}^{\dagger}(\mathbf{x}')],\end{aligned}$$

which leads to the basic property of antisymmetry and symmetry for Fermi and Bose statistics, respectively:

$$\mathcal{G}(\mathbf{x}, -\tau; \mathbf{x}', 0) = \mp\mathcal{G}(\mathbf{x}, -\tau+\beta; \mathbf{x}', 0), \quad 0 < \tau < \beta. \tag{18.10}$$

The above property has an immediate consequence on the Fourier series expansion of the Green function. In the interval $(-\beta, \beta)$ the Fourier expansion reads

$$\mathcal{G}(\tau) = \frac{1}{\beta}\sum_{n=-\infty}^{\infty} \mathrm{e}^{-\mathrm{i}\omega_n\tau}\mathcal{G}(\omega_n), \tag{18.11}$$

$$\mathcal{G}(\omega_n) = \frac{1}{2}\int_{-\beta}^{\beta} \mathrm{d}\tau\, \mathrm{e}^{\mathrm{i}\omega_n\tau}\mathcal{G}(\tau), \tag{18.12}$$

with $\omega_n = n\pi/\beta$, n running over the natural numbers and we have dropped the space arguments for the sake of simplicity. Only odd-(even-)integer frequences contribute for fermions (bosons). Indeed, thanks to the property (18.10), we have

$$\begin{aligned}\mathcal{G}(\omega_n) &= \frac{1}{2}\int_0^{\beta} \mathrm{d}\tau\, \mathrm{e}^{\mathrm{i}\omega_n\tau}\mathcal{G}(\tau) + \frac{1}{2}\int_{-\beta}^{0} \mathrm{d}\tau\, \mathrm{e}^{\mathrm{i}\omega_n\tau}\mathcal{G}(\tau) \\ &= \frac{1}{2}\int_0^{\beta} \mathrm{d}\tau\, \mathrm{e}^{\mathrm{i}\omega_n\tau}\mathcal{G}(\tau) \mp \frac{1}{2}\int_{-\beta}^{0} \mathrm{d}\tau\, \mathrm{e}^{\mathrm{i}\omega_n\tau}\mathcal{G}(\tau+\beta) \\ &= \frac{1}{2}\int_0^{\beta} \mathrm{d}\tau\, \mathrm{e}^{\mathrm{i}\omega_n\tau}\mathcal{G}(\tau) \mp \frac{1}{2}\int_0^{\beta} \mathrm{d}\tau\, \mathrm{e}^{\mathrm{i}\omega_n\tau}\mathcal{G}(\tau)\mathrm{e}^{-\mathrm{i}\omega_n\beta} \\ &= \frac{1 \mp \mathrm{e}^{-\mathrm{i}\omega_n\beta}}{2}\int_0^{\beta} \mathrm{d}\tau\, \mathrm{e}^{\mathrm{i}\omega_n\tau}\mathcal{G}(\tau) \\ &= \int_0^{\beta} \mathrm{d}\tau\, \mathrm{e}^{\mathrm{i}\omega_n\tau}\mathcal{G}(\tau).\end{aligned} \tag{18.13}$$

In the last expression only contributions of odd- and even-integer frequencies appear, i.e. the so-called Matusbara frequencies $\omega_n = (2n+1)\pi/\beta$ and $\omega_n = 2n\pi/\beta$, for Fermi and Bose statistics, respectively.

18.2 The thermal Green function for Fermi and Bose gases

As an important example, we now evaluate the Matsubara Green function $\mathcal{G}_0$ for Fermi and Bose gases with the Hamiltonian K given by the first term of Eq. (9.40). To this end we make use of the decomposition of the field operators (9.18), (9.19) in terms of the annihilation and creation operators with the choice of the plane waves (9.39) as wave functions. We have, for positive "times" $\tau > 0$,

$$\begin{aligned}
\mathcal{G}_{0,\alpha\beta}(\mathbf{x},\tau;\mathbf{x}',0) &= -\frac{1}{V}\sum_{\mathbf{k},\mathbf{k}'} e^{i(\mathbf{k}\cdot\mathbf{x}-\mathbf{k}'\cdot\mathbf{x}')}\mathrm{Tr}\big[e^{\beta\Omega-\beta K}e^{K\tau}a_{\mathbf{k},\alpha}e^{-K\tau}a^{\dagger}_{\mathbf{k}',\beta}\big] \\
&= -\frac{1}{V}\sum_{\mathbf{k},\mathbf{k}'} e^{i(\mathbf{k}\cdot\mathbf{x}-\mathbf{k}'\cdot\mathbf{x}')}e^{-\epsilon_{\mathbf{k}}\tau}\mathrm{Tr}\big[e^{\beta\Omega-\beta K}a_{\mathbf{k},\alpha}a^{\dagger}_{\mathbf{k}',\beta}\big] \\
&= -\frac{1}{V}\sum_{\mathbf{k},\mathbf{k}'} e^{i(\mathbf{k}\cdot\mathbf{x}-\mathbf{k}'\cdot\mathbf{x}')}e^{-\epsilon_{\mathbf{k}}\tau}(1\mp n_{\mathbf{k}})\delta_{\mathbf{k},\mathbf{k}'}\delta_{\alpha\beta} \\
&= -\delta_{\alpha\beta}\frac{1}{V}\sum_{\mathbf{k}} e^{i\mathbf{k}\cdot(\mathbf{x}-\mathbf{x}')}e^{-\epsilon_{\mathbf{k}}\tau}(1\mp n_{\mathbf{k}}) \\
&\equiv \delta_{\alpha\beta}\mathcal{G}_0(\mathbf{x},\tau;\mathbf{x}',0),
\end{aligned} \tag{18.14}$$

where the minus (plus) sign, inside the brackets, applies to fermions (bosons), we also used (9.51) and (9.52) and $\epsilon_{\mathbf{k}}$ has been defined in Eq. (9.41).

The translational invariance of the gas makes the Green function dependent only on the difference of the space arguments. This allows us to introduce the Fourier transform with respect to $\mathbf{x}-\mathbf{x}'$ as

$$\mathcal{G}_0(\mathbf{k},\tau) = \int d(\mathbf{x}-\mathbf{x}')e^{-i\mathbf{k}\cdot(\mathbf{x}-\mathbf{x}')}\mathcal{G}_0(\mathbf{x},\tau;\mathbf{x}',0), \tag{18.15}$$

$$\mathcal{G}_0(\mathbf{x},\tau;\mathbf{x}',0) = \frac{1}{V}\sum_{\mathbf{k}} e^{i\mathbf{k}\cdot(\mathbf{x}-\mathbf{x}')}\mathcal{G}_0(\mathbf{k},\tau). \tag{18.16}$$

A simple inspection of Eq. (18.14) shows that

$$\mathcal{G}_0(\mathbf{k},\tau) = -e^{-\epsilon_{\mathbf{k}}\tau}(1\mp n_{\mathbf{k}}), \qquad \tau > 0. \tag{18.17}$$

Finally, we can evaluate the Fourier transform with respect to the "time" τ according to the definition of Eq. (18.13) to read

$$\begin{aligned}
\mathcal{G}_0(\mathbf{k},\omega_n) &= -(1\mp n_{\mathbf{k}})\int_0^{\beta} d\tau e^{i\omega_n\tau}e^{-\epsilon_{\mathbf{k}}\tau} \\
&= (1\mp n_{\mathbf{k}})\frac{1-e^{(i\omega_n-\epsilon_{\mathbf{k}})\beta}}{i\omega_n-\epsilon_{\mathbf{k}}} \\
&= (1\mp n_{\mathbf{k}})\frac{1\pm e^{-\epsilon_{\mathbf{k}}\beta}}{i\omega_n-\epsilon_{\mathbf{k}}} \\
&= \frac{1}{i\omega_n-\epsilon_{\mathbf{k}}},
\end{aligned} \tag{18.18}$$

where we used the fact that $\exp(i\beta\omega_n) = \mp 1$ for Fermi and Bose statistics. According to the procedure of Eq. (18.6) ($\tau = 0$ and $\tau' = -0^+$), we obtain the one-particle reduced density matrix as an anti-Fourier transform

$$h_1(\mathbf{x}, \mathbf{x}') = \frac{1}{V} \sum_{\mathbf{k}} e^{i\mathbf{k}\cdot(\mathbf{x}-\mathbf{x}')} \frac{1}{\beta} \sum_{\omega_n} \frac{e^{i\omega_n 0^+}}{i\omega_n - \epsilon_{\mathbf{k}}}. \tag{18.19}$$

The connection with statistics is obtained by observing that the sum over the frequencies can be written in terms of the Fermi and Bose functions:

$$\frac{1}{\beta} \sum_{\omega_n} \frac{e^{i\omega_n 0^+}}{i\omega_n - \epsilon_{\mathbf{k}}} = \frac{1}{e^{\beta\epsilon_{\mathbf{k}}} \pm 1}. \tag{18.20}$$

This relation is proved in the answer to Problem 18.1. In the limit of zero kelvin, the sum over the frequencies can be replaced by an integral over the imaginary axis $\omega_n \to \tilde{\omega}$:

$$\frac{1}{\beta} \sum_{\omega_n} \cdots \to \int_{-\infty}^{\infty} \frac{d\tilde{\omega}}{2\pi} \cdots. \tag{18.21}$$

and Eq. (18.20) becomes

$$\begin{aligned} \frac{1}{\beta} \sum_{\omega_n} \frac{e^{i\omega_n 0^+}}{i\omega_n - \epsilon_{\mathbf{k}}} &\to \int_{-\infty}^{\infty} \frac{d\tilde{\omega}}{2\pi} \frac{e^{i\tilde{\omega}0^+}}{i\tilde{\omega} - \epsilon_{\mathbf{k}}} \\ &= -i \int_{-\infty}^{\infty} \frac{d\tilde{\omega}}{2\pi} \frac{e^{i\tilde{\omega}0^+}}{\tilde{\omega} + i\epsilon_{\mathbf{k}}} \\ &= \theta(-\epsilon_{\mathbf{k}}). \end{aligned} \tag{18.22}$$

The integral can be regularized for $i\tilde{\omega} < 0$ only, i.e. for the pole for $\epsilon_{\mathbf{k}} < 0$. We see that the result coincides with the zero kelvin limit of the Fermi and Bose gas distributions. For a Fermi gas it defines the discontinuity at the Fermi surface. For the Bose gas, the result is zero for any $\epsilon_{\mathbf{k}} \neq 0$ and it is not defined for $\epsilon_{\mathbf{k}} = 0$. Indeed, at $\epsilon_{\mathbf{k}} = 0$, the sum over the frequencies contains the singular term $\omega_n = 0$. The regularized integral only yields the zero occupation above the ground-state condensate.

Notice that for a Bose system in the presence of a condensate, as we have seen in Section 15.8, the creation and annihilation operators in and out of it are c-numbers. A special treatment is then required, particularly in the presence of interaction, which connects the particles in the condensate with those out of it. In the following we will refer either to the non-interacting Bose systems or to temperatures above the critical condensation temperature. Since, at low temperature, we will mainly consider specific fermion problems, for the treatment of the Green function for Bose systems in the presence of the condensate the reader is referred to the book of Abrikosov et al. (1963).

18.3 The connection with the time-dependent Green function at finite temperature

We have seen in Eq. (18.6) the connection between the reduced one-particle density matrix and the equal-time thermal Green function. As long as we are interested in equilibrium properties the connection provided by Eq. (18.6) is sufficient. However, if we want to investigate dynamical effects or transport properties such as the Boltzmann equation, we have to consider quantum-statistical averages of operators dependent on real time. Such averages are also expressed as time-dependent Green functions, for which, at $T \neq 0$, there is no perturbative expansion as for the Matsubara Green function, discussed in Section 18.5. However, as we shall see in this section, there is a direct connection between thermal Green functions and their time-dependent counterpart, so that a knowledge of the former is sufficient for obtaining the expression for the latter.

To show the above mentioned connection we need to take into account the time-dependence of the Heisenberg field operators:

$$\hat{\psi}_{\mathrm{H}}(\mathbf{x}, t) = \mathrm{e}^{\mathrm{i}Kt/\hbar}\hat{\psi}(\mathbf{x})\mathrm{e}^{-\mathrm{i}Kt/\hbar}, \quad \hat{\psi}^{\dagger}_{\mathrm{H}}(\mathbf{x}, t) = \mathrm{e}^{\mathrm{i}Kt/\hbar}\hat{\psi}^{\dagger}(\mathbf{x})\mathrm{e}^{-\mathrm{i}Kt/\hbar}, \tag{18.23}$$

in terms of which we define the *real-time*-dependent Green function

$$G(\mathbf{x}, t; \mathbf{x}', t') = -\mathrm{i}\mathrm{Tr}\big[\mathrm{e}^{\beta\Omega-\beta K}\mathrm{T}_t\big(\hat{\psi}_{\mathrm{H}}(\mathbf{x}, t)\hat{\psi}^{\dagger}_{\mathrm{H}}(\mathbf{x}', t')\big)\big], \tag{18.24}$$

and for the m-particle Green function

$$G^{2m}(x_1, \ldots, x_m; x'_1, \ldots, x'_m) = (-\mathrm{i})^m \langle \mathrm{T}_t \hat{\psi}(x_1)\cdots\hat{\psi}(x_m)\hat{\psi}^{\dagger}(x'_1)\cdots\hat{\psi}^{\dagger}(x'_m)\rangle, \tag{18.25}$$

where T_t indicates the time-ordering operator for real times and $x_i \equiv (\mathbf{x}_i, t_i)$, $i = 1, \ldots, m$ and $x'_i \equiv (\mathbf{x}'_i, t'_i)$, $i = 1, \ldots, m$. In this case the time-dependent one-particle reduced density matrix $h_1(\mathbf{x}, \mathbf{x}'; t)$ coincides with

$$h_1(\mathbf{x}, \mathbf{x}'; t) = \mp\mathrm{i}\lim_{t'\to t^+} G(\mathbf{x}, t; \mathbf{x}', t'). \tag{18.26}$$

The connection between the above time-dependent Green function and the Matsubara Green function $\mathcal{G}$ of Eq. (18.2) is best shown by using the so-called Lehmann decomposition, as done also for the linear response function in Eq. (10.14). We again introduce the resolution of the identity in terms of a complete set of exact Hamiltonian eigenstates to get

$$\begin{aligned} G(\mathbf{x}, t; \mathbf{x}', t') = &-\mathrm{i}\theta(t-t')\sum_{l,m}\mathrm{e}^{\beta\Omega}\langle l|\mathrm{e}^{-\beta K}\hat{\psi}_{\mathrm{H}}(\mathbf{x}, t)|m\rangle\langle m|\hat{\psi}^{\dagger}_{\mathrm{H}}(\mathbf{x}', t')|l\rangle \\ &\pm \mathrm{i}\theta(t'-t)\sum_{l,m}\mathrm{e}^{\beta\Omega}\langle l|\mathrm{e}^{-\beta K}\hat{\psi}^{\dagger}_{\mathrm{H}}(\mathbf{x}', t')|m\rangle\langle m|\hat{\psi}_{\mathrm{H}}(\mathbf{x}, t)|l\rangle. \end{aligned}$$

By exploiting the fact that the exponential factors can be evaluated exactly when acting on the eigenstates, we obtain

$$G(\mathbf{x},t;\mathbf{x}',t') = -\mathrm{i}\theta(t-t')\sum_{l,m} \mathrm{e}^{\beta\Omega}\mathrm{e}^{-\beta K_l}\mathrm{e}^{\mathrm{i}(t-t')\omega_{lm}}\langle l|\hat{\psi}(\mathbf{x})|m\rangle\langle m|\hat{\psi}^\dagger(\mathbf{x}')|l\rangle$$
$$\pm\mathrm{i}\theta(t'-t)\sum_{l,m} \mathrm{e}^{\beta\Omega}\mathrm{e}^{-\beta K_l}\mathrm{e}^{-\mathrm{i}(t-t')\omega_{lm}}\langle l|\hat{\psi}^\dagger(\mathbf{x}')|m\rangle\langle m|\hat{\psi}(\mathbf{x})|l\rangle, \quad (18.27)$$

where $\omega_{lm} = (K_l - K_m)/\hbar$. Finally, by interchanging the indices l and m in the second term, we obtain

$$G(\mathbf{x},t;\mathbf{x}',t') = -\mathrm{i}\theta(t-t')\sum_{l,m} \mathrm{e}^{\beta\Omega}\mathrm{e}^{-\beta K_l}\mathrm{e}^{\mathrm{i}(t-t')\omega_{lm}}\langle l|\hat{\psi}(\mathbf{x})|m\rangle\langle m|\hat{\psi}^\dagger(\mathbf{x}')|l\rangle$$
$$\pm\,\mathrm{i}\theta(t'-t)\sum_{l,m} \mathrm{e}^{\beta\Omega}\mathrm{e}^{-\beta K_m}\mathrm{e}^{\mathrm{i}(t-t')\omega_{lm}}\langle l|\hat{\psi}(\mathbf{x})|m\rangle\langle m|\hat{\psi}^\dagger(\mathbf{x}')|l\rangle.$$

We can take the Fourier transform with respect to $t - t'$ by using

$$\int_{-\infty}^{\infty} \mathrm{d}(t-t')\theta(\pm t \mp t')\mathrm{e}^{\mathrm{i}(t-t')(\omega\pm\mathrm{i}0^+ +\omega_{lm})} = \frac{\pm\mathrm{i}}{\omega+\omega_{lm}\pm\mathrm{i}0^+} = \frac{\pm\mathrm{i}}{\omega+\omega_{lm}} + \pi\delta(\omega+\omega_{lm}) \quad (18.28)$$

and

$$B_{lm}(\mathbf{x},\mathbf{x}') = \langle l|\hat{\psi}(\mathbf{x})|m\rangle\langle m|\hat{\psi}^\dagger(\mathbf{x}')|l\rangle \quad (18.29)$$

to get

$$G(\mathbf{x},\mathbf{x}';\omega) = \sum_{lm} \mathrm{e}^{\beta\Omega}B_{lm}(\mathbf{x},\mathbf{x}')\mathrm{e}^{-\beta K_l}\left[\frac{1}{\omega+\omega_{lm}+\mathrm{i}0^+} \pm \frac{\mathrm{e}^{\beta\hbar\omega_{lm}}}{\omega+\omega_{lm}-\mathrm{i}0^+}\right], \quad (18.30)$$

which is not analytic in terms of the complex variable ω. This is clear from the fact that the first and second terms on the right-hand side have poles lying on the lower and upper half of the complex plane, respectively. It is worthwhile noticing that $G(\mathbf{x},\mathbf{x}';\omega)$ and $\mathcal{G}(\mathbf{x},\mathbf{x}';\omega_n)$, being the Fourier transforms with respect to the real and fictitious times, respectively, have physical dimensions differing by the ratio of an energy to a frequency.

By performing a similar analysis for the thermal Green function, we obtain

$$\mathcal{G}(\mathbf{x},\tau;\mathbf{x}',\tau') = -\theta(\tau-\tau')\sum_{l,m} \mathrm{e}^{\beta\Omega}\mathrm{e}^{-\beta K_l}\mathrm{e}^{(\tau-\tau')\,\hbar\omega_{lm}}\langle l|\hat{\psi}(\mathbf{x})|m\rangle\langle m|\hat{\psi}^\dagger(\mathbf{x}')|l\rangle$$
$$\pm\,\theta(\tau'-\tau)\sum_{l,m} \mathrm{e}^{\beta\Omega}\mathrm{e}^{-\beta K_m}\mathrm{e}^{-(\tau-\tau')\,\hbar\omega_{lm}}\langle l|\hat{\psi}(\mathbf{x})|m\rangle\langle m|\hat{\psi}^\dagger(\mathbf{x}')|l\rangle.$$

By using

$$\int_0^\beta \mathrm{d}(\tau-\tau')\mathrm{e}^{(\mathrm{i}\omega_n+\hbar\omega_{lm})(\tau-\tau')} = \frac{\mp\mathrm{e}^{\beta\hbar\omega_{lm}}-1}{\mathrm{i}\omega_n+\hbar\omega_{lm}} \quad (18.31)$$

we get

$$\mathcal{G}(\mathbf{x},\mathbf{x}';\omega_n) = \sum_{lm} \mathrm{e}^{\beta\Omega}B_{lm}(\mathbf{x},\mathbf{x}')\mathrm{e}^{-\beta K_l}\frac{1\pm\mathrm{e}^{\beta\hbar\omega_{lm}}}{\mathrm{i}\omega_n+\hbar\omega_{lm}}. \quad (18.32)$$

As a function of ω_n, $\mathcal{G}(\mathbf{x}, \mathbf{x}'; \omega_n)$ can be regarded as the function $\mathcal{G}(\mathbf{x}, \mathbf{x}'; \omega)$ of the complex variable ω:

$$\mathcal{G}(\mathbf{x}, \mathbf{x}'; \omega) = \sum_{lm} \mathrm{e}^{\beta\Omega} B_{lm}(\mathbf{x}, \mathbf{x}') \mathrm{e}^{-\beta K_l} \frac{1 \pm \mathrm{e}^{\beta\hbar\omega_{lm}}}{\hbar\omega + \hbar\omega_{lm}} \tag{18.33}$$

evaluated on the sequence of points $\hbar\omega = \mathrm{i}\omega_n$ along the imaginary axis. Clearly, $\mathcal{G}(\mathbf{x}, \mathbf{x}'; \omega)$ is not analytic in the complex plane. However, it is possible to analytically continue $\mathcal{G}(\mathbf{x}, \mathbf{x}'; \omega_n)$ in the complex plane provided this is done separately for the upper and lower half-planes with the prescriptions $\mathrm{i}\omega_n \to \hbar\omega + \mathrm{i}0^+$ and $\mathrm{i}\omega_n \to \hbar\omega - \mathrm{i}0^+$, respectively. By indicating with $G^{\mathrm{R}}(\mathbf{x}, \mathbf{x}'; \omega)$ and $G^{\mathrm{A}}(\mathbf{x}, \mathbf{x}'; \omega)$ the retarded and advanced Green functions, being, respectively, the analytical continuation of $\mathcal{G}(\mathbf{x}, \mathbf{x}'; \omega_n) \to G^R(\mathbf{x}, \mathbf{x}'; \omega)/\hbar$ in the upper and lower half-planes, we obtain

$$G^{\mathrm{R}}(\mathbf{x}, \mathbf{x}'; \omega) = \sum_{lm} \mathrm{e}^{\beta\Omega} B_{lm}(\mathbf{x}, \mathbf{x}') \mathrm{e}^{-\beta K_l} \frac{1 \pm \mathrm{e}^{\beta\hbar\omega_{lm}}}{\omega + \omega_{lm} + \mathrm{i}0^+} \tag{18.34}$$

and

$$G^{\mathrm{A}}(\mathbf{x}, \mathbf{x}'; \omega) = \sum_{lm} \mathrm{e}^{\beta\Omega} B_{lm}(\mathbf{x}, \mathbf{x}') \mathrm{e}^{-\beta K_l} \frac{1 \pm \mathrm{e}^{\beta\hbar\omega_{lm}}}{\omega + \omega_{lm} - \mathrm{i}0^+}, \tag{18.35}$$

with $\mathrm{Re}\, G^{\mathrm{R}} = \mathrm{Re}\, G^{\mathrm{A}} = \mathrm{Re}\, G$ and $\mathrm{Im}\, G^{\mathrm{R}} = -\mathrm{Im}\, G^{\mathrm{A}}$.

It is instructive to derive the expression for the retarded and advanced Green functions as functions of time. For instance, by anti-Fourier transforming Eq. (18.34) we have

$$G^{\mathrm{R}}(\mathbf{x}, t; \mathbf{x}', t') = -\mathrm{i}\theta(t - t') \sum_{lm} \mathrm{e}^{\beta\Omega} B_{lm}(\mathbf{x}, \mathbf{x}') \mathrm{e}^{-\beta K_l} (1 \pm \mathrm{e}^{\beta\hbar\omega_{lm}}) \mathrm{e}^{\mathrm{i}\omega_{lm}(t-t')}, \tag{18.36}$$

where we used

$$\int_{-\infty}^{\infty} \frac{d\omega}{2\pi} \frac{\mathrm{e}^{-\mathrm{i}\omega t}}{\omega + \omega_{lm} + \mathrm{i}0^+} = -\mathrm{i}\theta(t - t') \mathrm{e}^{\mathrm{i}\omega_{lm}(t-t')}.$$

This is nothing but the Lehmann decomposition of

$$G^{\mathrm{R}}(\mathbf{x}, t; \mathbf{x}', t') = -\mathrm{i}\theta(t - t') \mathrm{Tr}\left[\mathrm{e}^{\beta\Omega - \beta K} \left[\hat{\psi}_{\mathrm{H}}(\mathbf{x}, t), \hat{\psi}^\dagger_{\mathrm{H}}(\mathbf{x}', t')\right]_\pm\right], \tag{18.37}$$

where $[\ldots, \ldots]_\pm$ indicates the anticommutator and the commutator, respectively. Similarly, for the advanced Green function we obtain

$$G^{\mathrm{A}}(\mathbf{x}, t; \mathbf{x}', t') = \mathrm{i}\theta(t' - t) \mathrm{Tr}\left[\mathrm{e}^{\beta\Omega - \beta K} \left[\hat{\psi}_{\mathrm{H}}(\mathbf{x}, t), \hat{\psi}^\dagger_{\mathrm{H}}(\mathbf{x}', t')\right]_\pm\right]. \tag{18.38}$$

The vanishing of G^{R} (G^{A}) for $t < t'$ ($t > t'$) explains the use of retarded and advanced denominations.

By comparing Eqs. (18.30) and (18.34), we finally obtain

$$\mathrm{Im}\, G(\mathbf{x}, \mathbf{x}'; \omega) = \frac{1 \mp \mathrm{e}^{-\beta\hbar\omega}}{1 \pm \mathrm{e}^{-\beta\hbar\omega}} \mathrm{Im}\, G^{\mathrm{R}}(\mathbf{x}, \mathbf{x}'; \omega), \tag{18.39}$$

which is the analog for Fermi and Bose field operators to the fluctuation–dissipation theorem (10.51), which was introduced for two generic observables $\mathcal{O}_A$ and $\mathcal{O}_B$. We see that the case of Bose statistics coincides with the case of the density fluctuations that are boson-like both for Fermi and Bose systems. Indeed, for Bose systems, the retarded Green function is the expectation value of the commutator of the field operators (see Eq. (18.37)) and has the same structure as the response function (10.8) multiplied by $\hbar$. The imaginary part of the Green function, as can be seen from Eq. (18.30),

$$\mathrm{Im}\, G(\mathbf{x}, \mathbf{x}'; \omega) = -\pi \sum_{lm} \mathrm{e}^{\beta\Omega} B_{lm}(\mathbf{x}, \mathbf{x}') \mathrm{e}^{-\beta K_l} \delta(\omega + \omega_{lm})(1 + \mathrm{e}^{\beta\hbar\omega_{lm}}), \tag{18.40}$$

coincides with minus the spectral representation (10.48) of the expectation value of the anti-commutator of the field operators and has the same structure as the fluctuations correlation function introduced in Eq. (10.45).

For Fermi operators, the roles of Im G and Im G^{R} are inverted.

18.4 The physical meaning of the poles of the Green function

The previous sections have presented the main properties of the thermal and time-dependent Green functions as well as their connections. It is the purpose of this section to elucidate the physical content of the Green function and its relation to the analytic structure derived previously. We emphasize that Eq. (18.39) shows how the Green function contains both the spectral information associated with the retarded G^{R} and the statistical one in the hyberbolic functions. Here we concentrate on the spectral part and, for simplicity, confine ourselves to the zero-kelvin limit for G, where the the statistical average is replaced by the average over the ground state in Eq. (18.27).

For $t > 0$, the sum over the index l is restricted to the ground state with N particles, whereas that over m runs over all states with $N + 1$ particles. Hence we have

$$\begin{aligned}
\hbar\omega_{0m} &= E_0(N) - \mu N - (E_m(N+1) - \mu(N+1)) \\
&= -(E_m(N+1) - E_0(N+1) + E_0(N+1) - E_0(N) - \mu) \\
&= -\epsilon_m,
\end{aligned}$$

where ϵ_m represents the excitation energy of the state $|m\rangle$ with respect to the ground state with $N + 1$ particles and we have identified $E_0(N + 1) - E_0(N)$ as the chemical potential for a state with N particles. Similarly, for $t < 0$, the sum over m runs over all states with $N - 1$ particles and we have

$$\begin{aligned}
\hbar\omega_{0m} &= E_0(N) - \mu N - (E_m(N-1) - \mu(N-1)) \\
&= -(E_m(N-1) - E_0(N-1) + E_0(N-1) - E_0(N) + \mu) \\
&= -\epsilon_m,
\end{aligned}$$

with now ϵ_m being the excitation energy of the state $|m\rangle$ with respect to the ground state with $N - 1$ particles and having identified $E_0(N) - E_0(N - 1)$ as the chemical potential for a state with $N - 1$ particles. In principle, the chemical potentials in the two cases differ, but the difference is of the order $1/N$ and is negligible in the thermodynamic limit. As a result, Eq. (18.30) becomes

$$G(\mathbf{x}, \mathbf{x}'; \omega) = \sum_m \left[\frac{B_{0m}(\mathbf{x}, \mathbf{x}')}{\omega - \epsilon_m/\hbar + \mathrm{i}0^+} \pm \frac{B_{m0}(\mathbf{x}, \mathbf{x}')}{\omega + \epsilon_m/\hbar - \mathrm{i}0^+} \right]. \tag{18.41}$$

We see that the Green function has singularities at the excitation energies. In particular, for positive excitation energies, corresponding to states with one more particle than the ground state, the singularities lie in the lower half of the complex plane of the variable ω. Conversely, for negative excitation energies, corresponding to states with one fewer particle than the ground state, the singularities are in the upper half of the complex plane. The definitions (18.34) and (18.35) of the retarded and advanced Green functions in terms of the $T = 0$ limit of the analytic continuation of the thermal Green function remain valid, with $\mathrm{Re}\, G^{\mathrm{R}} = \mathrm{Re}\, G^{\mathrm{A}} = \mathrm{Re}\, G$ and $\mathrm{Im}\, G^{\mathrm{R}} = -\mathrm{Im}\, G^{\mathrm{A}}$. In particular relation (18.39) reduces to

$$\mathrm{Im}\, G(\mathbf{x}, \mathbf{x}'; \omega) = \mathrm{sign}(\omega)\mathrm{Im}\, G^{\mathrm{R}}(\mathbf{x}, \mathbf{x}'; \omega). \tag{18.42}$$

By using the above equation together with Eq. (18.41) we find that the single-particle density of states in terms of the imaginary part of the Green function,

$$-\frac{1}{\pi}\mathrm{Im}\, G^{\mathrm{R}}(\mathbf{x}, \mathbf{x}; \omega) = \sum_m B_{0m}(\mathbf{x}, \mathbf{x})\delta(\omega - \epsilon_m/\hbar) \equiv \nu(\omega). \tag{18.43}$$

$\nu(\omega)$ is the generalization of the single-particle density of states (7.14) valid for a gas. The spectral decomposition introduces the weighting factor $B_{0m}(\mathbf{x}, \mathbf{x})$ for the excited state.

To proceed further, we consider the important case of translationally invariant systems. In this case, we can use the plane-wave decomposition of the field operators and evaluate their matrix elements (18.29) to read (notice that here $|l = 0\rangle \equiv |0\rangle$ indicates the ground state and not the vacuum as in Chapter 9)

$$B_{0m}(\mathbf{x}, \mathbf{x}') = \frac{1}{V}\sum_{\mathbf{k}} \mathrm{e}^{\mathrm{i}\mathbf{k}\cdot(\mathbf{x}-\mathbf{x}')}|\langle 0|a_{\mathbf{k}}|m\rangle|^2, \tag{18.44}$$

$$B_{m0}(\mathbf{x}, \mathbf{x}') = \frac{1}{V}\sum_{\mathbf{k}} \mathrm{e}^{\mathrm{i}\mathbf{k}\cdot(\mathbf{x}-\mathbf{x}')}\left|\langle 0|a^\dagger_{\mathbf{k}}|m\rangle\right|^2. \tag{18.45}$$

The proof of the above relations is shown in the answer to Problem 18.2. Their use allows us to take the Fourier transform of Eq. (18.41) with respect to the difference of the space arguments $\mathbf{x} - \mathbf{x}'$ to read

$$G(\mathbf{k}; \omega) = \sum_m \left[\frac{|\langle 0|a_{\mathbf{k}}|m\rangle|^2}{\omega - \epsilon_m/\hbar + \mathrm{i}0^+} \pm \frac{\left|\langle 0|a^\dagger_{\mathbf{k}}|m\rangle\right|^2}{\omega + \epsilon_m/\hbar - \mathrm{i}0^+} \right]. \tag{18.46}$$

It is useful at this point to examine the above expression in the case of the Fermi gas. The ground state $|0\rangle$ is obtained by filling all momentum single-particle states up to the Fermi momentum k_{F}. The states $|m\rangle$ appearing in Eq. (18.46) are the states differing from the ground state by the addition or removal of a particle with respect to the ground state. Hence the excitation energy of the state $|m\rangle$ is simply the energy of a free particle with wave vector $\mathbf{k}$, i.e., $\epsilon_m = \epsilon_\mathbf{k}$ so that

$$\sum_m \langle 0|a_\mathbf{k}|m\rangle\langle m|a^\dagger_\mathbf{k}|0\rangle = \theta(k - k_\mathrm{F})$$

and

$$\sum_m \langle 0|a^\dagger_\mathbf{k}|m\rangle\langle m|a_\mathbf{k}|0\rangle = 1 - \theta(k - k_\mathrm{F})$$

and the Green function for the gas reads

$$G^0(\mathbf{k};\omega) = \frac{\hbar}{\hbar\omega - \epsilon_\mathbf{k} + \mathrm{sign}(k - k_\mathrm{F})\mathrm{i}0^+}. \tag{18.47}$$

In the case of a Bose gas, the ground state is obtained by having all the particles in the zero momentum state. As a result, in order to excite the system, it is only possible to create a particle with momentum $\mathbf{k} \neq 0$:

$$\sum_m \langle 0|a_\mathbf{k}|m\rangle\langle m|a^\dagger_\mathbf{k}|0\rangle = 1$$

and

$$\sum_m \langle 0|a^\dagger_\mathbf{k}|m\rangle\langle m|a_\mathbf{k}|0\rangle = 0$$

and the Green function reads

$$G^0(\mathbf{k};\omega) = \frac{\hbar}{\hbar\omega - \epsilon_\mathbf{k} + \mathrm{i}0^+}. \tag{18.48}$$

In the case of an interacting Bose system, due to the depletion the ground state is no longer a fully condensate state and the proper description requires the introduction of the so-called anomalous Green functions (see Abrikosov et al. (1963)) connecting via the interaction the condensate to the depleted particles.

For a Fermi gas the Green function singularities are indeed simple poles located at the excitation energies $\epsilon_\mathbf{k}$ analogously to Eq. (18.18) for the thermal Green function. A key point necessary to arrive at Eq. (18.47) is the fact that the state $|m\rangle = a^\dagger_\mathbf{k}|0\rangle$ obtained by adding a particle to the ground state of the Fermi gas is also an eigenstate of the Hamiltonian as well as the state obtained by removing a particle $|m\rangle = a_\mathbf{k}|0\rangle$. This is no longer true in the general case when the interaction is present and the ground state cannot be built by filling the single particle momentum eigenstates anymore. In general, the states $|m\rangle = a^\dagger_\mathbf{k}|0\rangle$ and $|m\rangle = a_\mathbf{k}|0\rangle$ are a complicated superposition of the exact eigenstates and the form (18.46) can no longer be simplified. All we can say is that the Green function must have singularities in the upper and lower half planes depending whether the excitation energies are negative or positive. However, if the singularities of the Green function acquire the form of simple poles, we may attach a very transparent physical meaning to them. Let us assume, for instance, that in the lower half plane the closest singularity to the real axis is a simple pole located at $\epsilon_\mathbf{k} - \mathrm{i}\gamma_\mathbf{k}$ with residue $z(\mathbf{k})$. Near the singularity the Green function has then the form

$$G(\mathbf{k};\omega) = \frac{z(\mathbf{k})}{\hbar\omega - \epsilon_\mathbf{k} + \mathrm{i}\gamma_\mathbf{k}} + \text{incoherent}, \tag{18.49}$$

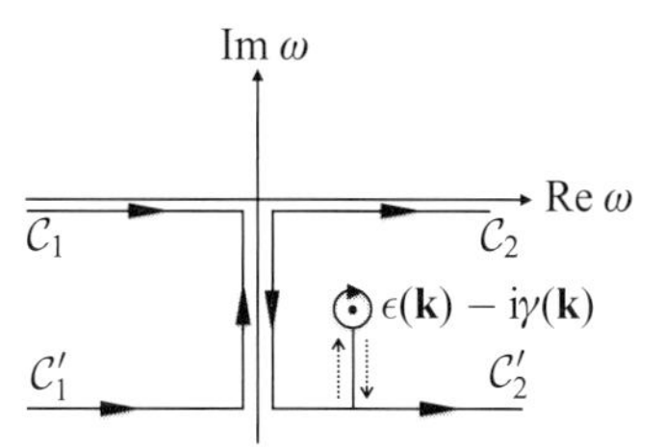

Fig. 18.1 Contours in the complex plane of the variable ω. The horizontal parts of the contours C_1' and C_2' are meant to be shifted to an arbitrarily large distance from the real axis.

where the "incoherent" part indicates a contribution to G which is not singular. Notice that $\epsilon_{\mathbf{k}}$ is not the free particle dispersion, but the real part of the location of the pole of the Green function at fixed wave vector $\mathbf{k}$. Similarly, $\gamma_{\mathbf{k}}$ indicates the imaginary part. The approximate polar structure of the Green function has a direct physical meaning. It expresses the fact that, even for an interacting system, the dynamics of the low-energy excitations can be described in terms of *independent* quasiparticles characterized by a energy dispersion $\epsilon_{\mathbf{k}}$ and lifetime $\hbar/\gamma_{\mathbf{k}}$. The residue $z(\mathbf{k})$ quantifies the relative importance of the quasiparticle contribution with respect to incoherent excitations. The non-interacting limit corresponds to $z(\mathbf{k}) = 1$, as we have seen. The validity of the approximate form (18.49) provides a microscopic justification of the quasiparticle concept introduced in the Landau theory of Fermi liquids in Chapter 12.

We proceed now to give a more mathematical discussion of the physical meaning of the poles of the Green function. Let us consider the Green function for $t > 0$. We must evaluate the integral

$$\int_{-\infty}^{\infty} \frac{\mathrm{d}\omega}{2\pi}\, \mathrm{e}^{-\mathrm{i}\omega t} G(\mathbf{k},\omega) = \int_{-\infty}^{0} \frac{\mathrm{d}\omega}{2\pi}\, \mathrm{e}^{-\mathrm{i}\omega t} G^{\mathrm{A}}(\mathbf{k},\omega) + \int_{0}^{\infty} \frac{\mathrm{d}\omega}{2\pi}\, \mathrm{e}^{-\mathrm{i}\omega t} G^{\mathrm{R}}(\mathbf{k},\omega),$$

where we used the fact that at $T = 0$ G coincides with G^{A} and G^{R} in the first and second integrals, respectively. Then we may deform the path of integration in the two integrals as shown in Fig. 18.1, i.e., we have

$$\int_{C_1} = \int_{C_1'}, \quad \int_{C_2} = \int_{C_2'}.$$

According to Eq. (18.46), all Green functions behave as ω^{-1} at large frequencies. Hence, because of the oscillating factor $\exp(-\mathrm{i}\omega t)$, the contributions from the horizontal parts of the contours C_1' and C_2' can be made arbitrarily small if their position is far enough from the real axis. Hence, because of the analyticity of G^{A}, the first integral reduces to the integral along the vertical part of C_1'. The second integral, since G^{R} has a pole, reduces to an integral along the vertical part of C_2' plus a contribution coming from the residue of G^{R}. We then have

$$\int_{-\infty}^{\infty} \frac{\mathrm{d}\omega}{2\pi}\, \mathrm{e}^{-\mathrm{i}\omega t} G(\mathbf{k},\omega) = \int_{0}^{-\mathrm{i}\infty} \frac{\mathrm{d}\omega}{2\pi}\, \mathrm{e}^{-\mathrm{i}\omega t} [G^{\mathrm{R}}(\mathbf{k},\omega) - G^{\mathrm{A}}(\mathbf{k},\omega)] - \mathrm{i}z(\mathbf{k})\mathrm{e}^{-\mathrm{i}\epsilon_{\mathbf{k}} t/\hbar}\mathrm{e}^{-\gamma_{\mathbf{k}} t/\hbar}. \tag{18.50}$$

By making the variable change $\mathrm{i}\omega = u$ in the first term on the right-hand side, we have

$$\int_{-\infty}^{\infty} \frac{\mathrm{d}\omega}{2\pi}\, \mathrm{e}^{-\mathrm{i}\omega t} G(\mathbf{k}, \omega)$$
$$= -\mathrm{i} \int_{0}^{\infty} \frac{\mathrm{d}u}{2\pi}\, \mathrm{e}^{-ut}[G^{\mathrm{R}}(\mathbf{k}, -\mathrm{i}u) - G^{\mathrm{A}}(\mathbf{k}, -\mathrm{i}u)] - \mathrm{i}z(\mathbf{k})\mathrm{e}^{-\mathrm{i}\epsilon_{\mathbf{k}}t/\hbar}\mathrm{e}^{-\gamma_{\mathbf{k}}t/\hbar}. \tag{18.51}$$

The region of large values of u does not contribute much to the integral due to the presence of the exponential factor. As a result, by considering that the contribution to the integral comes mostly from the region at small values of u, where the Green function behavior is dominated by the pole, and remembering that G^{R} and G^{A} differ, at $T = 0$, only by the sign of the imaginary part, we can approximate the integral over u as

$$-\mathrm{i} \int_{0}^{\infty} \frac{\mathrm{d}u}{2\pi}\, \mathrm{e}^{-ut} \left[G^{\mathrm{R}}(\mathbf{k}, -\mathrm{i}u) - G^{\mathrm{A}}(\mathbf{k}, -\mathrm{i}u)\right] \approx - \int_{0}^{\infty} \frac{\mathrm{d}u}{2\pi} \mathrm{e}^{-ut} \frac{2z(\mathbf{k})\gamma_{\mathbf{k}}/\hbar}{(\mathrm{i}u + \epsilon_{\mathbf{k}}/\hbar)^2 + \gamma_{\mathbf{k}}^2/\hbar^2}.$$

By neglecting u in the denominator, we finally arrive at

$$\int_{-\infty}^{\infty} \frac{\mathrm{d}\omega}{2\pi}\, \mathrm{e}^{-\mathrm{i}\omega t} G(\mathbf{k}, \omega) = -\frac{1}{\pi t} \frac{z(\mathbf{k})\hbar\gamma_{\mathbf{k}}}{\epsilon_{\mathbf{k}}^2 + \gamma_{\mathbf{k}}^2} - \mathrm{i}z(\mathbf{k})\mathrm{e}^{-\mathrm{i}\epsilon_{\mathbf{k}}t/\hbar}\mathrm{e}^{-\gamma_{\mathbf{k}}t/\hbar}. \tag{18.52}$$

The above is the central result of this section. It shows that for times $\epsilon_{\mathbf{k}}^{-1} \ll t \le \gamma_{\mathbf{k}}^{-1}$, the first term is negligible with respect to the second. The Green function as a function of time has the form of the wave function of a particle with energy $\epsilon_{\mathbf{k}}$ and lifetime $\gamma_{\mathbf{k}}^{-1}$. As far as the meaning of the residue $z(\mathbf{k})$, we reason as follows. We define in an arbitrary interacting Fermi system the Fermi momentum as $\epsilon_{\mathbf{k}_{\mathrm{F}}} = 0$, separating positive and negative energies. Then, from Eq. (18.26) we evaluate the particle number in terms of the Green function as

$$n_{\mathbf{k}} = -\mathrm{i} \int_{-\infty}^{\infty} \frac{\mathrm{d}\omega}{2\pi}\, \mathrm{e}^{\mathrm{i}\omega 0^+} G(\mathbf{k}, \omega). \tag{18.53}$$

Notice that we are considering infinitesimal negative time $t \to -0^+$. With respect to the previous case we have to deform the contours of integration in the upper complex plane of the variable ω in a way symmetrical with respect to that of Fig. 18.1. The result of the integral depends of the position of $\mathbf{k}$ with respect to $\mathbf{k}_{\mathrm{F}}$. First we split the integral for negative and positive frequencies. For $\mathbf{k}$ within the Fermi sphere, the singularity of the Green function is in the upper half plane. For $\mathbf{k}$ outside the Fermi sphere, there are no poles. As a result, one discovers that the occupation number $n_{\mathbf{k}}$ is discontinuous at k_{F} with

$$n_{\mathbf{k}_{\mathrm{F}-}} - n_{\mathbf{k}_{\mathrm{F}+}} = z(\mathbf{k}) \le 1, \tag{18.54}$$

where $\mathbf{k}_{\mathrm{F}\mp}$ indicates the limit $\mathbf{k} \to \mathbf{k}_{\mathrm{F}}$ from below and above, respectively. The upper bound comes from the Pauli exclusion principle. In the case of a Fermi gas, the jump goes to unity. As anticipated, the jump gives the fraction of the excitations of the interacting systems, which can be described in terms of effective free particles. This observation paved the way to the notion of the quasiparticle at the heart of the Landau Fermi liquid theory, which was the subject of Chapter 12.

As a final remark, we point out that at $T \neq 0$, G does not coincide with G^{R} and G^{A} at positive and negative excitation energies, and instead of Eq. (18.42), one has to refer to

Eq. (18.39). Nevertheless, the polar approximation (18.49) can be extended to the thermal Green function

$$\mathcal{G}(\mathbf{k}, \omega_n) = \frac{z(\mathbf{k})}{\mathrm{i}\omega_n - \epsilon_{\mathbf{k}} + \mathrm{i}\gamma_{\mathbf{k}}\mathrm{sgn}(\omega_n)} + \delta\mathcal{G}(\mathbf{k}, \omega_n), \tag{18.55}$$

$\delta\mathcal{G}(\mathbf{k}, \omega_n)$ indicating the regular part. For $\gamma_{\mathbf{k}} \ll \epsilon_{\mathbf{k}}$, by means of (18.20) one obtains the Fermi and Bose distribution functions, weighted by the residue $z(\mathbf{k})$, with the free particle dispersion relation substituted by the single-particle excitation energy from the ground state. When $\gamma_{\mathbf{k}}$ is not negligible, it produces an additional smearing of the distribution with respect to the one due to the temperature.

18.5 The perturbative expansion of the Matsubara Green function

In this section we show how to systematically obtain a perturbative expansion of the Green function. To this end we distinguish the *unpertubed* operator $K_0 = H_0 - \mu N$ from the full K. In the statistical average, the operator K always enters in an exponential function, so that we need to evaluate the difference with respect to the exponential of K_0. We define the "statistical S-matrix $\mathcal{S}$" as

$$\mathrm{e}^{-K\tau} = \mathrm{e}^{-K_0\tau}\mathcal{S}(\tau), \quad 0 < \tau \neq \beta \tag{18.56}$$

with the condition $\mathcal{S}(0) = 1$. $\mathcal{S}(\tau)$ will allow for a peculiar interaction representation of the field operators, which is adequate for a perturbative expansion. By differentiating both terms of the equation with respect to τ, one obtains

$$\frac{\partial\mathcal{S}(\tau)}{\partial\tau} = -\mathrm{e}^{K_0\tau}H_{\mathrm{i}}\mathrm{e}^{-K_0\tau}\mathcal{S}(\tau) \equiv -H_{\mathrm{i}}(\tau)\mathcal{S}(\tau), \tag{18.57}$$

where $H_i = K - K_0$ is the *interaction* Hamiltonian. The differential equation can be formally solved by iteration with the result

$$\mathcal{S}(\tau) = \mathrm{T}_\tau\mathrm{e}^{-\int_0^\tau \mathrm{d}\tau' H_{\mathrm{i}}(\tau')}. \tag{18.58}$$

The solution (18.58) is deceptively simple. The key point is the presence of the "time"-ordering operator in front of the exponential. The iterative solution naturally produces, order by order, terms which are "time"-ordered with the largest time on the left. Hence, to reproduce all such terms, one must first expand the exponential and then order the various operators according to their time arguments. It is this structure of the S-matrix that motivates the introduction of the "time"-ordered Green function.

The introduction of the S-matrix allows us to write the "time"-dependent field operators (18.3) in the *interaction* picture as

$$\hat{\psi}(\mathbf{x}, \tau) = \mathcal{S}^{-1}(\tau)\hat{\psi}_0(\mathbf{x}, \tau)\mathcal{S}(\tau), \quad \hat{\bar{\psi}}(\mathbf{x}, \tau) = \mathcal{S}^{-1}(\tau)\hat{\bar{\psi}}_0(\mathbf{x}, \tau)\mathcal{S}(\tau), \tag{18.59}$$

where the subscript 0 indicates the "time" dependence of the field operator with respect to K_0. To appreciate the usefulness of the "time"-ordering, let us consider the "time"-ordered product appearing into the definition of $\mathcal{G}$ in Eq. (18.2). By using Eqs. (18.56) and (18.59) we get, for $\tau > \tau'$,

$$\begin{aligned} e^{-\beta K}\hat{\psi}(\mathbf{x},\tau)\hat{\bar{\psi}}(\mathbf{x}',\tau') &= e^{-\beta K_0}\mathcal{S}(\beta)\mathcal{S}^{-1}(\tau)\hat{\psi}_0(\mathbf{x},\tau)\mathcal{S}(\tau)\mathcal{S}^{-1}(\tau')\hat{\bar{\psi}}_0(\mathbf{x}',\tau')\mathcal{S}(\tau') \\ &= e^{-\beta K_0}\mathcal{S}(\beta,\tau)\hat{\psi}_0(\mathbf{x},\tau)\mathcal{S}(\tau,\tau')\hat{\bar{\psi}}_0(\mathbf{x}',\tau')\mathcal{S}(\tau'), \end{aligned}$$

where for brevity $\mathcal{S}(\tau,\tau') = \mathcal{S}(\tau)\mathcal{S}^{-1}(\tau')$. The two "times" τ and τ' split the inteval $(0,\beta)$ into three pieces: $(0,\tau')$, (τ',τ) and (τ,β). The unification of the $\mathcal{S}$-matrix coming from the Boltzmann factor and from the "time" Heisenberg evolution of the field operators would not have been possible for the real-time Green function at finite temperature.

By using the unitarity of the S-matrix:

$$\mathcal{S}^{-1}(\tau,\tau') = \mathcal{S}(\tau',\tau), \tag{18.60}$$

the "time"-ordered product allows us to combine all the S-matrix terms into a single one to get

$$e^{-\beta K}\hat{\psi}(\mathbf{x},\tau)\hat{\bar{\psi}}(\mathbf{x}',\tau') = e^{-\beta K_0}\mathrm{T}_\tau(\hat{\psi}_0(\mathbf{x},\tau)\hat{\bar{\psi}}_0(\mathbf{x}',\tau')\mathcal{S}(\beta,0)). \tag{18.61}$$

When dealing with the perturbative expansion of $\mathcal{S}$ the "time"-ordering will indeed place the resulting string of the field operators in the right order. To better appreciate this point, consider, for instance, the first order term of the S-matrix:

$$\mathcal{S}^{(1)}(\beta) = -\int_0^\beta d\tau'' H_{\mathrm{i}}(\tau'') = \mathcal{S}^{(1)}(\beta,\tau) + \mathcal{S}^{(1)}(\tau,\tau') + \mathcal{S}^{(1)}(\tau',0).$$

For the given choice of "times" $\tau > \tau'$, the "time"-ordering places each of the above terms in the correct position with respect to the field operators.

By performing the same analysis for $\tau < \tau'$, one then concludes that the Green function defined in (18.2) with the average (18.5) may be written as

$$\begin{aligned} \mathcal{G}(\mathbf{x},\tau;\mathbf{x}',\tau') &= -\mathrm{Tr}[e^{\beta\Omega}e^{-\beta K_0}\mathrm{T}_\tau(\hat{\psi}_0(\mathbf{x},\tau)\hat{\bar{\psi}}_0(\mathbf{x}',\tau')\mathcal{S}(\beta,0))] \\ &= -\frac{\mathrm{Tr}[e^{-\beta K_0}\mathrm{T}_\tau(\hat{\psi}_0(\mathbf{x},\tau)\hat{\bar{\psi}}_0(\mathbf{x}',\tau')\mathcal{S}(\beta,0))]}{\mathrm{Tr}[e^{-\beta K_0}\mathcal{S}(\beta,0)]} \\ &\equiv -\left\langle \mathrm{T}_\tau\left(\hat{\psi}_0(\mathbf{x},\tau)\hat{\psi}_0^\dagger(\mathbf{x}',\tau')\mathcal{S}(\beta,0)\right)\right\rangle_0, \end{aligned} \tag{18.62}$$

where we have expressed the normalization factor in terms of the trace:

$$e^{-\beta\Omega} = \mathrm{Tr}[e^{-\beta K_0}\mathcal{S}(\beta,0)]. \tag{18.63}$$

Expressions (18.62) and (18.63) are the starting point of the perturbative expansion for the Green function and the thermodynamic potential, respectively. The statistical average is performed with respect to the unperturbed system as specified by the subscript 0. The interaction Hamiltonian appears via $\mathcal{S}(\beta,0)$. By expanding the exponential (18.58), the expression of the Green function and the partition function become an infinite series with each term being a statistical average of a string of field operators. The unperturbed Hamiltonian H_0 is chosen to be separable in the sense that the Hamiltonian contains only one-body terms, the statistical average of a string of field operators can be decomposed as the sum of terms, each being the product of statistical averages of only two field operators, i.e. the average of n field operators $\hat{\psi}$ and n operators $\bar{\hat{\psi}}$ can be decomposed into products of n averaged pairs with respect to the unperturbed system in all possible combinations. This is the statistical Wick theorem, which will be proven in the next section and is at the basis of the perturbative expansion in terms of independent $\mathcal{G}_0$.

The above factorization of the thermal averages of strings of field operators in terms of unperturbed Green functions is not possible for the real-time-ordered Green functions at finite temperature due to the presence of the real-time dependent Heisenberg exponentials and of the statistical Boltzmann factors. At $T = 0$, since there are no Boltzmann factors present, a perturbative expansion is possible for the real-time Green function. In this limit the Matsubara frequencies become continuous variables on the imaginary axis with respect to the real frequencies.

The perturbative expansion for the real-time-ordered Green function can then be directly obtained by the corresponding expansion of the thermal Green function with the replacement of the sum over Matsubara frequencies with the integral over a continuous frequency on the imaginary axis:

$$\frac{1}{\beta}\sum_{\omega_n} \to \frac{\hbar}{2\pi}\int_{-i\infty}^{i\infty} d\omega \tag{18.64}$$

and by a rotation in the complex plane.[1]

18.6 The statistical Wick theorem

The basic argument is the same as that used in deriving the relation (9.52). Because the Hamiltonian conserves the number of particles, the trace is performed over a set of states with a well defined number of particles. As a consequence, in order for a string to have an average different from zero, it must have an equal number of creation and annihilation operators. To prove the theorem, let us begin with the specific example of the average of a string of four "Heisenberg" field operators:

$$\mathrm{Tr}[e^{\beta\Omega}e^{-\beta K_0}\mathrm{T}_\tau(\bar{\hat{\psi}}(\mathbf{x}_1,\tau_1)\bar{\hat{\psi}}(\mathbf{x}_2,\tau_2)\hat{\psi}(\mathbf{x}_3,\tau_3)\hat{\psi}(\mathbf{x}_4,\tau_4))].$$

The basic assumption of dealing with a one-body Hamiltonian allows us to expand the field operators in terms of wave functions and creation and annihilation operators (see Eqs. (9.18) and (9.19)). By making the correspondence $it/\hbar \to \tau$, Eq. (9.51) leads to

$$a_{\kappa_i}(\tau) = e^{K\tau}a_{\kappa_i}e^{-K\tau} = e^{-\epsilon_{\kappa_i}\tau}a_{\kappa_i}, \quad \bar{a}_{\kappa_i}(\tau) = e^{K\tau}a^\dagger_{\kappa_i}e^{-K\tau} = e^{\epsilon_{\kappa_i}\tau}a_{\kappa_i}, \tag{18.65}$$

so we now have

$$\sum_{\kappa_1,\kappa_2,\kappa_3,\kappa_4} g^*_{\kappa_1}(\mathbf{x}_1)g^*_{\kappa_2}(\mathbf{x}_2)g_{\kappa_3}(\mathbf{x}_3)g_{\kappa_4}(\mathbf{x}_4) \times \mathrm{Tr}\left[e^{\beta\Omega}e^{-\beta K_0}\mathrm{T}_\tau\left(e^{\epsilon_{\kappa_1}\tau_1}a^\dagger_{\kappa_1}e^{\epsilon_{\kappa_2}\tau_2}a^\dagger_{\kappa_2}e^{-\epsilon_{\kappa_3}\tau_3}a_{\kappa_3}e^{-\epsilon_{\kappa_4}\tau_4}a_{\kappa_4}\right)\right]. \tag{18.66}$$

[1] Such a rotation explains the presence of the factors of i in the definition of the time-dependent Green function.

We have made use of Eq. (9.51) to make explicit the "time" dependence of the creation and annihilation operators. Let us consider that the "times" are $\tau_1 > \tau_2 > \tau_3 > \tau_4$. This fixes the order of the operators. The exponential "time"-dependent factors can then be taken out of the trace sign and we are left with the average of a string of four annihilation and creation operators. Our problem is then reduced to evaluating

$$\mathrm{Tr}\left[\mathrm{e}^{\beta\Omega}\mathrm{e}^{-\beta K_0}a^\dagger_{\kappa_1}a^\dagger_{\kappa_2}a_{\kappa_3}a_{\kappa_4}\right].$$

Now we move the operator $a^\dagger_{\kappa_1}$ one step to the right, taking into account that it anticommutes or commutes with $a^\dagger_{\kappa_2}$ depending on whether we are dealing with Fermi or Bose statistics:

$$\mp\mathrm{Tr}\left[\mathrm{e}^{\beta\Omega}\mathrm{e}^{-\beta K_0}a^\dagger_{\kappa_2}a^\dagger_{\kappa_1}a_{\kappa_3}a_{\kappa_4}\right].$$

In the next step, to move the operator $a^\dagger_{\kappa_1}$ further to the right, we must consider the anticommutator or commutator. The result is

$$\mp\mathrm{Tr}\left[\mathrm{e}^{\beta\Omega}\mathrm{e}^{-\beta K_0}a^\dagger_{\kappa_2}\left[a^\dagger_{\kappa_1},a_{\kappa_3}\right]_\pm a_{\kappa_4}\right]+\mathrm{Tr}\left[\mathrm{e}^{\beta\Omega}\mathrm{e}^{-\beta K_0}a^\dagger_{\kappa_2}a_{\kappa_3}a^\dagger_{\kappa_1}a_{\kappa_4}\right].$$

By using the commutation rules, we then get

$$\mp\delta_{\kappa_1\kappa_3}\mathrm{Tr}\left[\mathrm{e}^{\beta\Omega}\mathrm{e}^{-\beta K_0}a^\dagger_{\kappa_2}a_{\kappa_4}\right]+\mathrm{Tr}\left[\mathrm{e}^{\beta\Omega}\mathrm{e}^{-\beta K_0}a^\dagger_{\kappa_2}a_{\kappa_3}a^\dagger_{\kappa_1}a_{\kappa_4}\right].$$

To have the operator $a^\dagger_{\kappa_1}$ in the second term moved beyond a_{κ_4}, we must do as in the previous step with the result

$$\mp\delta_{\kappa_1\kappa_3}\mathrm{Tr}\left[\mathrm{e}^{\beta\Omega}\mathrm{e}^{-\beta K_0}a^\dagger_{\kappa_2}a_{\kappa_4}\right]+\delta_{\kappa_1\kappa_4}\mathrm{Tr}\left[\mathrm{e}^{\beta\Omega}\mathrm{e}^{-\beta K_0}a^\dagger_{\kappa_2}a_{\kappa_3}\right]\mp\mathrm{Tr}\left[\mathrm{e}^{\beta\Omega}\mathrm{e}^{-\beta K_0}a^\dagger_{\kappa_2}a_{\kappa_3}a_{\kappa_4}a^\dagger_{\kappa_1}\right].$$

In the last term, we can perform a cyclic permutation under the trace:

$$\mp\delta_{\kappa_1\kappa_3}\mathrm{Tr}\left[\mathrm{e}^{\beta\Omega}\mathrm{e}^{-\beta K_0}a^\dagger_{\kappa_2}a_{\kappa_4}\right]+\delta_{\kappa_1\kappa_4}\mathrm{Tr}\left[\mathrm{e}^{\beta\Omega}\mathrm{e}^{-\beta K_0}a^\dagger_{\kappa_2}a_{\kappa_3}\right]\mp\mathrm{Tr}\left[\mathrm{e}^{\beta\Omega}a^\dagger_{\kappa_1}\mathrm{e}^{-\beta K_0}a^\dagger_{\kappa_2}a_{\kappa_3}a_{\kappa_4}\right].$$

In the last term, before the operator $a^\dagger_{\kappa_1}$ we insert $\mathrm{e}^{-\beta K_0}\mathrm{e}^{\beta K_0}$ and use again the property (18.65):

$$\mp\delta_{\kappa_1\kappa_3}\mathrm{Tr}\left[\mathrm{e}^{\beta\Omega}\mathrm{e}^{-\beta K_0}a^\dagger_{\kappa_2}a_{\kappa_4}\right]+\delta_{\kappa_1\kappa_4}\mathrm{Tr}\left[\mathrm{e}^{\beta\Omega}\mathrm{e}^{-\beta K_0}a^\dagger_{\kappa_2}a_{\kappa_3}\right]\mp\mathrm{e}^{\beta\epsilon_{\kappa_1}}\mathrm{Tr}\left[\mathrm{e}^{\beta\Omega}\mathrm{e}^{-\beta K_0}a^\dagger_{\kappa_1}a^\dagger_{\kappa_2}a_{\kappa_3}a_{\kappa_4}\right].$$

We see that the last term, apart from the exponential $\mathrm{e}^{\beta\epsilon_{\kappa_1}}$, is the same term we started from. Hence it can be combined with it with the result

$$\begin{aligned}&\mathrm{Tr}\left[\mathrm{e}^{\beta\Omega}\mathrm{e}^{-\beta K_0}a^\dagger_{\kappa_1}a^\dagger_{\kappa_2}a_{\kappa_3}a_{\kappa_4}\right]\\&\quad=\mp\delta_{\kappa_1\kappa_3}n_{\kappa_1}\mathrm{Tr}\left[\mathrm{e}^{\beta\Omega}\mathrm{e}^{-\beta K_0}a^\dagger_{\kappa_2}a_{\kappa_4}\right]+\delta_{\kappa_1\kappa_4}n_{\kappa_1}\mathrm{Tr}\left[\mathrm{e}^{\beta\Omega}\mathrm{e}^{-\beta K_0}a^\dagger_{\kappa_2}a_{\kappa_3}\right],\end{aligned}\tag{18.67}$$

having used the fact that $(1\pm\mathrm{e}^{\beta\epsilon_{\kappa_1}})^{-1}=n_{\kappa_1}$. Equation (18.67) contains the core of the argument. It shows how an average of four operators has been reduced to the average of two operators. The latter has been evaluated explicitly in (9.52) so that we can write

$$\mathrm{Tr}\left[\mathrm{e}^{\beta\Omega}\mathrm{e}^{-\beta K_0}a^\dagger_{\kappa_1}a^\dagger_{\kappa_2}a_{\kappa_3}a_{\kappa_4}\right]=\mp\delta_{\kappa_1\kappa_3}n_{\kappa_1}\delta_{\kappa_2\kappa_4}n_{\kappa_2}+\delta_{\kappa_1\kappa_4}n_{\kappa_1}\delta_{\kappa_2\kappa_3}n_{\kappa_2}.\tag{18.68}$$

Equation (18.68) can be rewritten as

$$\langle a^\dagger_{\kappa_1}a^\dagger_{\kappa_2}a_{\kappa_3}a_{\kappa_4}\rangle_0=\mp\langle a^\dagger_{\kappa_1}a_{\kappa_3}\rangle_0\langle a^\dagger_{\kappa_2}a_{\kappa_4}\rangle_0+\langle a^\dagger_{\kappa_1}a_{\kappa_4}\rangle_0\langle a^\dagger_{\kappa_2}a_{\kappa_3}\rangle_0,\tag{18.69}$$

with the understanding that the average defined in Eq. (18.5) has been performed with respect to K_0. Hence the average of four operators is decomposed as the sum of terms, each being the product of two field operators. The appearance of minus signs is associated, in the case of Fermi statistics, with the number of permutations necessary to place the pair of operators which must be averaged.

The evaluation of the Matsubara Green function done for Fermi and Bose gases can be repeated in the general case of one-body Hamiltonians with the result, for $\tau > 0$, from Eq. (18.17)

$$\mathcal{G}_0(\mathbf{x}, \tau; \mathbf{x}', 0) = -\sum_\kappa g^*_\kappa(\mathbf{x}')g_\kappa(\mathbf{x})\mathrm{e}^{-\epsilon_\kappa\tau}(1 \mp n_\kappa) = -\sum_\kappa g^*_\kappa(\mathbf{x}')g_\kappa(\mathbf{x})\mathrm{e}^{-\epsilon_\kappa\tau}\langle a_\kappa a^\dagger_\kappa\rangle_0, \tag{18.70}$$

the expression for $\tau < 0$ is derivable from the symmetry relation (18.10). Hence, for the choice of "times" $\tau_1 > \tau_2 > \tau_3 > \tau_4$ we can conclude that

$$\begin{aligned}&\langle\hat{\bar{\psi}}(\mathbf{x}_1, \tau_1)\hat{\bar{\psi}}(\mathbf{x}_2, \tau_2)\hat{\psi}(\mathbf{x}_3, \tau_3)\hat{\psi}(\mathbf{x}_4, \tau_4)\rangle_0\\ &\quad = \mp\mathcal{G}_0(\mathbf{x}_3, \tau_3; \mathbf{x}_1, \tau_1)\mathcal{G}_0(\mathbf{x}_4, \tau_4; \mathbf{x}_2, \tau_2) + \mathcal{G}_0(\mathbf{x}_4, \tau_4; \mathbf{x}_1, \tau_1)\mathcal{G}_0(\mathbf{x}_3, \tau_3; \mathbf{x}_2, \tau_2),\end{aligned} \tag{18.71}$$

where the Green functions are those appropriate for the "time" arguments. It is clear that the above derivation can be repeated for any ordering of the "time" arguments, so that one can conclude that the same relation is valid also under the sign of the "time"-ordering operator

$$\begin{aligned}&\langle \mathrm{T}_\tau(\hat{\bar{\psi}}(\mathbf{x}_1, \tau_1)\hat{\bar{\psi}}(\mathbf{x}_2, \tau_2)\hat{\psi}(\mathbf{x}_3, \tau_3)\hat{\psi}(\mathbf{x}_4, \tau_4))\rangle_0\\ &\quad = \mp\mathcal{G}_0(\mathbf{x}_3, \tau_3; \mathbf{x}_1, \tau_1)\mathcal{G}_0(\mathbf{x}_4, \tau_4; \mathbf{x}_2, \tau_2) + \mathcal{G}_0(\mathbf{x}_4, \tau_4; \mathbf{x}_1, \tau_1)\mathcal{G}_0(\mathbf{x}_3, \tau_3; \mathbf{x}_2, \tau_2).\end{aligned} \tag{18.72}$$

This completes the proof of the statistical Wick theorem for strings of four field operators. Of course, the theorem is far more general and, at this stage, it must be obvious how to proceed in order to extend the validity to an arbitrary even numbers of field operators. For instance, with six field operators, the first step will be to reduce the average to a sum of averages of four field operators and to apply the result (18.71) to them.

18.7 Diagrammatic expansion of the one-particle thermal Green function and the thermodynamic potential

Before expanding the $\mathcal{S}$-matrix of Eq. (18.58) we need to integrate the interaction Hamiltonian, whose second quantized form is given by Eq. (9.37). With this spin-conserving Hamiltonian, we will drop the indices most of the time. It is, hence, convenient to introduce a symmetrized form of the interparticle potential, namely the four-point bare vertex, to take into account the direct and exchange terms:

$$\begin{aligned}\Gamma^0(x_1, x_2, x_3, x_4) \equiv\ &\delta(\tau_1 - \tau_2)\delta(x_1 - x_4)\delta(x_2 - x_3)V(\mathbf{x}_1 - \mathbf{x}_2)\\ &\mp\delta(\tau_1 - \tau_2)\delta(x_1 - x_3)\delta(x_2 - x_4)V(\mathbf{x}_1 - \mathbf{x}_2),\end{aligned} \tag{18.73}$$

$$\mathcal{G}_0(x,\tau;x',\tau') = x,\tau \longrightarrow x',\tau' \qquad \Gamma^0(x_1,x_2,x_3,x_4) =$$

(a) (b)

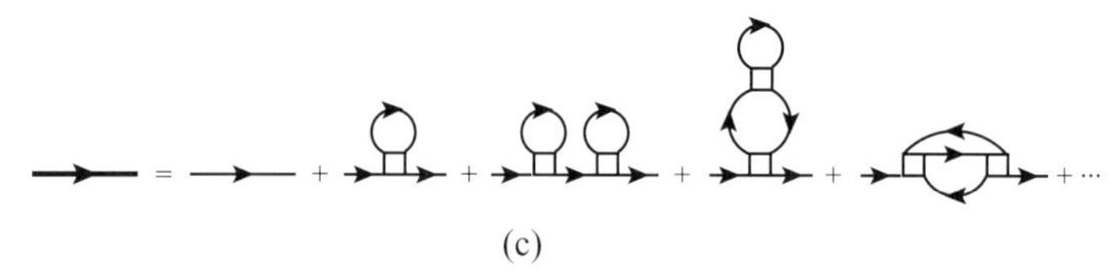

(c)

Fig. 18.2 (a) The non-interacting $\mathcal{G}_0$ Green function line. (b) The bare symmetrized interaction Γ^0. The dashed lines represent the four field operators of the building block Eq. (18.74). (c) Diagrammatic representation of the perturbative expansion of the single-particle Green function at second order.

where $x_i \equiv (\mathbf{x}_i, \tau_i)$. The minus and plus signs refer to Fermi and Bose statistics, respectively. In terms of Γ^0, the building block of the perturbation expansion reads

$$\int \mathrm{d}\tau_1 H_{\mathrm{i}}(\tau_1) = \frac{1}{4}\int \mathrm{d}x_1\, \mathrm{d}x_2\, \mathrm{d}x_3\, \mathrm{d}x_4 \Gamma^0(x_1,x_2,x_3,x_4)\hat{\bar{\psi}}(x_1)\hat{\bar{\psi}}(x_2)\hat{\psi}(x_3)\hat{\psi}(x_4). \quad (18.74)$$

We start from Eq. (18.62) for the perturbative expansion of the Green function. By considering the first order expression of the $\mathcal{S}$-matrix and the symmetrized form of the interaction (18.74), we get (see the second term in the diagrammatic expansion of Fig. 18.2(c)):

$$\delta\mathcal{G}(x,x') = (-1)^2 \int \mathrm{d}x_1 \mathrm{d}x_2 \mathrm{d}x_3 \mathrm{d}x_4 \frac{1}{4}\Gamma^0(x_1,x_2,x_3,x_4)$$
$$\times \langle \hat{\psi}(x)\hat{\bar{\psi}}(x_1)\hat{\bar{\psi}}(x_2)\hat{\psi}(x_3)\hat{\psi}(x_4)\hat{\bar{\psi}}(x')\rangle, \quad (18.75)$$

where one minus sign is due to the order (the first) in the perturbation expansion and the other to the definition of the Green function (see Eq. (18.2)). The next step amounts to writing the average of the string of field operators in terms of averages of pairs of field operators, indicated as *contractions*, by using the Wick theorem.

The perturbative expansion of the Green function may be graphically represented in terms of the so-called Feynman diagrams, as shown in Fig. 18.2. With the non-interacting or bare propagator $\mathcal{G}_0$ we associate an oriented line going from $\mathbf{x}\,\tau$ to $\mathbf{x}'\tau'$ (see Fig. 18.2 (a)). The symmetrized interaction connects these lines. The empty square of Fig. 18.2(b) is the graphic representation of Γ^0, from which the four fermion fields of the building block (18.74) originate.

It is important to emphasize that we must consider only *connected* diagrams, i.e. diagrams which are not made of independent subdiagrams. This is the theorem of the cancellation of the so-called *disconnected* diagrams. The origin of this theorem can be identified in the form of the perturbative expansion (18.62). There one must expand both the numerator and the denominator. After some reflection, one can convince oneself that the

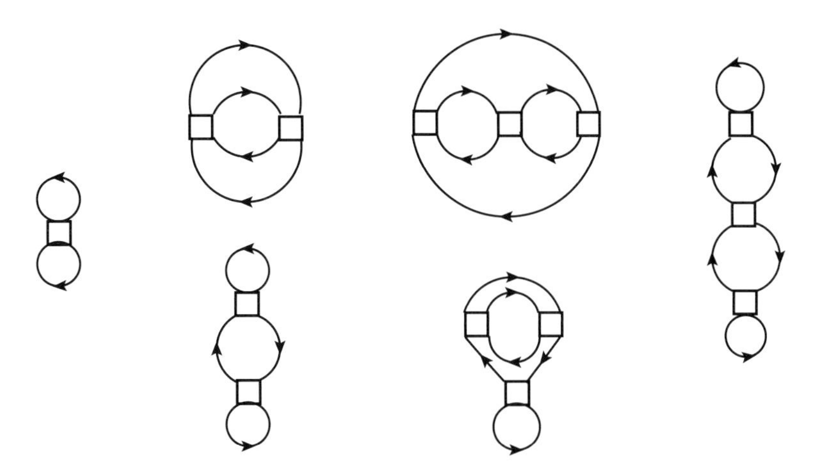

Fig. 18.5 Diagrams for the thermodynamical potential up to third order.

18.8 The Dyson equation

The Dyson equation expresses in very compact form the thermal one-particle Green function $\mathcal{G}$ in terms of the self-energy we are going to introduce.

The thermal Green function defined in Eq. (18.2) obeys an "equation of motion", which can be derived by exploiting the "Heisenberg form" of the field operators introduced in Eq. (18.3). Let us derive with respect to the fictitious time τ the Green function of Eq. (18.2). The derivative acts both on the theta functions and on the operator $\hat{\psi}(\mathbf{x}, \tau)$. The derivative of the theta function is the delta function $\delta(\tau - \tau')$, while the derivative of the field operator gives

$$-\frac{\partial}{\partial \tau}\hat{\psi}(\mathbf{x}, \tau) = [\hat{\psi}(\mathbf{x}, \tau), K], \tag{18.81}$$

which reduces to the standard Heisenberg equation of motion (9.43) with the replacement $\tau = \mathrm{i}t/\hbar$. The equation for $\mathcal{G}$ becomes

$$-\frac{\partial}{\partial \tau}\mathcal{G}(\mathbf{x}, \tau; \mathbf{x}', \tau') = \delta(\tau - \tau')\delta(\mathbf{x} - \mathbf{x}') - \langle \mathrm{T}_\tau[\hat{\psi}(\mathbf{x}, \tau), K]\bar{\hat{\psi}}(\mathbf{x}', \tau')\rangle. \tag{18.82}$$

Notice that the delta function $\delta(\mathbf{x} - \mathbf{x}')$ originates from terms with delta function $\delta(\tau - \tau')$, where one may use the commutation relations of the field operators at equal "times". A more explicit expression may be obtained by specifying the form of the Hamiltonian. Let us consider as a specific example the Hamiltonian K given by Eq. (9.37) with $K = K_0 + H_\mathrm{i}$, as done in the previous sections.

To begin with it is instructive to start from the case of the Fermi and Bose gases, where $H_\mathrm{i} = 0$. In this case the commutator appearing under the "time"-ordering operator in Eq. (18.82) can be easily evaluated and Eq. (18.82) becomes

$$-\frac{\partial}{\partial \tau}\mathcal{G}_0(\mathbf{x}, \tau; \mathbf{x}', \tau') = \delta(\tau - \tau')\delta(\mathbf{x} - \mathbf{x}') + \left(-\frac{\hbar^2\nabla_\mathbf{x}^2}{2m} - \mu\right)\mathcal{G}_0(\mathbf{x}, \tau; \mathbf{x}', \tau'). \tag{18.83}$$

The above is the Dyson equation for non-interacting particles. This is a non-homogeneous differential equation, which can be solved by Fourier transforming with respect to $\tau - \tau'$ and $\mathbf{x} - \mathbf{x}'$. By so doing one easily obtains $\mathcal{G}_0(\mathbf{k}, \omega_n)$, as already derived in Eq. (18.18).

The zero-kelvin time-dependent Green function obeys a similar equation, which, by the above mentioned replacement $\tau = \mathrm{i}t/\hbar$, may be written as

$$\left(\mathrm{i}\hbar\frac{\partial}{\partial t} + \frac{\hbar^2\nabla^2_{\mathbf{x}}}{2m} + \mu\right) G_0(\mathbf{x}, t; \mathbf{x}', t') = \delta(t - t')\delta(\mathbf{x} - \mathbf{x}'), \tag{18.84}$$

which shows that the Green function for non-interacting particles is the Green function of the Schrödinger differential operator. Indeed, if we denote with $g(\mathbf{x}', t')$ the wave function at time t' as function of $\mathbf{x}'$, then

$$g(\mathbf{x}, t) = \int \mathrm{d}\mathbf{x}' G_0(\mathbf{x}, t; \mathbf{x}', t') g(\mathbf{x}', t') \tag{18.85}$$

represents the wave function at time $t > t'$ as a function of $\mathbf{x}$. Hence the Green function describes the spreading in space and time of a particle which was initially localized. In this sense the Green function is also called the *propagator*.

Before addressing the Green function for interacting systems, it may be useful to consider the case of non-interacting particles in an external field $U(\mathbf{x})$. By evaluating the commutator as in Eq. (18.82), we see that Eq. (18.83) becomes

$$\left(-\frac{\partial}{\partial \tau} + \frac{\hbar^2\nabla^2_{\mathbf{x}}}{2m} + \mu - U(\mathbf{x})\right) \mathcal{G}(\mathbf{x}, \tau; \mathbf{x}', \tau') = \delta(\tau - \tau')\delta(\mathbf{x} - \mathbf{x}'), \tag{18.86}$$

which may be rewritten as

$$\mathcal{G}(\mathbf{x}, \tau; \mathbf{x}', \tau') = \mathcal{G}_0(\mathbf{x}, \tau; \mathbf{x}', \tau') + \int \mathrm{d}\tau'' \int \mathrm{d}\mathbf{x}'' \mathcal{G}_0(\mathbf{x}, \tau; \mathbf{x}'', \tau'') U(\mathbf{x}'') \mathcal{G}(\mathbf{x}'', \tau''; \mathbf{x}', \tau'). \tag{18.87}$$

This form lends itself to a perturbative solution by iteration. Indeed, by replacing $\mathcal{G}$ with $\mathcal{G}_0$ in the integral on the right-hand side, we obtain $\mathcal{G}^1$, which, in turn, can be inserted in the above integral to obtain $\mathcal{G}^2$:

$$\mathcal{G}^1(\mathbf{x}, \tau; \mathbf{x}', \tau') = \int \mathrm{d}\tau'' \int \mathrm{d}\mathbf{x}'' \mathcal{G}_0(\mathbf{x}, \tau; \mathbf{x}'', \tau'') U(\mathbf{x}'') \mathcal{G}_0(\mathbf{x}'', \tau''; \mathbf{x}', \tau'),$$

$$\mathcal{G}^2(\mathbf{x}, \tau; \mathbf{x}', \tau') = \int \mathrm{d}\tau'' \int \mathrm{d}\mathbf{x}'' \mathcal{G}_0(\mathbf{x}, \tau; \mathbf{x}'', \tau'') U(\mathbf{x}'') \mathcal{G}^1(\mathbf{x}'', \tau''; \mathbf{x}', \tau'),$$

and so on and so forth. The physical meaning of the expression for $\mathcal{G}^1$ is that a particle propagates freely from the point $\mathbf{x}'$ at "time" τ' until point $\mathbf{x}''$ at "time" τ'', when it is subject to the action of the potential. Then it propagates freely again from $\mathbf{x}''$ to $\mathbf{x}$. The intermediate point $\mathbf{x}''$ can be anywhere and can be reached at any "time" τ'', so that one must integrate over $\mathbf{x}''$ and τ''.

To make the expressions less cumbersome, the integration over the space and "time" variables can be regarded as a sum over *continuous* matrix indices.

Let us introduce the *self-energy* due to an external field as

$$\Sigma(\mathbf{x}, \tau; \mathbf{x}', \tau') = U(\mathbf{x})\delta(\mathbf{x} - \mathbf{x}')\delta(\tau - \tau'), \tag{18.88}$$

then the formal solution of the Dyson equation in compact matrix notation reads

$$\mathcal{G} = \mathcal{G}_0 + \mathcal{G}_0 \Sigma \mathcal{G}_0 + \mathcal{G}_0 \Sigma \mathcal{G}_0 \Sigma \mathcal{G}_0 + \cdots , \tag{18.89}$$

which can be summed as a geometric series to give

$$\mathcal{G} = (1 - \mathcal{G}_0 \Sigma)^{-1} \mathcal{G}_0,$$

so that

$$(1 - \mathcal{G}_0 \Sigma) \mathcal{G} = \mathcal{G}_0,$$

or

$$\left((\mathcal{G}_0)^{-1} - \Sigma\right) \mathcal{G} = 1. \tag{18.90}$$

We have then re-obtained the Dyson equation (18.86) provided we make the formal identification

$$(\mathcal{G}_0)^{-1}(\mathbf{x}, \tau; \mathbf{x}', \tau') = \delta(\mathbf{x} - \mathbf{x}')\delta(\tau - \tau') \left(-\frac{\partial}{\partial \tau} + \frac{\hbar^2 \nabla_{\mathbf{x}}^2}{2m} + \mu \right), \tag{18.91}$$

corresponding to Eq. (18.84).

For the relatively simple case of particles in an external field, the above discussion may appear excessive. However, the self-energy and its appearance in the Dyson equation in the form (18.90) is far more general than the simple case of a one-body potential. In general, in the presence of arbitrary interactions, we can formally define the self-energy via Eq. (18.90). A generic term of the perturbation expansion of $\mathcal{G}$ will contain a certain number of *propagators* $\mathcal{G}_0$ connecting pairs of space–"time" points. A *one-particle irreducible* self-energy is an expression which can be attached to two propagators $\mathcal{G}_0$ to produce a term in the expansion of $\mathcal{G}$ and cannot, in turn, be divided, by cutting a $\mathcal{G}_0$ line, in pieces which may be used as self-energy themselves. Once the self-energy is known, one may generate an infinite sequence of terms, as done in Eq. (18.89). It must be emphasized that the self-energy itself may be written as an infinite sequence of terms.

The self-energy is represented by a filled black circle, which can be inserted between two $\mathcal{G}_0$ lines (see Fig. 18.6(a)).

Before deriving the Dyson equation for a generic interaction system and providing a more precise definition of the self-energy, to appreciate the importance of the concept of the self-energy, let us assume that a self-energy $\Sigma(\mathbf{k}, \omega_n)$ is given for an interacting system. By analytical continuation $\mathrm{i}\omega_n \to \hbar\omega$ we get the corresponding retarded self energy $\Sigma^{\mathrm{R}}(\mathbf{k}, \omega)$ entering the equation for G^{R}

$$G^{\mathrm{R}}(\mathbf{k}, \omega) = \frac{\hbar}{\hbar\omega - \epsilon_{\mathbf{k}}^0 - \Sigma^{\mathrm{R}}(\mathbf{k}, \omega)}, \tag{18.92}$$

where we have performed the Fourier transform of the self-energy in exactly the same way as for the Green function.

We have denoted with $\epsilon_{\mathbf{k}}^0$ the free particle energy dispersion relation measured as usual with respect to the chemical potential. Let us suppose that at $\hbar\omega = \epsilon_{\mathbf{k}} - \mathrm{i}\gamma_{\mathbf{k}}$ the Green function has a pole (the position of the pole relative to the real axis is fixed by the analytic properties of G^{R}) such that

$$\epsilon_{\mathbf{k}} - \mathrm{i}\gamma_{\mathbf{k}} = \epsilon_{\mathbf{k}}^0 + \Sigma^{\mathrm{R}}(\mathbf{k}, \epsilon_{\mathbf{k}}/\hbar - \mathrm{i}\gamma_{\mathbf{k}}/\hbar). \tag{18.93}$$

We assume that the pole is close to the real axis and $\gamma_{\mathbf{k}} \ll \epsilon_{\mathbf{k}}$, as happens in the quasiparticle description of the Fermi liquid. Hence, by considering the real $\Sigma^{\mathrm{R}'}$ and imaginary $\Sigma^{\mathrm{R}''}$ parts of the self energy Σ^{R} we have

$$z_{\mathbf{k}} = \left(1 - \frac{\partial}{\partial \hbar\omega}\Sigma^{\mathrm{R}'}(\mathbf{k}, \omega)|_{\omega=\epsilon_{\mathbf{k}}/\hbar}\right)^{-1}, \tag{18.94}$$

$$\epsilon_{\mathbf{k}} = \epsilon_{\mathbf{k}}^0 + \Sigma^{\mathrm{R}'}(\mathbf{k}, \epsilon_{\mathbf{k}}/\hbar), \tag{18.95}$$

$$\gamma_{\mathbf{k}} = -\Sigma^{\mathrm{R}''}(\mathbf{k}, \epsilon_{\mathbf{k}}/\hbar). \tag{18.96}$$

From the above relations, one sees that the frequency dependence of the real part of the self-energy is responsible for the change of the distribution function in **k**-space with respect to the gas, whereas the real and imaginary parts yield the true energy dispersion and the particle relaxation time.

We are now ready to come back to Eq. (18.82) in the presence of an interaction among the particles. To evaluate the commutator in Eq. (18.82), we make use of the equality (9.44).

After evaluating the commutator containing the interaction Hamiltonian, we get the Dyson equation for the Green function:

$$-\frac{\partial}{\partial\tau}\mathcal{G}(\mathbf{x}, \tau; \mathbf{x}', \tau') = \delta(\tau - \tau')\delta(\mathbf{x} - \mathbf{x}') + \left(-\frac{\hbar^2\nabla_{\mathbf{x}}^2}{2m} - \mu\right)\mathcal{G}(x; x') \tag{18.97}$$

$$-\frac{1}{2}\int \mathrm{d}x_2\mathrm{d}x_3\mathrm{d}x_4\Gamma^0(x, x_2, x_3, x_4)\langle \mathrm{T}_\tau \bar{\hat{\psi}}(x_2)\hat{\psi}(x_3)\hat{\psi}(x_4)\bar{\hat{\psi}}(x')\rangle,$$

where the two-particle Green function $\mathcal{G}^{(4)}$ defined in Eq. (18.9) appears. $\mathcal{G}^{(4)}$, being a "time"-ordered product, can be evaluated in perturbation theory along the lines used to evaluate the one-particle Green function. Also in this case only connected diagrams appear, which now, however, are those with no parts without connection to one of the external points. In the absence of interaction, the two-particle Green function factorizes into the product of two one-particle Green functions:

$$\mathcal{G}^{(4)}(x_1, x_2, x_3, x_4) = \mathcal{G}_0(x_1, x_4)\mathcal{G}_0(x_2, x_3) \mp \mathcal{G}_0(x_1, x_3)\mathcal{G}_0(x_2, x_4), \tag{18.98}$$

where, as usual, the upper and lower signs apply to Fermi and Bose statistics. The terms on the right-hand side of Eq. (18.98) correspond to the first two diagrams of Fig. 18.6(b). In the presence of an interaction, terms will appear that connect the two one-particle Green function lines. This will happen thanks to the interaction term containing Γ^0, which has four field operators. To first order in the $\mathcal{S}$-matrix expansion, one obtains the second term in the expansion of $\mathcal{G}^{(4)}$:

$$\mathcal{G}^{(4)}(x_1, x_2, x_3, x_4) = \int \prod_{i=1,4} \mathrm{d}x_i'\mathcal{G}_0(x_1, x_1')\mathcal{G}_0(x_2, x_2')\Gamma^0(x_1', x_2', x_3', x_4')\mathcal{G}_0(x_3', x_3)\mathcal{G}_0(x_4', x_4), \tag{18.99}$$

which is the third diagram of Fig. 18.6(b) as an empty square with two incoming and two outgoing $\mathcal{G}_0$ lines. To go beyond the lowest order terms of Fig. 18.6(b), one may divide

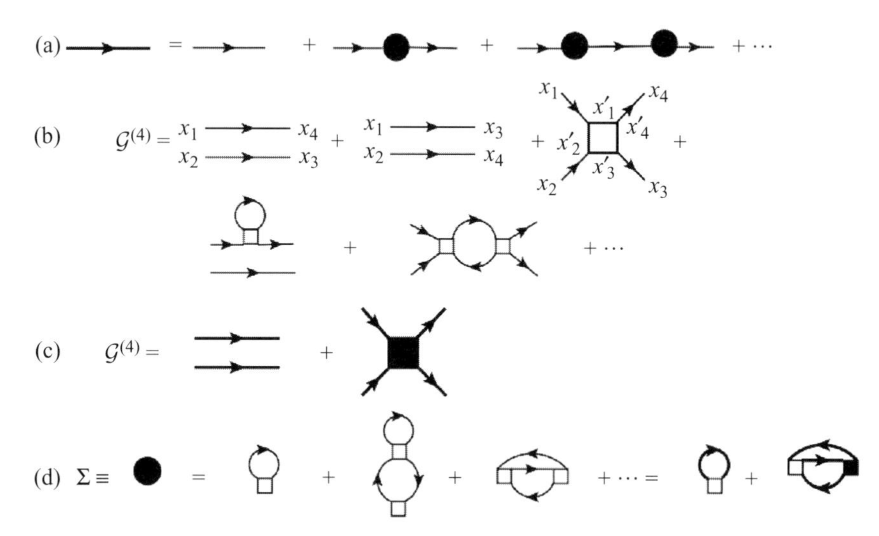

Fig. 18.6 (a) Expansion of the *full* (one-particle) Green function $\mathcal{G}$ (thick line) in terms of the non-interacting one $\mathcal{G}_0$ (thin line) and of the self-energy Σ (filled black circle). (b) Perturbative expansion of the two-particle Green function $\mathcal{G}^{(4)}$. The lowest order contribution is just that of two $\mathcal{G}_0$ lines propagating independently. The *bare* vertex part (empty square) introduces a connection between the two lines. The fourth diagram is a first order example of dressing a $\mathcal{G}_0$ line with a self-energy insertion. The fifth diagram is a first order example of vertex corrections by the connection of two one-particle Green functions with Γ^0. (c) The two-particle Green function $\mathcal{G}^{(4)}$ expressed in terms of the full $\mathcal{G}$ lines and the full vertex part (filled black square). (d) Perturbative expansion of the self-energy and its expression in terms of the full Green function and full vertex part. Diagrams (a) and (d) are a graphical representation of the Dyson equation.

the resulting connected terms into two classes. In the first class, we include all the terms which correspond to the expansion of the two one-particle Green functions, independently. The first example of this type of correction is shown by the fourth diagram of Fig. 18.6(b). These terms will be automatically included if we replace the $\mathcal{G}_0$ lines with the $\mathcal{G}$ lines, as shown in Fig. 18.6(a). Such an operation is called a *dressing* of the *bare* lines. Hence $\mathcal{G}$ is called the dressed or full line. In the second class, we include all the terms in which the two one-particle Green function lines are connected. The first example of these corrections is shown in the fifth diagram of Fig. 18.6(b). Such terms can be represented as the dressing of the first order diagram for $\mathcal{G}^{(4)}$ in Fig. 18.6 (b). The dressing operates on the bare Green function lines and on the bare vertex part Γ^0. The full Green function $\mathcal{G}^{(4)}$ is represented in Fig. 18.6(c), where the dressed vertex part is the filled square and will be denoted by $\Gamma^{(4)}$, and its expression reads

$$\begin{aligned}\mathcal{G}^{(4)}(x_1, x_2, x_3, x_4) &= \mathcal{G}(x_1, x_4)\mathcal{G}(x_2, x_3) \mp \mathcal{G}(x_1, x_3)\mathcal{G}(x_2, x_4) \\ &+ \int \prod_{i=1,4} \mathrm{d}x'_i \mathcal{G}(x_1, x'_1)\mathcal{G}(x_2, x'_2)\Gamma^{(4)}(x'_1, x'_2, x'_3, x'_4)\mathcal{G}(x'_3, x_3)\mathcal{G}(x'_4, x_4),\end{aligned} \tag{18.100}$$

i.e. $\Gamma^{(4)}$ is the one-particle irreducible contribution to $\mathcal{G}^{(4)}$. Its first perturbative diagrams are the third (bare vertex) and the fifth of Fig. 18.6(b) without the external legs.

Equation (18.97) can then be written as

$$-\frac{\partial}{\partial\tau}\mathcal{G}(x;x') = \delta(x-x') + \left(-\frac{\hbar^2\nabla_{\mathbf{x}}^2}{2m} - \mu\right)\mathcal{G}(x;x')$$
$$-\int \mathrm{d}x_2\mathrm{d}x_3\mathrm{d}x_4\Gamma^0(x,x_2,x_3,x_4)\mathcal{G}(x_3,x_2)\mathcal{G}(x_4,x')$$
$$-\int \mathrm{d}x_2\mathrm{d}x_3\mathrm{d}x_4 \prod_{i=1,4}\mathrm{d}x_i'\,\Gamma^0(x,x_2,x_3,x_4)\mathcal{G}(x_3,x_1')\mathcal{G}(x_4,x_2')$$
$$\times\,\Gamma^{(4)}(x_1',x_2',x_3',x_4')\mathcal{G}(x_3',x_2)\mathcal{G}(x_4',x'). \tag{18.101}$$

The vertex $\Gamma^{(4)}$ represents, therefore, the connected, one-particle irreducible part of $\mathcal{G}^{(4)}$ obtained with the truncation of the four external lines $\mathcal{G}$.

The above result can be generalized to define the vertex function $\Gamma^{(2m)}$ from the m-particle Green function $\mathcal{G}^{(2m)}$, Eq. (18.8), as the truncated connected irreducible part.

Equation (18.101) can be re-written in the form (18.90) as

$$\int \mathrm{d}x''\,\left((\mathcal{G}_0)^{-1}(x,x'') - \Sigma(x,x'')\right)\mathcal{G}(x'',x') = \delta(x-x'), \tag{18.102}$$

where the self-energy is given by

$$\Sigma(x,x'') = \int \mathrm{d}x_2\mathrm{d}x_3\mathcal{G}(x_3,x_2)\Gamma^0(x,x_2,x_3,x'') + \int \mathrm{d}x_2\mathrm{d}x_3\mathrm{d}x_4\prod_{i=1,4}\mathrm{d}x_i'\Gamma^0(x,x_2,x_3,x_4)$$
$$\times\,\mathcal{G}(x_3,x_1')\mathcal{G}(x_4,x_2')\mathcal{G}(x_3',x_2)\Gamma^{(4)}(x_1',x_2',x_3',x''). \tag{18.103}$$

Equations (18.102) and (18.103) are a central result of this chapter and are the heart of many-body theory. They correspond to diagram (a) and the re-summed diagrammatic form (d) in Fig. 18.6. The self-energy is a functional of the Green function and of the vertex part and the equation for $\mathcal{G}$ is not closed in itself. The problem is reduced then to the evaluation of the vertex part $\Gamma^{(4)}$, which in turn is connected to Green functions with $m > 2$. In general this task cannot be accomplished exactly and one must rely on an approximate solution in perturbation theory. By neglecting the second term in the self-energy (18.103), one obtains, for instance, the Hartree–Fock approximation discussed in Chapter 9. In this case, the self-energy is a functional of the Green function only. Later on in the book we will encounter several examples of approximate evaluations of the self-energy.

18.9 Thermal density and current response functions

In the same way that the time-dependent Green function can be obtained in terms of the analytical continuation of the thermal Green function, we may define a *thermal* density and current correlation function, whose operatorial form was given in Eqs. (9.29) and (9.45).

Since we are using the thermal Green function depending on the fictitious time τ, we begin by translating the continuity equation (10.32) in terms of the derivative with respect to τ. By remembering the position $t \to -i\hbar\tau$, we obtain for the number density $n(\mathbf{x}, \tau)$ and

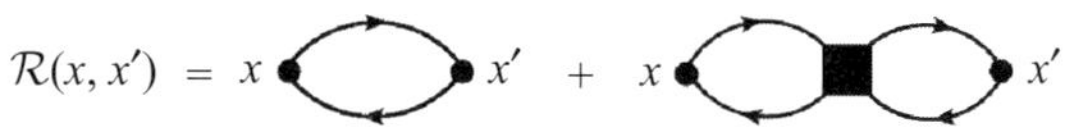

$$\mathcal{R}(x,x') = x \bullet\!\!\bigcirc\!\!\bullet x' + x \bullet\!\!\bigcirc\!\blacksquare\!\bigcirc\!\!\bullet x'$$

Fig. 18.7 Diagrammatic representation of the thermal Green function for the density–density correlation function known as bubble diagrams. The first diagram includes only the dressing of the Green function lines. In a non-interacting system, this diagram with the bare Green function lines yields the full density–density correlation function. In an interacting system, besides the so-called simple bubble, there appears a term containing the vertex part. The black dots $j_0^\mu(x)$ at the left and right extrema of the diagram represent the coupling of the density operator with an external field.

number current $\mathbf{j}(\mathbf{x},\tau)$:

$$\partial_\tau n(\mathbf{x},\tau) - \mathrm{i}\hbar\nabla_{\mathbf{x}}\cdot\mathbf{j}(\mathbf{x},\tau) \equiv \partial_\mu j^\mu(x) = 0, \tag{18.104}$$

where we used the four-vector notation of Chapter 10, $\partial_\mu = (\partial_\tau, -\mathrm{i}\hbar\nabla_{\mathbf{x}})$, $j^\mu(x) = \hat{\psi}^\dagger(x) j_0^\mu \hat{\psi}(x)$ with $j_0^\mu = (1, -\mathrm{i}\hbar\nabla/m)$ and $x = (\tau, \mathbf{x})$ and the fictitious time-dependent number four-current in accordance with the definition of the field operators in Eq. (18.3):

$$j^\mu(x) = \mathrm{e}^{K\tau} j^\mu(\mathbf{x})\mathrm{e}^{-K\tau}.$$

For later usage, we also give the continuity equation (18.104) in the momentum and frequency representation by taking into account the expressions (9.31) and (9.46):

$$-\mathrm{i}\omega_\nu n(\mathbf{q},\omega_\nu) + \hbar\mathbf{q}\mathbf{j}(\mathbf{q},\omega_\nu) \equiv q_\mu j^\mu(q) = 0, \quad j^\mu(q) = \sum_{\mathbf{k}} \hat{a}^\dagger_{\mathbf{k}-\mathbf{q}/2}\hat{a}_{\mathbf{k}+\mathbf{q}/2} j_0^\mu(k), \tag{18.105}$$

with $q_\mu = (\mathrm{i}\omega_\nu, -\hbar\mathbf{q})$ ω_ν being a Matsubara bosonic frequency (any physical observable has Bose-like commutation relations) and $j_0^\mu(\mathbf{k}) = (1, \mathbf{k}/m)$ the bare four-current vertex. The corresponding charge and electrical current densities are obtained by multiplication by $-e$ in the case of electrons.

In terms of the number four-current $j^\mu(x)$ we then define the thermal current–current correlation function

$$\mathcal{R}^{\mu\nu}(x,x') = -\langle T_\tau j^\mu(x) j^\nu(x')\rangle. \tag{18.106}$$

After Fourier transforming using bosonic Matsubara frequencies, we get the relation

$$\tilde{R}^{\mu\nu}(\omega) = \tilde{\mathcal{R}}^{\mu\nu}(\mathrm{i}\omega_\nu)|_{\mathrm{i}\omega_\nu\to\hbar\omega+\mathrm{i}0^+}, \tag{18.107}$$

giving the prescription on how to obtain the response function $\tilde{R}^{\mu\nu}(\omega)$ of Chapter 10 from the thermal one.

The knowledge of the two-particle Green function $\mathcal{G}^{(4)}$ allows us also to evaluate the linear density–density response function. For instance, by considering the definition (18.9) with pairwise coinciding space–"time" arguments one obtains the density–density thermal Green function

$$\mathcal{R}^{00} \equiv \mathcal{R}(x,x') = -\langle \mathrm{T}_\tau \bar{\hat{\psi}}(x) j_0^0(x)\hat{\psi}(x)\bar{\hat{\psi}}(x') j_0^0(x')\hat{\psi}(x')\rangle = \mathcal{G}^{(4)}(x,x',x',x). \tag{18.108}$$

Together with $\mathcal{G}$ and $\mathcal{G}^{(4)}$, $\mathcal{R}$ also can be represented diagrammatically, as shown in Fig. 18.7, obtained from Fig. 18.6(c) by closing the external lines with the two j_0^0 signing the position of this closure. The relevant diagram takes the form of the so-called *bubble*,

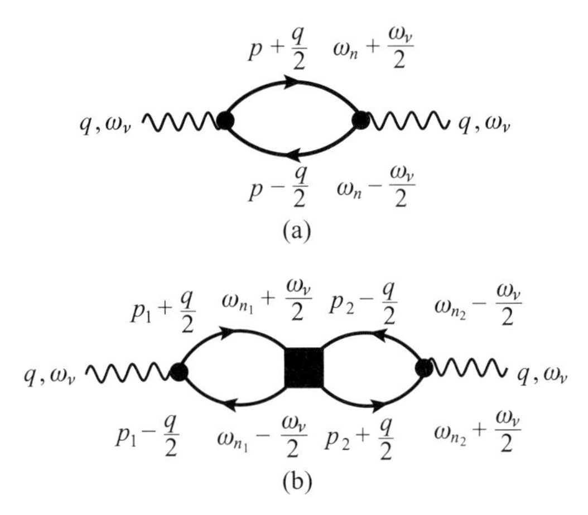

Fig. 18.8 The diagrams in the momentum–frequency representation of the density–density correlation function. The Matsubara frequencies are fermionic, $\epsilon_n = (2n+1)\pi/\beta$ for Green function propagators $\mathcal{G}$. In the wavy lines flows a bosonic frequency $\omega_\nu = 2\pi\nu/\beta$. (a) The simple bubble $\mathcal{R}(q)$. (b) The bubble with the vertex part insertion. Here the black dot is the vertex in the Fourier space j_0^0.

where the left and right extrema describe the coupling of the density operator to the external field. Clearly, this procedure can be generalized to any one-particle observable which has a second quantized expression with the proviso that the last equality in Eq. (18.108) is strictly valid for the density–density response function for which $j_0^0 = 1$; for instance j_0^i would appear explicitly in the current–current response function.

Whereas we will leave to the following chapters an explicit evaluation of diagrams, it may be useful at this point to write down the expression of the density–density diagrams of Figs. 18.8(a) and (b) in momentum and frequency space. We have

$$\tilde{\mathcal{R}}^{(a)}(q) = \int \mathrm{d}p\, \mathcal{G}(p+q/2)\mathcal{G}(p-q/2) \tag{18.109}$$

and

$$\begin{aligned}\tilde{\mathcal{R}}^{(b)}(q) = &-\int \mathrm{d}p_1 \int \mathrm{d}p_2\, \mathcal{G}(p_1+q/2)\mathcal{G}(p_1-q/2)\\ &\times \Gamma^{(4)}(p_1+q/2, p_2-q/2, q)\mathcal{G}(p_2+q/2)\mathcal{G}(p_2-q/2).\end{aligned} \tag{18.110}$$

The above equations will turn out very useful in Chapter 19 where we will discuss the microscopic foundations of the Landau theory of Fermi liquids.

18.10 The Ward identity

Charge conservation is expressed by the continuity equation (see Eq. (10.32)) and manifests in field theory as a relation connecting the Green function to the vertex associated with the external electromagnetic field. This relation, which is called the Ward identity, will be derived in this section.

To derive the Ward identity we introduce a density/current vertex function $\Lambda^{\mu}_{s's''}(x, x', x'')$ defined as

$$\Lambda^{\mu}_{s's''}(x, x', x'') = \langle T_\tau j^\mu(x)\hat{\psi}_{s'}(x')\hat{\bar{\psi}}_{s''}(x'')\rangle, \tag{18.111}$$

where T_τ is the "time"-ordering operator and the average, $\langle \dots \rangle$, is defined in Eq. (18.5) and we have introduced explicitly the spin dependence for generality.

We define the Fourier transform of Eq. (18.111) as

$$\Lambda^{\mu}_{s's''}(x, x', x'') = \int \mathrm{d}q\, \mathrm{d}p\, \mathrm{e}^{\mathrm{i}(p-q/2)(x'-x)/\hbar}\, \mathrm{e}^{\mathrm{i}(p+q/2)(x-x'')/\hbar}\, \Lambda^\mu(p, q). \tag{18.112}$$

For later use, we can define in general the mixed thermal Green function $\mathcal{G}^{(2m,l)}$ as the "time"-ordering average of l composite operators j^μ and $2m$ fields:

$$\begin{aligned} \mathcal{G}^{(2m,l)} &= \langle T_\tau j^{\mu_1}(x_1)\cdots j^{\mu_l}(x_l)\hat{\psi}(x'_1)\cdots\hat{\psi}(x'_m)\hat{\bar{\psi}}(x''_1)\cdots\hat{\bar{\psi}}(x''_m)\rangle, \\ \Lambda^\mu(x, x', x'') &\equiv \mathcal{G}^{(2,1)}(x, x', x''). \end{aligned} \tag{18.113}$$

The corresponding truncated vertices $\Gamma^{(2m,l)}$ are obtained by cutting the external $\mathcal{G}$ legs. To avoid cumbersome notation we have again dropped the spin indices in the general definition.

The vertex function (18.111) can be directly related to the response function by appropriate choice of the "time" and space arguments. For instance, for the density–density response function, with $\tau'' = \tau' + 0^+$ and $\mathbf{x}'' = \mathbf{x}'$,

$$\sum_s \Lambda^0_{ss}(x, x', x') = \mathcal{R}^{00}(x, x'), \tag{18.114}$$

or in Fourier space

$$\tilde{\mathcal{R}}^{00}(q) = \int \mathrm{d}p \sum_s \Lambda^0_{ss}(p, q). \tag{18.115}$$

The comparison between Eqs. (18.114) and (18.108) provides a connection between the density vertex Λ^0 and $\mathcal{G}^{(4)}$ with pairwise coinciding space–"time" arguments. Then, using Eq. (18.100), the vertex Λ^0 can be expressed in terms of $\Gamma^{(4)}$. The same connection can also be carried in Fourier space, as shown in Fig. 18.9, which is obtained from Fig. 18.8 by opening one loop.

The representation of the vertex Λ^0 given in Fig. 18.9 is generally valid for the vertex Λ^i with the bare vertex j^i_0 given by p^i/m in the first diagram and p^i_1/m in the integrated loop of the second diagram.

To get the equation of motion of Λ^μ, we notice that when the derivative ∂_μ acts on the right-hand side of Eq. (18.111), one obtains two contributions. One is due to the derivative acting on j^μ and gives zero due to the continuity equation (18.104). A second contribution comes from the "time"-derivative of the T_τ-product of Λ^0, which is expressed in terms of step functions with the "time" differences as its arguments. As a result one gets the eight

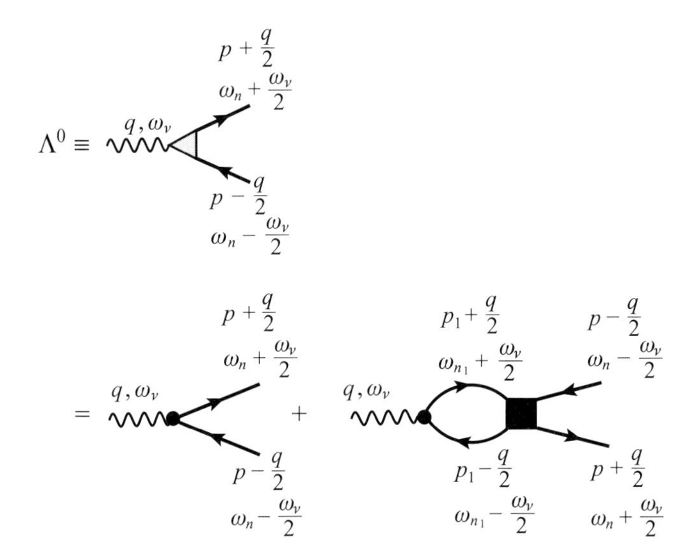

Fig. 18.9 Momentum and frequency representation of the vertex Λ^0 in terms of $\Gamma^{(4)}$. The dot represent the density vertex $j_0^0 = 1$.

terms containing the density j^0 only:

$$
\begin{aligned}
\partial_\mu \Lambda^\mu_{s's''}(x, x', x'') = {} & \langle \delta(\tau - \tau')\theta(\tau - \tau'') j^0(x) \hat{\psi}_{s'}(x') \hat{\bar{\psi}}_{s''}(x'') \rangle \\
& - \langle \delta(\tau - \tau')\theta(\tau'' - \tau) \hat{\bar{\psi}}_{s''}(x'') j^0(x) \hat{\psi}_{s'}(x') \rangle \\
& + \langle \delta(\tau'' - \tau)\theta(\tau - \tau') \hat{\bar{\psi}}_{s''}(x'') j^0(x) \hat{\psi}_{s'}(x') \rangle \\
& - \langle \delta(\tau'' - \tau)\theta(\tau' - \tau) \hat{\psi}_{s'}(x') \hat{\bar{\psi}}_{s''}(x'') j^0(x) \rangle \\
& - \langle \delta(\tau - \tau'')\theta(\tau - \tau') j^0(x) \hat{\bar{\psi}}_{s''}(x'') \hat{\psi}_{s'}(x') \rangle \\
& - \langle \delta(\tau' - \tau)\theta(\tau - \tau'') \hat{\psi}_{s'}(x') j^0(x) \hat{\bar{\psi}}_{s''}(x'') \rangle \\
& + \langle \delta(\tau - \tau'')\theta(\tau' - \tau) \hat{\psi}_{s'}(x') j^0(x) \hat{\bar{\psi}}_{s''}(x'') \rangle \\
& + \langle \delta(\tau' - \tau)\theta(\tau'' - \tau) \hat{\bar{\psi}}_{s''}(x'') \hat{\psi}_{s'}(x') j^0(x) \rangle .
\end{aligned}
$$

The above terms can be arranged by collecting together those with the delta and theta functions with the same "time" arguments

$$
\begin{aligned}
\partial_\mu \Lambda^\mu_{s's''}(x, x', x'') = {} & \delta(\tau - \tau')\theta(\tau - \tau'') \langle [j^0(x), \hat{\psi}_{s'}(x')] \hat{\bar{\psi}}_{s''}(x'') \rangle \\
& + \delta(\tau - \tau')\theta(\tau'' - \tau) \langle \hat{\bar{\psi}}_{s''}(x'') [\hat{\psi}_{s'}(x'), j^0(x)] \rangle \\
& + \delta(\tau - \tau'')\theta(\tau - \tau') \langle [\hat{\bar{\psi}}_{s''}(x''), j^0(x)] \hat{\psi}_{s'}(x') \rangle \\
& + \delta(\tau - \tau'')\theta(\tau' - \tau) \langle \hat{\psi}_{s'}(x') [j^0(x), \hat{\bar{\psi}}_{s''}(x'')] \rangle .
\end{aligned}
$$

By using the fact that at equal "times" the field operators satisfy the commutation relations (9.20) we obtain

$$[j^0(x), \hat{\psi}_{s'}(x')] = -\delta_{ss'}\delta(\mathbf{x}-\mathbf{x}')\hat{\psi}_s(\mathbf{x},\tau),$$
$$[\bar{\hat{\psi}}_{s''}(x''), j^0(x)] = -\delta_{ss'}\delta(\mathbf{x}-\mathbf{x}'')\bar{\hat{\psi}}_s(\mathbf{x},\tau).$$

From this, by recalling the definition of the thermal Green function (18.2), one gets the following Ward identity connecting the vertex Λ^μ to $\mathcal{G}$:

$$\partial_\mu \Lambda^\mu_{s's''}(x, x', x'') = \delta(x-x')\mathcal{G}_{s's''}(x, x'') - \delta(x-x'')\mathcal{G}_{s's''}(x', x). \quad (18.116)$$

In terms of the Fourier transforms with respect to the relative coordinates $x - x''$ and $x' - x$ (the arguments of the two Green functions on the right-hand side of Eq. (18.116)) the Ward identity becomes

$$\mathrm{i}\omega_\nu \Lambda^0_{s's''}(p, q) - q_i \Lambda^i_{s's''}(p, q) = \mathcal{G}_{s's''}(p - q/2) - \mathcal{G}_{s's''}(p + q/2). \quad (18.117)$$

By analytical continuation a similar Ward identity can be written for the real time-dependent Green function G with the replacements $\mathrm{i}\epsilon_n \to \hbar\epsilon$, $\mathrm{i}\omega_\nu \to \hbar\omega$:

$$\omega\Lambda^0_{s's''}(p, q) - q_i \Lambda^i_{s's''}(p, q) = G(p - q/2) - G(p + q/2). \quad (18.118)$$

Furthermore, we introduce the truncated vertex $\mathcal{T}^\mu \equiv \Gamma^{(2,1)}$ defined as

$$\Lambda^\mu(p, q) = \mathcal{G}(p + q/2)\mathcal{T}^\mu(p, q)\mathcal{G}(p - q/2), \quad (18.119)$$

whose graphical representation is easily obtained from Fig. 18.9 by cutting the external legs. For a non-interacting system one has $\mathcal{T}^\mu(p, q) = j_0^\mu = (1, \mathbf{p}/m)$.

In terms of $\mathcal{T}^\mu$ the Ward identity (18.117) reads

$$\mathrm{i}\omega_\nu \mathcal{T}^0(\mathbf{p}, \epsilon_n; \mathbf{q}, \omega_\nu) - q_i \mathcal{T}^i(\mathbf{p}, \epsilon_n; \mathbf{q}, \omega_\nu)$$
$$= \mathcal{G}^{-1}\left(\mathbf{p} + \frac{\mathbf{q}}{2}, \epsilon_n + \frac{\omega_\nu}{2}\right) - \mathcal{G}^{-1}\left(\mathbf{p} - \frac{\mathbf{q}}{2}, \epsilon_n - \frac{\omega_\nu}{2}\right), \quad (18.120)$$

where we have expanded the four-vector notation for clarity.

18.11 Problems

18.1 Prove the relation (18.20).

18.2 Prove the relations (18.44) and (18.45).

19 The microscopic foundations of Fermi liquids

In this chapter we provide a microscopic justification of the Landau Fermi-liquid phenomenological assumption. A detailed and exhaustive treatment of this issue can be found in the excellent book by Nozières (1964). Here we content ourselves in giving the main argument, which was presented by Landau (1958) himself. We will see only that the Landau phenomenological parameters can be expressed in terms of certain limits of the vertex part of the many-body formulation of the interacting Fermi gas problem. To this end we will mainly rely on the last sections of Chapter 18, where we summarized the general structure of the perturbation theory for the thermal Green and response functions. Whenever necessary we will also use the connection of the thermal Green function with the time-dependent Green function, as done in Section 18.3.

19.1 Scattering amplitudes

Let us begin by recalling the perturbative expansion for the momentum and energy dependent vertex part $\Gamma^{(4)}(p_1, p_2, q)$, which was introduced in Eq. (18.76). In Fig. 19.1(a), $\Gamma^{(4)}(p_1, p_2, q)$ is represented as a graphical resummation of diagrams, where the lowest order $\Gamma^0(p_1, p_2, q)$ term can be obtained from the interaction potential according to the prescription of Eq. (18.73). Diagrams (b), (c) and (d) of Fig. 19.1 represent first-order corrections to Γ^0. The key observation, due to Landau, is that diagram (d) becomes singular in the limit of small momentum and energy transfer. To appreciate this point, we notice that all diagrams contain a product of two thermal Green functions, as shown in Fig. 19.1. When $q \to 0$, the poles of the Green functions in diagram (d) merge, in contrast to what happens for diagrams (b) and (c). Such a combination of Green function lines is called anomalous. We may, then, collect all the diagrams which do not contain such an anomalous combination. Let us call such a combination $\Gamma^1(p_1, p_2) \equiv \Gamma^{\text{regular}}(p_1, p_2, q = 0)$, where we have set $q = 0$, since such a sum does not contain dangerous anomalous elements and it is *regular* in this limit. Once we have Γ^1, the full vertex part can be obtained by iterating the anomalous combination in a ladder-like fashion.

As a result we obtain an integral equation for the vertex part:

$$\Gamma^{(4)}(p_1, p_2, q) = \Gamma^1(p_1, p_2, q) + \int \mathrm{d}k\, \Gamma^1(k, p_1)\mathcal{G}(k)\mathcal{G}(k+q)\Gamma^{(4)}(k, p_2, q), \quad (19.1)$$

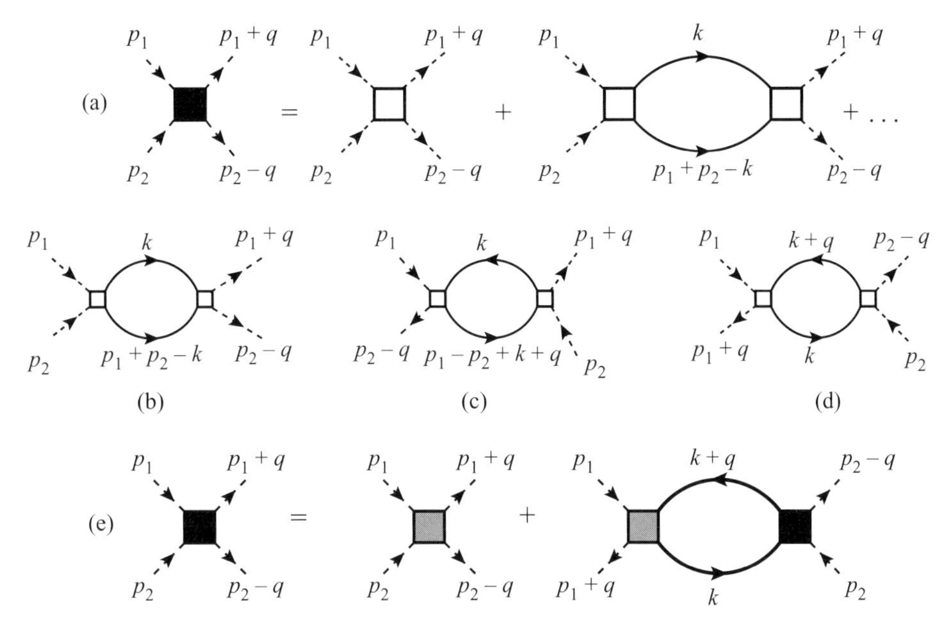

Fig. 19.1 (a) Perturbative expansion of the vertex part in momentum space. The filled black ($\Gamma^{(4)}$) and empty (Γ^0) boxes indicate the full and bare vertex parts. Here, for the sake of brevity, we use the four-momentum notation $p \equiv (\mathbf{p}, \omega_n)$. (b), (c) and (d) are lowest-order corrections to the vertex part. Notice the different values of the four-momenta flowing in the internal Green function lines. The diagram (d) contains a product of two Green functions, whose poles come close for small energy–momentum transfer. Such a product of two Green functions is called anomalous. (e) Diagrammatic representation of the integral equation for the vertex part. The filled gray box here represents the sum of all diagrams contributing to the vertex part, which are irreducible for cutting an anomalous Green functions product. Dashed Green function lines entering the vertex are only shown to indicate the four-momenta, but are not present in the vertex equation. Internal thick Green function lines indicate inclusion of self-energy corrections.

where the Green function has the form

$$\mathcal{G}(k) = \frac{z(\mathbf{k})}{\mathrm{i}\epsilon_n - \epsilon_{\mathbf{k}} + \mathrm{i}\gamma_{\mathbf{k}}} + \delta\mathcal{G}(k), \tag{19.2}$$

$\delta\mathcal{G}(k)$ indicating the regular part of the Green function. The position $\epsilon_{\mathbf{k}} - \mathrm{i}\gamma_{\mathbf{k}}$ and the residue $z_{\mathbf{k}}$ of the pole can be obtained from the self-energy, as indicated in Eqs. (18.94)–(18.96). To study the above integral equation we need to make more transparent the singular behavior of the product $\mathcal{G}(k)\mathcal{G}(k+q)$ at small q. To this end, we begin by considering the frequency sum in Eq. (19.1):

$$\frac{1}{\beta}\sum_{\epsilon_n} \mathcal{G}(\mathbf{k}, \epsilon_n)\mathcal{G}(\mathbf{k}+\mathbf{q}, \epsilon_n + \omega_\nu)F(\epsilon_n), \tag{19.3}$$

where ω_ν represents the transmitted Bosonic frequency according to Fig. 18.3(b) and $F(\epsilon_n)$ represents all the dependence on ϵ_n coming from the vertex parts Γ^1 and $\Gamma^{(4)}$. The frequency sum can be carried out with the technique explained in Appendix H. By using suitably

Eq. (H.1) and limiting ourselves to the most singular part of the Green function, we find

$$\frac{1}{2}\frac{1}{\mathrm{i}\omega_\nu - \epsilon_{\mathbf{k}+\mathbf{q}} + \epsilon_{\mathbf{k}}}\left(\tanh\left(\frac{\beta\epsilon_{\mathbf{k}+\mathbf{q}}}{2}\right)F(\epsilon_{\mathbf{k}+\mathbf{q}} - \mathrm{i}\omega_\nu) - \tanh\left(\frac{\beta\epsilon_{\mathbf{k}}}{2T}\right)F(\epsilon_{\mathbf{k}})\right), \quad (19.4)$$

under the reasonable assumption that the function F is regular close to the poles of $\mathcal{G}$ and $\gamma_{\mathbf{k}} \ll \epsilon_{\mathbf{k}}$. By expanding for small $\mathbf{q}$ and ω_ν, we get

$$\delta(\epsilon_{\mathbf{k}})F(0)\frac{\mathbf{v}\cdot\mathbf{q}}{\mathrm{i}\omega_\nu - \mathbf{v}\cdot\mathbf{q}}, \quad (19.5)$$

where, recalling that the Fermi-liquid theory is valid for $T \ll T_{\mathrm{F}}$, the delta function $\delta(\epsilon_{\mathbf{k}})$ is the zero-kelvin limit of the derivative of the Fermi function. Considering the result, the product of the two Green functions in Eq. (19.1) can be approximated as

$$L_k(q) \equiv \mathcal{G}(k)\mathcal{G}(k+q) = \beta z_{\mathbf{k}}^2\delta(\epsilon_n)\delta(\epsilon_{\mathbf{k}})\frac{\mathbf{v}\cdot\mathbf{q}}{\mathrm{i}\omega_\nu - \mathbf{v}\cdot\mathbf{q}} + L_k \equiv \Delta_k(q) + L_k. \quad (19.6)$$

Notice that we have set $q = 0$ in L_k, the non-singular part of $L_k(q)$. By inserting Eq. (19.6) into Eq. (19.1), we can write in compact form

$$\Gamma^{(4)} = \Gamma^1 + \Gamma^1\Delta_k(q)\Gamma^{(4)} + \Gamma^1 L_k\Gamma^{(4)}. \quad (19.7)$$

The next step in the Landau reasoning is based on the observation that the singular character of (19.6) implies two different limits for small momentum and energy transfer. First, we may consider the limit $\mathbf{q} = 0$ and $\omega_\nu \to 0$. This will be called the *dynamic* limit. The other limit amounts to set in agreement with the discussion about response functions in Section 10.2 $\omega_\nu = 0$ first and then sending $\mathbf{q}$ to zero. In this case we speak of the *static* limit. In the dynamic limit, the integral equation becomes

$$\Gamma^\omega = \Gamma^1 + \Gamma^1 L_k\Gamma^\omega, \quad \text{or} \quad \Gamma^\omega = (1 - \Gamma^1 L_k)^{-1}\Gamma^1, \quad (19.8)$$

where Γ^ω denotes the dynamic limit of the full vertex part. By using the expression of Γ^ω to replace Γ^1 in Eq. (19.7), we obtain

$$\Gamma^{(4)} = \Gamma^\omega + \Gamma^\omega\Delta_k(q)\Gamma^{(4)}. \quad (19.9)$$

By considering the static limit Γ^q of the vertex part and setting the external momenta $\mathbf{p}_1$ and $\mathbf{p}_2$ on the Fermi surface with $\omega_{n_1} = \omega_{n_2} = 0$, we finally obtain

$$\Gamma^q(\mathbf{p}_1, \mathbf{p}_2) = \Gamma^\omega(\mathbf{p}_1, \mathbf{p}_2) - z_{k_{\mathrm{F}}}^2\nu_{\mathrm{F}}^{\mathrm{i}}\int\frac{\mathrm{d}\Omega}{4\pi}\Gamma^\omega(\mathbf{p}_1, \mathbf{k})\Gamma^q(\mathbf{k}, \mathbf{p}_2), \quad (19.10)$$

where $\nu_{\mathrm{F}}^{\mathrm{i}}$ is the density of states of the excitations of the interacting system.

The above equation can be solved if spin rotational invariance is present. In this case we can introduce for Γ^q and Γ^ω, in terms of the spin singlet and triplet, the same decomposition (12.6) we have used for the phenomenological parameter f:

$$\Gamma^{q,\omega}(\mathbf{p}, \mathbf{p}', \boldsymbol{\sigma}, \boldsymbol{\sigma}') = \Gamma_{\mathrm{s}}^{q,\omega}(\mathbf{p}, \mathbf{p}')\mathbf{I} + \Gamma_{\mathrm{t}}^{q,\omega}(\mathbf{p}, \mathbf{p}')\boldsymbol{\sigma}\cdot\boldsymbol{\sigma}' \quad (19.11)$$

with the expansion in terms of Legendre polynomials

$$\Gamma_{\mathrm{s,t}}^{q,\omega}(\mathbf{p}, \mathbf{p}') = \sum_l \Gamma_{\mathrm{s,t}}^{q,\omega}(l)P_l(\cos(\hat{\mathbf{p}}\cdot\hat{\mathbf{p}}')). \quad (19.12)$$

As a result one finds

$$\Gamma^{q}_{\mathrm{s,t}}(l) = \frac{\Gamma^{\omega}_{\mathrm{s,t}}(l)}{1 + z^{2}_{k_{\mathrm{F}}} \nu^{i}_{\mathrm{F}} \Gamma^{\omega}_{\mathrm{s,t}}(l)/(2l+1)}. \qquad (19.13)$$

This is a central result of this section. It relates the two limits of the vertex part. The dynamical limit Γ^{ω} represents the interaction among the quasiparticles. The static limit has the physical meaning of the fully dressed scattering amplitude between two quasiparticles as a resummation of Γ^{ω} and determines the behavior of the static response function.

19.2 The static and dynamic limit of the truncated vertex function

We now refer to Section 18.10 and in particular to the truncated density/current vertex. In the absence of interaction, by using the expression of the Green function for the Fermi gas, we obtain $\mathcal{T}^{\mu} = j^{\mu}_{0}$, i.e. the truncated vertex reduces to the bare four-current vertex. In the general case, the identity (18.120) allows us to derive some useful relations, once we recall Eq. (18.95) giving the pole condition at $\hbar\epsilon = \epsilon_{\mathbf{k}}$:

$$\epsilon_{\mathbf{k}} - \epsilon^{0}_{\mathbf{k}} - \Sigma^{\mathrm{R}}(\mathbf{k}, \epsilon_{\mathbf{k}}/\hbar) = 0, \qquad (19.14)$$

where $\epsilon^{0}_{\mathbf{k}}$ is the energy dispersion in the absence of interaction. By performing the dynamic limit in the Ward identity (18.120) only the density vertex survives and one obtains

$$\mathcal{T}^{0}_{\omega} = \left(1 - \frac{\partial \Sigma^{\mathrm{R}'}}{\partial \hbar\epsilon}\right) = z^{-1}_{\mathbf{k}}, \qquad (19.15)$$

where the last equality follows by comparison with Eq. (18.94). In the above $\mathcal{T}^{0}_{\omega}$ denotes the dynamic limit of the corresponding quantity $\mathcal{T}^{0}$. The connection of the dynamic limit of the density vertex with the residue $z_{\mathbf{k}}$ of the Green function expresses the connection between the wave function renormalization and charge renormalization, well known in quantum electrodynamics implementing the charge conservation.

In the static limit, instead, the Ward identity (18.120) gives a relation for the current vertex

$$\mathcal{T}^{a}_{q} = v^{a} + \frac{\partial \Sigma^{\mathrm{R}'}}{\partial k_{a}}, \qquad a = x, y, z, \qquad (19.16)$$

where $\mathcal{T}^{a}_{q}$ denotes the static limit and $v^{a} = \partial\epsilon^{0}_{\mathbf{k}}/\partial k_{a} = k_{a}/m = j^{a}_{0}$. By differentiating the pole condition (19.14) with respect to k_{a}, we obtain

$$V^{a}\left(1 - \frac{\partial \Sigma^{\mathrm{R}'}}{\partial \hbar\epsilon}\right) = v^{a} + \frac{\partial \Sigma^{\mathrm{R}'}}{\partial k_{a}},$$

where $V^{a} = \partial\epsilon_{\mathbf{k}}/\partial k_{a} = k_{a}/m^{*}$. As a result we get

$$\mathcal{T}^{a}_{q} = V^{a}\left(1 - \frac{\partial \Sigma^{\mathrm{R}'}}{\partial \hbar\epsilon}\right) = \frac{V^{a}}{z_{\mathbf{k}}}. \qquad (19.17)$$

Finally, by noticing that in the dynamic limit $\Lambda^0 j_0^a = \Lambda^a$ (see Nozières (1964) Eq. (6.133)), we get

$$\mathcal{T}_\omega^a = \left(1 - \frac{\partial \Sigma^{R'}}{\partial \hbar \epsilon}\right) j_0^a = \frac{v^a}{z_{\mathbf{k}}}. \tag{19.18}$$

Since k_F is invariant in the Landau theory of a Fermi liquid, $V_F/v_F = m/m^*$. This relation together with Eqs. (19.15), (19.17) and (19.18) will be used in the next section to establish the Fermi liquid theory.

19.3 Response functions and Fermi liquid parameters

In terms of the truncated vertex $\mathcal{T}^\mu$, the response functions with the structure similar to Eqs. (18.109) and (18.110) of the previous chapter read

$$\tilde{\mathcal{R}}^{\mu\nu}(q) = -\int \mathrm{d}p \; j_0^\mu(p) \mathcal{T}^\nu(p + q/2, q) L_{p+q/2}(q), \tag{19.19}$$

with the truncated vertex expressed as

$$\mathcal{T}^\mu(p + q/2, q) = j_0^\mu(p) + \int \mathrm{d}p' \Gamma^{(4)}(p + q/2, p' + q/2) L_{p'+q/2}(q) j_0^\mu(p' + q/2). \tag{19.20}$$

In more compact notation we have

$$\tilde{\mathcal{R}}^{\mu\nu} = -j_0^\mu L_k(q) \mathcal{T}^\nu, \tag{19.21}$$

$$\mathcal{T}^\mu = j_0^\mu + j_0^\mu L_k(q) \Gamma^{(4)}. \tag{19.22}$$

It is direct to take the dynamic limit of (19.22):

$$\mathcal{T}_\omega^\mu = j_0^\mu + j_0^\mu L_k \Gamma^\omega. \tag{19.23}$$

By inserting (19.23) into (19.22), keeping in mind (19.6) and (19.9), we get

$$\mathcal{T}^\mu = \mathcal{T}_\omega^\mu + \mathcal{T}^\mu \Delta_k(q) \Gamma^\omega. \tag{19.24}$$

With some algebra this may be written also as

$$\mathcal{T}^\mu = \mathcal{T}_\omega^\mu + \mathcal{T}_\omega^\mu \Delta_k(q) \Gamma^{(4)}. \tag{19.25}$$

The static limit of Eq. (19.24) for the truncated vertex for $\mu = a$, after recalling Eq. (19.6) and the identities (19.17) and (19.18), becomes

$$V^a(\mathbf{k}) = v^a(\mathbf{k}) - \nu_F^{\mathrm{i}} z_{k_F}^2 \int \frac{\mathrm{d}\Omega_{\mathbf{k}'}}{4\pi} \Gamma^\omega(\mathbf{k}, \mathbf{k}') V^a(\mathbf{k}'). \tag{19.26}$$

This is nothing but the Landau equation (12.11) for the effective mass after the identification $\nu_F^{\mathrm{i}} = \nu_F^{\mathrm{L}}$ with V^a the quasiparticle velocity and the Landau parameter

$$F_1 = \nu_F^{\mathrm{i}} z_{k_F}^2 \Gamma_1^\omega, \tag{19.27}$$

Γ_1^ω being the p-wave component of the expansion of Γ^ω in Legendre polynomials. The identification (19.27) provides a microscopic expression for the Landau parameter F_1 and agrees with the interpretation pointed out after Eq. (19.13) of the dynamic limit of the vertex part as the interaction among the quasiparticles. Clearly, the establishing of the Landau Fermi-liquid theory is based on the assumption that the dynamic limit of the vertex part is regular, i.e., all the singular behavior of the theory arises from the merging of the poles of the Green functions in the small momentum and energy transfer limit.

The identification (19.27) is not restricted to the $l = 1$ component of the Legendre polynomial expansion, but it is general. Let us, as a further important example, examine the case of compressibility. Let us manipulate Eq. (19.21) with the definition (19.6) of $\Delta_k(q)$ and L_k. By writing $\mathcal{T}^\nu = \mathcal{T}^\nu - \mathcal{T}^\nu_\omega + \mathcal{T}^\nu_\omega$,

$$\tilde{\mathcal{R}}^{\mu\nu} = -j_0^\mu L_k \mathcal{T}^\nu_\omega - j_0^\mu L_k\big(\mathcal{T}^\nu - \mathcal{T}^\nu_\omega\big) - j_0^\mu \Delta_k(q)\mathcal{T}^\nu,$$

so that, by using (19.24),

$$= \tilde{\mathcal{R}}^{\mu\nu}_\omega - j_0^\mu L_k \Gamma^\omega \Delta_k(q)\mathcal{T}^\nu - j_0^\mu \Delta_k(q)\mathcal{T}^\nu$$

and (19.23)

$$= \tilde{\mathcal{R}}^{\mu\nu}_\omega - \mathcal{T}^\mu_\omega \Delta_k(q)\mathcal{T}^\nu$$

and expressing $\mathcal{T}^\nu$ by (19.25)

$$\begin{aligned} &= \tilde{\mathcal{R}}^{\mu\nu}_\omega - \mathcal{T}^\mu_\omega \Delta_k(q)\big(\mathcal{T}^\mu_\omega + \Gamma^{(4)}\Delta_k(q)\mathcal{T}^\mu_\omega\big) \\ &= \tilde{\mathcal{R}}^{\mu\nu}_\omega - \mathcal{T}^\mu_\omega \Delta_k(q)\mathcal{T}^\mu_\omega - \mathcal{T}^\mu_\omega \Delta_k(q)\Gamma^{(4)}\Delta_k(q)\mathcal{T}^\mu_\omega. \end{aligned} \tag{19.28}$$

The compressibility is the static limit of the above relation for $\mu = \nu = 0$. Since in the dynamic limit $\Delta_k(q) = 0$, the first term on the right-hand side of (19.28) represents the dynamic limit. The latter vanishes due to gauge invariance, as shown in Eq. (10.40) when dealing with the linear response to an electromagnetic field. Hence we have in compact notation

$$\kappa = \mathcal{T}^0_\omega z^2_{k_{\rm F}} \nu^{\rm i}_{\rm F} \mathcal{T}^0_\omega - \mathcal{T}^0_\omega z^2_{k_{\rm F}} \nu^{\rm i}_{\rm F} \Gamma^q_{\rm s} z^2_{k_{\rm F}} \nu^{\rm i}_{\rm F} \mathcal{T}^0_\omega. \tag{19.29}$$

By expressing the static limit $\Gamma^q_{\rm s}$ of the vertex part in terms of the dynamic one $\Gamma^\omega_{\rm s}$ with the help of Eq. (19.13) and recalling the identity (19.15), we finally obtain

$$\kappa = \nu^{\rm i}_{\rm F}\left(1 - \frac{\nu^{\rm i}_{\rm F} z^2_{k_{\rm F}} \Gamma^\omega_{\rm s}}{1 + \nu^{\rm i}_{\rm F} z^2_{k_{\rm F}} \Gamma^\omega_{\rm s}}\right) = \frac{\nu^{\rm i}_{\rm F}}{1 + \nu^{\rm i}_{\rm F} z^2_{k_{\rm F}} \Gamma^\omega_{\rm s}} = \frac{\nu^{\rm L}_{\rm F}}{1 + F_0}. \tag{19.30}$$

In the last equality we have identified the symmetric s-wave of the Landau parameter with the dynamic limit of the vertex part in the same way as in Eq. (19.27). The spin susceptibility can be treated along the same lines and as a result one finds the Landau expression (12.28) with the identification of the s-wave component of the function G with the triplet component of the vertex part.

20 The Luttinger liquid

20.1 The breakdown of the Fermi liquid in one dimension

For pure fermionic systems, and in the absence of symmetry breaking (e.g. magnetism, superconductivity, ...), two types of metallic phases are well established: (i) the "normal" Fermi liquid phase in $d = 3$ presented in Chapters 12 and 19 via the response and the Green functions techniques; and (ii) the "anomalous" Luttinger liquid phase in $d = 1$ originally formulated with the bosonization technique by which the fermionic operators are represented in terms of "bosonic" density operators (Luther and Peschel (1974); Haldane (1981)) with the exact solution (Mattis and Lieb (1965)) of the Tomonaga–Luttinger model (Tomonaga (1950); Luttinger (1963)) discussed in many reviews and books (see for instance the book by Giamarchi (2004)).

Here, to stress similarities and differences between the Fermi liquid and the Luttinger liquid, we will follow the alternative route of response and Green functions techniques (Dzyaloshinskii and Larkin (1973), Di Castro and Metzner (1991), Metzner et al. (1998)) and add a few comments on the renormalization group approach (Metzner and Di Castro (1993)).

As we have seen in Chapters 12 and 19, the Fermi liquid is described asymptotically by the free low-lying single-particle excitations, i.e., the quasiparticles with a zero-kelvin discontinuous occupation number in momentum space $n_{\mathbf{k}}$ (see Eq. (18.54)) at a well defined Fermi surface ($\mathbf{k} = \mathbf{k}_{\mathrm{F}}$):

$$|n_{\mathbf{k}_{\mathrm{F}>}} - n_{\mathbf{k}_{\mathrm{F}<}}| = z_{\mathbf{k}_{\mathrm{F}}} < 1, \tag{20.1}$$

where here $\mathbf{k}_{\mathrm{F}>}$ and $\mathbf{k}_{\mathrm{F}<}$ indicate the limit $\mathbf{k} \to \mathbf{k}_{\mathrm{F}}$ from above and below. This discontinuity still marks the Fermi surface in the interacting system. The finite reduction of the single-particle spectral weight $z_{\mathbf{k}_{\mathrm{F}}}$ with respect to the Fermi gas ($z_{\mathbf{k}_{\mathrm{F}}} = 1$) is given by the finite (i.e. non-critical) "wavefunction renormalization" (see Eq. (18.94)). The presence of the discontinuity at the Fermi surface, together with the Pauli principle, compel the inverse quasiparticle lifetime to be $\tau^{-1} \approx \max(T^2, \epsilon^2)$, where ϵ is the deviation of the energy of the quasiparticle from the Fermi energy. The energy uncertainty $\hbar\tau^{-1}$ due to the finite lifetime of a quasiparticle near the Fermi surface is small compared to its energy ϵ, and the quasiparticle concept is well defined. All the momentum transferring scattering processes become asymptotically ineffective, as derived phenomenologically in Chapter 12 and confirmed microscopically in Chapter 19. The expressions of the specific heat, the spin susceptibility and the compressibility or the Drude peak of a Fermi gas are still valid, but for finite multiplicative renormalizations due to the residual Hartree type interaction among the quasiparticles (see Eqs. (12.17), (12.24), (12.28), (12.45)).

The Fermi liquid breaks down in $d = 1$. The Fermi surface reduces to two Fermi points $\pm k_{\rm F}$ and the quasiparticle spectral weight as $k \to \pm k_{\rm F}$ is suppressed by the presence of an interaction-induced anomalous dimension η in $z_{\mathbf{k}_{\rm F}} \sim |k - k_{\rm F}|^{\eta}$. This non-Fermi-liquid behavior in $d = 1$ is named a Luttinger liquid. As we shall see in this chapter, the vanishing of $z_{k_{\rm F}}$ corresponds to the suppression of the low-lying single particle excitations with the single-particle density of states acquiring an anomalous behavior $\nu(\epsilon) \sim |\epsilon|^{\eta}$ (no fermionic quasiparticles at the Fermi surface). The low-energy behavior of the system is dominated by the charge and spin collective modes. These modes propagate with different velocities, leading to the so-called charge and spin separation, as opposed to the behavior of a normal Fermi liquid, where quasiparticle excitations simultaneously carry charge and spin. This Luttinger liquid behavior is obtained as the exact solution of the Tomonoga–Luttinger model, which can be carried out thanks to the existence of symmetries proper to fermions in $d = 1$ with forward scattering. In this case, in addition to the total charge and spin conservation valid also for a normal Fermi liquid, the particles moving in one direction cannot mix with the others going in the opposite direction and their total numbers are conserved separately. The Ward identity associated with this additional conservation law is the technical ingredient which allows for the exact solution of the Tomonaga–Luttinger model providing the Luttinger liquid behavior.

We shall end this chapter with a few comments on the use of the renormalization group for an interacting system with forward scattering in $d = 1$. The exponent η of the spectral weight corresponds to an interaction dependent scaling index for the wave function renormalization of the fermion field.

As in the case of critical phenomena, the existence of an anomalous scaling dimension manifests in perturbation theory in the appearance of logarithmic divergences when the system size is sent to infinity. In contrast to critical phenomena, these divergences are not the signal of a phase instability, but reflect the anomalous behavior of the single-particle Green function. These divergences can still be controlled by the methods of the renormalization group approach developed for critical phenomena. From this point of view, the one-dimensional interacting electron system provides a paradigmatic example of how the renormalization group approach is suitable for tackling both stable and unstable phases. In particular, for the present case, the program of the renormalization group approach can be fully carried out because of these additional symmetries, which allow for the exact solution of the Tomonaga–Luttinger model and guarantee the cancellation of singularities in perturbation theory to all orders yielding finite values of the response functions, as expected in a stable phase. All the divergences of the theory are absorbed in the wave function renormalization, which gives the anomalous behavior of the quasiparticle spectral weight.

20.2 The Tomonaga–Luttinger (TL) model

In one-dimensional systems the Fermi surface consists of two points $\alpha k_{\rm F}$, where $\alpha = \pm$ refer to right- and left-moving fermions, respectively. The peculiar Fermi surface leads

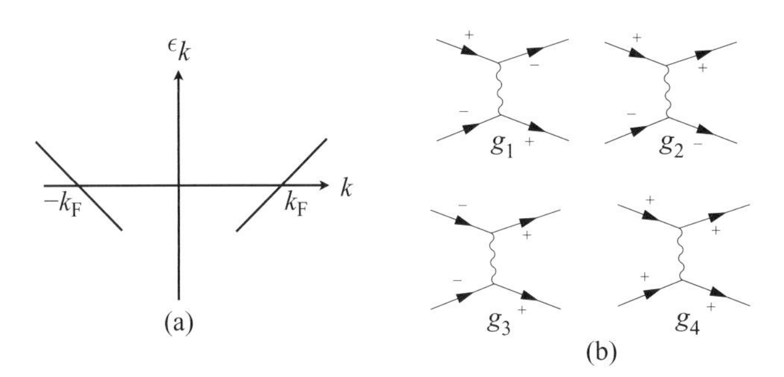

Fig. 20.1 (a) The energy dispersion around the two Fermi points. (b) The relevant scattering processes considered within the g-ology model. The plus and minus labels on the Green function lines denote the right and left branches of the energy dispersion.

us in general to consider a few scattering processes that are incorporated in the continuous low-energy model known as the "g-ology model" (Sólyom (1979)). As depicted in Fig. 20.1(b), this model includes the small momentum transfer scattering processes (g_2, where the two colliding particles reside at the opposite Fermi points, and g_4, where both particles are at the same Fermi point), the back scattering (g_1, where each particle is scattered at the opposite Fermi point) and Umklapp scattering (g_3, where two particles around the same Fermi point are scattered to opposite Fermi points). A treatement of the full g-ology model within a weak-coupling renormalization group approach is given in Sólyom (1979).

For the sake of simplicity, we will only consider the exactly soluble Tomonaga–Luttinger model with spectrum ($\epsilon(\mathbf{k}) = \alpha v_F k$) (see Fig. 20.1(a)) linearized around the Fermi points, a bandwidth cut–off Λ and the couplings g_4 and g_2 for the two forward scattering processes of Fig. 20.1(b). We also assume g_2 and g_4 positive, i.e. we do not consider the possible tendency towards a broken symmetry as would happen, for instance, with g_2 negative.

For a system of size L, the Hamiltonian $H = H_0 + H_2 + H_4$ (H_0 is the free particle Hamiltonian) is specified by

$$
\begin{aligned}
H_0 &= \sum_{|\mathbf{k}|<\Lambda,\alpha,\sigma} \alpha v_F \mathbf{k} a^+_{\alpha,\sigma}(\mathbf{k}) a_{\alpha,\sigma}(\mathbf{k}), \\
H_2 &= \frac{1}{L} \sum_{\mathbf{q}<\mathbf{q}_0} \sum_{\sigma\sigma'} g_2^{\sigma\sigma'} n_{+,\sigma}(\mathbf{q}) n_{-,\sigma'}(-\mathbf{q}), \\
H_4 &= \frac{1}{2L} \sum_{\mathbf{q}<\mathbf{q}_0} \sum_{\sigma\sigma'} g_4^{\sigma\sigma'} \sum_{\alpha} n_{\alpha,\sigma}(\mathbf{q}) n_{\alpha,\sigma'}(-\mathbf{q}).
\end{aligned}
\tag{20.2}
$$

Here $a_{\alpha\sigma}(\mathbf{k})$ and $a^\dagger_{\alpha\sigma}(\mathbf{k})$ are annihilation and creation operators for right-moving ($\alpha = +$) and left-moving ($\alpha = -$) particles with momentum $\mathbf{k}$ and spin projection $\sigma = \pm 1$. Henceforth, v_F indicates the Fermi velocity and we adopt the convention of measuring energies and momenta in units of inverse time, i.e., we take out a factor $\hbar$. We adopt the boldface notation for the momentum $\mathbf{k}$, measured relative to $\pm k_F$, to distinguish it from bivectors $k = (\epsilon, \mathbf{k})$, which include the energy variable ϵ. We have also introduced the operators $n_{\alpha,\sigma}(\mathbf{q})$ describing density fluctuations for a fixed Fermi point branch α and spin projection σ:

$$n_{\alpha,\sigma}(\mathbf{q}) = \sum_{\mathbf{k}} a^{+}_{\alpha,\sigma}(\mathbf{k}-\mathbf{q})a_{\alpha,\sigma}(\mathbf{k}), \tag{20.3}$$

with the restriction in the sum that both $\mathbf{k}$ and $\mathbf{k}-\mathbf{q}$ are in the band. Such a restriction on the sum over momenta will be understood in the following and recalled only when necessary. We also restrict the range of interaction by introducing a *momentum-transfer cutoff* $\mathbf{q}_0 \ll \Lambda$ to confine all the effects of the interaction around the Fermi points. Both cutoffs Λ and $\mathbf{q}_0$ are necessary to regularize, in the ultraviolet region, the divergency of some of the integrals in perturbation theory, but the infrared asymptotic behavior should not depend on the cutoff procedure adopted and the final exact solution should not depend on it.

Finally, the couplings may be spin dependent,

$$g_i^{\sigma\sigma'} = g_{i||}\delta_{\sigma,\sigma'} + g_{i\perp}\delta_{\sigma,-\sigma'}. \tag{20.4}$$

In one-dimensional systems with a linear spectrum and forward scattering, we find it convenient to introduce, for the total charge ($a = \mathrm{c}$) and spin ($a = \mathrm{s}$), bicurrent operators $j^{a\mu}(\mathbf{q}) = (n^a(\mathbf{q}), \mathbf{j}^a(\mathbf{q}))$ with $\mu = 0, 1$ defined by

$$j^{a0} \equiv n^a(\mathbf{q}) = \sum_{\alpha} n^a_{\alpha}(\mathbf{q}), \quad j^{a1} \equiv \mathbf{j}^a(\mathbf{q}) = \sum_{\alpha} \alpha n^a_{\alpha}(\mathbf{q}), \tag{20.5}$$

$$n^{\mathrm{c}}_{\alpha} = n_{\alpha\uparrow}(\mathbf{q}) + n_{\alpha\downarrow}(\mathbf{q}), \tag{20.6}$$

$$n^{\mathrm{s}}_{\alpha} = n_{\alpha\uparrow}(\mathbf{q}) - n_{\alpha\downarrow}(\mathbf{q}). \tag{20.7}$$

In terms of the bicurrent operators $j^{a\mu}(\mathbf{q})$, the interacting part of the Hamiltonian can be rewritten in the compact form

$$H_2 + H_4 = \frac{1}{2L}\sum_{\mathbf{q},a} j^{a\mu}(\mathbf{q}) g^{a,\mu\nu} j^{a\nu}(-\mathbf{q}), \tag{20.8}$$

where the repeated indices are summed and the couplings between densities ($\mu = 0$) and currents ($\mu = 1$) in the channel a are given by

$$g^{a,00} = \frac{1}{2}\left(g_4^a + g_2^a\right), \quad g^{a,11} = \frac{1}{2}\left(g_4^a - g_2^a\right), \quad g^{a,10} = g^{a,01} = 0. \tag{20.9}$$

For the charge and spin channels we have

$$g_i^{\mathrm{c}} = \frac{1}{2}(g_{i||} + g_{i\perp}), \quad g_i^{\mathrm{s}} = \frac{1}{2}(g_{i||} - g_{i\perp}). \tag{20.10}$$

20.2.1 Anomalous commutation rules for density operators

A basic feature of the TL model is that, not only is the total number of particles conserved, but also the right and left movers are conserved separately. Moreover, the right and left density operators, for fixed spin projection, satisfy unusual commutation relations.

To see this, let us take $\mathbf{q} > 0$ for definiteness and evaluate the commutator of the density operators for the right-moving particles. We drop for a moment the spin index for simplicity and, by taking care that all the involved momenta are within the band $(-\Lambda, \Lambda)$, compute

$$
\begin{aligned}
[n_+(\mathbf{q}), n_+(-\mathbf{q})] &= \sum_{-\Lambda+\mathbf{q}<\mathbf{k}<\Lambda} \sum_{-\Lambda<\mathbf{k}'<\Lambda-\mathbf{q}} [a^\dagger(\mathbf{k}-\mathbf{q})a(\mathbf{k}), a^\dagger(\mathbf{k}'+\mathbf{q})a(\mathbf{k}')] \\
&= \sum_{-\Lambda+\mathbf{q}<\mathbf{k}<\Lambda} \sum_{-\Lambda<\mathbf{k}'<\Lambda-\mathbf{q}} (\delta_{\mathbf{k},\mathbf{k}'+\mathbf{q}} a^\dagger(\mathbf{k}-\mathbf{q})a(\mathbf{k}') - \delta_{\mathbf{k}',\mathbf{k}-\mathbf{q}} a^\dagger(\mathbf{k}'+\mathbf{q})a(\mathbf{k})) \\
&= \sum_{-\Lambda+\mathbf{q}<\mathbf{k}<\Lambda} (a^\dagger(\mathbf{k}-\mathbf{q})a(\mathbf{k}-\mathbf{q}) - a^\dagger(\mathbf{k})a(\mathbf{k})) \\
&= \sum_{-\Lambda<\mathbf{k}<\Lambda-\mathbf{q}} a^\dagger(\mathbf{k})a(\mathbf{k}) - \sum_{-\Lambda+\mathbf{q}<\mathbf{k}<\Lambda} a^\dagger(\mathbf{k})a(\mathbf{k}),
\end{aligned}
$$

where to get the last line we have made the shift $\mathbf{k} \to \mathbf{k} + \mathbf{q}$. Since $\mathbf{q}_0 \ll \Lambda$ we observe that the most important contribution to the difference of the two sums in the last line comes from states far below the Fermi energy (see the first sum) or well above the Fermi energy (see the second sum). We may expect that in those regions the quantum fluctuations of the number operator $a^\dagger(\mathbf{k})a(\mathbf{k})$ are small and we may replace it with its ground-state value $\langle a^\dagger(\mathbf{k})a(\mathbf{k})\rangle = \theta(-\mathbf{k})$. The two sums differ only in the region $(-\Lambda \le \mathbf{k} \le -\Lambda + \mathbf{q})$. This, after replacing the sums with integrals in the standard way, leads to

$$
[n_+(\mathbf{q}), n_+(-\mathbf{q})] = \frac{L}{2\pi}\mathbf{q}. \tag{20.11}
$$

This result was first obtained in Tomonaga (1950), where it was emphasized that the commutation relation (20.11) holds with good accuracy if one assumes that excitations lie in the neighborhood of the Fermi level. For an ideal Fermi gas this is justified if the gas is not too excited. For an interacting Fermi gas one requires, as has been mentioned above, a long range interaction with a small momentum transfer cutoff.[1] By performing a similar calculation for left-moving particles, one obtains

$$
[n_-(\mathbf{q}), n_-(-\mathbf{q})] = -\frac{L}{2\pi}\mathbf{q}. \tag{20.12}
$$

The minus sign originates from the fact that for $\alpha = -1$, we have the ground-state value $\langle a^\dagger(\mathbf{k})a(\mathbf{k})\rangle = \theta(\mathbf{k})$.

Since left- and right-moving particle operators anticommute, their occupation numbers commute:

$$
[n_+(\mathbf{q}), n_-(-\mathbf{q})] = 0. \tag{20.13}
$$

Hence, it is easy to see that the total density $n(\mathbf{q}) = n_+(\mathbf{q}) + n_-(-\mathbf{q})$ satisfies

$$
[n(\mathbf{q}), n(-\mathbf{q})] = 0, \tag{20.14}
$$

i.e., the total density operators satisfy a standard commutation relation. We emphasize that the anomalous commutators arise from the peculiarity of one-dimensional systems

[1] We point out that in the model discussed in Luttinger (1963) with no bandwidth cutoff, where one assumes a filled Fermi sea with an infinite number of particles, the commutation relation (20.11) can be proven rigorously, as shown by Mattis and Lieb (1965).

of splitting the density operators into two parts for left- and right-moving particles. By reintroducing spin, we obtain in general

$$[n_{\alpha\sigma}(\mathbf{q}), n_{\alpha'\sigma'}(\mathbf{q}')] = \frac{\alpha L\mathbf{q}}{2\pi}\delta_{\alpha\alpha'}\delta_{\sigma\sigma'}\delta(\mathbf{q}+\mathbf{q}'), \tag{20.15}$$

which allows us to derive an anomalous commutator for the bicurrent (20.5) operators

$$[j^{a\mu}(\mathbf{q}), j^{a\nu}(\mathbf{q}')] = (1-\delta_{\mu\nu})\frac{2L\mathbf{q}}{\pi}\delta(\mathbf{q}+\mathbf{q}'), \tag{20.16}$$

$\delta_{\mu\nu}$ being the Kronecker symbol. Equation (20.16) will prove very useful to get the equations of motion for the problem at hand.

20.2.2 Equations of motion and response functions

In order to find the equation of motion for the density operators in the Heisenberg representation, we need to evaluate their commutator with the Hamiltonian.

Let us first examine what happens for the 1d gas with Hamiltonian H_0. We have

$$\begin{aligned} \mathrm{i}\partial_t n_{\alpha,\sigma}(\mathbf{q},t) &= [n_{\alpha,\sigma}(\mathbf{q},t), H_0] \\ &= \alpha v_{\mathrm{F}} \sum_{\mathbf{k}} \mathrm{e}^{\mathrm{i}H_0 t}\left(\mathbf{k}a^\dagger_{\alpha,\sigma}(\mathbf{k}-\mathbf{q})a_{\alpha,\sigma}(\mathbf{k})\right. \\ &\quad \left. -(\mathbf{k}-\mathbf{q})a^\dagger_{\alpha,\sigma}(\mathbf{k}-\mathbf{q})a_{\alpha,\sigma}(\mathbf{k})\right)\mathrm{e}^{-\mathrm{i}H_0 t} \\ &= \alpha v_{\mathrm{F}}\mathbf{q}n_{\alpha,\sigma}(\mathbf{q},t). \end{aligned} \tag{20.17}$$

This last result is a consequence of the linearity of the spectrum. This is manifest in the second line of Eq. (20.17), where the two terms in the sum over $\mathbf{k}$ contain the excitation $\epsilon(\mathbf{k}) = \alpha v_{\mathrm{F}}\mathbf{k}$ and $\epsilon(\mathbf{k}-\mathbf{q}) = \alpha v_{\mathrm{F}}(\mathbf{k}-\mathbf{q})$. The quadratic and higher order terms in the spectrum are, however, irrelevant in the renormalization group language as far as the low-lying asymptotic behavior of the system is concerned.

Equation (20.17) implies the continuity equation for charge and spin, in the Heisenberg representation, for *free* fermions:

$$\mathrm{i}\partial_t n^a(\mathbf{q},t) = \mathbf{q}v_{\mathrm{F}}\mathbf{j}^a(\mathbf{q},t). \tag{20.18}$$

By taking the commutator of the current $\mathbf{j}^a(\mathbf{q})$ with H_0 it is easy to see that

$$\mathrm{i}\partial_t \mathbf{j}^a(\mathbf{q},t) = \mathbf{q}v_{\mathrm{F}}n^a(\mathbf{q},t), \tag{20.19}$$

which also has the form of a conservation law. In this case the conserved quantity is the current $\mathbf{j}^a(\mathbf{q})$ and the charge plays the role of the associated current. This extra conservation law is a peculiar feature of the discrete nature of the Fermi manifold in one-dimensional systems and reflects the separate conservation laws for right- and left-moving particles. Indeed, by taking the sum and the difference of Eqs. (20.18) and (20.19), we may easily write separate continuity equations for left- and right-density operators.

In the interacting system, by using the commutator (20.16) and the form (20.8) for the interacting Hamiltonian, we get

$$[j^{a0}(\mathbf{q}), H_2+H_4] = \mathbf{q}\frac{g_4^a - g_2^a}{\pi}j^{a1}(\mathbf{q}),\ [j^{a1}(\mathbf{q}), H_2+H_4] = \mathbf{q}\frac{g_4^a + g_2^a}{\pi}j^{a0}(\mathbf{q}), \tag{20.20}$$

from which we have the conservation laws for *interacting* 1d fermions:

$$i\partial_t n^a(\mathbf{q},t) = \mathbf{q} v^a \mathbf{j}^a(\mathbf{q},t), \quad v^a = v_F + \frac{1}{\pi}\left(g_4^a - g_2^a\right), \tag{20.21}$$

$$i\partial_t \mathbf{j}^a(\mathbf{q},t) = \mathbf{q} \tilde{v}^a n^a(\mathbf{q},t), \quad \tilde{v}^a = v_F + \frac{1}{\pi}\left(g_4^a + g_2^a\right). \tag{20.22}$$

The second continuity equation reflects the additional conservation law of the 1d system with forward scattering processes g_2 and g_4. With respect to Eqs. (20.18) and (20.19), a different velocity appears in the continuity equation for $n^a(\mathbf{q})$ and $\mathbf{j}^a(\mathbf{q})$.

Equation (20.21) and (20.22) can also be written in terms of the densitites for each particle branch:

$$i\partial_t n_+^a(\mathbf{q},t) = \mathbf{q}\frac{v^a + \tilde{v}^a}{2} n_+^a(\mathbf{q},t) + \mathbf{q}\frac{\tilde{v}^a - v^a}{2} n_-^a(\mathbf{q},t), \tag{20.23}$$

$$i\partial_t n_-^a(\mathbf{q},t) = \mathbf{q}\frac{v^a - \tilde{v}^a}{2} n_+^a(\mathbf{q},t) - \mathbf{q}\frac{\tilde{v}^a + v^a}{2} n_-^a(\mathbf{q},t). \tag{20.24}$$

These equations will be useful when solving the Dyson equation for the Green function.

By combining Eqs. (20.21) and (20.22), one obtains a one-dimensional wave equation for $n^a(\mathbf{q})$:

$$\frac{\partial^2}{\partial t^2} n^a(\mathbf{x},t) = (u^a)^2 \frac{\partial^2}{\partial_{\mathbf{x}^2}} n^a(\mathbf{x},t), \tag{20.25}$$

in which the velocity u^a for the a-type of density excitation appears:

$$u^a = \sqrt{v^a \tilde{v}^a}. \tag{20.26}$$

Equation (20.25) describes undamped propagating harmonic oscillations of the density with frequencies $u^a|\mathbf{q}|$, implying that, in the 1d interacting system, charge and spin modes propagate with different velocities u^c and u^s. At finite temperature, the gas of charge and spin density excitations contribute to the energy of the system. By considering that both charge and spin excitations are bosonic gases with linear spectrum in one dimension, we easily find that their contribution to the specific heat is linear in temperature (see Problem 20.1):

$$c_V = \frac{\pi}{3}\frac{u^c + u^s}{u^c u^s} T. \tag{20.27}$$

We emphasize that, although we have a linear in T behavior as in the Fermi gas and the Fermi liquid, such a contribution here arises solely from the collective modes and not from the quasiparticles as in Fermi liquids.

The symmetry properties discussed so far allow us also to give the expressions for the response functions free of any singularity for the Luttinger liquid phase.

We introduce the bicurrent–bicurrent thermal "time-ordered" correlation function $\mathcal{R}^{a\mu\nu}$ (18.106) and consider its analytic continuation (18.107). Similarly to the relations of Chapter 10 between the density and current response (see Eqs. (10.34) and (10.35) and their consequences), by using the continuity equation (20.21) and the commutator (20.16), one obtains

$$\omega \tilde{R}^{a0\nu}(q) - v^a \mathbf{q} \tilde{R}^{a1\nu} = \delta_{\nu 1} \frac{2\mathbf{q}}{\pi}. \tag{20.28}$$

By itself the above relation following from the total conservation of particles is not sufficient to solve the problem. By using the additional continuity equation (20.22) one has the further relation

$$\omega \tilde{R}^{a1\nu}(q) - \tilde{v}^a \mathbf{q} \tilde{R}^{a0\nu} = \delta_{\nu 0} \frac{2\mathbf{q}}{\pi}. \tag{20.29}$$

By combining Eqs. (20.28) and (20.29), one can eliminate the current–current response function and obtain the density–density response function for both charge and spin ($a = (\mathrm{c}, \mathrm{s})$) explicitly:

$$\tilde{R}^{a00}(\omega, \mathbf{q}) = \frac{2}{\pi} \frac{v^a \mathbf{q}^2}{\omega^2 - (u^a)^2 \mathbf{q}^2}. \tag{20.30}$$

Hence, by Eq. (10.44) the interaction dependent compressibility κ_T, analogous to the formula (12.25) for the Fermi liquid, is

$$n^2 \kappa_T = -\lim_{\mathbf{q} \to 0} \tilde{R}^{c00}(0, \mathbf{q}) = 2/(\pi \tilde{v}^c). \tag{20.31}$$

Similarly, for the susceptibility,

$$\chi = -\lim_{\mathbf{q} \to 0} \tilde{R}^{s00}(0, \mathbf{q}) = 2/(\pi \tilde{v}^s). \tag{20.32}$$

From Eq. (10.43) the dynamical electrical conductivity, in the small $\mathbf{q}$ and ω region, becomes

$$\sigma(\omega, \mathbf{q}) = e^2 \frac{2}{\pi} v^{\mathrm{c}} \frac{\mathrm{i}\omega}{\omega^2 - (u^{\mathrm{c}} \mathbf{q})^2}. \tag{20.33}$$

The absorptive part in a homogeneous electric field ($\mathbf{q} \to 0$) reads

$$\sigma(\omega) = \lim_{\mathbf{q} \to 0} \mathrm{Re}\, \sigma(\omega, \mathbf{q}) = \mathrm{Re}\, 2e^2 v^{\mathrm{c}} \frac{\mathrm{i}}{\omega + \mathrm{i}0^+} = 2e^2 v^{\mathrm{c}} \delta(\omega), \tag{20.34}$$

as is usual for clean systems at $T = 0$, the absorptive factor is a delta peak at $\omega = 0$. v^a, $\tilde{v}^a$ are then related to observable quantities.

20.3 The Dyson equation, conservation laws and Ward identities

As we have seen in the previous sections, the peculiar symmetry properties of the 1d system render all the response functions finite, in spite of the singular perturbation theory. The same symmetries associated with the conservation laws (20.21) and (20.22) yield also additional Ward identities (WI) besides those discussed in Chapter 18. All together these WIs allow us to close the Dyson equation for the *interacting* single-particle Green function, thus showing the anomalous behavior of single-particle excitations of the Luttinger liquid. This is the only drastic change in the behavior of the system compared to the Fermi liquid.

The Dyson equation for the thermal Green function was derived in Chapter 18 (see Eqs. (18.101)–(18.103)). Hence, one could start from Eq. (18.103) for the self-energy and use the specific form of the interaction in Eq. (20.2). To derive the expression of the self-energy one must analyze the diagrammatic structure of the theory. Such an analysis is shown in Appendix L.

In the present case, the interaction is relatively simple, so we therefore prefer to derive the Dyson equation for the zero-kelvin single-particle Green function directly, starting from its very definition in terms of a time-ordered product of the Heinseberg operators for right- or left-moving particles:

$$G_{\alpha\sigma}(\mathbf{p}, t) = -\mathrm{i}\left\langle \mathrm{T}_t a_{\alpha\sigma}(\mathbf{p}, t) a^{+}_{\alpha\sigma}(\mathbf{p}, 0)\right\rangle . \tag{20.35}$$

The Green function in the frequency representation will be denoted by $G_{\alpha\sigma}(p)$ and is obtained as a Fourier transform of Eq. (20.35) with respect to the time arguments. The single-particle Green function for the free case, analogous to Eq. (18.47) is

$$G^0_{\alpha\sigma}(\epsilon, \mathbf{p}) = [\varepsilon - \alpha v_{\mathrm{F}}\mathbf{p} + \alpha \mathrm{sign}(\mathbf{p})\mathrm{i}0^{+}]^{-1} \tag{20.36}$$

with $z_{\mathbf{p}_{\mathrm{F}}} = 1$.

As was done in the derivation of the WIs in Section 18.10, we take the time derivative of the definition (20.35) and use the commutation relations for the fermion operator. After Fourier transforming we get the Dyson equation

$$(\epsilon - \alpha v_{\mathrm{F}}\mathbf{p})G_{\alpha\sigma}(p) = 1 + \mathrm{i}\sum_{a\alpha'}\int \frac{\mathrm{d}\mathbf{q}}{2\pi}\int_{-\infty}^{\infty}\frac{\mathrm{d}\omega}{2\pi} g^a_{\alpha'\alpha}\epsilon^a_\sigma \Lambda^a_{\alpha',\alpha\sigma}(p - q/2, q), \tag{20.37}$$

where the interaction inside each branch of the dispersion is $g^a_{++} = g^a_4$ and the interaction connecting the two branches is $g^a_{-+} = g^a_2$ and the three-leg vertex of the density of the branch α' is defined by

$$\Lambda^a_{\alpha',\alpha\sigma}(\mathbf{p} - \mathbf{q}/2, \mathbf{q}; t, t', t'') = -\left\langle \mathrm{T}_t n^a_{\alpha'}(\mathbf{q}, t) a_{\alpha\sigma}(\mathbf{p} - \mathbf{q}, t') a^{\dagger}_{\alpha\sigma}(\mathbf{p}, t'')\right\rangle . \tag{20.38}$$

Notice that if we use bicurrent operators we may define also the density/current vertex

$$\Lambda^{a\mu}_{\alpha\sigma}(\mathbf{p} - \mathbf{q}/2, \mathbf{q}; t, t', t'') = -\langle \mathrm{T}_t j^{a\mu}(\mathbf{q}, t) a_{\alpha\sigma}(\mathbf{p} - \mathbf{q}, t') a^{\dagger}_{\alpha\sigma}(\mathbf{p}, t'')\rangle . \tag{20.39}$$

The vertex (20.39) is the zero-kelvin limit of the vertex defined in Eq. (18.111). As was mentioned before, the remarkable feature of the problem at hand is the fact that the Dyson equation for $G_{\alpha\sigma}(\mathbf{p}, t)$ can be written in a closed form which can then be solved. Let us see the main steps of the derivation. The conservation laws (20.23) and (20.24) or (20.21) and (20.22) can be transformed into Ward identities for the vertex $\Lambda^{a\mu}_{\alpha\sigma}$ or $\Lambda^a_{\alpha',\alpha\sigma}$, respectively, along the lines of Section 18.10.

The conservation law of total charge and spin (20.21) connects the one-particle Green function $G_{\alpha\sigma}$ to the vertex $\Lambda^{a\mu}_{\alpha\sigma}$ for charge (spin) density ($\mu = 0$) and current ($\mu = 1$):

$$\omega \Lambda^{a0}_{\alpha,\sigma}(p, q) - v^a \mathbf{q}\Lambda^{a1}_{\alpha,\sigma}(p, q) = \varepsilon^a_\sigma [G_{\alpha,\sigma}(p - q/2) - G_{\alpha,\sigma}(p + q/2)], \tag{20.40}$$

with $\epsilon^a_\sigma = -1$ if $a = \mathrm{s}$ and $\sigma = \downarrow$ and $\epsilon^a_\sigma = +1$ otherwise. This WI (20.40) is valid for all systems with total conservation (see (18.118)). By itself it is, of course, not enough to solve

in terms of the Heisenberg fermion and density operators:

$$\begin{aligned}
&G^{(2m,l)}(\mathbf{k}_1,t_1,\ldots;\mathbf{k}'_1,t'_1,\ldots;\mathbf{k}''_1,t''_1,\ldots)\\
&=(-\mathrm{i})^{m+l}\langle \mathrm{T}_t a_{\alpha_1\sigma_1}(\mathbf{k}_1,t_1)\cdots a_{\alpha_m\sigma_m}(\mathbf{k}_m,t_{m1})a^+_{\alpha'_1\sigma'_1}(\mathbf{k}'_1,t'_1)\cdots a^+_{\alpha'_m\sigma'_m}(\mathbf{k}'_m,t'_m)\\
&\quad\times n_{\alpha''_1\sigma''_1}(\mathbf{k}''_1t''_1)\ldots n_{\alpha''_l\sigma''_l}(\mathbf{k}''_lt''_l)\rangle.
\end{aligned}\tag{20.57}$$

To avoid formulae with many indices, we have defined the Green functions $G^{(2m,l)}$ without the subscripts relative to branch and spin.

The diagrams of $G^{(2m,l)}$ can be classified according to the possibility of dividing them into diagrams which can or cannot be divided into disjoint parts by cutting a fermion line. The latter are called irreducible. The corresponding vertices $\Gamma^{(2m,l)}$ are obtained as the one-particle irreducible of $G^{(2m,l)}$ after truncating the external lines. $\Gamma^{(2)}$ coincides with G^{-1}.

The dimensional analysis of the TL model given above can be summarized by the canonical dimensions in inverse length: $[H]=1$, $[a(\mathbf{k})]=-1/2$ and therefore $[g_i]=0$, i.e. g_i is are marginal. Every integration over the bimomentum in a Feynman diagram behaves as $\mathbf{q}_0^2$ and every Green function line goes as $\mathbf{q}_0^{-1}$. This predicts that divergences can originate from the self-energy $\Gamma^{(2)}$ (see Problem 20.4), the density–density response (the polarization bubble) $\Gamma^{(0,2)}$ or the truncated three-leg density vertex $\Gamma^{(2,1)}$ ($\mathcal{T}^{a0}_{\alpha,\sigma}$, defined in Fig. L.4) and the effective truncated four-leg interaction $\Gamma^{(4)}$. All divergences in the Green functions (and vertices) with more than two particles will be a consequence of the divergences of the above Green functions.

According to the general framework of the field-theoretic renormalization group discussed in Chapter 17, we define renormalized vertex functions as

$$\overline{\Gamma}^{(2m,l)}(P;\overline{g};\lambda)=\lim_{\mathbf{q}_0\to\infty}Z^m(g,\lambda/\mathbf{q}_0)Z_n^l(g,\lambda/\mathbf{q}_0)\Gamma^{(2m,l)}(P;g,\mathbf{q}_0),\tag{20.58}$$

where we have introduced a wave function renormalization Z associated with the fermion fields, a density operator renormalization Z_n and $\overline{g}(g,\lambda/\mathbf{q}_0)$ stands for all renormalized couplings $\overline{g}_i$. The symbol P indicates a normalization point at the corresponding scale. A possible choice for the normalization point is that the vertex functions coincide with their bare values:

$$\overline{\Gamma}^{(2)}=\Gamma^{(2)0},\tag{20.59}$$

$$\overline{\Gamma}^{(4)}=g,\tag{20.60}$$

$$\overline{\Gamma}^{(2,1)}=1,\tag{20.61}$$

at

$$P^{(2)}_\lambda=(v_\mathrm{F}\lambda,\mathbf{0}),\tag{20.62}$$

$$P^{(4)}_\lambda=((3v_\mathrm{F}\lambda/2,\mathbf{0}),(-v_\mathrm{F}\lambda/2,\mathbf{0}),(v_\mathrm{F}\lambda/2,\mathbf{0}),(v_\mathrm{F}\lambda/2,\mathbf{0})),\tag{20.63}$$

$$P^{(2)}_\lambda=((v_\mathrm{F}\lambda,\mathbf{0});(v_\mathrm{F}\lambda/2,\mathbf{0}),(-v_\mathrm{F}\lambda/2,\mathbf{0})).\tag{20.64}$$

A change of the normalization point from λ to λ' implies

$$\overline{\Gamma}^{(2m,l)}(\overline{g}';\lambda')=z^m(g,\lambda'/\lambda)z_n^l(g,\lambda'/\lambda)\overline{\Gamma}^{(2m,l)}(\overline{g};\lambda'),\tag{20.65}$$

where

$$z(g, \lambda'/\lambda) = Z(g, \lambda'/\mathbf{q}_0)/Z(g, \lambda/\mathbf{q}_0), \tag{20.66}$$

$$z_n(g, \lambda'/\lambda) = Z_n(g, \lambda'/\mathbf{q}_0)/Z_n(g, \lambda/\mathbf{q}_0), \tag{20.67}$$

with the ultraviolet cutoff $\mathbf{q}_0$ large since we are dealing with the infrared divergences.

By an infinitesimal change of λ one gets the differential form of the renormalization group transformation:

$$[\lambda\partial_\lambda + \beta(\overline{g})\partial_{\overline{g}} - m\gamma(\overline{g}) - l\gamma_n(\overline{g})]\overline{\Gamma}^{(2m,l)}(\overline{g}, \lambda) = 0, \tag{20.68}$$

where

$$\beta(\overline{g}) = \left(\frac{\partial \overline{g'}}{\partial \lambda'/\lambda}\right)_{\lambda'=\lambda}, \tag{20.69}$$

$$\gamma(\overline{g}) = \left(\frac{\partial \ln z}{\partial \lambda'/\lambda}\right)_{\lambda'=\lambda}, \tag{20.70}$$

$$\gamma_n(\overline{g}) = \left(\frac{\partial \ln z_n}{\partial \lambda'/\lambda}\right)_{\lambda'=\lambda}. \tag{20.71}$$

At a fixed point $\beta(g*) = 0$, $\gamma(g*)$ would correspond to η, which is twice the anomalous dimension of the fermion field, and $\gamma_n(g*)$ to the anomalous dimension of the density vertex. By using the constraints imposed by the WIs associated with the conservation laws of the TL model, we can now draw some general conclusions about the renormalization of the TL model by recalling some results of its diagrammatic structure discussed in Appendix L.

First, the TL couplings g_2 and g_4 do not get renormalized and are the values of the points on the fixed-point line of the renormalization group equations $\beta(g) \equiv 0$. This result leads to the anomalous index $\eta = \gamma(g*) = \gamma(g)$ depending on the bare couplings, as in the exact solution Eq. (20.49).

Second, the density vertex $\Gamma^{(2,1)}$ via the WIs (20.40) and (20.41) or, more directly, its interaction irreducible truncated part $\mathcal{I}^{a0}_{\alpha,\sigma}$ (see Section L.3), via the WI (L.13), does not require additional renormalizations once G is renormalized, i.e. $Z_n = 1$.

Third, in the perturbative structure (Fig. L.1) of the truncated vertex $\Gamma^{a(4)}_{\alpha\alpha'}$, the block $(\mathcal{I}^{a0}_{\alpha,\sigma})^2 D^a_{\alpha\alpha'}$ repeats itself at each order. Since $D^a_{\alpha\alpha'}$ does not renormalize and the vertex $\mathcal{I}^{a0}_{\alpha,\sigma}$ renormalizes as G^{-1}, $\Gamma^{a(4)}_{\alpha\alpha'}$ renormalizes as G^{-2}. In conclusion, we have to consider the wave function renormalization only to make the theory finite.

Before evaluating $z(g, \lambda'/\lambda)$ and the index η, let us explain more extensively why the interaction does not renormalize. We first notice that because of the loop cancellation theorem (see Appendix L.2), diagrams with fermionic loops with more than two insertions (Fig. L.2) vanish. As a first consequence, discussed in Appendix L, the full interaction $D^a_{\alpha\alpha'}$ can only be formed by a geometric series summation in terms of the bare coupling $g^a_{\alpha\alpha'}$ and the full polarization bubble (see Section L.4). This last one, in turn, reduces to the bare polarization bubble (Fig. L.3(a), (b) and (c)), as is guaranteed by the WI (L.13) with the cancellation of the vertex $\mathcal{I}^{a0}_{\alpha,\sigma}$ renormalization with the self-energy corrections.

The absence of logarithmic singularities in the perturbative evaluation of the renormalized coupling occurs thanks to exact cancellation of diagrams, as shown in the answer to Problem 20.3. To all orders in perturbation theory, as mentioned above, such cancellation is, therefore, granted from the cancellation between self-energy and vertex corrections in the polarization bubble (see Eq. (L.16)) as required by the WI of Eq. (L.13), which is the result of the extra conservation law valid in the model. It is the same WI which guarantees that the density vertex renormalizes as G^{-1} and therefore $Z_n = 1$, i.e. no renormalization of the density vertex besides the wave function renormalization Z is required. Of course also $\Gamma^{(0,2)}$ is finite.

We now proceed to the calculation of the wave function renormalization Z which makes $Z\Gamma^{(2)}$ and $Z^2\Gamma^{(4)}$ finite. Evaluating Eq. (20.58) at the normalization point, Z turns out to be

$$Z(g, \lambda/\mathbf{q}_0) = \frac{\Gamma^{(2)0}(v_F\lambda, \mathbf{0})}{\Gamma^{(2)}((v_F\lambda, \mathbf{0}); g, \mathbf{q}_0)} = \frac{1}{1 - \Sigma(v_F\lambda, \mathbf{0})/v_F\lambda}. \tag{20.72}$$

Z can be evaluated explicitly along the same line as the calculation of the self energy corrections in the answer to Problem 20.4. There is no infrared singular term coming from the first order diagram (a) of Fig. 18.4 of the self-energy. The second order correction gives

$$Z((v_F\lambda, \mathbf{0}; g, \mathbf{q}_0) = 1 + \frac{g_2^2}{4\pi^2 v_F^2} \ln \frac{\lambda}{2\mathbf{q}_0}. \tag{20.73}$$

We notice that at this order only g_2 contributes, as shown in Problem 20.4. In Problem 20.5 we evaluate the second order correction to $\Gamma^{(4)}$ and we find

$$\Gamma_{+-}^{a(4)} = g_2^a \left(1 + \frac{g_2^2}{2\pi^2 v_F^2} \ln \frac{2v_F\mathbf{q}_0}{|\epsilon|}\right) = g_2^a Z^{-2}, \tag{20.74}$$

and at this order $\Gamma^{(4)} Z^2$ is finite.

Then

$$z(g, \lambda'/\lambda) = 1 + \frac{g_2^2}{4\pi^2 v_F^2} \ln \frac{\lambda'}{\lambda} \tag{20.75}$$

and via (20.70)

$$\gamma(\overline{g}) = \left(\frac{\partial \ln z}{\partial \lambda'/\lambda}\right)_{\lambda'=\lambda} = \frac{g_2^2}{4\pi^2 v_F^2} = \eta, \tag{20.76}$$

which coincides with the weak coupling limit of the exact solution (20.49). Further terms would reproduce the exact result order by order.

In this case the use of the renormalization group may appear redundant. The first lesson we learned, however, is to see explicitly how the renormalization group method obtains power laws out of the logarithmic singularities of the perturbation theory. Secondly, it makes clear how important it is to use symmetry properties to condition the proper choice of the relevant variables to be renormalized and the relation between them, in particular for the cases where it is difficult to make such a choice a priori.

20.5 Problems

20.1 Evaluate the specific heat associated with the charge and spin density modes.

20.2 Derive Eq. (20.44).

20.3 Show that the one-loop correction to g_2 are free from logarithmic divergences.

20.4 Show that the one-loop correction to the self-energy yields the correct perturbative expression for the anomalous exponent η.

20.5 Evaluate the two-loop corrections to the vertex $\Gamma^{a(4)}_{+-}$ due to the coupling g_2 only.

21 Quantum interference effects in disordered electron systems

This chapter provides an introduction to the theory of strongly disordered electron systems. As for the case of superconductivity, the understanding of transport and thermodynamical properties of dirty metals and semiconductors required the introduction of new concepts in solid-state physics.

In non-interacting systems the quantum interference effects yield corrections to the Drude–Boltzmann classical theory of electrical transport (developed in Chapter 11) and for sufficiently strong disorder the correction terms produce a transition to an insulating phase where the electrons are localized, named the Anderson localization (Anderson (1958)).

To emphasize the peculiarity of this transition, we recall that in a perfect crystal, according to Bloch theory, the electron states are extended plane waves modulated by the crystalline potential, grouped in finite energy bands. Within this picture, partial occupation of the bands leads to a metallic state, whereas insulators and semiconductors have an energy gap between the last fully occupied band and the first empty one. The classification of metallic and insulating states is done accordingly to the above filling procedure. The new aspect introduced by Anderson is the possibility of an insulating state arising from the localization of the electron wave function in a partially occupied band.

The effect of electron–electron interaction also introduces surprising effects. The resulting scenario is different from the single-particle picture and provides an effective Fermi liquid picture with frequency and momentum dependent Landau parameters. These parameters then become scale dependent and give rise to a complex flow of the renormalization group equations.

There exist several review articles which give an account of the problem from different viewpoints and at different stages of the historical development (Bergmann (1984); Lee and Ramakrishnan (1985); Altshuler and Aronov (1985); Finkelstein (1990); Belitz and Kirkpatrick (1994); Di Castro and Raimondi (2004)).

We will concentrate here on those aspects that we believe are fundamental for the understanding of the problem. Therefore we present, at first, the theory in a simple phenomenological way, then we introduce all the required technicalities necessary to proceed towards a microscopic theory.

The microscopic theory will be presented here using the standard language of many-body theory developed so far, rather than the field-theoretic non-linear sigma model introduced by Wegner (1979), Efetov et al. (1980) and Hikami (1981) for the non-interacting case and generalized by Finkelstein (1983, 1984) to include the interaction. These last papers were the real breakthrough on this topic. In the non-interacting case, we present a generalization (Gor'kov et al. (1979); Hikami et al. (1980); Altshuler et al. (1980b); Castellani et al. (1983)) to strong disorder of the impurity technique for a Fermi gas, as developed in the book by Abrikosov et al. (1963). In the interacting case, this problem is treated in terms of a disordered Fermi liquid theory (Altshuler and Aronov (1970); Altshuler et al. (1980a); Finkelstein (1983, 1984); Castellani et al. (1984a, b, 1986); Castellani and Di Castro (1986)).

21.1 Experimental evidence of anomalous disorder effects

Typical disordered systems show, at low temperature, strong anomalies with respect to the conceptual classical scenario of transport, as developed in Section 11.4 and Chapter 12.

Experimental realizations of electronic disordered systems are obtained in doped semiconductors like Si:P, Ge:Sb and amorphous alloys of a semiconductor and a metal such as Nb_x:Si_{1-x}, Al_x:Ge_{1-x}, Au_x:Ge_{1-x}. Above a critical value of doping and of metallic component concentration x_c, systems of both classes become metallic.

Coming from the metallic side, at a given amount of disorder for the temperature dependence of the conductivity of Si:P one measures

$$\sigma(T) = \sigma_0 + mT^n, \tag{21.1}$$

where the coefficient m can be both positive and negative and $n = 1/2$ (Rosenbaum et al. (1983)), contrary to the classical expectation that the conductivity should only increase by lowering the temperature and $n = 2$ according to the temperature corrections due to inelastic electron–electron interactions (see Eq. (12.1)). Even stronger anomalies are present in metallic films, with logarithmically singular temperature dependent corrections, as observed e.g. in PdAu films (Dolan and Osheroff (1979)):

$$\sigma(T) = \sigma(T_0) + m \ln T/T_0, \tag{21.2}$$

where m is positive and the resistivity increases by lowering the temperature.

By decreasing the P-concentration below a critical value, the Si:P system undergoes a metal–insulator transition at $T = 0$ in the sense that

$$\sigma_0 \approx (n - n_c)^\mu, \tag{21.3}$$

with the critical exponent $\mu = 1/2$ (Rosenbaum et al. (1980, 1983)). The value of μ for uncompensated Si:P is, however, uncertain and may depend on the identification of the critical concentration in the experimental data (Stupp et al. (1993))[1]. Compensated samples (Thomas et al. (1982)) and alloys (Hertel et al. (1983); Yamaguchi et al. (1983); Rohde and Micklitz (1987)) have $\mu = 1$.

According to classical theory, as a consequence of the collisions, the electrons undergo a classical random walk of step l and a *diffusive* motion, as we have seen in Chapter 11. Within the independent electron approximation, only one diffusion constant D controls the charge, spin and heat transport, leading to three Einstein relations for the electrical (11.40),

[1] A doped semiconductor is compensated when, besides the majority donor atoms, it contains also some minority acceptor atoms. In this way, some of the donor electrons are captured by the acceptor levels, by allowing the tunneling of a donor electron from an occupied level to an unoccupied one.

thermal (11.55) and spin (11.60) conductivity:

$$\sigma_0 = e^2 \left(\frac{\partial n}{\partial \mu}\right)_0 D, \quad \kappa_{\mathrm{E},0} = c_{V,0} D, \quad \sigma_{\mathrm{s},0} = \chi_0 D, \tag{21.4}$$

in which appears the same diffusion constant $D = v_{\mathrm{F}} 2\tau/d$ (Eq. (11.34)) and $(\partial n/\partial \mu)_0 = 2N_0$, $c_{V,0} = (2\pi^2/3)k_{\mathrm{B}}^2 N_0 T$ and $\chi_0 = (g\mu_{\mathrm{B}}/2)^2 2N_0$ are the thermodynamic density of states (7.28), the specific heat (7.37) and the spin susceptibility (7.32) of the electron gas, all expressed in terms of N_0, the d-dimensional generalization of the single-particle density of states per unit volume per spin of Eq. (7.14) evaluated at the Fermi energy, E_{F}:

$$N_0 \equiv \frac{1}{2}\nu_d(E_{\mathrm{F}}) = \frac{\Omega_d}{(2\pi\hbar)^d} m p_{\mathrm{F}}^{d-2} = \frac{1}{2}\frac{2\pi^{d/2}}{\Gamma(d/2)(2\pi\hbar)^d} 2^{\frac{d-2}{2}} m^{d/2} E_{\mathrm{F}}^{\frac{d-2}{2}}. \tag{21.5}$$

The subscript 0 in all the quantities emphasizes that the above expressions are valid for weak disorder and in the absence of electron–electron interaction, Ω_d is the solid angle in d dimensions.

At finite temperature, besides the impurities, deviations from the perfect lattice periodicity arise from the thermal vibrations of the ions (electron–phonon interactions) and, together with the inelastic electron–electron interactions, produce further diffusion of the electrons. At low temperature, we use the Matthiessen rule and sum the inverse of the scattering times for the various mechanisms obtaining for the conductivity σ:

$$\sigma(T) = \sigma(0) - aT^p, \tag{21.6}$$

where the coefficient a is positive (on lowering T the effect of the inelastic processes is decreasing) and the exponent $p \geq 2$, the value $p = 2$ corresponding to the case of electron–electron interaction. As already recalled above, these results are in contrast with the experimental findings of Eqs. (21.1) and (21.2).

In concluding this section, we introduce the conductance related to the conductivity by geometrical factors:

$$G = \frac{\sigma_0 \mathcal{S}}{\mathcal{L}} = \sigma_0 L^{d-2}, \tag{21.7}$$

where $\mathcal{S}$ and $\mathcal{L}$ are the cross section and length of the conductor to which we assign the typical size L. By using the explicit expression of the density of states, one may rewrite G as

$$G = \frac{2e^2}{h}\frac{\Omega_d}{(2\pi)^{d-1}d}\frac{2E_{\mathrm{F}}\tau}{\hbar}\left(\frac{p_{\mathrm{F}}L}{\hbar}\right)^{d-2} = \frac{2e^2}{h}\frac{\Omega_d}{(2\pi)^{d-1}d}\frac{p_{\mathrm{F}}l}{\hbar}\left(\frac{p_{\mathrm{F}}L}{\hbar}\right)^{d-2}, \tag{21.8}$$

which shows that, in the natural conductance units ($G_0 = 2e^2/h = (12.9 \times 10^3\Omega)^{-1}$), the value of $g \equiv G/G_0$ is controlled by the dimensionless parameters $p_{\mathrm{F}}l/\hbar = 2\pi l/\lambda_{\mathrm{F}}$ and $2\pi L/\lambda_{\mathrm{F}}$. Its inverse, the resistance r, is proportional to the amount of disorder present in the system and is small for good metals. In two dimensions, in particular, conductivity and conductance have the same physical dimensions and the ratio between the Fermi wave length and the mean free path l is the only parameter that controls the value of the conductivity. In the semiclassical limit, $\lambda_{\mathrm{F}} \ll l$, the Drude formula (11.33) predicts a high conductivity. The rate of collisions τ^{-1} is proportional to the impurity concentration (see (11.45)). By increasing the disorder in the semiclassical approach, one has that σ_0 diminishes, but remains finite. Ioffe and Regel (1960) stated the criterion that in the metallic phase the mean free path l cannot be smaller than the average interelectron distance proportional to $\hbar/p_F$, i.e., $p_F l/\hbar \geq 1$. Mott (1972) applied this criterion to the Drude conductivity, arguing that for $d \geq 2$ there is a minimum metallic conductivity, $\sigma_{0\min}$ when $l \approx \hbar/p_F$,

$$\sigma_{0\min} = \frac{2e^2}{h} \frac{\Omega_d}{(2\pi)^{d-1} d} \left(\frac{p_F}{\hbar}\right)^{d-2}, \tag{21.9}$$

which is universal (e^2/h) in $d = 2$. As a result, the metal–insulator transition, if any, should be a discontinuous one in the conductivity.

The vanishing of the conductivity (see Eq. (21.3)) implies that the transition is of second order, in contrast to the minimum metallic conductivity criterion. However, when $l \approx \hbar/p_F$ we are deeply in the quantum limit and, although the Ioffe–Regel criterion cannot be disputed, it cannot naively be applied to the semiclassical Drude formula. Indeed, one expects that corrections beyond the semiclassical approximations will strongly modify Eq. (11.33), opening the way to a new perspective on the metal–insulator transition.

On general grounds, for a disordered Fermi gas the only two characteristic energy scales are $\hbar\tau^{-1}$ related to the disorder and the Fermi energy E_F related to the kinetic energy. For weak disorder $E_F\tau/\hbar \propto g \gg 1$, hence, starting from the metallic phase, corrections to the Drude formula (11.33) will appear as powers in the small dimensionless parameter $1/g$, as we shall see in the next sections.

21.2 The Anderson transition and quantum interference

The discussion of the previous section suggests that disorder cannot be treated within a semiclassical approach. In 1958, Anderson started the field of *localization*, proposing that under certain circumstances diffusion may be completely suppressed (Anderson (1958)). He proposed a lattice model where the site energies are randomly distributed. When disorder is absent, a small hopping amplitude is enough to delocalize the electron states and form Bloch waves. However, at sufficiently strong disorder, the hopping processes may only spread an initially localized state over a finite distance, which defines the localization length ξ_0.

In fact, due to the random distribution of the site levels, when the disorder is strong, it is very unlikely for an electron to have a sufficiently near site with, at the same time, a compatible energy level to allow it to jump. In doped semiconductors, localized states are more likely to form in the tails of the bands originated by the dopant atoms. Mott (1967) argued that there must exist a critical energy E_c, called the *mobility edge*, which separates localized from extended states. When the Fermi energy E_F is below the mobility edge, the system is an insulator. When E_F passes through E_c, the system becomes metallic.

These concepts played an important role in shaping our modern view of the metal–insulator transition. A great impulse to the development of the field came, however, by the discovery of the phenomenon of *weak localization*.

As remarked at the end of the previous section, the standard theory of transport is based on a semiclassical approach, where in evaluating the probability for electron diffusion one

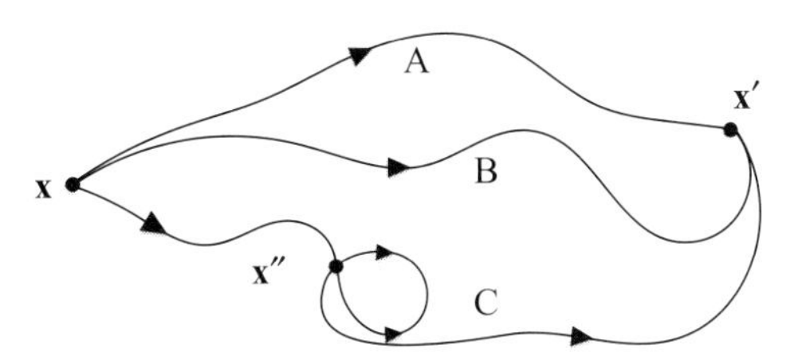

Fig. 21.1 Various paths for a particle to go from point $\mathbf{x}$ to point $\mathbf{x}'$. Paths A and B are non-self-intersecting paths. Path C has a self-intersection point at point $\mathbf{x}''$. Since the particle may go around the closed loop clockwise or counter-clockwise, path C, actually, identifies two distinct trajectories, which coincide outside the loop section of the path.

neglects the interference between the amplitudes corresponding to different trajectories and essentially reduces to a classical random walk with step l and diffusion constant D. This is indeed justified in many cases, where the semiclassical theory works. In fact, in a disorderd system, the phase difference for any two trajectories varies randomly. The situation changes, however, for *self-intersecting* trajectories, which come from closed loops. In this case, trajectories naturally come in pairs, depending on whether one goes around the loop clockwise or counter-clockwise, as shown in Fig. 21.1. The interference between these pairs of trajectories modifies the semiclassical result, giving rise to the so-called weak localization correction.

There is a simple argument to estimate the probability of having a self-intersecting trajectory. The electron motion is described by a classical diffusion process such that the average distance after a time t is, from Eq. (11.14), $\overline{\mathbf{r}^2} = \overline{x^2} + \overline{y^2} + \overline{z^2} = 2dDt$. The quantum nature of the electron, to a first approximation, may be introduced by considering a *tube* of trajectories of transverse size of the order of the Fermi wave length λ_{F}. In a time $\mathrm{d}t$, the volume spanned by the tube increases by

$$\mathrm{d}V_{\text{tube}} = \lambda_{\mathrm{F}}^{d-1} v_{\mathrm{F}} \mathrm{d}t. \tag{21.10}$$

Let us consider the ratio between the increase of the tube volume in time $\mathrm{d}t$ and the total volume $(Dt)^{d/2}$ generated by the diffusion process in the time t. The total probability for self-intersection may be estimated by integrating this ratio over time:

$$P \sim \int_{\tau}^{\tau_\phi} \frac{\lambda_{\mathrm{F}}^{d-1} v_{\mathrm{F}} \mathrm{d}t}{(Dt)^{d/2}}, \tag{21.11}$$

where the lower limit τ is the time above which the diffusive regime, after a few collisions, starts to set in. The upper limit, τ_ϕ, is the persistance time of the wave function phase coherence. The probability reads

$$P \sim \frac{v_{\mathrm{F}} \lambda_{\mathrm{F}}^{d-1}}{D^{d/2} \tau^{(d-2)/2}} \frac{2}{d-2} \left[\left(\frac{\tau}{\tau_\phi} \right)^{\frac{d-2}{2}} - 1 \right] \propto \frac{G_0}{\sigma_0 l^{d-2}} \frac{2}{d-2} \left[\left(\frac{\tau}{\tau_\phi} \right)^{\frac{d-2}{2}} - 1 \right]. \tag{21.12}$$

In general, *inelastic* processes at finite temperature make τ_ϕ a decreasing function of the temperature. At zero temperature, the upper limit of the time integration is provided by $\tau_L \equiv L^2/D$, the time necessary to diffuse over the entire system of size L. In this way, the probability of self-intersection acquires a scale dependence.

By assuming that the conductivity corrections are proportional to this probability, at $T = 0$ one obtains

$$\frac{\delta\sigma}{\sigma_0} \propto \frac{1}{g_0}\frac{1}{\epsilon}\left[\left(\frac{l}{L}\right)^{\epsilon} - 1\right] \tag{21.13}$$

where $\epsilon = d - 2$ and $g_0 \equiv G(l)/G_0$ is the conductance at the scale l in units of G_0. Equation (21.13) is valid at $T = 0$ and the inverse scattering time or the inverse mean free path plays the role of the ultraviolet cutoff, whereas τ_L^{-1} is the infrared cutoff. At $d = 2$, the conductivity correction is log-singular.

At finite temperature, when $\tau_\phi < \hbar/k_B T$, the infrared cutoff becomes temperature dependent and in $d = 2$ the correction becomes logarithmic in temperature, as it was observed in the disordered metallic films (PdAu). In $d = 3$ the correction is representative of the anomalous behavior proportional to $T^{1/2}$.

This opens the way to the scaling theory discussed in the next section. Indeed, since $\tau/\tau_L = l^2/L^2$, $\sigma(0)$ as function of L can be written as

$$\sigma_L(0) = \sigma_0\left\{1 + \frac{a_d}{g_0}\frac{1}{\epsilon}\left[\left(\frac{l}{L}\right)^{\epsilon} - 1\right]\right\} = \sigma_0 Z_\sigma\left(\frac{l}{L}, g_0\right), \tag{21.14}$$

with a_d a numerical factor depending on the space dimensionality and Z_σ is the renormalization of the conductivity depending on the length scale evaluated at first order in $1/g_0$.

As a final remark to this section, we point out that the weak localization phenomenon is sensitive to any perturbation that breaks the time reversal invariance. This can be understood from the above argument of the interference between time reversed trajectories, which are no longer equivalent or in phase. For instance, in the presence of a magnetic field, the two trajectories acquire a phase difference $\phi_1 - \phi_2 = (2e/\hbar c)\Phi_B$, proportional to the magnetic flux Φ_B, threading the surface delimited by the closed loop. At finite temperature the logarithmic singularity appears at time τ_ϕ. In order to cut off this singularity the magnetic field must be, at least, of the order of a flux quantum over a region whose size is of the order of the dephasing length $L_\phi = \sqrt{D\tau_\phi}$. This gives the condition $B \geq (h/ec)/L_\phi^2$. The weak localization is therefore depressed by the presence of the field because of the dephasing mechanism. The conductivity increases with the field, a phenomenon called anomalous magnetoresistance. It is also clear that other dephasing mechanisms, as for instance spin-flip scattering due to magnetic impurities with typical time τ_{spin}, become important when $\tau_{\text{spin}} < \tau_\phi$. In general, the presence of a time-reversal breaking suppresses the first interference correction in $1/g$. As we shall see, quantum interference appears again at the order $1/g^2$, as a result of the interference between two diffusion processes rather than between two time-reversed paths of single electrons.

21.3 The scaling theory of the metal–insulator transition

The starting point is the argument of Thouless concerning the evolution of the wave function as the system size is increased (Edwards and Thouless (1972); Licciardello and Thouless

(1975)). To fix the ideas, let us imagine that the system of size $2L$ is made up by combining blocks of size L. We pose the question whether electrons in one block can go through it and eventually pass into the next block or not. Electrons would correspondingly be in extended states and mobile or localized. Suppose we know the eigenstates for a block of size L with level spacing ΔE. Let δE be the energy brought about by the perturbation of joining the blocks together. The energy δE measures the sensitivity to a change in the boundary conditions. In a diffusive system this may be related to the uncertainty in the energy the electrons acquire during the time τ_L necessary to reach the boundary, $\delta E = \hbar D/L^2$. δE is also known as the Thouless energy. The level spacing, on the other hand is related to the inverse density of states, $\Delta E = 1/(N_0 L^d)$. By using the first of the Einstein relations (21.4), the ratio of the two energies gives

$$\frac{\delta E}{\Delta E} = N_0 L^d \frac{\hbar D}{L^2} = \frac{1}{2\pi} \frac{\sigma_0 L^{d-2}}{2e^2/h} = \frac{g(L)}{2\pi}, \tag{21.15}$$

where $g(L)$ is the conductance at the scale L in units of $2e^2/h$. If $g \ll 1$, the new eigenstates are not modified much by the assembling of the blocks. The blocks are almost independent and for the electrons reaching the boundary it will be difficult to find a level of the next block compatible with their energy. On the other hand, when $g \gg 1$, the new states are delocalized in all the blocks.

The scaling theory (Abrahams et al. (1979)) assumes that $g(L)$ is the only parameter which controls the evolution of the eigenstates when we rescale the system size. Mathematically, this is expressed by requiring that the conductance of a block of size $L' = bL$ is expressed in terms of the conductance of a block of size L by a function of L'/L and $g(L)$ only, i.e., $g(L') = f(L'/L, g(L))$. Its logarithmic derivative for $L' = L$, which defines the *beta* function of the corresponding renormalization group equation, depends on the scale L only through $g(L)$ itself:

$$\frac{\mathrm{d} \ln g(L)}{\mathrm{d} \ln L} = \beta(g(L)). \tag{21.16}$$

The vanishing of the beta function controls the scale-invariant limit, i.e., provides the fixed point of the trasformation g^*. In the case of the one-parameter equation the fixed point g^* coincides with the critical value g_c. Linearization of the transformation, starting from the fixed point, provides the scaling behavior of the physical quantities. The β-function is relatively well known on physical grounds in the two limits of a good metal, where Ohm's law is valid and σ is a constant, and in the strongly localized insulating regime, where the scale-dependent conductance falls off exponentially over a localization length ξ_0 as

$$g(L) = g_0 \mathrm{e}^{-L/\xi_0}. \tag{21.17}$$

One then gets immediately

$$\beta(g) = d - 2, \quad g \gg 1, \tag{21.18}$$

$$\beta(g) = \ln \frac{g}{g_0}, \quad g \ll 1, \tag{21.19}$$

where g_0 is the conductance at some initial microscopic scale l. Under reasonable assumptions, by a continuous interpolation between the two above limits, the β-function has the qualitative behavior shown in Fig. 21.2. A positive (negative) value for β means that upon

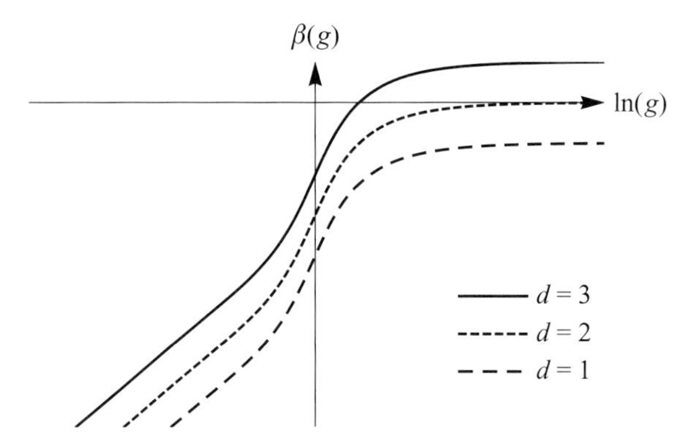

Fig. 21.2 Qualitative plot of the β-function in terms of $\ln g$ for different space dimensionality d. For $d > 2$, at the critical point, the slope of the function determines the critical index $x_g = 1/\nu$. In the extreme localized regime the slope is 1. Only when $\nu = 1$ do the two slopes coincide. Otherwise, for $\nu < 1$, the slope at the critical point is larger than that in the localized regime. For $d \leq 2$ there is no zero for the β-function.

increasing the system size, g increases (decreases) corresponding to a metallic (insulating) behavior.

A value g_c of g, zero of the β-function, signals an unstable fixed point for the flow of g. This represents a metal–insulator transition. One consequence of Eqs. (21.18) and (21.19) is that, for $d \leq 2$, the system is always an insulator at zero temperature and all states are localized. To obtain the behavior near the critical point $g = g_c$ and the associated critical index, we must linearize the β-function to get

$$\frac{\mathrm{d}(g - g_c)}{\mathrm{d}\ln L} = g_c \beta'(g_c)(g - g_c), \quad \beta'(g_c) = \left(\frac{\mathrm{d}\beta(g)}{\mathrm{d}g}\right)_{g=g_c}, \tag{21.20}$$

from which

$$g(L) - g_c = (g_0 - g_c)\left(\frac{L}{l}\right)^{x_g}, \tag{21.21}$$

where g_0 is the conductance at scale l and

$$x_g = g_c \beta'(g_c). \tag{21.22}$$

Here $g_0 - g_c$ is the variable controlling the deviation from the critical point and is the analog of $T - T_c$ in ordinary critical phenomena at finite temperature. By defining a correlation length, which coincides with the localization length in the insulating side, it diverges at criticality as

$$\xi \sim (g_0 - g_c)^{-\nu}. \tag{21.23}$$

By assuming that ξ is the only relevant length, it scales as $\xi' = \xi/(L/l)$ and one deduces $\nu = 1/x_g$. Furthermore, taking into account the relation between conductance and conductivity (21.7) on the scale $L \sim \xi$, the critical value for the conductivity is given by

$\sigma_c \sim \lim_{\xi\to\infty} g_c/\xi^{d-2} = 0$ and the approach to the critical point is given by

$$\delta\sigma \sim (g_0 - g_c)^{\mu} \sim \frac{\delta g(\xi)}{\xi^{d-2}} \sim \frac{(g_0 - g_c)}{\xi^{d-2}} \left(\frac{\xi}{l}\right)^{x_g} \sim (g_0 - g_c)^{1-\nu x_g+\nu(d-2)}. \quad (21.24)$$

The scaling relation between ν and μ derived by Wegner is (Wegner (1976))

$$\mu = (d-2)\nu \equiv \epsilon\nu. \quad (21.25)$$

In the metallic regime, where g is large, we can assume that the β-function can be expanded in a power series in $1/g$:

$$\beta(g) = d - 2 - \frac{a}{g} - \frac{b}{g^2} + \cdots. \quad (21.26)$$

Above two dimensions, if $a > 0$, one has a fixed point $g_c = a/\epsilon$ and $\nu = 1/\epsilon$ leading to $\mu = 1$. On the other hand, if $a = 0$, the fixed point is determined by the second order term, $g_c^2 = b/\epsilon$, which implies $\nu = 1/(2\epsilon)$ and $\mu = 1/2$. Finally, when $a < 0$, there is no fixed point at this order.

At $d = 2$ the scaling equation reduces to

$$\frac{\mathrm{d}g}{\mathrm{d}\ln L} = -\frac{a}{g}. \quad (21.27)$$

For $a > 0$ the system scales to an insulator, whereas for $a < 0$ it scales to a perfect conductor. This phenomenological theory does not allow for a metallic phase in $d = 2$.

The microscopic theory of a gas of fermions in the presence of disorder, as will be shown in the next section, reproduces the above picture by an explicit perturbative evaluation of the β function in the metallic regime.

21.4 The quantum theory of the disordered Fermi gas

In this section we develop the quantum theory of the disordered Fermi gas. We will see that the quasi-classical treatment of Chapter 11 emerges as a leading approximation in the expansion in the small parameter λ_F/l.

Non-interacting electrons in the presence of disorder are described by the following Hamiltonian:

$$H = \int \mathrm{d}\mathbf{x} \hat{\psi}_\sigma^\dagger(\mathbf{x}) \left[-\frac{\hbar^2\nabla^2}{2m} + U(\mathbf{x}) \right] \hat{\psi}_\sigma(\mathbf{x}), \quad (21.28)$$

where $U(\mathbf{x})$ is taken as a *Gaussian* random variable with zero mean and variance given by

$$\overline{U(\mathbf{x})} = 0, \quad \overline{U(\mathbf{x})U(\mathbf{x}')} = \overline{u^2}\delta(\mathbf{x} - \mathbf{x}'), \quad (21.29)$$

where $\overline{u^2}$ represents the strength of the disorder proportional to the concentration of impurities. Having in mind a perturbation expansion in the potential $U(\mathbf{x})$, we may use the Dyson equation Eq. (18.87), which can be solved by iteration, each step corresponding to a term of a given order in the potential $U(\mathbf{x})$. In terms of Feynman diagrams, we may depict

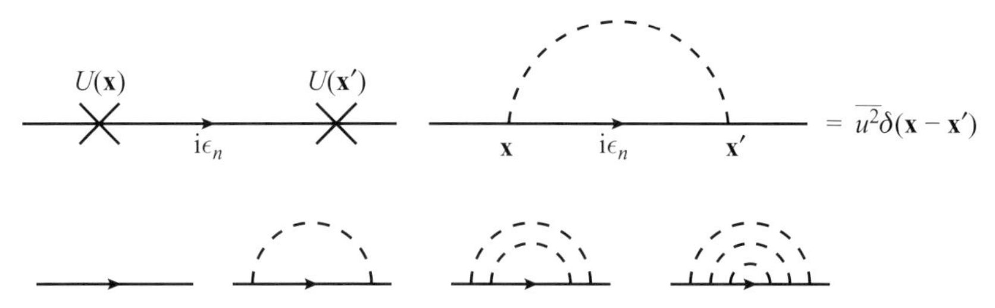

Fig. 21.3 Self-energy in the Born approximation before (top left) and after (top right) averaging over the impurity distribution. The dashed line represents the average of two impurity insertions. The Matsubara frequency does not change along the line because the scattering is elastic. At the bottom is shown the sequence of *rainbow* diagrams selected by the self-consistent solution for the Green function.

with a single line the Green function $\mathcal{G}_0$ of the *clean* Fermi gas when the potential is zero or a constant. A cross insertion on the line indicates a single iteration step in the solution of Eq. (18.87). Given the model potential of Eq. (21.29), only terms with an *even* number of cross insertions are different from zero. Hence, in the lowest approximation, the so-called Born approximation, we have the term shown in the top left of Fig. 21.3. The averaging over the impurity distribution restores the translational invariance and allows us to Fourier transform in momentum space. The lowest order impurity-average represented by the top right diagram of Fig. 21.3 plays the role of an effective self-energy and it is the starting point of the so-called impurity technique.

21.4.1 The self-consistent Born approximation

When the lowest Born approximation to the self-energy is made self-consistent, the first conserving approximation for the current–current response function is obtained. This corresponds to the semiclassical analysis of Drude.

The self-energy term in real space of the diagram shown in Fig. 21.3 has the expression

$$\Sigma^{(1)}(\mathbf{x}, t; \mathbf{x}', t') = \overline{u^2}\delta(\mathbf{x} - \mathbf{x}')\mathcal{G}_0(\mathbf{x}, t; \mathbf{x}', t'). \tag{21.30}$$

In momentum and energy Fourier space, Eq. (21.30) reads

$$\Sigma^{(1)}(\mathbf{p}, \epsilon_n) = \overline{u^2} \int_{\mathbf{p}'} \frac{1}{\mathrm{i}\epsilon_n - \mathbf{p}'^2/2m + \mu}, \quad \int_{\mathbf{p}'} \equiv \int \frac{\mathrm{d}\mathbf{p}'}{(2\pi\hbar)^d}. \tag{21.31}$$

The chosen disorder model does not allow a dependence on the momentum transfer upon scattering. For large values of $\mathbf{p}'$, the real part of the integral over $\mathbf{p}'$ diverges, but its value *does* not depend on energy ϵ_n and momentum $\mathbf{p}$. This divergence is a consequence of the simple model taken for the scattering potential. A more realistic model may cure the divergence by introducing a cutoff for the scattering processes with large momentum transfer and give rise to a finite contribution that can be absorbed into a redefinition of the chemical potential. In the following we ignore the real part of the self-energy with the understanding that it has been absorbed in a shift of the chemical potential. The main contribution to the energy dependence of the imaginary part of the integral comes from the values of $\mathbf{p}'$ close to the Fermi surface. To perform the integral, we pass from momentum to the energy variable $\xi_{\mathbf{p}} = \mathbf{p}^2/2m - \mu$:

$$\int_{\mathbf{p}} \cdots = N_0 \int_{-\mu}^{\infty} d\xi_{\mathbf{p}} \cdots \approx N_0 \int_{-\infty}^{\infty} d\xi_{\mathbf{p}} \cdots, \tag{21.32}$$

where the lower limit of integration is sent to minus infinity, since $\mu \approx E_{\mathrm{F}}$ is the biggest energy scale in the problem. By performing the integral, as shown in the answer to Problem 21.1, we obtain

$$\Sigma^{(1)}(\mathbf{p}, \epsilon_n) = -i\pi N_0 \overline{u^2} \mathrm{sign}(\epsilon_n) \equiv -i\frac{\hbar}{2\tau}\mathrm{sign}(\epsilon_n), \tag{21.33}$$

where we have introduced the elastic quasiparticle relaxation time $\tau^{-1} = 2\pi N_0 \overline{u^2}/\hbar$. This expression reproduces that of Eq. (11.45) if we identify the variance $\overline{u^2}$ with $n_i v_0^2$. Hence, the variance is chosen in such a way as to sum the scattering from all the impurities *incoherently*.

To proceed in the perturbative expansion, one replaces the above result into the Green function and computes the next iteration for the self-energy $\Sigma^{(2)}$. At the second iteration, the Green function has the same expression as before apart from the replacement $|\epsilon_n| \to |\epsilon_n| + i\hbar/2\tau$ provided by $\Sigma^{(1)}$. As shown in Problem 21.1, this replacement does not affect the value of the integral and the right-hand side of Eq. (21.33) yields a self-consistent solution of the Dyson equation for the Green function:

$$\mathcal{G}(\mathbf{p}, \epsilon_n) = \frac{1}{i\epsilon_n - \xi_{\mathbf{p}} + i(\hbar/2\tau)\mathrm{sign}(\epsilon_n)}. \tag{21.34}$$

The disorder-dressed retarded and advanced Green functions G^{R} and G^{A} are then

$$G^{\mathrm{R(A)}}(\mathbf{p}, \epsilon) = \frac{\hbar}{\hbar\epsilon - \xi_{\mathbf{p}} \pm i\hbar/2\tau}. \tag{21.35}$$

The single-particle density of states at the Fermi surface from Eq. (18.43) reads

$$\nu(0) = -\frac{2}{\pi}\mathrm{Im}\int_{\mathbf{p}} G^{\mathrm{R}}(\mathbf{p}, 0) = -\frac{1}{\pi}\int_{\mathbf{p}}(G^{\mathrm{R}} - G^{\mathrm{A}}) = 2N_0. \tag{21.36}$$

The single-particle density of states coincides with that of the free clean gas.

We stress that the consistency of the above approximation for evaluating integrals over momentum is based on the fact that the scale of variation of the integrand $(|\epsilon_n| + \hbar/2\tau)^{-1}$, set up by the position of the pole, is much smaller than the lower limit of integration, $-\mu$. This confirms the expression of the small expansion parameter discussed previously:

$$\frac{\hbar}{\tau} \ll \mu \approx E_{\mathrm{F}} = \frac{v_{\mathrm{F}} p_{\mathrm{F}}}{2}, \quad \Rightarrow p_{\mathrm{F}} l \gg \hbar, \tag{21.37}$$

where $l = v_{\mathrm{F}}\tau$ is the mean free path.

The *self-consistent* Born approximation effectively selects the subset of the so-called *rainbow* diagrams shown in the bottom of Fig. 21.3 which are characterized by the absence of *crossing* of impurity averaged lines. This sequence of independent scattering events leads, as we will see, to a ladder resummation of diagrams for the effective four-point vertex part $\Gamma^{(4)}$ given by the repetition of disorder lines.

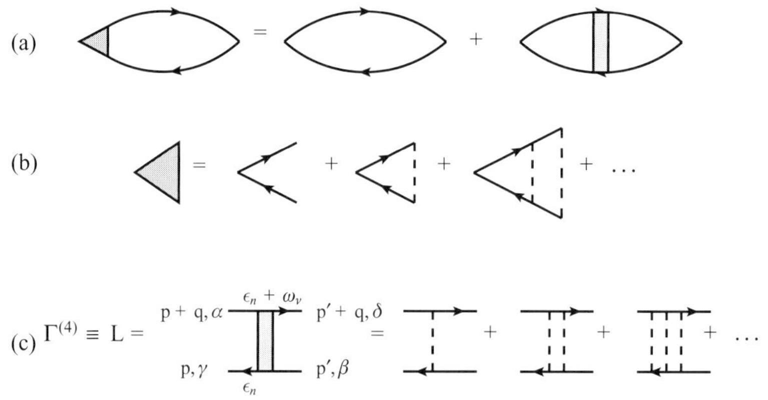

(a)

(b)

(c) $\Gamma^{(4)} \equiv L =$

Fig. 21.4 (a) Diagram for the correlation function. The gray triangle indicates the vertex part. (b) The vertex part is obtained as an infinite resummation of non-crossing impurity lines. (c) Ladder resummation shown as a gray rectangular box.

The replacement of the Green function dressed with the self-energy in the place of the bare one shares the same idea at the basis of the mean-field theory that we have discussed e.g. in the case of ferromagnets. The self-energy is like the effective magnetic field in spin systems or the effective potential in real gases. Through the self-energy we evaluate the Green function. In the present case a form for the self-energy is inserted into the Green function by requiring self-consistency. The guessed form corresponds to a Fermi gas with disordered-broadened levels, the broadening being inversely proportional to the elastic quasiparticle relaxation time induced by the disorder. This corresponds to having $z_{\mathbf{k}} = 1$ and $\gamma_{\mathbf{k}} = \hbar/2\tau$ in Eq. (18.52). As in all mean-field treatments, we obtain a solution which is not perturbatively related to the initial system. The clean Fermi gas is made of single-particle states which are momentum eigenstates. The disordered Fermi gas, in the semiclassical approximation, is made of single-particle states which are momentum eigenstates over a finite time τ only.

The technique of impurity averaging that we have applied to the diagrams for the Green function can be extended also to those for the evaluation of both the density–density and current–current response functions.

The calculation of the density–density response function at this approximation well reproduces the pole of the diffusive motion discussed in the semiclassical approximation in Chapter 11. At the same time it provides both the thermodynamic density of states $dn/d\mu$ (Eq. (10.22)) and the conductivity σ (Eq. (10.43)).

We begin with the thermal density–density response function (18.108) in the Born approximation. Its expression in terms of the bare $j_0^0 = 1$ and fully dressed truncated vertex $\mathcal{T}^0$ has been given in Eqs. (18.115) and (18.119) and reads

$$\tilde{\mathcal{R}}^{00}(\mathbf{q}, \omega_\nu) = 2e^2 \frac{1}{\beta} \sum_{\epsilon_n} \int_{\mathbf{p}} j_0^0 \mathcal{G}(p + q/2) \mathcal{T}^0(p, q) \mathcal{G}(p - q/2) . \qquad (21.38)$$

The dressed truncated vertex $\mathcal{T}^0$ is indicated by a gray triangle in Fig. 21.4. The Green functions appearing in the diagrams are those evaluated within the self-consistent Born approximation (21.34). The calculation of the vertex $\mathcal{T}^0$ requires the knowledge of the effective vertex $\Gamma^{(4)}$ as the sequence of ladder diagrams shown in Fig. 21.4(c). The sum of this sequence of ladder diagrams, to be called the ladder from now on, is obtained by solving the integral equation

$$L_{\mathbf{p},\mathbf{p}',\epsilon_n}(\mathbf{q},\omega_\nu) = L^{(0)} + L^{(0)} \int_{\mathbf{p}''} \mathcal{G}(\mathbf{p}''_+,\epsilon_+)\mathcal{G}(\mathbf{p}''_-,\epsilon_-)L_{\mathbf{p}'',\mathbf{p}',\epsilon}(\mathbf{q},\omega_\nu), \tag{21.39}$$

where $L^{(0)} = \overline{u^2} = \hbar/2\pi N_0\tau$ and $\mathbf{p}_\pm = \mathbf{p} \pm \mathbf{q}/2$, $\epsilon_\pm = \epsilon_n \pm \omega_\nu/2$. The ladder generally depends on three momenta and two energies. However, its actual dependence is only on $\mathbf{q}$ and ω_ν due to the simple model of the disorder potential, which does not transmit energy and momentum. This is the reason for the notation adopted. Later on we will drop the subscripts $\mathbf{p}$, $\mathbf{p}'$ and ϵ_n. The kernel of the integral equation (21.39) gives a non-vanishing contribution when the poles of the two Green functions lie on opposite sides of the real axis. To see this, let us consider the term with two impurity lines in the ladder sequence. Apart from $\overline{u^2}$ factors, it reads

$$\Pi(\mathbf{q},\epsilon_+,\epsilon_-) = \int_{\mathbf{p}} \mathcal{G}(\mathbf{p}_+,\epsilon_+)\,\mathcal{G}(\mathbf{p}_-,\epsilon_-). \tag{21.40}$$

To evaluate the integral we go to the energy variable introduced in Eq. (21.32). We have $\xi_{\mathbf{p}\pm\mathbf{q}/2} = \xi_{\mathbf{p}} \pm \mathbf{p}\cdot\mathbf{q}/2m + q^2/8m$. Since the integral is dominated by the contribution coming from the poles we set to p_{F} the absolute value of $\mathbf{p}$ in the scalar product with $\mathbf{q}$. The integral over the energy $\xi_{\mathbf{p}}$ may then be carried out by the residue method, as explained in the answer to Problem 21.2. Also we notice that the condition of having poles on opposite sides of the real axis implies a restriction on the frequencies, i.e., $\epsilon_+ > 0$ and $\epsilon_- < 0$. The integration over $\xi_{\mathbf{p}}$ (see Problem 21.2) gives

$$\Pi(\mathbf{q},\epsilon_+,\epsilon_-) = \theta(-\epsilon_+\epsilon_-)\frac{2\pi N_0\tau}{\hbar}\int \frac{\mathrm{d}\Omega_{\hat{\mathbf{p}}}}{\Omega_{\hat{\mathbf{p}}}}\frac{1}{1+|\omega_\nu|\tau/\hbar + \mathrm{i}lq\cos(\theta)/\hbar}, \tag{21.41}$$

where $\Omega_{\hat{\mathbf{p}}}$ is the solid angle, θ is the angle between $\mathbf{p}$ and $\mathbf{q}$ and with $\theta(x)$ we indicate the Heaviside function. Although the angular integral may be evaluated exactly, in the following we will be interested in the limit of small frequency and large wavelength $|\omega_\nu|\tau \ll \hbar$ and $lq \ll \hbar$, which defines the *diffusive*[2] transport regime. In this limit it is then convenient to expand the right-hand side of Eq. (21.41) and perform the angular integral. We get finally

$$\Pi(\mathbf{q},\epsilon_+,\epsilon_-) = \theta(-\epsilon_+\epsilon_-)\frac{2\pi N_0\tau}{\hbar}(1 - |\omega_\nu|\tau/\hbar - Dq^2\tau/\hbar^2), \tag{21.42}$$

where the diffusion coefficient is $D = v_{\mathrm{F}}l/d$ in agreement with (11.34), d being the dimensionality. As anticipated, the ladder does not depend on the momenta $\mathbf{p}$ and $\mathbf{p}'$ and the

[2] This expansion is sufficient in the low temperature regime when $k_{\mathrm{B}}T\tau \ll \hbar$. At higher temperature with $k_{\mathrm{B}}T\tau \gtrsim \hbar$ one must retain the full frequency and momentum dependence of $\Pi(\mathbf{q},\epsilon_+,\epsilon_-)$. This defines the quasi-ballistic regime.

integral equation becomes an algebraic one. The final solution reads

$$L(\mathbf{q},\omega_\nu) = \frac{L^{(0)}}{1 - L^{(0)}\Pi(\mathbf{q},\omega_\nu)} = \frac{\hbar^2}{2\pi N_0 \tau^2} \frac{\theta(-\epsilon_+\epsilon_-)}{|\omega_\nu| + Dq^2/\hbar}. \tag{21.43}$$

The above equation shows, after performing the analytical continuation $\omega_\nu \to -\mathrm{i}\hbar\omega$, how the diffusive pole emerges from the repeated elastic scattering by impurities. Classically it corresponds to a Brownian motion, as discussed in the semiclassical approximation of Chapter 11.

In terms of the ladder, the vertex $\mathcal{T}^0$ reads (see Fig. 21.4(b))

$$\begin{aligned}\mathcal{T}^0(\mathbf{p},\epsilon,\mathbf{q},\omega) &\equiv j_0^0 + \delta\mathcal{T}^0 \\ &= \left(1 + \int_{\mathbf{p}'} \mathcal{G}(\mathbf{p}'_+,\epsilon_+)\mathcal{G}(\mathbf{p}'_-,\epsilon_-) L_{\mathbf{p}',\mathbf{p},\epsilon_n}(\mathbf{q},\omega_\nu)\right) \\ &= \left(1 + \frac{2\pi N_0\tau}{\hbar} L(q,\omega_\nu)\right) = \left(1 + \frac{\theta(-\epsilon_+\epsilon_-)}{|\omega_\nu|\tau/\hbar + D\tau q^2/\hbar^2}\right).\end{aligned} \tag{21.44}$$

Because of the ladder, the small $\mathbf{q}$ and ω_ν region is dominating the integral over $\mathbf{p}'$ and, in getting the third step, we have set to zero both $\mathbf{q}$ and ω_ν in the Green functions. This integral is then done by using the residue theorem in the answer to Problem 21.3. One notices that in the zero-momentum limit the expression for $\mathcal{T}^0$ may be obtained directly from the Ward identity (18.120) and this implies that the present approximation is charge conserving.

Once the vertex has been calculated, the evaluation of the density–density response function can be completed. It is natural to split the latter into two parts corresponding to the two contributions in the vertex. The contribution, which does not contain the diffusive pole and does not have restrictions on the frequency range, reads

$$\begin{aligned}\tilde{\mathcal{R}}^{00}_{++} + \tilde{\mathcal{R}}^{00}_{--} &= 2e^2 \frac{1}{\beta} \sum_{\epsilon_n} \int_{\mathbf{p}} j_0^0 \mathcal{G}(\mathbf{p}_+,\epsilon_+)\, j_0^0 \mathcal{G}(\mathbf{p}_-,\epsilon_-) \\ &\approx 2e^2 \frac{1}{\beta} \sum_{\epsilon_n} \int_{\mathbf{p}} \mathcal{G}(\mathbf{p},\epsilon)^2 .\end{aligned}$$

By analytic continuation we transform the sum over Matsubara frequencies into an integral in the complex plane, where we have a cut along the real axis. By reasoning in the same way as in the answer to Problem 18.1, in terms of the retarded and advanced Green functions (21.35) we get

$$\tilde{\mathcal{R}}^{00}_{++} + \tilde{\mathcal{R}}^{00}_{--} = -2e^2 \int_{\mathbf{p}} \int_{-\infty}^{\infty} \frac{\mathrm{d}\epsilon}{2\pi\mathrm{i}} f(\hbar\epsilon)[(G^{\mathrm{R}}(\mathbf{p},\epsilon))^2 - (G^{\mathrm{A}}(\mathbf{p},\epsilon))^2].$$

In the zero-temperature limit ($f(x) \to \theta(-x)$) the integral over the energy is easily performed by considering that $(G^{\mathrm{R(A)}})^2 = -\partial_\epsilon G^{\mathrm{R(A)}}$ and one obtains

$$\begin{aligned}\tilde{\mathcal{R}}^{00}_{++} + \tilde{\mathcal{R}}^{00}_{--} &=_{T\to 0} 2e^2 \frac{1}{2\pi\mathrm{i}} \int_{\mathbf{p}} [G^{\mathrm{R}}(\mathbf{p},0) - G^{\mathrm{A}}(\mathbf{p},0)] \\ &= -2e^2 N_0.\end{aligned} \tag{21.45}$$

We remark that integrals over momentum of combinations of retarded and advanced Green functions can be handled exactly as the integrals over combinations of thermal Green functions $\mathcal{G}$ with energies with opposite sign in the way explained in the answer to Problem 21.3.

In obtaining (21.45), we have taken the *static* limit (frequency goes to zero first, and then momentum). Given the restriction in the frequencies for the second part in the vertex, one could naively think that in the small frequency and momentum limit this contribution would be vanishingly small. This, however, depends on the order the two limits are performed. Due to the presence of the diffusive pole in the ladder, one obtains for the second *dynamic* contribution

$$\begin{aligned}\tilde{\mathcal{R}}^{00}_{+-} &= 2e^2\frac{1}{\beta}\sum_{\epsilon_n}\int_{\mathbf{p}}\theta(-\epsilon_-\epsilon_+)\mathcal{G}(\mathbf{p}_+,\epsilon_+)\mathcal{G}(\mathbf{p}_-,\epsilon_-)\frac{1}{|\omega_\nu|\tau/\hbar + D\tau q^2/\hbar^2}\\ &\approx 2e^2 N_0\frac{|\omega_\nu|}{|\omega_\nu| + Dq^2/\hbar},\end{aligned} \tag{21.46}$$

where, due to the diffusive pole, one may set $\mathbf{q} = \mathbf{0}$ and $\omega_\nu = 0$ in the Green functions and perform the integral with the residue method. The factor $|\omega_\nu| = |2\pi\nu/\beta|$ in the numerator is reconstructed by performing the sum over ϵ_n given by the ν equal terms fixed by the $\theta(-\epsilon_-\epsilon_+)$ factor. By combining together Eqs. (21.45) and (21.46), one obtains the total density–density response function at leading order in $E_{\rm F}\tau/\hbar$ in complete agreement with the result of Eq. (11.39) based on a phenomenological approach:

$$\tilde{\mathcal{R}}^{00} = -2e^2 N_0\frac{Dq^2}{Dq^2/\hbar + |\omega_\nu|}. \tag{21.47}$$

We also note that, by taking the static limit, the compressibility is equal to the single-particle density of states. Both are given by the clean Fermi gas expression N_0.

At this order of approximation, the Drude electrical conductivity is obtained via the relation (10.43). Its direct evaluation by using the current–current response function in the Kubo formula (10.31),

$$\tilde{\mathcal{R}}^{ij}(\mathbf{q},\omega_\nu) = -2e^2\frac{1}{\beta}\sum_{\epsilon_n}\int_{\mathbf{p}} j^i_0(\mathbf{p})\mathcal{G}(\mathbf{p}_+,\epsilon_+)\mathcal{T}^j(\mathbf{p},\epsilon_n;\mathbf{q},\omega_\nu)\mathcal{G}(\mathbf{p}_-,\epsilon_-), \tag{21.48}$$

is actually much simpler than the corresponding calculation for the density since there are no ladder diagrams involved. Even the presence of a single impurity line in the bubble separates the integration over the momenta flowing in the current vertices. Due to the vectorial character of the current vertex, each integration averages to zero. In this approximation the current–current response function is simply given by a bubble with the disorder-dressed Green function (21.34). The explicit evaluation of the current–current response function is left as Problem 21.4.

21.4.2 Weak localization corrections

The leading approximation in the expansion in the small parameter $\hbar/(E_{\rm F}\tau)$ is obtained by considering diagrams without crossing of impurity lines. We begin our discussion of the

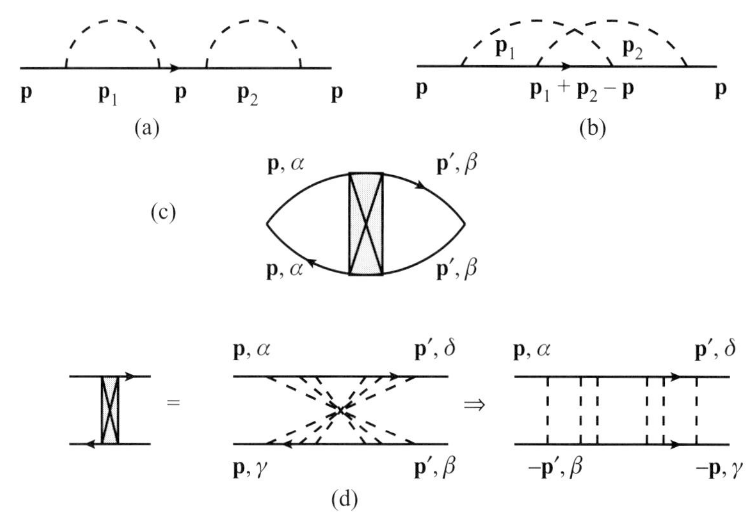

Fig. 21.5 (a) A diagram without crossing. (b) A diagram with crossing. (c) Diagram contributing to the next-to-leading correction to the current–current response function. The Greek indices label the spin of the Green function line. Notice that at the density vertex the spin is conserved. (d) Crossed ladder, denoted by a gray rectangular box with cross sign, expressed in term of the direct ladder. Notice how the spin indices are interchanged in the bottom Green function line. The relevant combination of spin indices for the evaluation of the conductivity is obtained by setting $\delta = \beta$ and $\gamma = \alpha$ and by summing over α and β.

next-to-leading corrections by showing how crossing of impurity lines increases the order of a diagram. A simple example is shown in Fig. 21.5(a) and (b), where both diagrams are of the same order in the impurity lines, but differ in powers of $\hbar/(E_F\tau)$. Let us estimate these diagrams. By recalling the self-energy expression (21.33) and the Green function (21.34), the diagram (a) reads

$$(a) = \left(\frac{\mathrm{i}\,\mathrm{sign}(\epsilon_n)}{2\tau/\hbar}\right)^2 \mathcal{G}(\mathbf{p}, \epsilon_n) \approx_{\epsilon_n \ll \hbar/\tau, \xi_{\mathbf{p}} \sim 0} \frac{\hbar^2}{\tau^2}\frac{\tau}{\hbar} = \frac{\hbar}{\tau}, \tag{21.49}$$

while the diagram (b) yields

$$\begin{aligned}
(b) &= \left(\frac{\hbar}{2\pi N_0 \tau}\right)^2 \int_{\mathbf{p}_2,\mathbf{p}_2} \mathcal{G}(\mathbf{p}_1, \epsilon_n)\mathcal{G}(\mathbf{p}_2, \epsilon_n)\mathcal{G}(\mathbf{p}_1 + \mathbf{p}_2 - \mathbf{p}, \epsilon_n) \\
&= \frac{\hbar^2}{(2\pi N_0 \tau)^2} \int_{\mathbf{p}_2,\mathbf{p}_2} \frac{1}{(\epsilon_n - \xi_1 + \mathrm{i}\delta_{\epsilon_n})(\epsilon_n - \xi_2 + \mathrm{i}\delta_{\epsilon_n})} \\
&\quad \times \frac{1}{\left(\epsilon_n - \xi_1 - \xi_2 - \xi + 2\mu + \frac{\mathbf{p}_1\cdot\mathbf{p}_2}{m} + \frac{\mathbf{p}\cdot(\mathbf{p}_1+\mathbf{p}_2)}{m} + \mathrm{i}\delta_{\epsilon_n}\right)} \\
&\approx \frac{\hbar^2}{\tau^2 \mu},
\end{aligned}$$

where for brevity $\delta_{\epsilon_n} = \mathrm{sign}(\epsilon_n)\,\hbar/2\tau$ and $\xi_1 \equiv \xi_{\mathbf{p}_1}$ and similarly for $\mathbf{p}_2$ and $\mathbf{p}$. The above result is obtained by noting that, due to the poles of the first two Green functions, we can set $\xi_1 = \xi_2 = 0$ in the third one, which then contributes order $1/\mu$. Clearly (b) is smaller than (a) by a factor $\hbar/\mu\tau$.

In general, to take diagrams with crossing of impurity lines becomes a very complicated problem. There is, however, a subset of diagrams, the so-called maximally crossed diagrams, which can be evaluated. These are more easily analyzed in the evaluation of the current–current response function, which does not have contributions from the ladder (with no crossing of impurity lines) diagrams (see Problem 21.4) and does not allow for mixed direct and crossed ladder terms.

The contribution of the maximally crossed diagrams to the current–current response function is shown in Fig. 21.5(c). The corresponding expression reads

$$\delta\tilde{\mathcal{R}}^{ij}(\mathbf{0},\omega) = -2e^2\frac{1}{\beta}\sum_{\epsilon_n}\int_{\mathbf{p},\mathbf{p}'} j_0^i(\mathbf{p})j_0^j(\mathbf{p}')\mathcal{G}(\mathbf{p},\epsilon_+)\mathcal{G}(\mathbf{p},\epsilon_-) \times L_c(\mathbf{p},\mathbf{p}',\omega_\nu)\mathcal{G}(\mathbf{p}',\epsilon_+)\mathcal{G}(\mathbf{p}',\epsilon_-), \tag{21.50}$$

where with $L_c(\mathbf{p},\mathbf{p}',\omega_\nu)$ we have indicated the sequence of the maximally crossed diagrams, which from now on will be called a crossed ladder, L_c. The evaluation of $L_c(\mathbf{p},\mathbf{p}',\omega_\nu)$ can be expressed in terms of the direct ladder by reversing one of the electron Green function lines, as shown in Fig. 21.5(d). This graphical operation is valid as long as time reversal invariance is a symmetry of the Hamiltonian. We stress that the invariance must apply to each Green function separately. Reversing both lines just gives the same diagram. In reversing, say, the bottom Green function line, the direction of the momentum is also reversed and we obtain

$$L_c(\mathbf{p},\mathbf{p}',\omega_\nu) = L(\mathbf{p}+\mathbf{p}',\omega_\nu) - \frac{\hbar}{2\pi N_0\tau}. \tag{21.51}$$

Hence, the crossed ladder has a diffusive form with respect to the combination $\mathbf{p}+\mathbf{p}'$. This means that the dominant contribution comes from backscattering processes with $\mathbf{p}' \sim -\mathbf{p}$. The direct ladder describes the propagation of a particle–hole pair. The crossed ladder, when one of the lines is reversed, describes the propagation of a particle–particle pair and the diffusive pole is with respect to the total incoming momentum of the pair. This is reminiscent of the interaction scattering channel relevant for superconductivity and for this reason the crossed ladder is also called the cooperon ladder. The direct ladder, on the other hand, is, in technical jargon, called the diffusion ladder. A key point to keep in mind about the difference between the diffuson and the cooperon is the total electric charge q of the pair. While for the diffuson $q = 0$, $q = 2e$ for the cooperon.

We are now ready to complete our derivation of the correction to the current response. The main aspect for the evaluation of the conductivity is that the presence of the cooperon makes $\mathbf{p} \sim -\mathbf{p}'$ and the two integrations are no longer independent, yielding a finite result. By performing the integrals as shown in the answer to Problem 21.3 and using the relation between the cooperon and the diffuson given by Eq. (21.51), Eq. (21.50) becomes

$$\delta\tilde{\mathcal{R}}^{ij}(\mathbf{0},\omega_\nu) = -2e^2\frac{1}{\beta}\sum_{\epsilon_n}\int_{\mathbf{p},\mathbf{p}'} j_0^i(\mathbf{p})^i j_0^j(\mathbf{p}')\mathcal{G}(\mathbf{p},\epsilon_+)\mathcal{G}(\mathbf{p},\epsilon_-)\mathcal{G}(\mathbf{p}',\epsilon_+)\mathcal{G}(\mathbf{p}',\epsilon_-) \times \frac{\hbar^2}{2\pi N_0\tau^2}\frac{\theta(-\epsilon_-\epsilon_+)}{|\omega_\nu| + D(\mathbf{p}+\mathbf{p}')^2/\hbar}.$$

Because of the theta function of the energies, ϵ_+ and ϵ_- have opposite signs. On the other hand, ω_ν because of the diffusive pole must be small. Hence, we may set $\epsilon_\pm = 0$ in the Green functions, which then coincide with the corresponding retarded and advanced Green functions. The sum over ϵ_n reproduces a factor ω_ν as in Eq. (21.46). Finally we perform a change of variable $\mathbf{Q} = \mathbf{p} + \mathbf{p}'$ and we get

$$\begin{aligned}\delta\tilde{\mathcal{R}}^{ij}(\mathbf{0}, \omega_\nu) &\approx -2e^2 \frac{|\omega_\nu|}{2\pi\hbar^4} \int_{\mathbf{p}} \int_{\mathbf{Q}} j_0^i(\mathbf{p}) j_0^j(\mathbf{Q}-\mathbf{p}) G^{\mathrm{R}}(\mathbf{p}, 0)\, G^{\mathrm{A}}(\mathbf{p}, 0) \\ &\quad \times \frac{\hbar^2}{2\pi N_0 \tau^2} \frac{1}{|\omega_\nu| + D\mathbf{Q}^2/\hbar} G^{\mathrm{R}}(\mathbf{Q}-\mathbf{p}, 0)\, G^{\mathrm{A}}(\mathbf{Q}-\mathbf{p}, 0) \\ &= |\omega_\nu| \frac{2e^2}{\pi} \int_{\mathbf{Q}} \frac{1}{|\omega_\nu|/D + \mathbf{Q}^2}.\end{aligned}$$

The logarithmic divergence of the above integral has to be cut off at large $\mathbf{Q}$ by the inverse mean free path, i.e. the distance beyond which diffusive motion starts to set in. At zero temperature, the small $\mathbf{Q}$ cutoff is instead provided by the inverse of some characteristic length scale, given by the system size L. At finite temperature, inelastic processes would provide a so-called dephasing length L_ϕ depending on T. From the above equation and Eq. (10.31), at $d = 2$ and $T = 0$, we get the correction to the conductivity

$$\delta\sigma = -\frac{e^2}{\pi^2\hbar} \ln\left(\frac{L}{l}\right) = -\sigma_0 2r \ln\left(\frac{L}{l}\right). \tag{21.52}$$

The above equation represents the *weak localization* correction and $r = (4\pi^2 N_0 D\hbar)^{-1} = (2\pi E_{\mathrm{F}}\tau/\hbar)^{-1}$ is the effective small disorder expansion parameter in the metallic regime. To make a connection with the phenomenological scaling theory, we note that the parameter r coincides with the inverse conductance in units of G_0 divided by 2π, $r = 1/(2\pi g)$. At first order in r the critical index of the conductivity is then $\mu = 1$.

As discussed in the first section of this chapter, the physical origin of the weak localization correction is due to quantum interference. To connect this idea to the diagrammatic description of the present section, we observe that in a Feynman diagram a Green function line describes the probability amplitude for going from one point to another. In the response functions, the two Green function lines represent the operation of taking the product of one amplitude with its complex conjugate in order to obtain the probability of the transfer process. Each probability amplitude is the sum over all possible paths so that interference terms may appear in the product. The leading approximation, by restricting itself to ladder diagrams, makes an *effective* selection of paths. When we average over the impurity configurations, by connecting two impurity insertions by a dashed line, both the upper and lower Green function lines are going through the same scattering center, i.e., we are dressing both Green functions with the same sequence of rainbow diagrams (see bottom line of Fig. 21.3). In conclusion, we are considering the product of the amplitude of a given path with its complex conjugate and, in this leading approximation, there is no interference. When considering, on the other hand, the maximally crossed diagrams, one observes that the upper Green function line goes through a sequence of scattering events which is exactly the opposite of the one followed by the lower line. This represents the interference between trajectories that are the time reversal of one another.[3]

So far we have established that in $d = 2$ the electrical conductivity acquires a logarithmic correction to σ_0 at order $\hbar/E_F\tau$. The correction has a negative sign and signals a slowing down of the electron diffusion. It is then legitimate to ask what this implies for the full momentum and frequency dependence of the density–density response function. At small momentum and frequency, due to the conservation laws, the density–density response function must retain the form (21.47) apart from possible renormalizations of the density of states N_0 and the diffusion coefficient D. We now show that the density of states is not renormalized and hence the renormalization of the diffusion coefficient is the only possible source of weak localization corrections to the density–density response function to produce a slowing down of the diffusion.

By inserting a factor 2 for the spin degeneracy in Eq. (18.115) for the density–density response function and taking advantage of the Ward identity (18.118) connecting the Green function and the reducible density vertex Λ^0, one gets

$$\begin{aligned}\tilde{\mathcal{R}}^{00}_{+-}(\mathbf{0},\omega) &= 2e^2\frac{1}{\beta}\sum_{-\omega_\nu/2<\epsilon_n<\omega_\nu/2}\int_{\mathbf{p}}\Lambda^0(\epsilon_n,\mathbf{p};\omega_\nu,\mathbf{0})\\ &\approx_{\omega\to 0} -2e^2\frac{1}{\pi}\int_{\mathbf{p}}\operatorname{Im}G^{\mathrm{R}}(\mathbf{p},0)\\ &= e^2\nu,\end{aligned} \tag{21.53}$$

where ν is the exact single-particle density of states (including spin) and we made use of the fact that the sign of the frequency determines whether the Green function is analytical in the upper (retarded, R) or in the lower (advanced, A) half of the complex plane as function of the frequency. Since by charge conservation $\tilde{\mathcal{R}}^{00}(0,\omega) = 0$ (see Eq. (10.40)) and that $\tilde{\mathcal{R}}^{00}_{++} + \tilde{\mathcal{R}}^{00}_{--} = -2e^2N_0$ (see Eq. (21.45)) is not affected by ladder insertions, the density of states is not renormalized, $\nu = 2N_0$.

When considering diagrams correcting the density–density response function, several of them cancel among themselves or are negligible in the small momentum and frequency limit. Such cancellations must necessarily occur to keep the density of states unrenormalized. The only diagrams which survive are those responsible for the renormalization of the diffusion coefficient and are evaluated in the answer to Problem 21.5. To first order in r, we get (see the answer to Problem 21.5 for details) as a correction to the dynamical contribution, $\tilde{\mathcal{R}}^{00}_{+-}$, of the density–density response Eq. (21.46):

$$\delta\tilde{\mathcal{R}}^{00}_{+-} = \frac{2|\omega_\nu|Dq^2}{(Dq^2+|\omega_\nu|)^2}\frac{e^2\hbar}{\pi}\int_{\mathbf{Q}}\frac{1}{D\mathbf{Q}^2/\hbar+|\omega_\nu|} \equiv -\frac{2e^2|\omega_\nu|N_0\mathbf{q}^2}{(Dq^2/\hbar+|\omega_\nu|)^2}\delta D(\omega_\nu), \tag{21.54}$$

where

$$\delta D(\omega_\nu) = -\frac{D\hbar}{\pi N_0}\int_{\mathbf{Q}}\frac{1}{D\mathbf{Q}^2/\hbar+|\omega_\nu|},\qquad \lim_{\omega_\nu\to 0}\delta D(\omega_\nu) = \frac{\delta\sigma}{2e^2N_0}$$

[3] One also notices that these trajectories are made by closed loops and always come in pairs, due to the fact the loop can be gone around clockwise or counter-clockwise.

agrees with the expression (21.52) for $\delta\sigma$ derived from the current–current response function. The above equation shows that the correction to $\tilde{\mathcal{R}}^{00}$, made of many different contributions, can, as anticipated, be absorbed into a renormalization of the diffusion coefficient $D_{\mathrm{R}} = D + \delta D$:

$$\tilde{\mathcal{R}}^{00}_{+-} = 2e^2 N_0 \frac{|\omega_\nu|}{|\omega_\nu| + D_{\mathrm{R}} q^2/\hbar}, \quad \tilde{\mathcal{R}}^{00} = -2e^2 N_0 \frac{D_{\mathrm{R}} q^2}{D_{\mathrm{R}} q^2/\hbar + |\omega_\nu|}, \tag{21.55}$$

and by virtue of Eq. (10.44) also the compressibility $\mathrm{d}n/\mathrm{d}\mu$ is not renormalized and coincides with $2N_0$.

Since only one renormalization ($D \to D_{\mathrm{R}}$) is required in the present non-interacting Fermi gas (to be contrasted with the interacting Fermi liquid to be discussed later on), the one-parameter scaling theory follows. Given the expression (21.52) for $\delta\sigma$, the group equation for the conductance g has an expansion in r of its β-function (Eq. (21.26)) with the coefficient $a = 1/\pi$. The critical index for the conductivity, at order ϵ in $d = 3$ is $\mu = 1$. The frequency cuts off the singularity in the diffusion ladder and acts in this transition as an external field in ordinary transitions. Its scaling index x_ω has the same value as the scaling dimension of the $D_{\mathrm{R}} q^2$, i.e., $x_\omega = \epsilon + 2 = d$, where ϵ is the scaling dimension of D_{R} because, at the critical fixed point, the conductivity $\sigma \sim g_{\mathrm{c}} \xi^{-\epsilon}$ and $\partial n/\partial \mu$ does not scale.

21.4.3 Weak localization in the presence of time reversal breaking and spin–orbit coupling

To complete the correspondence with the phenomenological scaling theory of Section 21.3, we discuss how the weak localization corrections are affected by changing the symmetry properties of the Hamiltonian.

An external magnetic field breaks the time reversal invariance and then destroys the quantum interference effect responsible for the weak localization correction. As a consequence of its total charge $q = 2e$, the cooperon is affected by the presence of a magnetic field, while the diffuson is not touched. In real space and time, the cooperon is obtained by transforming back the analytically continued diffusive pole in $\mathbf{Q}$ and ω of Eq. (21.43) to real space $\mathbf{r} - \mathbf{r}'$ and time $t - t'$ coordinates. The cooperon then describes the motion of the center of mass of the pair from $\mathbf{r}$ to $\mathbf{r}'$ in the time $t' - t$. The corresponding diffusion equation is given by Eq. (11.35) with delta-like sources:

$$\left(\frac{\partial}{\partial t} - D \nabla^2_{\mathbf{r}} \right) L^{t,t'}_{\mathrm{c}}(\mathbf{r}, \mathbf{r}') = \frac{\hbar}{2\pi N_0 \tau^2} \delta(t - t') \delta(\mathbf{r} - \mathbf{r}'). \tag{21.56}$$

By expressing the magnetic field in terms of a vector potential and by allowing the minimal coupling substitution as usual, $-\mathrm{i}\hbar\nabla \to -\mathrm{i}\hbar\nabla - q\mathbf{A}/c$, one obtains the real-space differential equation for the cooperon in the presence of the magnetic field $\mathbf{B}$. The problem for the cooperon becomes equivalent to that of a particle of charge $2e$ and mass $\hbar^2/2D$ in an external magnetic field, a problem discussed in Appendix F when dealing with the Landau diamagnetism. The diffusive motion becomes quantized with the effective cyclotron frequency $\omega_{\mathrm{B}} = 4eDB/\hbar^2 c$ and the lowest eigenvalue of the Laplacian operator corresponds to the zero-point energy. As a result the diffusive pole is shifted to finite frequency $\omega = \mathrm{i}D\mathbf{Q}^2/\hbar^2 \to \mathrm{i}/\tau_\mathrm{B}$ ($\tau_\mathrm{B} = \omega_\mathrm{B}^{-1}$) and kills the logarithmic singularity of Eq. (21.52). Experimentally, the progressive reduction of the interference effect due to an increasing field results in a *negative* magnetoresistance. Interference effects manifest to second order in the effective coupling r with the crossing of two direct ladders and the critical exponent gets the value $\mu = 1/2$.

Besides the magnetic field, there are other physical mechanisms that affect the electron propagation in a disordered medium. Time reversal invariance is also broken by magnetic impurities. As a result we expect that the diffusive pole in the cooperon is shifted also to finite frequency $\omega = \mathrm{i}D\mathbf{Q}^2/\hbar^2 \to \mathrm{i}D\mathbf{Q}^2/\hbar^2 + 1/\tau_\mathrm{spin}$, τ_spin being the spin relaxation time proportional to the inverse concentration of magnetic impurities. Again the critical exponent should get the value $\mu = 1/2$.

More complex is the case provided by the spin–orbit coupling originating from the non-relativistic limit of the Dirac equation:

$$H_\mathrm{so} = \int \mathrm{d}\mathbf{x}\hat{\psi}^\dagger_\alpha(\mathbf{x})\frac{\hbar}{4m^2c^2}\nabla U(\mathbf{x}) \times (-\mathrm{i}\hbar\nabla)\cdot \boldsymbol{\sigma}_{\alpha\beta}\hat{\psi}_\beta(\mathbf{x}). \tag{21.57}$$

This coupling, while producing spin flip with a relaxation time τ_so proportional to the inverse impurity concentration, preserves the time reversal invariance because both the spin and the momentum operators change sign upon time reversal.

When the spin is conserved, the response functions depend on the spin only via a degeneracy factor. When the spin–orbit Hamiltonian is present the spin along each Green function line of the response functions is no longer conserved. To see how the spin relaxation affects the weak localization correction, the cooperon spin structure has to be decomposed into singlet S_c and triplet T_c components. Of course only the latter is affected by the spin–orbit coupling.

In the response function the spin conservation at the vertices according to Fig. 21.4(b) and (c) requires the combination $\delta_{\alpha\gamma}\delta_{\beta\delta}$. To express this in terms of the singlet and triplet components we use the Fierz identities for combinations of Pauli matrices:

$$\delta_{\alpha\gamma}\delta_{\beta\delta} = \frac{1}{2}\delta_{\alpha\delta}\delta_{\beta\gamma} + \frac{1}{2}\boldsymbol{\sigma}_{\alpha\delta}\cdot\boldsymbol{\sigma}_{\beta\gamma} \equiv S_\mathrm{d}, \tag{21.58}$$

$$\boldsymbol{\sigma}_{\alpha\gamma}\cdot\boldsymbol{\sigma}_{\beta\delta} = \frac{3}{2}\delta_{\alpha\delta}\delta_{\beta\gamma} - \frac{1}{2}\boldsymbol{\sigma}_{\alpha\delta}\cdot\boldsymbol{\sigma}_{\beta\gamma} \equiv T_\mathrm{d}. \tag{21.59}$$

The first identity represents the combination which selects the singlet component S_d of the diffuson ladder in terms of the spin indices of the incoming (α, γ) and outgoing (β, δ) particle–hole pairs. The second one is the expression for the triplet component T_d.

To obtain the decomposition for the cooperon ladder it is enough to apply the time reversal operation on one of the Green function lines, say, the bottom one. This amounts to the replacement $\boldsymbol{\sigma}_{\beta\gamma} \to -\boldsymbol{\sigma}_{\beta\gamma}$ in the right-hand sides of Eqs. (21.58) and (21.59). Finally one obtains

$$\delta_{\alpha\gamma}\delta_{\beta\delta} = \frac{1}{2}\left(-S_\mathrm{c} + T_\mathrm{c}\right). \tag{21.60}$$

By using this decomposition in the evaluation of the diagram for the weak localization correction, one sees that the triplet, due to its multiplicity, is minus three times the singlet contribution. Hence, in the absence of spin–orbit coupling the weak localization correction is dominated by the triplet channel. By switching on the spin–orbit coupling the spin relaxation kills the singularity in the triplet channel and only the singlet contribution with opposite sign appears in the correction to σ at this order as an antilocalizing term.

21.5 A few remarks on the experiments

It is now time to compare the theoretical predictions obtained from the microscopic approach with some experiments. Even if the theoretical values of the critical index μ at order ϵ are not to be trusted for a strict comparison with the experiments, there are a number of facts which suggest that a proper description of the metal–insulator transition cannot be obtained without taking into account the effects of the electron–electron interaction. (1) In semiconductor–metal alloys, the metal–insulator transition is observed with a conductivity critical exponent $\mu = 1$. This is in agreement with the one-parameter scaling theory. On the other hand, tunneling measurements reveal that the single-particle density of states has strong anomalies. For instance, the single particle density of states of $Ge_{1-x}Au_x$ (McMillan and Mochel (1981)) is measured for different values of the Au concentration above and below the critical concentration for the metal–insulator transition. On the metallic side, the density of states shows a dip at the Fermi energy, which in the insulating regime, at lower Au concentrations, develops in a gap. In the case of the Nb_xSi_{1-x} alloy, the value of the density of states, as obtained from tunneling measurements, is plotted as a function of the Nb concentration, and goes to zero by approaching the critical concentration for the metal–insulator transition (Hertel et al. (1983)), while $\mathrm{d}n/\mathrm{d}\mu$ remains finite. In the one-particle picture both densities of states and $\mathrm{d}n/\mathrm{d}\mu$ cannot renormalize. In the Fermi liquid theory they are different and could respond differently to the disorder. (2) The presence of magnetic impurities, in the one-parameter scaling theory, should lead to a value $\mu = 1/2$. However, in metal films of Cu:Mn, one finds experimentally $\mu = 1$ (Okuma et al. (1988)). (3) In the amorphous alloy Si:Au, it has been observed (Nishida et al. (1984)) $\mu = 1$ both in the absence and in the presence of a magnetic field of 5 Tesla. This system has a strong spin–orbit coupling which, within the single-particle scheme, should switch from an incipient antilocalizing behavior in the absence of the magnetic field, to a transition with $\mu = 1/2$ in the presence of the field. (4) Similarly, a value $\mu = 1$, both with and without a magnetic field, has also been observed in $Al_{0.3}Ga_{0.7}As$:Si with a fine tuning of the electron concentration close to the metal–insulator transition, by using the photoconductivity effect (Katsumoto et al. (1987); Katsumoto (1988)). Also in this system, it is estimated that spin–orbit scattering is relevant. The localizing term must have a different origin with respect to the cooperon sensitive to the magnetic field. (5) Experiments in Si:P show that there is a strong enhancement at low temperature in Si:P of the electronic specific heat (Thomas et al. (1981)) and the spin susceptibility (Paalanen et al. (1986)), impossible to obtain in the one-particle picture.

Finally, we comment on the problem of the metal–insulator transition in the two-dimensional electron gas as in Si-MOSFET devices and semiconducting heterostructures (Dobrosavljević et al. (2012)). Although there is quite a rich experimental literature on this phenomenon, there is not yet a general consensus. In any case, from the point of view of the present discussion about the relevance of the electron–electron interaction in disordered systems, Si-MOSFET devices and semiconducting heterostructures are even a stronger case. In fact, if there is a metal–insulator transition, this is clearly beyond the conventional non-interacting scaling theory for which all states are localized in two dimensions for any value of the disorder.

The interplay between disorder and electron–electron interaction will account for some of the questions arisen above.

21.6 Electron–electron interaction in the presence of disorder

21.6.1 Specific heat, spin susceptibility and single particle density of states

In this section we describe the origin of the electron–electron interaction effects in disordered electron systems. Generally speaking, their physical origin can be traced back to the diffusive character of the electron motion and it is distinct from the interference mechanism analyzed in the case of the weak localization correction. However, the electron–electron interaction effects also yield logarithmic singularities in two spatial dimensions. In contrast to the non-interacting disordered Fermi gas, the presence of the interaction yields logarithmic singularities in the single-particle density of states and thermodynamic quantities, in addition to the electrical conductivity.

The electron–electron interaction is described by the Hamiltonian (see Eqs. (9.37) and (9.40)) for the real and momentum space representations, respectively:

$$H_{\text{int}} = \int \mathrm{d}\mathbf{x} \int \mathrm{d}\mathbf{x}' \frac{1}{2} V(\mathbf{x}-\mathbf{x}')\hat{\psi}^{\dagger}_{\alpha}(\mathbf{x})\hat{\psi}^{\dagger}_{\beta}(\mathbf{x}')\hat{\psi}_{\beta}(\mathbf{x}')\hat{\psi}_{\alpha}(\mathbf{x}) \tag{21.61}$$

$$= \frac{1}{2V}\sum_{\mathbf{k},\mathbf{k}',\mathbf{q}} V(\mathbf{q})\, a^{\dagger}_{\mathbf{k}+\mathbf{q},\alpha} a^{\dagger}_{\mathbf{k}'-\mathbf{q},\beta} a_{\mathbf{k}',\beta} a_{\mathbf{k},\alpha}, \tag{21.62}$$

where the sum over repeated spin indices is understood. The first order correction in the interaction of the thermodynamic potential is the quantum statistical average of the interaction Hamiltonian:

$$\Delta\Omega^{(1)} = \langle \hat{H}_{\text{int}} \rangle = \int \mathrm{d}\mathbf{x} \int \mathrm{d}\mathbf{x}' \frac{1}{2} V(\mathbf{x}-\mathbf{x}') h_{2,\alpha\beta\beta\alpha}(\mathbf{x},\mathbf{x}';\mathbf{x}',\mathbf{x}), \tag{21.63}$$

where the statistical average of the four field operators entering the expression of H_{int} has been written in terms of the two-particle reduced density matrix defined in Eq. (9.58). To lowest order in the interaction, the two-particle reduced density matrix can be evaluated by using the expression for a non-interacting Fermi system. Accordingly, we use the decomposition in terms of the one-particle reduced density matrices as in Eq. (9.63) and get

$$\Delta\Omega^{(1)} = \int \mathrm{d}\mathbf{x} \int \mathrm{d}\mathbf{x}' \frac{1}{2} V(\mathbf{x}-\mathbf{x}') \left(h_{1,\alpha\alpha}(\mathbf{x},\mathbf{x}) h_{1,\beta\beta}(\mathbf{x}',\mathbf{x}') - h_{1,\alpha\beta}(\mathbf{x},\mathbf{x}') h_{1,\beta\alpha}(\mathbf{x}',\mathbf{x}) \right), \tag{21.64}$$

where the first and second terms are the Hartree and exchange contributions.

Let us examine first the exchange contribution for which it is possible to derive an expression in terms of the equal-time density–density correlation function $S^{00}(\mathbf{x},t;\mathbf{x}'t)$. Since we have defined in Eq. (10.45) the correlation functions in terms of the fluctuations of the operators with respect to their average, we cannot have in the expression of the correlation function terms which factorize as far as the dependence on $\mathbf{x}$ and $\mathbf{x}'$ is concerned. This means that the Hartree term in Eq. (21.64) cannot be directly interpreted in terms of the density–density correlation function.[4] As a result, by writing the equal-time correlation function as an integral over frequency of its Fourier transform with respect to time, the contribution of the exchange (Fock) term, reads

$$\Delta\Omega^{(1)}_{\mathrm{exc}} = \int \mathrm{d}\mathbf{x}\, \mathrm{d}\mathbf{x}' V(\mathbf{x},\mathbf{x}') \int_{-\infty}^{\infty} \mathrm{d}\omega \tilde{S}^{00}(\mathbf{x},\mathbf{x}';\omega). \tag{21.65}$$

The above equation allows us to study the effects of the electron–electron interaction depending on the nature of the density fluctuations in the system. In order to estimate the behavior of Eq. (21.65), we assume a translationally invariant statically screened interaction such that it remains finite in the limit of vanishing momentum transfer. In a translationally invariant system, as in the Fermi gas, we then have

$$\begin{aligned} \Delta\Omega^{(1)}_{\mathrm{exc}} &= V \int_{\mathbf{q}} V(\hbar\mathbf{q}) \int_{-\infty}^{\infty} \mathrm{d}\omega \tilde{S}^{00}_{\mathbf{q}}(\omega) \\ &= -V \int_{\mathbf{q}} V(\hbar\mathbf{q}) \int_{-\infty}^{\infty} \mathrm{d}\omega\hbar \coth\left(\frac{\beta\hbar\omega}{2}\right) \mathrm{Im}\, \tilde{R}^{00}_{\mathbf{q}}(\omega), \end{aligned} \tag{21.66}$$

where V is the volume and $V(\hbar\mathbf{q})$ is the Fourier transform of the interaction. We have used the fluctuation–dissipation theorem of Eq. (10.51) in going from the first to the second line. Hence, the correction to the thermodynamic potential can be seen as the energy cost, due to the interaction, for a density fluctuation. Here the symbol $\int_{\mathbf{q}}$ indicates integration over the wave vector $\mathbf{q}$ measured in units of inverse length, as defined in Eq. (21.31).

Since in the Fermi gas, at small momentum and frequency, the response function is real (see Eq. (10.22)), the correction (21.66) to the thermodynamic potential yields a non-singular contribution. In a disordered Fermi gas the situation is very different. As explained in Section 11.4, due to scattering with the impurities, the electrons perform, at classical level, a Brownian motion. On average the translational invariance is recovered and the density–density response function gets the diffusive form of Eq. (11.39). This

[4] By anticipating the later discussion, the connection with the correlation function and the exchange term can also be understood by recalling the bubble structure of the thermal response function of Fig. 18.7, which is given by the product $\mathcal{G}(x,x')\mathcal{G}(x',x)$. The equal-time limit of such an expression can be written as the product of one-particle reduced density matrices by the use of Eq. (18.6) and has the structure of the exchange term in Eq. (21.64).

immediately leads, when used in Eq. (21.66), to a singular-in-temperature correction to the thermodynamic potential in two dimensions:

$$\begin{aligned}\Delta\Omega^{(1)}_{\rm exc} &= V\int_{\mathbf{q}} V(\hbar\mathbf{q})\int_{-\infty}^{\infty} d\omega\hbar\coth\left(\frac{\beta\hbar\omega}{2}\right)\frac{\omega N_0 Dq^2}{(Dq^2)^2+\omega^2}\\ &= -2N_0(N_0V(0))rV\int_0^{\infty} d\omega\hbar^2\omega b(\beta\hbar\omega)\ln(|\omega|\tau)\\ &= -\frac{\pi^2}{3}N_0V_1rV(k_{\rm B}T)^2\ln(k_{\rm B}T\tau/\hbar),\end{aligned} \tag{21.67}$$

where $b(x)$ is the Bose function and r the effective disorder expansion parameter introduced in Eq. (21.52) when discussing the weak localization correction. In going from the first to the second line we have performed the integral over the wave vector $\mathbf{q}$ and used the parity in ω of the integrand function. The integral over momentum is logarithmically singular at large momentum and we used as upper cutoff the inverse scattering time due to the diffusion approximation $Dq^2\tau \ll 1$ of Eq. (21.42). Furthermore, in the wave vector integration, due to the singularity at small wave vector and frequency of the diffusive pole of the density–density response function, we have safely neglected the wave vector dependence of the electron–electron interaction, which can be replaced by its zero-wave vector limit

$$V_1 = N_0V(\mathbf{q}\to\mathbf{0}). \tag{21.68}$$

The free single-particle density of states N_0 defined in Eq. (21.5) has been included in the definition of V_1 to make it dimensionless. Finally, the integral over frequency, via the change of variable $y=\beta\hbar\omega$ is reduced to the Bose integral $I_1^-(0)=\pi^2/6$ (see Eq. (E.7) of Appendix E).

The analysis of the Hartree contribution is more involved for the reasons mentioned above, but the final result, besides an obvious factor -2, contains a parameter $V_2 = N_0V(q\sim 2k_{\rm F})$, which involves large momentum transfer processes in contrast to the exchange contribution dominated by small momentum transfer. A more precise definition of V_2 will be given below in Eq. (21.79), when the connection between the Hartree and exchange terms with large and small momentum regions will become clear by using Feynman diagrams.

Equation (21.67) for the correction to the thermodynamic potential implies a correction to the specific heat:

$$\delta c_V = c_V(V_1-2V_2)r\ln(k_{\rm B}T\tau/\hbar), \tag{21.69}$$

where c_V is the specific heat (7.37) of the Fermi gas and the Hartree contribution has been included as well.

Together with the specific heat, the Pauli spin susceptibility also acquires a correction. To see how this arises we observe that the contribution to the specific heat can be seen as due to the number and spin density fluctutations which have become diffusive in the disordered Fermi gas. Such an observation relies on the fact that the combination V_1-2V_2 of the interaction couplings can be written as the sum of number density fluctuations proportional to $V_1-(1/2)V_2$ and spin-density fluctuations proportional to $(-1/2)V_2$.

The detailed explanation of this decomposition is postponed to the beginning of the next section.

We are ready to tackle the problem of the spin susceptibility, i.e. of the response to a magnetic field entering via a Zeeman term in the Hamiltonian as in Eq. (7.29). On physical grounds, we may expect that only spin fluctuations with total spin projection along the quantization axis different from zero are affected by the magnetic field. Our task reduces, then, to evaluating the correction to the thermodynamic potential (21.64) by confining ourselves to the spin fluctuations with spin projection different from zero. In order to do so, we need the expression of the spin-density response function. We have already seen that in a disordered system the response function for the density is diffusive and microscopically is obtained by the ladder resummation of Eq. (21.43). What remains to be done, then, is the derivation of the ladder in the presence of a Zeeman term in the Hamiltonian H_Z (see Eq. (7.29)). This is carried out in Problem 21.6 with the result

$$L_{jM}(q,\omega) = \frac{\hbar}{2\pi N_0 \tau^2} \frac{1}{Dq^2 - \mathrm{i}\omega - \mathrm{i}M\Delta/\hbar}, \tag{21.70}$$

where $j = 0, 1$ identifies the density and spin channels, respectively, while $M = 0, \pm 1$ labels the different total spin projections of the corresponding channel.

We are now in a position to obtain the desired correction to the thermodynamic potential starting from Eq. (21.67) and considering only the channels with $M = \pm 1$ with the result

$$\Delta\Omega_{\mathrm{B}}^{(1)} = V \frac{r N_0 V_2}{2} (\Delta)^2 \ln(k_{\mathrm{B}} T\tau/\hbar). \tag{21.71}$$

From Eq. (21.71) one finally gets, by differentianting twice with respect to the magnetic field, the correction to the spin susceptibility:

$$\chi = -\chi_0 2r V_2 \ln(k_{\mathrm{B}} T\tau/\hbar), \tag{21.72}$$

where we have introduced the non-interacting value $\chi_0 = 2N_0(g\mu_{\mathrm{B}}/2)^2$. No correction is found for the compressibility, since no explicit dependence on the Fermi energy (i.e. the chemical potential) appears in the correction to the thermodynamic potential. We will see later that this is not accidental, but it remains true also when considering higher orders in the interaction.

As anticipated, singular corrections arise also for the density of states at zero kelvin. This can be seen by considering the shift in the single-particle energy level to first order in the interaction. To this end we may use the expression (9.70) of the Hartree–Fock energy level derived in Section 9.8. The exchange-induced shift in the single-particle energy, at zero temperature, after projecting over the single-particle eigenfunctions $g^*_{\kappa_1}(\mathbf{x})$ and $g_{\kappa_1}(\mathbf{x}')$, reads

$$\delta\epsilon_{\kappa_1} = -\int \mathrm{d}\mathbf{x}\,\mathrm{d}\mathbf{x}'\, V(\mathbf{x},\mathbf{x}') \sum_{\epsilon_{\kappa_2}<0} g^*_{\kappa_1}(\mathbf{x}) g_{\kappa_1}(\mathbf{x}') g^*_{\kappa_2}(\mathbf{x}') g_{\kappa_2}(\mathbf{x}), \tag{21.73}$$

where the restriction $\epsilon_{\kappa_2} < 0$ means a summation over the occupied states, which have negative energy with respect to the chemical potential. As a consequence of the modification of the single-particle levels, the density of states also changes. By using the expression of the single-particle density of states for a generic set of quantum numbers κ and spin index α,

$$\nu(\epsilon) = \frac{1}{V}\sum_{\kappa_1,\alpha} \delta(\epsilon - \epsilon_{\kappa_1} - \delta\epsilon_{\kappa_1}), \tag{21.74}$$

to first order in the interaction, the variation of the density of states can be written as

$$\begin{aligned}\delta\nu(\epsilon) &= -\frac{1}{V}\sum_{\kappa_1,\alpha} \frac{\partial}{\partial\epsilon}\delta(\epsilon - \epsilon_{\kappa_1})\delta\epsilon_{\kappa_1} \\ &= \frac{\partial}{\partial\epsilon}\frac{2}{V}\int_{\epsilon}^{\infty} \mathrm{d}\omega\hbar \int \mathrm{d}\mathbf{x}\,\mathrm{d}\mathbf{x}' V(\mathbf{x},\mathbf{x}') \\ &\quad \times \sum_{\kappa_1\kappa_2} \delta(\epsilon - \epsilon_{\kappa_1})\delta(\epsilon - \hbar\omega - \epsilon_{\kappa_2}) g^*_{\kappa_1}(\mathbf{x}) g_{\kappa_1}(\mathbf{x}') g^*_{\kappa_2}(\mathbf{x}') g_{\kappa_2}(\mathbf{x}). \end{aligned} \tag{21.75}$$

The appearance of the second δ-function in Eq. (21.75) is associated with the extra energy integration over ω and also to the fact that the sum over the set of quantum numbers κ_2 is no longer restricted to the occupied states.

By expressing the density operators in terms of field operators and using for the latter the expansions (9.18) and (9.19), it is possible to show that (see Problem 21.7)

$$\begin{aligned}\mathrm{Im}\,\tilde{R}^{00}(\mathbf{x},\mathbf{x}';\omega) &= \frac{\pi}{2}\int_{-\infty}^{\infty} \mathrm{d}\epsilon \left[\tanh\left(\frac{\beta\epsilon}{2}\right) - \tanh\left(\beta\frac{\epsilon+\hbar\omega}{2}\right)\right] \\ &\quad \times \sum_{\kappa_1\kappa_2} \delta(\epsilon - \epsilon_{\kappa_1})\delta(\epsilon + \hbar\omega - \epsilon_{\kappa_2}) g^*_{\kappa_1}(\mathbf{x}) g_{\kappa_2}(\mathbf{x}) g^*_{\kappa_2}(\mathbf{x}') g_{\kappa_1}(\mathbf{x}'). \end{aligned} \tag{21.76}$$

By using the zero-kelvin limit of Eq. (21.76) together with (21.75), we get for the correction to the single-particle density of states

$$\begin{aligned}\delta\nu(\epsilon) &= \frac{1}{\pi}\hbar \int_{\mathbf{q}} V(\hbar\mathbf{q})\mathrm{Im}\frac{1}{\epsilon}\tilde{R}^{00}_{\mathbf{q}}(\epsilon/\hbar) \\ &= -\frac{1}{\pi}\hbar \int_{\mathbf{q}} V(\hbar\mathbf{q}) 2N_0 \frac{Dq^2}{(Dq^2)^2 + (\epsilon/\hbar)^2}. \end{aligned} \tag{21.77}$$

By performing the integral in two spatial dimensions and including the Hartree contribution, one finally arrives at

$$\delta\nu(\epsilon) = 2N_0(V_1 - 2V_2) r \ln|\epsilon\tau/\hbar|. \tag{21.78}$$

21.6.2 Decomposition of the interaction and effective couplings in the presence of disorder

In order to prepare the ground for the theory of the disordered Fermi liquid of the next section, we show how the density of states correction arises diagrammatically. Figure 21.6(a) represents the lowest order in the electron–electron interaction exchange diagram with disorder potential insertions. The average procedure leads to diagram (b) of the same figure. The diagram to be evaluated is diagram (c), where the Green function lines have the self-energy insertion and the interaction vertices are dressed by disorder as in Fig. 21.4(b). Diagram (d) corresponds to the Hartree term with dressing due to ladder insertions. The explicit evaluation of both diagrams is carried out in the answer to Problem 21.8 and agrees

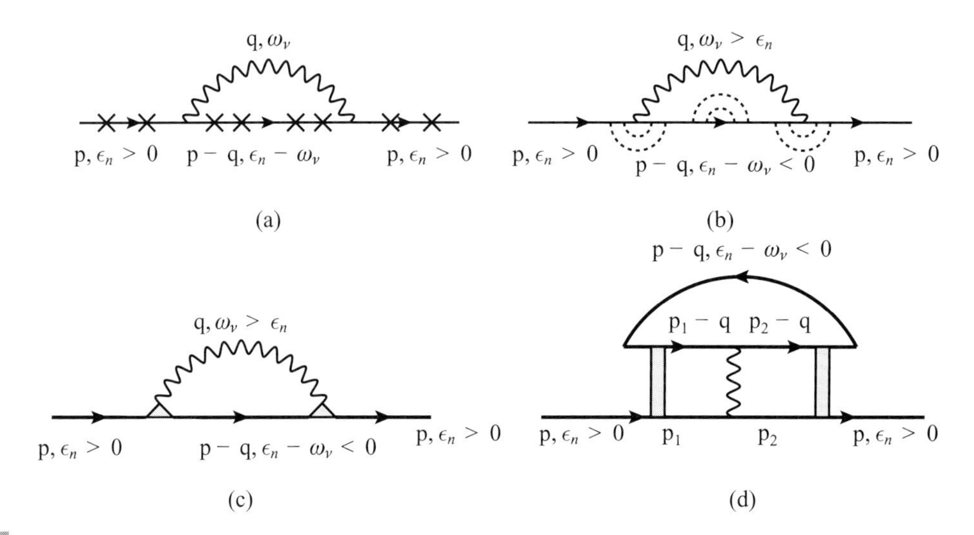

Fig. 21.6 Diagrams for the correction of the single-particle Green function at first order in the electron–electron interaction. (a) Exchange diagram to first order in the electron–electron interaction (wavy line) with disorder potential insertions (crosses) into the Green function lines. We have fixed the fermionic Matsubara frequency ϵ_n of the incoming electron to be positive. At this stage there is no restriction on the bosonic frequency ω_ν. (b) After performing the disorder average, the Green functions are dressed self-consistently with self-energy insertions as done in the bottom line of Fig. 21.3. (c) The resulting diagram contains the diffusive singular part of two vertices $\mathcal{T}^0$ of the type shown in Fig. 21.4. In the figure we emphasize that the dominant contribution arises when the Green functions entering each vertex have opposite sign for the Matsubara frequency, thus requiring a condition on the bosonic freqeuncy $\omega_\nu > 0$. (d) Hartree diagram. Notice that here the combination $\mathbf{p}_1 - \mathbf{p}_2$ flows in the interaction, meaning that are possible scattering events at arbitrary angle.

with the result (21.78). The key point to notice is that the insertion of ladders introduces diffusive poles (two in this case) and determines the singular behavior of the diagram. Hence, as shown in Problem 21.8, the integration over the transferred momentum $\mathbf{q}$ and frequency ω_ν is responsible for the emergence of logarithmic singularities. The role of the interaction is to allow the transfer of energy and hence to the possibility of ladder insertion between Green functions with opposite sign of the frequency. This is why such a correction is absent in the non-interacting case. The emergence of small and large momentum transfer is explained in Fig. 21.7, where the internal Green function line (that with momentum $\mathbf{p} - \mathbf{q}$) of the self-energy diagrams is cut in such a way as to produce the corresponding contribution to the four-leg scattering amplitudes. In case of the exchange diagram, the momentum of the diffusive pole is the same momentum transferred by the electron–electron interaction, selecting the small $\mathbf{q}$ region. In the Hartree case, instead, the momentum transferred by the electron–electron interaction is $\mathbf{p}_1 - \mathbf{p}_2$, where both $\mathbf{p}_1$ and $\mathbf{p}_2$ vary on the Fermi surface. The actual evaluation of the diagrams shows that the precise expression of V_2 is

$$V_2 = \frac{\int_{\mathbf{p}_1}\int_{\mathbf{p}_2} \mathcal{G}(p_1)\mathcal{G}(p_1 - q)\mathcal{G}(p_2)\mathcal{G}(p_2 - q)V(\mathbf{p}_1 - \mathbf{p}_2)}{\int_{\mathbf{p}_1} \mathcal{G}(p_1)\mathcal{G}(p_1 - q)\int_{\mathbf{p}_2} \mathcal{G}(p_2)\mathcal{G}(p_2 - q)}, \tag{21.79}$$

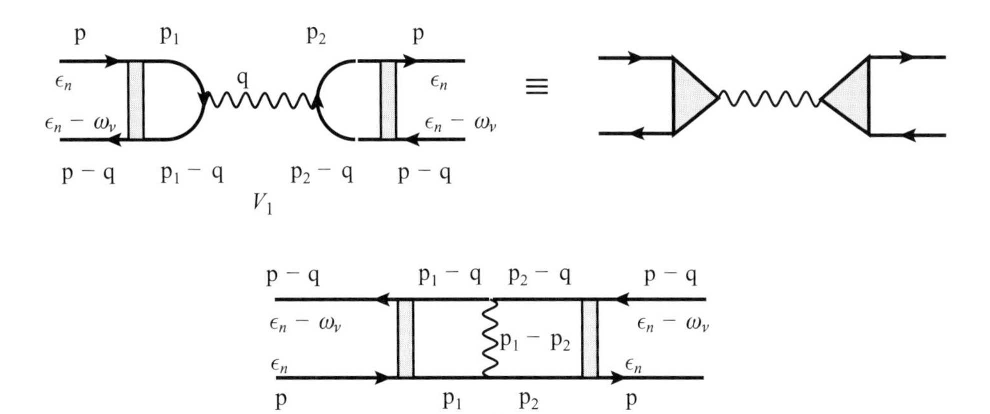

Fig. 21.7 Small, V_1, (top) and large, V_2, (bottom) momentum transfer scattering amplitudes.

showing that V_2 arises as an average of scattering events across the Fermi surface, since both $\mathbf{p}_1$ and $\mathbf{p}_2$ are on the Fermi surface due to the poles in the Green functions. Furthermore, in the evaluation of V_2 $\mathbf{q}$ can be set to zero due to the presence of the ladders. The rather involved expression of V_2 is not surprising considering the previous discussion about the correction of the thermodynamic potential.

By recalling that the corrections due to the combined effect of disorder and interaction involve typically small and large momentum transfer, we can rewrite the interaction term introduced in Eq. (9.40) as

$$H_{\text{int}} = \frac{1}{2VN_0} \sum_{\mathbf{k},\mathbf{k}',\mathbf{q}\sim 0,\alpha\beta}{}^{'} \left[V_1 a^{\dagger}_{\mathbf{k}+\mathbf{q},\alpha} a^{\dagger}_{\mathbf{k}'-\mathbf{q},\beta} a_{\mathbf{k}',\beta} a_{\mathbf{k},\alpha} + V_2 a^{\dagger}_{\mathbf{k}+\mathbf{q},\alpha} a^{\dagger}_{\mathbf{k}'-\mathbf{q},\beta} a_{\mathbf{k},\beta} a_{\mathbf{k}',\alpha} \right],$$

where the prime over the sum symbol indicates that the sum over $\mathbf{q}$ is limited to small values. We note that in the absence of spin-flip mechanisms, the total spin of the two colliding particles is a conserved quantity. By keeping fixed the momenta of the initial and final states in a scattering event, one realizes that different combinations of the two parameters V_1 and V_2 enter depending on the consideration of the singlet or triplet scattering channel. To see this explicitly, let us rewrite the interaction term by emphasizing the different spin structures of the small and large momentum transfer processes:

$$(V_1 \delta_{\alpha\delta}\delta_{\beta\gamma} - V_2 \delta_{\alpha\gamma}\delta_{\beta\delta}) a^{\dagger}_{\mathbf{k}+\mathbf{q},\alpha} a^{\dagger}_{\mathbf{k}'-\mathbf{q},\beta} a_{\mathbf{k}',\gamma} a_{\mathbf{k},\delta}.$$

By using the first Fierz identity (21.58), we write the interaction term as

$$V_1 \delta_{\alpha\delta}\delta_{\beta\gamma} - V_2 \delta_{\alpha\gamma}\delta_{\beta\delta} = \left(V_1 - \frac{1}{2} V_2 \right) \delta_{\alpha\delta}\delta_{\beta\gamma} - \frac{1}{2} V_2 \boldsymbol{\sigma}_{\alpha\delta} \cdot \boldsymbol{\sigma}_{\gamma\delta} \tag{21.80}$$

$$= \left(V_1 - \frac{1}{2} V_2 \right) \mathbf{I} - \frac{1}{2} V_2 \boldsymbol{\sigma} \cdot \boldsymbol{\sigma}', \tag{21.81}$$

where we have used the same form of the decomposition as introduced in Eq. (12.6) for the interaction function among the quasiparticles. We also note that the decomposition

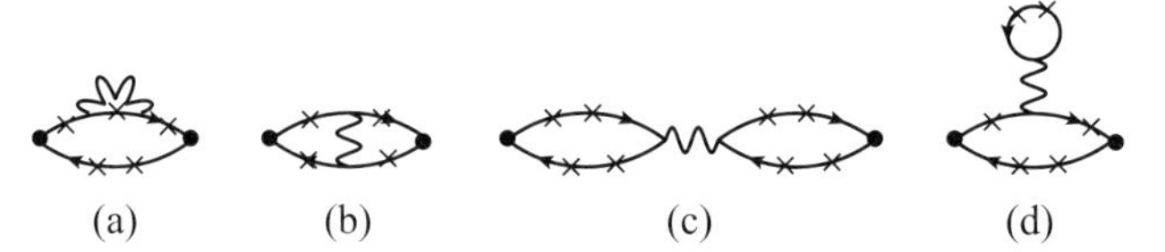

Fig. 21.8 Diagrams for the conductivity to the lowest order in the interaction in the presence of scattering by impurities. (a) and (b) exchange, (c) and (d) Hartree. Interaction is shown as a wiggly line. The dots are the current vertices $j_0^{\rm i} = p^{\rm i}/m$.

used in Eqs. (21.58) and (21.59) selects the *singlet* and *triplet* components with respect to the pair α and δ of *incoming* and the pair β and γ of *outgoing* indices, respectively. It is then convenient to introduce the *singlet* $V_{\rm s}$ and *triplet* $V_{\rm t}$ interaction parameters, which are nothing but the lowest order vertex part in the *static* limit, according to the terminology of Chapter 19,

$$V_{\rm s} = V_1 - \frac{1}{2}V_2, \quad V_{\rm t} = -\frac{1}{2}V_2. \tag{21.82}$$

21.6.3 Electrical conductivity

Finally, we consider the interaction corrections to the electrical conductivity. The physical origin is the same as for the equilibrium properties and density of states discussed so far, but their derivation is much more intricate. Basically, these corrections arise from the interference from two scattering processes. The scattering from impurities makes the electron density locally non-uniform. As a consequence, the Hartree–Fock potential dressed by disorder is non-uniform locally and adds up to the impurity potential as a source of random scattering. The interference between the scattering from impurities and from the impurity-modulated Hartree–Fock potential is the origin of the quantum interaction corrections to the electrical conductivity in disordered systems, i.e. the effective interaction becomes dynamically dressed by the diffusion. The quantitative evaluations of these corrections is best carried out in terms of diagrams, as done for the disordered Fermi gas. In Fig. 21.8 we show the diagrams contributing to the electrical conductivity to lowest order in the electron–electron interaction. Diagrams (a) and (d) are self-energy corrections and one must also include the specular diagrams with the self-energy insertion in the bottom Green function line. Diagrams (b) and (c) are vertex corrections, where the two Green function lines are connected by the interaction. The diagrams can also be divided in exchange ((a) and (b)) and Hartree ((c) and (d)) contributions. The latter have an extra fermionic loop and the associated factor of -2 with respect to the former.

As they stand the diagrams of Fig. 21.8 are assumed to include disorder insertions at arbitrary order as in the formal solution (18.89) of the Dyson equation (18.87). To obtain the effective diagrams, in analogy with the procedure followed for the density of states, we must perform the disorder average by dressing the diagrams with impurity dashed lines and making the ladder insertion. This procedure yields the effective exchange diagrams shown in Fig. 21.9. Similar diagrams can be obtained for the Hartree terms. The diagrams of

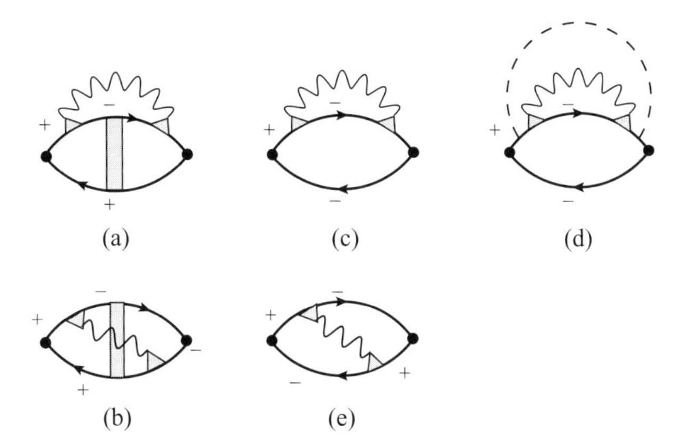

Fig. 21.9 Current–current response function diagrams after disorder averaging the exchange diagrams (a) and (b) of Fig. 21.8. Similar diagrams can be drawn for the Hartree contributions (c) and (d) of Fig. 21.8. The above diagrams are obtained first by dressing the interaction according to Fig. 21.7 and then inserting impurity lines in all possible ways without crossing. The insertion between Green functions with opposite sign of the energy results in a ladder, whereas only a single impurity line can be inserted between two advanced or two retarded Green functions.

Fig. 21.9 can be divided in two classes depending on whether impurity lines connect ((a) and (b)) or not ((c), (d) and (e)) the upper and lower Green function lines. In Appendix M.1, where we provide the detailed evaluation of the diagrams, it is shown that the three diagrams (c)–(e) cancel each other. Diagrams (c)–(e), as far as disorder dressing is concerned, affect the two Green function lines separately.

The evaluation of diagrams (a) and (b) reads

$$\delta\sigma = -\frac{2\sigma_0}{\pi d}\hbar \int_{\mathbf{q}} D\mathbf{q}^2 \int_{-\infty}^{\infty} \mathrm{d}\omega \frac{\partial}{\partial\omega}\left(\omega \coth\left(\frac{\beta\hbar\omega}{2}\right)\right) \mathrm{Im}\left[\frac{V(\mathbf{q})}{(D\mathbf{q}^2 - \mathrm{i}\omega)^3}\right] \tag{21.83}$$

where the dimensionality factor d comes from the angular integration over $\mathbf{q}$. As for the case of the evaluation of the density of states, the exchange contribution selects the small momentum transfer region due to the presence of the diffusive pole. We may then replace the potential $V(\mathbf{q})$ with its zero-momentum limit V_1 (see Eq. (21.68)). By considering also the disorder-averaged Hartree diagrams, one obtains a similar expression, with the amplitude V_1 replaced by V_2. In two dimensions, we then get the correction to the electrical conductivity:

$$\delta\sigma = -\frac{e^2}{\pi^2\hbar}(V_1 - 2V_2)\frac{1}{2}\ln\left(\frac{\hbar}{k_B T\tau}\right). \tag{21.84}$$

The above correction due to V_1, which has a physical origin distinct from the weak localization correction, has, however, the same localizing effect. The V_2 contribution comes with an opposite sign and is instead antilocalizing.

Before leaving the section, we mention that the self-energy diagram due to interaction can also be disorder dressed with maximally crossing impurity lines as shown in Fig. 21.10. Consideration of these diagrams requires the introduction of a further interaction constant besides V_1 and V_2. However, to keep our subsequent discussion simple, in the remaining

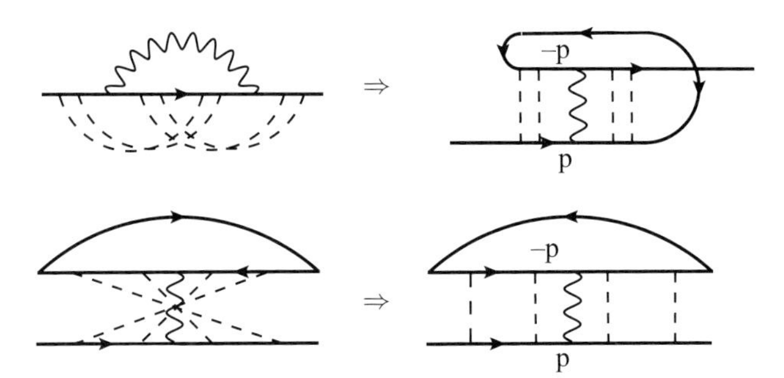

Fig. 21.10 Typical maximally crossed diagrams for the correction to the density of states due for exchange (top) and Hartree (bottom) contributions. Notice the combination of momenta of the propagating particle–particle pair typical of the cooperon channel.

part of this chapter we assume the presence of a small magnetic field in such a way that the cooperon ladder is not singular and hence diagrams of the type shown in Fig. 21.10 can be neglected.

21.7 Quantum theory of the disordered Fermi liquid

In this section we show that the corrections arising from the interplay of disorder and the interaction can be interpreted as a renormalized disordered Fermi liquid, where the Landau parameters, together with the electrical conductivity, acquire an anomalous behavior and are the set of running couplings flowing under the action of the renormalization group transformation discussed in the next section. The required renormalizations will be conditioned and controlled by means of the Ward identities derived for Fermi liquids in Chapters 18 and 19. For times $t > \tau$ (exchanged energy $\epsilon < \hbar/\tau$) the motion is diffusive and the disorder introduces retardation effects on the effective interactions. For exchanged energies $\epsilon > \hbar/\tau$, the disorder has no singular effect and the involved final dressing of the interaction produces the Landau theory of Fermi liquids of Chapter 19.

The first step in building a theory of the disordered Fermi liquid amounts to replace the lowest order vertex part $V_{1,2}$ with the full vertex part $\Gamma^q_{1,2} \equiv \Gamma_{1,2}$ so that the corresponding singlet and triplet amplitudes are

$$\Gamma_s = \Gamma_1 - \frac{1}{2}\Gamma_2, \quad \Gamma_t = -\frac{1}{2}\Gamma_2. \tag{21.85}$$

In Chapter 19, on the other hand, we have also shown that the Landau parameters are related to the *dynamic* limit of the vertex part (see e.g. Eqs. (19.27) and (19.30) for F_1 and F_0 respectively). The two limits of the vertex part are related by Eq. (19.13) for $l = 0$. This relation is written here in the form expressing the static limits Γ_s and Γ_t in terms of the Landau parameters F_0 and G_0 of the compressibility and the spin susceptibility:

$$z^2_{k_F}\nu^L_F\Gamma_s = \frac{F_0}{1+F_0}, \quad z^2_{k_F}\nu^L_F\Gamma_t = \frac{G_0}{1+G_0}. \tag{21.86}$$

We recall that the relation (19.13) has been derived by analyzing the singular part of the anomalous product of two Green functions as defined in Eq. (19.6). To develop a theory of disordered interacting electrons in terms of a *disordered* Fermi liquid, we have to make a key working hypothesis: the disorder has the main effect of renormalizing this singular part, which, instead of the form (19.6) of a *clean* Fermi gas, has the diffusive one appropriate for a *disordered* Fermi gas.

In terms of Γ_s and Γ_t the thermodynamic density of states $\left(\frac{\partial n}{\partial \mu}\right)^{L}$ (see Eq. (12.24)), the spin suceptibility χ^{L} (see Eq. (12.28)) and specific heat c_V^{L} (see Eq. (12.17)) read

$$\left(\frac{\partial n}{\partial \mu}\right)^{L} = \frac{\nu_F^L}{1+F_0} = Z_s^0 \frac{\nu_F^L}{\nu_F}\frac{\partial n}{\partial \mu}, \quad \chi^{L} = \chi \frac{\nu_F^L}{\nu_F}\frac{1}{1+G_0} = \frac{\nu_F^L}{\nu_F} Z_t^0 \chi, \quad c_V^L = \frac{\nu_F^L}{\nu_F} Z^0 c_V, \tag{21.87}$$

where $Z_s^0 = Z^0 - \nu_F^L z_{k_F}^2 \Gamma_s$, $Z_t^0 = Z^0 - \nu_F^L z_{k_F}^2 \Gamma_t$ and $Z^0 = 1$. The corresponding response functions have a common diffusive form (e.g. for the density–density Eq. (12.42)) with three different diffusion constants $D_{\rm ch} = (\nu_F/\nu_F^L)(1/Z_s^0)$, $D_{\rm spin} = (\nu_F/\nu_F^L)^2((1+G_1/3)/Z_t^0)$ and $D_{\rm Q} = (D/Z^0)(\nu_F/\nu_F^L)$ for charge, spin and heat, respectively.

From now on we first include the Fermi liquid wave function renormalization in the definition of the scattering amplitudes $z_{k_F}^2\Gamma_{s,t} \to \Gamma_{s,t}$. Secondly, for simplicity, we neglect the mass renormalization of the clean Fermi liquid $\nu_F^L = \nu_F \equiv \nu^L$ and we drop the factor $1 + G_1/3$ in the spin transport. The three diffusion constants then read

$$D_{\rm ch} = \frac{D}{Z_s^0}, \quad D_{\rm spin} = \frac{D}{Z_t^0}, \quad D_{\rm Q} = \frac{D}{Z^0}. \tag{21.88}$$

We begin our discussion of the interplay between interaction and disorder in the renormalization of the response functions, by considering the density–density response function. As in the case of the non-interacting theory of Eqs. (21.45) and (21.46), we split the response function into static and dynamic contributions. The effect of the interaction may be understood in terms of a skeleton perturbation theory as shown in Fig. 21.11. The first diagram (a), due to the ladder insertion, gives the dynamic contribution in the non-interacting case (see Eq. (21.46)). Its "dressing" due to the interaction is represented by diagram (b) with the three-point density vertex Λ_s (corresponding to the singlet interaction) and the diffusion ladder. This contribution, as will be explained soon, can be written in the form

$$\tilde{\mathcal{R}}_{(b)}^{00} = e^2 \nu^{L} \frac{|\omega_\nu| \zeta^2 \Lambda_s^2}{Dq^2/\hbar + |\omega_\nu| Z} \equiv \tilde{\mathcal{R}}_{(b)+-}^{00} \Lambda_s, \tag{21.89}$$

which generalizes Eq. (21.46). We now discuss it in detail. We begin with the ladder, which is the most important ingredient of the theory. The ladder, in fact, is the "effective" propagator of the diffusive mode, responsible for the singularities appearing in perturbation theory. It requires a *wave function* renormalization ζ, a frequency (effective external field)

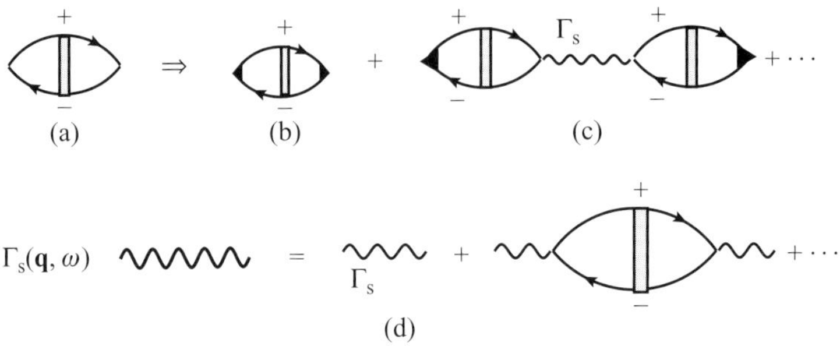

Fig. 21.11 Skeleton structure of the perturbative evaluation of the response function in the presence of interaction. The plus and minus sign indicate the sign of the energy flowing in the line. We recall that the ladder insertion is only possible between a pair of retarded and advanced Green functions. The first diagram (a) is the one used in the non-interacting theory. Diagram (b) represents how the interaction "dresses" (a). It leads to (i) a "dressing" of the vertex, here indicated as a black triangle, (ii) renormalization of the ladder. Interaction appears explicitly in diagram (c), which has to be considered together with all the other diagrams, indicated by the dots, obtained by the infinite resummation of the interaction. Depending on whether we consider the charge or spin response function one has the vertex $\Lambda_{\mathrm{s,t}}$ and interaction $\Gamma_{\mathrm{s,t}}$. For the energy response function, there is also a vertex Λ_{E}, but there is no infinite resummation of the interaction for the reasons explained in the text. (d) Dynamical resummation of the interaction.

renormalization Z, and a renormalization of the diffusion constant[5]

$$L(q,\omega) = \frac{\hbar^2}{\pi \nu^{\mathrm{L}} \tau^2} \frac{1}{Dq^2/\hbar + |\omega_\nu|} \rightarrow \frac{\hbar^2}{\pi \nu^{\mathrm{L}} \tau^2} \frac{\zeta^2}{Dq^2/\hbar + Z|\omega_\nu|}, \tag{21.90}$$

where now $\mathbf{q}$ and ω_ν have physical dimensions of momentum and energy, respectively. In Section M.2 it is shown that the interaction corrections to the ladder do not destroy its diffusive pole behavior and indeed lead to the renormalized form of Eq. (21.90). Furthermore, the logarithmically singular corrections to the single-particle density of states (see Eq. (21.78)), specific heat (see Eq. (21.69)) and conductivity (see Eq. (21.84)), which appear in two dimensions, can be absorbed into the above three renormalizations ζ, Z and D, respectively. Here, by exploiting the constraints given by the general conservation laws embodied in the Ward identities, we are able to directly express the ladder renormalization parameters in terms of physical quantities.

The definition of $\tilde{\mathcal{R}}^{00}_{(\mathrm{b})+-}$ in the last step of Eq. (21.89) defines the contribution of diagram (b) to $\tilde{\mathcal{R}}^{00}_{+-}$, which is the dynamic part of the response function ending with a retarded (+) and advanced (−) Green function. The vertex Λ_{s} is unity in the non-interacting case. In the interacting case, it represents the vertex which multiplies $\tilde{\mathcal{R}}^{00}_{+-}$ to give the total dynamic part of $\tilde{\mathcal{R}}^{00}$. Indeed, the interaction connects, via Λ_{s}, the $+-$ part to terms ending with two advanced ($--$) or two retarded ($++$) Green functions. The vertex Λ_{s} is irreducible for cutting a ladder propagator, otherwise its contribution would be included in the renormalization of the ladder.

[5] We use the same symbol as before for the interaction-renormalized diffusion constant to keep the notation simple. Everywhere in this section this is understood whenever we are dealing with the renormalized ladder.

Besides "dressing" the non-interacting diagram, interaction leads to new diagrams such as diagram (c) of Fig. 21.11. The contribution of this diagram to $\tilde{\mathcal{R}}^{00}$ can be obtained from the ladder (21.90) and by observing that, every time there are two Green function lines, the integration over momentum yields (see Problem 21.3) $2\pi N_0\tau/\hbar$ and summation over the restricted frequency range gives $|\omega_\nu|/2\pi$ (see the discussion following Eq. (21.46)):

$$\tilde{\mathcal{R}}^{00}_{(c)} = e^2\nu^{\mathrm{L}}\frac{|\omega_\nu|\zeta^2\Lambda_{\mathrm{s}}}{Dq^2/\hbar+|\omega_\nu|Z}\nu^{\mathrm{L}}\Gamma_{\mathrm{s}}\frac{|\omega_\nu|\zeta^2\Lambda_{\mathrm{s}}}{Dq^2/\hbar+|\omega_\nu|Z}. \tag{21.91}$$

By keeping in mind that the order in the expansion parameter r is determined by the number of integrations over the momenta flowing in the ladder propagator, we can replace, without changing the order in r, the scattering amplitude Γ_{s} by its screened form obtained by the infinite resummation (shown in Fig. 21.11(d)) and using Eq. (21.90):

$$\Gamma_{\mathrm{s}}(q,\omega_\nu) = \Gamma_{\mathrm{s}} + 2\Gamma_{\mathrm{s}}\frac{|\omega_\nu|}{2\pi}\frac{(\pi\nu^{\mathrm{L}}\tau)^2}{\hbar^2}L(q,\omega)\Gamma_{\mathrm{s}}(q,\omega_\nu) = \Gamma_{\mathrm{s}}\frac{Dq^2/\hbar+Z|\omega_\nu|}{Dq^2/\hbar+Z_{\mathrm{s}}|\omega_\nu|}. \tag{21.92}$$

$Z_{\mathrm{s}} = Z - \zeta^2\nu^{\mathrm{L}}\Gamma_{\mathrm{s}}$ is the expression dressed by the disorder of Z^0_{s} appearing in the Fermi-liquid compressibility. If we insert the dynamical amplitude (21.92) into the expression (21.91) for the first interaction correction and combine it with the contribution (21.89) of the first diagram, we arrive at the final form of the response function which resums all the infinite series of diagrams indicated by dots in Fig. 21.11:

$$\tilde{\mathcal{R}}^{00} = -e^2\frac{\partial n}{\partial\mu} + \tilde{\mathcal{R}}^{00}_{+-}\Lambda_{\mathrm{s}},\quad \tilde{\mathcal{R}}^{00}_{+-} = e^2\nu^{\mathrm{L}}\frac{|\omega_\nu|\zeta^2\Lambda_{\mathrm{s}}}{Dq^2/\hbar+|\omega_\nu|Z_{\mathrm{s}}}. \tag{21.93}$$

Global charge conservation (10.40) and the relation (21.53) for the full single-particle density of states ν applied to the two expressions in Eq. (21.93) allow us to eliminate the vertex Λ_{s} and to identify Z_{s} in terms of the compressibility and density of states:

$$\frac{\partial n}{\partial\mu} = \nu^{\mathrm{L}}\frac{\zeta^2\Lambda^2_{\mathrm{s}}}{Z_{\mathrm{s}}} = \nu\Lambda_{\mathrm{s}}, \tag{21.94}$$

$$\frac{\nu}{\nu^{\mathrm{L}}} = \frac{\zeta^2\Lambda_{\mathrm{s}}}{Z_{\mathrm{s}}}, \tag{21.95}$$

where the second equality in the first equation follows after insertion of Eq. (21.95). An immediate consequence of the global conservation (21.94) is that the density–density response function, at small $\mathbf{q}$ and ω_ν, has the usual diffusive form

$$\tilde{\mathcal{R}}^{00}(q,\omega) = -e^2\frac{\partial n}{\partial\mu}\frac{D_{\mathrm{ch}}q^2/\hbar}{D_{\mathrm{ch}}q^2/\hbar+|\omega_\nu|}, \tag{21.96}$$

where $D_{\mathrm{ch}} = D/Z_{\mathrm{s}}$ is the renormalized coefficient for charge diffusion.

From the perturbative calculation of Section M.2, we notice that ν/ν^{L} coincides with ζ. By assuming that this relation is valid at any order, it follows from Eq. (21.95) that $\zeta\Lambda_{\mathrm{s}} = Z_{\mathrm{s}}$, which when used in Eq. (21.94) gives

$$\frac{\partial n}{\partial\mu} = \nu^{\mathrm{L}}Z_{\mathrm{s}}. \tag{21.97}$$

Moreover, again perturbatively the compressibility has no logarithmic corrections, i.e., is given by the Fermi-liquid value in the absence of disorder with $Z_{\rm s} = \zeta\,\Lambda_{\rm s} = Z_{\rm s}^0$. This means that although both $\Lambda_{\rm s}$ and ζ have logarithmic corrections, the combination $\zeta\,\Lambda_{\rm s}$ does not.

In the presence of Coulomb long-range forces, to avoid double counting in the density–density response function, one has to subtract the statically screened long-range Coulomb Γ_0 from the full singlet scattering amplitude. This is because the screening of the Coulomb interaction includes already an infinite resummation of the density–density response function part which is irreducible for cutting a Coulomb interaction line:

$$\Gamma_0(q,\omega=0) = \frac{V_{\rm C}(q)\Lambda_{\rm s}^2}{1+V_{\rm C}(q)\partial n/\partial\mu} \to_{q\to 0} \frac{\Lambda_{\rm s}^2}{\partial n/\partial\mu}. \tag{21.98}$$

Hence, in the density–density response function we must operate the replacement $\Gamma_{\rm s} \to \Gamma_{\rm s} - \Gamma_0$ in $Z_{\rm s}$. As a consequence of Eq. (21.94) and Eq. (21.98), from $Z_{\rm s} = Z - \nu^{\rm L}\zeta^2(\Gamma_{\rm s} - \Gamma_0)$ we derive the constraint

$$Z = \nu^{\rm L}\zeta^2\Gamma_{\rm s}. \tag{21.99}$$

The analysis of the spin density–spin density and energy density–energy density response functions can be carried out in a way similar to that of the density–density response function. First we note that, as for $\Gamma_{\rm s}$, the retarded, due to the diffusion, scattering amplitude in the triplet channel reads

$$\Gamma_{\rm t}(q,\omega_\nu) = \Gamma_{\rm t} + 2\Gamma_{\rm t}\frac{|\omega_\nu|}{2\pi}\frac{(\pi\nu^{\rm L}\tau)^2}{\hbar^2}L(q,\omega_\nu)\Gamma_{\rm t}(q,\omega_\nu) = \Gamma_{\rm t}\frac{Dq^2/\hbar + Z|\omega_\nu|}{Dq^2/\hbar + Z_{\rm t}|\omega_\nu|}, \tag{21.100}$$

where $Z_{\rm t} = Z - \nu^{\rm L}\zeta^2\Gamma_{\rm t}$ is the expression dressed by the disorder of the Fermi-liquid renormalization for the spin susceptibility $Z_{\rm t}^0 = 1 - \nu^{\rm L}\Gamma_{\rm t}$. To analyze the spin density–spin density response function $\tilde{\chi}^{ab}(\mathbf{q},\omega_\nu)$, one introduces, in analogy with the density–density response function, the a-th component of the spin density ($a = 1, 2, 3$ determines the three components of the spin)

$$s^a(x) = \frac{g\mu_{\rm B}}{2}\hat{\psi}_\alpha^\dagger(x)\sigma_{\alpha\beta}^a\hat{\psi}_\beta(x), \tag{21.101}$$

and the associated spin–current density $\mathbf{j}^a$. Together they can be combined into a four-spin-current $j^{a,\mu} = (s^a, \mathbf{j}^a)$.

In the absence of the magnetic field, the calculation of the spin density–spin density response function would proceed exactly in the same way as for the density–density response function with the replacement $\Gamma_{\rm s} \to \Gamma_{\rm t}$. However, the study of the response function in the presence of the magnetic field will turn out to be useful in analyzing the flow of the group equations to be discussed in the next section.

In the continuity equation for the spin density, the magnetic field is a source term and produces the spin precession. In the presence of a magnetic field, chosen along the direction $a = 3$, the dynamical resummation of the skeleton structure, analog of Eq. (21.93) for $\tilde{\mathcal{R}}^{00}(q,\omega)$, with $\Gamma_{\rm s}$ replaced by $\Gamma_{\rm t}$ and $\partial n/\partial\mu$ or $\nu^{\rm L}$ by χ, gives

$$\tilde{\chi}^M_{+-}(\mathbf{q},\omega_\nu) = -\frac{|\omega_\nu|\chi^{\rm L}\zeta^2\Lambda_{\rm t}}{Dq^2/\hbar + |\omega_\nu|Z_{\rm t} - {\rm i}MZ_{\rm B}\Delta}, \quad \tilde{\chi}^M(q,\omega_\nu) = \chi(B) + \tilde{\chi}^M_{+-}(q,\omega_\nu)\Lambda_{\rm t}, \tag{21.102}$$

where we have introduced a renormalization factor, $Z_{\rm B}$, for the Zeeman energy Δ.

By proceeding as in the charge density case of Eq. (18.111), we introduce a vertex function

$$\Lambda^{a,\mu}_{\alpha\beta}(x,x',x'') = \langle T_{\rm t} j^{a,\mu}(x)\psi_\alpha(x')\psi^\dagger_\beta(x'')\rangle, \tag{21.103}$$

where the first upper index indicates which component of the spin density we are dealing with, while the second one, μ, distinguishes between density and current component. The Ward identity associated with $\Lambda^{a,\mu}$ reads (see Problem 21.9)

$$q_\mu\Lambda^{a,\mu}_{\alpha\beta}(p,q) = \frac{g\mu_{\rm B}}{2}\left[\mathcal{G}_\alpha(p-q/2)\sigma^a_{\alpha\beta} - \sigma^a_{\alpha\beta}\mathcal{G}_\beta(p+q/2)\right] + {\rm i}\epsilon_{ab3}\Delta\Lambda^{b,0}_{\alpha\beta}(p,q), \tag{21.104}$$

where ϵ_{ijk} is the full antisymmetric tensor and a source term appears.

We now consider the consequences of Eq. (21.104) in the limit $\mathbf{q}=\mathbf{0}$. Instead of working with the components $a=1,2$, it is convenient to switch to the circularly polarized ones defined as

$$\sigma^+ = \frac{1}{2}(\sigma^1 + {\rm i}\sigma^2), \quad \Lambda^{\uparrow\downarrow,0} = \frac{1}{2}(\Lambda^{1,0} + {\rm i}\Lambda^{2,0}), \tag{21.105}$$

$$\sigma^- = \frac{1}{2}(\sigma^1 - {\rm i}\sigma^2), \quad \Lambda^{\downarrow\uparrow,0} = \frac{1}{2}(\Lambda^{1,0} - {\rm i}\Lambda^{2,0}). \tag{21.106}$$

$\sigma^+ \equiv \sigma^{M=-1}$ and $\sigma^- \equiv \sigma^{M=1}$ are the standard raising and lowering spin operators and the vertices $\Lambda^{\uparrow\downarrow,0} \equiv \Lambda^{M=-1,0}$, $\Lambda^{\downarrow\uparrow,0} \equiv \Lambda^{M=1,0}$ correspond to the spin-density in the $M=\mp 1$ triplet channels, respectively, as shown in Fig. 21.12. By including the $M=0$ channel of the triplet corresponding to $\Lambda^{3,0}$, one gets the Ward identity (21.104) in the form (see Problem 21.10)

$$({\rm i}\omega_\nu + M\Delta)\Lambda^{M,0}_{\alpha\beta}(\epsilon_n,\omega_\nu) = \frac{g\mu_{\rm B}}{2}\left[\mathcal{G}_\alpha(\epsilon_-)\sigma^M_{\alpha\beta} - \sigma^M_{\alpha\beta}\mathcal{G}_\beta(\epsilon_+)\right], \tag{21.107}$$

where $\epsilon_\pm = \epsilon_n \pm \omega_\nu/2$ and σ^M are defined as in Eqs. (21.105) and (21.106) and we have dropped the explicit dependence on momentum $\mathbf{p}$ both in the vertex and in the Green functions.[6]

The analog of Eq. (10.40) for the spin density, in the limit $\mathbf{q}\to 0$, tells that in the channel $M=0$, where the spin is conserved, $\tilde{\chi}^{M=0}(\mathbf{0},\omega) = 0$. For the $M=\mp 1$ channel we rely on the Ward identity (21.107). To obtain $\tilde{\chi}^M$ we must multiply the vertex Λ^M by $g\mu_{\rm B}(\sigma^{-M})$, integrate over momentum and frequency and sum over spin variables (see Problem 21.11)

$$\tilde{\chi}^M(0,\omega_\nu) = \chi(B)\frac{M\Delta}{{\rm i}\omega_\nu + M\Delta}. \tag{21.108}$$

[6] Notice that the Green function gets a spin label in the presence of Zeeman coupling.

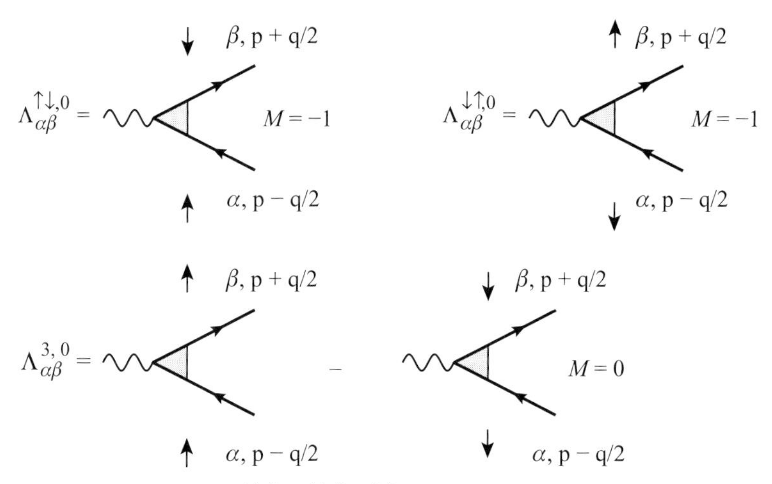

Fig. 21.12 The spin structure of the spin vertices $\Lambda_{\alpha\beta}^{\uparrow\downarrow,0}$, $\Lambda_{\alpha\beta}^{\downarrow\uparrow,0}$ $\Lambda_{\alpha\beta}^{3,0}$ controlling the channel $M = -1, M = 1$ and $M = 0$, respectively.

Indeed, in the limit of zero magnetic field Eq. (21.108) gives the total spin conservation by the vanishing of the response function at finite ω as $\mathbf{q}$ goes to zero, as for the $M = 0$ channel. By restricting the frequency integration range in analogy to Eq. (21.53), we select the dynamic contribution to the response function ending with retarded and advanced Green function lines (in analogy to Problem 21.11):

$$\tilde{\chi}_{+-}^{M}(0, \omega_\nu) = -\chi^{\mathrm{L}} \frac{\nu}{\nu^{\mathrm{L}}} \frac{\mathrm{i}\omega_\nu}{\mathrm{i}\omega_\nu + M\Delta}, \tag{21.109}$$

which in the zero-magnetic field limit gives the single-particle density of states but a constant.

We now impose the consequences (21.108)–(21.109) of the Ward identity on the skeleton structure (21.102) of the spin density–spin density response. The compatibility between the first condition (21.108) and the $\mathbf{q} = \mathbf{0}$ limit of Eq. (21.102) gives the analog of (21.94) and the identification of the renormalization of the external field Z_{B} as

$$\chi = \chi^{\mathrm{L}} \frac{\zeta^2 \Lambda_{\mathrm{t}}^2}{Z_{\mathrm{t}}}, \quad Z_{\mathrm{B}} = Z_{\mathrm{t}}. \tag{21.110}$$

The compatibility between Eqs. (21.109) and (21.102) now gives the renormalization of the single-particle density of states in analogy with (21.95):

$$\frac{\nu}{\nu^{\mathrm{L}}} = \frac{\zeta^2 \Lambda_{\mathrm{t}}}{Z_{\mathrm{t}}}. \tag{21.111}$$

By using again the perturbative result $\nu/\nu^{\mathrm{L}} = \zeta$, one obtains

$$\chi = \chi^{\mathrm{L}} Z_{\mathrm{t}}. \tag{21.112}$$

Finally, the small $\mathbf{q}$ and ω limit of the total response function has the form

$$\tilde{\chi}^{M}(q, \omega_\nu) = \chi \frac{D_{\mathrm{spin}} q^2/\hbar - \mathrm{i}M\Delta}{D_{\mathrm{spin}} q^2/\hbar - \mathrm{i}M\Delta + |\omega_\nu|}, \quad D_{\mathrm{spin}} = D/Z_{\mathrm{t}}, \tag{21.113}$$

which in the zero-magnetic field limit reproduces the standard diffusive form.

The energy–energy response function $\tilde{\chi}_{\mathrm{Q}}(\mathbf{q}, \omega_\nu)$ can also be decomposed analogously to Eq. (21.93), where $e^2\partial n/\partial\mu$ is replaced by $c_V T$, Z_{s} by Z and Λ_{s} by Λ_{Q}. In contrast to the density and spin response functions, in the energy response function the renormalization parameter does not require additional terms due to the interaction besides Z. The analog of diagram (c) of Fig. (21.11) is missing in the small frequency limit. This occurs since the interaction separates the integration at the two vertices in the response function diagrams. Due to the presence of the energy in the thermal vertex, each integration over the energy contributes at least a term with the second power of the frequency. Hence the terms due to the dynamical resummation of the interaction give rise to negligible contributions going with the fourth power of the frequency.

The analogous Ward identity gives the final form of the energy–energy response function in the standard diffusive form:

$$\tilde{\chi}_{\mathrm{Q}}(\mathbf{q}, \omega_\nu) = -c_V T \frac{D_{\mathrm{Q}} q^2/\hbar}{D_{\mathrm{Q}} q^2/\hbar + |\omega_\nu|}, \quad c_V = Z c_V^{\mathrm{L}}, \quad D_{\mathrm{Q}} = D/Z, \tag{21.114}$$

and the frequency renormalization Z of the ladder is identified with the specific heat renormalization.

In this way the general formulation of the effective renormalized Fermi-liquid theory is completed.

21.8 The renormalization group flow of the disordered Fermi liquid

In this section we discuss the physical picture emerging from the solution of the renormalization group equations, which can be derived by the perturbative analysis of the disordered Fermi liquid.

Before considering the explicit perturbative expressions, from the skeleton structure analysis given in the previous section, we can deduce various general consequences on the renormalization group flowing variables. Firstly, ζ^2 is always associated with either Γ_{s} and Γ_{t} and, if we do not consider explicitly the single-particle density of states, it will drop out of the renormalization group equations. Secondly, from the condition that the compressibility does not renormalize, Z_{s} does not renormalize. This means that we can consider either Z or Γ_{s} as the independent quantity to be renormalized. Finally, we are left with the Z renormalization associated with the specific heat, Z_{t} associated with the spin susceptibility and r, the disorder expansion parameter, associated with the electrical conductivity σ.

To evaluate the group equations explicitly and derive the physical scenario following from it, we need the first order (in r) expressions of the variation of the thermodynamical potential $\Delta\Omega$ and of the electrical conductivity σ for the skeleton structure theory.

We begin by recalling the perturbative expressions of the electrical conductivity (21.84), specific heat (21.69) and spin susceptibility (21.72). These expressions were derived to first order in the interaction. However, we have seen that we may relax this condition in two ways. Firstly, we may use the Fermi liquid scattering amplitudes. Secondly, we can make an infinite dynamical resummation (Eqs. (21.92), (21.100)).

The full expression for the correction to the thermodynamic potential is obtained by introducing in Eqs. (21.67) and (21.71) the singlet and triplet resummed scattering amplitudes and the renormalized ladder, including the magnetic-field renormalization $Z_{\mathrm{B}} = Z_{\mathrm{t}}$. In so doing, due to the presence of the dynamical resummation of the interaction, we use the standard trick of multiplying the interaction by a parameter $0 < \lambda < 1$ (see Eq. (18.80)). This is done by replacing Γ_{s} in front of L with $\lambda\Gamma_{\mathrm{s}}$ in Eq. (21.92). This procedure corresponds to the resummation depicted in Fig. 18.5. The same is for Γ_{t} in Eq. (21.100). For the magnetic-field independent part one then gets

$$\Delta\Omega = -\frac{1}{2\beta}\sum_{\omega_\nu}\int_{\mathbf{q}}\int_0^1 \mathrm{d}\lambda\left[\frac{\nu^{\mathrm{L}}\zeta^2\Gamma_{\mathrm{s}}|\omega_\nu|}{Dq^2/\hbar + (Z - \lambda\nu^{\mathrm{L}}\zeta^2\Gamma_{\mathrm{s}})|\omega_\nu|} + \frac{3\nu^{\mathrm{L}}\zeta^2\Gamma_{\mathrm{t}}|\omega_\nu|}{Dq^2/\hbar + (Z - \lambda\nu^{\mathrm{L}}\zeta^2\Gamma_{\mathrm{t}})|\omega_\nu|}\right], \tag{21.115}$$

while the field-dependent contribution reads

$$\Delta\Omega_{\mathrm{B}} = -\frac{1}{2\beta}\sum_{\omega_\nu}\int_{\mathbf{q}}\int_0^1 \mathrm{d}\lambda \times \sum_{M\pm 1}\left[\frac{\nu^{\mathrm{L}}\zeta^2\Gamma_{\mathrm{t}}|\omega_\nu|}{Dq^2/\hbar + (Z - \lambda\nu^{\mathrm{L}}\zeta^2\Gamma_{\mathrm{t}})|\omega_\nu| - \mathrm{i}M(Z - \nu^{\mathrm{L}}\zeta^2\Gamma_{\mathrm{t}})\Delta\,\mathrm{sign}(\omega_\nu)} - \frac{\nu^{\mathrm{L}}\zeta^2\Gamma_{\mathrm{t}}|\omega\nu|}{Dq^2/\hbar + (Z - \lambda\nu^{\mathrm{L}}\zeta^2\Gamma_{\mathrm{t}})|\omega_m|}\right]. \tag{21.116}$$

As a result, as shown in Problem 21.12, the corrections to the specific heat and to the spin susceptibility are

$$\delta c_V = \frac{1}{2}c_V^{\mathrm{L}} r(\nu^{\mathrm{L}}\zeta^2\Gamma_{\mathrm{s}} + 3\nu^{\mathrm{L}}\zeta^2\Gamma_{\mathrm{t}})\ln(k_{\mathrm{B}}T\tau/\hbar) = c_V^{\mathrm{L}}\delta Z, \tag{21.117}$$

$$\delta\chi = \chi^{\mathrm{L}} 2r\nu^{\mathrm{L}}\zeta^2\Gamma_{\mathrm{t}}\frac{Z_{\mathrm{t}}}{Z}\ln(k_{\mathrm{B}}T\tau/\hbar) = \chi^{\mathrm{L}}\delta Z_{\mathrm{t}}. \tag{21.118}$$

For the electrical conductivity, by inserting the dynamical scattering amplitudes in Eq. (21.83) one has

$$\begin{aligned}\frac{\delta\sigma}{\sigma} &= -\frac{2}{2\pi\mathrm{i}}\int_{-\infty}^{\infty}\mathrm{d}\omega\frac{\partial}{\partial\omega}\left(\omega\coth\left(\frac{\beta\omega}{2}\right)\right)\int_{\mathbf{q}}\frac{(Dq^2/\hbar)(\Gamma_{\mathrm{s}}(q,\omega) + 3\Gamma_{\mathrm{t}}(q,\omega))}{(Dq^2/\hbar - \mathrm{i}Z\omega)^3}\\ &= r\left[1 + \frac{Z_{\mathrm{s}}}{\nu^{\mathrm{L}}\zeta^2\Gamma_{\mathrm{s}}}\ln\frac{Z_{\mathrm{s}}}{Z} + 3\left(1 + \frac{Z_{\mathrm{t}}}{\nu^{\mathrm{L}}\zeta^2\Gamma_{\mathrm{t}}}\ln\frac{Z_{\mathrm{t}}}{Z}\right)\right]\ln(k_{\mathrm{B}}T\tau/\hbar)\\ &= -\frac{\delta r}{r},\end{aligned} \tag{21.119}$$

which coincides with $\delta D/D$ found from the ladder renormalization in Section M.2. Furthermore, for small Γ_{s} and Γ_{t} ($Z = 1$) one recovers the first order short-range interaction result of Eq. (21.84). In the case of Coulomb long-range forces, the constraint

Eq. (21.99) implies the disappearance of the second term of the singlet, i.e. the dressed singlet interaction acts as a localizing term exactly in the same way as the weak localization cooperon correction.

At zero kelvin the wave vector squared $(\mathbf{q}/\hbar)^2$ and frequency ω have l^{-2} (l mean free path) and τ^{-1} as ultraviolet cutoff Λ^2. By rescaling Λ with the parameter $\lambda^2 = (\Lambda')^2/\Lambda^2$, the running variables $(\mathbf{q}/\hbar)^2$ and ω/τ^{-1} are driven towards their infrared region. The logarithmic singularities in the running variables ($s = -\ln((\mathbf{q}/\hbar)^2/l^{-2}) \sim -\ln(\mathbf{q}/\hbar)^2/\tau^{-1})$ of the perturbation theory are replaced by the scaling parameter $s = -\ln\lambda^2$ ($s \to \infty$) of the flow equations of the renormalization group. At finite temperature the flow variable is $s = -\ln k_{\rm B}T\tau/\hbar$, so that $s \to \infty$ corresponds to the infrared limit. In the present case in getting the singularity we have in fact integrated the momentum variable in the entire region $0 < (\mathbf{q}/\hbar)^2 < l^{-2}$ and used the temperature $k_{\rm B}T$ as the infrared cutoff of the energy integration $(k_{\rm B}T\tau/\hbar) < \omega\tau < 1$.

By defining $\gamma_{\rm t} = -\nu^{\rm L}\zeta^2\Gamma_{\rm t}/Z$ ($Z_{\rm t} = Z(1+\gamma_{\rm t})$), from Eqs. (21.117) and (21.118) we have

$$\frac{{\rm d}Z}{{\rm d}s} = -\frac{r}{2}Z(1-3\gamma_{\rm t}), \tag{21.120}$$

$$\frac{{\rm d}Z_{\rm t}}{{\rm d}s} = 2rZ\gamma_{\rm t}(1+\gamma_{\rm t}). \tag{21.121}$$

By using the above equations, one obtains

$$\frac{{\rm d}\gamma_{\rm t}}{{\rm d}s} = \frac{r}{2}(1+\gamma_{\rm t})^2. \tag{21.122}$$

By writing the correction to the conductivity in Eq. (21.119) in terms of $\gamma_{\rm t}$ and confining ourselves to the long-range case, the dependence on Z drops out. One gets

$$\frac{{\rm d}r}{{\rm d}s} = r^2\left[1 + 3\left(1 - \frac{1+\gamma_{\rm t}}{\gamma_{\rm t}}\ln(1+\gamma_{\rm t})\right)\right]. \tag{21.123}$$

By resuming the weak-localization contribution, one would obtain a term identical to the singlet contribution (the first one in the square brackets). The two terms, although identical, have therefore a completely different origin. The presence of a magnetic field kills the weak-localization contribution and does not affect the singlet one. Moreover, the presence of the triplet interaction $\gamma_{\rm t}$ introduces an antilocalizing term.

Equations (21.122) and (21.123) together with Eq. (21.120) are the renormalization group equations at one-loop order for the problem of interacting disordered systems at $d = 2$, when there is no magnetic coupling in the system. In $d = 2+\epsilon$, the term coming from the bare dimension ϵ (Eq. (21.7)) of r, has to be added into Eq. (21.123):

$$\frac{{\rm d}r}{{\rm d}s} = -\epsilon\frac{r}{2} + r^2\left[1 + 3\left(1 - \frac{1+\gamma_{\rm t}}{\gamma_{\rm t}}\ln(1+\gamma_{\rm t})\right)\right]. \tag{21.124}$$

We now examine the consequences of the physical picture we have developed so far and briefly compare the results with experiments. To this end we discuss here in some detail the solution of the renormalization group equations. The first observation is that Eqs. (21.122) and (21.123) do not depend on Z. After solving for r and $\gamma_{\rm t}$ one may successively solve Eq. (21.120) for Z.

Let us consider first the case $d = 2$, i.e., $\epsilon = 0$. Equation (21.122) for γ_t implies a continuous growth. By integrating it between s_0 and s, one has

$$\frac{1}{1+\gamma_t(s)} = \frac{1}{1+\gamma_t(s_0)} - \frac{1}{2}\int_{s_0}^{s} \mathrm{d}s' r(s'), \tag{21.125}$$

from which one sees that γ_t diverges at a finite value, s_c, of the flow parameter:

$$1 = \frac{1}{2}(1+\gamma_t(s_0))\int_{s_0}^{s_c} \mathrm{d}s' r(s'). \tag{21.126}$$

Equation (21.123) for r says that after an initial increase for not too large $\gamma_t(s_0)$, the growth of γ_t makes the triplet antilocalizing contribution the dominant one. As a result, r goes through a maximum. In Fig. 21.13 we show the renormalization group flow in terms of the variables $r/(1+r)$ and $\gamma_t/(1+\gamma_t)$. For all the RG trajectories $\gamma_t = \infty$ at some finite value s_c, which depends on the initial values. Due to this, one cannot seriously trust the above equations quantitatively. Nevertheless, the appearance of a finite length scale for the spin fluctuations may indicate a self-formation of local magnetic moments on the same scale. Furthermore, the dominating antilocalizing effect of the triplet while r remains finite supports within this model the possibility of a metallic phase at low temperature associated with an enhancement of the spin susceptibility, in contrast with the non-interacting theory based on weak localization only.

We may finally notice that the inclusion of the cooperon contribution would modify the above group equations, without qualitative changes in the overall behavior.

In $d = 3$ one has a richer scenario depending on the initial value of the running variables. γ_t is again enhanced by the disorder. However, for a very small initial value of r, the term due to the bare dimension of r dominates the flow of the equations and there is the possibility of scaling to a conductor. For a large value of the initial value of r the quadratic term dominates and one expects a behavior similar to the one for $d = 2$. Hence, there must be a critical value $r_c(\epsilon, \gamma_{t0})$ depending on the initial value of γ_t, such that for the initial value of the disorder $r_0 < r_c$ the system scales to a conductor, while for $r_0 > r_c$, γ_t scales to infinity at a finite value s_0 of the scaling parameter. This critical line has been computed numerically and reported in Fig. 21.14. For large values of γ_t and small values of r it can be analytically controlled. In this regime, the product $r\gamma_t$ obeys the equation

$$\frac{\mathrm{d}(r\gamma_t)}{\mathrm{d}s} = \gamma_t\frac{\mathrm{d}r}{\mathrm{d}s} + r\frac{\mathrm{d}\gamma_t}{\mathrm{d}s} \approx \frac{r\gamma_t}{2}(r\gamma_t - \epsilon), \tag{21.127}$$

which has a fixed point for $r_c\gamma_{t,c} = \epsilon$, which is the asymptotic expression for a critical line in the $r - \gamma_t$ plane. Close to this critical line, for large values of γ_t, the approximate solution of Eq. (21.124) reads

$$r(s) = r(s_0)\mathrm{e}^{-\epsilon(s-s_0)/2}. \tag{21.128}$$

After substituing it into Eq. (21.122), one has

$$\gamma_t^{-1}(s) = \gamma_t^{-1}(s_0) - \frac{r(s_0)}{\epsilon} + \frac{r(s_0)}{\epsilon}\mathrm{e}^{-\epsilon(s-s_0)/2}. \tag{21.129}$$

One immediately sees from the above equations that for low disorder, $r(s_0) < \epsilon/\gamma_t(s_0)$, as $s \to \infty$, $r(s)$ scales to zero and $\gamma_t(s)$ scales to a finite value. In the high-disorder regime,

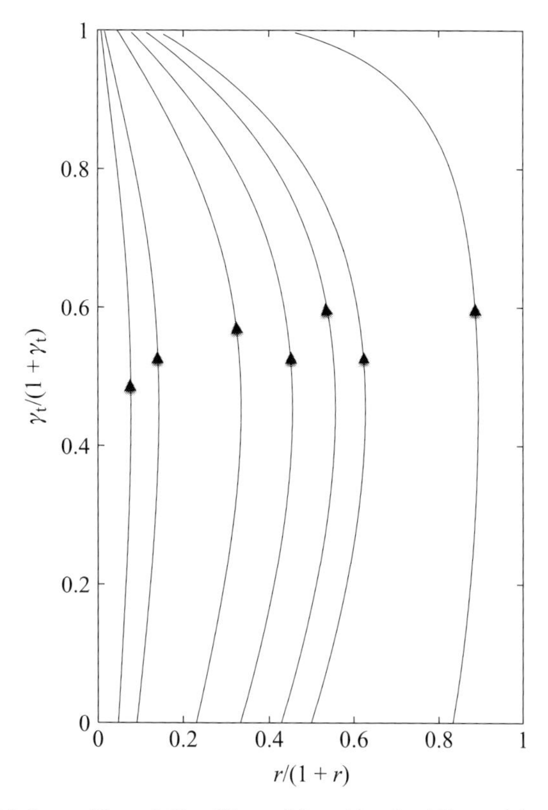

Fig. 21.13 The RG flow, for $d = 2$, in terms of the variable $r/(1+r)$ (x-axis) and $\gamma_t/(1+\gamma_t)$ (y-axis). In the figure the flow lines start on the x-axis with $\gamma_t = 0$. For all of them $\gamma_t = \infty$ at some finite value s_c, which depends on the initial values.

$r(s_0) > \epsilon/\gamma_t(s_0)$, γ_t diverges at a finite value of s as in the two-dimensional case and $r(s_c)$ stays finite. Since γ_t cannot become negative in the Fermi liquid, s cannot become larger than the value s_c for which the right-hand side of Eq. (21.129) vanishes. In Fig. 21.14 we report the flow obtained by numerically integrating the RG equations. We note, however, that the strong-coupling runaway flow requires us to go beyond the one-loop approximation presented here, leaving open the issue whether this proposed scenario is realized or not.[7]

Near the critical line for disorder less than r_c, by converting to a length scale via $s \sim 2\ln(L/l)$, one has that, while the resistance on the scale $L, r(L) = r(l)(L/l)^{-\epsilon}$ vanishes as $L \to \infty$, the conductivity

$$\sigma(L) = r_c^{-1}(L)L^{-\epsilon} \tag{21.130}$$

stays finite.

[7] An approximate treatment of the two-loop correction is possible, but its discussion is well outside the scope of this section. We refer the reader to the review by Belitz and Kirkpatrick (1994). This problem is still under investigation, see e.g. Punnoose and Finkel'stein (2002).

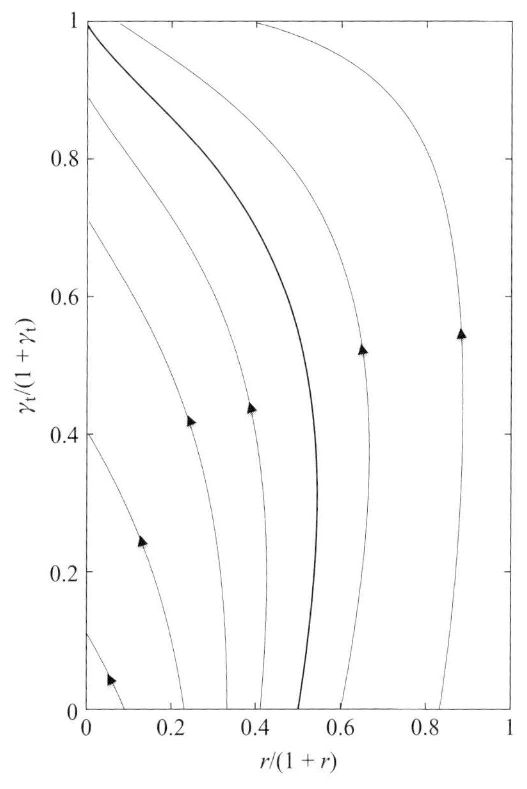

Fig. 21.14 The RG flow, in $d = 3$, in terms of the variable $r/(1+r)$ and $\gamma_t/(1+\gamma_t)$. In the figure the flow lines start on the x-axis with $\gamma_t = 0$. On the x-axis, there is a value $r/(1+r) \sim 0.5$ below which the RG flows to a state with zero r and a finite value of γ_t. The critical line originating from this value of r is shown by a thicker line. At large values of γ_t this critical line is well described by the approximate formula given in the text. For initial larger values of r, the RG flow is qualitatively similar to the two-dimensional case.

The equation for Z, in addition, gives now

$$\frac{\mathrm{d}Z}{\mathrm{d}s} \sim \frac{3}{2}\epsilon Z, \Rightarrow Z \sim L^{3\epsilon}. \tag{21.131}$$

In contrast to the finite value of σ, according to Eqs. (21.113) and (21.114) the spin and heat diffusion constants vanish, due to the divergenece of Z_t and Z. Higher order correction terms could modify this result leading to a vanishing conductivity. Alternatively, we can hypothesize that the critical line does not characterize the metal–insulator transition. The strong spin fluctuations which are the relevant physical effect associated with $D_s \to 0$ ($\chi \to \infty$) lead instead to an instability line before the localization takes place. In this case the system, before reaching the instability, due to the presence of the self-generated local moments acting as magnetic impurities, should make a crossover to one of the universality classes with magnetic couplings, which will be discussed soon. In any case, the theory predicts a low temperature enhancement of the specific heat and spin susceptibility, the latter being generally stronger, as observed in Si:P.

In the presence of any mechanism which inhibits the spin fluctuation enhancement, the localizing term in Eq. (21.124) for r dominates and one has a metal–insulator transition with $r_c \sim \mathcal{O}(\epsilon)$ with critical index for the conductivity $\mu = 1$.

In the presence of a magnetic field the ladders in the triplet channel with projection $M = \pm 1$ are not infrared singular. From Eq. (21.115) for the correction to the thermodynamic potential, one sees immediately that the spin susceptibility is no longer singular, i.e., Z_t does not renormalize. As a consequence, eliminating the contribution of the triplet components with $M = \pm 1$ in Eq. (21.120) for Z one obtains

$$\frac{dZ}{ds} = -\frac{r}{2} Z(1 - \gamma_t). \qquad (21.132)$$

Furthermore, using the relation $dZ_t/ds = d(Z(1 + \gamma_t))/ds = 0$, the equation for γ_t reads

$$\frac{d\gamma_t}{ds} = \frac{r}{2}\left(1 - \gamma_t^2\right). \qquad (21.133)$$

The above equations have a fixed point $\gamma_t^* = 1$ with a constant Z. It is now direct to obtain the equation for the parameter r. After suppressing the $M = \pm 1$ triplet channel contributions in Eq. (21.124) and by using the fixed-point condition $\gamma_t^* = 1$, one obtains

$$\frac{dr}{ds} = -\epsilon \frac{r}{2} + (2 - 2\ln 2) r^2, \qquad (21.134)$$

which has a fixed point $r^* = \epsilon/(2(2 - 2\ln 2))$ and gives $\mu = 1$ as in the non-magnetic impurity case for the non-interacting system. The thermal diffusion constant also vanishes.

In the case of magnetic impurities, spin is no longer conserved. Due to the spin relaxation time τ_{spin}, all triplet channels are massive, and γ_t drops out in the equations for Z and r, which now read

$$\frac{dr}{ds} = -\epsilon \frac{r}{2} + r^2, \qquad (21.135)$$

$$\frac{dZ}{ds} = -\frac{r}{2} Z. \qquad (21.136)$$

The equation for r gives the fixed point $r^* = \epsilon/2$ and the conductivity scaling exponent is again $\mu = 1$. The spin diffusion vanishes at the transition as D. By using the fixed point value for r in the equation for Z, one obtains

$$Z \approx e^{-(\epsilon/4)\ln s} \rightarrow \approx T^{\epsilon/4}. \qquad (21.137)$$

As $T \to 0$, the specific heat vanishes more rapidly than linearly (see Eq. (21.114)). Notice that when the contribution of the pure localization is also taken into account nothing happens for the magnetic impurities case. In fact, since magnetic impurities break time reversal invariance, there is no longer the interference effect responsible for the weak localization correction. Technically, this manifests with the appearance of a spin relaxation time in the diffusive pole of the cooperon ladder.

In the spin–orbit case, the situation is somehow intemediate. Time reversal invariance is still present, but spin is not conserved and only the singlets in the particle–particle and particle–hole ladder channel are left. As has been discussed when dealing with the non-interacting case, whereas the cooperon ladder in the triplet channel is no longer diffusive, the singlet one still yields a singular contribution. Furthermore, the triplet and singlet channels contribute with opposite signs. Hence, the standard weak localization correction is antilocalizing. It must be subtracted to the localizing term due to interaction in the singlet channel resulting in a global localizing term. The equation for r reads

$$\frac{\mathrm{d}r}{\mathrm{d}s} = -\frac{\epsilon}{2}r + \frac{r^2}{2}. \tag{21.138}$$

The critical index for the conductivity is the same as the one for the magnetic impurities $\mu = 1$, even though the fixed point value and the approach to it differ, giving $Z \sim T^{\epsilon/2}$. The thermal diffusion constant $D_{\mathrm{Q}} = D/Z$ stays finite. In the experiments discussed in Subsection 21.5 a value of $\mu = 1$ is observed, both in the absence and presence of a magnetic field, as in the present result of combined disorder and interaction effects. The magnetic field simply controls the antilocalizing contribution and only changes the approach to the fixed point, as observed experimentally (Katsumoto et al. (1987)).

From a theoretical point of view, we finally comment that in order to perform a quantitative analysis in the presence of both diffuson and cooperon channels requires, however, the inclusion of the interaction amplitude V_3 in the Cooper channel. We would see that all the universality classes discussed so far, share the same conductivity exponent $\mu = 1$, but differ as far as the behavior of Z (and hence of the specific heat) is concerned. To the best of our knowledge there are no experiments available to check this last point.

In conclusion, in Section 21.5 a few facts from the experiments were indicated which could not be dealt with in the non-interacting theory and would require the introduction of the interaction, which has been discussed here as the disordered Fermi liquid.

In two dimensions the most relevant result is the possibility of a metallic phase, which is observed to be suppressed by the magnetic field. However, a full account of this phase both theoretically and experimentally is not yet reached. All the theories realized so far give rise to a perfect metal with infinite conductivity.

In three dimensions, a bona fide metal–insulator transition is obtained whenever there is a magnetic coupling in the system and most of the puzzles met while discussing the non-interacting case have found plausible explanations. In the general case, with no magnetic coupling present, although the strong enhancement for the specific heat and spin susceptibility predicted by the theory appear also in some experiments, a deeper understanding is clearly needed and further theoretical and experimental work is required with particular emphasis on the formation of strong spin fluctuations to clear the scenario of the metal–insulator transition.

21.9 Problems

21.1 Evaluate the first order self-energy of the expression (21.31).

21.2 Evaluate the $\xi_{\mathbf{p}}$-integral of Eq. (21.40).

21.3 Derive a formula for $(m, n) \equiv \int_{\mathbf{p}} \mathcal{G}^m(\mathbf{p}, \epsilon_+)\mathcal{G}^n(\mathbf{p}, \epsilon_-)$ with the condition $\epsilon_+ > 0$ and $\epsilon_- < 0$.

21.4 Evaluate the electrical conductivity at leading order in the parameter $\hbar/E_F\tau$, by using the Kubo formula (10.31).

21.5 Evaluate the next-to-leading correction to the density–density response function.

21.6 Derive the ladder in the presence of a Zeeman term in the Hamiltonian.

21.7 Derive the relation (21.76).

21.8 Evaluate the density-of-states diagrams (c) and (d) of Fig. 21.6.

21.9 Derive the Ward identity (21.104).

21.10 Derive Eq. (21.107).

21.11 Derive Eq. (21.109).

21.12 Evaluate the expressions (21.115) and (21.116).

Appendix A The central limit theorem

Let $\{x_i\}$ be a set of N independent random variables, whose probability distribution $p(x_i)$ satisfies

$$\int_{-\infty}^{\infty} \mathrm{d}x_i\, p(x_i) = 1, \quad \int_{-\infty}^{\infty} \mathrm{d}x_i\, p(x_i) x_i = 0, \quad \int_{-\infty}^{\infty} \mathrm{d}x_i\, p(x_i) x_i^2 = \sigma^2, \; i = 1, \ldots, N, \tag{A.1}$$

where we have assumed zero mean for simplicity. Given the normalization condition, we assume the existence of the Fourier transform $\hat{p}(q_i)$ of the probability distribution

$$\hat{p}(q_i) = \int_{-\infty}^{\infty} \mathrm{d}x_i \mathrm{e}^{-\mathrm{i}q_i x_i} p(x_i), \tag{A.2}$$

in terms of which one can obtain the moments as

$$\frac{\mathrm{d}^n \hat{p}(q_i)}{\mathrm{d}q_i^n}|_{q_i=0} = (-\mathrm{i})^n \int_{-\infty}^{\infty} \mathrm{d}x_i\, p(x_i) x_i^n. \tag{A.3}$$

By virtue of the independence of the random variables, the joint probability distribution is simply the product of the distributions of all the variables $\mathrm{P}(\{x_i\}) = p(x_1) \cdots p(x_N)$. Let us consider now the variable x defined as

$$x = \frac{1}{\sqrt{N}} \sum_{i=1}^{N} x_i. \tag{A.4}$$

The probability distribution $\mathcal{P}(x)$ reads

$$\mathcal{P}(x) = \int_{-\infty}^{\infty} \mathrm{d}x_1 \cdots \int_{-\infty}^{\infty} \mathrm{d}x_N \mathrm{P}(\{x_i\}) \delta\left(x - \frac{1}{\sqrt{N}} \sum_{i=1}^{N} x_i\right). \tag{A.5}$$

The delta function can be expressed as a Fourier integral:

$$\delta\left(x - \frac{1}{\sqrt{N}} \sum_{i=1}^{N} x_i\right) = \int_{-\infty}^{\infty} \frac{\mathrm{d}\lambda}{2\pi} \mathrm{e}^{\mathrm{i}\lambda\left(x - \frac{1}{\sqrt{N}} \sum_{i=1}^{N} x_i\right)}. \tag{A.6}$$

The Fourier transform of $\mathcal{P}(x)$ then reads

$$\hat{\mathcal{P}}(q) = \int_{-\infty}^{\infty} \mathrm{d}x \mathrm{e}^{-\mathrm{i}qx} \int_{-\infty}^{\infty} \mathrm{d}x_1 \cdots \int_{-\infty}^{\infty} \mathrm{d}x_N \mathrm{P}(\{x_i\}) \int_{-\infty}^{\infty} \frac{\mathrm{d}\lambda}{2\pi} \mathrm{e}^{\mathrm{i}\lambda\left(x - \frac{1}{\sqrt{N}} \sum_{i=1}^{N} x_i\right)}. \tag{A.7}$$

After integrating over x one obtains $q = \lambda$ and the repeated integration over $x_1, \ldots, x_N$ yields

$$\hat{\mathcal{P}}(q) = \hat{p}^N\left(\frac{q}{\sqrt{N}}\right) = \mathrm{e}^{N \ln \hat{p}(q/\sqrt{N})}. \tag{A.8}$$

The function $\hat{p}(q/\sqrt{N})$ can be Taylor expanded:

$$\hat{p}(q/\sqrt{N}) = 1 - \frac{\sigma^2}{2N} q^2 + \cdots , \tag{A.9}$$

where the dots indicate terms going like N^{-l} with $l \geq 2$, which will eventually be negligible in the limit of large N. Hence, at large N, one has

$$\hat{\mathcal{P}}(q) \approx \mathrm{e}^{-\sigma^2 q^2/2}. \tag{A.10}$$

The probability distribution $\mathcal{P}(x)$, being the antitransform of a Gaussian, is still a Gaussian with variance σ^2:

$$\hat{\mathcal{P}}(x) = \frac{1}{\sqrt{2\pi\sigma^2}} \mathrm{e}^{-x^2/2\sigma^2}. \tag{A.11}$$

Hence, in the limit of large N the sum of N identically distributed, independent random variables with finite variance has a Gaussian distribution probability, provided the sum is normalized with the factor $1/\sqrt{N}$.

Appendix B Some useful properties of the Euler Gamma function

By integrating the definition (3.57) by parts we have the recurrence formula

$$\begin{aligned}\Gamma(z) &= t^{z-1}\mathrm{e}^{-t}|_{\infty}^{0} + (z-1)\int_0^{\infty} \mathrm{d}t\mathrm{e}^{-t}t^{z-2} \\ &= (z-1)\Gamma(z-1).\end{aligned} \tag{B.1}$$

For $z = n$ integer, the Gamma function is just the factorial

$$\Gamma(n) = (n-1)\Gamma(n-1) = \cdots = (n-1)!. \tag{B.2}$$

For half-integer z, we first consider $z = 1/2$:

$$\begin{aligned}\Gamma(1/2) &= \int_0^{\infty} \mathrm{d}t\frac{\mathrm{e}^{-t}}{\sqrt{t}} \\ &= 2\int_0^{\infty} \mathrm{d}x\mathrm{e}^{-x^2} \\ &= \sqrt{\pi}\end{aligned} \tag{B.3}$$

and then use the recurrence formula.

For large values of z, one can evaluate the integral in (3.57) approximatly with the saddle point method. First we rewrite the Gamma function as

$$\Gamma(z) = \int_0^{\infty} \mathrm{d}t\ \mathrm{e}^{-f(t,z)},$$

where

$$f(t,z) = t - (z-1)\ln t.$$

As a function of t, $f(t,z)$ has a minimum in $t^* = z-1$, since

$$f'(t,z) = 1 - \frac{z-1}{t}, \quad f''(t,z) = \frac{z-1}{t^2},$$

where the prime indicates the derivative with respect to t. Hence

$$\begin{aligned}\Gamma(z) &= \int_0^\infty \mathrm{d}t\ \mathrm{e}^{-f(t^*,z)-(1/2)f''(t^*,z)(t-t^*)^2} \\ &= \mathrm{e}^{-(z-1)+(z-1)\ln(z-1)} \int_0^\infty \mathrm{d}t\ \mathrm{e}^{-(t-t^*)^2/(2(z-1))} \\ &\approx (z-1)^{z-1}\mathrm{e}^{-(z-1)} \int_{-\infty}^\infty \mathrm{d}t\ \mathrm{e}^{-(t-t^*)^2/(2(z-1))} \\ &= (z-1)^{z-1}\mathrm{e}^{-z+1}\sqrt{2\pi(z-1)} \\ &\approx z^z\mathrm{e}^{-z}\sqrt{2\pi z}. \end{aligned} \tag{B.4}$$

It may interesting to notice that in Eq. (B.4), the factor under the square root gives the contribution due to the deviation of the value of the function with respect to the minimum value, whereas all other factors originate from the estimate of the integral at the saddle-point level. Finally, for large integer z, Eq. (B.4) yields the Stirling formula

$$n! \approx \sqrt{2\pi n}n^n\mathrm{e}^{-n}, \quad n \to \infty. \tag{B.5}$$

Appendix C Proof of the second theorem of Yang and Lee

We recall that all the roots of the grand partition function lie off the positive real axis. As the region R is always free of roots we take a circle of radius σ centered at $z = \eta$, where η is on the intersection of R and the positive real axis. By introducing the new variable $y = z - \eta$, we rewrite (5.2) as

$$\mathcal{Z}(z) = \prod_{i=1}^{N_{\max}} \left(1 - \frac{y}{y_i}\right) \frac{y_i}{z_i}, \tag{C.1}$$

from which

$$\frac{\ln \mathcal{Z}}{V} = \frac{1}{V} \sum_{i=1}^{N_{\max}} \left[\ln\left(1 - \frac{y}{y_i}\right) + \ln\left(\frac{y_i}{z_i}\right)\right]. \tag{C.2}$$

Since the roots y_i lie outside the circle, i.e., $|y_i| > \sigma$, for any y within the circle $|y/y_i| < 1$. We may then expand the logarithm in (C.2) and get

$$\ln\left(1 - \frac{y}{y_i}\right) = -\sum_{l=1}^{\infty} \frac{1}{l} \left(\frac{y}{y_i}\right)^l,$$

which yields

$$\frac{\ln \mathcal{Z}}{V} = \sum_{l=0}^{\infty} b_l(V) y^l \tag{C.3}$$

with

$$b_0(V) = \frac{1}{V} \sum_{i=1}^{M} \ln\left(\frac{y_i}{z_i}\right), \tag{C.4}$$

$$b_l(V) = -\frac{1}{V} \sum_{i=1}^{M} \frac{1}{l} \left(\frac{1}{y_i}\right)^l. \tag{C.5}$$

Equation (C.3) states that, for any value of V, the function $\ln \mathcal{Z}/V$ is a power series in y, uniformly converging in the circle $|y| < \sigma$. By using the result of the first theorem, we have that, for any y,

$$\lim_{V \to \infty} \frac{\ln \mathcal{Z}}{V} = \phi_0(y).$$

Then by setting $y = 0$ in (C.3) we obtain

$$\phi_0(0) = \lim_{V \to \infty} b_0(V). \tag{C.6}$$

From the expression (C.5), if $\epsilon < \sigma/2$,

$$\sum_{l=2}^{\infty} b_l(V)\epsilon^l \leq A \sum_{l=2}^{\infty} \left(\frac{\epsilon}{\sigma}\right)^l = A\left\{\frac{1}{1-\epsilon/\sigma} - 1 - \frac{\epsilon}{\sigma}\right\} = A\frac{\epsilon^2/\sigma^2}{1-\epsilon/\sigma} \leq 2A'\epsilon^2, \quad \text{(C.7)}$$

where $A' = A/\sigma^2$ and A is a constant larger than $M(V)/V$. It is then required that $M(V)$ does not grow faster than V. Now we can write

$$\epsilon b_1(V) = \frac{\ln \mathcal{Z}(\epsilon)}{V} - b_0(V) - \sum_{l=2}^{\infty} b_l(V)\epsilon^l. \quad \text{(C.8)}$$

Since $0 < \epsilon < \sigma/2$, we can take ϵ as small as we wish. Equation (C.7) tells us that the third term on the right-hand side of (C.8) is smaller than $2A'\epsilon^2$. Since both $b_0(v)$ and $\ln \mathcal{Z}/V$ have limits as $V \to \infty$, by taking a pair of volumes V and W large enough, we may satisfy

$$\left|\frac{\ln \mathcal{Z}}{V} - \frac{\ln \mathcal{Z}}{W}\right| < \epsilon^2$$

and

$$|b_0(V) - b_0(W)| < \epsilon^2,$$

from which it follows, via (C.8), that

$$|\epsilon b_1(V) - \epsilon b_1(W)| < 2(A'+1)\epsilon^2 \quad \text{(C.9)}$$

and by the Cauchy convergence test there must exist the limit

$$\lim_{V\to\infty} b_1(V) = b_1. \quad \text{(C.10)}$$

By iterating the reasoning, one can prove for any l, that

$$\lim_{V\to\infty} b_l(V) = b_l. \quad \text{(C.11)}$$

Putting together (C.11), (C.3) and the first theorem, we get

$$\phi_0(y) = \sum_{l=0}^{\infty} b_l y^l, \quad \text{(C.12)}$$

with the power series uniformly converging within the circle $|y| < \sigma$. Hence the series in (C.12) is an analytic function of y.

Appendix D The most probable distribution for the quantum gases

We show here how the equilibrium Fermi–Dirac and Bose–Einstein distributions arise as the most probable ones.

The Boltzmann procedure of discretizing the phase space Γ can be applied also in the quantum case. The coarse graining process is carried out by taking the volume element ω_l as the number of states labelled with an index i, all having energy values ϵ_i equal or near to a given energy level ϵ_l. The occupation n_l of the l-th cell is given by

$$n_l = \sum_{i \in (l\text{-th cell})} n_i ,$$

where n_i is the occupation of the i-th level of the l-th cell. For a Bose system, given a set of n_l, the problem is to distribute n_l indistinguishable objects among ω_l states with no restriction of the occupation of each state. Remember that once the occupation of $\omega_l - 1$ states is chosen, the last one is fixed. This leads to

$$V_\Gamma(\{n_l\}) = \prod_l \frac{(\omega_l - 1 + n_l)!}{n_l!(\omega_l - 1)!}. \tag{D.1}$$

The combinatorial factor $N!/\prod_l n_l!$ of Eq. (3.12) is now absent since for indistinguishable objects there is one way only to distribute N objects for a given set of n_l. By considering the variation of the logarithm of $V_\Gamma(\{n_l\})$ with the constraints (3.11), where now the sum is extended to all coarse grained cells, and considering both ω_l and $n_l \gg 1$, we obtain

$$\delta n_l \left[\ln(\omega_l - 1 + n_l) - \ln n_l - \alpha - \beta\epsilon_l\right] = 0, \tag{D.2}$$

where again the Lagrange multipliers α and β have to be determined by implementing the constraints. Then the Bose–Einstein distribution follows for the average occupation, $\bar{n}_l/\omega_l$, of each state i belonging to the cell:

$$\frac{\bar{n}_l}{\omega_l} = \frac{1}{\mathrm{e}^{\alpha+\beta\epsilon_l} - 1}. \tag{D.3}$$

When the system is diluted,

$$n_l \ll \omega_l, \tag{D.4}$$

the classical limit is recovered by observing that

$$\frac{(\omega_l - 1 + n_l)!}{(\omega_l - 1)!} = \omega_l\,(\omega_l + 1)\cdots(\omega_l - 1 + n_l) \approx \omega_l\,\omega_l\cdots\omega_l = \omega_l^{n_l}. \tag{D.5}$$

Substituting Eq. (D.5) in Eq. (D.1), one reproduces the formula (3.12) with no $N!$ from the beginning.

In the case of fermions, the Pauli exclusion principle tells us that in each state we can accommodate no more than one particle. Then, if we have ω_l states available for n_l particles, the problem is to count the ways we can choose n_l objects out of ω_l. This leads to

$$V_\Gamma(\{n_l\}) = \prod_l \frac{\omega_l!}{n_l!(\omega_l - n_l)!}. \tag{D.6}$$

By considering the variation with respect to the occupation numbers, we get

$$\delta n_l \left[\ln(\omega_l - n_l) - \ln n_l - \alpha - \beta\epsilon_l\right] = 0, \tag{D.7}$$

from which the Fermi–Dirac distribution function for the average occupation of the states, $\bar{n}_l/\omega_l$, in the cell follows:

$$\frac{\bar{n}_l}{\omega_l} = \frac{1}{\mathrm{e}^{\alpha+\beta\epsilon_l} + 1}. \tag{D.8}$$

Notice that the average occupation number $\langle n_i \rangle$, derived in the main text, coincides with the average occupation per state $\bar{n}_l/\omega_l$ given by the ratio between the most probable occupation of the l-th cell and the number of states in the cell with $\epsilon_i \approx \epsilon_n$ for all the states in the cell. With this last identification, the expression of the entropy (7.7) corresponds to $S = k_\mathrm{B} \ln V_\Gamma(\{\bar{n}_l\})$, where $V_\Gamma(\{\bar{n}_l\})$ is given by Eqs. (D.1) and (D.6), respectively.

E

Appendix E Fermi–Dirac and Bose–Einstein integrals

In dealing with the statistics of the quantum gases we encounter the Fermi–Dirac and Bose–Einstein integrals, $I_\alpha^\pm(x)$ (7.18). In this appendix we provide a number of properties of these integrals as well as their evaluation in various ranges of the variable $x = \beta\mu$. For more details and further discussion we refer to the book by Dingle (1973) and references therein.

Let us begin with their definition:

$$I_\alpha^\pm(x) = \int_0^\infty \mathrm{d}y \, \frac{y^\alpha}{\mathrm{e}^{y-x} \pm 1}, \tag{E.1}$$

where α is called the order of the integral. In general α is complex, but we limit ourselves to real values. In practice, we will only need integer and half-integer values. By observing that $(\mathrm{e}^x + 1)^{-1} = (\mathrm{e}^x - 1)^{-1} - 2(\mathrm{e}^{2x} - 1)^{-1}$, we get

$$I_\alpha^+(x) = I_\alpha^-(x) - 2^{-\alpha} I_\alpha^-(2x). \tag{E.2}$$

For $x \leq 0$, we may use the geometric series expansion to get

$$\begin{aligned} I_\alpha^-(x) &= \int_0^\infty \mathrm{d}y \, y^\alpha \sum_{k=0}^\infty \mathrm{e}^{-(k+1)(y-x)} \\ &= \sum_{k=1}^\infty \frac{\mathrm{e}^{kx}}{k^{\alpha+1}} \int_0^\infty \mathrm{d}s \, s^\alpha \mathrm{e}^{-s} \\ &= \mathrm{Li}_{\alpha+1}(\mathrm{e}^x)\Gamma(\alpha + 1), \end{aligned} \tag{E.3}$$

where the function $\Gamma(z)$ was discussed in Appendix B and the polylogarithm $\mathrm{Li}_s(z)$ is defined as

$$\mathrm{Li}_s(z) = \sum_{k=1}^\infty \frac{z^k}{k^s}, \tag{E.4}$$

s being the order of the polylogarithm. Notice that for $s = 1$, $\mathrm{Li}_1(z) = -\ln(1 - z)$. The polylogarithm obeys the recurrence property

$$\frac{\mathrm{dLi}_s(z)}{\mathrm{d}z} = \frac{1}{z}\mathrm{Li}_{s-1}(z). \tag{E.5}$$

By using Eq. (E.2), one has

$$I_\alpha^+(x) = -\mathrm{Li}_{\alpha+1}(-\mathrm{e}^x)\Gamma(\alpha + 1). \tag{E.6}$$

Since in most software packages presently available the polylogarithm and its analytical continuation in the complex plane can be computed with high precision, for practical purposes we could end our discussion here. However, for the physical discussion of the properties of the quantum gases, it is useful to have analytical expressions, even though approximate, of the integrals $I_\alpha^\pm$.

When $x \to 0^-$, or the fugacity $z = \mathrm{e}^x \to 1^-$, we have

$$I_\alpha^-(0) = \Gamma(\alpha+1)\zeta_\mathrm{R}(\alpha+1), \tag{E.7}$$

$$I_\alpha^+(0) = \Gamma(\alpha+1)(1-2^{-\alpha})\zeta_\mathrm{R}(\alpha+1) \tag{E.8}$$

where we have introduced the Riemann zeta function $\zeta_\mathrm{R}(s)$:

$$\zeta_\mathrm{R}(s) = \sum_{k=1}^{\infty} \frac{1}{k^s}. \tag{E.9}$$

In the following we will mostly need the zeta function at integer and half-integer values. We list below some of them:

$$\zeta_\mathrm{R}(3/2) \approx 2.612, \tag{E.10}$$

$$\zeta_\mathrm{R}(2) = \frac{\pi^2}{6} \approx 1.645, \tag{E.11}$$

$$\zeta_\mathrm{R}(5/2) \approx 1.342, \tag{E.12}$$

$$\zeta_\mathrm{R}(3) \approx 1.202, \tag{E.13}$$

$$\zeta_\mathrm{R}(7/2) \approx 1.127, \tag{E.14}$$

$$\zeta_\mathrm{R}(4) = \frac{\pi^4}{90} \approx 1.0823. \tag{E.15}$$

For $\alpha = -1/2$ the Bose–Einstein integral diverges when $x \to 0^-$. In the theory of the Bose degenerate gas it is useful to quantify such a divergence. For x small and negative, by expanding the exponential in the denominator, since most of the integral value comes from the lower limit of integration, we have

$$\begin{aligned} I_{-1/2}^- &\approx \int_0^\infty \mathrm{d}y\, \frac{1}{\sqrt{y}} \frac{1}{y+|x|} \\ &= 2\int_0^\infty \mathrm{d}t\, \frac{1}{t^2+|x|} \\ &= \frac{\pi}{\sqrt{|x|}}. \end{aligned} \tag{E.16}$$

In the limit of large positive x it is possible to derive an asymptotic expansion for the Fermi–Dirac integral. We closely follow the derivation by Dingle, confining ourselves to the case of half-integer $\alpha = 1/2 + m$, $m \geq -1$. After integration by parts we have

$$I_{1/2+m}^+(x) = \frac{1}{3/2+m} \int_0^\infty \mathrm{d}y\, \frac{y^{m+3/2}}{(\mathrm{e}^{y-x}+1)(\mathrm{e}^{x-y}+1)}. \tag{E.17}$$

Due to the half-integer power in the numerator, the above integral can be written as the real part of the integral over the full real axis:

$$I_{1/2+m}^+(x) = \frac{1}{3/2+m} \mathrm{Re} \int_{-\infty}^\infty \mathrm{d}y\, \frac{y^{m+3/2}}{(\mathrm{e}^{y-x}+1)(\mathrm{e}^{x-y}+1)}, \tag{E.18}$$

since $\sqrt{-|y|} = \pm i\sqrt{|y|}$, the sign depending on the convention for making the square root definite. Once the integral is over the full real axis we can make the change of variable $w = y - x$ to take advantage of the symmetry of the denominator around $w = 0$:

$$I^{+}_{1/2+m}(x) = \frac{1}{3/2+m} \mathrm{Re} \int_{-\infty}^{\infty} \mathrm{d}w \, \frac{(x+w)^{m+3/2}}{(\mathrm{e}^{w}+1)(\mathrm{e}^{-w}+1)}. \tag{E.19}$$

In order to find an asymptotic expansion for large values of x, we exploit the fact that the most important contribution of the integral comes from the values around $w = 0$ due to the fast varying function $((\mathrm{e}^{w}+1)(\mathrm{e}^{-w}+1))^{-1}$. Then, we expand the power in the numerator as

$$(x+w)^{m+3/2} = x^{m+3/2} \left(1 + (m+3/2)\frac{w}{x} + \frac{(m+3/2)(m+1/2)}{2!}\frac{w^2}{x^2} + \cdots \right). \tag{E.20}$$

We now replace the above expansion in Eq. (E.19) and integrate term by term. Since only even powers survive, we get

$$I^{+}_{1/2+m}(x) = \frac{x^{m+3/2}}{m+3/2} \left(1 + \frac{(m+3/2)(m+1/2)}{2!} 4I_1^{+}(0)\frac{1}{x^2} + \cdots \right), \tag{E.21}$$

where we have used

$$\int_{-\infty}^{\infty} \mathrm{d}w \, \frac{w^2}{(\mathrm{e}^{w}+1)(\mathrm{e}^{-w}+1)} = 4I_1^{+}(0), \tag{E.22}$$

which follows by reducing integration to positive w and integration by parts. The observant reader may notice that the expansion (E.20) is not valid over the entire range of integration since the series converges only for $|w| < x$. However, for large x, the function $((\mathrm{e}^{w}+1)(\mathrm{e}^{-w}+1))^{-1}$ makes the integral negligible before w reaches x. It is also clear that taking the real part becomes redundant since potential imaginary terms are negligible at large x. We refer to Dingle's book (1973) for a general discussion about the form of the asymptotic expansion for arbitrary α and complex x.

In the theory of the degenerate Fermi gas we especially need the Fermi–Dirac integrals for $\alpha = 1/2$ and $\alpha = 3/2$. By using Eqs. (E.8) and (E.11) we get

$$I^{+}_{1/2}(x) = \frac{2}{3}x^{3/2} \left(1 + \frac{3\pi^2}{24}\frac{1}{x^2} + \cdots \right), \tag{E.23}$$

$$I^{+}_{3/2}(x) = \frac{2}{5}x^{5/2} \left(1 + \frac{15\pi^2}{24}\frac{1}{x^2} + \cdots \right). \tag{E.24}$$

Appendix F The Fermi gas in a uniform magnetic field: Landau diamagnetism

In this appendix we show that a Fermi gas has a diamagnetic orbital susceptibility which is one third of the Pauli spin susceptibility of Eq. (7.32). This is known as Landau diamagnetism.

By adopting the Landau gauge to describe a static and uniform magnetic field along the z axis

$$\mathbf{A} = (0, xB, 0), \tag{F.1}$$

the Hamiltonian for the particles of the Fermi gas reads

$$H = \frac{1}{2m}\left[p_x^2 + \left(p_y + \frac{eB}{c}x\right)^2 + p_z^2\right]. \tag{F.2}$$

Since H does not depend on y and z, we look for a solution of the form

$$\psi(x, y, z) = e^{ik_y y} e^{ik_z z}\phi(x). \tag{F.3}$$

The equation for $\phi(x)$ becomes

$$\frac{1}{2m}\left[p_x^2 + \left(\hbar k_y + \frac{eB}{c}x\right)^2\right]\phi(x) = \left(E - \frac{1}{2m}\hbar^2 k_z^2\right)\phi(x), \tag{F.4}$$

which corresponds to an harmonic oscillator with frequency $\omega_c = (eB/mc)$ (the cyclotron frequency) centered at $x_0 = -(\hbar c/eB)k_y$. The energy levels, also called Landau levels, are then given by

$$\epsilon(k_z, n) = \frac{\hbar^2 k_z^2}{2m} + \hbar\omega_c\left(\frac{1}{2} + n\right). \tag{F.5}$$

To determine the degeneracy of the Landau levels, we consider the allowed values for k_y and k_z and use the standard periodic boundary conditions. For the latter quantum number we get $k_z = (2\pi/L)n_z$, $n_z = 0, 1, 2, \ldots$. For k_y we note that the distance Δx_0 between two possible centers for the oscillator is related to the difference between two successive k_y, $\Delta x_0 = (\hbar c/eB)\Delta k_y = (\hbar c/eB)(2\pi/L)$. As a result along the x axis we can accommodate $g = L/\Delta x_0 = L^2(eB/hc)$ centers. This number represents the energy degeneracy for each oscillator level and physically is the magnetic flux through the surface $L^2 = A$ measured in units of the flux quantum hc/e.

To evaluate the magnetic susceptibility we need the second derivative of the thermodynamic grand canonical potential with respect to the magnetic field. We have for the grand canonical potential

$$\Omega = -k_B T \sum_{k_z} 2g \sum_{n=0}^{\infty} \ln\left[1 + \exp\left(\frac{\mu_z - \hbar\omega_c(1/2+n)}{k_B T}\right)\right], \tag{F.6}$$

where $\mu_z = \mu - \frac{\hbar^2 k_z^2}{2m}$. By using the Poisson summation formula

$$\sum_{k=0}^{\infty} F(k) = \int_{-1/2}^{\infty} dx\, F(x)\left(1 + 2\sum_{n=1}^{\infty}\cos(2\pi n x)\right) \tag{F.7}$$

we transform Eq. (F.6) into an integral:

$$\Omega = -k_B T \sum_{k_z} \hbar\omega_c L^2 \nu_2 \int_{-1/2}^{\infty} dx \ln\left[1 + \exp\left(\frac{\mu_z - \hbar\omega_c(1/2+x)}{k_B T}\right)\right]$$
$$\times\left(1 + \sum_{k=-\infty}^{\infty}\exp(i2\pi k x)\right),$$

where $\nu_2 = 2m/(2\pi\hbar^2)$ is the density of states in two dimensions and in the sum the term $k = 0$ is not included. By making the change of variable $\varepsilon = \hbar\omega_c(x + 1/2)$, one sees that the magnetic field only appears in the oscillating terms. As a result we are only interested in the magnetic field dependent part of the grand canonical potential:

$$\Delta\Omega = -k_B T L^2 \nu_2 \sum_{k_z}\sum_{k=-\infty}^{\infty}(-1)^k \int_0^{\infty} d\varepsilon \ln\left[1 + \exp\left(\frac{\mu_z - \varepsilon}{k_B T}\right)\right]\exp\left(\frac{i2\pi k}{\hbar\omega_c}\varepsilon\right). \tag{F.8}$$

After an integration by parts we get

$$\Delta\Omega = -L^2\nu_2 \sum_{k_z}\sum_{k=-\infty}^{\infty}(-1)^k(-i)\frac{\hbar\omega_c}{2\pi k}\left[k_B T \ln\left[1 + \exp\left(\frac{\mu_z}{k_B T}\right)\right]\right.$$
$$\left. + \int_0^{\infty} d\varepsilon \frac{\exp\left(\frac{2\pi k i}{\hbar\omega_c}\varepsilon\right)}{1 + \exp\left(\frac{\varepsilon - \mu_z}{k_B T}\right)}\right]$$
$$= -L^2\nu_2 \sum_{k_z}\sum_{k=-\infty}^{\infty}(-1)^k(-i)\frac{\hbar\omega_c}{2\pi k}\int_0^{\infty} d\varepsilon \frac{\exp\left(\frac{2\pi k i}{\hbar\omega_c}\varepsilon\right)}{1 + \exp\left(\frac{\varepsilon - \mu_z}{k_B T}\right)}, \tag{F.9}$$

where the last step follows from the fact that the sum over k of the first term in the square bracket vanishes since positve and negative terms cancel each other. We see that the resulting expression has an explicit factor linear in the magnetic field apart from the dependence on the oscillating factor in the integral. This leads to an oscillatory behavior of the susceptibility as a function of the field strength. We limit our discussion to the zero field limit for the susceptibility, i.e. to terms in Ω up to second order in the field. By further integration by parts we get

$$\Delta\Omega = -L^2\nu_2 \sum_{k_z} \sum_{k=-\infty}^{\infty} (-1)^k (-\mathrm{i})^2 \left(\frac{\hbar\omega_c}{2\pi k}\right)^2 \times \left[\frac{1}{1+\exp\left(\frac{-\mu_z}{k_B T}\right)} - \frac{1}{k_B T}\int_0^\infty \mathrm{d}\varepsilon \frac{\exp\left(\frac{2\pi k \mathrm{i}}{\hbar\omega_c}\varepsilon\right)\exp\left(\frac{\varepsilon-\mu_z}{k_B T}\right)}{\left(1+\exp\left(\frac{\varepsilon-\mu_z}{k_B T}\right)\right)^2} \right]. \tag{F.10}$$

We now notice that the second term in square brackets can be further integrated by parts yielding at each successive integration a further power of the magnetic field. Hence for the purpose of evaluating the magnetic susceptibility it is enough to keep only the first term. By considering the zero-temperature limit, after transforming the sum over k_z to an integral we arrive at

$$\begin{aligned}\chi_L &= \nu(E_F) 4 \left(\frac{\hbar e}{2\pi mc}\right)^2 \sum_{k=1}^{\infty} (-1)^k \frac{1}{k^2} \\ &= \nu(E_F) 4 \frac{\mu_B^2}{\pi^2}\left(-\frac{\pi^2}{12}\right) \\ &= -\frac{1}{3}\nu(E_F)\mu_B^2. \end{aligned} \tag{F.11}$$

Appendix G Ising and gas-lattice models

In the lattice model description, we divide the volume V into $\mathcal{N}$ small cells v_0 which can accommodate only one particle. Introducing the occupation numbers t_i for the cell whose center is at lattice site i,

$$t_i = \begin{cases} 1, & \text{occupied} \\ 0, & \text{empty} \end{cases} \quad \text{for all } i, \quad \sum_{i=1}^{\mathcal{N}} t_i = N, \tag{G.1}$$

the interaction is given by

$$U = -\frac{1}{2} \sum_{ij} \phi_a(|\mathbf{r}_i - \mathbf{r}_j|) t_i t_j. \tag{G.2}$$

The interaction corresponds only to the attractive part ($\phi_a > 0$), since the hard-core contribution is taken into account by the introduction of the cells with volume v_0. The grand canonical partition function for the gas-lattice model is

$$\mathcal{Z}(T, V, \mu) = \sum_{N=0}^{\infty} \frac{e^{\beta\mu N} v_0^N}{\lambda^{3N}} \sum_{\{t_i\}:\sum_i t_i = N} e^{(1/2)\beta \sum_{ij} \phi_a(|\mathbf{r}_i - \mathbf{r}_j|) t_i t_j}, \tag{G.3}$$

where for each N the sum over the configurations $\{t_i\}$ is conditioned by $\sum_i t_i = N$, μ is the chemical potential and λ the thermal wave length due to the kinetic part of the Hamiltonian. Notice that the factor v_0^N keeps track of the change from the continuous integration over the N space position vectors to the discrete summation over the cells. By observing that in Eq. (G.3) we have the sum over all possible values of N, we can rewrite the partition function as a sum over all possible sets $\{t_i\}$ without restriction:

$$\mathcal{Z}(T, V, \mu) = \sum_{\{t_i\}} e^{\sum_i t_i(\beta\mu - \ln(\lambda^3/v_0)) + (1/2)\beta \sum_{ij} \phi_a(|\mathbf{r}_i - \mathbf{r}_j|) t_i t_j}. \tag{G.4}$$

The canonical partition function for the Ising model with uniform $h_i = h$ reads instead

$$Z(T, h) = \sum_{\{\sigma_i\}} e^{\beta \sum_i h\sigma_i + (\beta/2) \sum_{ij} J_{ij} \sigma_i \sigma_j}, \tag{G.5}$$

with the magnetization per spin given by

$$m = \frac{1}{N} \sum_{i=1}^{N} \langle \sigma_i \rangle = k_B T \frac{\partial \ln Z}{\partial h}. \tag{G.6}$$

By making the position

$$t_i = \frac{1}{2}(1 - \sigma_i) \tag{G.7}$$

and the identification

$$\phi_a(|\mathbf{r}_i - \mathbf{r}_j|) = 4J_{ij}, \quad \tilde{\phi}(0) = \sum_j \phi_a(|\mathbf{r}_i - \mathbf{r}_j|) = 4\sum_j J_{ij} = 4\tilde{J}_0, \tag{G.8}$$

$$\mu = -2h - \frac{1}{2}\tilde{\phi}(0) + \frac{1}{\beta}\ln(\lambda^3/v_0), \tag{G.9}$$

we obtain the relation between the two partition functions:

$$\mathcal{Z} = \mathrm{e}^{\mathcal{N}(\beta\mu - \ln(\lambda^3)/2/v_0)}\mathrm{e}^{\mathcal{N}\beta\tilde{\phi}(0)/8}Z. \tag{G.10}$$

We see that, at constant T, variations with respect to h in the Ising model corresponds to variations with respect to μ in the gas-lattice. From Eq. (G.10) follows the connection between the grand potential of the gas-lattice and the free energy of the Ising model:

$$PV = \mathcal{N}(\mu - (1/\beta)\ln(\lambda^3/v_0))/2 + \mathcal{N}\tilde{\phi}(0)/8 - F. \tag{G.11}$$

By using Eq. (G.9), the above equation becomes

$$P = \frac{1}{v_0}\left[-h - \frac{1}{8}\tilde{\phi}(0) - v_0 f\right], \quad f = F/V, \tag{G.12}$$

$$\begin{aligned} n \equiv \frac{\langle N \rangle}{V} &= \left(\frac{\partial P}{\partial \mu}\right)_\beta = \left(\frac{\partial P}{\partial h}\right)_\beta \left(\frac{\partial h}{\partial \mu}\right)_\beta \\ &= \frac{1}{2}\left(\frac{1}{v_0} + \left(\frac{\partial f}{\partial h}\right)_\beta\right) \\ &= \frac{1}{2v_0}(1 - m), \end{aligned} \tag{G.13}$$

where $m = \sum_{i=1}^{\mathcal{N}} \langle \sigma_i \rangle / \mathcal{N}$ is the average magnetization per spin.

H

Appendix H Sum over discrete Matsubara frequencies

In this appendix we provide a few technical details on how to perform the sum over the Matsubara frequencies involving more than one Green function. The basic idea is the same as in Problem 18.1.

Consider the case of a sum over fermionic frequencies involving two Green functions, as in the case of Eq. (19.3). We define the quantity

$$\Pi(\omega_\nu) = \frac{1}{\beta}\sum_{\epsilon_n} \mathcal{G}_{\kappa_1}(\mathrm{i}\epsilon_n)\mathcal{G}_{\kappa_2}(\mathrm{i}(\epsilon_n+\omega_\nu)), \tag{H.1}$$

where $\epsilon_n = (2n+1)\pi/\beta$ and $\omega_\nu = 2\nu\pi/\beta$ and having defined the Green function as in Eq. (18.18):

$$\mathcal{G}_\kappa(\mathrm{i}\epsilon_n) = (\mathrm{i}\epsilon_n - \epsilon_\kappa)^{-1}. \tag{H.2}$$

We may expand the product of the two Green functions in partial fractions:

$$\begin{aligned}
\Pi(\mathrm{i}\omega_\nu) &= \frac{1}{\beta}\sum_{\epsilon_n} \frac{1}{\mathrm{i}\epsilon_n - \epsilon_{\kappa_1}}\frac{1}{\mathrm{i}(\epsilon_n+\omega_\nu)-\epsilon_{\kappa_2}} \\
&= \frac{1}{\beta}\sum_{\epsilon_n} \frac{1}{\mathrm{i}\omega_\nu - \epsilon_{\kappa_2}+\epsilon_{\kappa_1}}\left(\frac{1}{\mathrm{i}\epsilon_n - \epsilon_{\kappa_1}} - \frac{1}{\mathrm{i}(\epsilon_n+\omega_\nu)-\epsilon_{\kappa_2}}\right) \\
&= \frac{1}{\mathrm{i}\omega_\nu - \epsilon_{\kappa_2}+\epsilon_{\kappa_1}}(f(\epsilon_{\kappa_1}) - f(\epsilon_{\kappa_2}-\mathrm{i}\omega_\nu)) \\
&= \frac{1}{2}\frac{1}{\mathrm{i}\omega_\nu - \epsilon_{\kappa_2}+\epsilon_{\kappa_1}}\left(\tanh\left(\beta\frac{\epsilon_{\kappa_2}-\mathrm{i}\omega_\nu}{2}\right) - \tanh\left(\beta\frac{\epsilon_{\kappa_1}}{2}\right)\right),
\end{aligned}$$

where we used the relation (18.20) and the identity $f(z) = (1-\tanh(\beta z/2))/2$. We notice that the shift $\mathrm{i}\omega_\nu$ can be dropped since $\exp(\beta\mathrm{i}\omega_\nu) = 1$. The quantity $\Pi(\mathrm{i}\omega_\nu)$ can be analytically continued to real frequencies as $\mathrm{i}\omega_\nu \to \hbar\omega \pm \mathrm{i}0^+$ yielding the corresponding retarded $\Pi^{\mathrm{R}}(\omega)$ and advanced $\Pi^{\mathrm{A}}(\omega)$ satisfying $\mathrm{Re}\,\Pi^{\mathrm{R}}(\omega) = \mathrm{Re}\,\Pi^{\mathrm{A}}(\omega)$, $\mathrm{Im}\,\Pi^{\mathrm{R}}(\omega) = -\mathrm{Im}\,\Pi^{\mathrm{A}}(\omega)$ with

$$\mathrm{Im}\,\Pi^{\mathrm{R}}(\omega) = \frac{\pi}{2}\int_{-\infty}^{\infty} \mathrm{d}\epsilon\left(\tanh\left(\frac{\beta\epsilon}{2}\right) - \tanh\left(\frac{\beta(\epsilon+\hbar\omega)}{2}\right)\right)\delta(\epsilon-\epsilon_{\kappa_1})\delta(\epsilon+\hbar\omega-\epsilon_{\kappa_2}). \tag{H.3}$$

From the above one sees that $\mathrm{Im}\,\Pi^{\mathrm{R}}(-\omega) = -\mathrm{Im}\,\Pi^{\mathrm{R}}(\omega)$ with the interchange of ϵ_{κ_1} and ϵ_{κ_2}.

I Appendix I Two-fluid hydrodynamics: a few hints

The starting point is Eqs. (15.4) and (15.5) which define the two-fluid model. The conservation laws for the particle mass, momentum and entropy densities read

$$\frac{\partial \rho}{\partial t} = -\nabla \cdot \mathbf{J}, \tag{I.1}$$

$$\frac{\partial \mathbf{J}}{\partial t} = -\nabla P, \tag{I.2}$$

$$\frac{\partial s}{\partial t} = -\nabla \cdot s\mathbf{v}_{\mathrm{n}}. \tag{I.3}$$

The last equation tells us that the entropy density $s = S/V$ is carried by the normal component. We assume for simplicity that $c_{\mathrm{p}} \approx c_{\mathrm{v}}$ and there are no external forces so that only the gradient of the pressure enters the conservation law for the momentum density. The standard sound equation is derived by taking the derivative of the first equation with respect to time and then inserting in it the second equation, to read, in terms of compressibility,

$$\frac{\partial^2 \rho}{\partial t^2} = c_1^2 \nabla^2 \rho, \quad c_1^2 = \frac{1}{\rho \kappa_T}. \tag{I.4}$$

Within the assumption of keeping only linear terms with respect to the velocities $\mathbf{v}_{\mathrm{n}}$ and $\mathbf{v}_{\mathrm{s}}$, after taking the derivative of Eq. (I.3) with respect to time, we get

$$\frac{\partial^2 s}{\partial t^2} = -\nabla \cdot s \frac{\partial \mathbf{v}_{\mathrm{n}}}{\partial t}. \tag{I.5}$$

In obtaining the above equation, we have discarded the time derivative of the entropy density, because we are confining to first order variations with respect to equilibrium. The latter is defined by the vanishing of the velocities $\mathbf{v}_{\mathrm{n}}$ and $\mathbf{v}_{\mathrm{s}}$. Since the momentum density $\mathbf{J}$ involves both normal and superfluid velocities, in order for Eq. (I.2) to be useful, we need an extra equation for the time derivative of the velocities. To this end we follow an argument due to Anderson (1966). We observe that, in accordance with the discussion of the symmetry of a complex order parameter done in Section 15.6, number and phase are dynamically conjugate variables, i.e.

$$[\hat{N}, \hat{\theta}] = \mathrm{i}. \tag{I.6}$$

In the representation $\hat{N} = N$ and $\hat{\theta} = -\mathrm{i}\partial_N$, the equation of motion for the phase operator $\hat{\theta}$ reads

$$\hbar \partial_t \hat{\theta} = \mathrm{i}[\hat{H}, \hat{\theta}] = -\frac{\partial \hat{H}}{\partial N}. \tag{I.7}$$

By taking the statistical average of such an equation, we have that the dynamics of the phase of the order parameter satisfies

$$\frac{\partial \theta}{\partial t} = -\frac{1}{\hbar}\mu, \tag{I.8}$$

from which, by using the irrotational character of the superfluid flow with $\mathbf{v}_s = \hbar \nabla \theta / m$,

$$\frac{\partial \mathbf{v}_s}{\partial t} = -\frac{1}{m}\nabla \mu. \tag{I.9}$$

Finally, the thermodynamical relation $dG = -SdT + VdP = Nd\mu$ gives

$$d\mu = -\frac{s}{n}dT + \frac{1}{n}dP \tag{I.10}$$

and one gets

$$\frac{\partial \mathbf{v}_s}{\partial t} = \frac{s}{\rho}\nabla T - \frac{1}{\rho}\nabla P. \tag{I.11}$$

Notice that in equilibrium, Eq. (I.11) yields the thermomechanical effect, where a pressure gradient is induced by a temperature gradient according to $\nabla P = s\nabla T$. This is the origin of the fountain effect discussed in the main text.

Equation (I.11), when used together with Eq. (I.2), yields the time derivative of the normal velocity:

$$\frac{\partial \mathbf{v}_n}{\partial t} = -\frac{s\rho_s}{\rho_n \rho}\nabla T - \frac{1}{\rho}\nabla P. \tag{I.12}$$

At constant pressure and with linear terms in the velocity, upon inserting Eq. (I.12) into (I.5) we get the second sound equation for entropy propagation:

$$\frac{\partial^2 s}{\partial t^2} = c_2^2 \nabla^2 s, \; c_2^2 = \frac{\rho_s}{\rho_n}\frac{Ts^2}{\rho c_P}. \tag{I.13}$$

If we add a damping term in the right-hand side of Eq. (I.3), $D_2 \nabla^2 s$, the equation determining the dispersion relation of the second sound mode reads

$$\omega^2 + i\omega k^2 D_2 - c_2^2 k^2 = 0. \tag{I.14}$$

J

Appendix J The Cooper problem in the theory of superconductivity

We consider the problem of two identical fermions of mass m interacting through a potential $V(r)$ in the presence of a filled Fermi sea, which has the only purpose of forbidding scattering of the two particles in states with $k < k_{\mathrm{F}}$. Does there exist a bound state with energy less than twice the Fermi energy, corresponding to the minimum kinetic energy of the two independent particles?

The interaction does not depend on the spin, $V(\mathbf{x}_1 - \mathbf{x}_2)$, and the antisymmetric wave function is of the form $\Psi(\mathbf{x}_1, \mathbf{x}_2; \sigma_1, \sigma_2) = \chi(\mathbf{x}_1 - \mathbf{x}_2)\phi(\sigma_1, \sigma_2)$. The Schrödinger equation then reads

$$\left[-\frac{\hbar^2}{2m}\left(\nabla_1^2 + \nabla_2^2\right) + V(\mathbf{x}_1 - \mathbf{x}_2)\right]\chi(\mathbf{x}_1 - \mathbf{x}_2) = E\chi(\mathbf{x}_1 - \mathbf{x}_2), \tag{J.1}$$

where $\chi(\mathbf{x}_1 - \mathbf{x}_2)$ is a symmetric function of $\mathbf{x}_1 - \mathbf{x}_2$ when $\phi(\sigma_1, \sigma_2)$ is antisymmetric in σ_1 and σ_2 as in the spin singlet state. Vice versa, $\chi(\mathbf{x}_1 - \mathbf{x}_2)$ is an odd function of $\mathbf{x}_1 - \mathbf{x}_2$ when $\phi(\sigma_1, \sigma_2)$ is symmetric as in the spin triplet state. Let us introduce the relative and center-of-mass variables in real and momentum space:

$$\mathbf{x} = \mathbf{x}_1 - \mathbf{x}_2,\ \mathbf{X} = \frac{1}{2}(\mathbf{x}_1 + \mathbf{x}_2),\ \mathbf{k} = \frac{1}{2}(\mathbf{k}_1 - \mathbf{k}_2),\ \mathbf{K} = \mathbf{k}_1 + \mathbf{k}_2. \tag{J.2}$$

In the Fourier transformed space Eq. (J.1) reads

$$\left(\frac{\hbar^2 K^2}{4m} + \frac{\hbar^2 k^2}{m} - E\right)\chi(\mathbf{K}, \mathbf{k}) = -{\sum_{\mathbf{k}'}}' V(|\mathbf{k} - \mathbf{k}'|)\chi(\mathbf{K}, \mathbf{k}'), \tag{J.3}$$

where the prime symbol indicates that the sum is restricted to the permitted states above the Fermi sea and $\mathbf{K} = \mathbf{K}'$ due to momentum conservation. For zero center-of-mass momentum $\mathbf{K} = 0$, the number of states available for scattering is maximum and the possibility of forming a bound state with energy $E < 2E_{\mathrm{F}} = \hbar^2 k_{\mathrm{F}}^2/m$ is maximum.

It is worthwhile noticing that in the absence of the Fermi sea, Eq. (J.3) would have a bound state solution for $E < 0$. If the potential $V(\mathbf{x})$, with no restriction in the sum over $\mathbf{k}'$, were uniformly repulsive, there would not be any bound state. Even in the case of attractive negative V, there might not be a solution for too weak a potential.

Let us go back to the restricted case. The given interaction V can be decomposed in terms of Legendre polynomials $P_l(\cos(\theta))$,

$$V(|\mathbf{k} - \mathbf{k}'|) \equiv V(k, k', \cos(\theta)) = \sum_l (2l+1)V_l(k, k')P_l(\cos(\theta)), \tag{J.4}$$

where θ is the angle between $\mathbf{k}$ and $\mathbf{k}'$ and the coefficients V_l are given by

$$V_l(k, k') = \int \frac{d\Omega}{4\pi} V(|\mathbf{k} - \mathbf{k}'|) P_l(\cos(\theta)). \tag{J.5}$$

Ω is the solid angle and the angular momentum l is even for a spin singlet and odd for a spin triplet.

At $\mathbf{K} = \mathbf{K}' = 0$, substitute the decomposition Eq. (J.4) into Eq. (J.3) and search for solutions with given angular momentum l and projection m:

$$\chi(0, \mathbf{k}) = \chi_l(k) Y_{lm}(\hat{\mathbf{k}}). \tag{J.6}$$

$Y_{lm}(\hat{\mathbf{k}})$ is a spherical harmonic and $\chi_l(k)$ is the radial function. Recalling the property of the Legendre polynomials in terms of the spherical harmonics:

$$P_l(\cos(\theta)) = \frac{4\pi}{2l+1} \sum_{m=-l}^{l} Y_{lm}(\hat{\mathbf{k}})^* Y_{lm}(\hat{\mathbf{k}}'),$$

Eq. (J.4) becomes

$$\left(\frac{\hbar^2 k^2}{m} - E\right) \chi_l(k) = -\frac{1}{(2\pi)^3} 4\pi \int dk' \, (k')^2 V_l(k, k') \chi_l(k'), \tag{J.7}$$

where the sum over $\mathbf{k}'$ has been replaced by the continuous integration per unit volume, Eq. (7.10), and the integration is limited by the Fermi sea. We consider the solution of Eq. (J.7) in the simple case of an attractive constant interaction in a shell $\hbar\omega_c \ll E_F$ above the Fermi surface:

$$V_l(k, k') = -V_l < 0, \quad \text{for} \quad E_F < \frac{\hbar^2 k^2}{2m}, \frac{\hbar^2 k'^2}{2m} < E_F + \hbar\omega_c \text{ and } V_l(k, k') = 0 \text{ otherwise.}$$

$\chi_l(k)$ has the simple form

$$\chi_l(k) = \frac{C}{\hbar^2 k^2/m - E}$$

with a constant C. Equation (J.7) reduces to

$$\frac{1}{V_l} = \frac{\nu(E_F)}{4} \int_{2E_F}^{2E_F + 2\hbar\omega_c} \frac{d\epsilon}{\epsilon - E} = \frac{\nu(E_F)}{4} \ln \frac{2E_F + 2\hbar\omega_c - E}{2E_F - E},$$

with the density of states $\nu(E_F)$, Eq. (7.14) evaluated at the Fermi energy. The binding energy is $E_b = 2E_F - E = 2\hbar\omega_c/(\exp(4/\nu(E_F)V_l) - 1)$.

Contrary to the isolated pairs, in the presence of a Fermi sea any weak attractive interaction produces a bound state. The Fermi sea is unstable with respect to pair formation. The N-body equivalent is the BCS instability. Which l-component of the interaction dominates will depend on the type of interaction. Finally, we recall that the expression for the relative motion of the two particles in the real space $\chi_l(\mathbf{x})$ at small $\mathbf{x}$ vanishes as x^l (see Problem 15.9). This means that whenever the potential is repulsive at short distances the s-wave pairing is hindered and $l \geq 1$ pair states are more likely to occur.

K

Appendix K Superconductive fluctuation phenomena

In this appendix we derive the Landau–Ginzburg equations for superconductivity, starting from the microscopic BCS theory, and discuss their application to the phenomenon of paraconductivity, first discussed by Aslamazov and Larkin in 1968. For a comprehensive discussion of the theory of superconductive fluctuations phenomena and their experimental observation, we refer to the book by Larkin and Varlamov (2005). Here we confine ourselves to the main aspects of the microscopic derivation.

K.1 The derivation of the Landau–Ginzburg equations

The starting point is the equation of motion Eq. (15.196) for the density matrix K defined in Eq. (15.156):

$$i\hbar\partial_t K = [M, K], \tag{K.1}$$

where the matrix M has been introduced in Eq. (15.165). In the following, for simplicity, we neglect the Hartree–Fock contribution to the matrix M. Close to T_c, the pair potential $\mu_{\mathbf{k}}$ is small and can be treated as a perturbation. We then split the matrix M into two parts, $M^{(0)}$ and $M^{(1)}$, corresponding to the non-interacting Fermi gas and the pairing potential, respectively. To carry out a perturbative evaluation with respect to $\mu_{\mathbf{k}}$, we rewrite Eq. (K.1) as

$$i\hbar\partial_t K = \left[M^{(0)}, K\right] + \left[M^{(1)}, K\right], \tag{K.2}$$

which, after expanding $K = \sum_n K^{(n)}$ and equating order by order, can be formally solved to give

$$K^{(n)}(t) = -\frac{i}{\hbar}\int_{-\infty}^{t} dt' \, \exp\left(-\frac{i}{\hbar}M^{(0)}(t-t')\right)\left[M^{(1)}(t'), K^{(n-1)}(t')\right] \times \exp\left(\frac{i}{\hbar}M^{(0)}(t-t')\right). \tag{K.3}$$

One sees that the approximation of order $K^{(n-1)}$ for K generates the successive order $K^{(n)}$. The general structure of the matrix K and of $M^{(0)}$ and $M^{(1)}$ is given by

$$K = \begin{pmatrix} h & \chi \\ \chi* & 1-h \end{pmatrix}, \quad M^{(0)} = \begin{pmatrix} \epsilon & 0 \\ 0 & -\epsilon \end{pmatrix}, \quad M^{(1)} = \begin{pmatrix} 0 & \mu \\ \mu* & 0 \end{pmatrix}, \tag{K.4}$$

where h, χ, ϵ and μ are all operators and can be represented in terms of their matrix elements either in momentum or real space. For simplicity, we are confining ourselves to just one block, since there are no spin-dependent interactions.

To first order in the pairing potential, there is only a contribution to the off-diagonal part, χ, of the generalized density matrix:

$$\chi^{(1)}_{\mathbf{kk}'}(t) = -\frac{\mathrm{i}}{\hbar}\int_{-\infty}^{t} \mathrm{d}t' \, \exp\left(-\frac{\mathrm{i}}{\hbar}(\epsilon_{\mathbf{k}} + \epsilon_{\mathbf{k}'})(t-t')\right)\mu_{\mathbf{kk}'}(t')(1 - f_{\mathbf{k}} - f_{\mathbf{k}'}), \quad \text{(K.5)}$$

where $h_{\mathbf{k}} = f_{\mathbf{k}} = f(\epsilon_{\mathbf{k}})$, $f(x)$ being the Fermi function. The above equation has the structure of a convolution product. The Fourier transform of the first factor reads

$$\left(-\frac{\mathrm{i}}{\hbar}\theta(t)\exp\left(-\frac{\mathrm{i}}{\hbar}(\epsilon_{\mathbf{k}} + \epsilon_{\mathbf{k}'})t\right)\right) = \frac{1}{\hbar\omega - (\epsilon_{\mathbf{k}} + \epsilon_{\mathbf{k}'}) + \mathrm{i}0^+}. \quad \text{(K.6)}$$

To proceed further, it is convenient to consider the values of the frequency ω on the imaginary axis according to the prescription $\hbar\omega \to \mathrm{i}\omega_\nu$, where $\omega_\nu = 2\pi\nu/\beta$ is a bosonic Matsubara frequency, introduced in Chapter 18. The Fourier transform of Eq. (K.5), evaluated at Matsubara frequencies, reads

$$\chi^{(1)}_{\mathbf{kk}'}(\omega_\nu) = \mu_{\mathbf{kk}'}(\omega_\nu)\frac{1 - f_{\mathbf{k}} - f_{\mathbf{k}'}}{\mathrm{i}\omega_\nu - (\epsilon_{\mathbf{k}} + \epsilon_{\mathbf{k}'})}. \quad \text{(K.7)}$$

For the time being, we confine to the first order in the pairing potential. This is equivalent to the Gaussian approximation in the fluctuations. The effect of higher order terms will be considered later. By recalling the property, with $\epsilon_n = (2n+1)\pi/\beta$ as a fermionic Matsubara frequency,

$$\frac{1}{2} - f_{\mathbf{k}} = \frac{1}{\beta}\sum_{\epsilon_n}\frac{1}{\epsilon_{\mathbf{k}} - \mathrm{i}\epsilon_n}, \quad \text{(K.8)}$$

we rewrite Eq. (K.7) as

$$\chi^{(1)}_{\mathbf{kk}'}(\omega_\nu) = -\mu_{\mathbf{kk}'}(\omega_\nu)\frac{1}{\beta}\sum_{\epsilon_n}\mathcal{G}(\mathbf{k},\epsilon_n)\mathcal{G}(\mathbf{k}',\omega_\nu - \epsilon_n), \quad \mathcal{G}(\mathbf{k},\epsilon_n) = \frac{1}{\mathrm{i}\epsilon_n - \epsilon_k}, \quad \text{(K.9)}$$

where $\mathcal{G}(\mathbf{k},\epsilon_n)$ is the thermal Green function for the Fermi gas (see Eq. (18.18)). The space and frequency dependent pairing function is obtained by the pairing potential at coinciding space points:

$$\Delta(\mathbf{x}',\omega_\nu) = \int_{\mathbf{k}}\int_{\mathbf{k}'} \mathrm{d}\mathbf{x}' \, \mathrm{e}^{\mathrm{i}\mathbf{k}\cdot\mathbf{x}'}\mu_{\mathbf{kk}'}(\omega_\nu)\mathrm{e}^{-\mathrm{i}\mathbf{k}'\cdot\mathbf{x}'}. \quad \text{(K.10)}$$

The self-consistency equation is obtained by requiring that the pairing function is expressed in terms of the off-diagonal part of the density matrix:

$$\Delta(\mathbf{x},\omega_\nu) = -V\chi^{(1)}(\mathbf{x},\mathbf{x};\omega_\nu) = -V\int_{\mathbf{k}}\int_{\mathbf{k}'}\chi^{(1)}_{\mathbf{kk}'}(\omega_\nu)\mathrm{e}^{i(\mathbf{k}-\mathbf{k}')\cdot\mathbf{x}}. \quad \text{(K.11)}$$

By using the real space form of the thermal Green function

$$\mathcal{G}(\mathbf{x},\epsilon_n) = \int_{\mathbf{k}}\mathcal{G}(\mathbf{k},\epsilon_n)\mathrm{e}^{\mathrm{i}\mathbf{k}\cdot\mathbf{x}}, \quad \text{(K.12)}$$

we write (K.11) as

$$\Delta(\mathbf{x},\omega_\nu) = \int d\mathbf{x}'\ \Delta(\mathbf{x}',\omega_\nu)\frac{1}{\beta}\sum_{\epsilon_n}\mathcal{G}(\mathbf{x}-\mathbf{x}',\epsilon_n)\mathcal{G}(\mathbf{x}'-\mathbf{x},\omega_\nu-\epsilon_n). \tag{K.13}$$

This can be written in the well known form in terms of the Cooper channel bubble $\Pi^{\mathrm{C}}(\mathbf{q},\omega_\nu)$

$$\Delta(\mathbf{q},\omega_\nu) = \Delta(\mathbf{q},\omega_\nu)\Pi^{\mathrm{C}}(\mathbf{q},\omega_\nu), \tag{K.14}$$

where

$$\Pi^{\mathrm{C}}(\mathbf{q},\omega_\nu) = V\frac{1}{\beta}\sum_{\epsilon_n}\int_{\mathbf{k}}\mathcal{G}(\mathbf{k},\epsilon_n)\mathcal{G}(\mathbf{k}+\mathbf{q},\omega_\nu-\epsilon_n). \tag{K.15}$$

The strategy is clear. We expand $\Pi^{\mathrm{C}}(\mathbf{q},\omega_\nu)$ at small $\mathbf{q}$ and ω_ν, insert the resulting expression into (K.14), analytically continue to real frequencies, and Fourier transform back to real space and time. The zero momentum and frequency of Eq. (K.14) is nothing but the self-consistency equation (15.189) of the BCS theory, so that

$$\Pi(0,0) = VN_0\ln\left(\frac{\omega_{\mathrm{D}}e^{I}}{2T}\right). \tag{K.16}$$

To analyze the momentum and frequency dependence, we transform the momentum integral in Eq. (K.15) by passing to the energy variable $\xi_{\mathbf{k}} = \mathbf{k}^2/2m - \mu$:

$$\int_{\mathbf{k}}\cdots = N_0\int_{-\infty}^{\infty}d\xi_{\mathbf{k}},$$

to read

$$\int_{-\infty}^{\infty}d\xi_{\mathbf{k}}\frac{1}{i(\omega_\nu-\epsilon_n)-\xi_{\mathbf{k}+\mathbf{q}}}\frac{1}{i\epsilon_n-\xi_{\mathbf{k}}} = -2\pi\ \mathrm{sign}(\epsilon_n)\frac{\theta(\epsilon_n(\epsilon_n-\omega_\nu))}{\omega_\nu-2\epsilon_n-i\mathbf{v}_{\mathrm{F}}\cdot\mathbf{q}}. \tag{K.17}$$

After expanding for small $\mathbf{q}$ and ω_ν, which means assuming slow variations in space and time of $\Delta(\mathbf{x},t)$ with respect to λ_{F} and $\hbar/(k_{\mathrm{B}}T)$, we get

$$\Pi^{\mathrm{C}}(\mathbf{q},\omega_\nu) = VN_0\ln\left(\frac{\omega_{\mathrm{D}}e^{I}}{2T}\right)+\omega_\nu\frac{4\pi VN_0}{\beta}\sum_{\epsilon_n>0}\frac{1}{(2\epsilon_n)^2}-q^2v_{\mathrm{F}}^2\frac{4\pi VN_0}{3\beta}\sum_{\epsilon_n>0}\frac{1}{(2\epsilon_n)^3}. \tag{K.18}$$

The frequency sums can be done as

$$\frac{1}{\beta}\sum_{\epsilon_n>0}\frac{1}{(2\epsilon_n)^2} = \frac{\beta}{4\pi^2}\sum_{n=0}^{\infty}\frac{1}{(2n+1)^2} = \frac{\beta}{4}\frac{1}{8}, \tag{K.19}$$

$$\frac{1}{\beta}\sum_{\epsilon_n>0}\frac{1}{(2\epsilon_n)^3} = \frac{\beta}{8\pi^3}\sum_{n=0}^{\infty}\frac{1}{(2n+1)^3} = \frac{\beta^2}{8}\frac{7\zeta_{\mathrm{R}}(3)}{8}, \tag{K.20}$$

$$\Pi(\mathbf{q},\omega_\nu) = VN_0\ln\left(\frac{\omega_{\mathrm{D}}e^{I}}{2T}\right)+\frac{\pi VN_0}{k_{\mathrm{B}}T}\omega_\nu - VN_0\xi_{0,\mathrm{LG}}^2q^2, \tag{K.21}$$

where we have defined

$$\xi_{0,\mathrm{LG}} = \sqrt{\frac{7\zeta_{\mathrm{R}}(3)\hbar^2 v_{\mathrm{F}}^2}{48\pi^2 k_{\mathrm{B}}^2 T_{\mathrm{c}}^2}}. \tag{K.22}$$

As a result, within the linear approximation, the time-dependent equation for the pairing function assumes the Landau–Ginzburg form

$$\left(\frac{T - T_{\mathrm{c}}}{T_{\mathrm{c}}} - \xi_{0,\mathrm{LG}}^2 \nabla^2\right) \Delta(\mathbf{x}, t) = -\frac{\hbar\pi}{8k_{\mathrm{B}}T} \partial_t \Delta(\mathbf{x}, t). \tag{K.23}$$

This equation is sufficient for the evaluation of the paraconductivity performed in the next section.

In the remaining part of this section, we show how to obtain the higher order terms. By using Eq. (K.3), it not difficult to convince ourselves that odd order terms contribute to the off-diagonal part of the density matrix, whereas even order terms appear in the diagonal part. The first correction to the diagonal density matrix reads

$$\begin{aligned} h^{(2)}_{\mathbf{k}\mathbf{k}'}(t) = &-\frac{\mathrm{i}}{\hbar}\int_{-\infty}^{t} \mathrm{d}t' \, \exp\left(-\frac{\mathrm{i}}{\hbar}(\epsilon_{\mathbf{k}} - \epsilon_{\mathbf{k}'})(t - t')\right) \\ &\times \left(\chi^{(1)}_{\mathbf{k}\mathbf{k}_1}(t')\mu^*_{\mathbf{k}_1\mathbf{k}'}(t') - \mu_{\mathbf{k}\mathbf{k}_1}(t')\chi^{(1)*}_{\mathbf{k}_1\mathbf{k}'}(t')\right), \end{aligned} \tag{K.24}$$

and the third order correction to the off-diagonal part is

$$\begin{aligned} \chi^{(3)}_{\mathbf{k}\mathbf{k}'}(t) = &-\frac{\mathrm{i}}{\hbar}\int_{-\infty}^{t} \mathrm{d}t' \, \exp\left(-\frac{\mathrm{i}}{\hbar}(\epsilon_{\mathbf{k}} + \epsilon_{\mathbf{k}'})(t - t')\right) \\ &\times \left(h^{(2)}_{\mathbf{k}\mathbf{k}_1}(t')\mu_{\mathbf{k}_1\mathbf{k}'}(t') + \mu_{\mathbf{k}\mathbf{k}_1}(t')h^{(2)}_{\mathbf{k}_1\mathbf{k}'}(t')\right). \end{aligned} \tag{K.25}$$

Using again the assumption of slow space and time variations of the pairing function, we approximate, after inserting Eq. (K.5) in (K.24) and Eq. (K.24) in (K.25), the above equations become

$$h^{(2)}_{\mathbf{k}\mathbf{k}'}(t) \approx -\mu_{\mathbf{k}\mathbf{k}_1}(t)\mu^*_{\mathbf{k}_1\mathbf{k}'}(t)\frac{1}{\beta}\sum_{\epsilon_n} \mathcal{G}(\mathbf{k}, \epsilon_n)\mathcal{G}(\mathbf{k}_1, -\epsilon_n)\mathcal{G}(\mathbf{k}', \epsilon_n) \tag{K.26}$$

and

$$\chi^{(3)}_{\mathbf{k}\mathbf{k}'}(t) \approx -\mu_{\mathbf{k}\mathbf{k}_2}(t)\mu^*_{\mathbf{k}_2\mathbf{k}_1}(t)\mu_{\mathbf{k}_1\mathbf{k}'}(t)\frac{1}{\beta}\sum_{\epsilon_n} \mathcal{G}(\mathbf{k}, \epsilon_n)\mathcal{G}(\mathbf{k}_2, -\epsilon_n)\mathcal{G}(\mathbf{k}_1, \epsilon_n)\mathcal{G}(\mathbf{k}', -\epsilon_n). \tag{K.27}$$

As a result the self-consistency equation takes the form

$$\begin{aligned} \Delta(\mathbf{x}, t) = &V\int \mathrm{d}\mathbf{x}' \int \mathrm{d}t' \, \Delta(\mathbf{x}', t')\mathcal{G}(\mathbf{x} - \mathbf{x}', t - t')\mathcal{G}(\mathbf{x}' - \mathbf{x}, t - t') \\ &-V\int \mathrm{d}\mathbf{x}_1 \int \mathrm{d}\mathbf{x}_2 \int \mathrm{d}\mathbf{x}_3 \, \Delta(\mathbf{x}_1, t)\Delta^*(\mathbf{x}_2, t)\Delta(\mathbf{x}_3, t) \\ &\times\frac{1}{\beta}\sum_{\epsilon_n} \mathcal{G}(\mathbf{x} - \mathbf{x}_1, \epsilon_n)\mathcal{G}(\mathbf{x}_1 - \mathbf{x}_2, -\epsilon_n)\mathcal{G}(\mathbf{x}_2 - \mathbf{x}_3, \epsilon_n)\mathcal{G}(\mathbf{x}_3 - \mathbf{x}, -\epsilon_n). \end{aligned} \tag{K.28}$$

The linear terms have been already discussed. The third order term can be obtained by replacing $\mathbf{x}_1 = \mathbf{x}_2 = \mathbf{x}_3 \approx \mathbf{x}$ and the resulting coefficient reads

$$\frac{1}{\beta}\sum_{\epsilon_n} N_0 \int_{-\infty}^{\infty} \mathrm{d}\xi_k \mathcal{G}(k,\epsilon_n)\mathcal{G}(k,-\epsilon_n)\mathcal{G}(k,\epsilon_n)\mathcal{G}(k,-\epsilon_n) = \frac{1}{\beta}\sum_{\epsilon_n}\frac{\pi N_0}{2|\epsilon_n|^3} = N_0\frac{7\zeta_{\mathrm{R}}(3)}{8\pi^2 k_{\mathrm{B}}^2 T^2}, \tag{K.29}$$

which, when added to the linear terms in Eq. (K.23), yields the equation for the pairing function close to T_{c}:

$$\left(\frac{T-T_{\mathrm{c}}}{T_{\mathrm{c}}} - \xi_{0,\mathrm{LG}}^2\nabla^2 + \frac{7\zeta_{\mathrm{R}}(3)}{8\pi^2 k_{\mathrm{B}}^2 T_{\mathrm{c}}^2}|\Delta(\mathbf{x},t)|^2\right)\Delta(\mathbf{x},t) = -\frac{\hbar\pi}{8k_{\mathrm{B}}T_{\mathrm{c}}}\partial_t\Delta(\mathbf{x},t). \tag{K.30}$$

The order parameter $\psi(\mathbf{x},t)$ introduced in Section 15.7 is normalized in such a way that its modulus squared coincides with the density n. We then define

$$\psi(\mathbf{x},t) = \sqrt{\frac{n7\zeta_{\mathrm{R}}(3)}{8\pi^2 k_{\mathrm{B}}^2 T_{\mathrm{c}}^2}}\Delta(\mathbf{x},t), \tag{K.31}$$

so that the equation for $\psi(\mathbf{x},t)$ reads

$$\left(\frac{T-T_{\mathrm{c}}}{T_{\mathrm{c}}} - \xi_{0,\mathrm{LG}}^2\nabla^2 + \frac{1}{n}|\psi(\mathbf{x},t)|^2\right)\psi(\mathbf{x},t) = -\frac{\hbar\pi}{8k_{\mathrm{B}}T_{\mathrm{c}}}\partial_t\psi(\mathbf{x},t). \tag{K.32}$$

To get the explicit expression of the coefficients a, b and c of the Landau–Ginzburg theory of Section 15.7 we must multiply Eq. (K.32) by a suitable factor. By requiring that the coefficient of the term with the Laplacian corresponds to that of a particle of mass $2m$, we must multiply Eq. (K.32) by $\hbar^2/(4m\xi_{0,\mathrm{LG}}^2)$, and we obtain, after recalling Eqs. (K.22) and (K.23),

$$c = \frac{\hbar^2}{4m},\ a'T_{\mathrm{c}} = \frac{c}{\xi_{0,\mathrm{LG}}^2},\ b = \frac{a'T_{\mathrm{c}}}{n},\ \gamma = \frac{\pi\hbar}{8k_{\mathrm{B}}T_{\mathrm{c}}}a'T_{\mathrm{c}},\ \xi_{0,\mathrm{LG}}^2 = \frac{7\zeta_{\mathrm{R}}(3)\hbar^2 v_{\mathrm{F}}^2}{48\pi^2 k_{\mathrm{B}}^2 T_{\mathrm{c}}^2}, \tag{K.33}$$

which are the microscopic expressions of the Landau–Ginzburg coefficients.

K.2 The Aslamazov–Larkin paraconductivity

To a good approximation, from Eq. (15.129) when we are not too close to T_{c}, we may consider the superconductive fluctuations as a gas of independent particles with charge $-2e$ and mass $2m$ with dispersion relation $\epsilon(\mathbf{k}) = a + ck^2$ and velocity $\mathbf{v}_{\mathbf{k}} = \hbar^{-1}\partial_{\mathbf{k}}\epsilon(\mathbf{k}) = 2\hbar^{-1}c\mathbf{k}$. By using $c = \hbar^2/4m$, we have the velocity of a particle of mass $2m$ and momentum $\hbar\mathbf{k}$. Then we write a Boltzmann transport equation similar to Eq. (11.47) for the electron gas:

$$-2e\mathbf{E}\cdot\frac{1}{\hbar}\partial_{\mathbf{k}}f(\mathbf{k}) = -\frac{2}{\tau_{\mathbf{k}}}(\delta f(\mathbf{k}) - \langle\delta f(\mathbf{k}')\rangle), \tag{K.34}$$

where $\tau_{\mathbf{k}}$ was introduced in Eq. (15.132). Notice the factor of 2, due to the fact that the modulus squared $\langle|\psi(\mathbf{k},t)|^2\rangle$ decays twice as fast.

The current due to the fluctuations can be written, in analogy with Eq. (11.49) for the electron gas, as

$$\mathbf{j} = -2e \int \frac{\mathrm{d}^3 k}{(2\pi)^3} \mathbf{v}_{\mathbf{k}} \delta f(\mathbf{k}). \tag{K.35}$$

In the left-hand side of Eq. (K.34), we may use the equilibrium distribution for the fluctuation:

$$f^{(0)}(\mathbf{k}) = \frac{k_{\mathrm{B}} T}{a + ck^2},$$

from which

$$\frac{1}{\hbar} \partial_{\mathbf{k}} f^{(0)}(\mathbf{k}) = -\frac{k_{\mathrm{B}} T}{(a + ck^2)^2} \mathbf{v}(\mathbf{k}). \tag{K.36}$$

Taking into account Eq. (K.34), the current becomes

$$\begin{aligned}
\mathbf{j} &= (-2e)^2 \int \frac{\mathrm{d}^3 k}{(2\pi)^3} \frac{\tau_{\mathbf{k}}}{2} \frac{k_{\mathrm{B}} T}{(a + ck^2)^2} \mathbf{v}(\mathbf{k}) \mathbf{v}(\mathbf{k}) \cdot \mathbf{E} \\
&= (-2e)^2 \frac{2c^2 \gamma k_{\mathrm{B}} T}{3\hbar^2} \int \frac{\mathrm{d}^3 k}{(2\pi)^3} \frac{k^2}{(a + ck^2)^3} \mathbf{E}.
\end{aligned}$$

By performing the integral,[1] we obtain, after recalling that $c/a' T_{\mathrm{c}} = \xi_{0,\mathrm{LG}}^2$ and the result $\gamma = a' T_{\mathrm{c}} \pi \hbar / (8 k_{\mathrm{B}} T_{\mathrm{c}})$ (see Eq. (K.33)), the famous Aslamazov–Larkin contribution to the paraconductivity

$$\sigma_{\mathrm{AL}} = \frac{e^2}{32 \hbar \xi_{0,\mathrm{LG}}} \left(\frac{T_{\mathrm{c}}}{T - T_{\mathrm{c}}} \right)^{1/2}. \tag{K.37}$$

In a film of thickness d, σ_{AL} reads

$$\sigma_{\mathrm{AL}} = \frac{e^2}{32 \hbar \xi_{0,\mathrm{LG}}} \frac{2\xi(T)}{d} \left(\frac{T_{\mathrm{c}}}{T - T_{\mathrm{c}}} \right)^{1/2}. \tag{K.38}$$

Close to the critical point, still in the normal metallic phase, the superconductive fluctuations produce an additional channel for electrical conduction. To estimate the size of the correction in $d = 3$ we recall that the Drude conductivity (11.33) is controlled by the parameter $l k_{\mathrm{F}} \sim \tau E_{\mathrm{F}} / \hbar$, $\sigma = (e^2 n \tau)/m \sim (e^2/\hbar)(1/\lambda_{\mathrm{F}})(\tau E_{\mathrm{F}}/\hbar)$ and the coherence distance $\xi_{0,\mathrm{LG}} \sim \lambda_{\mathrm{F}} (E_{\mathrm{F}} / k_{\mathrm{B}} T_{\mathrm{c}})$. The ratio of the Aslamazov–Larkin paraconductivity to the Drude conductivity reads

$$\frac{\sigma_{\mathrm{AL}}}{\sigma} \sim \frac{1}{k_{\mathrm{F}} l} \frac{k_{\mathrm{B}} T_{\mathrm{c}}}{E_{\mathrm{F}}} \left(\frac{T_{\mathrm{c}}}{T - T_{\mathrm{c}}} \right)^{1/2}. \tag{K.39}$$

Good metals with low-T_{c} superconductivity have $k_{\mathrm{F}} l$ large and $k_{\mathrm{B}} T_{\mathrm{c}} / \epsilon_{\mathrm{F}}$ small, and only very close to the critical temperature does the effect become important. The situation may change either in the so-called dirty superconductors, where the parameter $k_{\mathrm{F}} l$ and $\xi_{0,\mathrm{LG,d}} \sim \sqrt{l \xi_{0,\mathrm{LG}}}$

[1] Note that $\int_0^\infty \mathrm{d}x\, x^4 (1 + x^2)^{-3} = 3\pi/16$.

become smaller[2] or in restricted dimensionality as in thin films, where $\xi_{0,\mathrm{LG}}$ is replaced by the film tickness $d \ll \xi_{0,\mathrm{LG}}$ and the divergence of σ_{AL} is stronger, see Eq. (K.38). The phenomenon of paraconductivity has been observed experimentally in numerous cases. A further reason for the amplification of the paraconductivity, where fluctuation effects become in general relevant, comes from a reduction of the coherence distance as in high-T_{c} superconductors.

[2] The expression for $\xi_{0,\mathrm{LG,d}}$ in dirty superconductors can be understood physically in the following way. Equation (K.32), when restricted to the linear terms at $T = T_{\mathrm{c}}$, has the form of a diffusion equation with the combination $\xi_{0,\mathrm{LG}}^2(8k_{\mathrm{B}}T_{\mathrm{c}}/\pi\hbar)$ playing the role of the effective diffusion coefficient. The dirty limit expression is then obtained by replacing such a combination with the diffusion coefficient $\xi_{0,\mathrm{LG,d}}^2(8k_{\mathrm{B}}T_{\mathrm{c}}/\pi\hbar) \equiv D = v_{\mathrm{F}}^2\tau/d$ of Eq. (11.34) for the dirty Fermi gas.

Appendix L Diagrammatic aspects of the exact solution of the Tomonaga–Luttinger model

L.1 General features

The simplicity of the exact solution for the single-particle Green function, due to the additional Ward identities associated with left- and right-particle conservation, has a counterpart in the diagrammatic structure of the theory.

To make contact with the general theory developed in Chapter 18, we begin by analyzing the four-leg vertex function $\Gamma^{(4)}$, which appears in the expression of the self-energy (18.103) in Fourier space:

$$\Sigma_{\alpha\sigma}(p) = \sum_a \int_{p'} g^a_{\alpha\alpha'} G_{\alpha'\sigma'}(p') + \int_{p_1,p_2} g^a_{\alpha\alpha'} \Gamma^{a(4)}_{\alpha'\alpha}(p, p_1 + p_2 - p, p_1 - p) \\ \times G_{\alpha'\sigma'}(p_1) G_{\alpha'\sigma'}(p_2) G_{\alpha\sigma'}(p_1 + p_2 - p), \qquad \text{(L.1)}$$

and whose diagrammatic representation is shown in Fig. 18.6(d). Here we have replaced the empty box Γ^0 with the bare interaction $g^a_{\alpha\alpha'}$ ($g^a_{\alpha\alpha} = g^a_4$ and $g^a_{\alpha\bar{\alpha}} = g^a_2$) and the full four-point interaction vertex $\Gamma^{(4)}$ with $\Gamma^{a(4)}_{\alpha\alpha'}$ connecting particles in the branch α and α' and interacting in the charge ($a =$ c) or spin ($a =$ s) channel. In general, the renormalized full box $\Gamma^{a(4)}_{\alpha\alpha'}$ cannot be determined exactly and one must rely on some suitable truncation of its perturbative expansion. In the present case, it turns out that the four-leg vertex $\Gamma^{a(4)}_{\alpha\alpha'}$ can be expressed in terms of the effective interaction $D^a_{\alpha\alpha'}$, which will be defined in the following, and of the three-leg density vertex $\mathcal{I}^{a0}_{\alpha,\sigma}$, which is obtained from $\Lambda^{a0}_{\alpha\sigma}$ in Eq. (20.39) as the irreducible part for cutting an interaction line of the truncated vertex $\mathcal{T}^{a\mu}_{\alpha,\sigma}$ for $\mu = 0$ (see Eq. (18.119)), as is explained in Section L.3. The Dyson equation (L.1) assumes eventually the form depicted in Fig. L.3(d). As we will show in this appendix, the diagrammatic expansion of the four-leg vertex $\Gamma^{a(4)}_{\alpha\alpha}$ must have the form shown in Fig. L.1. One sees that, at each order, a vertex $\mathcal{I}^{a0}_{\alpha,\sigma}$ and an integration over two propagators G are added. (Remember that the vertex $\mathcal{I}^{a0}_{\alpha,\sigma}$ is irreducible with respect to cutting an interaction line.) The WI associated with the conservation law for left- and right-particles allows us to eliminate the density vertex $\mathcal{I}^{a0}_{\alpha,\sigma}$ in favor of G (see Eq. (L.13)) and to obtain a closed equation of motion (see Eq. (20.44)). Indeed, a remarkable aspect of the diagrammatic structure of the theory, on which the form of $\Gamma^{a(4)}_{\alpha\alpha'}$ is based, is the vanishing of diagrams containing a fermion loop with more than two fermion Green function lines, which will be shown in the next section. Examples of such diagrams are shown in Fig. L.2. This means that the diagrams contributing to $\Gamma^{a(4)}_{\alpha\alpha'}$ can only contain loops with two fermion lines or *bubbles* or repetition of bubbles and the three-leg density vertex according to the form shown in Fig. L.1. Moreover, the effective interaction $D^a_{\alpha\alpha'}$ is a geometric series of bubbles as shown in Eq. (L.14) and

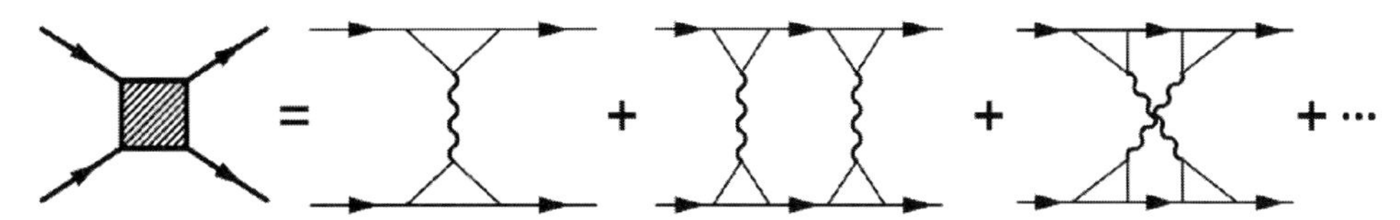

Fig. L.1 Perturbative corrections to the four-point vertex $\Gamma^{(4)}$ (hatched square): the vertex $\mathcal{I}^a_{\alpha,\sigma}$ (empty triangle) and the effective interaction $D^a_{\alpha\alpha'}$ (wiggly thick line) always appear in the combination $(\mathcal{I}^{0a}_{\alpha,\sigma})^2 D^a_{\alpha\alpha}$.

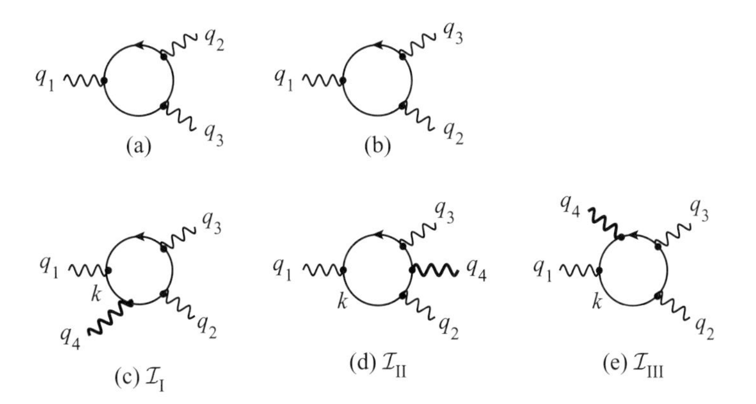

Fig. L.2 Fermionic loops with three, (a) and (b), and four, (c), (d), (e), insertions. Diagrams (a) and (b) correspond to the arrangements (q_1, q_2, q_3) and (q_1, q_3, q_2), respectively. Diagrams (c), (d) and (e) are obtained from diagram (b) by an additional insertion (thicker wiggly line) with bimomentum q_4 and correspond to the expressions indicated as I, II and III, respectively.

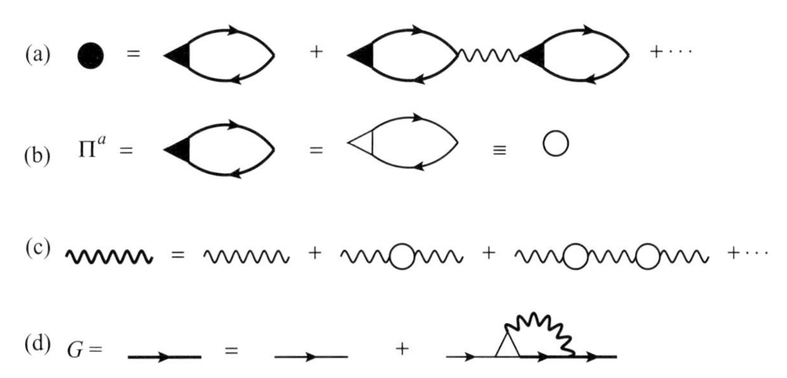

Fig. L.3 (a) Diagrammatic representation for the bicurrent–bicurrent correlation function. (b) The irreducible polarization Π coincides with the non-interacting one due to the WI Eq. (L.13). (c) Dyson equation for the dressed interaction in terms of the polarization Π. (d) Equation for the Green function in terms of the dressed interaction and the irreducible vertex $\mathcal{I}$.

diagrammatically in Fig. L.3(c). A series of bubbles also describes the response functions shown in Fig. L.3(a). Their exact expression was derived in Eq. (20.30) as a consequence of the conservation laws. This appears diagramatically in the fact that the bubble has the same expression as for the non-interacting system, a fact which is emphasized in Fig. L.3(b). As a result, the effective interaction $D^a_{\alpha\alpha'}$ is known.

After these preliminary remarks about the diagrammatic structure of the theory, we move to the next section, where we discuss the origin of the disapperance of fermion loops with more than two vertices, and to Section L.3 where we show the WI for the vertex $\mathcal{I}^{a0}_{\alpha,\sigma}$. In Section L.4 we derive the expression of the resummed interaction, whereas in the final Section L.5 we provide a few technical details on the solution of the Dyson equation for the Green function.

L.2 Theorem of loop cancellation

To understand the origin of loop cancellation we notice that, in general, a closed fermion loop can be seen as the the expectation value of a string of density operators with the number of density operators equal to the number of fermion lines. The above mentioned feature can then be stated as the fact that the connected diagram of the expectation value of a string of density operators can be made of, at most, two fermion lines.

Let us consider, for definiteness, the case of a fermion loop with four density operators (see Fig. L.2) with bimomenta q_i, $i = 1, \dots, 4$ obeying the conservation rule $\sum_{i=1}^{4} q_i = 0$. We adopt the convention that the density momenta entering the loop are positive and those leaving are negative. A generic diagram for a loop with four density operator insertions has the expression

$$I(\{q_i\}) = \int_k G_k G_{k+q_1} G_{k+q_1+q_2} G_{k+q_1+q_2+q_3},$$

where the notation $\int_k$ stands for both momentum and frequency integration. As the Fermi velocity is constant, all the possible bicurrent density insertions do not depend on the momenta which appear as arguments in the Green function lines. Hence, they only contribute a constant factor to the diagram and are irrelevant as far as the proof is concerned. The full set of diagrams for fixed external momenta $\{q_i\}$ can be obtained by considering all non-equivalent ways of arranging the density operator insertions in the loop. The key idea of the proof relies on the observation that diagrams with four insertions can be obtained from those with three insertions by considering all possible ways of making an additional insertion. When making the additional insertion one uses the property valid for the non-interacting Green function:

$$G_{k-q/2} G_{k+q/2} = \frac{1}{q}(G_{k-q/2} - G_{k+q/2}), \tag{L.2}$$

where $q \equiv \omega - \alpha v_F \mathbf{q}$ for α particles.

For instance, when considering diagrams with three insertions, there are two possible diagrams corresponding to the two non-equivalent ways of inserting three external densities into a loop. In a clockwise order we have the arrangements $(q_1\, q_2\, q_3)$ and $(q_1\, q_3\, q_2)$. Let us consider the latter and enumerate the ways we can make an additional insertion in all possible ways. We obtain three diagrams, labelled as I_{I}, I_{II} and I_{III} corresponding to the arrangements $(q_1\, q_3\, q_2\, q_4)$, $(q_1\, q_3\, q_4\, q_2)$ and $(q_1\, q_4\, q_3\, q_2)$, respectively. Here $q_4 = -(q_1 + q_2 + q_3)$ by energy–momentum conservation. The expressions for the corresponding diagrams read

$$\mathrm{I}_{\mathrm{I}} = \int_k G_k G_{k+q_1} G_{k+q_1+q_3} G_{k+q_1+q_2+q_3}, \tag{L.3}$$

$$\mathrm{I}_{\mathrm{II}} = \int_k G_k G_{k+q_1} G_{k+q_1+q_2} G_{k-q_2}, \tag{L.4}$$

$$\mathrm{I}_{\mathrm{III}} = \int_k G_k G_{k+q_1} G_{k-q_2-q_3} G_{k-q_2}. \tag{L.5}$$

By using the property (L.2), at each additional q_4-insertion we get

$$\mathrm{I}_{\mathrm{I}} = \frac{1}{q_1+q_2+q_3} \int_k G_{k+q_1} G_{k+q_1+q_3} \left(G_k - G_{k+q_1+q_2+q_3}\right), \tag{L.6}$$

$$\mathrm{I}_{\mathrm{II}} = \frac{1}{q_1+q_2+q_3} \int_k G_k G_{k+q_1} \left(G_{k-q_2} - G_{k+q_1+q_3}\right), \tag{L.7}$$

$$\mathrm{I}_{\mathrm{III}} = \frac{1}{q_1+q_2+q_3} \int_k G_k G_{k-q_2} \left(G_{k-q_2-q_3} - G_{k+q_1}\right). \tag{L.8}$$

Then it is immediate to verify that the first term in Eq. (L.6) cancels the second one in Eq. (L.7). In a similar way, the first term in Eq. (L.7) cancels the second one in Eq. (L.8). Finally, after making the shift $k \to k + q_1 + q_2 + q_3$, the first term in Eq. (L.8) cancels with the second one in Eq. (L.6). Hence the sum of all diagrams with four insertions obtained from a given diagram with three insertions cancel among themselves. Then the sum of all diagrams with four insertions cancel. The proof can be easily generalized by considering a given diagram with $n-1$ insertions and considering all diagrams with n obtained from the given one by making the additional insertion in all possible ways. As before, the expression of each diagram with n insertions so obtained can be split into two contributions and the cancellation of the diagrams among themselves occurs via the cancellation of pairs of contributions from diagrams by making the additional insertion in consecutive positions.

L.3 The irreducible vertex $\mathcal{I}^{a\mu}_{\alpha,\sigma}$

We have seen that the irreducible vertex $\mathcal{I}^{a0}_{\alpha,\sigma}$ is one of the elements on which the diagrammatic structure of the theory is built. In this section we show how to obtain it by means of the Ward identities.

As done in Section 18.10, we begin by introducing the truncated vertices $\mathcal{T}^{a\mu}_{\alpha,\sigma}(k,q)$ which, as in Eq. (18.119), are obtained from $\Lambda^{a\mu}_{\alpha,\sigma}$ defined in Eq. (20.39) by cutting the external fermion lines (see Fig. L.4(a) and (b), respectively). In this sense, the truncated vertex $\mathcal{T}^{a\mu}_{\alpha,\sigma}(k,q)$ is irreducible by cutting a Green function line. As shown in Fig. L.4(c), the vertex $\mathcal{T}^{a\mu}_{\alpha,\sigma}$ can be further expressed in terms of the vertex $\mathcal{I}^{a\mu}_{\alpha,\sigma}$, which is defined as the *irreducible part* of $\mathcal{T}^{a\mu}_{\alpha,\sigma}$ for cutting an interaction line. This gives

$$\mathcal{T}^{a\mu}_{\alpha\sigma} = \mathcal{I}^{a\mu}_{\alpha\sigma} + \mathcal{R}^{a\mu\nu} g^{a,\nu\lambda} \mathcal{I}^{a\lambda}_{\alpha\sigma}, \tag{L.9}$$

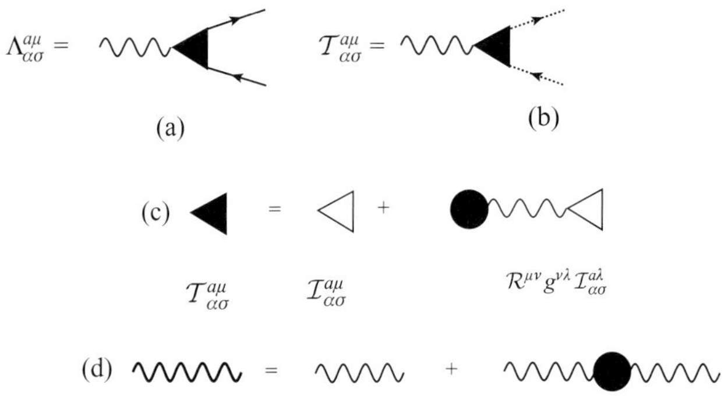

$\Lambda^{a\mu}_{\alpha\sigma} =$ (a) $\mathcal{T}^{a\mu}_{\alpha\sigma} =$ (b)

(c) $\mathcal{T}^{a\mu}_{\alpha\sigma}$ = $\mathcal{I}^{a\mu}_{\alpha\sigma}$ + $\mathcal{R}^{\mu\nu} g^{\nu\lambda} \mathcal{I}^{a\lambda}_{\alpha\sigma}$

(d) = +

Fig. L.4 (a) The reducible vertex Λ. (b) The associated vertex $\mathcal{T}$ obtained by truncating the external Green function lines. (c) The vertex $\mathcal{T}$ expressed in terms of the vertex $\mathcal{I}$, which is irreducible for cutting an interaction wiggly line. (d) The fully dressed interaction expressed in terms of the density–density response function $\mathcal{R}^{00}$.

where $\mathcal{R}^{a\mu\nu}$ is the bicurrent–bicurrent time-ordered correlation function. As a result of the WIs (20.40) and (20.41), the vertex $\mathcal{I}^{a\mu}_{\alpha\sigma}$ satisfies

$$\omega \mathcal{I}^{a0}_{\alpha,\sigma}(p,q) - v_{\rm F}\mathbf{q}\mathcal{I}^{a1}_{\alpha,\sigma}(p,q) = \varepsilon^a_\sigma \left[G^{-1}_{\alpha,\sigma}(p+q/2) - G^{-1}_{\alpha,\sigma}(p-q/2)\right], \quad \text{(L.10)}$$

$$\omega \mathcal{I}^{a1}_{\alpha,\sigma}(p,q) - v_{\rm F}\mathbf{q}\mathcal{I}^{a0}_{\alpha,\sigma}(p,q) = \alpha\varepsilon^a_\sigma \left[G^{-1}_{\alpha,\sigma}(p+q/2) - G^{-1}_{\alpha,\sigma}(p-q/2)\right]. \quad \text{(L.11)}$$

Notice that $v_{\rm F}$, instead of v^a and $\tilde{v}^a$, appears in the above WIs for $\mathcal{I}^{a\mu}_{\alpha\sigma}$. Then

$$\mathcal{I}^{a1}_{\alpha,\sigma}(p,q) = \alpha \mathcal{I}^{a0}_{\alpha,\sigma}(p,q) \quad \text{(L.12)}$$

and

$$(\omega - \alpha v_{\rm F}\mathbf{q})\mathcal{I}^{a0}_{\alpha,\sigma}(p,q) = \varepsilon^a_\sigma \left[G^{-1}_{\alpha,\sigma}(p+q/2) - G^{-1}_{\alpha,\sigma}(p-q/2)\right]. \quad \text{(L.13)}$$

Hence we have expressed the irreducible density vertex $\mathcal{I}^{a0}_{\alpha,\sigma}$ in terms of G only, as was mentioned in Section L.1.

L.4 The effective interaction $D^a_{\alpha\alpha'}$

In analogy with the skeleton diagrammatic structure of quantum electrodynamics, where the role of the interaction is played by the photon propagator, any Green or vertex function can be expressed in terms of the propagator G, the irreducible charge (spin) vertex $\mathcal{I}^{a0}_{\alpha,\sigma}$ and the effective interaction $D^a_{\alpha\beta}$, obtained as a resummation of the couplings g^a (analogous to the photon propagator in quantum electrodynamics) via the polarization bubble Π^a (see Fig. L.3(c)):

$$D^a = g^a + g^a \Pi^a D^a, \quad \text{(L.14)}$$

where D^a, g^a and Π^a are all matrices with respect to the right and left branch indices. Specifically,

$$g^a = \begin{pmatrix} g^a_4 & g^a_2 \\ g^a_2 & g^a_4 \end{pmatrix}, \quad D^a = \begin{pmatrix} D^a_{++} & D^a_{+-} \\ D^a_{-+} & D^a_{--} \end{pmatrix}.$$

$D^a_{\alpha\alpha'}$ can be obtained as the summation of the bubbles Π^a for each coupling g_a, according to Fig. L.3(c). Π^a represents the one-interaction line irreducible contribution to the effective interaction $D^a_{\alpha\beta}$. The remarkable consequence of the additional conservation law of the TL model is the fact that the irreducible bubble Π^a coincides with the bubble Π^a_0 of the non-interacting system, as depicted in Fig. L.3(b). This corresponds to the fact that the vertex and self-energy corrections to the renormalized bubbles indeed cancel each other due to Eq. (L.13), as depicted in Fig. L.3(b). We recall the general formula (19.19) for the structure the density–density response function, which, by taking into account Eq. (19.6), is expressed in terms of the truncated vertex $\mathcal{T}^{a\mu}_{\alpha\sigma}$ and the single-particle Green function $G_{\alpha,\sigma}$. Since, by definition, Π^a is irreducibile for cutting an interaction line, we must replace in Eq. (19.19) the truncated vertex $\mathcal{T}^{a\mu}_{\alpha\sigma}$ with the one irreducible for cutting an interaction line $\mathcal{I}^{a0}_{\alpha,\sigma}$ to read

$$\Pi^a_\alpha(q) = \mathrm{i}\sum_\sigma \epsilon^a_\sigma \sum_{|\mathbf{k}|<\Lambda} \int_{-\infty}^{\infty} \frac{d\epsilon}{2\pi} G_{\alpha,\sigma}(p+q/2)G_{\alpha,\sigma}(p-q/2)\mathcal{I}^{a0}_{\alpha,\sigma}. \tag{L.15}$$

After using the WI (L.13), we get (notice that by symmetry $\Pi^a_{\alpha-\alpha} = 0$)

$$\Pi^a_{\alpha\alpha}(q) = \mathrm{i}\sum_\sigma \frac{1}{\omega - \alpha v_{\mathrm{F}}\mathbf{q}} \sum_{|\mathbf{k}|<\Lambda} \int_{-\infty}^{\infty} \frac{d\epsilon}{2\pi} \left(G_{\alpha,\sigma}(p+q/2) - G_{\alpha,\sigma}(p-q/2)\right). \tag{L.16}$$

The integral over frequency can be expressed in terms of the occupation number in **k**-space of the interacting system:

$$\begin{aligned}
\Pi^a_{\alpha\alpha}(q) &= \sum_\sigma \frac{1}{\omega - \alpha v_{\mathrm{F}}\mathbf{q}} \frac{1}{2\pi} \int_{-\Lambda}^{\Lambda} d\mathbf{k}\,(n_\alpha(\mathbf{k}-\mathbf{q}/2) - n_\alpha(\mathbf{k}+\mathbf{q}/2)) \\
&= \sum_\sigma \frac{1}{\omega - \alpha v_{\mathrm{F}}\mathbf{q}} \frac{1}{2\pi} \left(\int_{-\Lambda-\mathbf{q}/2}^{\Lambda-\mathbf{q}/2} d\mathbf{k}\, n_\alpha(\mathbf{k}) - \int_{-\Lambda+\mathbf{q}/2}^{\Lambda+\mathbf{q}/2} d\mathbf{k}\, n_\alpha(\mathbf{k})\right) \\
&= \sum_\sigma \frac{1}{\omega - \alpha v_{\mathrm{F}}\mathbf{q}} \frac{1}{2\pi} \left(\int_{-\Lambda-\mathbf{q}/2}^{-\Lambda+\mathbf{q}/2} d\mathbf{k}\, n_\alpha(\mathbf{k}) - \int_{\Lambda-\mathbf{q}/2}^{\Lambda+\mathbf{q}/2} d\mathbf{k}\, n_\alpha(\mathbf{k})\right) \\
&= \frac{1}{\pi} \frac{\alpha\mathbf{q}}{\omega - \alpha v_{\mathrm{F}}\mathbf{q}},
\end{aligned} \tag{L.17}$$

where the last result follows from the fact that in the relevant region of integration, which is far from the Fermi points, we can use the non-interacting value $n_\alpha(\mathbf{k} \sim -\alpha\Lambda) = 1$ and $n_\alpha(\mathbf{k} \sim \alpha\Lambda) = 0$. The dressed interaction $D^a_{\alpha\alpha}$, which is relevant for the following, after some algebra from Eq. (L.14), reads

$$\begin{aligned}
D^a_{\alpha\beta}(\mathbf{q},\omega) &= \frac{\omega^2 - v_{\mathrm{F}}^2 q^2}{\omega^2 - (u^a)^2 q^2} \left(g^a_4 + \frac{\left(g^a_4\right)^2 - \left(g^a_2\right)^2}{\pi} \frac{\alpha\mathbf{q}}{\omega + \alpha v_{\mathrm{F}}\mathbf{q}}\right) \\
&= \pi(\omega - \alpha v_{\mathrm{F}}\mathbf{q}) \left(\frac{(u^a - \alpha v_{\mathrm{F}}\mathrm{sign}\mathbf{q})(\eta^a + \theta(\alpha\mathbf{q}))}{\omega - \alpha u^a|\mathbf{q}|}\right. \\
&\qquad \left. + \frac{(u^a + \alpha v_{\mathrm{F}}\mathrm{sign}\mathbf{q})(\eta^a + \theta(-\alpha\mathbf{q}))}{\omega + \alpha u^a|\mathbf{q}|}\right),
\end{aligned} \tag{L.18}$$

where the parameter η^a is defined in Eq. (20.49). Notice that the expression (L.18) coincides with that obtained in Eq. (20.45) of the main text and expresses the fermion effective interaction mediated by the density and spin modes.

L.5 Solution of the Dyson equation for the fermion Green function

Notice that the convolution product in Eq. (20.44) has become a direct product after the Fourier transform in Eq. (20.47). We see that the Green function satisfies the symmetry property $G_{-\alpha\sigma}(\mathbf{x},t) = G_{\alpha\sigma}(-\mathbf{x},t)$. This is apparent in the form of the differential operator in Eq. (20.46) and it is also a property of the function $L^a_{-\alpha-\alpha}(\mathbf{x},t) = L^a_{\alpha\alpha}(-\mathbf{x},t)$. Hence in the following we take $\alpha = +1$ for the sake of definiteness and drop the lower indices for the Green function and $L^a_{\alpha\alpha}$.

By introducing variables $s = \mathbf{x} + v_\mathrm{F} t$ and $r = \mathbf{x} - v_\mathrm{F} t$, Eq. (20.46) can be reduced to a simple differential equation:

$$2v_\mathrm{F}\partial_s G(s,r) = -\mathrm{i}\delta((s+r)/2)\delta((s-r)/2v_\mathrm{F}) + L(s,r)G(s,r) \tag{L.19}$$

with solution

$$G(s,r) = h(r)G_0(s,r)\exp\left(\frac{1}{2v_\mathrm{F}}\int_r^s \mathrm{d}\tilde{s}\, L(\tilde{s},r)\right), \tag{L.20}$$

where $G_0(s,r) = G_0(\mathbf{x},t) = (1/2\pi)[\mathbf{x} - v_\mathrm{F} t + \mathrm{i}0^+\mathrm{sign}(t)]^{-1}$ is the non-interacting Green function and the smooth function $h(r) = h(\mathbf{x} - v_\mathrm{F} t)$ must satifiy the boundary condition $h(0) = 1$.

To appreciate the meaning of this solution we need to make explicit the dependence on $\mathbf{x}$ and t of $L(\mathbf{x},t)$ by evaluating Eq. (20.47). The integral over frequency can be performed by shifting the poles of $L(\mathbf{q},\omega)$ with $\omega \to \omega + \mathrm{i}\,\mathrm{sign}(\omega)0^+$ and using

$$\int_{-\infty}^{\infty} \frac{\mathrm{d}\omega}{2\pi} \mathrm{e}^{-\mathrm{i}\omega t} \frac{1}{\omega - u^a|\mathbf{q}| + \mathrm{i}\,\mathrm{sign}(\omega)0^+} = -\mathrm{i}\,\mathrm{sign}(t)\mathrm{e}^{-\mathrm{i}\,\mathrm{sign}(t)u^a|\mathbf{q}|t}.$$

The remaining integral over the momentum is divergent at large $\mathbf{q}$. This can be cured by remembering that the model was introduced with a cutoff $\mathbf{q}_0$ in momentum space. To make the integral easier to handle we introduce here a cutoff via an exponential function $\mathrm{e}^{-\gamma_0|\mathbf{q}|}$. We then obtain

$$L(\mathbf{x},t) = \frac{1}{2}\sum_a \left(\frac{(u^a - v_\mathrm{F})(\eta^a + 1)}{\mathbf{x} - u^a t + \mathrm{i}\,\mathrm{sign}(t)\gamma_0} - \frac{(u^a + v_\mathrm{F})\eta^a}{\mathbf{x} + u^a t - \mathrm{i}\,\mathrm{sign}(t)\gamma_0}\right). \tag{L.21}$$

The exponential factor in the solution of the Green function becomes

$$\begin{aligned}
&\exp\left(\frac{1}{2v_\mathrm{F}}\int_r^s \mathrm{d}\tilde{s}\, L(\tilde{s},r)\right) \\
&= \prod_a \left(\frac{\mathbf{x} - v_\mathrm{F} t + \mathrm{i}\,\mathrm{sign}(t)\gamma_0}{\mathbf{x} - u^a t + \mathrm{i}\,\mathrm{sign}(t)\gamma_0}\right)^{(\eta^a+1)/2} \left(\frac{\mathbf{x} - v_\mathrm{F} t - \mathrm{i}\,\mathrm{sign}(t)\gamma_0}{\mathbf{x} + u^a t - \mathrm{i}\,\mathrm{sign}(t)\gamma_0}\right)^{\eta^a/2}.
\end{aligned} \tag{L.22}$$

We see that all factors in the above expression share the same analytical properties of the non-interacting Green function as function of t in the complex plane. The poles of G_0 are indeed $t = \mathbf{x}/v_F + \mathrm{i}\,\mathrm{sign}(t)0^+$ and hence G_0 is analytic in the lower right and upper left quadrants of the complex plane. The factor $\mathbf{x} - v_F t - \mathrm{i}\,\mathrm{sign}(t)\gamma_0$ has the wrong analytical properties. We notice that this factor depends only on the combination $\mathbf{x} - v_F t$. By looking at the solution (L.20) we may eliminate such an unwanted factor by suitably choosing the function $h(r)$. A simple choice is $h(r) = \prod_a \gamma_0^2/(\gamma_0^2 + r^2)^{\eta^a/2}$. Hence we get finally for the Green function the expression

$$G(\mathbf{x},t) = \frac{\gamma_0^{\eta^c+\eta^s}}{2\pi} \frac{1}{x - v_F t + \mathrm{i}\,\mathrm{sign}(t)0^+} \prod_a \left(\frac{x - v_F t + \mathrm{i}\,\mathrm{sign}(t)\gamma_0}{x - u^a t + \mathrm{i}\,\mathrm{sign}(t)\gamma_0} \right)^{1/2} \times \left(\frac{1}{x - u^a t + \mathrm{i}\,\mathrm{sign}(t)\gamma_0} \frac{1}{x + u^a t - \mathrm{i}\,\mathrm{sign}(t)\gamma_0} \right)^{\eta^a/2}. \tag{L.23}$$

M

Appendix M Details on the theory of the disordered Fermi liquid

M.1 Quantum interaction corrections to the conductivity

In this appendix, we show how to obtain the correction to the electrical conductivity due to the combined effect of disorder and interaction and that the logarithmic corrections found for the physical quantities can be absorbed into a renormalization of the parameters characterizing the ladder propagator. This identification is the formal basis of the renormalizability of the effective field theory, whose physical meaning is discussed in the text via the Ward identities. To avoid cumbersome factors of $\hbar$, we measure energies in units of inverse time.

The diagrams contributing to electrical conductivity to lowest order in the interaction are shown in Fig. 21.8. Diagrams (a) and (d) are obtained by inserting a self-energy correction into the Green function and one has to consider also the symmetric ones with the self-energy insertion in the bottom electron line. Diagrams (b) and (c) are due to vertex corrections. We begin our discussion with diagrams (a) and (b). The extension to diagrams (c) and (d) is straightforward. The expression for diagram (a) reads

$$R^{ij}_{(\mathrm{a})}(\mathbf{x},\mathbf{x}';\Omega_\nu) = 2\frac{1}{\beta}\sum_{\epsilon_n}\int \mathrm{d}\mathbf{x}_1\mathrm{d}\mathbf{x}_2 j^i_0(\mathbf{x})j^j_0(\mathbf{x}')\frac{1}{\beta}\sum_{\omega_m}\mathcal{V}(\mathbf{x}_1,\mathbf{x}_2;\omega_\mathrm{m})\mathcal{G}(\mathbf{x},\mathbf{x}_1;\epsilon_n+\Omega_\nu)$$
$$\times\mathcal{G}(\mathbf{x}_1,\mathbf{x}_2;\epsilon_n+\Omega_\nu-\omega_\mathrm{m})\mathcal{G}(\mathbf{x}_2,\mathbf{x}';\epsilon_n+\Omega_\nu)\mathcal{G}(\mathbf{x}',\mathbf{x};\epsilon_n). \quad \text{(M.1)}$$

Here we allow for a retarded interaction depending on the Matsubara frequency $\mathcal{V}(\mathbf{x}_1,\mathbf{x}_2;\omega_\mathrm{m})$. By analytical continuation $\mathrm{i}\omega_\mathrm{m}\to\omega\pm\mathrm{i}0^+$, $\mathcal{V}(\mathbf{x}_1,\mathbf{x}_2;\omega_\mathrm{m})\to V^{\mathrm{R(A)}}(\mathbf{r}_1,\mathbf{r}_2;\omega)$. To this expression one has to add the one corresponding to having the interaction line in the bottom electron line. $j^i_0(\mathbf{r})$ is the real-space representation of the current vertices. We do not need to write here its explicit expression since at the end, after the impurity average, we recover translational invariance and go back to the momentum representation. Diagram (b) is instead given by

$$R^{ij}_{(\mathrm{b})}(\mathbf{x},\mathbf{x}';\Omega_\nu) = 2\frac{1}{\beta}\sum_{\epsilon_n}\int \mathrm{d}\mathbf{x}_1\mathrm{d}\mathbf{x}_2 j^i_0(\mathbf{x})j^j_0(\mathbf{x}')\mathcal{G}(\mathbf{x},\mathbf{x}_1;\epsilon_n+\Omega_\nu)\mathcal{G}(\mathbf{x}_2,\mathbf{x};\epsilon_n)$$
$$\times\frac{1}{\beta}\sum_{\omega_\mathrm{m}}\mathcal{G}(\mathbf{x}_1,\mathbf{x}';\epsilon_n+\Omega_\nu-\omega_\mathrm{m})\mathcal{G}(\mathbf{x}',\mathbf{x}_2;\epsilon_n-\omega_\mathrm{m})\mathcal{V}(\mathbf{x}_1,\mathbf{x}_2;\omega_\mathrm{m}). \quad \text{(M.2)}$$

In both Eqs. (M.1) and (M.2), the Matsubara frequency sums are transformed to contour integrals in the complex plane by means of the standard manipulation illustrated in Appendix H. Then, one gets for diagram (a) (including the other diagram with the two electron lines interchanged)

$$
\begin{aligned}
R^{ij}_{(\mathrm{a})}(\mathbf{x},\mathbf{x}';\Omega) = -2\int \mathrm{d}\mathbf{x}_1\mathrm{d}\mathbf{x}_2\, j^i_0(\mathbf{x})j^j_0(\mathbf{x}') &\int_{-\infty}^{\infty}\frac{\mathrm{d}\epsilon}{2\pi\mathrm{i}}\int_{-\infty}^{\infty}\frac{\mathrm{d}\omega}{2\pi\mathrm{i}}\\
\text{(1)}\qquad &\times\big[f(\epsilon-\Omega)b(\omega)\big(V^{\mathrm{R}}_\omega - V^{\mathrm{A}}_\omega\big)G^{\mathrm{R}}_\epsilon G^{\mathrm{R}}_{\epsilon-\omega}G^{\mathrm{R}}_\epsilon\big(G^{\mathrm{R}}_{\epsilon-\Omega}-G^{\mathrm{A}}_{\epsilon-\Omega}\big)\\
\text{(2)}\qquad &\quad +f(\epsilon-\Omega)f(\omega-\epsilon)V^{\mathrm{R}}_\omega G^{\mathrm{R}}_\epsilon\big(G^{\mathrm{R}}_{\epsilon-\omega}-G^{\mathrm{A}}_{\epsilon-\omega}\big)G^{\mathrm{R}}_\epsilon\big(G^{\mathrm{R}}_{\epsilon-\Omega}-G^{\mathrm{A}}_{\epsilon-\Omega}\big)\\
\text{(3)}\qquad &\quad +f(\epsilon)b(\omega)\big(V^{\mathrm{R}}_\omega - V^{\mathrm{A}}_\omega\big)G^{\mathrm{R}}_\epsilon G^{\mathrm{R}}_{\epsilon-\omega}G^{\mathrm{R}}_\epsilon G^{\mathrm{A}}_{\epsilon-\Omega}\\
\text{(4)}\qquad &\quad +f(\epsilon)f(\omega-\epsilon)V^{\mathrm{R}}_\omega G^{\mathrm{R}}_\epsilon\big(G^{\mathrm{R}}_{\epsilon-\omega}-G^{\mathrm{A}}_{\epsilon-\omega}\big)G^{\mathrm{R}}_\epsilon G^{\mathrm{A}}_{\epsilon-\Omega}\\
\text{(5)}\qquad &\quad -f(\epsilon)b(\omega)\big(V^{\mathrm{R}}_\omega - V^{\mathrm{A}}_\omega\big)G^{\mathrm{A}}_\epsilon G^{\mathrm{A}}_{\epsilon-\omega}G^{\mathrm{A}}_\epsilon G^{\mathrm{A}}_{\epsilon-\Omega}\\
\text{(6)}\qquad &\quad -f(\epsilon)f(\omega-\epsilon)V^{\mathrm{A}}_\omega G^{\mathrm{A}}_\epsilon\big(G^{\mathrm{R}}_{\epsilon-\omega}-G^{\mathrm{A}}_{\epsilon-\omega}\big)G^{\mathrm{A}}_\epsilon G^{\mathrm{A}}_{\epsilon-\Omega}\\
\text{(7)}\qquad &\quad +f(\epsilon-\Omega)b(\omega)\big(V^{\mathrm{R}}_\omega - V^{\mathrm{A}}_\omega\big)G^{\mathrm{R}}_\epsilon G^{\mathrm{R}}_{\epsilon-\omega-\Omega}G^{\mathrm{R}}_{\epsilon-\Omega}G^{\mathrm{R}}_{\epsilon-\Omega}\\
\text{(8)}\qquad &\quad +f(\epsilon-\Omega)f(\omega-\epsilon+\Omega)V^{\mathrm{R}}_\omega G^{\mathrm{R}}_\epsilon G^{\mathrm{R}}_{\epsilon-\Omega}\big(G^{\mathrm{R}}_{\epsilon-\omega-\Omega}-G^{\mathrm{A}}_{\epsilon-\omega-\Omega}\big)G^{\mathrm{R}}_{\epsilon-\Omega}\\
\text{(9)}\qquad &\quad -f(\epsilon-\Omega)b(\omega)\big(V^{\mathrm{R}}_\omega - V^{\mathrm{A}}_\omega\big)G^{\mathrm{R}}_\epsilon G^{\mathrm{A}}_{\epsilon-\Omega}G^{\mathrm{A}}_{\epsilon-\omega-\Omega}G^{\mathrm{A}}_{\epsilon-\Omega}\\
\text{(10)}\qquad &\quad -f(\epsilon-\Omega)f(\omega-\epsilon+\Omega)V^{\mathrm{A}}_\omega G^{\mathrm{R}}_\epsilon G^{\mathrm{A}}_{\epsilon-\Omega}\big(G^{\mathrm{R}}_{\epsilon-\omega-\Omega}-G^{\mathrm{A}}_{\epsilon-\omega}\big)G^{\mathrm{A}}_{\epsilon-\Omega}\\
\text{(11)}\qquad &\quad +f(\epsilon)b(\omega)\big(V^{\mathrm{R}}_\omega - V^{\mathrm{A}}_\omega\big)\big(G^{\mathrm{R}}_\epsilon - G^{\mathrm{A}}_\epsilon\big)G^{\mathrm{A}}_{\epsilon-\Omega}G^{\mathrm{A}}_{\epsilon-\omega-\Omega}G^{\mathrm{A}}_{\epsilon-\Omega}\\
\text{(12)}\qquad &\quad +f(\epsilon)f(\omega-\epsilon+\Omega)V^{\mathrm{A}}_\omega\big(G^{\mathrm{R}}_\epsilon - G^{\mathrm{A}}_\epsilon\big)G^{\mathrm{A}}_{\epsilon-\Omega}\big(G^{\mathrm{R}}_{\epsilon-\omega-\Omega}-G^{\mathrm{A}}_{\epsilon-\omega-\Omega}\big)G^{\mathrm{A}}_{\epsilon-\Omega}\big].
\end{aligned}
\tag{M.3}
$$

We have enumerated the various terms for later easier reference. Diagram (b) gives

$$
\begin{aligned}
R^{ij}_{(\mathrm{b})}(\mathbf{x},\mathbf{x}';\Omega) = -2\int \mathrm{d}\mathbf{x}_1\mathrm{d}\mathbf{x}_2\, j^i_0(\mathbf{x})j^j_0(\mathbf{x}') &\int_{-\infty}^{\infty}\frac{\mathrm{d}\epsilon}{2\pi\mathrm{i}}\int_{-\infty}^{\infty}\frac{\mathrm{d}\omega}{2\pi\mathrm{i}}\\
\text{(1)}\qquad &\times\big[f(\epsilon)b(\omega)\big(V^{\mathrm{R}}_\omega - V^{\mathrm{A}}_\omega\big)G^{\mathrm{R}}_{\epsilon+\Omega}G^{\mathrm{R}}_\epsilon G^{\mathrm{R}}_{\epsilon+\Omega-\omega}G^{\mathrm{R}}_{\epsilon-\omega}\\
\text{(2)}\qquad &\quad +f(\epsilon)f(\omega-\epsilon)V^{\mathrm{R}}_\omega G^{\mathrm{R}}_{\epsilon+\Omega}G^{\mathrm{R}}_\epsilon G^{\mathrm{R}}_{\epsilon+\Omega-\omega}\big(G^{\mathrm{R}}_{\epsilon-\omega}-G^{\mathrm{A}}_{\epsilon-\omega}\big)\\
\text{(3)}\qquad &\quad +f(\epsilon)f(\omega-\epsilon-\Omega)V^{\mathrm{R}}_\omega G^{\mathrm{R}}_{\epsilon+\Omega}G^{\mathrm{R}}_\epsilon\big(G^{\mathrm{R}}_{\epsilon+\Omega-\omega}-G^{\mathrm{A}}_{\epsilon+\Omega-\omega}\big)G^{\mathrm{A}}_{\epsilon-\omega}\\
\text{(4)}\qquad &\quad -f(\epsilon)b(\omega)\big(V^{\mathrm{R}}_\omega - V^{\mathrm{A}}_\omega\big)G^{\mathrm{R}}_{\epsilon+\Omega}G^{\mathrm{A}}_\epsilon G^{\mathrm{R}}_{\epsilon+\Omega-\omega}G^{\mathrm{A}}_{\epsilon-\omega}\\
\text{(5)}\qquad &\quad -f(\epsilon)f(\omega-\epsilon)V^{\mathrm{A}}_\omega G^{\mathrm{R}}_{\epsilon+\Omega}G^{\mathrm{A}}_\epsilon G^{\mathrm{R}}_{\epsilon+\Omega-\omega}\big(G^{\mathrm{R}}_{\epsilon-\omega}-G^{\mathrm{A}}_{\epsilon-\omega}\big)\\
\text{(6)}\qquad &\quad -f(\epsilon)f(\omega-\epsilon-\Omega)V^{\mathrm{R}}_\omega G^{\mathrm{R}}_{\epsilon+\Omega}G^{\mathrm{A}}_\epsilon\big(G^{\mathrm{R}}_{\epsilon+\Omega-\omega}-G^{\mathrm{A}}_{\epsilon+\Omega-\omega}\big)G^{\mathrm{A}}_{\epsilon-\omega}\\
\text{(7)}\qquad &\quad +f(\epsilon+\Omega)b(\omega)\big(V^{\mathrm{R}}_\omega - V^{\mathrm{A}}_\omega\big)G^{\mathrm{R}}_{\epsilon+\Omega}G^{\mathrm{A}}_\epsilon G^{\mathrm{R}}_{\epsilon+\Omega-\omega}G^{\mathrm{A}}_{\epsilon-\omega}\\
\text{(8)}\qquad &\quad +f(\epsilon+\Omega)f(\omega-\epsilon)V^{\mathrm{A}}_\omega G^{\mathrm{R}}_{\epsilon+\Omega}G^{\mathrm{A}}_\epsilon G^{\mathrm{R}}_{\epsilon+\Omega-\omega}\big(G^{\mathrm{R}}_{\epsilon-\omega}-G^{\mathrm{A}}_{\epsilon-\omega}\big)\\
\text{(9)}\qquad &\quad +f(\epsilon+\Omega)f(\omega-\epsilon-\Omega)V^{\mathrm{R}}_\omega G^{\mathrm{R}}_{\epsilon+\Omega}G^{\mathrm{A}}_\epsilon\big(G^{\mathrm{R}}_{\epsilon+\Omega-\omega}-G^{\mathrm{A}}_{\epsilon+\Omega-\omega}\big)G^{\mathrm{A}}_{\epsilon-\omega}\\
\text{(10)}\qquad &\quad -f(\epsilon+\Omega)b(\omega)\big(V^{\mathrm{R}}_\omega - V^{\mathrm{A}}_\omega\big)G^{\mathrm{A}}_{\epsilon+\Omega}G^{\mathrm{A}}_\epsilon G^{\mathrm{A}}_{\epsilon+\Omega-\omega}G^{\mathrm{A}}_{\epsilon-\omega}\\
\text{(11)}\qquad &\quad -f(\epsilon+\Omega)f(\omega-\epsilon)V^{\mathrm{A}}_\omega G^{\mathrm{A}}_{\epsilon+\Omega}G^{\mathrm{A}}_\epsilon G^{\mathrm{R}}_{\epsilon+\Omega-\omega}\big(G^{\mathrm{R}}_{\epsilon-\omega}-G^{\mathrm{A}}_{\epsilon-\omega}\big)\\
\text{(12)}\qquad &\quad -f(\epsilon+\Omega)f(\omega-\epsilon-\Omega)V^{\mathrm{A}}_\omega G^{\mathrm{A}}_{\epsilon+\Omega}G^{\mathrm{A}}_\epsilon\big(G^{\mathrm{R}}_{\epsilon+\Omega-\omega}-G^{\mathrm{A}}_{\epsilon+\Omega-\omega}\big)G^{\mathrm{A}}_{\epsilon-\omega}\big].
\end{aligned}
\tag{M.4}
$$

Since we are discussing the frequency structure, in Eqs. (M.3) and (M.4) we have dropped the explicit dependence on space coordinates. The Green functions are presented in the same order as in Eqs. (M.1) and (M.2) where the space dependence is shown. In Eqs. (M.3) and (M.4) we may now perform the impurity average. Firstly, averaging impurity pairs belonging to the same Green function line implies the replacement of the Green function with its self-consistent Born approximation expression of Eq. (21.35). Secondly, we have to perform the average of impurity pairs belonging to different Green function lines. This can be performed by arranging the Green functions on the sides of a square, the corners of the square being the current vertices and the interaction vertices. Such a square of Green functions has also occurred in the evaluation of the weak localization correction of the density–density correlation function in Problem 21.5. If two Green functions adjacent to the same vertex have different imaginary parts (one retarded and one advanced), then one inserts a ladder between them. This typically happens at an interaction vertex. At a current vertex, on the other hand, even though the insertion of a ladder would be possible when the Green functions have opposite imaginary parts, the vectorial nature of the vertex forbids the insertion of the ladder. One can also insert impurity lines connecting Green functions which are on opposite sides of the square. When inserting ladders, wherever possible, at the leading order in the expansion parameter, we neglect, as a rule, all the diagrams in which a crossing of impurity lines occurs. Depending on the sequence of retarded and advanced Green functions around the sides of the square, we may insert two or three ladders. For instance, terms with four retarded Green functions do not allow the insertion of any ladder and, furthermore, after integrating over the momentum of the Green functions, give zero, since all poles lie on the same side of the real axis. In Eq. (M.3), the terms that allow two or three ladder insertions are the second, fourth, sixth, eighth, tenth, and twelfth:

$$
\begin{aligned}
R^{ij}_{(a)}(\mathbf{0},\Omega) = & -2\int_{\mathbf{q}}\int_{-\infty}^{\infty}\frac{\mathrm{d}\epsilon}{2\pi\mathrm{i}}\int_{-\infty}^{\infty}\frac{\mathrm{d}\omega}{2\pi\mathrm{i}} \\
(2)\qquad & \times\Big[f(\epsilon-\Omega)f(\omega-\epsilon)V^{\mathrm{R}}_{\omega}(\mathbf{q})\overline{j^i_0 j^j_0\, G^{\mathrm{R}}_{\epsilon}\big(G^{\mathrm{R}}_{\epsilon-\omega}-G^{\mathrm{A}}_{\epsilon-\omega}\big)G^{\mathrm{R}}_{\epsilon}\big(G^{\mathrm{R}}_{\epsilon-\Omega}-G^{\mathrm{A}}_{\epsilon-\Omega}\big)} \\
(4)\qquad & +f(\epsilon)f(\omega-\epsilon)V^{\mathrm{R}}_{\omega}(\mathbf{q})\overline{j^i_0 j^j_0\, G^{\mathrm{R}}_{\epsilon}\big(G^{\mathrm{R}}_{\epsilon-\omega}-G^{\mathrm{A}}_{\epsilon-\omega}\big)G^{\mathrm{R}}_{\epsilon}G^{\mathrm{A}}_{\epsilon-\Omega}} \\
(6)\qquad & -f(\epsilon)f(\omega-\epsilon)V^{\mathrm{A}}_{\omega}(\mathbf{q})\overline{j^i_0 j^j_0\, G^{\mathrm{A}}_{\epsilon}\big(G^{\mathrm{R}}_{\epsilon-\omega}-G^{\mathrm{A}}_{\epsilon-\omega}\big)G^{\mathrm{A}}_{\epsilon}G^{\mathrm{A}}_{\epsilon-\Omega}} \\
(8)\qquad & +f(\epsilon-\Omega)f(\omega-\epsilon+\Omega)V^{\mathrm{R}}_{\omega}(\mathbf{q})\overline{j^i_0 j^j_0\, G^{\mathrm{R}}_{\epsilon}G^{\mathrm{R}}_{\epsilon-\Omega}\big(G^{\mathrm{R}}_{\epsilon-\omega-\Omega}-G^{\mathrm{A}}_{\epsilon-\omega-\Omega}\big)G^{\mathrm{R}}_{\epsilon-\Omega}} \\
(10)\qquad & -f(\epsilon-\Omega)f(\omega-\epsilon+\Omega)V^{\mathrm{A}}_{\omega}\overline{j^i_0 j^j_0\, G^{\mathrm{R}}_{\epsilon}G^{\mathrm{A}}_{\epsilon-\Omega}\big(G^{\mathrm{R}}_{\epsilon-\omega-\Omega}-G^{\mathrm{A}}_{\epsilon-\omega}\big)G^{\mathrm{A}}_{\epsilon-\Omega}} \\
(12)\qquad & +f(\epsilon)f(\omega-\epsilon+\Omega)V^{\mathrm{A}}_{\omega}(\mathbf{q})\overline{j^i_0 j^j_0\big(G^{\mathrm{R}}_{\epsilon}-G^{\mathrm{A}}_{\epsilon}\big)G^{\mathrm{A}}_{\epsilon-\Omega}\big(G^{\mathrm{R}}_{\epsilon-\omega-\Omega}-G^{\mathrm{A}}_{\epsilon-\omega-\Omega}\big)G^{\mathrm{A}}_{\epsilon-\Omega}}\Big].
\end{aligned}
\tag{M.5}
$$

In the same way, from Eq. (M.4) we pick up the third and eleventh terms:

$$
\begin{aligned}
R^{ij}_{(b)}(\mathbf{0},\Omega) = & -2\sum_{\mathbf{q}}\int_{-\infty}^{\infty}\frac{\mathrm{d}\epsilon}{2\pi\mathrm{i}}\int_{-\infty}^{\infty}\frac{\mathrm{d}\omega}{2\pi\mathrm{i}} \\
(3)\qquad & \times\Big[f(\epsilon)f(\omega-\epsilon-\Omega)V^{\mathrm{R}}_{\omega}(\mathbf{q})\overline{j^i_0 j^j_0\, G^{\mathrm{R}}_{\epsilon+\Omega}G^{\mathrm{R}}_{\epsilon}\big(G^{\mathrm{R}}_{\epsilon+\Omega-\omega}-G^{\mathrm{A}}_{\epsilon+\Omega-\omega}\big)G^{\mathrm{A}}_{\epsilon-\omega}} \\
(11)\qquad & -f(\epsilon+\Omega)f(\omega-\epsilon)V^{\mathrm{A}}_{\omega}(\mathbf{q})\overline{j^i_0 j^j_0\, G^{\mathrm{A}}_{\epsilon+\Omega}G^{\mathrm{A}}_{\epsilon}G^{\mathrm{R}}_{\epsilon+\Omega-\omega}\big(G^{\mathrm{R}}_{\epsilon-\omega}-G^{\mathrm{A}}_{\epsilon-\omega}\big)}\Big].
\end{aligned}
\tag{M.6}
$$

The bar over the products of Green functions indicates the impurity average. Notice that in Eqs. (M.5) and (M.6) the current vertices appear under the impurity average bar. After the average and the restoration of translational invariance, integrations over momenta appear. The integrations over the *slow* momenta (because of the diffusive pole) which enter the interaction and the ladders are performed at the end, while the integrations over the *fast* momenta (because of the pole at the Fermi surface) entering the Green functions are performed with the help of the residue theorem as explained in the answer to Problem 21.3. To this end, all frequencies in the Green functions can be set to zero in the leading order in the diffusive-regime expansion $\omega\tau \ll 1$. As a result, the impurity average of the product of Green functions does not depend on the energy. Since the ladders depend only on the *slow* frequencies ω and Ω, we can perform the ϵ-integration at once by using the useful identity

$$\int_{-\infty}^{\infty} \mathrm{d}\epsilon\; f(\epsilon)\, f(\omega-\epsilon) = \frac{\omega}{2}\left(\coth\left(\frac{\beta\omega}{2}\right) - 1\right) \equiv F(\omega). \tag{M.7}$$

We now consider the impurity average explicitly. To illustrate the procedure, let us consider the first term in Eq. (M.5). It contains four products of four Green functions each. The first product $G^{\mathrm{R}}_{\epsilon}G^{\mathrm{R}}_{\epsilon-\omega}G^{\mathrm{R}}_{\epsilon}G^{\mathrm{R}}_{\epsilon-\Omega}$ gives zero upon averaging. The term $G^{\mathrm{R}}_{\epsilon}G^{\mathrm{R}}_{\epsilon-\omega}G^{\mathrm{R}}_{\epsilon}G^{\mathrm{A}}_{\epsilon-\Omega}$ vanishes because of the vector nature of the current vertices. There remain the terms $G^{\mathrm{R}}_{\epsilon}G^{\mathrm{A}}_{\epsilon-\omega}G^{\mathrm{R}}_{\epsilon}G^{\mathrm{R}}_{\epsilon-\Omega}$ and $G^{\mathrm{R}}_{\epsilon}G^{\mathrm{A}}_{\epsilon-\omega}G^{\mathrm{R}}_{\epsilon}G^{\mathrm{A}}_{\epsilon-\Omega}$. Upon averaging, the first term gives rise to two diagrams: one with two and one with three ladders. The latter corresponds to diagram (a) of Fig. 21.9. The first diagram with two ladders, instead, must be combined with the two-ladder diagrams originating from the term $G^{\mathrm{R}}_{\epsilon}G^{\mathrm{A}}_{\epsilon-\omega}G^{\mathrm{R}}_{\epsilon}G^{\mathrm{A}}_{\epsilon-\Omega}$. All together the two-ladder diagrams are shown as (c) and (d) in Fig. 21.9. By following this line of reasoning, one obtains all the diagrams of Fig. 21.9, including those obtained by interchanging the top and bottom Green function lines. One key point to notice is that the diagrams with two ladders cancel each other, i.e., the sum of (c), (d) and (e). This cancellation is shown in detail in Altshuler et al. (1980b). To see it, let us consider the terms in Eqs. (M.3) and (M.4) which contain the retarded interaction V^{R}_{ω} (a similar analysis can be done for the terms containing V^{A}_{ω}). By using Eq. (M.7), we get from Eqs. (M.5) and (M.6)

$$\begin{aligned} -2\int_{\mathbf{q}}\int_{-\infty}^{\infty}\frac{\mathrm{d}\omega}{2\pi\mathrm{i}}V^{\mathrm{R}}_{\omega}\Big(F(\omega-\Omega)\Big[&\overline{j^i_0 j^j_0 G^{\mathrm{R}}_{\epsilon}G^{\mathrm{A}}_{\epsilon-\omega}G^{\mathrm{R}}_{\epsilon}G^{\mathrm{A}}_{\epsilon-\Omega}} - \overline{j^i_0 j^j_0 G^{\mathrm{R}}_{\epsilon}G^{\mathrm{A}}_{\epsilon-\omega}G^{\mathrm{R}}_{\epsilon}G^{\mathrm{R}}_{\epsilon-\Omega}}\Big] \\ &-F(\omega)\overline{j^i_0 j^j_0 G^{\mathrm{R}}_{\epsilon}G^{\mathrm{A}}_{\epsilon-\omega}G^{\mathrm{R}}_{\epsilon}G^{\mathrm{A}}_{\epsilon-\Omega}} - F(\omega)\overline{j^i_0 j^j_0 G^{\mathrm{R}}_{\epsilon}G^{\mathrm{A}}_{\epsilon-\omega}G^{\mathrm{R}}_{\epsilon}G^{\mathrm{R}}_{\epsilon-\Omega}} \\ &-F(\omega-\Omega)\overline{j^i_0 j^j_0 G^{\mathrm{R}}_{\epsilon+\Omega}G^{\mathrm{R}}_{\epsilon}G^{\mathrm{A}}_{\epsilon+\Omega-\omega}G^{\mathrm{A}}_{\epsilon-\omega}}\Big), \end{aligned}$$

which is readily seen to vanish by considering the following averages:

$$\overline{j^i_0 j^j_0 G^{\mathrm{R}}_{\epsilon}G^{\mathrm{A}}_{\epsilon-\omega}G^{\mathrm{R}}_{\epsilon}G^{\mathrm{R}}_{\epsilon-\Omega}} = -\delta_{ij}\frac{v_{\mathrm{F}}^2}{d}2\pi N_0\tau^3\left(\frac{1}{2\pi N_0\tau^2}\frac{1}{Dq^2-\mathrm{i}\omega}\right)^2,$$

$$\overline{j^i_0 j^j_0 G^{\mathrm{R}}_{\epsilon}G^{\mathrm{A}}_{\epsilon-\omega}G^{\mathrm{R}}_{\epsilon}G^{\mathrm{A}}_{\epsilon-\Omega}} = \delta_{ij}\frac{v_{\mathrm{F}}^2}{d}2\pi N_0\tau^3\left(\frac{1}{2\pi N_0\tau^2}\frac{1}{Dq^2-\mathrm{i}\omega}\right)^2,$$

$$\overline{j^i_0 j^j_0 G^{\mathrm{R}}_{\epsilon+\Omega}G^{\mathrm{R}}_{\epsilon}G^{\mathrm{A}}_{\epsilon+\Omega-\omega}G^{\mathrm{A}}_{\epsilon-\omega}} = \delta_{ij}\frac{v_{\mathrm{F}}^2}{d}4\pi N_0\tau^3\left(\frac{1}{2\pi N_0\tau^2}\frac{1}{Dq^2-\mathrm{i}\omega}\right)^2.$$

Let us now turn our attention to the diagrams with three ladders. The impurity average needed for diagram (a) of Fig. 21.9 is

$$
\begin{aligned}
&\overline{j_0^i j_0^j G_\epsilon^R G_{\epsilon-\omega}^A G_\epsilon^R G_{\epsilon-\Omega}^R} \\
&= \int_{\mathbf{p}} j_0^i G_\epsilon^R(\mathbf{p}) G_{\epsilon-\omega}^A(\mathbf{p}) G_{\epsilon-\Omega}^R(\mathbf{p}-\mathbf{q}) \\
&\times \int_{\mathbf{p}'} j_0^j G_\epsilon^R(\mathbf{p}') G_{\epsilon-\omega}^A(\mathbf{p}') G_{\epsilon-\Omega}^R(\mathbf{p}'-\mathbf{q}) (\mathcal{T}^0(\mathbf{q},\omega))^2 L(\mathbf{q},\omega-\Omega) \\
&= \frac{(-4\pi e N_0 D\tau^2 q^i)(-4\pi e N_0 D\tau^2 q^j)}{2\pi N_0 \tau^4} \frac{1}{(D\mathbf{q}^2 - \mathrm{i}\omega)^2 (D\mathbf{q}^2 - \mathrm{i}(\omega-\Omega))} \\
&= 4\pi\sigma_0 \frac{Dq^i q^j}{(D\mathbf{q}^2 - \mathrm{i}\omega)^2 (D\mathbf{q}^2 - \mathrm{i}(\omega-\Omega))},
\end{aligned}
\tag{M.8}
$$

where the factors in round brackets, in the third line, arise from the integration of the three Green functions with a current vertex. Notice that in diagram (b) of Fig. 21.9, the two integrations over products of three Green functions produce an opposite sign. This gives an overall minus sign for diagram (b) with respect to (a). In the last line $\sigma_0 = 2e^2 N_0 D$ is the Drude conductivity. By collecting in Eqs. (M.5) (first, third, fourth and sixth lines) and (M.6) all the terms giving rise to diagrams with three ladders we get

$$
\begin{aligned}
R^{ij}(\mathbf{0},\Omega) = -4\sigma_0 \int_{\mathbf{q}} \int_{-\infty}^{\infty} \frac{\mathrm{d}\omega}{2\pi}
&\times \Bigg[F(\omega-\Omega) V_\omega^R(\mathbf{q}) \frac{Dq^i q^j}{(D\mathbf{q}^2 - \mathrm{i}\omega)^2 (D\mathbf{q}^2 - \mathrm{i}(\omega-\Omega))} \\
&+ F(\omega)\ V_\omega^A(\mathbf{q}) \frac{Dq^i q^j}{(D\mathbf{q}^2 + \mathrm{i}\omega)^2 (D\mathbf{q}^2 + \mathrm{i}(\omega-\Omega))} \\
&+ F(\omega)\ V_\omega^R(\mathbf{q}) \frac{Dq^i q^j}{(D\mathbf{q}^2 - \mathrm{i}\omega)^2 (D\mathbf{q}^2 - \mathrm{i}(\omega+\Omega))} \\
&+ F(\omega+\Omega) V_\omega^A(\mathbf{q}) \frac{Dq^i q^j}{(D\mathbf{q}^2 + \mathrm{i}\omega)^2 (D\mathbf{q}^2 + \mathrm{i}(\omega+\Omega))} \\
&- F(\omega-\Omega) V_\omega^R(\mathbf{q}) \frac{Dq^i q^j}{(D\mathbf{q}^2 - \mathrm{i}\omega)^2 (D\mathbf{q}^2 - \mathrm{i}(\omega+\Omega))} \\
&- F(\omega-\Omega) V_\omega^R(\mathbf{q}) \frac{Dq^i q^j}{(D\mathbf{q}^2 - \mathrm{i}\omega)^2 (D\mathbf{q}^2 - \mathrm{i}(\omega-\Omega))} \\
&- F(\omega+\Omega) V_\omega^A(\mathbf{q}) \frac{Dq^i q^j}{(D\mathbf{q}^2 + \mathrm{i}\omega)^2 (D\mathbf{q}^2 + \mathrm{i}(\omega+\Omega))} \\
&- F(\omega+\Omega) V_\omega^A(\mathbf{q}) \frac{Dq^i q^j}{(D\mathbf{q}^2 + \mathrm{i}\omega)^2 (D\mathbf{q}^2 + \mathrm{i}(\omega-\Omega))} \Bigg]
\end{aligned}
\tag{M.9}
$$

Notice that the first term cancels with the sixth and the fourth with the seventh. By diving by $-\mathrm{i}\Omega$ and sending Ω to zero we get Eq. (21.83) quoted in the text.

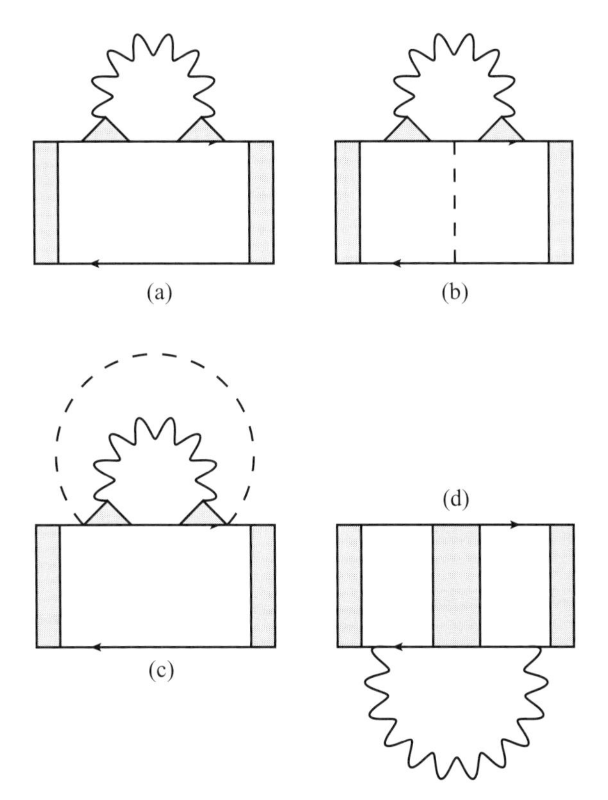

Fig. M.1 Diagrams for the ladder self-energy. A similar set of diagrams is generated by the interchange of the interaction line (wiggly line) between top and bottom Green function lines.

M.2 The ladder renormalization in the presence of electron–electron interaction

Let us assume that the ladder, in the presence of interaction, gets renormalized as

$$L(q,\omega) = \frac{1}{2\pi N_0\tau^2}\frac{1}{Dq^2+|\omega_\nu|} \to \frac{1}{2\pi N_0\tau^2}\frac{\zeta^2}{D_{\mathrm{R}}q^2+Z|\omega_\nu|}, \tag{M.10}$$

where ζ, D_{R} and Z represent the effective wave function renormalization, the renormalized diffusion coefficient and the renormalization of the frequency. By expanding,

$$\frac{\zeta^2}{D_{\mathrm{R}}q^2+Z|\omega_\nu|} = \frac{1}{Dq^2+|\omega_\nu|} + \frac{(2\delta\zeta-\delta D/D)Dq^2+(2\delta\zeta-\delta Z)|\omega_\nu|}{(Dq^2+|\omega_\nu|)^2} \equiv \frac{\Sigma_{\mathrm{L}}(q,\omega)}{(Dq^2+|\omega_\nu|)^2} \tag{M.11}$$

and the last equation defines the ladder self-energy. The diagrams for the self-energy are shown in Fig. M.1. The first step is the integration over the *fast* momenta running within the Green functions. This integration amounts to the evaluation of several integrals of products of the type $(G^{\mathrm{R}})^m(G^{\mathrm{A}})^n$, whose results are given in the answer to Problem 21.3. For diagrams (a)–(c), in the small k, Ω_{m}, q and ω_ν limit, we obtain

$$I_{\mathrm{abc}} = (2\pi N_0\tau)^2 2\pi N_0\tau^4[D(q^2+k^2)+|\omega_\nu+\Omega_{\mathrm{m}}|-2\mathbf{q}\cdot\mathbf{k}], \tag{M.12}$$

where the first factor $(2\pi N_0\tau)^2$ represents the two integrations over the two Green functions at the interaction vertices. The rest gives the integration over the remaining Green functions. In a similar way, integration over Green functions gives for diagram (d)

$$I_{\mathrm{d}} = -(2\pi N_0\tau^2)^3. \tag{M.13}$$

The diagrams (a), (b), (c) shown in Fig. M.1 finally yield

$$\Sigma_{\mathrm{L,abc}} = -2\frac{1}{\beta}\sum_{\epsilon_n+\omega_\nu<\Omega_{\mathrm{m}}}\int_{\mathbf{k}}\frac{I_{\mathrm{abc}}(\mathbf{k},\mathbf{q},\Omega_{\mathrm{m}},\omega_\nu)\mathcal{V}(\mathbf{k},\Omega_{\mathrm{m}})}{(2\pi N_0\tau^2)^3(Dk^2+|\Omega_{\mathrm{m}}|)^2} \tag{M.14}$$

and diagram (d) gives

$$\Sigma_{\mathrm{L,d}} = 2\frac{1}{\beta}\sum_{\epsilon_n<\Omega_{\mathrm{m}}}\int_{\mathbf{k}}\frac{I_{\mathrm{d}}\mathcal{V}(\mathbf{k},\Omega_{\mathrm{m}})}{(2\pi N_0\tau^2)^3(D(\mathbf{k}+\mathbf{q})^2+|\Omega_{\mathrm{m}}+\omega_\nu|)}. \tag{M.15}$$

In Eqs. (M.14) and (M.15) we have used Matsubara frequencies. The relative minus sign comes from the integration over the *fast* momenta. The factor of 2 is due to the fact that there is another set of diagrams generated by interchanging the interaction line between the two electron lines. One may check that the sum of Eqs. (M.14) and (M.15) vanishes in the limit $q=0$ and $\omega=0$. For small, but finite, external momentum and frequency, we rewrite Eq. (M.14) in the form

$$\begin{aligned}\Sigma_{\mathrm{L,abc}} = &-2\frac{1}{\beta}\sum_{\epsilon_n+\omega_\nu<\Omega_{\mathrm{m}}}\int_{\mathbf{k}}\frac{\mathcal{V}(\mathbf{k},\Omega_{\mathrm{m}})}{Dk^2+|\Omega_{\mathrm{m}}|}\\ &-2\frac{1}{\beta}\sum_{\epsilon_n+\omega_\nu<\Omega_{\mathrm{m}}}\int_{\mathbf{k}}\frac{(Dq^2+|\omega_\nu|)\mathcal{V}(\mathbf{k},\Omega_{\mathrm{m}})}{(Dk^2+|\Omega_{\mathrm{m}}|)^2}\end{aligned} \tag{M.16}$$

and Eq. (M.15) as

$$\begin{aligned}\Sigma_{\mathrm{L,d}} = &\,2\frac{1}{\beta}\sum_{\epsilon_n<\Omega_{\mathrm{m}}<\epsilon_n+\omega_\nu}\int_{\mathbf{k}}\frac{\mathcal{V}(\mathbf{k},\Omega_{\mathrm{m}})}{D(\mathbf{k}+\mathbf{q})^2+|\Omega_{\mathrm{m}}+\omega_\nu|}\\ &+2\frac{1}{\beta}\sum_{\epsilon_n+\omega_\nu<\Omega_{\mathrm{m}}}\int_{\mathbf{k}}\left[\frac{\mathcal{V}(\mathbf{k},\Omega_{\mathrm{m}})}{D\mathbf{k}^2+|\Omega_{\mathrm{m}}|}+\frac{\mathcal{V}(\mathbf{k},\Omega_{\mathrm{m}})}{D(\mathbf{k}+\mathbf{q})^2+|\Omega_{\mathrm{m}}+\omega_\nu)|}-\frac{\mathcal{V}(\mathbf{k},\Omega_{\mathrm{m}})}{D\mathbf{k}^2+|\Omega_{\mathrm{m}}|}\right].\end{aligned} \tag{M.17}$$

The first term in the square brackets of Eq. (M.17) cancels with the first term in Eq. (M.16). Let us analyze the first term of Eq. (M.17), which due to the frequency restriction becomes proportional to ω_ν. $\mathbf{q}$ and ω_ν can therefore be put to zero in performing the integrals. By transforming the Matsubara sum into an integral in the complex plane and analytically continuing $\epsilon \to -\mathrm{i}\epsilon$ and $\epsilon + \omega \to -\mathrm{i}(\epsilon + \omega)$, it reads

$$\begin{aligned}
\Sigma^1_{\mathrm{L,d}} &= 2\frac{1}{\beta} \sum_{\epsilon_n < \Omega_\mathrm{m} < \epsilon_n + \omega_\nu} \int_\mathbf{k} \frac{\mathcal{V}(\mathbf{k}, \Omega_\mathrm{m})}{D(\mathbf{k}+\mathbf{q})^2 + |\Omega_\mathrm{m} + \omega_\nu|} \\
&\approx -\frac{2}{2\pi \mathrm{i}} \int_{-\infty}^{\infty} \mathrm{d}\Omega \, [f(\Omega - \epsilon - \omega) - f(\Omega - \epsilon)] \int_\mathbf{k} \frac{V^\mathrm{R}(0,0)}{D\mathbf{k}^2 - \mathrm{i}(\Omega + \omega)} \\
&= -\frac{2\omega}{2\pi \mathrm{i}} \int_\mathbf{k} \frac{V^\mathrm{R}(0,0)}{D\mathbf{k}^2 - \mathrm{i}(\epsilon + \omega)} \\
&\Rightarrow \mathrm{i}\omega \frac{V_1 - 2V_2}{4\pi^2 D} \ln\left(\frac{1}{\epsilon\tau}\right) \\
&= -\mathrm{i}\omega \, \mathrm{I}_3.
\end{aligned} \tag{M.18}$$

having used $f(\Omega - \epsilon - \omega) - f(\Omega - \epsilon) \approx \omega \, \partial_\Omega f(\Omega - \epsilon)$. In the last line we have also included the contribution of the Hartree diagrams. The difference between the second and third terms in the square brackets of Eq. (M.17) may be expanded in powers of q and ω_ν. The lowest order term reads

$$\begin{aligned}
\Sigma^2_{\mathrm{L,d}} &= 2\frac{1}{\beta} \sum_{\epsilon_n + \omega_\nu < \Omega_\mathrm{m}} \int_\mathbf{k} \mathcal{V}(\mathbf{k}, \Omega_\mathrm{m}) \left[\frac{2D^2 q^2 \mathbf{k}^2}{(D\mathbf{k}^2 + |\Omega_\mathrm{m}|)^3} - \frac{(Dq^2 + |\omega_\nu|)}{(D\mathbf{k}^2 + |\Omega_\mathrm{m}|)^2} \right] \\
&= -\frac{2}{2\pi \mathrm{i}} \int_{-\infty}^{\infty} \mathrm{d}\Omega f(\Omega - \epsilon - \omega) \int_\mathbf{k} \left[\frac{Dq^2 D\mathbf{k}^2 V^\mathrm{R}(0,0)}{(D\mathbf{k}^2 - \mathrm{i}\Omega)^3} - \frac{(Dq^2 - \mathrm{i}\omega) V^\mathrm{R}(0,0)}{(D\mathbf{k}^2 - \mathrm{i}\Omega)^2} \right] \\
&\Rightarrow Dq^2 \frac{V_1 - 2V_2}{4\pi^2 D} \ln\left(\frac{1}{\epsilon\tau}\right) - (Dq^2 - \mathrm{i}\omega) \frac{V_1 - 2V_2}{4\pi^2 D} \ln\left(\frac{1}{\epsilon\tau}\right) \\
&= Dq^2 \, \mathrm{I}_2 - (Dq^2 - \mathrm{i}\omega) \mathrm{I}_1.
\end{aligned} \tag{M.19}$$

Finally, the second term in Eq. (M.16) reads

$$\begin{aligned}
\Sigma_{\mathrm{L,abc}} &= -(Dq^2 + |\omega_\nu|) 2\frac{1}{\beta} \sum_{\epsilon_n + \omega_\nu < \Omega_\mathrm{m}} \int_\mathbf{k} \frac{\mathcal{V}(\mathbf{k}, \Omega_\mathrm{m})}{(Dk^2 + |\Omega_\mathrm{m}|)^2} \\
&= -(Dq^2 - \mathrm{i}\omega) \frac{2}{2\pi \mathrm{i}} \int_{-\infty}^{\infty} \mathrm{d}\Omega f(\Omega - \epsilon - \omega) \int_\mathbf{k} \frac{V^\mathrm{R}(0,0)}{(Dk^2 - \mathrm{i}\Omega)^2} \\
&\Rightarrow -(Dq^2 - \mathrm{i}\omega) \frac{V_1 - 2V_2}{4\pi^2 D} \ln\left(\frac{1}{\epsilon\tau}\right) \\
&= -(Dq^2 - \mathrm{i}\omega) \, \mathrm{I}_1.
\end{aligned} \tag{M.20}$$

One sees that the by inserting the self-energy results (M.18)–(M.20) into Eq. (M.11) one gets

$$\zeta = 1 - \mathrm{I}_1, \tag{M.21}$$

$$\frac{D_\mathrm{R}}{D} = 1 - \mathrm{I}_2, \tag{M.22}$$

$$Z = 1 - \mathrm{I}_3. \tag{M.23}$$

In the main text the renormalized diffusion coefficient D_{R} will be renamed D. We conclude this appendix by remarking that the above renormalizations coincide with the perturbative corrections of the single-particle density of states, Eq. (21.78), conductivity, Eq. (21.84) and specific heat, Eq. (21.69), satisfying at this order the Ward identities identifications.

N Appendix N Answers to problems

Chapter 1

1.1 We note that

$$\frac{\left(\frac{\partial U}{\partial V}\right)_P}{\left(\frac{\partial U}{\partial P}\right)_V} = \frac{c_P}{c_V}\frac{\left(\frac{\partial T}{\partial V}\right)_P}{\left(\frac{\partial T}{\partial P}\right)_V},$$

as can easily be seen by applying the rule of composite differentiation to both numerator and denominator.

Hence the slope for an adiabatic process reads from (1.8) and (1.9)

$$\left(\frac{\partial P}{\partial V}\right)_{\text{adia}} = -\frac{P}{\left(\frac{\partial U}{\partial P}\right)_V} + \frac{c_P}{c_V}\left(\frac{\partial P}{\partial V}\right)_T < \frac{c_P}{c_V}\left(\frac{\partial P}{\partial V}\right)_T < 0,$$

since the internal energy is an increasing function of the pressure for the first inequality and $(\mathrm{d}P/\mathrm{d}V)_T < 0$ and $c_P > c_V$ by the stability conditions (1.51) and (1.52). Upon increasing the pressure, the negative work done on the system implies an increase of the internal energy by (1.6) with $Q = 0$.

1.2 Without loss of generality, let us assume $Q_2' < Q_2$. Then, we imagine a combined cycle of R' and the reverse of R. Since the heat balance at θ_1 is zero, the work done in the combined cycle is

$$W = Q_2 - Q_2' > 0,$$

which contradicts the Kelvin statement of the second principle.

The second point is shown by combining an irreversible cycle acting between T_1 and T_2 with two Carnot cycles acting between T_0 and T_1 and between T_0 and T_2, respectively. The resulting work is equal to the heat absorbed in the two cycles at temperature T_0 and must be negative to avoid contradiction with the second principle.

1.3 Start from

$$c_V = \frac{T}{V}\left(\frac{\partial S}{\partial T}\right)_V$$

and notice that

$$\left(\frac{\partial S}{\partial T}\right)_V = \frac{\partial(S,V)}{\partial(T,V)}\frac{\partial(T,P)}{\partial(T,P)} = \left(\frac{\partial S}{\partial T}\right)_P - \left(\frac{\partial S}{\partial P}\right)_T\left(\frac{\partial P}{\partial V}\right)_T\left(\frac{\partial V}{\partial T}\right)_P.$$

By using the Maxwell relation (1.29), we get

$$c_V = \frac{T}{V}\left(\frac{\partial S}{\partial T}\right)_P + \frac{T}{V}\left(\frac{\partial V}{\partial T}\right)_P^2 \left(\frac{\partial P}{\partial V}\right)_T .$$

By recalling (1.50) and

$$c_P = \frac{T}{V}\left(\frac{\partial S}{\partial T}\right)_P$$

we obtain Eq. (1.52).

1.4 Let us begin by denoting with $S(0, X_{\mathrm{i}})$ and $S(0, X_{\mathrm{f}})$ the limiting values of the two curves of Fig. 1.3 by considering the possibility that they may be different from one another. As a function of temperature and the parameter X we have

$$\mathrm{d}S = \left(\frac{\partial S}{\partial T}\right)_X \mathrm{d}T + \left(\frac{\partial S}{\partial X}\right)_T \mathrm{d}X,$$

implying for the entropy change along the two curves

$$S(T_{\mathrm{i}}, X_{\mathrm{f}}) - S(0, X_{\mathrm{f}}) = \int_0^{T_{\mathrm{i}}} \mathrm{d}T \frac{V c_{X_{\mathrm{f}}}}{T},$$

$$S(T_{\mathrm{f}}, X_{\mathrm{i}}) - S(0, X_{\mathrm{i}}) = \int_0^{T_{\mathrm{f}}} \mathrm{d}T \frac{V c_{X_{\mathrm{i}}}}{T},$$

where T_{i} and T_{f} are the initial and final temperatures and $c_{X_{\mathrm{i}}}$ and $c_{X_{\mathrm{f}}}$ the specific heat along the two curves at a constant value of X. By definition of isoentropic transformation $S(T_{\mathrm{i}}, X_{\mathrm{f}}) = S(T_{\mathrm{f}}, X_{\mathrm{i}})$ and one obtains

$$S(0, X_{\mathrm{f}}) - S(0, X_{\mathrm{i}}) = \int_0^{T_{\mathrm{f}}} \mathrm{d}T \frac{V c_{X_{\mathrm{f}}}}{T} - \int_0^{T_{\mathrm{i}}} \mathrm{d}T \frac{V c_{X_{\mathrm{i}}}}{T}.$$

The above equation allows us to see the logical equivalence between the impossibility of attaining absolute zero and the vanishing in the zero-temperature limit of the entropy change for an isothermic reversible transformation.

If one could reach absolute zero, it should be possible to find T_{i} such that $T_{\mathrm{f}} = 0$. Then, since $c_{X_{\mathrm{i}}} > 0$ for thermodynamic stability, one would have $S(0, X_{\mathrm{f}}) - S(0, X_{\mathrm{i}}) < 0$, which contradicts Eq. (1.18). This shows that the unattainability formulation implies the vanishing of the entropy change. Vice versa, by assuming $S(0, X_{\mathrm{f}}) - S(0, X_{\mathrm{i}}) < 0$ in Eq. (1.18), there should exist a value T_{i} such that the second integral in the relation above is equal to $S(0, X_{\mathrm{f}}) - S(0, X_{\mathrm{i}})$. Then, necessarily, T_{f} must vanish, implying that absolute zero would be attainable.

Chapter 2

2.1 The time reversed process of the collision $(\mathbf{p}_1, \mathbf{p}_2 \to \mathbf{p}_1', \mathbf{p}_2')$ is obtained by reversing the momenta and exchanging initial and final states. Time reversal invariance then requires

$$W(-\mathbf{p}_1, -\mathbf{p}_2; -\mathbf{p}_1', -\mathbf{p}_2') = W(\mathbf{p}_1', \mathbf{p}_2'; \mathbf{p}_1, \mathbf{p}_2).$$

Invariance under space reversal or parity yields the additional constraint

$$W(-\mathbf{p}_1, -\mathbf{p}_2; -\mathbf{p}_1', -\mathbf{p}_2') = W(\mathbf{p}_1, \mathbf{p}_2; \mathbf{p}_1', \mathbf{p}_2').$$

By combining both relations one obtains

$$W(\mathbf{p}_1', \mathbf{p}_2'; \mathbf{p}_1, \mathbf{p}_2) = W(\mathbf{p}_1, \mathbf{p}_2; \mathbf{p}_1', \mathbf{p}_2'),$$

which is known as the principle of detailed balance.

2.2 We need to find the extrema of

$$\widetilde{\mathrm{H}} = \int \mathrm{d}^3 p \, f(\mathbf{p}) \left[\ln f(\mathbf{p}) + \alpha + \beta \frac{p^2}{2m} \right] - \alpha \frac{N}{V} - \beta \frac{E}{V},$$

where α and β are Lagrange multipliers. By performing a variation of $f(\mathbf{p})$ we get

$$\frac{\delta \widetilde{\mathrm{H}}}{\delta f(\mathbf{p})} = \ln f(\mathbf{p}) + 1 + \alpha + \beta \frac{p^2}{2m} = 0$$

from which $f(\mathbf{p}) \propto \mathrm{e}^{-\beta p^2/2m}$ in agreement with Eq. (2.43). By performing the second variation of $\widetilde{\mathrm{H}}$

$$\frac{\delta^2 \widetilde{\mathrm{H}}}{\delta f^2(\mathbf{p})} = \frac{1}{f} > 0,$$

which shows that the Maxwell–Boltzmann distribution is a minimum.

2.3 Due to the given assumptions, which may be justified by isotropy of space, we have

$$g(v_x) g(v_y) g(v_z) = G\left(v_x^2 + v_y^2 + v_z^2\right),$$

where G is a function of the modulus squared of the total velocity. To solve the above functional equation we take the logarithm of both sides:

$$\ln g(v_x) + \ln g(v_y) + \ln g(v_z) = \ln G\left(v_x^2 + v_y^2 + v_z^2\right)$$

and differentiate with respect to v_x:

$$\frac{1}{2v_x} \frac{g'}{g} = \frac{G'}{G}.$$

Whereas the left hand side depends only on v_x, the right hand side is a function of v_x, v_y and v_z. Hence compatibility requires both members of the equation to be equal to a constant:

$$\frac{1}{2v_x} \frac{g'}{g} = -b, \qquad \frac{G'}{G} = -b.$$

After integrating we get

$$g(v_x) = -b v_x^2 + a', \qquad G(v^2) = -b v^2 + a,$$

and we must have $a' = a/3$.

2.4 At room temperature the molecular velocity for oxygen is about $v \sim 4 \times 10^2 \mathrm{m\,s^{-1}}$. The particle travels undisturbed over a mean free path $l \sim 0.9 \times 10^{-6}\mathrm{m}$ for a typical time $\tau \sim 0.2 \times 10^{-8}\mathrm{s}$. Hence, we may consider a one-dimensional motion of the particle as composed of independent steps x_k with average values $\langle x_k \rangle = 0$ and $\langle x_k^2 \rangle = l^2$, each step being covered in a time τ. The fact that $\langle x_k \rangle = 0$ corresponds physically to the fact that, as a consequence of collisions, the particle may bounce forward and backward with the same average probability. After N steps the distance covered is the sum $\sum_{k=1}^{N} x_k$. Hence the problem of the average of the distance covered reduces to the average of N independent variables. As a result the mean square distance reads $l_N = \sqrt{\langle \sum_{k,k'=1}^{N} x_k x_{k'} \rangle} = \sqrt{N l^2}$. The time to cover such a distance will be $t = N\tau = l_N^2 \tau / l^2 = l_N^2 / vl$. For $l_N = 1\mathrm{m}$ this gives $t \sim 2 \times 10^3\mathrm{s}$, much longer than the fraction of a second for an undisturbed motion. This is typical of the Brownian or diffusive motion which will be discussed in Chapter 11. Here we have seen that, in the context of the kinetic theory for a Maxwell–Boltzmann distribution the diffusion coefficient is $D \sim vl$.

2.5 We integrate Eq. (2.18) with respect to $\mathbf{p}_1$. For the right-hand side we get

$$\int_1 \int_2 \int_{1'} \int_{2'} W_{\text{class}}(\mathbf{p}_{1'}, \mathbf{p}_{2'}; \mathbf{p}_1, \mathbf{p}_2)(f_{1'} f_{2'} - f_1 f_2) = 0$$

since now the integral is symmetric with respect to the two pairs of initial and final momenta. Also

$$\int_1 \mathbf{F} \cdot \frac{\partial f_1}{\partial \mathbf{p}_1} = 0$$

since the distribution function vanishes at infinity. Finally, by introducing the mass density and the associated current

$$\rho(\mathbf{q}, t) = m \int_1 f_1, \quad \mathbf{j}(\mathbf{q}, t) = \int_1 \mathbf{p}_1 f_1,$$

we obtain the equation of continuity

$$\frac{\partial \rho}{\partial t} + \frac{\partial}{\partial \mathbf{q}} \cdot \mathbf{j}(\mathbf{q}, t) = 0.$$

2.6 We write $f_1 = f_1^0 + \delta f_1$, f_1^0 being the equilibrium distribution function. We then get

$$f_{1'} f_{2'} - f_1 f_2 = \delta f_{1'} f_{2'}^0 + \delta f_{2'} f_{1'}^0 - \delta f_1 f_2^0 - \delta f_2 f_1^0,$$

since $f_{1'}^0 f_{2'}^0 = f_1^0 f_2^0$. By writing $\delta f_i = f_i^0 \beta \chi_i$, the linearized collision integral reads

$$I[\chi] = \beta \int_2 \int_{1'} \int_{2'} W_{\text{class}}(\mathbf{p}_1', \mathbf{p}_2'; \mathbf{p}_1, \mathbf{p}_2) f_1^0 f_2^0 (\chi_{1'} + \chi_{2'} - \chi_1 - \chi_2).$$

2.7 Due to the smallness of the mean free path, the molecules of the gas undergo many collisions in a region of space where the temperature is approximately constant. This means that the gas reaches *local equilibrium* with a Maxwell–Boltzmann distribution at the local temperature $T(\mathbf{q})$. To linear order in the temperature gradient, the linearized Boltzmann equation reads

$$\frac{\mathbf{p}_1}{m} \cdot \frac{\partial T(\mathbf{q})}{\partial \mathbf{q}} \frac{\partial f_1^0}{\partial T} = I[\chi],$$

where we have used the linearized collision integral of Problem 2.6 and f_1^0 is the equilibrium distribution function in the absence of the temperature gradient. At constant pressure, the temperature derivative reads

$$\frac{\partial f_1^0}{\partial T} = f_1^0 \beta k_\mathrm{B} \left[\beta \frac{p_1^2}{2m} - \frac{5}{2} \right],$$

so that the linearized Boltzmann equation reads

$$\frac{\mathbf{p}_1}{m} \cdot \frac{\partial (k_\mathrm{B} T(\mathbf{q}))}{\partial \mathbf{q}} \left[\beta \frac{p_1^2}{2m} - \frac{5}{2} \right] = \frac{a^2}{4} \int \mathrm{d}^3 p_2 \mathrm{d}\Omega \frac{|\mathbf{p}_1 - \mathbf{p}_2|}{m} f_2^0 (\chi_{1'} + \chi_{2'} - \chi_1 - \chi_2),$$

where we have used the expression (2.19) for the scattering kernel and the hard spheres result $\mathrm{d}\sigma/\mathrm{d}\Omega = a^2/4$. We notice that the integral on the right-hand side is of order $\sim (N/V) a^2 \chi \sim l^{-1} \chi$ whereas the left-hand side goes like $k_\mathrm{B} T(\mathbf{q})/L_\mathrm{c}$, L_c being the characteristic length scale over which the temperature varies. This suggests that one may look for a solution in powers of the small parameter l/L_c. We then make the ansatz

$$\chi_\mathbf{p} = \mathbf{g} \cdot \frac{\partial (k_\mathrm{B} T(\mathbf{q}))}{\partial \mathbf{q}}, \qquad \mathbf{g} = A \frac{\mathbf{p}}{m} \left[\beta \frac{p^2}{2m} - \frac{5}{2} \right] + \cdots$$

where the dots indicate component terms that are relevant only in higher orders in l/L_c. By neglecting these terms the coefficient A can be determined by multiplying both terms by $f_1^0 \mathbf{p}_1 (\beta p_1^2/2m - (5/2))$ and integrating over $\mathbf{p}_1$ to read

$$A = \frac{\int \mathrm{d}^3 p_1 \, f_1^0 p_1^2 [\beta \frac{p_1^2}{2m} - \frac{5}{2}]^2}{\frac{\beta^2 a^2}{64 m^2} \int \mathrm{d}^3 p_1 \mathrm{d}^3 p_2 \mathrm{d}\Omega \frac{|\mathbf{p}_1 - \mathbf{p}_2|}{m} f_1^0 f_2^0 \left(p_{1'}^2 \mathbf{p}_{1'} + p_{2'}^2 \mathbf{p}_{2'} - p_1^2 \mathbf{p}_1 - p_2^2 \mathbf{p}_2 \right)^2} = -\frac{15}{8} \frac{l}{\overline{v}},$$

where $\overline{v} = 2\sqrt{2 k_\mathrm{B} T}/\sqrt{\pi m}$. In performing the integral in the denominator, we have used the center-of-mass and relative velocities defined in (2.46).

The above approximate solution can be systematically improved by carrying on the expansion in powers of l/L_c, as first shown by Enskog and Chapman (see for instance Chapman and Cowling (1970)).

2.8 By using the result of Problem 2.7 we have

$$\mathbf{j}_Q = \int \mathrm{d}^3 p \, \frac{\mathbf{p}}{m} \frac{p^2}{2m} f^0(\mathbf{p}) \beta \mathbf{g}$$

from which we derive the expression for the thermal conductivity defined by $\mathbf{j}_Q = -\kappa \partial T / \partial \mathbf{q}$ as

$$\kappa = -\frac{k_\mathrm{B}}{3} A \int \mathrm{d}^3 p \, \frac{p^2}{m^2} \frac{\beta p^2}{2m} f^0(\mathbf{p}) \left(\frac{\beta p^2}{2m} - \frac{5}{2} \right) = -\frac{N}{V} k_\mathrm{B} A \frac{5\pi}{16} \overline{v}^2 = c_P \frac{15\pi}{64} l \overline{v},$$

where we have used the fact the specific heat at constant pressure reads $c_P = (5/2)(N/V)k_B$. We may then write

$$\mathbf{j}_Q = -\frac{15\pi}{64} l\bar{v} \frac{\partial(c_P T)}{\partial \mathbf{q}},$$

expressing the fact that the thermal energy current as in a diffusive system is proportional to minus the gradient of the enthalpy of the gas $c_P T$. This confirms the estimate derived in Problem 2.4 for the diffusion coefficient of a gas $D \sim l\bar{v}$.

Chapter 3

3.1 By taking the divergence of the phase velocity

$$\text{div} \cdot \mathbf{v} \equiv \sum_{i=1}^{6N} \frac{\partial \dot{x}_i}{\partial x_i} = \sum_{i=1}^{3N} \left[\frac{\partial \dot{q}_i}{\partial q_i} + \frac{\partial \dot{p}_i}{\partial p_i} \right],$$

which, because of Eq. (3.1), becomes

$$\text{div} \cdot \mathbf{v} = \left[\frac{\partial}{\partial q_i} \frac{\partial H}{\partial p_i} - \frac{\partial}{\partial p_i} \frac{\partial H}{\partial q_i} \right] = 0.$$

3.2 For simplicity we consider the case of N_1 particles in a volume V_1 and N_2 particles in a volume V_2 with $V_1 + V_2 = V$. On average we expect that the fraction of particles in a given volume corresponds to the fraction of the given volume with respect to the total volume. Thus the probability of finding a particle in the volume V_1 is given by the fraction V_1/V and analogously for the corresponding probability of finding a particle in the volume V_2. By composition of probabilities, the probability of having N_1 particles in a volume V_1 and N_2 in V_2 is given by

$$P(N_1, N_2) = \frac{N!}{N_1! N_2!} \left(\frac{V_1}{V} \right)^{N_1} \left(\frac{V_2}{V} \right)^{N_2},$$

where the binomial coefficient gives the number of possible ways of putting N_1 particles in V_1 and N_2 in V_2. The probability $P(N_1, N_2)$ is indeed normalized to one.

The most probable distribution is obtained by minimizing $\ln P$ after using the Stirling formula (B.5) derived in Appendix B. As was expected, the ratio of the most probable values $\bar{N}_1$, $\bar{N}_2$ is equal to the ratio of the volumes $\bar{N}_1/N = V_1/V$.

If we further assume $V_1 = V_2 = V/2$ then

$$P(N_1, N_2) = \frac{N!}{N_1!(N - N_1)!} \frac{1}{2^N}.$$

All combinations of N_1 and N_2 are possible microscopic configurations but their statistical weight is extremely low for macroscopic values of N of the order of the Avogadro number.

The deviation x from the most probable configuration $N_1 = N_2 = N/2$ can be defined as $N_1 = N(1 + x)/2$ and $N_2 = N(1 - x)/2$. By using the Stirling formula

$N! \approx (N/\mathrm{e})^N \sqrt{2\pi N}$, the formula for $P(x)$ is easily obtained:

$$P(x) = \sqrt{\frac{2}{\pi N}}(1+x)^{-N(1+x)/2}(1-x)^{-N(1-x)/2}.$$

For large N and small x, as we expect, we can expand $P(x)$ up to quadratic terms:

$$P(x) = \sqrt{\frac{2}{\pi N}}\mathrm{e}^{-Nx^2/2}.$$

From this Gaussian form, considering x as a continuous variable, we easily obtain

$$\langle x^2 \rangle = \frac{1}{N}.$$

For instance, for a gas at ambient pressure ($P \sim 10^5$ Pa) and temperature ($T \sim 290$ K), the number per cm^3 is $\sim 10^{19}$ and one should observe a relative deviation of density of the order of $\sim 10^{-9}$. As already remarked, macroscopic deviations (as for instance $N_1 = N$ and $N_2 = 0$), even if possible, are statistically unobservable.

Chapter 4

4.1 Let us divide, ideally, the cylinder into two parts: the first from the ground up to the height z and the second from there to infinity. Let us denote with $N_\pm$ the numbers of the particles contained in the upper and lower parts, respectively, and $N_+ + N_- = N$. The free energy for the two parts is readily obtained by using Eq. (4.8) as

$$F_\pm = -k_B T N_\pm \ln\left(\frac{I_\pm}{N_\pm \lambda^3}\right) - k_B T N_\pm,$$

where the $I_\pm$ represents the contribution of the integral over the space variables (x, y, z') in the volumes $V_\pm$ corresponding to $z' > z$ and $0 < z' < z$, respectively,

$$I_+ = A\int_z^\infty \mathrm{d}z' \mathrm{e}^{-\beta m g z'}, \quad I_- = A\int_0^z \mathrm{d}z' \mathrm{e}^{-\beta m g z'}.$$

In statistical equilibrium, at the interface between the two parts, there is exact compensation between the pressure exercised by the two portions of the gas. Both pressures describe the rate of change of the free energy when the volume is varied by an infinitesimal amount. Let us imagine changing z by the amount $\mathrm{d}z$, so that the lower part increases its volume by $A\mathrm{d}z$, while the upper part decreases its own by an equal amount. As a result

$$P_\pm = -\frac{1}{A}\frac{\partial F_\pm}{\partial z}.$$

By imposing the condition $P_+ + P_- = 0$ we can solve for N_+ and N_- with the condition $N_+ + N_- = N$. After some simple algebra one finds

$$N_+ = N\beta m g \int_0^z \mathrm{d}z' \mathrm{e}^{-\beta m g z'},$$

which leads to the *barometric formula*

$$P(z) = \frac{Nmg}{A} e^{-\beta mgz}. \tag{N.1}$$

4.2 By introducing center-of-mass and relative coordinates

$$\mathbf{r} = \frac{m_a \mathbf{r}_a + m_b \mathbf{r}_b}{m_a + m_b}, \qquad \boldsymbol{\rho} = \mathbf{r}_a - \mathbf{r}_b,$$

the Hamiltonian for a molecule becomes

$$H = \frac{\mathbf{p}^2}{2M} + \frac{\mathbf{q}^2}{2\mu} + U(|\boldsymbol{\rho}|),$$

where $\mathbf{p}$ and $\mathbf{q}$ are conjugated momenta to $\mathbf{r}$ and $\boldsymbol{\rho}$ with total mass $M = m_a + m_b$ and reduced mass $\mu^{-1} = m_a^{-1} + m_b^{-1}$. Clearly the first term of the Hamiltonian is nothing but the kinetic energy of the center-of-mass motion and yields a contribution to the specific heat as that of a perfect monoatomic gas, $(3/2)Nk_B$. As for the other terms in the Hamiltonian, we may exploit the spherical symmetry of the potential. In spherical coordinates the Hamiltonian reads

$$H = \frac{p_\rho^2}{2I} + \frac{1}{2I}\left(p_\theta^2 + \frac{p_\phi^2}{\sin^2(\theta)}\right) + U(\rho) \equiv \frac{p_\rho^2}{2I} + \frac{\mathbf{L}^2}{2I} + U(\rho), \tag{N.2}$$

where $I = \mu\rho^2$ is the moment of inertia of the relative motion and $\mathbf{L}$ is the angular momentum. To obtain the partition function of the molecule, one must integrate over the spherical cordinates (ρ, θ, ϕ) and their conjugated momenta $(p_\rho, p_\theta, p_\phi)$. The integration of the Boltzmann factor of the angular variables yields

$$\int_0^\pi d\theta \int_0^{2\pi} d\phi \int_{-\infty}^{\infty} dp_\theta \int_{-\infty}^{\infty} dp_\phi e^{-\beta \mathbf{L}^2/2I} = 8\pi^2 k_B T \mu\rho^2.$$

The last result is nothing but the spherical surface at a distance ρ times the kinetic energy contribution of the motion along the surface. Hence from the rotational degrees of freedom of the molecule there is a contribution Nk_B to the specific heat. We are then left with the vibrational motion associated with the coordinate ρ, which is the interatomic distance. Its Hamiltonian is given by the first and third terms in Eq. (N.2). Under the general physical assumption that the interatomic potential can be expanded around the equilibrium configation of the molecule and by confining to the harmonic approximation, the vibrational motion reduces to a simple harmonic oscillator. Notice that, in principle, the coordinate of the harmonic oscillator is confined within a distance whose size goes with the inverse third power of the volume where the molecules are contained. However, since at large distances the Boltzmann factor decays exponentially, one can safely relax the limits of integration to infinity. As a result, by a direct application of the equipartition theorem we obtain the contribution to the specific heat as Nk_B. Overall, then, one obtains $(7/2)Nk_B$ for a gas of biatomic molecules. However, at room temperature, the measured specific heat of a biatomic gas is only $(5/2)Nk_B$. This discrepancy can only be explained by considering quantum mechanics. What turns out is that, at room temperature, the vibrational motion appears to be *frozen*, so that it cannot deliver its contribution to the total energy. This freezing of a degree of freedom is a general feature of quantum statistical mechanics, to be developed later on in this book. Here we content ourselves to noticing that the freezing occurs whenever the available thermal energy $\sim k_B T$ is small compared with the energy separation between the ground states and the first excited state of the spectrum associated with the degree of freedom at stake.

4.3 By taking the derivative of Eq. (4.27) with respect to μ, one gets

$$\beta \sum_{N=0}^{\infty} e^{\beta\mu N} Z(T, V, N) e^{\beta\Omega(T,V,\mu)} \left(N + \frac{\partial \Omega}{\partial \mu} \right) = 0,$$

which gives the first of the relations (4.28). Next we differentiate Eq. (4.27) with respect to T. One gets

$$\beta \sum_{N=0}^{\infty} e^{\beta\mu N} e^{\beta\Omega(T,V,\mu)} \left[-\frac{\mu N + \Omega(T, V, \mu)}{k_B T^2} Z(T, V, N) + \frac{\partial Z(T, V, N)}{\partial T} + \beta Z(T, V, N) \frac{\partial \Omega}{\partial T} \right] = 0.$$

By observing that

$$\frac{\partial Z(T, V, N)}{\partial T} = \frac{1}{k_B T^2} \int d\Gamma \, e^{-\beta H} H$$

and recalling that $\Omega = U - TS - \mu N$, one obtains the second of the relations (4.28).
Finally, by differentiating with respect to V one obtains

$$\left(\frac{\partial Z}{\partial V} \right)_{T,N} + Z\beta \left(\frac{\partial \Omega}{\partial V} \right)_{T,\mu} = 0$$

and the last of the relations (4.28) follows.

4.4 We have

$$\begin{aligned} \left\langle x_i \frac{\partial H}{\partial x_i} \right\rangle &= \frac{1}{Z} \int d\Gamma x_i \frac{\partial H}{\partial x_i} e^{-\beta H} \\ &= \frac{1}{\beta} \frac{1}{Z} \int d\Gamma e^{-\beta H} \\ &= k_B T, \end{aligned}$$

where we made an integration by parts and neglected the contribution from infinity.

4.5 By using the equipartition theorem the energy reads $U = 3Nk_B T$ and the specific heat is $C = 3Nk_B$.

Chapter 6

6.1 For two distinguishable particles, the wave function is the product of orthonormal single-particle wave functions $g_\kappa(\mathbf{x})$, $g^{(2)}_{\kappa\kappa'}(\mathbf{x}_1, \mathbf{x}_2) = g_\kappa(\mathbf{x}_1) g_{\kappa'}(\mathbf{x}_2)$. For Bose (Fermi) particles, the normalized wave function $g^{(2)}_{\kappa\kappa'}(\mathbf{x}_1, \mathbf{x}_2)$ is given by the symmetric (antisymmetric) combination (6.1). The mean square distance among the two particles is given by

$$\left\langle (\mathbf{x}_1 - \mathbf{x}_2)^2 \right\rangle = \left\langle \mathbf{x}_1^2 \right\rangle + \left\langle \mathbf{x}_2^2 \right\rangle - 2\langle \mathbf{x}_1 \rangle \langle \mathbf{x}_2 \rangle,$$

where the average is taken with respect to the wave function $g^{(2)}_{\kappa\kappa'}(\mathbf{x}_1, \mathbf{x}_2)$ with the appropriate symmetry.

For distinguishable particles

$$\left\langle (\mathbf{x}_1 - \mathbf{x}_2)^2 \right\rangle_{\mathrm{D}} = \left\langle \mathbf{x}_1^2 \right\rangle_{\kappa\kappa} + \left\langle \mathbf{x}_2^2 \right\rangle_{\kappa'\kappa'} - 2\langle \mathbf{x}_1 \rangle_{\kappa\kappa} \langle \mathbf{x}_2 \rangle_{\kappa'\kappa'},$$

where the subscripts indicate the average is over the corresponding single-particle state $g_\kappa(\mathbf{x})$.

For Bose (Fermi) particles, with $\kappa \neq \kappa'$,

$$\begin{aligned} \left\langle (\mathbf{x}_1 - \mathbf{x}_2)^2 \right\rangle_{\mathrm{B,F}} &= \left\langle (\mathbf{x}_1 - \mathbf{x}_2)^2 \right\rangle_{\mathrm{D}} \mp 2\langle \mathbf{x}_1 \rangle_{\kappa\kappa'} \langle \mathbf{x}_2 \rangle_{\kappa'\kappa} \\ &= \left\langle (\mathbf{x}_1 - \mathbf{x}_2)^2 \right\rangle_{\mathrm{D}} \mp 2|\langle \mathbf{x} \rangle_{\kappa\kappa'}|^2, \end{aligned} \tag{N.3}$$

where we have used the orthogonality of the single-particle wave functions. The contraction and repulsion, on average, of the Bose and Fermi particles depend on the overlap between the wave functions $g_\kappa(\mathbf{x})$ and $g_{\kappa'}(\mathbf{x})$.

Chapter 7

7.1 Let us first evaluate the density of states:

$$\nu(\epsilon) = 2\frac{V_d \Omega_d}{(2\pi\hbar)^d} \int \mathrm{d}\epsilon\ \delta(\epsilon - cp) = 2\frac{V_d \Omega_d}{(2\pi\hbar)^d c^d} \epsilon^{d-1},$$

where V_d is volume in d space dimensions and Ω_d the corresponding solid angle introduced in Eq. (3.58). By using the above density of states, at zero kelvin, we get

$$N = 2\frac{V_d \Omega_d}{(2\pi\hbar)^d c^d} \frac{\epsilon_{\mathrm{F}}^d}{d}, \qquad U = 2\frac{V_d \Omega_d}{(2\pi\hbar)^d c^d} \frac{\epsilon_{\mathrm{F}}^{d+1}}{d+1}.$$

By taking the ratio U/N, one obtains Eq. (7.97).

7.2 Instead of the energy spectrum (7.9), one has

$$\epsilon(p_x, p_y, n_z) = \frac{1}{2m}\left(p_x^2 + p_y^2\right) + \epsilon_z(n_z), \tag{N.4}$$

where n_z is the quantum number associated with the motion along the z-axis along which the gravitational field is directed. As the Hamiltonian is separable, the motion along the z axis satisfies the equation

$$-\frac{\hbar^2}{2m}\frac{\mathrm{d}^2\varphi(z)}{\mathrm{d}z^2} + mgz\varphi(z) = \epsilon_z(n_z)\varphi(z). \tag{N.5}$$

It is useful to introduce reduced variables for z and ϵ in terms of units of length and energy defined by

$$\ell = \left(\frac{\hbar^2}{2m} \frac{1}{mg} \right)^{1/3}, \quad \epsilon_\ell = mg\ell, \tag{N.6}$$

so that Eq. (N.5) gets the form $\varphi'' - (z - \epsilon_z(n_z))\varphi = 0$, whose solution can be written in terms of the Airy function $\mathrm{Ai}(x)$ (see, for instance, Abramowitz and Stegun (1965))

$$\varphi(z) = \mathrm{Ai}(z - \epsilon_z(n_z)), \quad \mathrm{Ai}(x) = \frac{1}{\pi} \int_0^\infty \mathrm{d}t \cos \left(xt + \frac{t^3}{3} \right). \tag{N.7}$$

The value of the eigenvalues ϵ_z is determined by imposing the vanishing of the wave function:

$$\mathrm{Ai}(-\epsilon_z(n_z)) = 0. \tag{N.8}$$

In the thermodynamic limit, the most important contribution comes from the eigenvalues with quantum number $n_z \gg 1$, hence to determine the form of the eigenvalue from Eq. (N.8) we can use the asymptotic expression of the Airy function:

$$\mathrm{Ai}(x) \approx \frac{\sin\left(\frac{2}{3}|x|^{3/2} + \frac{\pi}{4}\right)}{\sqrt{\pi}|x|^{1/4}}, \quad x \to -\infty. \tag{N.9}$$

By using the form (N.9) into (N.8) we obtain

$$\epsilon_z(n_z) = \epsilon_\ell \left(\frac{3}{2} \pi n_z \right)^{2/3}, \quad n_z \gg 1, \tag{N.10}$$

where we have restored the physical units (notice the factor ϵ_ℓ). By transforming the sum over the quantum numbers p_x, p_y and n_z into integrals we obtain

$$\begin{aligned} \nu(E) &= \frac{2A}{(2\pi\hbar)^2} \int \mathrm{d}p_x \mathrm{d}p_y \mathrm{d}n_z \, \delta \left(E - \epsilon_\ell \left(\frac{3}{2}\pi n_z \right)^{2/3} - \frac{1}{2m}(p_x^2 + p_y^2) \right) \\ &= \frac{2}{3} \frac{(2m)^{3/2} A}{2\pi^2 \hbar^3 mg} E^{3/2}. \end{aligned} \tag{N.11}$$

The expression (N.11) compared to the case without the gravity field has a simple interpretation. The quantity $(2/3)E/mg$ plays the role of an effective cylinder height. Particles with energy E cannot, classically, exceed this height. Indeed, the single-particle phase space available, classically, up to energy E reads

$$\mu(E) = 2 \int \mathrm{d}^3 r \int \mathrm{d}^3 p \, \theta(E - H(\mathbf{p}, \mathbf{r})), \tag{N.12}$$

where the integration over the momentum space is extended to $\mathbf{R}^3$ as usual and that over the configuration space to the available volume. The number $N(E)$ of quantum states in the region $\mu(E)$ is obtained by dividing by h^3, so that the density of states is estimated as

$$\nu(E) \approx \frac{1}{h^3} \frac{\mathrm{d}\mu(E)}{\mathrm{d}E} = \int_0^\infty \mathrm{d}z \, \frac{2Am^{3/2}}{\sqrt{2}\pi^2\hbar^3} \sqrt{E - mgz} \, \theta(E - mgz), \tag{N.13}$$

which agrees with (N.11). By using the general formula (7.25) with the density of states (N.11) one obtains the Fermi energy as

$$E_{\rm F} = \left(\frac{15}{4} N \frac{\sqrt{2}\pi^2\hbar^3 mg}{2m^{3/2} A} \right)^{2/5} . \tag{N.14}$$

Finally, to obtain the local density, one notices that the integrand in Eq. (N.13) represents the density of states per unit length along the z-axis. Consistently, the local density is obtained by integrating it up to the Fermi energy. We get

$$\rho(z) = \frac{2}{3} \frac{2m^{3/2}}{\sqrt{2}\pi^2\hbar^3} (E_{\rm F} - mgz)^{3/2} . \tag{N.15}$$

It is left as a further exercise to the reader to show that the local pressure $p(z)$ goes like $\rho(z)^{5/3}$.

7.3 We first calculate the total number of states $N(\epsilon)$ from 0 up to the energy $\epsilon - \epsilon_0$ in the continuum limit where the $\sum_{n_x,n_y,n_z}$ is replaced by the integral in $dn_x dn_y dn_z$.
Introducing the new variables $\epsilon_x = n_x\hbar\omega_x$, $\epsilon_y = n_y\hbar\omega_y$ and $\epsilon_z = n_z\hbar\omega_z$,

$$\begin{aligned} N(\epsilon) &= \frac{1}{(\hbar\bar{\omega})^3} \int_0^{\epsilon-\epsilon_0} d\epsilon_z \int_0^{\epsilon-\epsilon_0-\epsilon_z} d\epsilon_y \int_0^{\epsilon-\epsilon_0-\epsilon_z-\epsilon_y} d\epsilon_x \\ &= \frac{1}{(\hbar\omega)^3}(\epsilon-\epsilon_0)^3,\ \bar{\omega} = (\omega_x\omega_y\omega_z)^{1/3}. \end{aligned} \tag{N.16}$$

By taking the derivative of the last equation with respect to $\epsilon - \epsilon_0$, we obtain Eq. (7.55). In two dimensions $\tilde{\nu}(\epsilon) \approx \epsilon$ and it is constant in one dimension.

7.4 By using the density of states (7.55) we have

$$E_{\rm F} = \epsilon_0 + 3^{1/3}\hbar\omega. \tag{N.17}$$

7.5 To determine this cusp one has to look at the derivative of the second term of Eq. (7.52), which can be obtained by looking at the expansion at small values of the chemical potential of $\mathrm{Li}_{-\alpha}(z)$ in Eq. (E.16) of Appendix E. We get

$$\begin{aligned} \frac{\partial}{\partial T}\Delta c_V &\equiv -\frac{\partial}{\partial T}\langle N\rangle k_{\rm B} \frac{9}{4}\frac{\mathrm{Li}_{3/2}(z)}{\mathrm{Li}_{1/2}(z)} \\ &\approx -\frac{\partial}{\partial T}\langle N\rangle k_{\rm B} \frac{9}{2}\frac{\zeta_{\rm R}(3/2)\Gamma(3/2)}{\pi}(-\ln z)^{1/2} \\ &= -\frac{27}{4}\left(\frac{\zeta_{\rm R}(3/2)\Gamma(3/2)}{\pi}\right)^2 \frac{\langle N\rangle k_{\rm B}}{T_{\rm c}} = -3.6658\frac{\langle N\rangle k_{\rm B}}{T_{\rm c}}. \end{aligned} \tag{N.18}$$

Chapter 8

8.1 The hard-core potential (8.3) eliminates from the partition function the configurations where at least two particles have equal values of their space coordinates. In one space dimension this becomes a very stringent condition. Suppose, for instance, that the system of N particles finds itself in a configuration where two particles have coordinates such that $x_1 < x_2$. Clearly, it is impossible that during the dynamical evolution the system may reach a configuration where $x_1 > x_2$, because in order to do so it would have to pass through the configuration $x_1 = x_2$. As a result the set of all the possible configurations naturally splits into disjointed sets. One set may specified as

$$0 < x_1 < x_2 < \cdots < x_N < L,$$

where L is the length of the segment where the particles are confined. Any other possible set can be obtained by a permutation of the N particles coordinates. The integral over all possible configurations reduces then to the sum of the integrals over each set of configurations specified by the above equation or by one obtained by a permutation. For reasons of symmetry all integrals are the same and their number is $N!$, i.e. the number of the permutations of N identical objects. As a result the partition function reads

$$\begin{aligned}
Z &= \frac{N!}{N!\lambda^N} \int_{(N-1)r_0}^{L} \mathrm{d}x_N \cdots \int_{r_0}^{x_3 - r_0} \mathrm{d}x_2 \int_0^{x_2 - r_0} \mathrm{d}x_1 \\
&= \frac{1}{\lambda^N} \int_0^{L-(N-1)r_0} \mathrm{d}y_N \cdots \int_0^{y_3} \mathrm{d}y_2 \int_0^{y_2} \mathrm{d}y_1 \\
&= \frac{1}{\lambda^N} \int_0^{L-(N-1)r_0} \mathrm{d}y_N \cdots \int_0^{y_3} \mathrm{d}y_2 y_2 \\
&= \frac{1}{\lambda^N} \int_0^{L-(N-1)r_0} \mathrm{d}y_N \cdots \int_0^{y_4} \mathrm{d}y_3 \frac{1}{2} y_3^2 \\
&= \frac{1}{\lambda^N} \int_0^{L-(N-1)r_0} \mathrm{d}y_N \cdots \int_0^{y_5} \mathrm{d}y_4 \frac{1}{2 \times 3} y_4^3 \\
&= \frac{1}{\lambda^N} \frac{(L - (N-1)r_0)^N}{N!}.
\end{aligned}$$

Hence it has the form of the partition function of a perfect gas where the "volume" L has been replaced by $L - (N-1)r_0$, meaning that for each particle or "rod" the allowed volume has been decreased by the volume occupied by all the other particles.

8.2 By introducing $f(r) = \exp(-\beta\phi(r)) - 1$, we may write

$$\mathrm{e}^{-\beta \sum_{i<j} \phi(|\mathbf{x}_i - \mathbf{x}_j|)} = \prod_{i<j} (1 + f(|\mathbf{x}_i - \mathbf{x}_j|)).$$

The pressure reads

$$\frac{PV}{Nk_{\mathrm{B}}T} = 1 + V \frac{\partial}{\partial V} \frac{1}{N} \ln \frac{1}{V^N} \int \mathrm{d}^{3N}x \prod_{i<j} (1 + f_{ij}),$$

where $f_{ij} = f(|\mathbf{x}_i - \mathbf{x}_j|)$. By keeping only terms linear in f_{ij} we split the integral over the coordinates into two parts. The first part, which comes from the terms to zero order in f_{ij}, yields, by integration, a factor V^N. The second part, corresponding to terms to first order in f_{ij}, is the sum of $N(N-1)/2$ contributions, which are the number of possible pairs of particles. In each of such terms, $N-2$ coordinates can be integrated, yielding a factor V^{N-2} so that

$$\frac{PV}{Nk_{\mathrm{B}}T} = 1 + V\frac{\partial}{\partial V}\frac{1}{N}\ln\left[1 + \frac{N(N-1)}{2V^2}\int \mathrm{d}^3x_i\mathrm{d}^3x_j f_{ij}\right].$$

By exploiting the fact that f_{ij} depends only on the difference of the coordinates, we arrive at

$$\frac{PV}{Nk_{\mathrm{B}}T} = 1 - \frac{N}{2V}\int \mathrm{d}^3 r f(r).$$

In the case of the hard-core potential (8.3), we easily get

$$\frac{PV}{Nk_{\mathrm{B}}T} = 1 + \frac{2\pi}{3}\frac{Nr_0^3}{V}.$$

8.3 By introducing the variable magnetization per spin $\tilde{m} = \sum_{i=1}^{N}\sigma_i/N$, the Hamiltonian (8.60) becomes

$$H = -\frac{JN}{2}\tilde{m}^2 - hN\tilde{m}, \tag{N.19}$$

where $\tilde{m}$ takes $2N+1$ possible values between -1 and 1. The strategy is then to write the partition function as a sum over all possible values of $\tilde{m}$. In so doing we must take care of the fact that each given value of $\tilde{m}$ can be obtained in a finite number of ways $C(\tilde{m})$. Notice that the number N_+ of spin up and the number N_- of spin down, at a given value of $\tilde{m}$ is $N_+ = N(1+\tilde{m})/2$ and $N_- = N(1-\tilde{m})/2$, $C(\tilde{m})$ then reads

$$C(\tilde{m}) = \frac{N!}{(N(1+\tilde{m})/2)!\,(N(1-\tilde{m})/2)!}.$$

As a result the partition function reads

$$Z = \sum_{\tilde{m}=-1}^{1} \mathrm{e}^{-N\left(-\beta J\tilde{m}^2/2 - \beta h\tilde{m} - \frac{1}{N}\ln C(\tilde{m})\right)} \equiv \sum_{\tilde{m}=-1}^{1} \mathrm{e}^{-N\beta f(\tilde{m})}. \tag{N.20}$$

By observing that the sum in (N.20) is always greater than its maximum addendum and smaller than a sum with all addenda equal to the maximum one, in the large N limit, the free energy per site reads, after using the Stirling formula for the factorials,

$$f(m) = -\frac{J}{2}m^2 - hm + \frac{1}{2\beta}\ln(1-m^2) + \frac{m}{2\beta}\ln\frac{1+m}{1-m} + \frac{1}{\beta}\ln 2, \tag{N.21}$$

where m is determined by requiring that it is the value at which $f(m)$ has a maximum. By taking the first derivative with respect to m, one obtains

$$f'(m) = -Jm - h + \frac{1}{2\beta}\ln\frac{1+m}{1-m} = 0, \tag{N.22}$$

which, after recalling that $(1/2)\ln(1+x)/(1-x)$ is the inverse hyperbolic tangent, reproduces Eq. (8.59).

Chapter 9

9.1 We evaluate

$$\begin{aligned}\langle\kappa_1,\ldots,\kappa_N|\kappa_1,\ldots,\kappa_N\rangle &= c^2\langle\kappa_N|\cdots\langle\kappa_1|\Lambda^2|\kappa_1\rangle\cdots|\kappa_N\rangle\\ &= c^2\langle\kappa_N|\cdots\langle\kappa_1|\Lambda|\kappa_1\rangle\cdots|\kappa_N\rangle\\ &= \frac{c^2}{N!}\sum_p\lambda_p\langle\kappa_N|\cdots\langle\kappa_1|U_p|\kappa_1\rangle\cdots|\kappa_N\rangle\\ &= \frac{c^2}{N!}\sum_p\lambda_p\langle\kappa_N|\cdots\langle\kappa_1|\,|\kappa_{p_1}\rangle\cdots|\kappa_{p_N}\rangle,\end{aligned}$$

where $p_1,\ldots,p_N$ indicates a permutation of the numbers $1,\ldots,N$. The state $|\kappa_1\rangle\ldots|\kappa_N\rangle$ means that the first particle is in the single-particle state κ_1, the second in κ_2 and so on. After the permutation generated by U_p, the first particle is in the state κ_{p_1} and so on and so forth. For fermions, due to the exclusion principle, each state can be occupied by at most one particle. This means that in the sum only the identical permutation survives and then $c^2 = N!$.

In the boson case, a single-particle state can be occupied by more than one particle. Suppose, for instance, that there are n_{κ_1} particles in the state κ_1. Hence all permutations of these n_{κ_1} particles among themselves do not change the state. The number of these permutations is evidently $n_{\kappa_1}!$. Hence in the sum over the permutations all the terms which can be obtained by the original disposition of the particles by those permutations of particles will survive, where only particles in the same state are interchanged among themselves. The total number of these permutations is $n_{\kappa_1}!\ldots n_{\kappa_N}!$. This leads to the expression $c^2 = N!/\prod_i n_{\kappa_i}!$.

9.2 We consider the action of $a^\dagger_{\kappa_i}a_{\kappa_j}$ on a basis vector $|\ldots,n_{\kappa_i},\ldots,n_{\kappa_j},\ldots\rangle$. We get for $i\neq j$

$$\begin{aligned}&a^\dagger_{\kappa_i}a_{\kappa_j}|\ldots,n_{\kappa_i},\ldots,n_{\kappa_j},\ldots\rangle\\ &\quad= a^\dagger_{\kappa_i}\sqrt{n_{\kappa_j}}|\ldots,n_{\kappa_i},\ldots,n_{\kappa_j}-1,\ldots\rangle\\ &\quad= \sqrt{n_{\kappa_i}+1}\sqrt{n_{\kappa_j}}|\ldots,n_{\kappa_i}+1,\ldots,n_{\kappa_j}-1,\ldots\rangle.\end{aligned}$$

Clearly, by reversing the order of i and j one obtains the same result. On the other hand, for $i=j$, instead

$$\begin{aligned}&a^\dagger_{\kappa_i}a_{\kappa_i}|\ldots,n_{\kappa_i},\ldots\rangle\\ &\quad= a^\dagger_{\kappa_i}\sqrt{n_{\kappa_i}}|\ldots,\ldots,n_{\kappa_i}-1,\ldots\rangle\\ &\quad= n_{\kappa_i}|\ldots,n_{\kappa_i},\ldots,\ldots\rangle,\end{aligned}$$

and for the reverse order

$$
\begin{aligned}
& a_{\kappa_i} a^{\dagger}_{\kappa_i} | \ldots, n_{\kappa_i}, \ldots \rangle \\
& \quad = a_{\kappa_i} \sqrt{n_{\kappa_i} + 1} | \ldots, \ldots, n_{\kappa_i} + 1, \ldots \rangle \\
& \quad = \left(n_{\kappa_i} + 1\right) | \ldots, n_{\kappa_i}, \ldots, \ldots \rangle .
\end{aligned}
$$

By subtracting the last two results one obtains the last of (9.10). In a similar way one can prove the other two relations of (9.10).

9.3 From (9.9), we have

$$
|n\rangle = \frac{\left(a^{\dagger}_{\kappa_i}\right)}{\sqrt{n}} |n-1\rangle .
$$

By iteration one finds

$$
|n\rangle = \frac{\left(a^{\dagger}_{\kappa_i}\right)}{\sqrt{n}} |n-1\rangle = \frac{\left(a^{\dagger}_{\kappa_i}\right)^2}{\sqrt{n(n-1)}} |n-2\rangle = \cdots .
$$

9.4 We may proceed as for bosons in Problem 9.3. We confine to the case of $i = j$. We have

$$
\begin{aligned}
& a^{\dagger}_{\kappa_i} a_{\kappa_i} | \ldots, n_{\kappa_i}, \ldots \rangle \\
& \quad = (-1)^{P_i} a^{\dagger}_{\kappa_i} n_{\kappa_i} | \ldots, \ldots, n_{\kappa_i} - 1, \ldots \rangle \\
& \quad = n_{\kappa_i} (2 - n_{\kappa_i}) | \ldots, n_{\kappa_i}, \ldots, \ldots \rangle ,
\end{aligned}
$$

and for the reverse order

$$
\begin{aligned}
& a_{\kappa_i} a^{\dagger}_{\kappa_i} | \ldots, n_{\kappa_i}, \ldots \rangle \\
& \quad = (-1)^{P_i} a_{\kappa_i} (1 - n_{\kappa_i}) | \ldots, \ldots, n_{\kappa_i} + 1, \ldots \rangle \\
& \quad = \left(1 - n^2_{\kappa_i}\right) | \ldots, n_{\kappa_i}, \ldots, \ldots \rangle .
\end{aligned}
$$

By summing the two terms and taking into account that either $n_{\kappa_i} = 0$ or 1, we obtain the last relation (9.17).

9.5 For instance the first relation is obtained by

$$
\begin{aligned}
[\hat{\psi}(\mathbf{x}), \hat{\psi}^{\dagger}(\mathbf{x}')] &= \sum_{\kappa_1, \kappa_2} \langle \mathbf{x} | \kappa_1 \rangle \langle \kappa_2 | \mathbf{x}' \rangle \left[a_{\kappa_1}, a^{\dagger}_{\kappa_2} \right] \\
&= \sum_{\kappa_1, \kappa_2} \langle \mathbf{x} | \kappa_1 \rangle \langle \kappa_2 | \mathbf{x}' \rangle \delta_{\kappa_1, \kappa_2} \\
&= \sum_{\kappa_1} \langle \mathbf{x} | \kappa_1 \rangle \langle \kappa_1 | \mathbf{x}' \rangle \\
&= \langle \mathbf{x} | \mathbf{x}' \rangle \\
&= \delta(\mathbf{x} - \mathbf{x}') .
\end{aligned}
$$

For the others, one proceeds in a similar way.

Chapter 10

10.1 Let us consider the Lagrangian

$$\mathcal{L} = \frac{1}{2}m\mathbf{v}^2 + \frac{q}{c}\mathbf{v}\cdot\mathbf{A}. \tag{N.23}$$

By the Eulero–Lagrange equation we obtain

$$m\frac{\mathrm{d}v_i}{\mathrm{d}t} + \frac{q}{c}v_j\frac{\partial A_i}{\partial x_j} = \frac{q}{c}v_j\frac{\partial A_j}{\partial x_i}.$$

After collecting together the two terms with the vector potential, one obtains the expression of the Lorentz force.

The Legendre transformation with respect to $\mathbf{v}$ is defined by

$$\mathbf{p} = \frac{\delta\mathcal{L}}{\delta\mathbf{v}},\ \ H = \mathbf{p}\cdot\mathbf{v} - \mathcal{L}$$

which leads to

$$H = \frac{1}{2m}\left(\mathbf{p} - \frac{q}{c}\mathbf{A}\right)^2. \tag{N.24}$$

10.2 Recall that $[\hat{p}_i, f] = -\mathrm{i}\partial_i f$, hence

$$[H, f] = \sum_i \frac{1}{2m}\left[\hat{p}_i^2, f\right] = \sum_i \frac{1}{2m}\left(-\partial_i^2 f - 2\mathrm{i}(\partial_i f)\hat{p}_i\right),$$

so that

$$\left[f^*, [H, f]\right] = \sum_i \frac{1}{2m}2(\partial_i f)(\partial_i f^*).$$

By using $f = \sum_i e^{\mathrm{i}\mathbf{k}\cdot\mathbf{x}_i}$ one obtains (10.58).

10.3 Let $\mathcal{P}$ be the momentum conjugated to $\mathcal{O}$. Hamilton's equations then give

$$\dot{\mathcal{O}} = \frac{\partial H_0}{\partial\mathcal{P}},$$
$$\dot{\mathcal{P}} = -\frac{\partial H_0}{\partial\mathcal{O}} + \lambda.$$

By integration by parts

$$-\int_0^\tau \mathrm{d}t\dot{\lambda}\mathcal{O} = -\left(\lambda_\tau\mathcal{O}_\tau - \lambda_0\mathcal{O}_0\right) + \int_0^\tau \mathrm{d}t\lambda\dot{\mathcal{O}}.$$

The last term can be rewritten by using the Hamilton equations

$$\begin{aligned}\int_0^\tau \mathrm{d}t\lambda\dot{\mathcal{O}} &= \int_0^\tau \mathrm{d}t\dot{\mathcal{O}}\left(\frac{\partial H_0}{\partial\mathcal{O}} + \dot{\mathcal{P}}\right)\\ &= \int_0^\tau \mathrm{d}t\left(\dot{\mathcal{O}}\frac{\partial H_0}{\partial\mathcal{O}} + \dot{\mathcal{P}}\frac{\partial H_0}{\partial\mathcal{P}}\right)\\ &= H_0(\tau) - H_0(0)\end{aligned}$$

which leads to the desired result.

Chapter 11

11.1 By using the translational invariance for the density–density correlation function, $f_{\mathbf{q}}^{nn}$, we have

$$f_{\mathbf{q}} = \langle \delta n_{\mathbf{q}} \delta n_{-\mathbf{q}} \rangle$$
$$= \int \mathrm{d}(\mathbf{x} - \mathbf{x}') \mathrm{e}^{-\mathrm{i}\mathbf{q}(\mathbf{x}-\mathbf{x}')} \langle \delta n(\mathbf{x}) \delta n(\mathbf{x}') \rangle$$
$$= \frac{1}{V} \int \mathrm{d}\mathbf{x}\, \mathrm{d}\mathbf{x}' \mathrm{e}^{-\mathrm{i}\mathbf{q}(\mathbf{x}-\mathbf{x}')} \langle \delta n(\mathbf{x}) \delta n(\mathbf{x}') \rangle. \tag{N.25}$$

Notice that in going from the second to the third line, we used the fact that $\langle \delta n(\mathbf{x}) \delta n(\mathbf{x}') \rangle$ depends only on the difference of the space variables. In the static limit then

$$\lim_{\mathbf{q}\to 0} f_{\mathbf{q}}^{nn} = \frac{1}{V} \langle (\Delta N)^2 \rangle, \tag{N.26}$$

since, at fixed volume,

$$\Delta N = \int \mathrm{d}\mathbf{x} \delta n(\mathbf{x}).$$

By recalling Eq. (4.39) expressing the fluctuations of particle number in terms of the isothermal compressibility κ_T, $\langle (\Delta N)^2 \rangle = N^2 k_{\mathrm{B}} T \kappa_T / V$, we obtain

$$\lim_{\mathbf{q}\to 0} f_{\mathbf{q}}^{nn} = n^2 k_{\mathrm{B}} T \kappa_T = k_{\mathrm{B}} T \frac{\partial n}{\partial \mu}, \tag{N.27}$$

where in the last step we used the expression (4.35) relating the the isothermal compressibility to the thermodynamic density of states. By using this result in (11.38) one verifies the classical fluctuation–dissipation theorem Eq. (10.55).

11.2 The scattering probability of Eq. (11.43) can be obtained directly by solving the Schrödinger equation in the presence of the impurity potential (11.42). Since this derivation is instructive, we show it here. The Schrödinger equation reads

$$-\frac{\hbar^2}{2m} \nabla^2 \psi(\mathbf{r}) + v_0 \delta(\mathbf{r}) \psi(\mathbf{0}) = E \psi(\mathbf{r}).$$

We express the energy in terms of the initial momentum $E = p^2/2m$ and rewrite the above equation in the form of an inhomogeneous Helmholtz equation:

$$(\nabla^2 + k^2) \psi(\mathbf{r}) = \frac{2m v_0}{\hbar^2} \psi(\mathbf{0}) \delta(\mathbf{r}), \tag{N.28}$$

where $k = p/\hbar$. Being non-homogeneous, the solution of Eq. (N.28) may be written as the sum of a solution of the associated homogeneous equation and a particular solution. According to the conditions fixed by the scattering problem, we take as the homogeneous solution the incident plane wave with momentum $\mathbf{k}$. The particular solution can be obtained by Fourier transforming Eq. (N.28):

$$\psi_{\mathbf{q}} = \frac{2mv_0}{\hbar^2}\psi(\mathbf{0})\frac{1}{k^2 - q^2}, \quad \psi_{\mathbf{q}} = \int d\mathbf{r} e^{-i\mathbf{q}\cdot\mathbf{r}}\psi(\mathbf{r}).$$

In the anti-Fourier transform we must regularize the integral in order to have a solution, corresponding to an outgoing wave

$$\int \frac{d^3 q}{(2\pi)^3} e^{i\mathbf{q}\cdot\mathbf{r}} \frac{1}{k^2 - q^2 + i0^+} = -\frac{e^{ikr}}{4\pi r}.$$

Hence the general solution reads

$$\psi(\mathbf{r}) = \frac{1}{\sqrt{V}} e^{i\mathbf{k}\cdot\mathbf{r}} - \frac{mv_0}{2\pi\hbar^2}\psi(\mathbf{0})\frac{e^{ikr}}{r} \approx \frac{1}{\sqrt{V}} e^{i\mathbf{k}\cdot\mathbf{r}} - \frac{mv_0}{2\pi\hbar^2\sqrt{V}}\frac{e^{ikr}}{r}, \tag{N.29}$$

the last equality being valid at the level of the Born approximation, where $\psi(\mathbf{0}) = 1/\sqrt{V}$. In Eq. (N.29), the plane wave is the incident part of the wave function, whereas the spherical wave is the outgoing part. From (N.29) we derive the scattering amplitude

$$f = -\frac{m}{2\pi\hbar^2\sqrt{V}} v_0. \tag{N.30}$$

In scattering theory the total cross section is given by integration over the solid angle of the modulus squared of the scattering amplitude:

$$\sigma_{\mathrm{T}} = 4\pi|f|^2 = \frac{m^2 v_0^2}{\pi\hbar^4 V} = 4\pi a_{\mathrm{s}}^2, \tag{N.31}$$

where we have introduced the scattering length a_{s} in the Born approximation. The scattering rate is given by multiplying the cross section by the velocity and the number of impurities N_{i}:

$$\frac{1}{\tau} = N_{\mathrm{i}}\sigma_T \frac{p}{m} = \frac{mpN_{\mathrm{i}}v_0^2}{\pi\hbar^4 V} = \frac{2\pi}{\hbar}\frac{\nu(\epsilon_{\mathbf{p}})}{s} n_{\mathrm{i}} v_0^2, \tag{N.32}$$

which reproduces Eq. (11.45) of the main text.

11.3 In the case of a classical gas interacting with fixed random impurities, Eq. (11.48) remains valid with the replacement of the Fermi distribution function with the Maxwell–Boltzmann one:

$$f_0(\mathbf{p}) = \frac{N}{V}\left(\frac{\beta}{2\pi m}\right)^{3/2} e^{-\beta p^2/2m}.$$

The expression for the electrical current leading to the result (11.49) also stays valid up to the momentum integration with two changes. There is no factor of two associated with the spin degeneracy and the factor $\hbar^3$ in the denominator is absorbed into the distribution function due to the different normalization in the classical and quantum cases, $f/h^3 \to f$. As a result we have

$$\mathbf{j} = -e^2\tau \int dp^3 \frac{\mathbf{p}}{m} \mathbf{E}\cdot\mathbf{v}_{\mathbf{p}} \frac{\partial f_0}{\partial \epsilon_{\mathbf{p}}}.$$

Performing the momentum integral leads to

$$\mathbf{j} = e^2 \left(\frac{N}{V}\right) \frac{\tau}{m} \mathbf{E},$$

which is again the Drude formula (11.33) with $n = N/V$. It is illuminating to recall the expression of the isothermal compressibility (1.50) for a perfect gas:

$$\kappa_T = \frac{1}{nk_\mathrm{B}T}$$

and the relation (4.35) connecting κ_T and the thermodynamic density of states $(\partial n/\partial\mu)_T$ yielding

$$n = k_\mathrm{B}T \left(\frac{\partial n}{\partial \mu}\right)_T .$$

This last relation gives then the Einstein relation (11.40) with $D = k_\mathrm{B}T\tau/m$, a result similar to Eq. (11.7).

Chapter 12

12.1 Consider the collision integral for fermions appearing in the right-hand-side of Eq. (6.3) with the scattering kernel defined by Eq. (2.20). The scattering rate for a quasiparticle of momentum $\mathbf{p}_1$ and energy $\epsilon_1 = \epsilon(\mathbf{p}_1)$ reads

$$\tau^{-1} = \frac{2\pi}{\hbar} \int \mathrm{d}^3 p_2 \int \frac{\mathrm{d}^3 p_1'}{(2\pi\hbar)^3} \int \frac{\mathrm{d}^3 p_2'}{(2\pi\hbar)^3} \delta^{(4)}_{1+2=1'+2'} f_2(1 - f_{1'})(1 - f_{2'})|U(\mathbf{p}_1 - \mathbf{p}_1')|^2, \tag{N.33}$$

where the conservation of energy and momentum is ensured by the delta function

$$\delta^{(4)}_{1+2=1'+2'} = \delta(\epsilon(\mathbf{p}_1) + \epsilon(\mathbf{p}_2) - \epsilon(\mathbf{p}_1') - \epsilon(\mathbf{p}_2'))\delta^{(3)}(\mathbf{p}_1 + \mathbf{p}_2 - \mathbf{p}_1' - \mathbf{p}_2'), \tag{N.34}$$

and

$$f_2 \equiv f(\epsilon(\mathbf{p}_2)) = (\exp(\beta(\epsilon(\mathbf{p}_2) - \mu)) + 1)^{-1}$$

and similarly for $f_{1'}$, $f_{2'}$. It is useful to define the energy and momentum transferred in the collision:

$$\omega = \epsilon(\mathbf{p}_1) - \epsilon(\mathbf{p}_1'), \quad \mathbf{q} = \mathbf{p}_1 - \mathbf{p}_1'. \tag{N.35}$$

The scattering rate (N.33), after using the conservation of momentum, becomes

$$\begin{aligned} \tau^{-1} = &\frac{2\pi}{\hbar} \int_{-\infty}^{\infty} \mathrm{d}\omega \int \frac{\mathrm{d}^3 p_2}{(2\pi\hbar)^3} \int \frac{\mathrm{d}^3 q}{(2\pi\hbar)^3} f(\epsilon(\mathbf{p}_2)) \\ &\times (1 - f(\epsilon(\mathbf{p}_1) - \omega))(1 - f(\epsilon(\mathbf{p}_2) + \omega))|U(\mathbf{q})|^2 \\ &\times \delta(\omega - \epsilon(\mathbf{p}_1) + \epsilon(\mathbf{p}_1 - \mathbf{q}))\delta(\omega - \epsilon(\mathbf{p}_2 + \mathbf{q}) + \epsilon(\mathbf{p}_2)). \end{aligned} \tag{N.36}$$

The integral over $\mathbf{p}_2$ can be decomposed into an integral over the energy $\epsilon_2 \equiv \epsilon(\mathbf{p}_2) - \mu$ and over the solid angle of $\hat{\mathbf{p}}_2$ by using the density of states per unit volume at the Fermi surface $\nu_F/2$. Hence, with $\epsilon_1 \equiv \epsilon(\mathbf{p}_1) - \mu$, we get

$$\tau^{-1} = \frac{\pi \nu_F}{\hbar} \int_{-\infty}^{\infty} d\omega \int_{-\mu}^{\infty} d\epsilon_2 f(\epsilon_2)(1 - f(\epsilon_1 - \omega))(1 - f(\epsilon_2 + \omega))$$
$$\times \int_0^{\infty} \frac{q^2 dq}{(2\pi\hbar)^3} \int \frac{d\Omega_{\hat{\mathbf{p}}_2}}{4\pi} \int d\Omega_{\hat{\mathbf{q}}} |U(\mathbf{q})|^2 \delta(\omega - \mathbf{p}_1 \cdot \mathbf{q}/m)\delta(\omega - \mathbf{p}_2 \cdot \mathbf{q}/m).$$

The integrals over the solid angles yield

$$\int \frac{d\Omega_{\hat{\mathbf{p}}_2}}{4\pi} \int d\Omega_{\hat{\mathbf{q}}} \delta(\omega - \mathbf{p}_1 \cdot \mathbf{q}/m)\delta(\omega - \mathbf{p}_2 \cdot \mathbf{q}/m) = \frac{\pi m^2}{p_1 p_2 q^2} \approx \frac{\pi}{v_F^2 q^2},$$

where the last step follows from the fact that at low temperatures the integrals over the energies are peaked around the Fermi surface. The integral over the transferred momentum q depends on the form of the interaction $U(\mathbf{q})$. It is useful to define a constant with the dimensions of a volume:

$$V_0 = \frac{\pi}{v_F^2} \int_0^{\infty} \frac{dq}{(2\pi\hbar)^3} |U(\mathbf{q})|^2. \tag{N.37}$$

The inverse scattering time, by recalling Eq. (7.25) to express the density of states in terms of the ratio between the density and the Fermi energy, becomes

$$\tau^{-1} = \frac{3\pi}{2\hbar\epsilon_F} n V_0 \int_{-\infty}^{\infty} d\omega \int_{-\mu}^{\infty} d\epsilon_2 f(\epsilon_2)(1 - f(\epsilon_1 - \omega))(1 - f(\epsilon_2 + \omega)). \tag{N.38}$$

Hence, in the absence of singular interaction as a function of the transferred momentum $\mathbf{q}$, the inverse scattering time is controlled by the integrals over the energies of the Fermi functions appearing in Eq. (N.38). Let us first consider the zero-kelvin case. The Fermi functions imply $0 < \omega < \epsilon_1$ and $-\omega < \epsilon_2 < 0$. This gives

$$\tau^{-1} = \frac{3\pi}{4\hbar} n V_0 \frac{\epsilon_1^2}{2E_F} \equiv \frac{a}{\hbar E_F} \epsilon_1^2. \tag{N.39}$$

At low temperature, we set $\epsilon_1 = 0$ and observe that, with $\mu \gg T$,

$$\int_{-\mu}^{\infty} d\epsilon_2 f(\epsilon_2)(1 - f(\epsilon_2 + \omega)) = \frac{\omega}{1 - e^{-\omega/T}}$$

and

$$\int_{-\infty}^{\infty} d\omega (1 - f(\epsilon_1 - \omega)) \frac{\omega}{1 - e^{-\omega/T}} = \frac{\pi}{4} T^2.$$

For the inverse scattering time or quasiparticle decay rate we get then

$$\tau^{-1} = \frac{3\pi}{2\hbar} n V_0 \frac{\pi T^2}{4E_F} \equiv \frac{b}{\hbar E_F} T^2. \tag{N.40}$$

In the case of electrons in metals, we can assume a screened interaction

$$U(\mathbf{q}) = \frac{4\pi\hbar^2 e^2}{q^2 + q_{TF}^2}, \tag{N.41}$$

where the Thomas–Fermi wave vector, $q_{\rm TF}$, is related to the screening length $l_s = \hbar/q_{\rm TF}$, so that we obtain

$$V_0 = \frac{\pi}{2} l_{\rm s}^3 \left(\frac{e^2}{\hbar v_{\rm F}}\right)^2 .$$

For the inverse scattering time we get the expression

$$\tau^{-1} = \frac{3\pi^2}{4\hbar\epsilon_{\rm F}} \left(\frac{1}{a_0 k_{\rm F}}\right)^2 \left(\frac{l_{\rm s}}{l}\right)^3 \int_{-\infty}^{\infty} {\rm d}\omega \int_{-\mu}^{\infty} {\rm d}\epsilon_2 f(\epsilon_2)(1 - f(\epsilon_1 - \omega))(1 - f(\epsilon_2 + \omega)), \tag{N.42}$$

where $a_0 = \hbar^2/(me^2)$ is the Bohr radius. For electrons in a metal, both $l_{\rm s} \sim l \sim n^{-1/3}$ and $a_0 k_{\rm F} \sim 1$, so that the dimensionless constants in front of the integrals in Eq. (N.42) yield a factor of order unity.

12.2 Set the whole system in motion with velocity $\delta\mathbf{p}/m$, then the quasiparticles of momentum $\mathbf{p}$ by Galilean invariance increase their energy by $\delta\epsilon = \mathbf{p} \cdot \delta\mathbf{p}/m$ and at the same time there is a variation of their distribution, so that

$$\mathbf{p} \cdot \frac{\delta\mathbf{p}}{m} = \nabla_{\mathbf{p}}\epsilon(\mathbf{p}) \cdot \delta\mathbf{p} + \sum_{\sigma'} \int \frac{{\rm d}^3 p'}{(2\pi\hbar)^3} f(\mathbf{p}, \mathbf{p}'; \sigma, \sigma')\delta n_{\sigma'}(\mathbf{p}'),$$

with only the first term on the right-hand side appearing for the Fermi gas. On the other hand $\delta n_{\sigma'}(\mathbf{p}') = -\nabla_{\mathbf{p}'} n_{\sigma'}(\mathbf{p}') \cdot \delta\mathbf{p} = -\hat{\mathbf{p}}'\delta(p' - p_F) \cdot \delta\mathbf{p}$ and Eq. (12.11) follows.

Chapter 13

13.1 In order to understand the origin of the second order transition in the presence of the magnetic field we make use of Fig. N.1 where the graphical solution of Eq. (13.31) is given. In general, for say $h > 0$, there is at least one solution. This is the case for the dotted curve in Fig. N.1. For the dashed curve there are three solutions, of which two are minima and one is maximum, as shown in Fig. N.2. Clearly the two minima are no longer equivalent. In the $b - a - h$ plane these two minima coexist on a surface which at $h = 0$ intersects the $b - a$ plane along the first order transition line. A specular surface exists in the half space with $h < 0$. The transition from one to three solutions occurs when the first derivative of the thermodynamic potential has an inflection point at a given value of the magnetic field. This requires both its first and second derivative to vanish, or, in terms of the thermodynamic potential, the vanishing of the second and third derivatives. This is the case of the full line in Fig. N.1. As a result the lines of second order critical points are obtained by the following

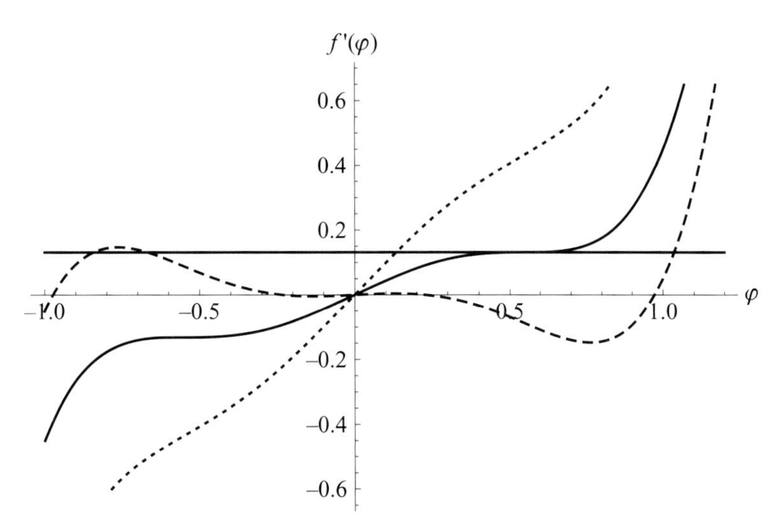

Fig. N.1 Plot of the first derivative of the thermodynamic potential for three different values of $a = 1, a = 0.25$ and $a = 1/20$ with $b = -d_0 = -1$. The horizontal line is fixed at the value of the external magnetic field. The solutions are then the intersections of the curves with the horizontal line.

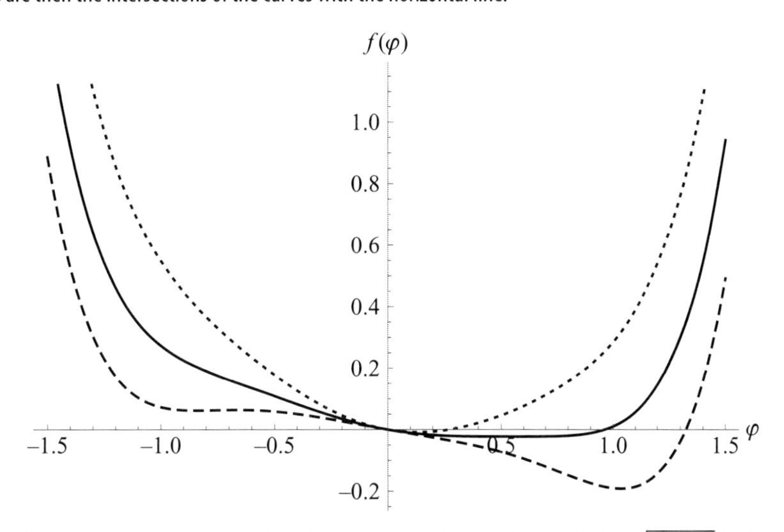

Fig. N.2 Plot of the thermodynamic potential including the magnetic field term at $h = (6/25)\sqrt{(3/10)}$ for three different values of a: $a = 1, a = 0.25$ and $a = 1/20$ with $b = -d_0 = -1$.

conditions:

$$\frac{\partial f}{\partial \varphi} = h, \tag{N.43}$$

$$\frac{\partial^2 f}{\partial \varphi^2} = 0, \tag{N.44}$$

$$\frac{\partial^3 f}{\partial \varphi^3} = 0. \tag{N.45}$$

By using the third condition

$$\varphi(6b + 20d_0\varphi^2) = 0, \quad \Rightarrow \varphi_0^2 = -\frac{3}{10}\frac{b}{d_0}. \tag{N.46}$$

This result can be used in the second condition to get a:

$$a = \frac{9}{20}\frac{b^2}{d_0}. \tag{N.47}$$

Finally one can use both (N.46) and (N.47) in the first condition to get h:

$$h = \frac{6}{25}\frac{b^2}{d_0}\sqrt{\frac{-3b}{10d_0}} \sim |b|^{5/2}. \tag{N.48}$$

The last two equations give in parametric form, as functions of the parameter b, the lines of critical points. For $b = -d_0 = 1$ one obtains the values of a and h used in the figures.

Chapter 15

15.1 Equation (15.7) is easily obtained by replacing $N_\mathrm{r}! = N_\mathrm{r}^{N_\mathrm{r}}\mathrm{e}^{-N_\mathrm{r}}$ and performing the derivative of F_r with respect to N_r:

$$\frac{\partial F_\mathrm{r}}{\partial N_\mathrm{r}} = \ln \frac{N_\mathrm{r}}{\frac{V}{(2\pi\hbar)^3}\int \mathrm{d}^3 p \mathrm{e}^{-\beta\epsilon_\mathrm{r}(\mathbf{p})}} = 0.$$

Equation (15.8) follows by noticing

$$\frac{\partial^2}{\partial T^2}(\mathrm{e}^{-\Delta_0/k_\mathrm{B}T}T^{3/2}) = \mathrm{e}^{-\Delta_0/k_\mathrm{B}T}\frac{1}{T^{1/2}}\left(\frac{3}{4} + \frac{\Delta_0}{k_\mathrm{B}T} + \frac{\Delta_0^2}{(k_\mathrm{B}T)^2}\right).$$

15.2 The symmetry is due to Bose–Einstein statistics. As far as reality is concerned, we note that if ϕ is a solution of the Schrödinger equation, ϕ^* is also a solution with the same energy. Hence, either $\mathrm{i}\phi$ is real or $\phi + \phi^*$ is real. We can then always choose the wave function to be real.

The requirement of a positive wave function implies the absence of nodes, which cannot be avoided for fermions. If the wave function had a node as a function of one of its arguments, then the wave function would change sign. We may then find a new wave function $|\phi|$ which has the same energy as the original one. The energy is given by

$$E = \frac{\int \mathrm{d}\mathbf{x}_1 \cdots \mathrm{d}\mathbf{x}_N \left(\sum_i (\nabla_i\phi)^2/2m + \sum_{i<j} V_{ij}\phi^2\right)}{\int \mathrm{d}\mathbf{x}_1 \cdots \mathrm{d}\mathbf{x}_N \phi^2}.$$

A wave function with lower energy can then be found by smoothing the cusp in $|\phi|$. This will decrease the gradient term in the energy, but it will not affect the potential term very much.

15.3 We consider **B** directed along the z-axis. Because the superconductor is infinite along the y–z plane, the magnetic field depends only on x. Hence we have

$$\partial_x^2 B_z(x) = \lambda^{-2} B_z(x) \Rightarrow B_z(x) = A\mathrm{e}^{x/\lambda} + B\mathrm{e}^{-x/\lambda},$$

with A and B constants to be fixed by the boundary conditions. The boundary conditions are $B_z(0) = B$ and $B_z(\infty) = 0$ leading to $B_z(x) = B\mathrm{e}^{-x/\lambda}$.

15.4 From (15.56) we have

$$\ln\frac{1}{z} = (N+1)f_{N+1}\left(\frac{V}{N+1}\right) - Nf_N\left(\frac{V}{N}\right), \quad f_N\left(\frac{V}{N}\right) = \frac{1}{N}\ln\frac{\Omega_N}{N!}.$$

In the thermodynamic limit $N \to \infty$, one has

$$\ln\frac{1}{z} = f(v) - v\frac{\partial f(v)}{\partial v}.$$

By using the virial expansion of Problem 8.2, one gets

$$f(v) = \ln v - \frac{2\pi}{3}\frac{d^3}{v},$$

where $d = 2.6$ Å is the diameter of the atoms. Hence the fraction of condensate from (15.58) reads

$$\frac{n_\sigma}{N} = \frac{1}{zv} = \mathrm{e}^{-1-\frac{4\pi}{3}\frac{d^3}{v}}.$$

By using $\overline{d} = 3.6$ Å one gets the estimate 0.08.

15.5 The field operator can expanded by using the single-particle wave functions $\phi_l(\mathbf{x})$:

$$\hat{\psi}(\mathbf{x}) = \sum_l \phi_l(\mathbf{x}) a_l,$$

where $l = 0$ corresponds to the single-particle wave function used in the ansatz (15.76). The ground state then can be written as

$$|\mathrm{GS}\rangle = \left(a_0^\dagger\right)^N |0\rangle,$$

where $|0\rangle$ is the vacuum state. By defining

$$h(\mathbf{x}) = -\frac{\hbar^2}{2m}\nabla^2 + V_{\mathrm{ext}}(\mathbf{x}),$$

we may write the Hartree–Fock energy as

$$\begin{aligned} W_0 = &\int \mathrm{d}\mathbf{x} \sum_{l_1,l_2} \phi_{l_1}^*(\mathbf{x}) h(\mathbf{x}) \phi_{l_2}(\mathbf{x}) \langle \mathrm{GS}| a_{l_1}^\dagger a_{l_2} |\mathrm{GS}\rangle \\ &+ \frac{1}{2}\int \mathrm{d}\mathbf{x} \int \mathrm{d}\mathbf{x}' \sum_{l_1,l_2,l_3,l_4} \phi_{l_1}^*(\mathbf{x}) \phi_{l_2}^*(\mathbf{x}') V(\mathbf{x}-\mathbf{x}') \phi_{l_3}(\mathbf{x}') \phi_{l_4}(\mathbf{x}) \\ &\times \langle \mathrm{GS}| a_{l_1}^\dagger a_{l_2}^\dagger a_{l_3} a_{l_4} |\mathrm{GS}\rangle \end{aligned}$$

and the problem is reduced to the evaluation of the expectation values of the operators $a^\dagger_{l_1} a_{l_2}$ and $a^\dagger_{l_1} a^\dagger_{l_2} a_{l_3} a_{l_4}$ in the ground state $|\mathrm{GS}\rangle$. Since in the ground state only the single-particle state $l = 0$ is occupied, all the indices l_i must be equal to zero. Then we have

$$\langle \mathrm{GS}|a^\dagger_0 a_0|\mathrm{GS}\rangle = N,$$
$$\langle \mathrm{GS}|a^\dagger_0 a^\dagger_0 a_0 a_0|\mathrm{GS}\rangle = N^2.$$

As a result the Hartree–Fock energy becomes

$$W_0 = N \int \mathrm{d}\mathbf{x} \sum_{l_1,l_2} \phi^*_0(\mathbf{x}) h(\mathbf{x}) \phi_0(\mathbf{x}) + \frac{N^2}{2} \int \mathrm{d}\mathbf{x} \int \mathrm{d}\mathbf{x}' \phi^*_0(\mathbf{x}) \phi^*_0(\mathbf{x}') V(\mathbf{x}-\mathbf{x}') \phi_0(\mathbf{x}') \phi_0(\mathbf{x}).$$

Next we make a variation of W_0 with respect to $\phi^*_0(\mathbf{x})$ and $\phi_0(\mathbf{x})$, obtaining

$$\delta W_0 = N \int \mathrm{d}\mathbf{x} \sum_{l_1,l_2} \delta\phi^*_0(\mathbf{x}) h(\mathbf{x}) \phi_0(\mathbf{x}) + \mathrm{c.c.} + N^2 \int \mathrm{d}\mathbf{x} \int \mathrm{d}\mathbf{x}' \delta\phi^*_0(\mathbf{x}) \phi^*_0(\mathbf{x}') V(\mathbf{x}-\mathbf{x}') \phi_0(\mathbf{x}') \phi_0(\mathbf{x}) + \mathrm{c.c.},$$

where the factor $1/2$ has been compensated by the fact that one can make a variation both with respect to $\phi^*_0(\mathbf{x})$ and $\phi^*_0(\mathbf{x}')$ and interchanging $\mathbf{x}$ with $\mathbf{x}'$ in one of the two terms. By requiring W_0 to be an extremum within the class of normalized wave functions one gets the Gross–Pitaevskii equation.

15.6 Fourier transforming the eigenvalue equation (15.44) and using the explicit BCS expressions (15.149) and (15.170), we obtain

$$h^2_{\mathbf{k}}(\phi(\mathbf{k}) + \phi(-\mathbf{k})) + \sum_{\mathbf{k}'} \phi(\mathbf{k}') \chi^*_{\mathbf{k}'} \chi_{\mathbf{k}} = \lambda \phi(\mathbf{k}), \tag{N.49}$$

and remembering that $\phi(\mathbf{k}) = \phi(-\mathbf{k})$, we have

$$\sum_{\mathbf{k}'} \phi(\mathbf{k}') \frac{\chi^*_{\mathbf{k}'} \chi_{\mathbf{k}}}{\lambda - 2h^2_{\mathbf{k}}} = \phi(\mathbf{k}) \tag{N.50}$$

which, after multiplying both members by $\chi^*_{\mathbf{k}}$ and summing over $\mathbf{k}$, reduces to

$$1 = \sum_{\mathbf{k}} \frac{|\chi_{\mathbf{k}}|^2}{\lambda - 2h^2_{\mathbf{k}}}. \tag{N.51}$$

From the constraint Eq. (15.168), since for fermions $h^2_{\mathbf{k}} \leq h_{\mathbf{k}}$, as long as the gap parameter is finite, we have:

$$\lim_{N\to\infty} \sum_{\mathbf{k}} |\chi_{\mathbf{k}}|^2 = \frac{V}{(2\pi)^3} \int \mathrm{d}\mathbf{k} |\chi_{\mathbf{k}}|^2 = \alpha N \tag{N.52}$$

with finite $\alpha < 1$. $h_{\mathbf{k}}$ is the microscopic occupation number of the state $\mathbf{k}$ and is therefore negligible compared to N in the limit of $N \to \infty$. In this limit (N.49) admits an eigenvalue

$$\lambda = \sum_{\mathbf{k}} |\chi_{\mathbf{k}}|^2 = \alpha N \tag{N.53}$$

and in the same limit no other isolated eigenvalue exists. As suggested in Section 15.5.2, where we discussed the condensation criterion, the Fourier component of the eigenfunction corresponding to this maximum eigenvalue Eq. (15.184) is

$$\phi(\mathbf{k}) = \frac{\chi_{\mathbf{k}}}{(\alpha N)^{1/2}}, \tag{N.54}$$

i.e. it coincides but for the factor $(\alpha N)^{1/2}$ with the function which characterized the ODLRO property of h_2 as in Eq. (15.46).

15.7 By means of the above identity, the gap equation (15.188) reads

$$\frac{1}{N_0 g} = \int_0^{\omega_D} d\epsilon_{\mathbf{k}} \beta^{-1} \sum_{l=0}^{\infty} \frac{4}{\epsilon_{\mathbf{k}}^2 + \overline{\Delta}^2 + (2l+1)^2 \pi^2 \beta^{-2}}.$$

We then expand the denominator in the sum

$$\frac{4}{\epsilon_{\mathbf{k}}^2 + (2l+1)^2 \pi^2 \beta^{-2}} - \frac{4\overline{\Delta}^2}{\left(\epsilon_{\mathbf{k}}^2 + (2l+1)^2 \pi^2 \beta^{-2}\right)^2} + \cdots .$$

The first term can be written again as before by using the identity backwards so that the gap equation becomes

$$\frac{1}{N_0 g} = \int_0^{\omega_D} d\epsilon_{\mathbf{k}} \left(\frac{1}{\epsilon_{\mathbf{k}}} \tanh\left(\frac{\beta \epsilon_{\mathbf{k}}}{2}\right) - \beta^{-1} \sum_{l=0}^{\infty} \frac{4\overline{\Delta}^2}{\left(\epsilon_{\mathbf{k}}^2 + (2l+1)^2 \pi^2 \beta^{-2}\right)^2} \right).$$

In the second term we interchange the $\epsilon_{\mathbf{k}}$ integration with the sum and send the upper limit to infinity since the integral converges fast:

$$\int_0^{\infty} d\epsilon_{\mathbf{k}} \frac{1}{\left(\epsilon_{\mathbf{k}}^2 + (2l+1)^2 \pi^2 \beta^{-2}\right)^2} = \frac{\pi}{4(2l+1)^3 \pi^3 \beta^{-3}}.$$

The summation can be easily performed by noticing that the sum over the odd integers can be written as the sum over all the integers minus the even integers

$$\sum_{l=0}^{\infty} \frac{1}{(2l+1)^3} = \sum_{l=1}^{\infty} \left(\frac{1}{l^3} - \frac{1}{(2l)^3} \right) = \frac{7}{8} \sum_{l=1}^{\infty} \frac{1}{l^3} = \frac{7}{8} \zeta_R(3).$$

By collecting all the results the gap equation reads

$$\frac{1}{N_0 g} = \int_0^{\omega_D} d\epsilon_{\mathbf{k}} \frac{1}{\epsilon_{\mathbf{k}}} \tanh\left(\frac{\beta \epsilon_{\mathbf{k}}}{2}\right) - \frac{7\zeta_R(3)\beta^2}{8\pi^2} \overline{\Delta}^2.$$

By recalling the evaluation of the integrals on the right-hand side of Eq. (15.189) we may rewrite it as

$$\ln \frac{\beta_c \omega_D 4 e^{\gamma_E}}{2\pi} = \ln \frac{\beta \omega_D 4 e^{\gamma_E}}{2\pi} - \frac{7\zeta_R(3)\beta^2}{8\pi^2} \overline{\Delta}^2,$$

where the left-hand side term $(N_0 g)^{-1}$ has been given in terms of the critical value $\beta_c = (k_B T_c)^{-1}$. Finally we arrive then at the form

$$\ln \frac{T}{T_c} = -\frac{7\zeta_R(3)\beta^2}{8\pi^2}\overline{\Delta}^2.$$

Close to T_c,

$$\ln \frac{T}{T_c} \approx \frac{T - T_c}{T_c},$$

which leads to an expression for the temperature dependence of the gap:

$$\overline{\Delta}^2 = \frac{8\pi^2 k_B^2 T_c^2}{7\zeta_R(3)} \frac{T_c - T}{T_c}.$$

15.8 It is convenient to define

$$v_{\mathbf{k}} = \frac{1}{1 + k^2/k_0^2}$$

and start from Eq. (15.163) expressing $\mu_{\mathbf{k}}$:

$$\begin{aligned} \mu_{\mathbf{k}} &= \sum_{\mathbf{k}'} V(\mathbf{k}, \mathbf{k}')\chi_{\mathbf{k}'} \\ &= -g \sum_{\mathbf{k}'} v_{\mathbf{k}} v_{\mathbf{k}'} \chi_{\mathbf{k}'} \\ &= g v_{\mathbf{k}} \sum_{\mathbf{k}'} \frac{v_{\mathbf{k}'}}{2E_{\mathbf{k}'}} \mu_{\mathbf{k}'}, \end{aligned}$$

where we used the second of Eq. (15.170). By making the ansatz $\mu_{\mathbf{k}} = \Delta v_{\mathbf{k}}$, we obtain

$$\frac{1}{g} = \sum_{\mathbf{k}} \frac{v_{\mathbf{k}}^2}{2E_{\mathbf{k}}}.$$

A similar analysis can be done for the equation for the single bound pair (15.201). We have

$$\begin{aligned} \chi_{\mathbf{k}} &= -\frac{1}{2\epsilon_{\mathbf{k}}} \sum_{\mathbf{k}'} V(\mathbf{k}, \mathbf{k}')\chi_{\mathbf{k}'} \\ &= \frac{g v_{\mathbf{k}}}{2\epsilon_{\mathbf{k}}} \sum_{\mathbf{k}'} v_{\mathbf{k}'} \chi_{\mathbf{k}'}. \end{aligned}$$

After multiplying both sides of the last line by $v_{\mathbf{k}}$ and summing over $\mathbf{k}$, we get

$$\frac{1}{g} = \sum_{\mathbf{k}} \frac{v_{\mathbf{k}}2}{2\epsilon_{\mathbf{k}}},$$

which has the same form as the gap equation with $E_{\mathbf{k}}$ replaced by $\epsilon_{\mathbf{k}}$.

15.9 In the Schrödinger equation (J.1) of Appendix J, after separating the center-of-mass motion in terms of the variable $\mathbf{X}$ from the relative motion in terms of the variable $\mathbf{x}$ of Eq. (J.2), write the Laplacian ∇_x^2 in spherical coordinates:

$$\frac{\hbar^2}{m}\left(-\frac{1}{x^2}\frac{\partial}{\partial x}\left(x^2\frac{\partial\chi}{\partial x}\right)+\frac{\mathbf{l}^2}{x^2}\chi\right)+V(\mathbf{x})\chi=E\chi,$$

where $\mathbf{l}^2$ is the square angular momentum operator in spherical coordinates. Look for a solution of the form

$$\chi(\mathbf{x})=\chi(x)Y_{lm}(\hat{\mathbf{x}}).$$

If the l-attractive component of the potential is such that

$$\lim_{x\to 0}x^2V_l(x)=0,$$

remembering that the the spherical harmonics are egenfuntions of $\mathbf{l}^2$ with eigenvalue $l(l+1)$, at small x the Schrödinger equation reads

$$\frac{\partial}{\partial x}\left(x^2\frac{\partial\chi}{\partial x}\right)-l(l+1)\chi=0$$

with solution $\chi(x)=x^l$.

15.10 The gap equation for e.g. the up spin reads

$$\mu_{\mathbf{k}\uparrow}=3g\sum_{\mathbf{k}'}\hat{\mathbf{k}}\cdot\hat{\mathbf{k}}'\frac{\mu_{\mathbf{k}'\uparrow}}{2E_{\mathbf{k}'\uparrow}}.$$

By inserting the ansatz (15.213) we get

$$\begin{aligned}
&\Delta\sqrt{3/2}(\hat{k}_x-\mathrm{i}\hat{k}_y)\\
&=3g\sum_{\mathbf{k}'}(\hat{k}_x\hat{k}'_x+\hat{k}_y\hat{k}'_y+\hat{k}_z\hat{k}'_z)\frac{\sqrt{3/2}(\hat{k}'_x-\mathrm{i}\hat{k}'_y)\Delta}{2\sqrt{\epsilon^2_{\mathbf{k}'}+\Delta^2(3/2)(\hat{k}'^2_x+\hat{k}'^2_y)}}\\
&=gN_0\Delta\sqrt{3/2}(\hat{k}_x-\mathrm{i}\hat{k}_y)3\left\langle\hat{k}'^2_x\left(\ln\frac{2\omega_D}{\Delta}+\ln\sqrt{\frac{2}{3}}-\frac{1}{2}\ln\left(\hat{k}'^2_x+\hat{k}'^2_y\right)\right)\right\rangle
\end{aligned}$$

where $\langle\dots\rangle$ is the solid angle integration over $\hat{\mathbf{k}}'$ and the integration over the absolute value of $\mathbf{k}'$ has been performed as in Eq. (15.181). By observing that

$$-\frac{3}{2}\int_0^{2\pi}\int_0^{\pi}\frac{\mathrm{d}\phi\mathrm{d}\theta}{4\pi}\sin\theta\cos^2(\phi)\sin^2(\theta)\ln\sin^2(\theta)=0.14,$$

we get for the gap equation

$$\frac{1}{gN_0}=\ln\left(0.94\frac{2\omega_\mathrm{D}}{\Delta}\right),$$

where $\exp(0.14)/\sqrt{3/2}=0.94$.

Chapter 16

16.1 We have

$$h_1(\mathbf{x}, \mathbf{x}') = |\varphi|^2 \left\langle e^{i(\theta(\mathbf{x})-\theta(\mathbf{x}'))} \right\rangle. \tag{N.55}$$

To perform the average, we expand the exponential by using the cumulant expansion defined in Eqs. (17.34)–(17.36):

$$\langle e^x \rangle = \exp\left(\sum_{n=0}^{\infty} \frac{1}{n!} \langle x^n \rangle_{\text{cum}} \right).$$

The first non-vanishing cumulant is for $n = 2$:

$$\frac{1}{2}\langle (\theta(\mathbf{x}) - \theta(\mathbf{x}'))^2 \rangle = \int \frac{d^d k}{(2\pi)^d} \left\langle \theta_{\mathbf{k}}^2 \right\rangle (1 - \cos(\mathbf{k} \cdot (\mathbf{x} - \mathbf{x}'))). \tag{N.56}$$

By using Eqs. (16.33) and (16.34) to perform the average of the phase $\theta_{\mathbf{k}}$, we get

$$\frac{1}{2}\langle (\theta(\mathbf{x}) - \theta(\mathbf{x}'))^2 \rangle = \int \frac{d^d k}{(2\pi)^d} \frac{2m^2 k_B T}{\hbar^2 \rho_s} \frac{1 - \cos(\mathbf{k} \cdot (\mathbf{x} - \mathbf{x}'))}{k^2}. \tag{N.57}$$

By performing the integral in $d = 3$ we get

$$\frac{1}{2}\langle (\theta(\mathbf{x}) - \theta(\mathbf{x}'))^2 \rangle = \frac{1}{\pi^2}\left(\xi \Lambda_{\text{uv}} - \pi \frac{\xi}{r} \right), \tag{N.58}$$

where $r = |\mathbf{x} - \mathbf{x}'|$ and Λ_{uv} is an ultraviolet cutoff and we expressed the superfluid density in terms of the correlation length according to the relation (16.36). As a result, in $d = 3$ in the long-distance behavior the reduced one-particle density matrix approaches a constant:

$$h_1(\mathbf{x}, \mathbf{x}') = |\psi|^2 e^{-\frac{1}{2\pi^2}\left(\xi \Lambda_{\text{uv}} - \pi \frac{\xi}{r}\right)} \to |\psi|^2 e^{-\frac{1}{\pi^2}\xi \Lambda_{\text{uv}}}. \tag{N.59}$$

The situation is different in $d = 2$, where the cumulant integral gives

$$\frac{1}{2}\langle (\theta(\mathbf{x}) - \theta(\mathbf{x}'))^2 \rangle = \frac{2k_B T m^2}{\hbar^2 \rho_s} \int_0^{\Lambda_{\text{uv}}} \frac{dk}{2\pi} \frac{1 - J_0(kr)}{k}, \tag{N.60}$$

where, after the angular integration, $J_0(x)$ is a Bessel function of the first kind. At high k the Bessel function is a slowly decaying oscillating function, so that the integral may be estimated with logarithmic accuracy as

$$\frac{1}{2}\langle (\theta(\mathbf{x}) - \theta(\mathbf{x}'))^2 \rangle = \frac{k_B T m^2}{\pi \hbar^2 \rho_s} \ln \Lambda_{\text{uv}} r. \tag{N.61}$$

As a result the long-distance behavior of h_1 reads

$$h_1(\mathbf{x}, \mathbf{x}') = |\psi|^2 r^{-s}, \quad s = \frac{k_B T m^2}{\pi \hbar^2 \rho_s}, \tag{N.62}$$

i.e. the off-diagonal elements of the one-particle reduced density matrix decay as a power law.

In $d = 1$ the integral can be performed exactly:

$$\frac{1}{2}\langle(\theta(\mathbf{x}) - \theta(\mathbf{x}'))^2\rangle = \frac{k_B T m^2}{\hbar^2 \rho_s} r, \tag{N.63}$$

leading to a decaying exponential of h_1:

$$h_1(\mathbf{x}, \mathbf{x}') = |\psi|^2 \mathrm{e}^{-\frac{k_B T m^2}{\hbar^2 \rho_s} r}. \tag{N.64}$$

Chapter 17

17.1 By definition ($c = 1$)

$$G(\mathbf{x}) = \int \frac{\mathrm{d}^d k}{(2\pi)^d} \frac{\mathrm{e}^{\mathrm{i}\mathbf{k}\cdot\mathbf{x}}}{t + k^2} \to \int \frac{\mathrm{d}^d k}{(2\pi)^d} \frac{\mathrm{e}^{\mathrm{i}\mathbf{k}\cdot\mathbf{x}}}{k^2}.$$

By means of the identity

$$\frac{1}{\alpha} = \int_0^\infty \mathrm{d}w \mathrm{e}^{-\alpha w}$$

we rewrite the propagator as

$$\begin{aligned} G(\mathbf{x}) &= \int \frac{\mathrm{d}^d k}{(2\pi)^d} \mathrm{e}^{\mathrm{i}\mathbf{k}\cdot\mathbf{x}} \int_0^\infty \mathrm{d}w\, \mathrm{e}^{-k^2 w} \\ &= \int_0^\infty \mathrm{d}w \prod_{i=1,d} \left(\int_{-\infty}^\infty \frac{\mathrm{d}k_i}{2\pi} \mathrm{e}^{\mathrm{i}k_i x_i} \mathrm{e}^{-w k_i^2} \right) \\ &= \int_0^\infty \mathrm{d}w \left(\frac{\pi}{w}\right)^{d/2} \frac{1}{(2\pi)^d} \mathrm{e}^{-|\mathbf{x}|^2/4w}, \end{aligned}$$

where the integrals over each component of the wave vector are of the Gaussian type as in Eq. (14.1). By performing the change of variable $s = 1/w$ one obtains

$$G(\mathbf{x}) = \frac{1}{2^d \pi^{d/2}} \int_0^\infty \mathrm{d}s s^{d/2-2} \mathrm{e}^{-|\mathbf{x}|^2 s/4} = \frac{\Gamma(d/2-1)}{4\pi^{d/2}|\mathbf{x}|^{d-2}},$$

$\Gamma(d/2 - 1)$ being the Euler Gamma function. In particular in $d = 4$

$$G(\mathbf{x}) = \frac{1}{4\pi^2|\mathbf{x}|^2} = \frac{2K_4}{|\mathbf{x}|^2}.$$

17.2 By considering the diagrams of Fig. 17.8, one has

$$\begin{aligned} \Gamma^{(2)}(\mathbf{k}) &= t + k^2 + 2(n+2)u \int_{|\mathbf{q}|<\Lambda} \frac{\mathrm{d}^d q}{(2\pi)^d} G(\mathbf{q}) \\ &\quad -32(n+2)u^2 \int_{|\mathbf{q}|<\Lambda} \frac{\mathrm{d}^d q}{(2\pi)^d} \int_{|\mathbf{p}|<\Lambda} \frac{\mathrm{d}^d p}{(2\pi)^d} G(\mathbf{q})G(\mathbf{p})G(-\mathbf{q}-\mathbf{p}-\mathbf{k}), \end{aligned}$$

where one sees that the second order term in u contains two propagators and an integration in addition to those of the first order. The multiplicity factors in front of the expression of each diagram are obtained with the same reasoning as Subsection 17.3.2.

17.3 From the definition of γ_t in Eq. (17.87), one has

$$\gamma_t(\mathbf{k}) = 1 - 2(n+2)u \int_{|\mathbf{q}|<\Lambda} \frac{\mathrm{d}^4 q}{(2\pi)^4} G^2(\mathbf{q}),$$

where the derivative acts under the integral on the propagator $\partial_t G = -G^2$. The integrand depends only on the absolute value of $\mathbf{q}$ so that the integral can be performed by going to hyperspherical coordinates:

$$K_4 \int_0^{\Lambda} \mathrm{d}q \frac{1}{(t+q^2)^2} = K_4 \frac{1}{2} \ln \frac{\Lambda^2}{t} + \cdots,$$

where the dots indicate terms less singular when $\Lambda \to \infty$ or $t \to 0$. As a result

$$\gamma_t(\mathbf{k}) = 1 - (n+2)uK_4 \left(\ln \frac{\Lambda^2}{t} + \cdots \right).$$

17.4 From Problem 17.2, we have

$$\gamma_\varphi = 1 - 32(n+2)u^2 \frac{\partial}{\partial k^2} \int_{|\mathbf{q}|<\Lambda} \frac{\mathrm{d}^d q}{(2\pi)^d} \int_{|\mathbf{p}|<\Lambda} \frac{\mathrm{d}^d p}{(2\pi)^d} G(\mathbf{q})G(\mathbf{p})G(-\mathbf{q}-\mathbf{p}-\mathbf{k}).$$

The double integral is formally divergent as Λ^2. However, the derivative with respect to k^2 lowers the divergence by two, as may be seen by expanding at small $\mathbf{k}$ the third propagator $G(-\mathbf{q}-\mathbf{p}-\mathbf{k})$. With logarithmic accuracy, it is actually more convenient to subtract the $\mathbf{k} = \mathbf{0}$ value, relax the constraint over the integration and expand to order k^2, as done in Eq. (17.74). By using $\exp(\mathrm{i}\mathbf{k} \cdot \mathbf{x}) - 1 \approx (\mathrm{i}\mathbf{k} \cdot \mathbf{x})^2/2$, one then has

$$\begin{aligned} \gamma_\varphi &= 1 - 32(n+2)u^2 \frac{\partial}{\partial k^2} \int \mathrm{d}^4 x \frac{1}{2} (-\mathrm{i}\mathbf{k} \cdot \mathbf{x})^2 G^3(\mathbf{x}) \\ &= 1 + 32(n+2)u^2 \frac{1}{8} \int \mathrm{d}^4 x |\mathbf{x}|^2 G^3(\mathbf{x}) \\ &= 1 + 32(n+2)u^2 \frac{1}{8} (2K_4)^3 K_4 (2\pi)^4 \int_0^\infty \mathrm{d}x \frac{1}{x}, \end{aligned} \tag{N.65}$$

where we have used the expression of $G(\mathbf{x})$ for $d = 4$ obtained in Problem 17.1 and used hyperspherical coordinates for the integral over $\mathbf{x}$. The last integral over the absolute value of $\mathbf{x}$ is logarithmically divergent as expected and needs to be regularized. The condition over the relevant wave vectors $t^{1/2} < q < \Lambda$ translates into $\Lambda^{-1} < x < t^{-1/2}$, yielding

$$\gamma_\varphi = 1 + 32(n+2)u^2 (K_4)^4 (2\pi)^4 \ln \frac{\Lambda}{t^{1/2}} + \cdots,$$

where the dots indicate terms less singular.

17.5 From Fig. 17.8, the expression for $\Gamma^{(4)}(\mathbf{k}_1, \mathbf{k}_2, \mathbf{k}_3, \mathbf{k}_4)$ up to second order in u is given by

$$\Gamma^{(4)}(\mathbf{k}_1, \mathbf{k}_2, \mathbf{k}_3, \mathbf{k}_4) = u - 4(n+8)u^2 \int_{|\mathbf{q}|<\Lambda} \frac{\mathrm{d}^d q}{(2\pi)^d} G(\mathbf{q})G(-\mathbf{q} - \mathbf{k}_1 - \mathbf{k}_2).$$

After dividing by u, we get

$$\gamma_u = 1 - 4(n+8)u \int_{|\mathbf{q}|<\Lambda} \frac{\mathrm{d}^d q}{(2\pi)^d} G^2(\mathbf{q}),$$

where we have set to zero the external wave vectors, having in mind the normalization point (17.90). We may finally perform the integral in $d = 4$, obtaining

$$\gamma_u = 1 - 4(n+8)uK_4 \ln \frac{\Lambda}{t^{1/2}} + \cdots,$$

where, as in the other equations, the dots indicate less singular terms.

Chapter 18

18.1 We begin by observing that the functions

$$f(z) = \frac{1}{e^{\beta z} + 1}, \quad b(z) = \frac{1}{e^{\beta z} - 1} \tag{N.66}$$

as functions of the complex variable z have simple poles on the imaginary axis at the points $z = \mathrm{i}\omega_n$ with residue $\mp 1/\beta$ for odd and even frequences, respectively. By using the Cauchy theorem, the sum over the discrete frequencies in Eq. (18.20) can be written as a contour integral along a path $\mathcal{C}$ encircling all the poles on the imaginary axis:

$$\frac{1}{\beta} \sum_{\omega_n} \frac{e^{\mathrm{i}\omega_n 0^+}}{\mathrm{i}\omega_n - \epsilon_{\mathbf{k}}} = -\frac{1}{2\pi \mathrm{i}} \oint_{\mathcal{C}} \mathrm{d}z \frac{e^{z0^+}}{e^{\beta z} \pm 1} \frac{1}{z - \epsilon_{\mathbf{k}}}. \tag{N.67}$$

The function

$$\frac{1}{z - \epsilon_{\mathbf{k}}}$$

must be understood as the analytic continuation of the sequence

$$\frac{1}{\mathrm{i}\omega_n - \epsilon_{\mathbf{k}}}.$$

Such an analytic continuation cannot be done in the same way over the entire complex plane, but we must distinguish the analytic continuation in the upper and lower half planes. We then introduce the retarded and advanced Green functions defined in Eqs. (18.34) and (18.35), which in the case of the gases have the simple form

$$G^{\mathrm{R,A}}(\mathbf{k}, \omega) = \frac{1}{\omega - \epsilon_{\mathbf{k}}/\hbar \pm \mathrm{i}0^+}. \tag{N.68}$$

The path $\mathcal{C}$ can be split into two paths $\mathcal{C}_\pm$ going around the poles in the upper ($\mathcal{C}_+$) and lower ($\mathcal{C}_-$) half-planes. A problem arises in the case of bosons, since the pole at the origin is not included in either path. Hence for bosons one must add to the two paths above a small circular path around the origin. Then we may deform the integration paths so that they both include the real axis and a semicircle closing at infinity in the upper and lower half-planes, respectively. One then realizes that for bosons the inclusion of the pole at the origin amounts to considering the real frequency integrals in the principal value sense. The integrals of $\mathcal{C}_\pm$ over the real axis are taken with opposite orientation of the path in such a way that they combine to

$$\frac{1}{\beta}\sum_{\omega_n}\frac{\mathrm{e}^{\mathrm{i}\omega_n 0^+}}{\mathrm{i}\omega_n - \epsilon_{\mathbf{k}}} = -\frac{1}{2\pi\mathrm{i}}\int_{-\infty}^{\infty}\mathrm{d}\omega\frac{\mathrm{e}^{\omega 0^+}}{\mathrm{e}^{\beta\omega}\pm 1}(G^{\mathrm{R}}(\mathbf{k},\omega) - G^{\mathrm{A}}(\mathbf{k},\omega)). \quad \text{(N.69)}$$

Notice that the existence of the integral for negative values of ω is guaranteed by the convergence factor $\mathrm{e}^{\omega 0^+}$. By observing that

$$G^{\mathrm{R}}(\mathbf{k},\omega) - G^{\mathrm{A}}(\mathbf{k},\omega) = -2\pi\mathrm{i}\delta(\omega - \epsilon_{\mathbf{k}}/\hbar)$$

one concludes the proof.

We remark that, in fact, the explicit form of the Green functions for the gases has been used only in the last step. Hence we further conclude that Eq. (N.69) is actually valid in general.

18.2 By assuming translational invariance, we can decompose the field operators

$$\hat{\psi}(\mathbf{x}) = \frac{1}{\sqrt{V}}\sum_{\mathbf{k}}\mathrm{e}^{\mathrm{i}\mathbf{k}\cdot\mathbf{x}}a_{\mathbf{k}}, \quad \hat{\psi}^\dagger(\mathbf{x}) = \frac{1}{\sqrt{V}}\sum_{\mathbf{k}}\mathrm{e}^{-\mathrm{i}\mathbf{k}\cdot\mathbf{x}}a^\dagger_{\mathbf{k}}$$

and obtain

$$B_{0m} = \sum_{\mathbf{k},\mathbf{k}'}\mathrm{e}^{\mathrm{i}\mathbf{k}\cdot\mathbf{x}}\mathrm{e}^{-\mathrm{i}\mathbf{k}'\cdot\mathbf{x}'}\langle 0|a_{\mathbf{k}}|m\rangle\langle m|a^\dagger_{\mathbf{k}'}|0\rangle.$$

The state $|m\rangle$ differs from the ground state $|0\rangle$ by the presence of one more particle with wave vector $\mathbf{k}$. By translational invariance, the momentum operator commutes with the Hamiltonian and the eigenstates can be chosen to be also momentum eigenstates. Hence the state $|m\rangle$ has momentum eigenvalue $\hbar\mathbf{k}$. Hence $\mathbf{k}' = \mathbf{k}$ since the same state $|m\rangle$ enters both matrix elements. As a result

$$B_{0m} = \sum_{\mathbf{k}}\mathrm{e}^{\mathrm{i}\mathbf{k}\cdot(\mathbf{x}-\mathbf{x}')}\langle 0|a_{\mathbf{k}}|m\rangle\langle m|a^\dagger_{\mathbf{k}}|0\rangle,$$

which is Eq. (18.44). Equation (18.45) can be proved similarly.

Chapter 20

20.1 Since charge and spin density modes are bosons, their energy reads

$$U = \sum_a\int_{-\infty}^{\infty}\frac{\mathrm{d}q}{2\pi}\frac{u^a|q|}{\mathrm{e}^{\beta u^a|q|}-1} = \sum_a\frac{T^2}{\pi u^a}I_1^-(0) = \frac{\pi T^2}{6u^a},$$

which then leads to Eq. (20.27) by taking the derivative with respect to T.

20.2 Let us rewrite the system (20.42)–(20.43) in the form

$$\begin{pmatrix} \omega - \mathbf{q}\frac{v^a+\tilde{v}^a}{2} & -\mathbf{q}\frac{\tilde{v}^a-v^a}{2} \\ \mathbf{q}\frac{\tilde{v}^a-v^a}{2} & \omega + \mathbf{q}\frac{v^a+\tilde{v}^a}{2} \end{pmatrix} \begin{pmatrix} \Lambda_{+,\alpha} \\ \Lambda_{-,\alpha} \end{pmatrix} = \begin{pmatrix} \delta_{+,\alpha} \\ \delta_{-,\alpha} \end{pmatrix} \varepsilon^a_\sigma [G_{\alpha,\sigma}(p_-) - G_{\alpha,\sigma}(p_+)], \tag{N.70}$$

from which it is clear that by sending $\mathbf{q} \to -\mathbf{q}$ and $\alpha \to -\alpha$, one has $\Lambda_{+,+} \to \Lambda_{-,-}$ and $\Lambda_{-,+} \to \Lambda_{+,-}$. We then have

$$\varepsilon^a_\sigma \begin{pmatrix} \Lambda_{+,\alpha} \\ \Lambda_{-,\alpha} \end{pmatrix} = \frac{1}{\det} \begin{pmatrix} \omega + \mathbf{q}\frac{v^a+\tilde{v}^a}{2} & \mathbf{q}\frac{\tilde{v}^a-v^a}{2} \\ -\mathbf{q}\frac{\tilde{v}^a-v^a}{2} & \omega - \mathbf{q}\frac{v^a+\tilde{v}^a}{2} \end{pmatrix} \begin{pmatrix} \delta_{+,\alpha} \\ \delta_{-,\alpha}, \end{pmatrix} [G_{\alpha,\sigma}(p_-) - G_{\alpha,\sigma}(p_+)], \tag{N.71}$$

where

$$\det = \omega^2 - \mathbf{q}^2 \left[\left(\frac{v^a + \tilde{v}^a}{2} \right)^2 - \left(\frac{v^a - \tilde{v}^a}{2} \right)^2 \right] = \omega^2 - (u^a)^2 \mathbf{q}^2, \tag{N.72}$$

after noting that

$$\frac{v^a + \tilde{v}^a}{2} = v_{\mathrm{F}} + \frac{g^a_4}{\pi}, \quad \frac{v^a - \tilde{v}^a}{2} = -\frac{g^a_2}{\pi}. \tag{N.73}$$

According to Eq. (20.37), for the branch $\alpha = +$ we need

$$\begin{aligned} &\sum_a g^a_{+,+} \varepsilon^a_\sigma \Lambda_{+,+} + g^a_{-,+} \varepsilon^a_\sigma \Lambda_{-,+} \\ &= \sum_a \frac{1}{\det} \left[g^a_4 \left(\omega + \mathbf{q}\frac{v^a + \tilde{v}^a}{2} \right) - g^a_2 \left(\mathbf{q}\frac{\tilde{v}^a - v^a}{2} \right) \right] \Delta G \\ &= \sum_a \frac{1}{\omega^2 - (u^a)^2 \mathbf{q}^2} \left[g^a_4 \left(\omega + \mathbf{q} (v_{\mathrm{F}} + g^a_4/\pi) \right) - \mathbf{q} \left(g^a_2 \right)^2 / \pi \right] \Delta G \\ &\equiv L_{++}(\mathbf{q}, \omega) \Delta G, \end{aligned} \tag{N.74}$$

where for brevity $\Delta G \equiv [G_{\alpha,\sigma}(p_-) - G_{\alpha,\sigma}(p_+)]$. By inserting this last result into (20.37), Eq. (20.44) follows.

20.3 At one-loop order, there are two diagrams (see Fig. N.3) for the corrections to g_2:

$$\begin{aligned} (\delta g_2)^{(1)} &= -\mathrm{i} g_2^2 \int_p G_+(p) G_-(p_1 + p_2 - p), \\ (\delta g_2)^{(2)} &= -\mathrm{i} g_2^2 \int_p G_+(p) G_-(p + p_2 - p_1), \end{aligned}$$

where the four bimomenta are $(p_1 + q/2, p_2 - q/2)$ for the incoming particles and $(p_1 - q/2, p_2 + q/2)$ for the outgoing particles. Each bimomentum is $p = (\epsilon, \mathbf{p})$.

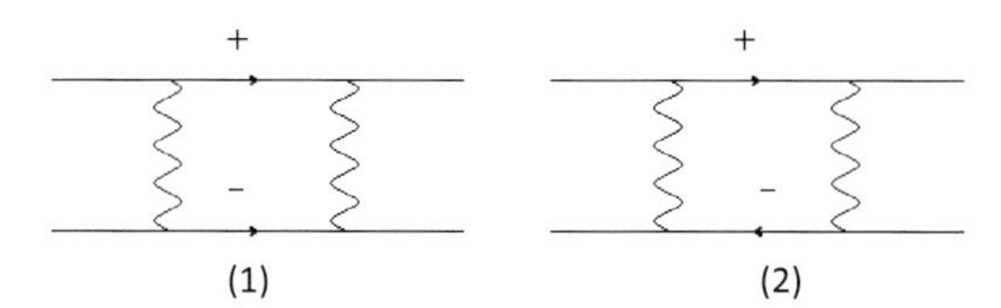

Fig. N.3 The one-loop corrections to the coupling g_2. The + and − signs on the fermion line label the energy branches.

Let us set $\mathbf{q} = \mathbf{p}_1 = \mathbf{p}_2 = 0$ and evaluate the first diagram with $\omega = \epsilon_1 + \epsilon_2$:

$$
\begin{aligned}
(\delta g_2)^{(1)} &= -\mathrm{i} g_2^2 \int_{\mathbf{p}} \int_{\epsilon} \frac{1}{\epsilon - v_{\mathrm{F}}\mathbf{p} + \mathrm{i}\,\mathrm{sgn}(\mathbf{p})} \frac{1}{\omega - \epsilon - v_{\mathrm{F}}\mathbf{p} - \mathrm{i}\,\mathrm{sgn}(\mathbf{p})} \\
&= -\mathrm{i} g_2^2 \int_{\mathbf{p}} \int_{\epsilon} \frac{1}{\omega - 2v_{\mathrm{F}}\mathbf{p}} \left(\frac{1}{\epsilon - v_{\mathrm{F}}\mathbf{p} + \mathrm{i}\,\mathrm{sgn}(\mathbf{p})} - \frac{1}{\epsilon - \omega + v_{\mathrm{F}}\mathbf{p} + \mathrm{i}\,\mathrm{sgn}(\mathbf{p})} \right) \\
&= g_2^2 \int_{\mathbf{p}} \frac{1}{\omega - 2v_{\mathrm{F}}\mathbf{p}} (\theta(-\mathbf{p}) - \theta(\mathbf{p})) \\
&= \frac{g_2^2}{2\pi v_{\mathrm{F}}} \ln \frac{v_{\mathrm{F}}\Lambda}{\omega}.
\end{aligned}
$$

For the second diagram we get, with $\omega = \epsilon_2 - \epsilon_1$,

$$
\begin{aligned}
(\delta g_2)^{(2)} &= -\mathrm{i} g_2^2 \int_{\mathbf{p}} \int_{\epsilon} \frac{1}{\epsilon - v_{\mathrm{F}}\mathbf{p} + \mathrm{i}\,\mathrm{sgn}(\mathbf{p})} \frac{1}{\omega + \epsilon + v_{\mathrm{F}}\mathbf{p} + \mathrm{i}\,\mathrm{sgn}(\mathbf{p})} \\
&= -\mathrm{i} g_2^2 \int_{\mathbf{p}} \int_{\epsilon} \frac{1}{\omega + 2v_{\mathrm{F}}\mathbf{p}} \left(\frac{1}{\epsilon - v_{\mathrm{F}}\mathbf{p} + \mathrm{i}\,\mathrm{sgn}(\mathbf{p})} - \frac{1}{\epsilon + \omega + v_{\mathrm{F}}\mathbf{p} + \mathrm{i}\,\mathrm{sgn}(\mathbf{p})} \right) \\
&= g_2^2 \int_{\mathbf{p}} \frac{1}{\omega + 2v_{\mathrm{F}}\mathbf{p}} (\theta(-\mathbf{p}) - \theta(\mathbf{p})) \\
&= -\frac{g_2^2}{2\pi v_{\mathrm{F}}} \ln \frac{v_{\mathrm{F}}\Lambda}{\omega}.
\end{aligned}
$$

Hence the logarithmic dependence on the cutoff cancels between the two diagrams.

20.4 The one-loop diagram for the self-energy (see Fig. N.4) with the interaction g_2 reads (we assume spin-independent couplings)

$$
\Sigma(p) = -2g_2^2 \int_q \int_{p_1} G_+(p+q)G_-(p_1)G_-(p_1 - q),
$$

where the bimomenta are $q = (\omega, \mathbf{q})$, $p = (\epsilon, \mathbf{p})$ and $p_1 = (\epsilon_1, \mathbf{p}_1)$. Notice that the integration over the bimomentum p_1 is equivalent to the bubble evaluation of Eq. (L.17), so that we get

$$
\Sigma(p) = -g_2^2 \int_q \frac{\mathrm{i}}{\pi} \frac{\mathbf{q}}{\omega + v_{\mathrm{F}}\mathbf{q} + \mathrm{i}\,\mathrm{sgn}\omega} \frac{1}{\epsilon + \omega - v_{\mathrm{F}}(\mathbf{p}+\mathbf{q}) + \mathrm{i}\,\mathrm{sgn}(\mathbf{p}+\mathbf{q})}.
$$

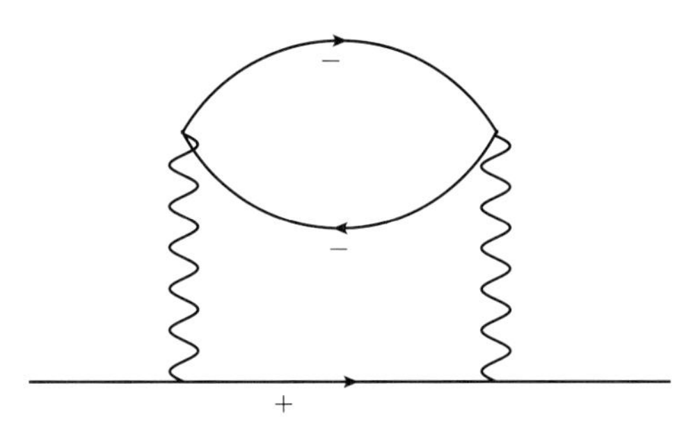

Fig. N.4 The one-loop self-energy correction due to the coupling g_2. The $+$ and $-$ signs on the fermion line label the energy branches.

By decomposing into partial fractions and setting $\mathbf{p} = 0$ we get

$$\Sigma(p) = -\frac{\mathrm{i}g_2^2}{4\pi^3}\int_{\mathbf{q}}\int_{\omega}\frac{\mathbf{q}}{\epsilon - 2v_{\mathrm{F}}\mathbf{q}}\left(\frac{1}{\omega + v_{\mathrm{F}}\mathbf{q} + \mathrm{i}\,\mathrm{sgn}\omega} - \frac{1}{\epsilon + \omega - v_{\mathrm{F}}\mathbf{q} + \mathrm{i}\,\mathrm{sgn}(\mathbf{q})}\right)$$
$$= \frac{g_2^2}{2\pi^2}\int_{\mathbf{q}}\frac{\mathbf{q}}{\epsilon - 2v_{\mathrm{F}}\mathbf{q}}(\theta(\mathbf{q}) - \theta(-\mathbf{q})) = \frac{g_2^2}{4\pi^2 v_{\mathrm{F}}^2}\epsilon\left(\ln\frac{\epsilon}{2v_{\mathrm{F}}\mathbf{q}_0} - \mathrm{i}\pi\right).$$

By taking the derivative with respect to ϵ, we get

$$\frac{\partial\Sigma(\epsilon)}{\partial\epsilon} = -\frac{g_2^2}{8\pi^2 v_{\mathrm{F}}^2}\int_{-2v_{\mathrm{F}}\mathbf{q}_0}^{2v_{\mathrm{F}}\mathbf{q}_0}\mathrm{d}y\frac{y}{(y-\epsilon)^2}(\theta(y) - \theta(-y))$$
$$= -\frac{g_2^2}{4\pi^2 v_{\mathrm{F}}^2}\ln\frac{2v_{\mathrm{F}}\mathbf{q}_0}{|\epsilon|}.$$

Hence the contribution to the wave function renormalization Z reads

$$Z = (1 - \partial_\epsilon\Sigma(\epsilon))^{-1} \approx 1 + \partial_\epsilon\Sigma(\epsilon) = 1 - \frac{g_2^2}{4\pi^2 v_{\mathrm{F}}}\ln\frac{2v_{\mathrm{F}}\mathbf{q}_0}{|\epsilon|}. \tag{N.75}$$

By recalling that $\gamma_0 = \mathbf{q}_0^{-1}$, the above result coincides with (20.55) with η given in the weak-coupling limit provided by (20.49). We finally notice that the corresponding diagram with the g_4 coupling has the expression

$$\Sigma(p) = -2g_4^2\int_q\int_{p_1} G_+(p+q)G_+(p_1)G_+(p_1 - q).$$

This diagram, however, does not contribute since the integration over the bimomentum p_1 yields a bubble with poles on the same side of the real axis of G_{p+q} and the integration over the frequency ω vanishes.

20.5 There are two diagrams. The first is obtained from the insertion of the coupling g_2 in the "+" internal Green function line of Figure N.4. The second one is obtained similarly by interchanging the role of the two branches. The expression for the first diagram reads

$$\left(\delta\Gamma_{+-}^{a(4)}\right)^{(1)} = -2g_2^3 \int_q \int_{p_1} G_+^2(p+q)G_-(p_1)G_-(p_1-q)$$
$$= -2g_2^3 \int_q \int_{p_1} (-\partial_\epsilon G_+(p+q))\, G_-(p_1)G_-(p_1-q)$$
$$= -g_2 \partial_\epsilon \Sigma(p),$$

where we have used the identity $G_+^2(\mathbf{p}+\mathbf{q}, \epsilon+\omega) = -\partial_\epsilon G(\mathbf{p}+\mathbf{q}, \epsilon+\omega)$, valid for non-interacting Green functions. The second diagram gives an identical contribution so that the total contribution reads

$$\delta\Gamma_{+-}^{a(4)} = \frac{g_2^3}{2\pi^2 v_F} \ln \frac{2v_F^2 \mathbf{q}_0}{|\epsilon|}. \tag{N.76}$$

Chapter 21

21.1 After transforming the integral over momenta into an integral over the energy $\xi_\mathbf{p}$ (see Eq. (21.32)), we write the first correction to the imaginary self-energy as

$$N_0\overline{u^2}\mathrm{Im}\int_{-\infty}^{\infty} \mathrm{d}\xi_\mathbf{p} \frac{1}{\mathrm{i}\epsilon_n - \xi_\mathbf{p}} = -N_0\overline{u^2}\int_{-\infty}^{\infty} \mathrm{d}\xi_\mathbf{p} \frac{\epsilon_n}{\epsilon_n^2 + \xi_\mathbf{p}^2} = -\pi N_0 \overline{u^2}\mathrm{sign}(\epsilon_n),$$

where the last step follows from the change of variable $\xi_\mathbf{p} \to \xi_\mathbf{p}/|\epsilon_n|$. Notice that the final result depends only on the sign of the energy ϵ_n.

21.2 In the energy region $\epsilon_+ > 0$ and $\epsilon_- < 0$ we have

$$\int_\mathbf{p} \mathcal{G}(\mathbf{p}_+, \epsilon_+)\mathcal{G}(\mathbf{p}_-, \epsilon_-)$$
$$= \int \frac{\mathrm{d}\Omega_{\hat{\mathbf{p}}}}{\Omega_{\hat{\mathbf{p}}}} N_0 \int_{-\infty}^{\infty} \mathrm{d}\xi_\mathbf{p} \frac{1}{\mathrm{i}\epsilon_+ - \xi_\mathbf{p} - \mathbf{p}\cdot\mathbf{q}/2m + \mathrm{i}\hbar/2\tau} \frac{1}{\mathrm{i}\epsilon_- - \xi_\mathbf{p} + \mathbf{p}\cdot\mathbf{q}/2m - \mathrm{i}\hbar/2\tau}$$
$$= \int \frac{\mathrm{d}\Omega_{\hat{\mathbf{p}}}}{\Omega_{\hat{\mathbf{p}}}} N_0 \frac{2\pi\mathrm{i}}{\mathrm{i}(\epsilon_+ - \epsilon_-) - v_F\hat{\mathbf{p}}\cdot\mathbf{q} + \mathrm{i}\hbar/\tau},$$

where we have neglected terms of order q^2 in the Green functions compared to terms of order $p_F q$. The above result leads immediately to Eq. (21.41) by multiplying both numerator and denominator by $-\mathrm{i}\tau/\hbar$.

21.3 By using the residue theorem and the formula for the residue of poles we get the useful formula

$$(m,n) = (-1)^m \mathrm{i}^{n+m} 2\pi N_0 \frac{n(n+1)\cdots(n+m-2)}{(m-1)!}\left(\frac{\tau}{\hbar}\right)^{n+m-1}. \tag{N.77}$$

The most frequent cases are

$$(1,1) = 2\pi N_0 \frac{\tau}{\hbar}; \qquad (2,2) = 2\; 2\pi N_0 \left(\frac{\tau}{\hbar}\right)^3; \qquad (1,2) = \mathrm{i}\; 2\pi N_0 \left(\frac{\tau}{\hbar}\right)^2;$$
$$(2,1) = -\mathrm{i}\; 2\pi N_0 \left(\frac{\tau}{\hbar}\right)^2; \quad (1,3) = -\; 2\pi N_0 \left(\frac{\tau}{\hbar}\right)^3; \quad (3,1) = -\; 2\pi N_0 \left(\frac{\tau}{\hbar}\right)^3;$$
$$(1,4) = -\mathrm{i}\; 2\pi N_0 \left(\frac{\tau}{\hbar}\right)^4; \quad (4,1) = \mathrm{i}\; 2\pi N_0 \left(\frac{\tau}{\hbar}\right)^4; \quad (2,4) = -4\; 2\pi N_0 \left(\frac{\tau}{\hbar}\right)^5;$$
$$(4,2) = -4\; 2\pi N_0 \left(\frac{\tau}{\hbar}\right)^5; \quad (3,3) = 6\; 2\pi N_0 \left(\frac{\tau}{\hbar}\right)^5.$$

Integrals containing scalar products are evaluated as

$$\int_{\mathbf{p}} (\mathbf{p}\cdot\mathbf{q})^2 \mathcal{G}^m(\mathbf{p},\epsilon_+)\mathcal{G}^n(\mathbf{p},\epsilon_-) = \frac{p_{\mathrm{F}}^2 q^2}{d} \int_{\mathbf{p}} \mathcal{G}^m(\mathbf{p},\epsilon_+)\mathcal{G}^n(\mathbf{p},\epsilon_-).$$

21.4 The expression of the current–current response function (21.48) has a minus sign difference with respect to the corresponding expression for the density–density response due to the different signs in front of the perturbing Hamiltonians of Eqs. (10.23) and (10.24). We have included the factor $1/c$ in front of the current–current response function in the vector potential, since this factor only appears in the connection between the vector potential and the electric field.

In contrast to the case of the density–density response, the current vertex $\mathcal{T}^{\mathrm{i}} = j_0^{\mathrm{i}}$ remains unrenormalized. In fact, the expression corresponding to (21.40) contains a vector vertex j_0^{i} which makes the integral vanish upon angular integration. As a result there are no ladder diagrams contributing to the current–current response function. The only contribution at this order of approximation is a simple bubble diagram with the Green function (21.34). The only delicate point is the correct handling of the small frequency limit, we need to divide by ω according to Eq. (10.31). With a small, but finite $|\omega_\nu|$, we separate the sum over the frequencies into two parts. The first corresponds to $\epsilon_+\epsilon_- > 0$ and the second to $\epsilon_+\epsilon_- < 0$. The two parts can be handled as the static (21.45) and dynamic (21.46) contributions of the density–density response, respectively. After a few manipulations, by recalling the angle average $\langle \hat{\mathbf{p}}^i \hat{\mathbf{p}}^j \rangle = 1/d$ and the relation $n = 2N_0 E_{\mathrm{F}}/d$, we get

$$\begin{aligned}\tilde{\mathcal{R}}^{ij}(\mathbf{0},\omega_\nu) &= -2e^2 \int_{\mathbf{p}} j_0^i(\mathbf{p}) j_0^j(\mathbf{p}) \left[\frac{1}{2\pi \mathrm{i}}(G^{\mathrm{R}}(\mathbf{p},0) - G^{\mathrm{A}}(\mathbf{p},0)) \right.\\ &\qquad \left. + |\omega_\nu| \frac{1}{2\pi} G^{\mathrm{R}}(\mathbf{p},0)\, G^{\mathrm{A}}(\mathbf{p},0)\right] \\ &= \frac{e^2 n}{m} - |\omega_\nu| \frac{e^2 n \tau}{\hbar m}. \end{aligned} \tag{N.78}$$

The first term, by recalling Eq. (9.48), cancels exactly with the diamagnetic contribution. The remaining term, after analytical continuation $\omega_\nu \to -\mathrm{i}\hbar\omega$ by using Eq. (10.31), reproduces the Drude formula (11.33) for the electrical conductivity.

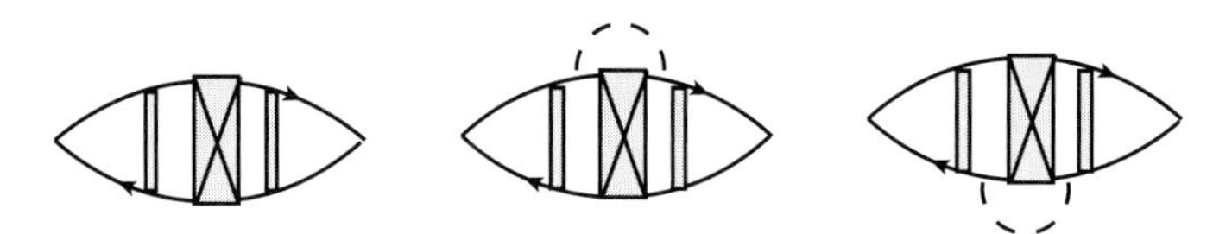

Fig. N.5 Diagram contributing to the next-to-leading correction to the density–density response function.

21.5 At first order in r, many diagrams have to be taken into account. Most of them cancel each other and we are left with those shown in Fig. N.5. The expression for the first diagram reads

$$\delta\tilde{\mathcal{R}}^{00}_{+-} = \frac{2\,|\omega_\nu|}{2\pi\hbar^4}\int_{\mathbf{p},\mathbf{p}'}(\delta\mathcal{T}^0(\mathbf{q},\omega_\nu))^2 G^{\mathrm{R}}(\mathbf{p}_+)\;G^{\mathrm{A}}(\mathbf{p}_-)$$
$$\times\, L_{\mathrm{c}}(\mathbf{p},\mathbf{p}',\omega_\nu)G^{\mathrm{R}}(\mathbf{p}'_+)G^{\mathrm{A}}(\mathbf{p}'_-). \qquad \text{(N.79)}$$

In Eq. (N.79) we have already performed the sum over the frequency ϵ_n, which gives rise to the factor ω_ν in front. In the Green functions, consistently with the approximations used up to now, we have set the frequency to zero and only the momentum argument is explicitly shown. The two vertex corrections $\delta\mathcal{T}^0$ at the extreme left and extreme right describe the direct ladders appearing in the diagram and are given by Eq. (21.44). To evaluate the integrals over the momenta $\mathbf{p}$ and $\mathbf{p}'$, we note that the cooperon ladder gives a big contribution to the integral when $\mathbf{p}+\mathbf{p}'$ is small. It is then convenient to introduce the variable $\mathbf{Q}=\mathbf{p}+\mathbf{p}'$ and set $\mathbf{Q}=0$ everywhere except in the cooperon ladder. We are then left with an integral over the cooperon ladder and a second over a product of four Green functions. However, this is not yet the full story. It turns out that there exist further diagrams (the second and the third) that may be obtained by simply *decorating* with an extra impurity line the first diagram of Fig. N.5. Such a decoration only adds two Green functions and an extra summation over a fast momentum. To proceed we have then to integrate the Green functions in the diagrams of Fig. N.5 according to the expression

$$I_{\mathrm{H}} = \frac{1}{\hbar^4}\int_{\mathbf{p}} G^{\mathrm{R}}(\mathbf{p}_+)G^{\mathrm{A}}(-\mathbf{p}_+)G^{\mathrm{R}}(-\mathbf{p}_-)G^{\mathrm{A}}(\mathbf{p}_-)$$
$$+\frac{\hbar}{2\pi N_0\tau}\frac{1}{\hbar^6}\int_{\mathbf{p}} G^{\mathrm{R}}(\mathbf{p}_+)G^{\mathrm{R}}(-\mathbf{p}_-)G^{\mathrm{A}}(\mathbf{p}_-)\int_{\mathbf{p}'} G^{\mathrm{R}}(\mathbf{p}'_+)G^{\mathrm{A}}(-\mathbf{p}'_+)G^{\mathrm{R}}(-\mathbf{p}'_-)$$
$$+\frac{\hbar}{2\pi N_0\tau}\frac{1}{\hbar^6}\int_{\mathbf{p}} G^{\mathrm{R}}(\mathbf{p}_+)G^{\mathrm{A}}(-\mathbf{p}_+)G^{\mathrm{A}}(\mathbf{p}_-)\int_{\mathbf{p}'} G^{\mathrm{A}}(-\mathbf{p}'_+)G^{\mathrm{A}}(\mathbf{p}'_-)G^{\mathrm{R}}(-\mathbf{p}'_-). \qquad \text{(N.80)}$$

Notice the factor $\frac{\hbar}{2\pi N_0\tau}$ for the extra impurity line. Since we are interested in the small momentum limit, we make an expansion in powers of $\mathbf{q}$. After a straightforward but lengthy calculation, one gets

$$I_{\mathrm{H}} = 4\pi\left(\frac{\tau}{\hbar}\right)^4 N_0 D\mathbf{q}^2. \qquad \text{(N.81)}$$

Finally, from Eq. (21.44) and the expression (21.51) for the crossed ladder, one gets the expression (21.54) reported in the text.

21.6 In the presence of a magnetic field, electron energies are changed by the Zeeman energy (see Eq. (7.29)), so that the Green function reads

$$\mathcal{G}_{\alpha\gamma}(\mathbf{p}, \epsilon_n) = \delta_{\alpha\gamma}\left[\mathrm{i}\epsilon_n - \xi_{\mathbf{p}} - \alpha\Delta/2 + \frac{\mathrm{i}\hbar}{2\tau}\mathrm{sign}(\epsilon_n)\right]^{-1}. \tag{N.82}$$

The spin projection takes values $\alpha = \pm 1$. One sees that the Zeeman energy $\alpha\Delta/2$ enters the Green function as an energy shift. This allows us immediately to get the ladder in the presence of magnetic Zeeman coupling by recalling Eq. (21.40). The energy argument of the ladder is the energy difference $\epsilon_+ - \epsilon_-$ and it is shifted by the difference of the Zeeman energies of the two Green functions (particle–hole) pair. Hence the ladder associated with the propagation of the two Green functions $\mathcal{G}_{\alpha\gamma}(\mathbf{p}_+, \epsilon_+)$ and $\mathcal{G}_{\beta\delta}(\mathbf{p}_-, \epsilon_-)$ acquires a spin dependence

$$L_{\alpha\beta\gamma\delta}(q, \omega_\nu) = \frac{\hbar^2}{2\pi N_0\tau^2}\frac{\delta_{\alpha\gamma}\delta_{\beta\delta}}{Dq^2/\hbar + |\omega_\nu| - \mathrm{i}(\alpha-\beta)\Delta/2}. \tag{N.83}$$

To understand the meaning of this expression, we notice that the ladder describes the propagation of the composite operator $\hat{\psi}_\alpha(\mathbf{x})\hat{\psi}_\delta^\dagger(\mathbf{x})$ which, being bilinear in the spinor fields, must behave under rotations as a scalar or a vector. Indeed, the two combinations $\hat{\psi}_1(\mathbf{x})\hat{\psi}_1^\dagger(\mathbf{x}) + \hat{\psi}_{-1}(\mathbf{x})\hat{\psi}_{-1}^\dagger(\mathbf{x})$ and $\hat{\psi}_1(\mathbf{x})\hat{\psi}_1^\dagger(\mathbf{x}) - \hat{\psi}_{-1}(\mathbf{x})\hat{\psi}_{-1}^\dagger(\mathbf{x})$ describe the propagation of the total number density and spin density component $M = 0$ along the z-axis, respectively. The remaining two possible combinations $\hat{\psi}_1(\mathbf{x})\hat{\psi}_{-1}^\dagger(\mathbf{x})$ and $\hat{\psi}_{-1}(\mathbf{x})\hat{\psi}_1^\dagger(\mathbf{x})$ correspond to the components with projection along the quantization axis $M = \pm 1$. As a result we can write the ladder in terms of the total spin j and its projection M of the particle–hole pair described by the composite operator $\hat{\psi}_\alpha(\mathbf{x})\hat{\psi}_\delta^\dagger(\mathbf{x})$

$$L_{jM}(q, \omega_\nu) = \frac{\hbar}{2\pi N_0\tau^2}\frac{1}{Dq^2/\hbar + |\omega_\nu| - \mathrm{i}M\Delta}. \tag{N.84}$$

21.7 By writing the density operator as

$$\hat{\psi}_\alpha^\dagger(\mathbf{x}, t)\hat{\psi}_\alpha(\mathbf{x}, t) = \sum_{\kappa_1,\kappa_2} \mathrm{e}^{\mathrm{i}(\epsilon_{\kappa_1}-\epsilon_{\kappa_2})t} g_{\alpha\kappa_1}^*(\mathbf{x}) g_{\alpha\kappa_2}(\mathbf{x}) a_{\alpha\kappa_1}^\dagger a_{\alpha\kappa_2}$$

the density–density response function becomes

$$R^{00}(\mathbf{x}, t; \mathbf{x}', t') = -\frac{\mathrm{i}}{\hbar}\theta(t-t')\sum_{\kappa_1,\kappa_2}\mathrm{e}^{-\mathrm{i}(\epsilon_{\kappa_2}-\epsilon_{\kappa_1})(t-t')} g_{\alpha\kappa_1}^*(\mathbf{x}) g_{\alpha\kappa_2}(\mathbf{x})$$
$$\times\, g_{\beta\kappa_2}^*(\mathbf{x}') g_{\beta\kappa_1}(\mathbf{x}')(f(\epsilon_{\kappa_1}) - f(\epsilon_{\kappa_2})),$$

where $f(x)$ is the Fermi function. After Fourier transforming with respect to $t - t'$ and taking the imaginary part, one obtains (21.76).

21.8 The diagram of Fig. 21.6, by recalling the expression for the vertex part (21.44), reads ($p = (\mathbf{p}, \epsilon_n)$ and $q = (\mathbf{q}, \omega_\nu)$)

$$\delta\mathcal{G}(p) = -\mathcal{G}^2(p)\frac{1}{\beta}\sum_\nu\int_{\mathbf{q}}\mathcal{G}(p-q)V(q)(\delta\mathcal{T}^0(p,q))^2, \;\; \delta\mathcal{T}^0(p,q) = \frac{\hbar/\tau}{|\omega_\nu| + Dq^2/\hbar}. \tag{N.85}$$

The density of states, after Eq. (18.43), is obtained by

$$\delta\nu(\epsilon) = -\frac{2}{\pi}\mathrm{Im}\left(\int_{\mathbf{p}} \delta\mathcal{G}(p)\right)|_{\mathrm{i}\epsilon_n=\hbar\epsilon+\mathrm{i}0^+}. \tag{N.86}$$

The integration over the momentum $\mathbf{p}$ can be done by setting $\mathbf{q} = 0$, $\omega_\nu = 0$ in the Green functions because of the diffusive pole. Then by recalling the results of Problem 21.3 we have

$$\int_{\mathbf{p}} \mathcal{G}^2(p)\mathcal{G}(p-q)|_{q=0} \equiv (2,1) = -\mathrm{i}2\pi N_0\left(\frac{\tau}{\hbar}\right)^2. \tag{N.87}$$

Hence, by integrating over momentum Eq. (N.85) and using (N.87), we get

$$\int_{\mathbf{p}} \delta\mathcal{G}(p) = 2\pi\mathrm{i}N_0\frac{1}{\beta}\sum_{\nu>n+1}\int_{\mathbf{q}} V\mathbf{q})\frac{1}{(|\omega_\nu| + Dq^2/\hbar)^2}. \tag{N.88}$$

Since the diffusive pole makes the small transferred momentum region dominant, we set $\mathbf{q} = 0$ in the interaction $V(\mathbf{q})$ and, after performing the $\mathbf{q}$-integral in two dimensions (with the change of variable $s = Dq^2/\hbar$), we arrive at

$$\int_{\mathbf{p}} \delta\mathcal{G}(p) = 2\pi\mathrm{i}N_0 V(\mathbf{0})\frac{\hbar}{4\pi D}\frac{1}{\beta}\sum_{\nu>n+1}\frac{1}{|\omega_\nu|} = \mathrm{i}\frac{N_0 V(\mathbf{0})}{4\pi D}\ln\left(\frac{\hbar}{\epsilon_n\tau}\right), \tag{N.89}$$

where the sum over the Matsubara frequency is logarithmically singular and can be approximated by an integral according to the replacement (18.21). The lower limit is provided by the condition $\omega_\nu > \epsilon_n$ due to the Green functions having Matsubara frequencies of opposite sign. The upper limit is determined by the diffusive approximation $|\omega_\nu\tau| \ll \hbar$ according to the expansion in Eq. (21.42). Finally, the insertion of the result (N.89) into (N.86) leads to the expression (21.78) (see the term with V_1) after taking into account also the Hartree diagram (see Fig. 21.6 (d)), which selects the events with large scattering angle associated with V_2.

21.9 We may follow the derivation of Section 18.10 with the spin vertices (21.101) replacing the density vertices of Eq. (18.105). For instance, in evaluating the commutators as in Eq. (18.116) with the spin vertex replacing the density vertex, one obtains

$$[s^a(x), \hat{\psi}_\alpha(x')] = -\frac{g\mu_{\mathrm{B}}}{2}\sigma^a_{\alpha\gamma}\delta(\mathbf{x}-\mathbf{x}')\hat{\psi}_\gamma(\mathbf{x},\tau).$$

In addition the presence of the magnetic field introduces a source term in the continuity equation for the spin density since

$$\partial_\tau s^a(\mathbf{x},\tau) = -\mathrm{i}\epsilon_{ab3}\Delta s^b(\mathbf{x},\tau).$$

This explains the last term in Eq. (21.104).

21.10 In the limit $\mathbf{q} = \mathbf{0}$, the explicit Ward identities for $a = 1, 2, 3$ read

$$\begin{aligned}
\mathrm{i}\omega_\nu\Lambda^1_{\alpha\beta} - \mathrm{i}\Delta\Lambda^2_{\alpha\beta} &= \mathcal{G}_\alpha(\epsilon_n - \omega_\nu/2)\sigma^1_{\alpha\beta} - \sigma^1_{\alpha\beta}G_\beta(\epsilon_n + \omega_\nu/2),\\
\mathrm{i}\omega_\nu\Lambda^2_{\alpha\beta} + \mathrm{i}\Delta\Lambda^1_{\alpha\beta} &= \mathcal{G}_\alpha(\epsilon_n - \omega_\nu/2)\sigma^2_{\alpha\beta} - \sigma^2_{\alpha\beta}G_\beta(\epsilon_n + \omega_\nu/2),\\
\mathrm{i}\omega_\nu\Lambda^3_{\alpha\beta} &= \mathcal{G}_\alpha(\epsilon_n - \omega_\nu/2)\sigma^3_{\alpha\beta} - \sigma^3_{\alpha\beta}G_\beta(\epsilon_n + \omega_\nu/2).
\end{aligned}$$

By summing the first equation with $\mp$i times the second one, we get

$$(\mathrm{i}\omega_\nu - \Delta)(\Lambda^1_{\alpha\beta} + \mathrm{i}\Lambda^2_{\alpha\beta}) = \mathcal{G}_\alpha(\epsilon_n - \omega_\nu/2)2\sigma^+_{\alpha\beta} - 2\sigma^+_{\alpha\beta}G_\beta(\epsilon_n + \omega_\nu/2),$$
$$(\mathrm{i}\omega_\nu + \Delta)(\Lambda^1_{\alpha\beta} - \mathrm{i}\Lambda^2_{\alpha\beta}) = \mathcal{G}_\alpha(\epsilon_n - \omega_\nu/2)2\sigma^-_{\alpha\beta} - 2\sigma^-_{\alpha\beta}G_\beta(\epsilon_n + \omega_\nu/2),$$
$$\mathrm{i}\omega_\nu\Lambda^3_{\alpha\beta} = \mathcal{G}_\alpha(\epsilon_n - \omega_\nu/2)\sigma^3_{\alpha\beta} - \sigma^3_{\alpha\beta}G_\beta(\epsilon_n + \omega_\nu/2).$$

The first two equations correspond to Eq. (21.107) for $M = \mp 1$, respectively.

21.11 The spin density–spin density response function $\tilde{\chi}^{ab}$ is related to the spin vertex $\Lambda^{a,0}$ by a relation similar to Eq. (18.115) relating the density–density response function and the density vertex. Specifically, we have

$$\tilde{\chi}^{ab}(q) = \frac{g\mu_\mathrm{B}}{2}\int_p \sum_{\alpha\beta} \Lambda^{a,0}_{\alpha\beta}(p,q)\sigma^b_{\beta\alpha}.$$

In considering $\tilde{\chi}^{M=-1}$, we must use the vertex $\Lambda^{\uparrow\downarrow}$ and the matrix $2\sigma^-$, obtaining

$$(\mathrm{i}\omega_\nu - \Delta)\tilde{\chi}^{M=-1}(q) = \frac{g\mu_\mathrm{B}}{2}\int_p \sum_{\alpha\beta} \Lambda^{\uparrow\downarrow}_{\alpha\beta}(p,q)2\sigma^-_{\beta\alpha}.$$

By using the Ward identity (21.107) and recalling that $\sigma^\pm\sigma^\mp = (\sigma^0 \pm \sigma^3)/2$, we get

$$\begin{aligned}(\mathrm{i}\omega_\nu - \Delta)\tilde{\chi}^{M=-1}(q) &= 2\left(\frac{g\mu_\mathrm{B}}{2}\right)^2 \int_p (\mathcal{G}_\uparrow(\epsilon_n - \omega_\nu/2) - \mathcal{G}_\downarrow(\epsilon_n + \omega_\nu/2)) \\ &= -2\left(\frac{g\mu_\mathrm{B}}{2}\right)^2 (n_\uparrow - n_\downarrow) \\ &= -g\mu_\mathrm{B}\chi^\mathrm{L} B \\ &= -\chi^\mathrm{L}\Delta,\end{aligned}$$

which gives Eq. (21.108) for $M = -1$. The corresponding equation for $M = 1$ is obtained by considering $\Lambda^{\downarrow\uparrow}$ combined with $2\sigma^+$.

21.12 To evaluate Eq. (21.115), we begin by considering the expression

$$\frac{1}{\beta}\sum_\nu V \int_\mathbf{q} \frac{|\omega_\nu|}{Z^\lambda_\mathrm{s}|\omega_\nu| + Dq^2/\hbar},$$

where $Z^\lambda_\mathrm{s} = Z - \lambda\nu^\mathrm{L}\zeta^2\Gamma_\mathrm{s}$. Due to the diffusion approximation the sum over the frequency is restricted $|\omega_\nu| < \hbar/\tau$. We relax this constraint by introducing a cutoff function

$$\left(\frac{\hbar\tau^{-1}}{\hbar\tau^{-1} + |\omega_\nu|}\right)^2$$

and, by analytic continuation $\omega_\nu \to -\mathrm{i}\hbar\omega$ transform the sum into an integral over the complex plane according to the technique of Appendix H. One then gets

$$\int_{-\infty}^{\infty} \frac{\mathrm{d}\omega}{\pi} b(\hbar\beta\omega) V \int_\mathbf{q} \mathrm{Im}\left[\left(\frac{\tau^{-1}}{\tau^{-1} - \mathrm{i}\omega}\right)^2 \frac{-\mathrm{i}\omega}{Dq^2/\hbar^2 - \mathrm{i}Z^\lambda_\mathrm{s}\omega}\right].$$

After performing the integral over the momentum, one obtains

$$N_0 r \int_{-\tau^{-1}}^{\tau^{-1}} \mathrm{d}\omega\, b(\beta\hbar\omega)\hbar^2\omega \ln\left(\tau Z_{\mathrm{s}}^{\lambda}\omega\right),$$

where r is the disorder expansion parameter and the cutoff function has been dropped by limiting the integral. By using $b(-x) = -(1 + b(x))$, reducing the integral over the positive frequency range and making the change of variable $y = \beta\hbar\omega$, we get

$$\begin{aligned}
&\approx 2(k_{\mathrm{B}}T)^2 r N_0 \int_0^{\tau^{-1}} \mathrm{d}y y b(y) \left[\ln(\tau k_{\mathrm{B}}T/\hbar) + \ln Z_{\mathrm{s}}^{\lambda}\right] \\
&\approx 2(k_{\mathrm{B}}T)^2 r N_0 I_1^-(0) \ln(\tau k_{\mathrm{B}}T/\hbar) \\
&= \frac{\pi^2}{3}(k_{\mathrm{B}}T)^2 r N_0 \ln(\tau k_{\mathrm{B}}T/\hbar),
\end{aligned}$$

where we dropped the term containing $\ln Z_{\mathrm{s}}^{\lambda}$ which is less singular in T. In the above $I_1^-(0) = \pi^2/6$ is the Bose integral of Appendix E. All the dependence on the parameter λ has disappeared from the most singular term and the integration over λ becomes trivial. The use of the above result in the two terms of (21.115), leads immediately to the expression for the specific heat in Eq. (21.117) after differentiating twice with respect to the temperature.

In order to evaluate the expression (21.116), we notice that performing the momentum integration and combining the two terms leads to

$$\begin{aligned}
&\frac{1}{\beta}\sum_{\nu} V \int_{\mathbf{q}} \sum_{M=\mp1} \left[\frac{|\omega_\nu|}{Z_{\mathrm{t}}^{\lambda}|\omega_\nu| + Dq^2/\hbar - \mathrm{i}MZ_{\mathrm{t}}\Delta} - \frac{|\omega_\nu|}{Z_{\mathrm{t}}^{\lambda}|\omega_\nu| + Dq^2/\hbar}\right] \\
&= V\pi N_0 r \frac{1}{\beta}\sum_{\nu} |\omega_\nu| \left[\ln\left(\frac{\left(Z_{\mathrm{t}}^{\lambda}|\omega_\nu|\right)^2}{\left(Z_{\mathrm{t}}^{\lambda}|\omega_\nu|\right)^2 + (Z_{\mathrm{t}}\Delta)^2}\right)\right] \\
&\approx -V\pi N_0 r \frac{1}{\beta}\sum_{\nu} \frac{(Z_{\mathrm{t}}\Delta)^2}{\left(Z_{\mathrm{t}}^{\lambda}\right)^2} \frac{1}{|\omega_\nu|} \\
&= VN_0 r \frac{(Z_{\mathrm{t}}\Delta)^2}{\left(Z_{\mathrm{t}}^{\lambda}\right)^2} \ln(\tau k_{\mathrm{B}}T/\hbar).
\end{aligned} \tag{N.90}$$

Integration over λ gives

$$\int_0^1 \mathrm{d}\lambda \frac{1}{\left(Z_{\mathrm{t}}^{\lambda}\right)^2} = \frac{1}{ZZ_{\mathrm{t}}}.$$

As a result

$$\Delta\Omega_{\mathrm{B}} = -Vr\chi^{\mathrm{L}}B^2\nu^{\mathrm{L}}\zeta^2\Gamma_{\mathrm{t}}\frac{Z_{\mathrm{t}}}{Z}\ln(\tau k_{\mathrm{B}}T/\hbar),$$

from which one obtains the spin susceptibility correction of Eq. (21.118).

References

Abanov, Ar., Chubukov, A. V. and Schmalian, J. 2003. Quantum-critical theory of the spin-fermion model and its application to cuprates: normal state analysis. *Adv. Phys.*, **52**, 119.

Abbamonte, P., Rusydi, A., Smadici, S., Gu, G. D., Sawatzky, G. A. and Feng, D. L. 2005. Spatially modulated 'Mottness' in $La_{2-x}Ba_xCuO_4$. *Nature Physics*, **1**, 155.

Abel, W. R., Anderson, A. C. and Wheatley, J. C. 1966. Propagation of zero sound in liquid He^3 at low temperatures. *Phys. Rev. Lett.*, **17**, 74.

Abrahams, E., Anderson, P. W., Licciardello, D. C. and Ramakrishnan, T. V. 1979. Scaling theory of localization: absence of quantum diffusion in two dimensions. *Phys. Rev. Lett.*, **42**, 673.

Abramowitz, M. and Stegun, I. A. 1965. *Handbook of Mathematical Functions with Formulas, Graphs, and Mathematical Tables*. New York: Dover.

Abrikosov, A. A. 1957. On the magnetic properties of superconductors of the second group. *Sov. Phys. JETP*, **5**, 1174.

Abrikosov, A. A., Gorkov, L. P. and Dzyaloshinski, I. E. 1963. *Methods of Quantum Field Theory in Statistical Physics*. Englewood Cliffs, New Jersey: Prentice-Hall, Inc.

Ahlers, G. 1968. Thermal conductivity of He I near the superfluid transition. *Phys. Rev. Lett.*, **21**, 1159.

Akhmetov, D. G. 2009. *Vortex Rings*. Berlin & Heidelberg: Springer.

Allen, J. F. and Misener, A. D. 1938. Flow of liquid helium II. *Nature*, **141**, 75.

Altshuler, B. L. and Aronov, A. G. 1970. Contribution to the theory of disordered metals in strongly doped semiconductors. *Zh. Eksp. Teor. Fiz*, **77**, 2028 (*JETP*, **50**, 968).

Altshuler, B. L. and Aronov, A. G. 1985. Electron–electron interaction in disordered conductors. In Pollak, M. and Efros, A. L. (eds.), *Electron–Electron Interactions in Disordered Systems*. Amsterdam: North-Holland.

Altshuler, B. L., Aronov, A. G. and Lee, P. A. 1980a. Interaction effects in disordered Fermi systems in two dimensions. *Phys. Rev. Lett.*, **44**, 1288.

Altshuler, B. L., Khmel'nitzkii, D., Larkin, A. I. and Lee, P. A. 1980b. Magnetoresistance and Hall effect in a disordered two-dimensional electron gas. *Phys. Rev. B*, **22**, 5142.

Alvesalo, T. A., Anufriyev, Yu. D., Collan, H. K., Lounasmaa, O. V. and Wennerström, P. 1973. Evidence for superfluidity in the newly found phases of ^{3}He. *Phys. Rev. Lett.*, **30**, 962.

Amit, D. J. and Martín-Mayor, V. 2005. *Field Theory, the Renormalization Group, and Critical Phenomena*. Singapore: World Scientific.

Andergassen, S., Caprara, S., Di Castro, C. and Grilli, M. 2001. Anomalous isotopic effect near the charge-ordering quantum criticality. *Phys. Rev. Lett.*, **87**, 056401.

Anderson, M. H., Ensher, J. R., Matthews, M. R., Wieman, C. E. and Cornell, E. A. 1995. Observation of Bose–Einstein condensation in dilute atomic vapor. *Science*, **269**, 198.

Anderson, P. W. 1958. Absence of diffusion in certain random lattices. *Phys. Rev.*, **109**, 1492.

Anderson, P. W. 1966. Considerations on the flow of superfluid helium. *Rev. Mod. Phys.*, **38**, 298.

Anderson, P. W. 2007. Is there glue in cuprate superconductors? *Science*, **316**, 1705.

Anderson, P. W. and Brinkman, W. F. 1973. Anisotropic superfluidity in ^{3}He: a possible interpretation of its stability as a spin-fluctuation effect. *Phys. Rev. Lett.*, **30**, 1108.

Anderson, P. W. and Morel, P. 1961. Generalized Bardeen–Cooper–Schrieffer states and the proposed low-temperature phase of liquid He^3. *Phys. Rev.*, **123**, 1911.

Ando, Y., Komiya, S., Segawa, K., Ono, S. and Kurita, Y. 2004. Electronic phase diagram of high-T_c cuprate superconductors from a mapping of the in-plane resistivity curvature. *Phys. Rev. Lett.*, **93**, 267001.

Andronikashvili, E. L. 1946. A direct observation of two kinds of motion in helium II. *J. Phys. (USSR)*, **10**, 201.

Aslamazov, L. G. and Larkin, A. I. 1968. The influence of fluctuation pairing of electrons on the conductivity of normal metal. *Phys. Lett. A*, **26**, 238.

Atkins, K. R. 1959. *Liquid Helium*. Cambridge: Cambridge University Press.

Balian, R. and Werthamer, N. R. 1963. Superconductivity with pairs in a relative *p* Wave. *Phys. Rev.*, **131**, 1553.

Bardeen, J., Cooper, L. N. and Schrieffer, J. R. 1957. Theory of superconductivity. *Phys. Rev.*, **108**, 1175.

Baym, G. and Pethick, C. 1978. Low temperature properties of dilute solutions of 3He in superfluid 4He. Page 123 of: Bennemann, K. H. and Ketterson, J. B. (eds.), *The Physics of Liquid and Solid Helium*. New York: Wiley.

Baym, G. and Pethick, C. 1991. *Landau Fermi-liquid Theory: Concepts and Applications*. New York: Wiley.

Bednorz, J. G., Müller, K. A. and Takashige, M. 1987. Superconductivity in alkaline earth-substituted La_2CuO_{4-y}. *Science*, **236**, 73.

Belitz, D. and Kirkpatrick, T. R. 1994. The Anderson–Mott transition. *Rev. Mod. Phys.*, **66**, 261.

Benettin, G., Di Castro, C., Jona-Lasinio, G., Peliti, L. and Stella, A. 1977. On the equivalence of different renormalization groups. In Lévy, M. and Mitter, P. (eds.), *New Developments in Quantum Theory and Statistical Mechanics*. New York: Plenum Press.

Bergmann, G. 1984. Weak localization in thin films: a time-of-flight experiment with conduction electrons. *Physics Reports*, **107**, 1.

Bernoulli, D. and Bernoulli, J. 2005. *Hydrodynamics and Hydraulics*. Mineola (N.Y.): Dover.

Biondi, M. A., Garfunkel, M. P. and Mc Coubrey, A. O. 1957a. Microwave measurements of the energy gap in superconducting aluminum. *Phys. Rev.*, **108**, 495.

Biondi, M. A., Forrester, A. T. and Garfunkel, M. P. 1957b. Millimeter wave studies of superconducting tin. *Phys. Rev.*, **108**, 497.

Bloch, I., Dalibard, J. and Zwerger, W. 2008. Many-body physics with ultracold gases. *Rev. Mod. Phys.*, **80**, 885.

Blume, M., Emery, V. J. and Griffiths, R. B. 1971. Ising model for the λ transition and phase separation in ^{3}He–^{4}He mixtures. *Phys. Rev. A*, **4**, 1071.

Bogoliubov, N. N. 1947. On the theory of superfluidity. *J. Phys.*, **11**, 23.

Bogoliubov, N. N. and Shirkov, D. V. 1959. *Introduction to the Theory of Quantized Fields*. New York: Wiley-Interscience.

Boltzmann, L. 1964. *Lectures on Gas Theory*. New York: Dover.

Bonch-Bruevich, V. L. and Tyablikov, S. V. 1962. *The Green Function Method in Statistical Physics*. Amsterdam: North-Holland Publishing Company.

Bose, S. N. 1924. Plancks Gesetz und Lichtquantenhypothese. *Z. Phys.*, **26**, 178.

Bradley, C. C., Sackett, C. A., Tollett, J. J. and Hulet, R. G. 1995. Evidence of Bose–Einstein condensation in an atomic gas with attractive interactions. *Phys. Rev. Lett.*, **75**, 1687.

Brezin, E., Le Guillou, J. C. and Zinn-Justin, J. 1973. Approach to scaling in renormalized perturbation theory. *Phys. Rev. D*, **8**, 2418.

Brooks, J. S. and Donnelly, R. J. 1977. The calculated thermodynamic properties of superfluid helium-4. *J. Phys. Chem. Ref. Data*, **6**, 51.

Brueckner, K. A., Soda, T., Anderson, P. W. and Morel, P. 1960. Level structure of nuclear matter and liquid He^3. *Phys. Rev.*, **118**, 1442.

Byers, N. and Yang, C. N. 1961. Theoretical considerations concerning quantized magnetic flux in superconducting cylinders. *Phys. Rev. Lett.*, **7**, 46.

Callen, H. B. and Welton, T. A. 1951. Irreversibility and generalized noise. *Phys. Rev.*, **83**, 34.

Campisi, M., Hänggi, P. and Talkner, P. 2011. Quantum fluctuation relations: foundations and applications. *Rev. Mod. Phys.*, **83**, 771.

Castellani, C. and Di Castro, C. 1986. Effective Landau theory for disordered interacting electron systems: specific-heat behavior. *Phys. Rev. B*, **34**, 5935.

Cassandro, M. and Gallavotti, G. 1975. The Lavoisier law and the critical point. *Nuovo Cimento B*, **25**, 691.

Cassandro, M. and Jona-Lasinio, G. 1978. Critical point behaviour and probability theory. *Advances in Physics*, **27**, 913.

Castellani, C., Di Castro, C., Forgacs, G. and Tabet, E. 1983. Gauge invariance and the multiplicative renormalisation group in the Anderson transition. *J. Phys. C Solid State Physics*, **16**, 159.

Castellani, C., Di Castro, C., Lee, P. A. and Ma, M. 1984a. Interaction-driven metal–insulator transitions in disordered fermion systems. *Phys. Rev. B*, **30**, 527.

Castellani, C., Di Castro, C., Lee, P. A., Ma, M., Sorella, S. and Tabet, E. 1984b. Spin fluctuations in disordered interacting electrons. *Phys. Rev. B*, **30**, 1596.

Castellani, C., Di Castro, C., Lee, P. A., Ma, M., Sorella, S. and Tabet, E. 1986. Enhancement of the spin susceptibility in disordered interacting electrons and the metal–insulator transition. *Phys. Rev. B*, **33**, 6169.

Castellani, C., Di Castro, C. and Grilli, M. 1995. Singular quasiparticle scattering in the proximity of charge instabilities. *Phys. Rev. Lett.*, **75**, 4650.

Castellani, C., Di Castro, C., Pistolesi, F. and Strinati, G. C. 1997. Infrared behavior of interacting bosons at zero temperature. *Phys. Rev. Lett.*, **78**, 1612.

Castellani, C., Di Castro, C. and Lee, P. A. 1998. Metallic phase and metal–insulator transition in two-dimensional electronic systems. *Phys. Rev. B*, **57**, 9381.

Chaikin, P. M. and Lubensky, T. C. 1995. *Principles of Condensed Matter Physics*. Cambridge: Cambridge University Press.

Chang, J., Blackburn, E., Holmes, A. T., et al. 2012. Direct observation of competition between superconductivity and charge density wave order in $YBa_2Cu_3O_{6.67}$. *Nature Physics*, **8**, 871.

Chapman, S. and Cowling, T. G. 1970. *The Mathematical Theory of Non-uniform Gases*. Cambridge: Cambridge University Press.

Chin, C., Grimm, R., Julienne, P. and Tiesinga, E. 2010. Feshbach resonances in ultracold gases. *Rev. Mod. Phys.*, **82**, 1225.

Chu, S. 1998. Nobel Lecture: The manipulation of neutral particles. *Rev. Mod. Phys.*, **70**, 685.

Clausius, R. 1857. Über die Art der Bewegung, welche wir Wärme nennen. *Ann. Phys.*, **176**, 353.

Clausius, R. 1858. Über die mittlere Länge der Wege, welche bei Molecularbewegung gasförmigen Körper von den einzelnen Molecülen zurücklegen werden, nebst anderen Bemerkungen über der mechanischen Wärmetheorie. *Ann. Phys.*, **181**, 239.

Cohen, D. and Imry, Y. 2012. Straightforward quantum-mechanical derivation of the Crooks fluctuation theorem and the Jarzynski equality. *Phys. Rev. E*, **86**, 011111.

Cohen-Tannoudji, C. N. 1998. Nobel Lecture: Manipulating atoms with photons. *Rev. Mod. Phys.*, **70**, 707.

Comin, R., Frano, A., Yee, M. M., et al. 2014. Charge order driven by Fermi-arc instability in $Bi_2Sr_{2?x}La_xCuO_{6+\delta}$. *Science*, **343**, 390.

Corak, W. S., Goodman, B. B., Satterthwaite, C. B. and Wexler, A. 1956. Atomic heats of normal and superconducting vanadium. *Phys. Rev.*, **102**, 656.

Crooks, G. E. 1999. Entropy production fluctuation theorem and the nonequilibrium work relation for free energy differences. *Phys. Rev. E*, **60**, 2721.

da Silva Neto, E. H., Aynajian, P., Frano, A., et al. 2014. Ubiquitous interplay between charge ordering and high-temperature superconductivity in cuprates. *Science*, **343**, 393.

Dalfovo, F., Giorgini, S., Pitaevskii, L. P. and Stringari, S. 1999. Theory of Bose–Einstein condensation in trapped gases. *Rev. Mod. Phys.*, **71**, 463.

Dash, J. G. and Taylor, R. D. 1957. Hydrodynamics of oscillating disks in viscous fluids: density and viscosity of normal fluid in pure He^4 from 1.2 K to the lambda point. *Phys. Rev.*, **105**, 7.

Davis, K. B., Mewes, M. O., Andrews, M. R., et al. 1995. Bose–Einstein condensation in a gas of sodium atoms. *Phys. Rev. Lett.*, **75**, 3969.

De Gennes, P. G. 1966. *Superconductivity of Metals and Alloys*. New York: W. A. Benjamin.

de Groot, S. R., Hooman, G. J. and Seldam, C. A. T. 1950. On the Bose–Einstein condensation. *Proc. R. Soc. Lond. A*, **203**, 266.

Deaver, B. S. and Fairbank, W. M. 1961. Experimental evidence for quantized flux in superconducting cylinders. *Phys. Rev. Lett.*, **7**, 43.

Debye, P. 1912. Zur Theorie der spezifischen Wärmen. *Ann. Phys.*, **344**, 789.

DeMarco, B. and Jin, D. S. 1999. Onset of Fermi degeneracy in a trapped atomic gas. *Science*, **285**, 1703.

DeMarco, B., Papp, S. B. and Jin, D. S. 2001. Pauli blocking of collisions in a quantum degenerate atomic Fermi gas. *Phys. Rev. Lett.*, **86**, 5409.

Di Castro, C. 1972. The multiplicative renormalization group and the critical behavior in $d = 4 - \epsilon$ dimensions. *Lett. Nuovo Cimento*, **5**, 69.

Di Castro, C. and Jona-Lasinio, G. 1969. On the microscopic foundation of scaling laws. *Phys. Lett. A*, **29**, 322.

Di Castro, C. and Metzner, W. 1991. Ward identities and the β function in the Luttinger liquid. *Phys. Rev. Lett.*, **67**, 3852.

Di Castro, C. and Raimondi, R. 2004. Disordered electron systems. In Giuliani, G. F. and Vignale, G. (eds.), *The Electron Liquid Paradigm in Condensed Matter Physics*. Amsterdam: IOS Press.

Dingle, R. B. 1973. *Asymptotic Expansions: Their Derivation and Interpretation*. London and New York: Academic Press.

Dirac, P. A. M. 1926. On the theory of quantum mechanics. *Proc. R. Soc. Lond. A*, **112**, 661.

Dobrosavljević, V., Trivedi, N. and Valles, J. M. Jr. (eds.). 2012. *Conductor–Insulator Quantum Phase Transitions*. Oxford: Oxford University Press.

Dolan, G. J. and Osheroff, D. D. 1979. Nonmetallic conduction in thin metal films at low temperatures. *Phys. Rev. Lett.*, **43**, 721.

Domb, C. and Green, M. S. (eds.). 1976. *Phase Transitions and Critical Phenomena: Volume 6*. San Diego: Academic Press Inc.

Drude, P. 1900. Zur Elektronentheorie der Metalle. *Ann. Phys.*, **306**, 566.

Dzyaloshinskii, I. E. and Larkin, A. I. 1973. Correlation functions for a one-dimensional Fermi system with long-range interaction (Tomonaga model). *Zh. Eksp. Teor. Fiz.*, **65**, 411 (*Sov. Phys. JETP*, **38**, 202 (1974)).

Edwards, J. T. and Thouless, D. J. 1972. Numerical studies of localization in disordered systems. *J. Phys. C*, **5**, 807.

Efetov, K. B., Larkin, A. I. and Khmel'nitskii, D. E. 1980. Interaction of diffusion modes in the theory of localization. *Zh. Eksp. Teor. Fiz.*, **79**, 1120 (*JETP*, **52**, 568).

Ehrenfest, P. 1933. Phasenumwandlungen im ueblichen und erweiterten Sinn, classifiziert nach den entsprechenden Singularitaeten des thermodynamischen Potentiales. *Proceedings Koninklijke Akademie van Wetenschappen*, **36**, 153.

Einstein, A. 1905. Über die von der molekularkinetischen Theorie der Wärme geforderte Bewegung von in ruhenden Flüssigkeiten suspendierten Teilchen. *Ann. Phys.*, **322**, 549.

Einstein, A. 1907. Berichtigung zu meiner Arbeit: die Plancksche Theorie der Strahlung. *Ann. Phys.*, **327**, 800.

Einstein, A. 1910. Theorie der Opaleszenz von homogenen Flüssigkeiten und Flüssigkeitsgemischen in der Nähe des kritischen Zustandes. *Ann. Phys.*, **338**, 1275.

Einstein, A. 1924. Quantentheorie des einatomigen idealen Gases / Quantum theory of ideal monoatomic gases. *Sitz. Ber. Preuss. Akad. Wiss.*, **22**, 261.

Einstein, A. 1925. Quantentheorie des einatomigen idealen Gases, 2. *Sitz. Preuss. Akad. Wiss.*, **1**, 3.

Emery, V. J. and Kivelson, S. A. 1993. Frustrated electronic phase separation and high-temperature superconductors. *Physica C*, **209**, 597.

Emery, V. J. and Sessler, A. M. 1960. Possible phase transition in Liquid He^3. *Phys. Rev.*, **119**, 43.

Esposito, M., Harbola, U. and Mukamel, S. 2009. Nonequilibrium fluctuations, fluctuation theorems and counting statistics in quantum systems. *Rev. Mod. Phys.*, **81**, 1665.

Fauqué, B., Sidis, Y., Hinkov, V., et al. 2006. Magnetic order in the pseudogap phase of high-T_C superconductors. *Phys. Rev. Lett.*, **96**, 197001.

Fermi, E. 1926a. Sulla quantizzazione del gas perfetto monoatomico. *Rendiconti della R. Accademia Nazionale dei Lincei*, **3**, 145.

Fermi, E. 1926b. Zur Quantelung des idealen einatomigen Gases. *Z. Phys.*, **36**, 902.

Ferrell, R. A., Menyhàrd, N., Schmidt, H., Schwabl, F. and Szépfalusy, P. 1968. Fluctuations and lambda phase transition in liquid helium. *Ann. Phys.*, **47**, 565.

Ferrenberg, A. M. and Landau, D. P. 1991. Critical behavior of the three-dimensional Ising model: a high-resolution Monte Carlo study. *Phys. Rev. B*, **44**, 5081.

Ferrier-Barbut, I., Delehaye, M., Laurent, S., et al. 2014. A mixture of Bose and Fermi superfluids. *Science*, **345**, 1035.

Fetter, A. L. and Walecka, J. D. 1971. *Quantum Theory of Many-particle Systems*. New York: McGraw-Hill.

Feynman, R. P. 1953a. Atomic theory of liquid helium near absolute zero. *Phys. Rev.*, **91**, 1301.

Feynman, R. P. 1953b. Atomic theory of the λ transition in helium. *Phys. Rev.*, **91**, 1291.

Feynman, R. P. 1954. Atomic theory of the two-fluid model of liquid helium. *Phys. Rev.*, **94**, 262.

Feynman, R. P. 1972. *Statistical Mechanics*. Reading, Mass.: W. A. Benjamin.

Feynman, R. P. and Cohen, M. 1956. Energy spectrum of the excitations in liquid helium. *Phys. Rev.*, **102**, 1189.

Finkelstein, A. M. 1983. Influence of Coulomb interaction on the properties of disordered metals. *Zh. Eksp. Teor. Fiz.*, **84**, 168 (*Sov. Phys. JETP*, **57**, 97).

Finkelstein, A. M. 1984. Weak localization and Coulomb interaction in disordered systems. *Z. Phys. B*, **56**, 189.

Finkelstein, A. M. 1990. Electron liquid in disordered conductors. *Sov. Sci. Rev.*, **14**, 1.

Fisher, M. E. 1967. The theory of equilibrium critical phenomena. *Rep. Prog. Phys*, **30**, 615.

Fokker, A. D. 1914. Die mittlere Energie rotierender elektrischer Dipole im Strahlungsfeld. *Ann. Phys.*, **348**, 810.

Gao, L., Xue, Y. Y., Chen, F., et al. 1994. Superconductivity up to 164 K in $HgBa_2Ca_{m-1}Cu_mO_{2m+2+\delta}$ ($m = 1$, 2, and 3) under quasihydrostatic pressures. *Phys. Rev. B*, **50**, 4260.

Gavoret, J. and Nozières, P. 1964. Structure of the perturbation expansion for the Bose liquid at zero temperature. *Ann. Phys.*, **28**(3), 349.

Gell-Mann, M. and Low, F. E. 1954. Quantum electrodynamics at small distances. *Phys. Rev.*, **95**, 1300.

Ghiringhelli, G., Le Tacon, M., Minola, M., et al. 2012. Long-range incommensurate charge fluctuations in (Y,Nd)Ba_2Cu3O_{6+x}. *Science*, **337**, 821.

Giamarchi, T. 2004. *Quantum Physics in One Dimension*. New York: Oxford University Press.

Gibbs, J. W. 1902. *Elementary Principles in Statistical Mechanics*. New York: Charles Scribner's Sons.

Giorgini, S., Pitaevskii, L. P. and Stringari, S. 2008. Theory of ultracold Fermi gases. *Rev. Mod. Phys.*, **80**, 1215.

Glover, R. E. 1967. Ideal resistive transition of a superconductor. *Phys. Lett. A*, **25**, 542.

Glover, R. E. and Tinkham, M. 1957. Conductivity of superconducting films for photon energies between 0.3 and $40kT_c$. *Phys. Rev.*, **108**, 243.

Goldner, L. S. and Ahlers, G. 1992. Superfluid fraction of ^{4}He very close to T_λ. *Phys. Rev. B*, **45**, 13129.

Gor'kov, L. P., Larkin, A. I. and Khmel'nitskii, D. E. 1979. Particle conductivity in a two-dimensional random potential. *JETP Lett.*, **30**, 228.

Graf, E. H., Lee, D. M. and Reppy, John D. 1967. Phase separation and the superfluid transition in liquid ^{3}He-^{4}He mixtures. *Phys. Rev. Lett.*, **19**, 417.

Griffiths, R. B. 1970. Thermodynamics near the two-fluid critical mixing point in ^{3}He-^{4}He. *Phys. Rev. Lett.*, **24**, 715.

Griffiths, R. B. and Pearce, P. A. 1978. Position-space renormalization-group transformations: some proofs and some problems. *Phys. Rev. Lett.*, **41**, 917.

Gross, E. P. 1961. Structure of a quantized vortex in boson systems. *Nuovo Cimento*, **20**, 454.

Haldane, F. D. M. 1981. 'Luttinger liquid theory' of one-dimensional quantum fluids. I. Properties of the Luttinger model and their extension to the general 1D interacting spinless Fermi gas. *J. Phys. C Solid State Phys.*, **14**, 2585.

Halperin, B. I. and Hohenberg, P. C. 1969a. Hydrodynamic theory of spin waves. *Phys. Rev.*, **188**, 898.

Halperin, B. I. and Hohenberg, P. C. 1969b. Scaling laws for dynamic critical phenomena. *Phys. Rev.*, **177**, 952.

Hebard, A. F., Rosseinsky, M. J., Haddon, R. C., et al. 1991. Superconductivity at 18 K in potassium-doped C_{60}. *Nature*, **350**, 600.

Heisenberg, W. 1926. Mehrkörperproblem und Resonanz in der Quantenmechanik. *Z. Phys.*, **38**, 411.

Henshaw, D. G. and Woods, A. D. B. 1961. Modes of atomic motions in liquid helium by inelastic scattering of neutrons. *Phys. Rev.*, **121**, 1266.

Hertel, G., Bishop, D. J., Spencer, E. G., Rowell, J. M. and Dynes, R. C. 1983. Tunneling and transport measurements at the metal-insulator transition of amorphous Nb: Si. *Phys. Rev. Lett.*, **50**, 743.

Hewson, A. C. 1997. *The Kondo Problem to Heavy Fermions*. Cambridge: Cambridge University Press.

Hikami, S. 1981. Anderson localization in a nonlinear σ-model representation. *Phys. Rev. B*, **24**, 2671.

Hikami, S., Larkin, A. I. and Nagaoka, Y. 1980. Spin–orbit interaction and magnetoresistance in the two dimensional random system. *Prog. Theor. Phys.*, **63**(2), 707.

Hohenberg, P. C. 1967. Existence of long-range order in one and two dimensions. *Phys. Rev.*, **158**, 383.

Hohenberg, P. C. and Halperin, B. I. 1977. Theory of dynamic critical phenomena. *Rev. Mod. Phys.*, **49**, 435.

Hohenberg, P. C. and Martin, P. C. 1965. Microscopic theory of superfluid helium. *Ann. Phys.*, **34**, 291.

Holm, C. and Janke, W. 1993. Critical exponents of the classical three-dimensional Heisenberg model: a single-cluster Monte Carlo study. *Phys. Rev. B*, **48**, 936.

Hor, P. H., Gao, L., Meng, R. L., et al. 1987. High-pressure study of the new Y-Ba-Cu-O superconducting compound system. *Phys. Rev. Lett.*, **58**, 911.

Huang, K. 1963. *Statistical Mechanics*. New York: J. Wiley & Sons.

Hubbard, J. 1963. Electron correlations in narrow energy bands. *Proc. Roy. Soc. A*, **276**, 238.

Ioffe, A. F. and Regel, A. R. 1960. Non-crystalline, amorphous and liquid electronic semiconductors. *Prog. Semicond.*, **4**, 237.

Ishida, K., Nakaii, Y. and Hosono, H. 2009. To what extent iron-pnictide new superconductors have been clarified: a progress report. *J. Phys. Soc. Jpn.*, **78**, 062001.

Ito, T., Takenaka, K. and Uchida, S. 1993. Systematic deviation from T-linear behavior in the in-plane resistivity of $YBa_2Cu_3O_{7-y}$: evidence for dominant spin scattering. *Phys. Rev. Lett.*, **70**, 3995.

Jaklevic, R. C., Lambe, J., Silver, A. H. and Mercereau, J. E. 1964. Quantum interference effects in Josephson tunneling. *Phys. Rev. Lett.*, **12**, 159.

Jarzynski, C. 1997. Nonequilibrium equality for free energy differences. *Phys. Rev. Lett.*, **78**, 2690.

Jasnow, D. and Wortis, M. 1968. High-temperature critical indices for the classical anisotropic Heisenberg model. *Phys. Rev.*, **176**, 739.

Johnson, J. B. 1928. Thermal agitation of electricity in conductors. *Phys. Rev.*, **32**, 97.

Jona-Lasinio, G. 1973. Generalized renormalization transformations. In Lundquist, B. and Lundquist, S. (eds.), *Collective Properties of Physical Systems*. New York: Academic Press,.

Josephson, B. D. 1964. Coupled superconductors. *Rev. Mod. Phys.*, **36**, 216.

Josephson, B. D. 1966. Relation between the superfluid density and order parameter for superfluid He near T_c. *Phys. Lett.*, **21**(6), 608–9.

Kac, M. 1959. *Probability and Related Topics in the Physical Sciences*. New York: Interscience.

Kadanoff, L. P. 1966. Scaling laws for Ising models near T_c. *Physics*, **2**, 263.

Kadanoff, L. P., Götze, W., Hamblen, D., et al. 1967. Static phenomena near critical points: theory and experiment. *Rev. Mod. Phys.*, **39**, 395.

Kamerling-Onnes, H. 1911. Further experiments with liquid helium. C. On the change of electric resistance of pure metals at very low temperatures etc. IV. The resistance of pure mercury at helium temperatures. *KNAW, Proceedings*, **13**, 1274.

Kamihara, Y., Watanabe, T., Hirano, M. and Osono, H. 2008. Iron-based layered superconductors La[$O_{1-x}F_x$]FeAs ($x = 0.05 \div 0.12$) with $T_c = 26$. *J. Am. Chem. Soc.*, **130**, 3296.

Kaminski, A., Rosenkranz, S., Fretwell, H. M., et al. 2002. Spontaneous breaking of time-reversal symmetry in the pseudogap state of a high-T_c superconductor. *Nature*, **416**, 610.

Kapitza, P. L. 1938. Viscosity of liquid helium below the λ-point. *Nature*, **141**, 74.

Katsumoto, S. 1988. The metal–insulator transition in a persistent photoconductor. In Ando, T. and Fukuyama, H. (eds.), *Anderson Localization. Proceedings of the International Symposium*, Tokyo, Japan, 1987. Berlin and New York: Springer.

Katsumoto, S., Komori, F., Sano, N. and Kobayashi, S.-I. 1987. Fine tuning of metal–insulator transition in $Al_{0.3}Ga_{0.7}As$ using persistent photoconductivity. *J. Phys. Soc. Jpn.*, **56**, 2259.

Keesom, W. H. and Clusius, K. 1932. Über die spezifische Wärme des flüssigen Heliums. *Proc. R. Acad. Amsterdam*, **35**.

Keesom, W. H. and MacWood, G. E. 1938. The viscosity of liquid helium. *Physica*, **5**, 737.

Keesom, W. H. and van den Ende, J. N. 1930. The specific heat of substances at the temperature obtained with the aid of helium. II Measurements of the atomic heats of lead and of bismath. *Proc. R. Acad. Amsterdam*, **33**, 243.

Kerrisk, J. F. and Keller, W. E. 1967. Thermal conductivity of liquid helium I. *Bull. Am. Phys. Soc. Ser. II*, **12**, 550.

Ketterle, W. and Zwierlein, M. W. 2008. Making, probing and understanding ultracold Fermi gases. In Inguscio, M., Ketterle, W. and Salomon, C. (eds.), *Ultra-cold Fermi Gases*. International School of Physics Enrico Fermi. Amsterdam: IOS Press.

Ketterle, W., Durfee, D. S. and Stamper-Kurn, D. M. 1999. Making, probing and understanding Bose–Einstein condensates. In Inguscio, M., Stringari, S. and Wieman, C. (eds.), *Bose–Einstein Condensation in Atomic Gases*. International School of Physics Enrico Fermi. Amsterdam: IOS Press.

Khalatnikov, I. M. 1965. *An Introduction to the Theory of Superfluidity*. New York, Amsterdam: W. A. Benjamin.

Khinchin, A. I. 1949. *Mathematical Foundations of Statistical Mechanics*. New York: Dover.

Khorana, B. M. and Chandrasekhar, B. S. 1967. AC Josephson effect in superfluid helium. *Phys. Rev. Lett.*, **18**, 230.

Kordyuk, A. A. 2012. Iron-based superconductors: magnetism, superconductivity, and electronic structure. *Low Temp. Phys.*, **38**, 888.

Krönig, A. 1856. Grundzüge einer Theorie der Gase. *Annalen der Physik und Chemie*, **99**, 315.

Lamb, H. 1945. *Hydrodynamics*. New York: Dover.

Landau, L. D. 1937a. Theory of phase transformations. I. *Zh. Eksp. Teor. Fiz.*, **7**, 19 (*Sov. Phys. JETP*, **11**, 26 (1937)).

Landau, L. D. 1937b. Theory of phase transformations. II. *Zh. Eksp. Teor. Fiz.*, **7**, 627 (*Sov. Phys. JETP*, **11**, 545 (1937)).

Landau, L. D. 1941. The theory of superfluidity on Helium II. *Zh. Eksp. Teor. Fiz.*, **11**, 542 (*J. Phys. USSR*, **5**, 71 (1941)).

Landau, L. D. 1947. On the Theory of Superfluidity of Helium II. *J. Phys. (USSR)*, **11**, 91.

Landau, L. D. 1956. The theory of a Fermi liquid. *Zh. Eksp. Teor. Fiz.*, **30**, 1058 (*Sov. Phys. JETP*, **3**, 920 (1957)).

Landau, L. D. 1957. Oscillations in a Fermi liquid. *Zh. Eksp. Teor. Fiz.*, **32**, 59 (*Sov. Phys. JETP*, **5**, 101 (1957)).

Landau, L. D. 1958. On the theory of the Fermi liquid. *Zh. Eksp. Teor. Fiz.*, **35**, 97 (*Sov. Phys. JETP*, **8**, 70 (1959)).

Landau, L. D. and Lifshitz, E. M. 1959. *Statistical Physics*. London: Pergamon Press.

Lanford, O. E. 1975. *Dynamical Systems, Theory and Applications*, Lecture Notes in physics, vol. 38. Berlin: Springer.

Langevin, P. 1908. Sur la théorie du mouvement brownien. *C.R. Acad. Sci. (Paris)*, **146**, 530.

Larkin, A. I. and Varlamov, A. 2005. *Theory of Fluctuations in Superconductors*. Oxford: Oxford University Press.

LeBoeuf, D., Doiron-Leyraud, N., Levallois, J., et al. 2007. Electron pockets in the Fermi surface of hole-doped high-T_c superconductors. *Nature*, **450**, 533.

Lee, D. M. and Leggett, A. J. 2011. Superfluid ^{3}He – the early days. *J. Low Temp. Phys.*, **164**, 140.

Lee, P. A. and Ramakrishnan, T. V. 1985. Disordered electronic systems. *Rev. Mod. Phys.*, **57**, 287.

Lee, P. A., Nagaosa, N. and Wen, X. G. 2006. Doping a Mott insulator: physics of high-temperature superconductivity. *Rev. Mod. Phys.*, **78**, 17.

Lee, T. D. and Yang, C. N. 1952. Statistical theory of equations of state and phase transitions. II. Lattice gas and Ising model. *Phys. Rev.*, **87**, 410.

Leggett, A. J. 1965. Theory of a superfluid Fermi liquid. I. General formalism and static properties. *Phys. Rev.*, **140**, A1869.

Leggett, A. J. 1980. Diatomic molecules and Cooper pairs. In Pfôkalski, A. and Przystawa, J. A. (eds.), *Modern Trends in the Theory of Condensed Matter*. Lecture Notes in Physics, vol. 115. Berlin: Springer.

Leggett, A. J. 2006. *Quantum Liquids*. Oxford: Oxford University Press.

Leggett, A. J. 1975. A theoretical description of the new phases of liquid ^{3}He. *Rev. Mod. Phys.*, **47**, 331.

Leggett, A. J. 2001. Bose–Einstein condensation in the alkali gases: some fundamental concepts. *Rev. Mod. Phys.*, **73**, 307.

Licciardello, D. C. and Thouless, D. J. 1975. Constancy of minimum metallic conductivity in two dimensions. *Phys. Rev. Lett.*, **35**, 1475.

Lipa, J. A., Swanson, D. R., Nissen, J. A., Chui, T. C. P. and Israelsson, U. E. 1996. Heat capacity and thermal relaxation of bulk helium very near the lambda point. *Phys. Rev. Lett.*, **76**, 944.

Livingston, J. D. 1963. Magnetic properties of superconducting lead-base alloys. *Phys. Rev.*, **129**, 1943.

London, F. 1938. The λ-phenomenon of liquid helium and the Bose–Einstein degeneracy. *Nature*, **141**, 643.

London, F. 1959. *Superfluids II*. New York: Wiley.

London, F. and London, H. 1935. The electromagnetic equations of the supraconductor. *Proc. R. Soc. Lond. A*, **149**, 71.

Löw, U., Emery, V. J., Fabricius, K. and Kivelson, S. A. 1994. Study of an Ising model with competing long- and short-range interactions. *Phys. Rev. Lett.*, **72**, 1918.

Luther, A. and Peschel, I. 1974. Single-particle states, Kohn anomaly and pairing fluctuations in one dimension. *Phys. Rev. B*, **9**, 2911–2919.

Luttinger, J. M. 1963. An exactly soluble model of a many-fermion system. *J. Math. Phys.*, **4**, 1154.

Ma, S. K. 1976. *Modern Theory of Critical Phenomena*. London: Benjamin.

Machida, K. 1989. Magnetism in La_2CuO_4 based compounds. *Physica C*, **158**, 192.

Mahan, G. D. 2000. *Many-particle Physics*. New York: Kluwer/Plenum.

Matsubara, T. 1955. A new approach to quantum-statistical mechanics. *Prog. Theor. Phys.*, **14**, 351.

Mattis, D. C. and Lieb, E. H. 1965. Exact solution of a many fermion system and its associated boson field. *J. Math. Phys.*, **6**, 304.

Maxwell, J. C. 1860a. Illustrations of the dynamical theory of gases. Part II. *Philos. Magazine*, **20**, 21.

Maxwell, J. C. 1860b. Illustrations of the dynamical theory of gases. Part I. On the motions and collisions of perfectly elastic spheres. *Philos. Magazine*, **19**, 19.

Maxwell, J. C. 1875. On the dynamical evidence of the molecular constitution of Bodies. *Nature*, **11**, 357.

McMillan, W. L. and Mochel, J. 1981. Electron tunneling experiments on amorphous $Ge_{1-x}Au_x$. *Phys. Rev. Lett.*, **46**, 556.

Meissner, W. and Ochsenfeld, R. 1933. Ein neuer Effekt bei Eintritt der Supraleitfähigkeit. *Naturwissenschaften*, **21**, 787.

Mermin, N. D. and Wagner, H. 1966. Absence of ferromagnetism or anti ferromagnetism in one- or two-dimensional isotropic Heisenberg models. *Phys. Rev. Lett.*, **17**, 1133.

Metzner, W. and Di Castro, C. 1993. Conservation laws and correlation functions in the Luttinger liquid. *Phys. Rev. B*, **47**, 16107.

Metzner, W., Castellani, C. and Di Castro, C. 1998. Fermi systems with strong forward scattering. *Adv. Phys.*, **47**, 317.

Migdal, A. A. 1969. A diagram technique near the Curie point and the second order phase transition in a Bose liquid. *Sov. Phys. JETP*, **28**, 1036.

Mott, N. F. 1967. Electrons in disordered structures. *Adv. Phys.*, **16**, 49.

Mott, N. F. 1972. Conduction in non-crystalline systems IX. The minimum metallic conductivity. *Philos. Magazine*, **26**, 1015.

Nakamura, Y. and Uchida, S. 1993. Anisotropic transport properties of single crystal $La_{2-x}Sr_xCuO_4$: evidence for the dimensional crossover. *Phys. Rev. B*, **47**, 8369R.

Nernst, W. H. 1906. Über die Berechnung chemischer Gleichgewichte aus thermischen Messungen. *Nachrichten von der Gesellschaft Wissenschaften zu Göttingen Matematisch-Physikalische Klasse*, 1.

Nishida, N., Furubayashi, T., Yamaguchi, M., Morigaki, K. and Ishimoto, H. 1985. Metal–insulator transition in the amorphous $Si_{1-x}Au_x$ system with a strong spin–orbit interaction. *Solid State Electronics*, **28**, 81.

Nozières, P. 1964. *Theory of Interacting Fermi Systems*. New York: W. A. Benjamin.

Nozières, P. and Pines, D. 1966. *The Theory of Quantum Liquids*. New York: W. A. Benjamin.

Nozières, P. and Schmitt-Rink, S. 1985. Bose condensation in an attractive fermion gas: from weak to strong coupling superconductivity. *J. L. Temp. Phys.*, **59**, 195.

Nyquist, H. 1928. Thermal agitation of electric charge in conductors. *Phys. Rev.*, **32**, 110.

Okuma, S., Komori, F. and Kobayashi, S. 1988. The metal–insulator transition in disordered metals. In Ando, T. and Fukuyama, H. (eds.), *Anderson Localization. Proceedings of the International Symposium*. Berlin: Springer.

Onsager, L. 1931a. Reciprocal relations in irreversible processes. I. *Phys. Rev.*, **37**, 405.

Onsager, L. 1931b. Reciprocal relations in irreversible processes. II. *Phys. Rev.*, **38**, 2265.

Ornstein, L. S. and Zernike, F. 1914. Accidental deviations of density and opalescence at the critical point of a single substance. *KNAW, Proceedings*, **17**, 793.

Osheroff, D. D., Richardson, R. C. and Lee, D. M. 1972a. Evidence for a new phase of solid He^3. *Phys. Rev. Lett.*, **28**, 885.

Osheroff, D. D., Gully, W. J., Richardson, R. C. and Lee, D. M. 1972b. New magnetic phenomena in liquid He^3 below 3 mK. *Phys. Rev. Lett.*, **29**, 920.

Paalanen, M. A., Sachdev, S., Bhatt, R. N. and Ruckenstein, A. E. 1986. Spin dynamics of nearly localized electrons. *Phys. Rev. Lett.*, **57**, 2061.

Patashinskij, A. Z. and Pokrovskij, V. L. 1966. Behavior of ordered systems near the transition point. *Sov. Phys. JETP*, **23**, 292.

Patashinskij, A. Z. and Pokrovskij, V. L. 1979. *Fluctuations Theory of Phase Transitions*. Oxford: Pergamon Press.

Paulson, D. N., Kojima, H. and Wheatley, J. C. 1974. Profound effect of a magnetic field on the phase diagram of superfluid ^3He. *Phys. Rev. Lett.*, **32**, 1098.

Peliti, L. 2011. *Statistical Mechanics in a Nutshell*. Princeton: Princeton University Press.

Penrose, O. and Onsager, L. 1956. Bose–Einstein condensation and liquid helium. *Phys. Rev.*, **104**, 576.

Perali, A., Pieri, P., Pisani, L. and Strinati, G. C. 2004. BCS-BEC crossover at finite temperature for superfluid trapped Fermi atoms. *Phys. Rev. Lett.*, **92**, 220404.

Perrin, J. B. 1913. *Les Atoms*. Paris: Librairie Felix Alcan.

Pethick, C. J. and Smith, H. 2008. *Bose–Einstein Condensation in Dilute Gases*. 2nd edn. Cambridge: Cambridge University Press.

Phillips, N. E. 1959. Heat capacity of aluminum between 0.1 K and 4.0K. *Phys. Rev.*, **114**, 676.

Phillips, W. D. 1998. Nobel Lecture: Laser cooling and trapping of neutral atoms. *Rev. Mod. Phys.*, **70**, 721.

Pippard, A. B. 1955. Trapped flux in superconductors. *Phil. Trans. R. Soc. Lond. A*, **248**, 97.

Pippard, A. B. 1957. *Elements of Classical Thermodynamics: for Advanced Students of Physics*. Cambridge: Cambridge University Press.

Pitaevskii, L. P. 1959. On the superfluidity of liquid ^3He. *Zh. Eksp. Teor. Fiz.*, **37**, 1794 (*Sov. Phys. JETP*, **10**, 1267 (1960)).

Pitaevskii, L. P. 1961. Vortex lines in imperfect Bose gas. *Zh. Eksp. Teor. Fiz.*, **40**, 646 (*Sov. Phys. JETP*, **13**, 451 (1961)).

Pitaevskii, L. P. and Stringari, S. 2003. *Bose–Einstein Condensation*. Oxford: Clarendon Press.

Planck, M. K. E. L. 1900a. Über eine Verbesserung der Wienschen Spectralgleichung. *Verhandlungen der Deutschen Physikalischen Gesellschaft*, **2**, 202.

Planck, M. K. E. L. 1900b. Zur Theorie des Gesetzes der Energieverteilung im Normalspectrum. *Verhandlungen der Deutschen Physikalischen Gesellschaft*, **2**, 237.

Planck, M. K. E. L. 1917. Über einen Satz der statischen Dynamik und seine Erweiterung in der Quantentheorie. *Sitzungberichte der Preussischen Akadademie der Wissenschaften*, 324.

Polyakov, A. M. 1968. Microscopic description of critical phenomena. *Zh. Eksp. Teor. Fiz.*, **55**, 1026 (*Sov. Phys. JETP*, **28**, 533 (1969)).

Polyakov, A. M. 1969. Properties of long and short range correlations in the critical region. *Zh. Eksp. Teor. Fiz.*, **57**, 271 (*Sov. Phys. JETP*, **30**, 151 (1970)).

Punnoose, A. and Finkel'stein, A. M. 2002. Dilute electron gas near the metal–insulator transition: role of valleys in silicon inversion layers. *Phys. Rev. Lett.*, **88**, 016802.

Qian, Z., Vignale, G. and Marinescu, D. C. 2004. Spin mass of an electron liquid. *Phys. Rev. Lett.*, **93**, 106601.

Raffa, F., Ohno, T., Mali, M., et al. 1998. Isotope dependence of the spin gap in $YBa_2Cu_4O_8$ as determined by Cu NQR relaxation. *Phys. Rev. Lett.*, **81**, 5912.

Rammer, J. 2007. *Quantum Field Theory of Non-equilibrium States*. Cambridge: Cambridge University Press.

Randeria, M. and Trivedi, N. 1998. Pairing correlations above T_c and pseudogaps in underdoped cuprates. *J. Phys. Chem. Solids*, **59**, 1754.

Rayfield, G. W. and Reif, F. 1964. Quantized vortex rings in superfluid helium. *Phys. Rev.*, **136**, A1194.

Richards, P. L. and Anderson, P. W. 1965. Observation of the analog of the AC Josephson effect in superfluid helium. *Phys. Rev. Lett.*, **14**, 540.

Riedel, E. K. and Wegner, F. J. 1974. Effective critical and tricritical exponents. *Phys. Rev. B*, **9**, 294.

Rohde, M. and Micklitz, H. 1987. Indication of universal behavior of Hall conductivity near the metal–insulator transition in disordered systems. *Phys. Rev. B*, **36**, 7572.

Rosenbaum, T. F., Andres, K., Thomas, G. A. and Bhatt, R. N. 1980. Sharp metal–insulator transition in a random solid. *Phys. Rev. Lett.*, **45**, 1723.

Rosenbaum, T. F., Milligan, R. F., Paalanen, M. A., Thomas, G. A., Bhatt, R. N. and Lin, W. 1983. Metal–insulator transition in a doped semiconductor. *Phys. Rev. B*, **27**, 7509.

Rubio Temprano, D., Mesot, J., Janssen, S., et al. 2000. Large isotope effect on the pseudogap in the high-temperature superconductor $HoBa2Cu_4O_8$. *Phys. Rev. Lett.*, **84**, 1990.

Schreck, F., Khaykovich, L., Corwin, K. L., et al. 2001. Quasipure Bose–Einstein condensate immersed in a Fermi sea. *Phys. Rev. Lett.*, **87**, 080403.

Schrieffer, J. R. 1999. *Theory of Superconductivity*. Reading, Mass.: Pegasus Books.

Schrödinger, E. 1968. *Statistical Thermodynamics*. Cambridge: Cambridge University Press.

Schroer, B. 1973. Theory of critical phenomena based on the normal-product formalism. *Phys. Rev. B*, **8**, 4200.

Sebastian, S. E., Harrison, N., Palm, E., et al. 2008. A multi-component Fermi surface in the vortex state of an underdoped high-T_c superconductor. *Nature*, **454**, 200.

Seibold, G., Grilli, M. and Lorenzana, J. 2012. Stripes in cuprate superconductors: excitations and dynamic dichotomy. *Physica C*, **481**, 132.

Serin, B., Reynolds, C. A. and Nesbitt, L. B. 1950. Mass dependence of the superconducting transition temperature of mercury. *Phys. Rev.*, **80**, 761.

Shankar, R. 1994. Renormalization-group approach to interacting fermions. *Rev. Mod. Phys.*, **66**, 129.

Shapiro, S., Janus, A. R. and Holly, S. 1964. Effect of microwaves on Josephson currents in superconducting tunneling. *Rev. Mod. Phys.*, **36**, 223.

Simon, M. E. and Varma, C. M. 2002. Detection and implications of a time-reversal breaking state in underdoped cuprates. *Phys. Rev. Lett.*, **89**, 247003.

Sólyom, J. 1979. The Fermi gas model of one-dimensional conductors. *Adv. Phys.*, **28**, 201.

Stanley, H. E. 1987. *Introduction to Phase Transitions and Critical Phenomena*. Oxford: Oxford University Press.

Stupp, H., Hornung, M., Lakner, M., Madel, O. and Löhneysen, H. V. 1993. Possible solution of the conductivity exponent puzzle for the metal-insulator transition in heavily doped uncompensated semiconductors. *Phys. Rev. Lett.*, **71**, 2634.

Tallon, J. L. and Loram, J. W. 2001. The doping dependence of T^* – what is the real high-T_c phase diagram? *Physica C: Superconductivity*, **349**, 53.

Thomas, G. A., Ootuka, Y., Kobayashi, S. and Sasaki, W. 1981. Comparison of the specific heat and conductivity of Si: P. *Phys. Rev. B*, **24**, 4886.

Thomas, G. A., Ootuka, Y., Katsumoto, S., Kobayashi, S. and Sasaki, W. 1982. Evidence for localization effects in compensated semiconductors. *Phys. Rev. B*, **25**, 4288.

Thompson, C. J. 1972. *Mathematical Statistical Mechanics*. Princeton: Princeton University Press.

Timusk, T. 2003. The mysterious pseudogap in high temperature superconductors: an infrared view. *Solid State Comm.*, **127**(5), 337.

Timusk, T. and Statt, B. 1999. The pseudogap in high-temperature superconductors: an experimental survey. *Rep. Prog. Phys.*, **62**, 61.

Tinkham, M. 1975. *Introduction to Superconductivity*. New York: McGraw-Hill.

Tino, G. M., Cataliotti, F. S., Cornell, E. A., Fort, C., Inguscio, M. and Prevedelli, M. 1999. Towards quantum degeneracy of bosonic and fermionic potassium atoms. In Inguscio, M., Stringari, S. and Wieman, C. (eds.), *Bose–Einstein Condensation in Atomic Gases*, Amsterdam: IOP Press.

Tisza, L. 1938. Transport phenomena in helium II. *Nature*, **141**, 913.

Tomonaga, S. 1950. Remarks on Bloch's method of sound waves applied to many-fermion problems. *Prog. Theor. Phys.*, **5**, 544.

Tranquada, J. M. 2012. Cuprates get orders to charge. *Science*, **337**, 811.

Tranquada, J. M., Sternlieb, B. J., Axe, J. D., Nakamura, Y. and Uchida, S. 1995. Evidence for stripe correlations of spins and holes in copper oxide superconductors. *Nature*, **375**, 561.

Tranquada, J. M., Axe, J. D., Ichikawa, N., Nakamura, Y., Uchida, S. and Nachumi, B. 1996. Neutron-scattering study of stripe-phase order of holes and spins in $La_{1.48}Nd_{0.4}Sr_{0.12}CuO_4$. *Phys. Rev. B*, **54**, 7489.

Tranquada, J. M., Axe, J. D., Ichikawa, N., Moodenbaugh, A. R., Nakamura, Y. and Uchida, S. 1997. Coexistence of and competition between, superconductivity and charge-stripe order in $La_{1.6-x}Nd_{0.4}Sr_xCuO_4$. *Phys. Rev. Lett.*, **78**, 338.

Truscott, A. G., Strecker, K. E., McAlexander, W. I., Partridge, G. B. and Hulet, R. G. 2001. Observation of Fermi pressure in a gas of trapped atoms. *Science*, **291**, 2570.

Tsuei, C. C. and Kirtley, J. R. 2000. Pairing symmetry in cuprate superconductors. *Rev. Mod. Phys.*, **72**, 969.

Valatin, J. G. 1962. Superconducting electron and nucleon systems. In *Lectures in Theoretical Physics, Boulder, Colorado (1961)*, vol. 4. New York: Interscience.

van der Waals, J. D. 1873. *Over de Continuiteit van den Gasen Vloeistoftoestand*. Leiden: A. W. Suthoff.

Van Hove, L. 1954. Time-dependent correlations between spins and neutron scattering in ferromagnetic crystals. *Phys. Rev.*, **95**, 1374.

Varma, C. M. 1993. Towards a theory of the marginal Fermi-liquid state. *J. Phys. Chem. Solids*, **54**, 1081.

Varma, C. M. 1999. Pseudogap phase and the quantum-critical point in copper-oxide metals. *Phys. Rev. Lett.*, **83**, 3538.

Vojta, M. 2009. Lattice symmetry breaking in cuprate superconductors: stripes, nematics and superconductivity. *Adv. Phys*, **58**, 699.

Vollhardt, D. 1998. Pair correlations in superfluid helium 3. In Kresin, V. (ed.), *Pair Correlations in Many-Fermion Systems*. New York: Plenum Press.

Wang, Y. and Chubukov, A. 2014. Charge-density-wave order with momentum $(2Q, 0)$ and $(0, 2Q)$ within the spin-fermion model: continuous and discrete symmetry breaking, preemptive composite order and relation to pseudogap in hole-doped cuprates. *Phys. Rev. B*, **90**, 035149.

Webb, R. A., Greytak, T. J., Johnson, R. T. and Wheatley, J. C. 1973. Observation of a second-order phase transition and its associated $P - T$ phase diagram in liquid He^3. *Phys. Rev. Lett.*, **30**, 210.

Wegner, F. 1976. Electrons in disordered systems. Scaling near the mobility edge. *Zeit. für Phys. B*, **25**, 327.

Wegner, F. 1979. The mobility edge problem: continuous symmetry and a conjecture. *Zeit. für Phys. B*, **35**, 307.

Wegner, F. J. and Riedel, E. K. 1973. Logarithmic corrections to the molecular-field behavior of critical and tricritical systems. *Phys. Rev. B*, **7**, 248.

Wheatley, J. C. 1966. *Quantum Fluids: Proceedings of the Sussex University Symposium 16–20 August 1965*. Amsterdam: North-Holland.

Wheatley, J. C. 1975. Experimental properties of superfluid ^3He. *Rev. Mod. Phys.*, **47**, 415.

Widom, B. 1974. The critical point and scaling theory. *Physica*, **73**, 107–118.

Wigner, E. P. 1957. Relativistic invariance and quantum phenomena. *Rev. Mod. Phys.*, **29**, 255.

Williams, G. V. M., Pringle, D. J. and Tallon, J. L. 2000. Contrasting oxygen and copper isotope effects in $YBa_2Cu_4O_8$ superconducting and normal states. *Phys. Rev. B*, **61**, R9257.

Wilson, K. G. 1971a. Renormalization group and critical phenomena. I. Renormalization group and the Kadanoff scaling picture. *Phys. Rev. B*, **4**, 3174.

Wilson, K. G. 1971b. Renormalization group and critical phenomena. II. Phase-space cell analysis of critical behavior. *Phys. Rev. B*, **4**, 3184.

Wilson, K. G. 1972. Feynman-graph expansion for critical exponents. *Phys. Rev. Lett.*, **28**, 548.

Wilson, K. G. and Fisher, M. E. 1972. Critical exponents in 3.99 dimensions. *Phys. Rev. Lett.*, **28**, 240.

Wilson, K. G. and Kogut, J. 1974. The renormalization group and the ϵ expansion. *Phys. Rep.*, **12**, 75.

Wu, M. K., Ashburn, J. R., Torng, C. J., et al. 1987. Superconductivity at 93 K in a new mixed-phase Y-Ba-Cu-O compound system at ambient pressure. *Phys. Rev. Lett.*, **58**, 908.

Wu, T., Mayaffre, H., Kramer, S., et al. 2011. Magnetic-field-induced charge-stripe order in the high-temperature superconductor $YBa_2Cu_3O_y$. *Nature*, **477**, 191.

Yamaguchi, M., Nishida, N., Furubayashi, T., Morigaki, K., Ishimoto, H. and Ono, K. 1983. Metal–nonmetal transition and superconductivity in amorphous Si_{1-x} Au_x System. *Physica B+C*, **117-118**, 694.

Yamase, H., Oganesyan, V. and Metzner, W. 2005. Mean-field theory for symmetry-breaking Fermi surface deformations on a square lattice. *Phys. Rev. B*, **72**, 035114.

Yang, C. N. 1962. Concept of off-diagonal long-range order and the quantum phases of liquid He and of superconductors. *Rev. Mod. Phys.*, **34**, 694.

Yang, C. N. and Lee, T. D. 1952. Statistical theory of equations of state and phase transitions. I. Theory of condensation. *Phys. Rev.*, **87**, 404.

Zaanen, J. and Gunnarsson, O. 1989. Charged magnetic domain lines and the magnetism of high-T_c oxides. *Phys. Rev. B*, **40**, 7391.

Zemansky, M. W. 1968. *Heat and Thermodynamics*. New York: MacGraw-Hill.

Index